Green Energy to Sustainability

Green Energy to Sustainability

Strategies for Global Industries

Edited by

Alain A. Vertès
Sloan Fellow, London Business School
UK

and

Managing Director of NxR Biotechnologies, Basel Switzerland

Nasib Qureshi
United States Department of Agriculture
National Center for Agricultural Utilization Research
Peoria, USA

and

University of Illinois at Urbana-Champaign, USA

Hans P. Blaschek
Department of Food Science and Human Nutrition
University of Illinois at Urbana-Champaign
USA

Hideaki Yukawa
Utilization of Carbon Dioxide Institute CO. Ltd.
Tokyo, Japan

This edition first published 2020

Registered Offices
John Wiley & Sons, Inc., 111 River Street, Hoboken, NJ 07030, USA
John Wiley & Sons Ltd, The Atrium, Southern Gate, Chichester, West Sussex, PO19 8SQ, UK

Editorial Office
The Atrium, Southern Gate, Chichester, West Sussex, PO19 8SQ, UK

For details of our global editorial offices, customer services, and more information about Wiley products visit us at www.wiley.com.

Wiley also publishes its books in a variety of electronic formats and by print-on-demand. Some content that appears in standard print versions of this book may not be available in other formats.

Library of Congress Cataloging-in-Publication Data applied for
HB ISBN: 9781119152026

Cover Design: Wiley
Cover Images: WMM DNA © 3divan/Shutterstock,
Power plan using renewable solar energy © Gencho Petkov/Shutterstock,
Oilseed Rape © TheDman/Getty Images

Set in 9.5/12.5pt STIXTwoText by SPi Global, Chennai, India

Printed and bound by CPI Group (UK) Ltd, Croydon, CR0 4YY

10 9 8 7 6 5 4 3 2 1

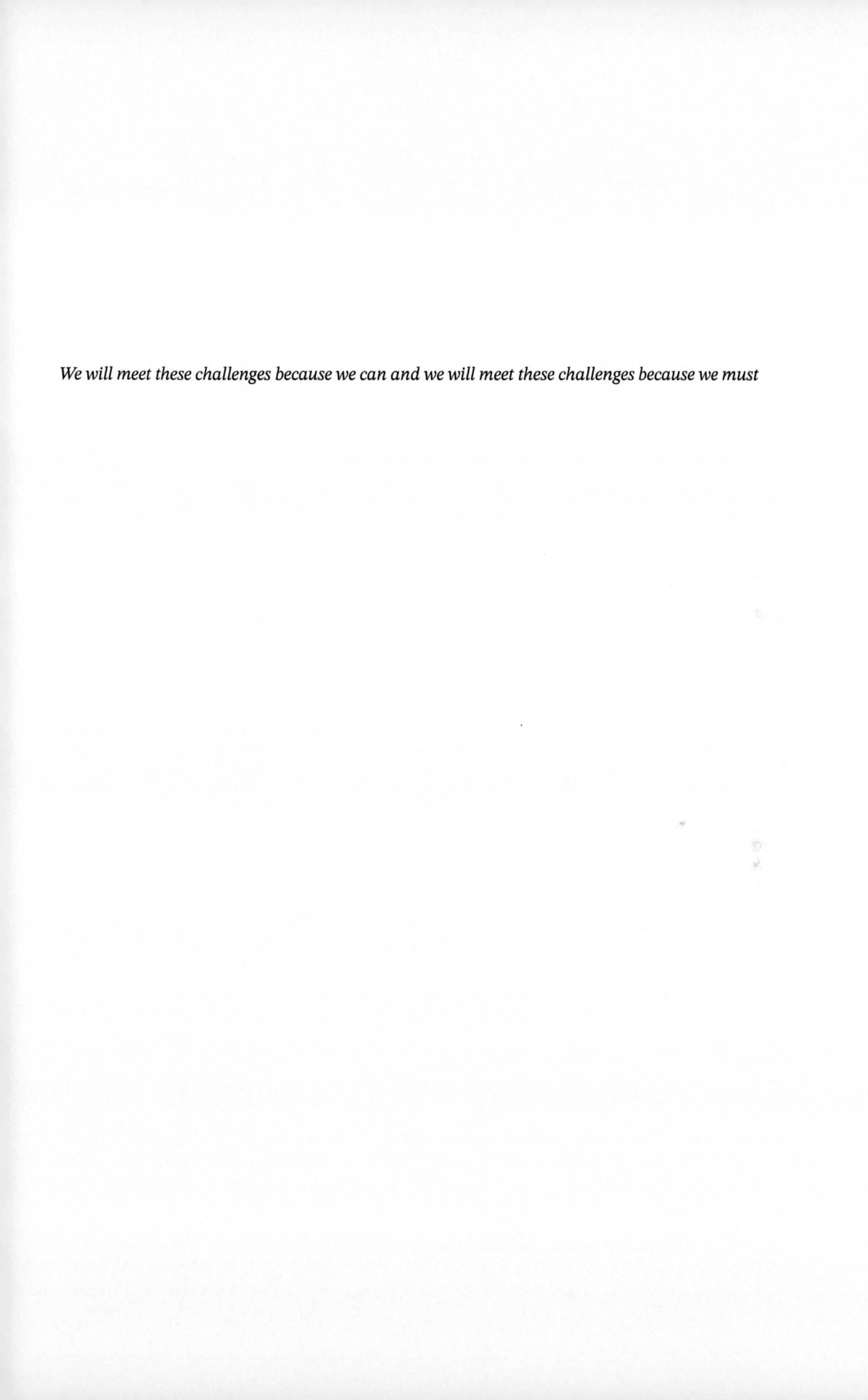

We will meet these challenges because we can and we will meet these challenges because we must

Contents

About the Editors

Dr. Alain A. Vertès is Managing Director at NxR Biotechnologies, a boutique global consulting firm based in Basel, Switzerland, where he advises clients on strategy, business development, in/out-licensing, entrepreneurship and investment. He brings to his role extensive experience in the pharmaceutical and industrial biotechnology sectors, in Europe, North America and Asia and in different functions including research, manufacturing, contract research, and strategic alliances. Dr. Vertès received his M.Sc. degree from the University of Illinois at Urbana-Champaign, his Ph.D. from the University of Lille Flandres Artois while conducting research at the Institut Pasteur in Paris, he has completed his post-doctoral training at Mitsubishi Petrochemical Company in Japan, and is a Sloan Fellow from London Business School (MBA/M.Sc.). Dr Vertès is a lead editor of several science and strategy books in the fields of regenerative medicine and sustainable chemistry.

Dr. Nasib Qureshi is a fellow in American Institute of Chemical Engineering (AIChE), the Society for Industrial Microbiology & Biotechnology (SIMB), and American Institute of Chemists (AIC). He is a Chemical/Biochemical Engineer by training with dual PhD degrees [one in Biochemical/Biological Engineering (University of Nebraska, Lincoln, NE, United States) and the other in Fermentation Technology (Institute of Chemical Technology, Bombay, India)]. He has a long association with the University of Illinois at Urbana Champaign (Illinois, USA, since 1987) and presently holds an Adjunct Professor's appointment. Currently, he is working for the United States Department of Agriculture (Agricultural Research Service) as a Research Chemical Engineer. His research focuses on developing novel bioprocess technologies for biofuels production. He has over 300 authoritative papers, chapters, review articles, patents, and conference presentations. He was President of American Institute of Chemical Engineers (AIChE), Central Illinois Section, 2008 & 2009, and American Chemical Society (ACS), Illinois Heartland Section, 2008. He was also an Advisory Board Member for the Society for Biological Engineering (SBE, USA). Dr. Qureshi is "Editor in Chief" for the World J. Microbiology & Biotechnology. He has edited several books on biofuels and biorefineries and stem cells in regenerative medicine (John Wiley & Elsevier). Dr. Qureshi has received many awards including from the World J. Microbiology & Biotechnology, American Chemical Society, United States Department of Agriculture, and University of Nebraska (Lincoln, NE, USA). His expertise in bioprocess/biochemical engineering and biofuels arena is widely and internationally sought.

Dr. Hans P. Blaschek is Professor Emeritus at the University of Illinois at Urbana-Champaign (UIUC). He served as Assistant Dean in the College of Agricultural Consumer and Environmental Sciences (ACES), Director of the Center for Advanced Bioenergy Research (CABER) and the Integrated Bioprocessing Research Laboratory (IBRL) at the University of Illinois. The IBRL is a $32M intermediate level pilot facility designed for the scale up of bench level processes and unit operations, with a particular focus on scaling biofuels and bioproducts technologies for commercialization. The mission of IBRL is to facilitate bioenergy-related activities as it relates to research, teaching and outreach. CABER and IBRL instituted a successful professional science MS degree program focused on numerous disciplinary areas including biofuels and bioprocessing. Dr. Blaschek served as theme leader of the Molecular Bioengineering of Biomass Conversion Theme of the Institute for Genomic Biology at UIUC and was interim Department Head of the Department of Food Science and Human Nutrition. He was the Co-Founder and CSO of TetraVitae Biosciences (TVB) focused on commercialization of the acetone - butanol fermentation building on technologies developed in his laboratory at the University of Illinois. Dr. Blaschek's research interests involve the genetic manipulation of microorganisms for biotechnological applications, examination of biomass as substrates for fermentation to value-added products, development of an integrated fermentation system for bioproducts production and recovery.

Dr. Hideaki Yukawa is Founder and CEO at Utilization of Carbon Dioxide Institute (UCDI), an R&D-focused venture company out of the University of Tokyo. Specializing in molecular biology, microbial applications and enzyme chemistry, Dr. Yukawa earned his doctorate from the University of Tokyo and climbed the ranks at Mitsubishi Chemical Corporation to the level of Research Fellow. He oversaw the establishment of the Microbiology Research Group at the Research Institute of Innovative Technology for the Earth (RITE), out of which he founded the Green Earth Institute (GEI) Company to stamp the role of renewables in industry. His current responsibilities at UCDI are anchored upon the pursuit of a low-carbon future society and advancement of biotechnology-based solutions to global food security issues. Dr. Yukawa's contributions have been recognized through awards such as the Japan Bioindustry Award for a bioprocess for production of biochemicals using nonlytic bacterial cells, the Tsukuba Foundation for Chemical and Biotechnology Award for establishment of recombinant DNA technology for coryneform bacteria and development of bioprocesses thereof, the Japan Society for Bioscience, Biotechnology and Agrochemistry (JSBBA) Award for achievements in technological research, and the Grand

Prize at the Nikkei Global Environmental Awards for bioethanol production from mixed sugars by genetically engineered *Corynebacterium glutamicum*. The first Japanese national to receive a Fellowship Award (for achievements in the field of applied microorganisms) from The Society of Industrial Microbiology and Biotechnology (SIMB), Dr. Yukawa has authored many scientific works, including papers, books and patents, and has mentored many scientists globally.

List of Contributors

Mairi J. Black
Department of Science, Technology
Engineering & Public Policy (STEaPP)
University College London
London
UK

Aiduan Li Borrion
Department of Civil, Environmental and
Geomatic Engineering
University College London
London
UK

Danilo S. Braz
University of Campinas (UNICAMP)
School of Chemical Engineering
Campinas
Brazil

Terri L. Butler
University of Washington
Buerk Center for Entrepreneurship
Seattle
USA

Daniel de Castro Assumpcao
University of Campinas (UNICAMP)
School of Chemical Engineering
Campinas
Brazil

Tunc Catal
Department of Molecular Biology and Genetics
Uskudar University
Istanbul
Turkey

Ziyu Dai
Chemical & Biological Process Development
Group, Pacific Northwest National Laboratory
Richland
USA

Jeyapraksh Damaraja
Division of Chemistry
Faculty of Science and Humanities
Sree Sowdambika College of Engineering
Aruppukottai,Tamil Nadu
India

Stephen DeVito
Toxics Release Inventory Program
United States Environmental Protection
Agency
Washington
USA

Carlos Hernández Díaz-Ambrona
Polytechnic University of Madrid
School of Agriculture
Spain

Miriam Felkers
Research Group Climate Change and Security
Institute of Geography
University of Hamburg
Germany

Rebecca Froese
Research Group Climate Change and Security
Institute of Geography
University of Hamburg
Germany

Sandra D. Gaona
Toxics Release Inventory Program
United States Environmental Protection Agency
Washington
USA

Patrick C. Hallenbeck
Département de Microbiologie
Infectiologie et Immunologie
Université de Montréal
Québec
Canada

Marcelo Hamaguchi
Life Sciences Research Center
Department of Biology
United States Air Force Academy
Valmet
Araucária
Brazil

John Hay
Department of Biological Systems Engineering
University of Nebraska–Lincoln
USA

Aurelia Karina Hillary
Centre for Environmental Policy
Imperial College London, Prince's Garden
South Kensington Campus
London
UK

Haibo Huang
Department of Food Science and Technology
Virginia Polytechnic Institute and State University
Blacksburg
USA

Stephen Hughes
Applied DNA Sciences
Stony Brook
USA

N. J. Ianno
Department of Electrical Engineering
University of Nebraska-Lincoln
USA

Lonnie O. Ingram
Department of Microbiology and Cell Science
University of Florida
Gainesville
USA

Qing Jin
Department of Food Science and Technology
Virginia Polytechnic Institute and State University
Blacksburg
USA

Marjorie A. Jones
Department of Chemistry
Illinois State University
Normal
USA

Ankita Juneja
Agricultural and Biological Engineering
University of Illinois Urbana-Champaign
Urbana
USA

Halil Kavakli
Departments of Chemical and Biological Engineering and Molecular Biology and Genetics
Koc University
Istanbul
Turkey

Cheryl Keenan
Eastern Research Group Inc.
Lexington
USA

Takuro Kobayashi
Center for Material Cycles and Waste Management Research
National Institute for Environmental Studies
Tsukuba
Japan

Gopalakrishnan Kumar
Department of Environmental Engineering
Daegu University
Gyeongsan
Republic of Korea

Institute of Chemistry
Bioscience and Environmental Engineering
Faculty of Science and Technology
University of Stavanger Stavanger
Norway

C.Q. Lan
Department of Chemical and Biological Engineering
University of Ottawa
Ottawa
Canada

Freeman Lan
Department of Chemical and Biological Engineering
University of Ottawa
Ottawa
Canada

Carolina Zampol Lazaro
Département de Microbiologie
Infectiologie et Immunologie
Université de Montréal
Montréal
Canada

Xueqin Lu
Department of Civil and Environmental Engineering
Graduate School of Engineering
Tohoku University
Sendai
Japan

Yanpin Lu
Washington State University, Bioproducts, Sciences, and Engineering Laboratory
Department of Biological Systems Engineering
Richland
USA

Emiliano Maletta
Bioenergy Crops
United Kingdom

Polytechnic University of Madrid
School of Agriculture
Spain

Adriano P. Mariano
University of Campinas (UNICAMP)
School of Chemical Engineering
Campinas
Brazil

Onesmus Mwaboje
Centre for Environmental Policy
Imperial College London
London
UK

Seiji Nakagame
The Faculty of Applied Bioscience
Kanagawa Institute of Technology
Atsugi
Japan

Licheng Peng
Department of Environmental Science
Hainan University
Haikou
China

Sivagurunathan Periyasamy
Center for Materials Cycles and Waste
Management Research
National Institute for Environmental Studies
Tsukuba
Japan

Nasib Qureshi
United States Department of Agriculture
Agricultural Research Service
National Center for Agricultural Utilization
Research
Bioenergy Research Unit
Peoria
USA

Lawrence Reichle
Abt Associates Inc.
Cambridge
USA

Jose Dilcio Rocha
Embrapa Agroenergy – The Brazilian
Agricultural Research Corporation (Embrapa)
Brasília
Brazil

Emrah Sagir
Département de Microbiologie
Infectiologie et Immunologie
Université de Montréal
Montréal
Canada

Ganesh Dattatraya Saratale
Department of Food Science and
Biotechnology
Dongguk University-Seoul
Goyang-si
Republic of Korea

Rijuta Ganesh Saratale
Research Institute of Biotechnology and
Medical Converged Science
Dongguk University-Seoul
Goyang-si
Republic of Korea

Jürgen Scheffran
Research Group Climate Change and Security
Institute of Geography
University of Hamburg
Germany

K. T. Shanmugan
Department of Microbiology and Cell Science
University of Florida
Gainesville
USA

Sutha Shobana
Department of Chemistry and Research Centre
Aditanar College of Arts and Science
Tiruchendur
India

Vijay Singh
Agricultural and biological Engineering
University of Illinois Urbana-Champaign
USA

Pedro F Souza Filho
Swedish Centre for Resource Recovery
University of Borås
Borås
Sweden

Lianghu Su
Nanjing Institute of Environmental Sciences of
the Ministry of Environmental Protection
Nanjing
PR China

Mohammed Taherzadeh
Swedish Centre for Resource Recovery
University of Borås
Borås
Sweden

Praveen Vadlani
Saivera Bio LLC
Puttaparthi
India

Cyril Vallet
Abt Associates Inc.
Cambridge
USA

Alain A. Vertès
Sloan Fellow
London Business School
UK

Henrique C. A. Venturelli
University of Campinas (UNICAMP)
School of Chemical Engineering
Campinas
Brazil

Elmar Mateo Villota
Washington State University, Bioproducts
Sciences, and Engineering Laboratory
Department of Biological Systems Engineering
Richland
USA

Jianhui Wang
Shanghai Key Lab for Urban Ecological
Processes and Eco-Restoration
School of Ecological and Environmental
Sciences
East China Normal University
Shanghai
PR China

Xiaohui Wang
Shanghai Key Lab for Urban Ecological
Processes and Eco-Restoration
School of Ecological and Environmental
Sciences
East China Normal University
Shanghai
PR China

Kaiqin Xu
Center for Material Cycles and Waste
Management Research
National Institute for Environmental Studies
Tsukuba
Japan

Asutosh T. Yagnik
AdSidera Ltd.
London, UK

Institute for Institute for Strategy
Resilience & Security
University College London
UK

Bin Yang
Washington State University, Bioproducts
Sciences, and Engineering Laboratory
Department of Biological Systems Engineering
Richland
USA

Lorraine P. Yomano
Department of Microbiology and Cell Science
University of Florida
Gainesville
USA

Sean W. York
Department of Microbiology and Cell Science
University of Florida
Gainesville
USA

Libing Zhang
Washington State University, Bioproducts
Sciences, and Engineering Laboratory
Department of Biological Systems Engineering
Richland
USA

Youcai Zhao
The State Key Laboratory of Pollution Control and Resource Reuse
Tongji University
Shanghai
PR China

Gunagyin Zhen
School of Ecological and Environmental Sciences
East China Normal University
Shanghai
PR China

Shaojuan Zheng
Shanghai Key Lab for Urban Ecological Processes and Eco-Restoration
School of Ecological and Environmental Sciences
East China Normal University
Shanghai
PR China

Zhongxiang Zhi
Shanghai Key Lab for Urban Ecological Processes and Eco-Restoration
School of Ecological and Environmental Sciences
East China Normal University
Shanghai
PR China

Foreword

Mahatma Gandhi is quoted as saying "The difference between what we do and what we are capable of doing is more than enough to solve the world's problems." This perhaps has never been more true than when we consider Society's greatest existential challenge of changing to a sustainable manner of living.

"Green Energy to Sustainability" is a volume that should be read by every person that cares about the future whether they are business people, policy makers, consumers, parents, activists, students, or mere scientists and engineers. This book is an essential resource for designing tomorrow to be better than today. With all of the valuable contributions by the leading figures on the topic of green energy, one would think that this is the most compelling aspect of the book, but I believe that it's not.

The *most* powerful benefit of this book, is how clearly it articulates what is possible *today*. What this book presents is not some unrealistic theory, distant vision, or science fiction. In the pages there are solutions that are well-demonstrated at various stages of developement and are available to be implemented at scale. It is the scale that will make the impact. It is the scale that is a matter of will. It is the scale that will determine how serious we are about taking the necessary actions with the urgency required in order to address the climate crisis and the related bio/geochemical cycle crises the planet is facing.

There is not a lack of scientific imagination that is a roadblock. It is not a lack of technical and engineering ingenuity that is an obstacle. It is not that we are waiting on new discoveries and new inventions. This book demonstrates that fact in page after page.

If the decision-makers, thought-leaders, capital investors, activists and influencers are set on mobilizing toward a sustainable future, this excellent volume has provided them with the information they need. In the war on unsustainability, the scientific and technical ammunition is there in abundance. But as has been said many times and in many ways, 'the best battle plans do not withstand the first encounter with the enemy' and so the dedicated metaphorical soldiers in the form of scientific innovators will be there to adapt and adjust to each unforeseen circumstance for every step forward until we achieve the world that our progeny deserve.

We will meet these challenges because we can and we will meet these challenges because we must.

Paul T. Anastas
New Haven, USA

Preface

Global warming (climate change) in the 2020's is marked by an inflexion point in biodiversity decline that is already translating into the premises of a mass extinction in numerous branches of the tree of life and notably in the extinction of a countless number of species of mammals, birds, insects, and fish as well as in dramatic global changes in plant and tree populations. All these consequences already require that ancestral agricultural practices and permanent vegetal cover adapt to changes in local climate given increased temperatures and more frequent severe heat waves, longer droughts, and very large scale fires. Ultimately the consequences of climate change will coalesce to significantly and durably impact the global food chain as illustrated by the impact that would result in a dramatic loss in the populations of crop-pollinating insects, for example, thus requiring adaptation to the new conditions. The last massive global extinction occurred 66 million years ago during the Cretaceous–Paleogene era. All these changes, as well as the threat of a significant rise of sea levels and the threat of an increasing desertification of whole regions that now constitute fertile lands, will greatly impact not only human quality of life but also the current status quo of economic activities. This may further translate into disequilibria and the exacerbation of the need to access vital resources as basic, but as precious, as water or hospitable lands. The accumulation of greenhouse gases in the atmosphere that drives the global warming experienced by the system Earth in the current geological era is absolutely unambiguously anthropogenic by nature. The current episode of climate change was initiated as early as the nineteenth century with the Industrial Revolution. It was fuelled by the exponential rise in fossil energy use to power the fast increasing demand for the cheap energy required to sustain economic growth at a heretofore unparalleled rapid rate and to bring mankind, in only a few generations, from a predominantly rural and artisanal era to a predominantly city-dwelling and industrial era accompanied by diminution of poverty and dramatic improvements in hygiene and healthcare; this was a period during which the quality of life of human populations in what constitutes currently developed countries, to take only this prism of analysis, dramatically rapidly improved as demonstrated by rises in life expectancies between 1850 and 2020 in various European countries. Changes in greenhouse gas concentrations that led to global climate change were long left unnoticed, thanks notably to the inertia of the Earth as a physical system. However, the wide-reaching impacts of climate change are now beyond question and global mass actions as well as a deep change in the global economic model are urgently needed to mitigate the worst consequences of climate change; this is hard to do because the inertia of the system Earth will result in any action taking a long time at the human scale to translate into practical positive changes observable by the naked eye. What is more, the thawing of the permafrost and the consequent release of its immense quantities of methane that have been sequestered for immemorial times represent the threat of a 'climatic event horizon' when anthropogenic climate change in the present geological era will enter a positive reinforcement loop and become totally out of human control, assuming it is not so already.

Climate change first and foremost constitutes a complex but burning political issue. It is a political issue because its mitigation requires fundamental changes at many levels, and notably at societal and industrial levels, and particularly at the level of the national energy mix. Profound changes are invariably painful. Profound changes impact vested interests that slow down needed changes. As a result, appropriate political agendas need to be set to minimize the social impacts of the necessary changes in lifestyles and fossil fuel consumption. This extreme challenge in efficiently dealing with climate change cannot be better exemplified than by the example of coal-fired power stations, which although they constitute a totally obsolete method of energy production remain in use in several jurisdictions given the need to recoup on their capital expenditures or to maintain mining industry gains, or to avoid the growing pain that inevitably accompanies a changing economy. Here, the political agenda is how to redirect the workforce and the various economic actors deeply dedicated to the coal-to-energy value chain and accompany them through the turmoil of change.

Climate change is a 'tragedy of the atmospheric commons'. The current macroeconomic model has evolved to maximally leverage comparative advantages, with the underlying assumption that the cost of atmospheric carbon dioxide disposal is nil. While it is convenient, because it avoids political complications and facilitates global trade by subsidizing the transportation industry in the form of free atmospheric commons, thus enabling to maximize the synergies of comparative advantages in the path to economic globalization, this model has for two centuries neglected the social cost of carbon. It is this assumption that the acknowledgement of the reality of climate change challenges. It is this assumption that made possible to economically transport at the antipodes commodity goods and particularly agricultural ones in spite it being possible to efficiently produce the very same goods locally; integrating a variable to capture the social cost of carbon could very well tip the balance in the opposite direction of the comparative advantages of today. It is this assumption that now needs to be urgently fixed. The current imbalance in global carbon budgets accumulated over the course of two centuries and embodied by the adverse consequences of global climate change calls for a global correction on a par with the geological imperative posed by the challenge of mounting atmospheric CO_2 concentrations. Permanent growth without recycling is not possible. The good news is that waste recycling has in a virtuous circle increasingly attracted attention in G20 countries, and the trend is bound to expand widely as the true costs of de novo production are increasingly integrated into the prices of goods. Here, CO_2 recycling constitutes another variable to integrate, with technological solutions being developed to achieve not only the mitigation of atmospheric CO_2, but also its valorization and recycling into valuable products either by chemical or by biotechnological means and by direct photosynthesis-mediated capture. Biochar obtained by the pyrolysis of biomass notably represents a very attractive method for sequestering or recycling $CO_{2,}$ with an estimated potential of fixing more than 10% of the current anthropogenic emissions of this greenhouse gas. Recycling alone, however, is unlikely to suffice, and additional changes will be necessary. This is where biotechnology has a major role to play to leverage the full potential of photosynthesis and biomass for sustainable energy and commodity chemicals production to enable the production of the goods necessary for modern life, including not only bioethanol or biodiesel but also sustainable chemical building blocks to complement conventional petrochemical processing. The coming of age of the technology of photovoltaic power backed up by the electricity generation potential of nuclear energy, which is still required for a few more decades but with the Chernobyl and Fukushima catastrophies serving as warnings, is enabling electricity to become a 'universal energy currency'. Notably, the potential of photovoltaic power would be decupled by deregulating and decentralizing energy production thereby enabling off-grid and on-grid electricity management. Here, on-grid power integrates into national total factor productivity, while off-grid power permits the harnessing of individual investments.

A lesson from history is that energy transitions take decades. This notwithstanding, a new transition is now urgently required to avoid the threat of experiencing a requiem for the natural world we have today and to which humans and their societies have adapted for thousands of years. The litmus test is whether the industrial world can evolve and achieve the synthesis of global trade with local development while maximizing recycling and minimizing environmental impact. This would represent the coming of age of a 'hypermodern-capitalism' characterized by accelerated changes to sense and respond to trends and competition with the objective of maximizing economic and social outcomes by optimizing the production function in terms of both local and global parameters in parallel. This might call for a revisit of the balance between regulation and deregulation on the one hand, and of corporate sustainability responsibility coupled with individual sustainability responsibility on the other.

The book *Green Energy to Sustainability: Strategies for Global Industries* is structured as four parts to revisit the green energy business and highlight critical potential and strategic directions for its future development.

In Part I, elements of the structure of the energy business are described that complement or provide critical updates to the chapters published in the first volume *Biomass to Biofuels: Strategies for Global Industries* that was published a decade ago. Whereas multiple parameters have evolved in this timeframe and notably the increasing rates of burning of fossil fuels and the increasing severity of the impacts of global warming, what has remained as a constant is the link between energy use and gross domestic product (GDP) growth. The opening chapter in this section thus examines in detail the relationship between economic growth and the global energy demand, with this link being further explored in terms of greenhouse gas emissions and the conditions that need to be met to achieve the appropriate energy transition towards a decarbonization of the economy. Strategic and operational considerations are also visited to propose an implementation roadmap as well as measures to monitor progress along the trajectory of the greening of the global economy via integrated assessment and decision-making in transition processes. The critical question of the national energy mix is subsequently explored through the lens of the Japanese economy post-Fukushima, analysing particularly the measurable impacts of two complementary strategies, namely the *National Energy Strategy* and the *National Energy and Environment Strategy for Technological Innovation Towards 2050*, implemented by the Japanese government to sustain economic growth in an environmentally-conscious approach. This analysis is particularly instructive as the policies at play aim at enhancing energy efficiency and clean energy deployment with the leading philosophy of meeting in parallel economic growth objectives and goals of greenhouse gases emission in a context of the shutdown of nuclear plants. The deployment and green energy technology in China and in emerging countries is discussed in detail with an emphasis on Africa, South East Asia, and South America, since although these jurisdictions are committed to advancing investments in green energy, they are confronted with very different driving forces and hurdles, perhaps best exemplified by the contrast between China's massive carbon footprint to power the phenomenal economic growth it experienced in the past two decades and sub-Saharan Africa's still emerging economic development. Through the lens of energy needs, energy mix, and energy policies, intrinsic roadmaps for greening of these economies are delineated. In a forward-looking chapter, the development of solar energy generation technologies and global production capacities are reviewed, progressing from the current state-of-the-art of photovoltaic cell science to global solar energy production capacity, with the journey highlighting advantages of grid integration versus point-of-production use. The production of sustainable aviation fuel is discussed in a separate chapter in which recent trends and the driving forces behind the growing biojet fuel market, as well as opportunities for implementing green chemistry principles and practices are highlighted. Transportation is further discussed through the lens of the environmental impact of pollution and pollution-prevention in the automotive industry and particularly in the automotive manufacturing industry, which has a complex upstream and downstream supply chain. The lessons learned from the manufacturing of combustion engine-powered vehicles are important lessons for a transition to fully electric vehicles and notably for very large-scale battery manufacturing. This section closes on an assessment of the global demand for biofuels and biotechnology-derived commodity chemicals.

Part II is centred on the technologies and practices of chemicals and transportation fuels from biomass. In the opening chapter of this section, the technologies enabling the production of chemicals from biomass are highlighted, particularly focusing on platform chemicals that serve as intermediaries in the synthesis of more complex compounds and notably for manufacturing a range of chemicals and materials including a plethora of plastics, the use of which permeate virtually every aspect of modern life. Genetic engineering techniques to develop performing microbial workhorses to complement petrochemistry by biotechnology-enabled C2-C6 chemistry are reviewed, with a focus on metabolic engineering techniques to develop superior microbial strains characterized by increased yields, increased ranges of substrates, and reduced sensitivities to growth inhibitors as well as reduced by-product formation. Scientific advances in the industrial-scale production of biofuels from seaweeds and microalgae are summarized in a technical chapter in which the main species that are being developed as biofuels producers are detailed. The major hurdles for reaching cost-effective biofuels generation from microalgae and seaweeds are subsequently discussed and forthcoming milestones in technology development to reach economic feasibility are proposed. Advanced fermentation technologies to achieve the conversion of biomass to ethanol by organisms other than yeasts are furthermore exemplified by the case of recombinant *Escherichia coli* though an analysis of the relative advantages of this facultative anaerobe as a novel production workhorse as well as hurdles to its industrial-scale implementation. The potential of strictly anaerobic bacteria is in turn discussed through revisiting biotechnological strains of the genus *Clostridium* and their process engineering for enabling cost-effective energy generation. In this chapter both the history of the use of these organisms for acetone/butanol/ethanol (ABE) production and the economic modelling of the ABE fermentation are integrated. The latest advances in the production of fuel ethanol production from lignocellulosic biomass using recombinant yeasts are presented in parallel, including the liquefaction and gasification of lignocellulosic materials, with the view of identifying technological opportunities to reduce the production costs of cellulosic ethanol. Advances in the catalysis of biomass and its saccharification are explored in a chapter where the latest progress made in biomass pre-treatment and hydrolysis technologies, including dilute sulfuric acid, sodium hydroxide, ammonia, and combined chemical as well as enzymatic treatments are discussed. The question whether generic pre-treatments exist that can be applied to most biomass sources, or whether tailor-designed pre-treatments are needed, is addressed particularly via a review of the various growth inhibitors that are generated during the pre-treatment and hydrolysis steps of biomass and potential technical solutions to mitigate their negative economic impacts. Starting from this review, a path to develop more robust cultures using microbial genetic techniques or adaptation techniques is suggested. To ensure that the net sustainable value of manufacturing processes for renewable chemicals is positive, life cycle assessments of biofuels and green commodity chemicals are proposed, focusing on biofuels. Recommendations to optimize the life cycle of such sustainable chemicals appear in the watermark of the chapter, including notably water usage, fossil fuel usage, and pesticide usage as appropriate life cycle assessments requiring an ongoing understanding of the supply chain, systems boundaries, displacement effects, direct and indirect effects, as well as the broader impacts that these effects have on wider sustainability issues.

Part III constitutes a brief overview of sustainable technologies to produce useful gases and the market potential of these gases. The biotechnological production of fuel hydrogen and its market deployment is first discussed, given that hydrogen represents an important option fuel for a clean automotive industry. Moreover, the deployment of biogas production technologies and production capacity in emerging countries is described. Remarkably, in emerging countries these technologies can essentially be deployed in greenfield projects without having to retrofit existing infrastructures albeit this sometimes implies conditions of limited general total factor productivity. Nonetheless, proactive efforts have been deployed and significant manufacturing capacity is increasingly being validated in many of these countries. Details of hydrogen production by algae are presented to identify the areas of greatest innovation need. Here, innovating cost-effective photoreactor designs appear to be the most critical success factor. Methane represents another gas with tremendous economic potential and notably as a renewable fuel. Approaches to enhance methane production from microalgae biomass are thus also reviewed, including pre-treatment methods and anaerobic co-digestion processes.

In Part IV, perspectives are presented to integrate what has been learnt from the preceding chapters and to highlight the potential and challenges of a sustainable economy. A first critical consideration is how to implement the biorefinery vision to displace and to complement where conventional petrochemistry makes sense. Integrated biorefineries for the production of bioethanol, biodiesel, and other commodity chemicals are revisited in terms of their key technologies and processes as well as the portfolio of products that make them economically competitive. Primary raw materials of the biorefinery of the future are lignocellulosic crops. These materials are reviewed in light of an economic lens of analysis and how they, or their underlying agricultural practices, could be enhanced in order to maximize their economic efficiency for bioenergy generation. Exploring downstream the bioenergy value chain, industrial waste valorization applied to the case of biofuels is also revisited, using food waste recycling as a key example of the potential of recycling technologies applied to high volumes. In the specific case of agricultural waste, cost-effective pre-treatment and hydrolysis methods remain particularly critical to convert these into sugar monomers. The food manufacturing sector is used here to illustrate the various positive impacts of the deployment of pollution prevention, sustainable energy generation, and other sustainable development strategies. However, the implementation of a sustainable economy enabled by green energy requires appropriate vectors for financial investment. An issue of large volume biomass-to-energy or biomass-to-chemicals projects is that they typically require high initial capital expenditures, have relatively long periods of break-even returns, and bear various limitation such as the seasonality of primary raw materials or supply risks; in addition, the competition with established petrochemical processes remains fierce. Financing strategies for sustainable energy production plants could include green banking and sustainable finance methods that are characterized by their factoring in environment risk and social burdens. Corporate social responsibility and corporate sustainability responsibility are furthermore discussed as powerful forces of change, the influence of which goes beyond the impact of public advocacy in changing business models or the increasing deployment of cash-generating waste recycling programmes, in that they already permeate management practices of today in leading companies given their measurable impacts on the bottom line of businesses. The closing chapter explores the transition of the industrial world that was inherited from the Industrial Revolution and is thus characterized by compounding carbon budgets to a new economy where the production function and economic comparative advantages on the one hand include the social cost of carbon in the production and transport of goods, and on the other deploys radical innovation to optimize operations via novel practices ranging from artificial intelligence to run smart cities, to electric cars with electricity as a universal high quality energy currency, generated and available on-grid as a component of total factor productivity as well as off-grid as a marker of both individual sustainability responsibility and of the re-localizing of the production function. This renewed impact of local production in the context of globalized trade could pave the way to a hyper-modern capitalism in which the *magic hand of the market* senses, responds, and swiftly adapts to maximize the production function while maintaining locally high quality of life and the global climate.

Alain A. Vertés
Basel, Switzerland

Nasib Qureshi
Peoria, IL, USA

Hans P. Blaschek
Urbana-Champaign, IL, USA

Hideaki Yukawa
Tokyo, Japan

Part I

Structure of the Energy Business

1

Economic Growth and the Global Energy Demand

Jürgen Scheffran[1], *Miriam Felkers*[1] *and Rebecca Froese*[1,2]

[1] *Research Group Climate Change and Security, Institute of Geography, Center for Earth System Research and Sustainability, University of Hamburg, 20144 Hamburg, Germany*
[2] *Research Group Landuse Conflicts, Institute of Environmental Sciences and Peace Academy Rhineland Palatinate, University of Koblenz-Landau, 76829 Landau, Germany*

CHAPTER MENU

1.1 Historical Context and Relationship Between Energy and Development

Energy is the driving force of all natural processes and a condition for the development of life on earth as well as for human society and its economic growth. From its origin, the character of energy is expressed by conceptions rooted in both the natural-physical world and in the human-social world. According to the Greek term, *energeia* (activity, reality) is an 'effecting force' that brings about the transition from possibility to reality. Aristotle saw *energeia* in connection with activity, action, and power. Later, Wilhelm von Humboldt established the relationship between energy and conscious human activity, and Leibniz spoke of the 'living force' (*vis viva*) which is preserved (Mittelstraß 1984).

In the course of the physical confinement of the energy concept and the distinction from the concept of force, energy has been regarded as an observable physical quantity to measure the motion of material bodies and as an acting force to implement the transition from possibility to reality. The theorem on the conservation and conversion of energy formulated in the second half of the nineteenth century proved that it is possible to convert different forms of energy (mechanical, elastic,

Green Energy to Sustainability: Strategies for Global Industries, First Edition.
Edited by Alain A. Vertès, Nasib Qureshi, Hans P. Blaschek and Hideaki Yukawa.

magnetic, electrical, chemical, heat, gravitational, and nuclear energy). While the total quantity of energy is not lost, but retained in each sequence of conversions, the quality and usability of energy declines with each stage. Following basic laws of thermodynamics, energy flows from a state of higher order to one of lower order where its entropy is increasing. Against this gradient, living organisms can maintain a dynamic equilibrium of flows by a continuous energy supply from the outside as provided by sunlight and food, or by control over other energy sources. Without this influx of energy, the order of living organisms disintegrates, and ultimately, they die.

The physical definition of energy as the ability to perform work offers a direct link into the human world. The potential work contained in natural processes is converted into actual work, affecting human values and shaping human interactions. Work is often associated with planned human activity directed towards goals which can be the satisfaction of needs or the maximization of profit and income. As an essential component of work, energy is becoming a factor of economic production, a measure of society's performance, a prerequisite for prosperity and satisfaction of needs, but also a condition for growth, power, and violence. The term 'power' has a wide range of meanings, from physical power to political power, also as an important instrument for empowering people (Scheffran 2002).

For human beings, energy appears in two forms: energy as a driver of the natural metabolism; and energy in the form of technical mechanisms that convert the energy available in nature to make it usable for human needs. Energy production and consumption is ambivalent: on the one hand, it is a prerequisite for human prosperity, on the other it affects the social and natural environment, with sometimes considerable risks for humans and nature. In the course of history, humans have succeeded in accessing ever greater sources of energy for their own usage to serve their own interests and develop societal structures. In long historical perspectives technological waves in agriculture and industrialization have been instrumental in shaping the modern developed world. Four historical periods can be distinguished regarding the forms of energy use (Sieferle 1981):

1. In the *solar energy system* of hunters and gatherers, people derive as much energy from their ecological niches as these can deliver sustainably, keeping the material cycles closed.
2. The sun was also the only source of energy in the *modelled energy system of agrarian societies* after the so-called 'Neolithic revolution' 10 000 years ago, when the energy flow was used in a much more focused way. This applied to the fields of agriculture, forestry, cattle farming and fishing which largely relied on the muscle power of humans and animals as well as the use of plant and animal biomass and, to a lesser degree, on water and wind power. In Europe, in the eighteenth and nineteenth centuries, the limits of the energetic capacity were reached by a growing human population and its rising demands.
3. *The fossil energy system of the industrial society:* Compared to the traditional solar energy system, the exploitation of fossil sources of energy (coal, petroleum, natural gas) accumulated during millions of years of Earth history, allowed humanity to multiply its energy consumption (EC) by a factor of more than 16, from 26 EJ in 1850 to 432 EJ in 2000 (Smil 2010; EIA 2016). This unique treasure of nature enabled the rapid growth of the industrial revolution but may be irretrievably exhausted in a few decades to centuries. The wasteful use of energy mobilized vast amounts of resources, ploughing through the surface of the Earth. It accumulated products of all kinds, released novel chemical substances, changed and destroyed ecosystems, exterminated a number of species, and polluted the world's atmosphere, oceans, and soils (Sieferle 1981).
4. *Post-industrial sustainable development in the Anthropocene*: Ultimately humanity affected the whole Earth system on a scale justifying a new geological epoch, the Anthropocene (Crutzen and Stoermer 2000). Humans are exceeding their footprint and population grows at a speed and intensity that is not sustainable and tends to collide with the planetary boundaries and carrying

capacities of nature (Rockstrom et al. 2009). The restraints on conventional energy supplies and the limited absorption capacity of the earth's atmosphere for emissions from the combustion of fossil energy sources foster the need for a transformation of the industrial production and lifestyle. The debate on the 'limits to growth' in the 1970s (Meadows et al. 1972), and on sustainable development in the 1980s and 1990s (WCED 1987) has triggered a wave of Earth politics (von Weizsäcker 1994) to protect the planet's integrity, notably the Rio Summit, the Sustainable Development Goals (SDGs), the quest for a sustainable energy transition and a 'Great Transformation' towards a decarbonization of the energy system.

The current situation in the energy sector is characterized by a growing energy demand combined with declining reserves, North–South gaps in energy supply, environmental risks and climate change, and geopolitical conflict potential due the dependency on fossil and nuclear energy sources. Given the steady increase in energy consumption due to population and consumption growth, the limits of the availability of cheap non-renewable primary energy carriers such as oil, natural gas and uranium are foreseeable, which are expected to last for a couple of decades at the expected usage rates. Although large coal reserves are still available, their risks and costs could undermine their competitiveness. During the twenty-first century, high demands for energy will encounter various problems of fossil and nuclear energy supplies, widening the gap between human needs and the supply of affordable and responsible energy.

Decoupling is understood as creating more economic wealth per resource unit, thus diminishing the input-to-output ratio. There is some evidence that global resource consumption has risen more slowly than gross domestic product (GDP), as world GDP grew by a factor of 23 in the twentieth century and resource use rose by a factor of eight (UNEP 2011). It is subject to debate whether it is possible to decouple economic growth from energy consumption. Energy intensity (energy used per unit of GDP) has declined in all developed and large developing countries, mainly due to technology, changes in economic structure, the mix of energy sources, and changes in the participation of inputs such as capital and labour used (IPCC 2014a,b). Newman (2017) suggests that the 'decoupling of fossil fuels from economic growth has not been imaginable for most of the industrial era but is now underway'.

Moreau and Vuille (2018) found that almost all members of the European Union have decoupled since 2005 as measured by a 'steady decline in energy intensity, the ratio of final energy consumption and gross domestic product. Economic growth, energy efficiency and structural changes have all contributed to changes in energy intensity of economic activities at the national level'. The authors compare the energy intensities with and without embodied energy in trade (which has reached 81% of final energy consumption in economic activities) and show that 'decoupling is more virtual than actual' (Moreau and Vuille 2018). Sorrell and Ockwell (2010) summarize the controversy: 'Orthodox economics assumes that the rebound effects from energy efficiency improvements are small and that decoupling is both feasible and relatively cheap. In contrast, ecological economics suggests that rebound effects can be large and that decoupling is both difficult and expensive. At present, there is insufficient evidence to demonstrate which perspective is correct'. For the future, 'resource productivity increases as high as 5–10-fold are already available' (von Weizsäcker et al. 2014).

While some reduction of coupling between economic growth and energy consumption is underway, largely due to efficiency gains in industrialized countries, it is outstripped by population growth and the attempt of the Global South to catch up in development. In the Western industrialized countries, the standard of living has so far been associated with comparatively higher energy efficiency and energy consumption per capita than in the Global South that has lower economic output per energy unit (IPCC 2014a). In some countries, available energy flows are below the minimum physical requirements of human existence, and insufficient to meet basic human needs. Developing countries pursuing industrialization are striving for an expansive energy policy, which often has little regard for ecological requirements. In order to leave this critical pathway, developed countries will need to establish technology and knowledge transfer to developing and emerging countries to enable a leapfrogging of these economies into low-carbon energy systems. Sustainable development, based on renewable energy sources, would not only be well aligned with the SDGs, but also benefit the developing and emerging economies as they are still not facing retrofitting costs from older carbon-based energy infrastructures (Scheffran and Froese 2016).

1.2 Conceptual Framework for Pathways of Energy Use

To analyse the role of energy in economic production and consumption, as well as possible risks related to environmental change, an integrated framework of decision-making and management of energy pathways is presented, driven by investments into energy pathways, inducing intended and unintended consequences evaluated in terms of benefits, costs and risks of economic growth (Figure 1.1). Key variables are the invested costs C into the production factors of energy generation E, capital K and labour (population) P to produce economic wealth W and utility U through distribution and market processes, and technical-economic parameters, such as the cost per energy unit, energy productivity and efficiency of conversion from primary energy to end use. A fraction of produced income is consumed and the remainder is re-invested. Energy pathways also generate environmental impacts (including climate change), indicated by emissions G and the associated risks R. Strategies seek to find investment and allocation pathways that achieve the 'best' net value gain of energy use $V(E) = U(E) - C(E) - R(E)$, in common units for these functions.

In the following we will consider key relationships and data to highlight the role of energy in this framework. We focus in particular on the impact chain in Figure 1.1 that determines greenhouse gas (GHG) emissions G, as a function of population P, economic wealth W (here measured by GDP) and energy use E. The ratio between these factors has been represented in the Kaya identity (Kaya 1990):

$$G = (G/E)\,(E/W)\,(W/P)\,P \tag{1.1}$$

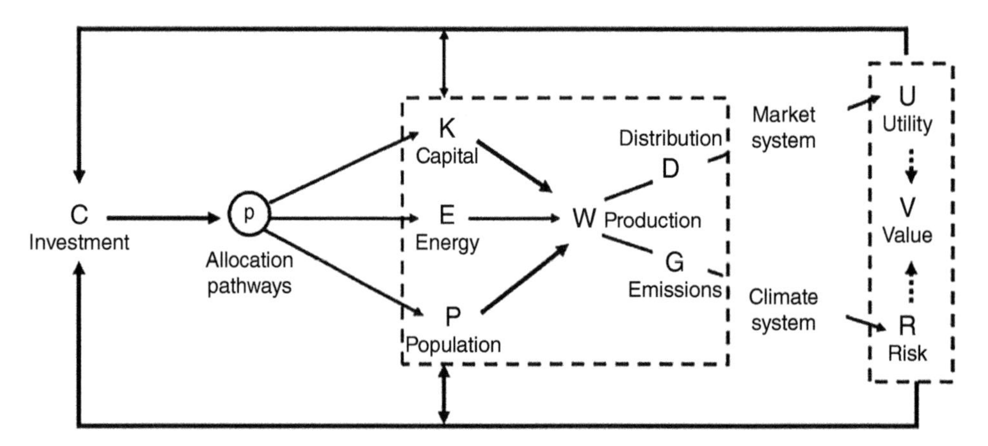

Figure 1.1 Conceptual framework for integrated energy assessment. Source: The authors, adapted and modified from Scheffran 2008a.

Besides population P, the Kaya factors are labour productivity W/P (GDP per capita), energy intensity E/W (energy use per GDP) and emission intensity G/E (emissions per energy unit). The Kaya identity goes back to the IPAT identity[1] by Ehrlich and Holdren (1971) and establishes a common framework for decomposing overall changes in GHG emissions into the socio-economic drivers. This allows to analyse the prevailing historical circumstances as well as predictions of future developments influencing the four factors, population, GDP, energy intensity and carbon intensity, each for a given time period (e.g. one year) and for particular actors who can be firms, consumers or nation states. For continuous time, these variables turn into flow variables. The following analysis reviews the development of these factors over time.

1.3 World Population Trends and Prospects

One key driver of economic growth, energy use and GHG emissions, is population growth. According to the UN 2015 Population Revision, the world population reached 7.35 billion in 2015 (UN 2015). Figure 1.2 illustrates the global population development from 1750 to 2015 and future prospects up to 2100 (Kremer 1993; UN 2015). Only about 200 years ago, around 1815, the global population approached the 1 billion mark for the first time in history. Since then, population has increased more than sevenfold. Along with the agricultural and industrial revolution and the introduction of vaccination and other medical and sanitation improvements in the second half of the eighteenth century, child life expectancy as well as general living conditions and healthcare increased strongly, leading to a constant increase in population (Cleland 2013). While it took more than an additional century to reach 2 billion in the early 1930s, population size is expected to

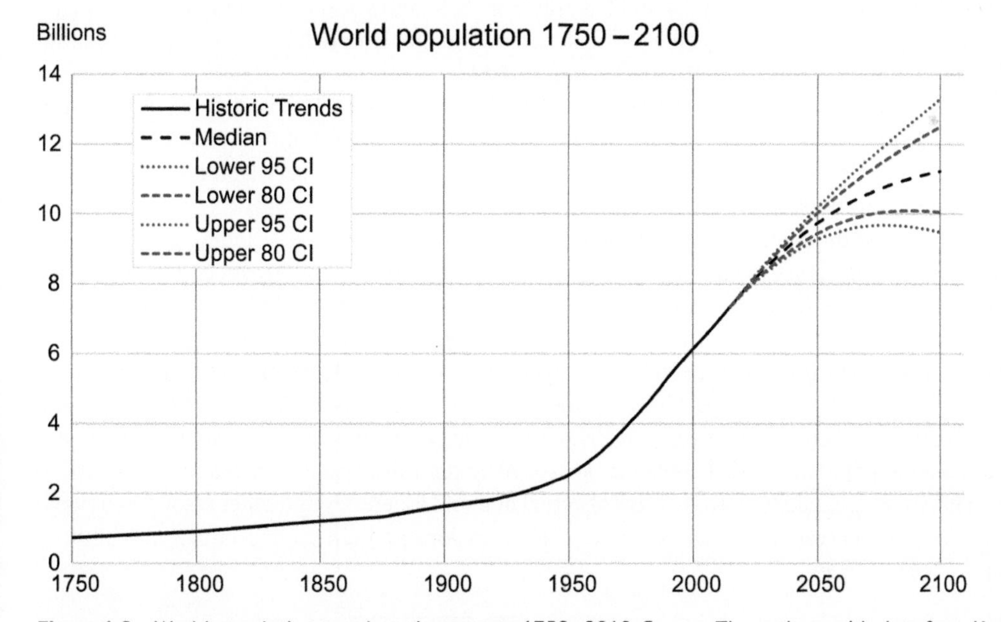

Figure 1.2 World population trends and prospects 1750–2010. Source: The authors with data from Kremer (1993) (data before 1950) and UN (2015) (after 1950).

1 The IPAT identity decomposes an impact (I), such as total GHG emissions, into population (P), affluence (A), e.g. income per capita, and technology (T), e.g. emission intensity of production and consumption.

exceed 8.5 billion in 2030, of which the last billion is expected to be added in just a little more than one decade (2017–2030). This corresponds to an increase of more than 6 billion within one century. The highest global population growth has been observed in the 1960s, when rates peaked at 2% per year. These high growth rates were due to decreasing death rates, particularly among infants and children triggered by the mass application of modern preventive health measures in many developing countries in Africa, Latin America and Asia as well as better transport logistics to relieve local food shortages (Cleland 2013).

While the global population grew rather slowly up until the 1940s, an exponential increase in population is observed when the global population has increased by 87% from 3.7 to 6.9 billion between 1970 and 2010 (UN 2015; Kremer 1993). Since then, growth rates have been decreasing due to overall declining fertility rates. Recent rates were at 1.24% per year in 2005 and 1.18% per year in 2015, adding about 83 million people annually (UN 2015). Due to this important demographic transition, societies have been moving away from high fertility levels and mortality rates, to a period of declining mortality rates.

Even though growth rates are declining, predictions indicate a further increase in the worldwide population, reaching 9.7 billion by 2050. During the second half of the present century, population growth rates are expected to moderately decrease while the total population size is still expected to increase, likely reaching 11.2 billion by the end of the twenty-first century (Figure 1.2, black-dotted curve). The presented growth numbers correspond to medium projections that assume a decline of fertility rates in countries of present high birth rates as well as slight increases of fertility rates in countries with birth rates smaller than two children per woman (UN 2015). However, as projections always imply a certain degree of uncertainty, Figure 1.2 also illustrates the corresponding 80% and 95% confidence intervals (CI) to depict the degree of uncertainty of the presented data. With a 95% degree of confidence it can be said that the global population will generally increase compared to present values and reach between 8.4 and 8.6 billion in 2030 and 9.5 and 13.3 billion in 2100. The scenarios, introduced later in this chapter, will assume the median predictions with a population development from 7.1 billion in 2013 to 9 billion in 2040.

Currently, 60% of the global population lives in Asia, 16% in Africa, 10% in Europe, 9% in Latin America and the Caribbean, and the remaining 5% in North America and Oceania. With more than 1 billion people respectively (UN 2015), China (1.4 billion) and India (1.3 billion) are the two most populous countries, together comprising more than one fourth of the world population. These numbers underline the significance of the population development in the developing and emerging countries for the trajectory of global population development. In the end, it will depend on these trends, while trends in industrial countries will play a rather minor role due to their smaller total population sizes and small changes in growth rates. In general, future population growth will be highly dependent on future fertility. In 2015, already 83 countries had growth rates below replacement fertility while global life expectancy is projected to rise from 70 years in 2010–2015 to 77 years in 2045–2050 and eventually to 83 years in 2095–2100. The predictions of discontinued growth and a further ageing population are a challenge for social systems and also drives planet Earth to the limits of its carrying capacity, both in terms of actual demand for energy and other resources as well as regarding human residuals such as waste, GHG emissions and other environmental pollution and their consequences.

1.4 Gross Domestic Product (GDP) and Economic Growth

Besides increasing population, global economic growth is a second major driver of energy consumption and global emissions. Figure 1.3 shows the development of global GDP based on

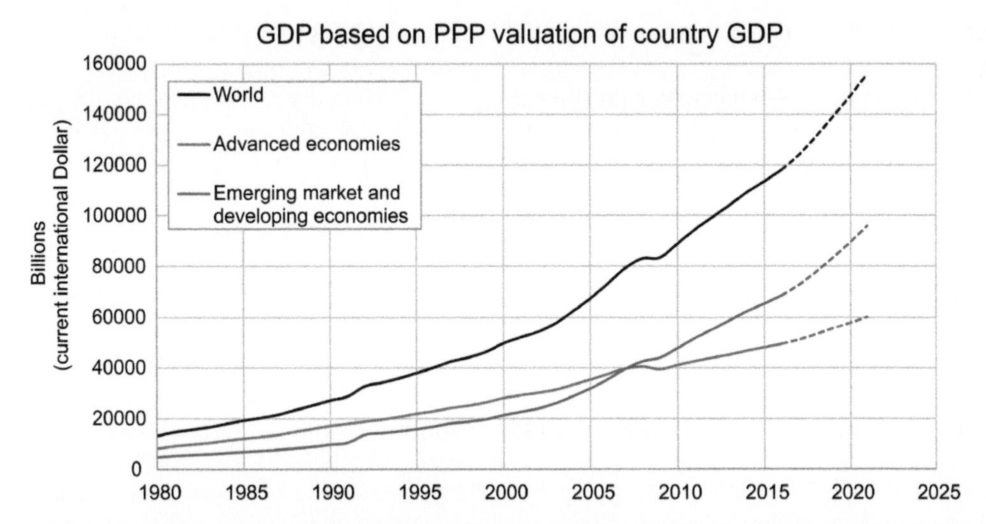

Figure 1.3 Gross Domestic Product (GDP) based on Purchasing Power Parity (PPP) valuation of country GDP for the World (black), Advanced Economies (green) and Emerging markets and developing economies (orange) from 1980 until 2016 and correspondent forecast until 2021. Source: The authors with data from IMF 2016.

purchasing power parity (PPP) as well as a breakdown into the two categories of advanced economies (green) and emerging markets and developing economies (orange) (IMF 2016). The graph indicates a general exponential increase in global GDP including some minor fluctuations, with the recession of 2008/2009 being the strongest and most prominent one. While North America and Europe shaped the global economic development since the industrial revolution until about 2007, this leading role has been overtaken by the emerging and developing economies as of 2007.

Additionally, these economies experienced a rather weak decrease in growth during the economic recession of 2008/2009, while the advanced economies were strongly impacted, with some economies even experiencing negative growth rates as it can be seen in Figure 1.4.

This figure also indicates that GDP growth experiences quite regular alterations over scales of several years. However, while the difference in growth between the advanced and emerging and developing economies did not exceed 2% up to the year 2000, this difference strongly increased afterwards. Still following similar trends and alterations, the growth rate in emerging and developing economies as of 2000 has been 4–8% higher than the growth rate in advanced economies (Figure 1.4) (IEA 2015b). Projections indicate a continuation of these global increasing trends and the general smaller growth rates in advanced economies and much higher growth rates in emerging and developing economies (Table 1.1).

Global trends in per capita GDP and GHG emissions vary dramatically by region (IPCC 2014a). While the global final energy consumption share is dominated by 78.3% of fossil fuels, renewables have only a share of about 19%, followed by nuclear power with about 2.5% on a global average (REN21 2016) (the role of renewables is further discussed in later sections). Economic growth was strongest in Asia with an average of 5% per annum over the 1970–2010 period, while the Middle East and Africa, and the Economies in Transition (Eastern Europe and part of former Soviet Union) saw setbacks in growth related to the changing price of oil and the collapse of the centrally planned economies, respectively (IPCC 2014a).

Looking at larger decadal trends and neglecting smaller fluctuations, real GDP grew by an average of 4.9% per year in emerging and developing economies, compared to 2.1% per year in advanced economies in 1990–2013 (see Table 1.1). Predictions indicate that world GDP is expected to rise from

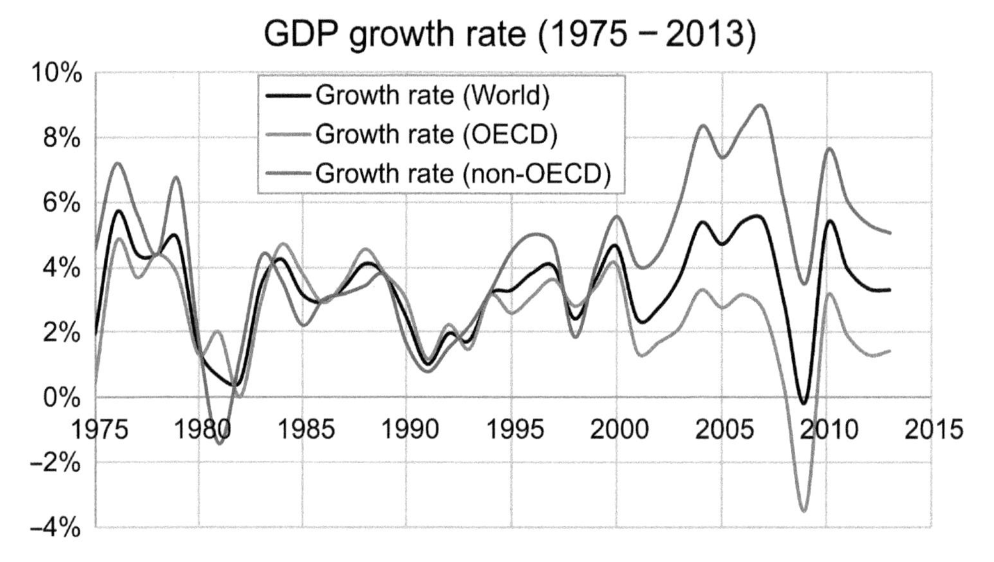

Figure 1.4 Growth rates in per cent of Gross Domestic Product from 1975 to 2013. Source: The authors with data from (IEA 2015a).

Table 1.1 Real GDP growth by region (compound average annual growth rates for the given periods) calculated based on GDP expressed in year-2014 dollars in PPP terms.

	1990–2013 (%)	2013–2020 (%)	2020–2030 (%)	2030–2040 (%)	2013–2040 (%)
World	**3.4**	**3.7**	**3.8**	**3.1**	**3.5**
Advanced economies	**2.1**	**2.2**	**1.9**	**1.7**	**1.9**
United States	2.5	2.5	2.0	2.0	2.1
Europe	1.8	1.9	1.8	1.6	1.7
Emerging and developing economies	**4.9**	**4.9**	**5.0**	**3.8**	**4.5**
Russia	0.7	0.2	3.1	2.7	2.2
China	9.9	6.4	5.3	3.1	4.8
India	6.5	7.5	7.0	5.3	6.5
Brazil	3.1	1.4	3.8	3.3	3.0

Source: Adapted from IEA 2015b.

3.4% in 1990–2013 up to 3.8% in 2020–2030 and to drop to 3.1% in 2030–2040. This trend is largely influenced by the emerging and developing economies that are expected to keep the current growth rate of 4.9% until 2020 and slightly rise to 5.0% in 2020–2030 before their growth rate is predicted to drop to 3.8% in 2030–2040 (IEA 2015b). Meanwhile, the growth rate of the advanced economies is expected to slightly increase from 2.1% in 1990–2013 to 2.2% in 2013–2020 before it is expected to continuously decrease to 1.9% in 2020–2030 and 1.7% in 2030–2040 (Figure 1.4). These long-term rates are based on estimations from the Organisation for Economic Co-operation and Development (OECD), the International Monetary Fund (IMF) and World Bank and depend on demographic and productivity trends, macroeconomic conditions and the pace of technological change. A principal tool to produce energy projections is the World Energy Model (WEM) which is a large-scale and data-intensive simulation tool that covers the entire global energy system and replicates how energy markets work. It consists of three main modules covering final energy consumption (including industry, transport, buildings, agriculture, and non-energy use), fossil fuel and bioenergy supply, and energy transformation (including power and heat generation, oil refining and other transformation). The primary model outputs are energy demand and supply by fuel, investment needs and CO_2 emissions for 25 distinct regions, 13 of which are individual countries (IEA 2015b).

1.5 Global Energy Development

Policy and technology together have been driving forces in transforming the energy sector in recent decades. They continue to play a central role in meeting the world's growing energy demands while addressing concerns for energy security, costs, and energy-related environmental impacts (Mattick et al. 2010). Energy conversion in the machines and information processors of the capital stock is a key driver for the growth of modern economies (Kümmel and Lindenberger 2014). While recent technology developments, markets, and energy-related events have demonstrated their capacity to influence global energy systems, determined action is needed to actively transform energy supply and end use (IEA 2014a,b).

To understand patterns in global energy development, it is important to look at historical structural changes as they have been described in the previous sections. The primary drivers of global energy consumption are population and economic growth as well as other accompanying factors (EIA 2016). The data provided in this chapter are based on the International Energy Outlook (IEO) 2016 that assesses the global energy markets and provides an energy outlook through to 2040. The reference case projection is a business-as-usual (BAU) trend estimated by the US Energy Information Administration (EIA) and based on existing government laws and regulations. In general, the IEO 2016 reflects the effects of current policies within the elaborated projections and assumes that current laws and regulations are maintained. These baselines are used for analysing international energy markets (see Section 1.8). Nevertheless, energy market projections are subject to great uncertainty because of future technology, demography, and resource developments that cannot be well foreseen. A good example for this is the rapid decrease in costs for solar power. While the average cost of solar cells in 1977 was around $76 per Watt, it went down to only $0.26 per Watt in 2016 (Bloomberg 2016). A more detailed description of the assumptions, methods, and findings can be found in the IEO 2016 (EIA 2016).

The IEO 2016 reference case forecasts a rising level in global energy demand over the 28-year period from 2012 to 2040. Figure 1.5 depicts the trends and prospects of worldwide consumption of marketed energy from 1990 to 2040 stated in exajoule (EJ). The projections differentiate between OECD members and non-members in order to take into account different economic and population developments. Total global energy consumption expands from 579 EJ in 2012 to 860 EJ in 2040 – a rise of 48% (see Figure 1.5 and Table 1.2). In the last two decades, a substantial share of the global increase in energy consumption occurs among developing countries along with population and economic growth (EIA 2016). According to the EIA, non-OECD energy use surpassed the OECD total in 2007 for the first time (EIA 2016). These data correspond well to the GDP development shown in Figure 1.3, which also indicates that developing and emerging economies have taken over the leading role in economic development.

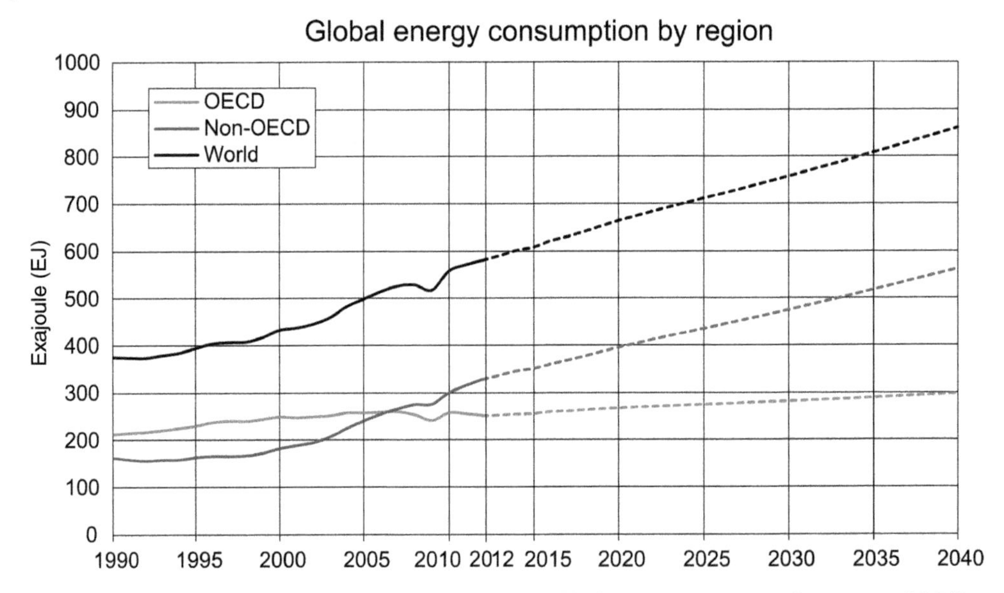

Figure 1.5 Global energy consumption from 1990 until 2012 and correspondent forecast until 2040. Source: The authors with data from (EIA 2016).

Table 1.2 World total primary energy consumption by region for selected past and projected future years, as well as average annual percent change for the reference case, 2011–2040, numbers in EJ.

	2011	2012	2020	2030	2040	Average annual percent change, 2012–2040 (%)
World	**540.5**	**549.3**	**628.9**	**717.7**	**815.0**	**1.4**
OECD	**242.0**	**228.4**	**253.9**	**267.2**	**282.1**	**0.6**
United States	96.8	94.4	100.8	102.9	105.7	0.4
Europe	82.0	81.4	84.9	90.3	95.5	0.6
Non-OECD	**298.6**	**310.8**	**375.0**	**450.5**	**532.8**	**1.9**
Russia	30.9	32.1	33.2	35.1	34.5	0.3
China	109.4	115.0	147.3	170.4	190.1	1.8
India	25.0	26.2	32.8	44.9	62.3	3.2
Brazil	14.8	15.2	16.3	20.0	24.3	1.7

Notes: Energy totals include net imports of coal coke and electricity generated from biomass in the United States. Totals may not equal the sum of components due to independent rounding. The electricity portion of the national fuel consumption values consists of generation for domestic use plus an adjustment for electricity trade based on a fuel's share of total generation in the exporting country.
Source: The authors based on data from EIA 2016.

Table 1.2 presents a detailed listing on energy consumption in selected OECD and non-OECD countries and country groupings. In the IEO 2016 reference case, energy consumption of OECD members is predicted to slightly increase at an average rate of 0.6% per year from 2012 to 2040 (see Table 1.2). Energy consumption of non-OECD members, especially in China, India, and Brazil, is predicted to increase significantly at an average rate of 1.9% per year within the same period.

Mattick et al. (2010) state that industrialized economies are becoming continuously less energy intensive. While energy consumption per capita in today's OECD member countries increased rapidly with the beginning of the Industrial Revolution, it stabilized over time through a rise in productivity (Mattick et al. 2010). In terms of economic added value, energy consumption grows as countries develop and enhance their living standards (EIA 2016).

Furthermore, the economy of developed countries is shifting increasingly toward services while goods and resources are imported from other, mostly emerging and developing countries. While doing so, developing countries migrate away from agriculture and toward manufacturing accompanied by an increased demand for appliances and transportation equipment as well as the growing production of goods. Eventually, this leads to a rapid increase in energy demand in developing countries while at the same time decreasing 'service economies' that rely on imports of industrial goods (EIA 2016).

Rühl et al. (2012) offer a closer look at changes in the composition of fuel sources over time. Fuels substituted each other (as in coal replacing wood, oil replacing coal, and gas replacing oil), but the most striking features are the extent to which the diversification of fuel supplies increases and its correspondence with technological diversification (Smil 2010). The role of coal increased significantly during the Industrial Revolution (Wrigley 2013; Mattick et al. 2010) with technological innovations such as the steam engine, railways systems and the arising electricity grid leading to a high demand for coal. Following the Industrial Revolution, the continued expansion of fossil fuels not only affected manufacturing, but also agriculture, transportation, and services (Mattick et al. 2010). In the first half of the twentieth century, other energy sources gained in importance (e.g. crude oil) although coal remained the main source of energy (Rühl et al. 2012).

Overall, the twentieth century is characterized by a high diversification of energy sources, which typifies a time of intense technological evolution. In the middle of the century, crude oil became the dominating source of commercial energy growth, but this dogma was challenged by the oil price shocks of the 1970s (Hartshorn 1993). While the share of coal in the energy mix remained substantial, nuclear power and natural gas replaced crude oil in electricity generation, while oil became more important in transport services (Rühl et al. 2012). Moreover, concerns about energy security, the cumulative effects of fossil fuel emissions on the environment, and sustained high world oil prices in the long term supported an increase of non-fossil renewable energy sources (EIA 2016). However, energy consumption has been increasing among all energy sources and is expected to increase further through 2040 (EIA 2016) as is shown in Figure 1.6. At an average rate of 2.6% per year, renewable energy is the world's fastest-growing energy source according to the IEO 2016; an increase that can be ascribed to various governmental policies and incentives (EIA 2016). Nuclear energy use increases by 2.3% per year, and natural gas use (the least carbon-intensive fossil fuel) increases by 1.9% per year (see Figure 1.6).

According to the IEO 2016 reference case, fossil fuels continue to play the main role of the world's energy demand (see Figure 1.6). In 2040, liquid fuels, natural gas, and coal will account for 78% of total global energy consumption (EIA 2016). Moreover, coal is expected to be the world's slowest growing energy source at an average rate of 0.6% per year. Total coal consumption is expected to account for 178 EJ in 2020 and 190 EJ in 2040. The coal use in China is predicted to reach 93 EJ in 2025 before declining to 88 EJ in 2040 (EIA 2016).

The share of renewables in the twentieth century results mainly from combustible biomass and wastes (9.5% in 1990). According to the Intergovernmental Panel on Climate Change (IPCC) with data from the International Energy Agency (IEA), combustible biomass and waste contributed approximately 10% of primary energy consumption in the period from 1971 to 2004 (IPCC 2007). More than 80% of it was used for traditional fuels for cooking and heating in developing countries

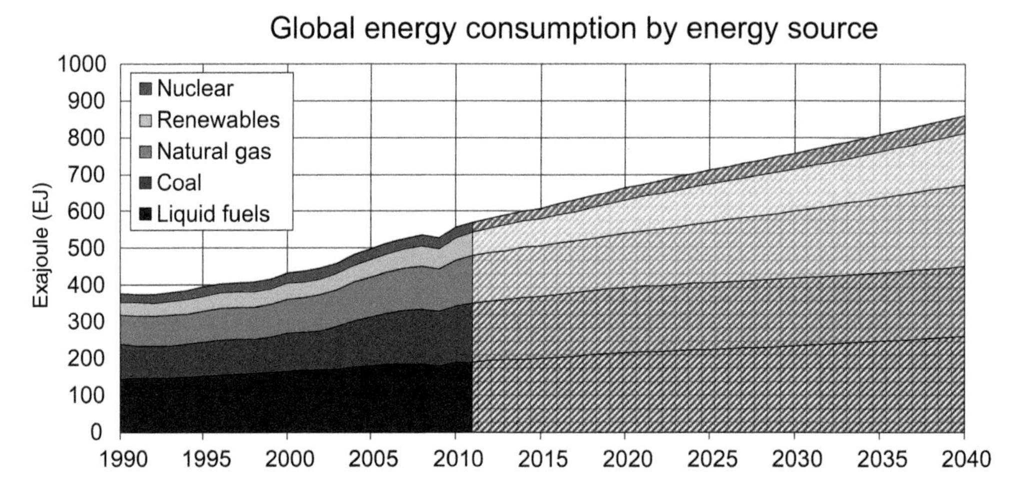

Figure 1.6 Global energy consumption by energy source from 1990 until 2012 and correspondent forecast (BAU) until 2040. Source: The authors with data from EIA 2016.

(IPCC 2007). In the IEO 2016 reference case, the share of renewables continues to increase significantly, led by renewable electricity generation (including hydropower). Hydropower and wind are the two largest contributors in renewable generation and are expected to account for more than 60% of the total increment from 2012 to 2040 (EIA 2016).

The world energy use is projected to rise from about 9 billion tonnes (Btoe) of oil equivalent today to 15 to 21 Btoe per year by the time the world's population has risen from 6 billion to around 12 billion people in the twenty-second century. Scenarios for global energy consumption vary with alternative assumptions of population growth and standards of living. Large amounts of energy will be required to fuel economic growth, increase standards of living, and to lift developing nations out of poverty (Brown et al. 2011). Despite the debate on energetic limits to economic growth, even high growth in per capita income over the next 20 years is not necessarily constrained by resource availability (Rühl et al. 2012). While energy scarcity and poverty may impose restraints on human growth, the availability of easily moveable, cheap fuels, requiring the use of all energy sources, will be important to allow the developing world to make the transition to a stable population with a decent standard of living (Sheffield 1998). One scenario combines the incremental increase in annual commercial energy use per capita and a corresponding decline in population growth rate (Gilland 1988), comparing the potential, indigenous energy sources with energy demands of the developing region to see whether supply might be able to cope with demand, without massive energy imports.

1.6 Global Emissions of Greenhouse Gases

Compared to the pre-industrial era, anthropogenic GHG emissions, driven largely by population and economic growth, have been growing exponentially over the past century as can be seen in Figure 1.7 (IPCC 2014b). Annual CO_2 emissions increased from near to zero to approximately 33 Gigatons (Gt)CO_2 in 2013, and were estimated to be 36.2 GtCO_2 in 2015 (UNEP 2016). These atmospheric concentrations of carbon dioxide (CO_2), methane (CH_4) and nitrous oxide (N_2O) have been unprecedented in at least the last 800 000 years (IPCC 2014b). During the period from 2000

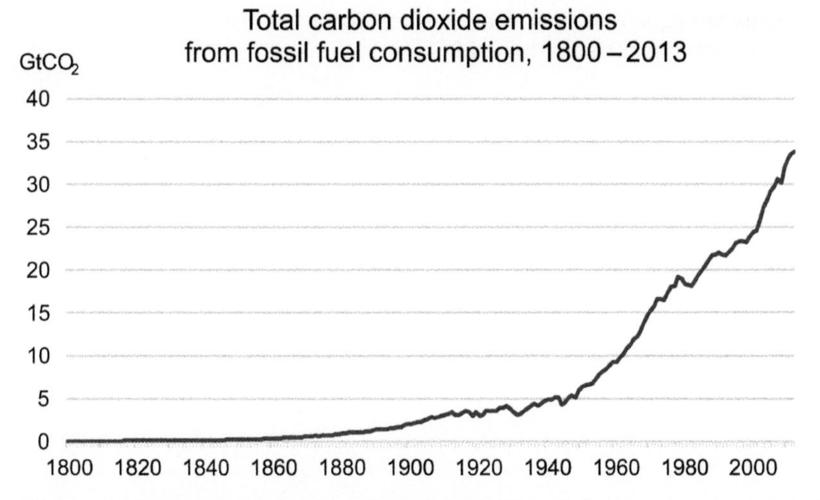

Figure 1.7 Total global carbon dioxide emissions from fossil fuel consumption between 1800 and 2013. Source: The authors with data from (CDIAC 2016).

to 2010 the rate of increase of global GHG emissions was even higher (2.2% per year) than during the period 1970–2000 (1.3% per year). After a rapid increase especially in 2010 and 2011 (3.5% per year), the rise of GHG emissions slowed down to 1.8% growth rate per year between 2012 and 2013 (UNEP 2016).

By sector, transportation and energy production dominated the observed global trend. In 2010, 11% of global anthropogenic greenhouse gases have been emitted by agriculture, contributing mainly CH_4 and N_2O from domestic livestock and rice cultivation (IEA 2015a). Industrial processes not related to energy production caused 7% of global GHG emissions, contributing mainly fluorinated gases and N_2O (IEA 2015a). However, the largest source of anthropogenic GHG emissions has been the production of energy, which contributes mainly to CO_2 emissions (IEA 2015a). In 2013, two-thirds of global CO_2 emissions were produced through electricity and heat generation (42%) as well as transport (23%) (IPCC 2014a). Worldwide, electricity and heat generation relies heavily on coal, which is known to be the most carbon-intensive fossil fuel (IEA 2015a). Coal and oil were each responsible for approximately 40% of global CO_2 emissions between the late 1980s and the early 2000s (IEA 2015a). Moreover, worldwide power sector emissions have tripled since 1970, while emissions from transport have doubled (IPCC 2014a).

Figure 1.8 depicts the emission developments between 1980 and 2014, distinguishing between industrialized OECD economies, and non-OECD countries most of which are developing and emerging economies. As it has been identified in previous graphics on economic development and energy consumption, the trend of developing economies surpassing developed countries can also be identified in the emission developments. In 2005, for the first time in history, emissions from developing and emerging economies have been higher than in industrialized economies. Developing and emerging economies only experienced a slight slowdown in emissions during the economic crisis in 2009 but continued with rising emissions afterwards. Meanwhile, emissions in OECD countries started to decline during the recession in 2009 and continued to stabilize and weakly decline afterwards (see Figure 1.8). This development indicates that the coupling of emissions and the economy is stronger in developing countries, while developed economies experience a slight decoupling due to increasing shares of renewable energies and enhanced energy efficiency and corresponding lower carbon intensities.

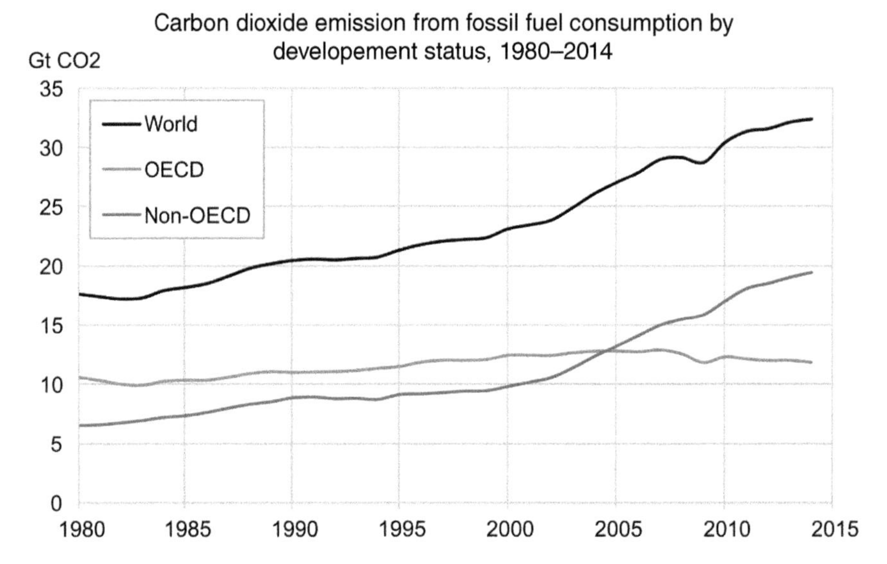

Figure 1.8 Global carbon dioxide emissions from fossil fuel consumption between 1980 and 2014 by development status, in billion tons (Gigatons). Source: The authors with data from IEA 2015a.

1.7 Linkages Between Kaya Factors

As already outlined in Section 1.2, the Kaya identity provides a common framework to determine GHG Emissions G as a function of the socio-economic drivers population P, labour productivity W/P (GDP per capita), energy intensity E/W (energy use per GDP) and emission intensity G/E (emissions per energy unit). The following subsections review the development of these factors over time.

1.7.1 Per Capita Energy and Growth

As for all organisms, the growth of the human population is limited by energy and other resources. Together with the growth of the human population, the economy has grown exponentially and now affects climate, ecosystem processes, and biodiversity, far exceeding those of any other species. Since a permanent growth in the use of energy and materials is not sustainable, solutions are needed to find a balance between energy use and decent standards of living for everybody, at a population size much higher than today. A key factor is the possible coupling between annual energy use per capita and the population growth rate for each region; and the consequences of such a connection to stabilize the world's population (Gilland 1988). For instance, energy facilitates increases in the standard of living and changes in the social conditions, such as higher female education and employment rates that are believed to influence the fertility rate in the demographic transition. This phenomenon can be observed across OECD countries, in which fertility rates declined strongly from 2.7 children per women in 1970 below replacement levels of 1.7 in 2014 (OECD 2016). At the same time, fertility rates in developing and emerging economies remain well above the replacement level (OECD 2016; UN 2015).

To understand contemporary and future trends, it is important to analyse historical patterns and relationships between energy use with population growth and social development (see Figures 1.9 and 1.10). From the timelines of human population and energy, the use of energy per capita and

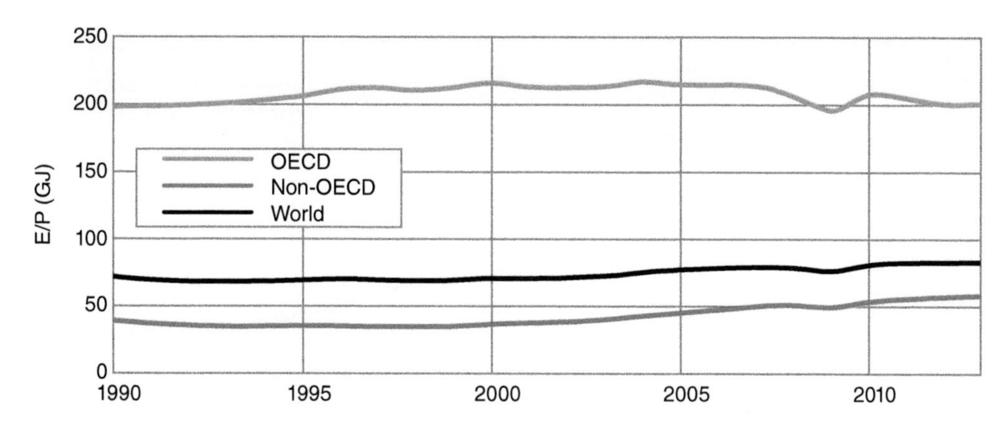

Figure 1.9 Per capita energy between 1990 and 2013.

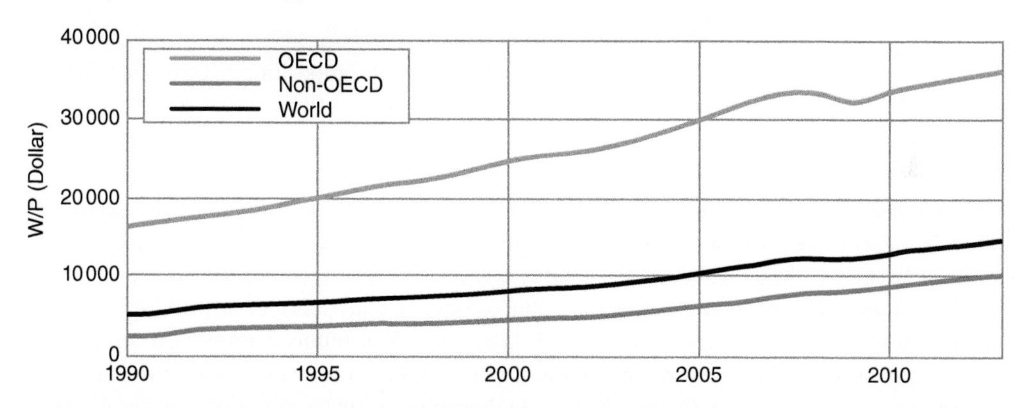

Figure 1.10 Per capita wealth between 1990 and 2013. Source: The authors with data from IEA 2015a.

the correlation between per capita energy consumption and human well-being can be determined. Since current choices are affecting the management of future pathways of human population and energy, an analytical understanding of the historical co-evolutionary patterns is important (Mattick et al. 2010). Apparently there is a large discrepancy and no convergence of energy and wealth per capita between OECD and non-OECD countries.

1.7.2 Energy Demand and Economic Production

Energy is an essential factor in both economic production and consumption where energy supply usually follows the rising demand for energy. It is widely held that there is a positive scaling relationship between per capita energy use and per capita GDP over time, both across nations and within nations (IEA 2015a; Mattick et al. 2010; EIA 2016). As large amounts of energy will be required to fuel economic growth, increase standards of living, and diminish poverty, a positive relationship between energy use and economic growth can be expected, as long as the adverse effects or the scarcity of energy beyond thresholds of planetary boundaries do not exclude further growth, e.g. due to the risks of climate change. However, energy intensity (E/W) has been declining by more than 50% for the world between 1990 and 2013 which is observed almost equally for OECD and non-OECD countries (Figure 1.11). This observation can partly be ascribed to energy savings and efficiency improvements.

Drawing on evidence from the last two centuries of industrialization, the declining energy intensity of economic output continues to rely on increased productivity of the fuel mix, accelerating

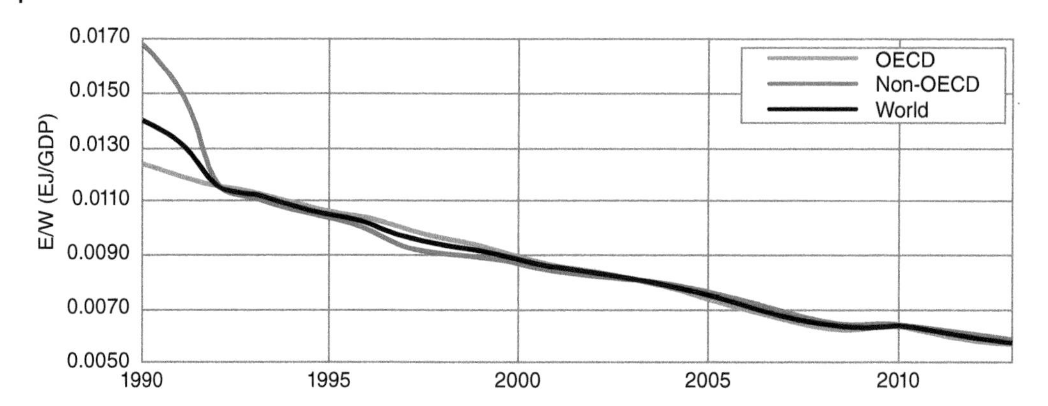

Figure 1.11 Energy intensity between 1990 and 2013 (GDP unit in US Dollar). Source: The authors with data from IEA 2015a.

technical progress of production and convergence of the sectoral composition of economies which will reduce the reliance on any single energy resource (Brown et al. 2011).

1.7.2.1 Energy as a Production Factor

In the neoclassical theory, capital and labour are the main production factors, with technical progress being a primary source of growth (Barro and Sala-i-Martin 2003) while energy and other natural resources are often neglected. The energy crises of the 1970s, demonstrated the 'limits to growth' and induced an economic crisis, providing a concrete experience of how a rise in energy prices could adversely affect economic growth. The crisis has induced numerous studies on natural resources and energy in economic growth, some of which have emphasized the need for an interdisciplinary perspective (Kümmel 1980; Eichhorn et al. 1992). Some studies emphasized the productive power of energy and used it as a factor in the production function (Renshaw 1981; Hudson and Jorgenson 1974).

Economic growth is measured through macroeconomic production functions with capital K, labour P and primary energy flow (installed power) E as production factors. Besides other production functions, such as the LINEX and constant elasticity of substitution (CES) production functions, most prominent is the Cobb–Douglas production function for economic output

$$W = A\,K^{\alpha}\,P^{\beta}\,E^{\gamma}$$

with productivity scale A which is a function of changing technology over time (Warr and Ayres 2012). The productive power and efficiency of each production factor are measured by the output elasticities or substitution exponents α, β, γ that satisfy $\alpha + \beta + \gamma = 1$. These elasticities correspond to the weights with which relative changes of the production factors contribute to the relative change of output W, such as the percentage increase in value added with a one-percent increase in energy input.

Standard economic theory regards capital and labour as the main factors of production that satisfy the 'cost-share theorem' stating that in economic equilibrium conditions the output elasticity of a production factor, which measures its productive power, is equal to the factor's share in total factor cost. In an economy in the neoclassical equilibrium, output elasticities equal factor cost shares.

Although energy conversion in the machines of the capital stock has been the basis of industrial growth, energy plays only a minor role in standard theories of economic growth. Due to its small cost share of about 5%, energy is often neglected altogether (Kümmel et al. 2010). By adding energy as a third production factor, physical constraints on substitution among the factors arise. The inclusion of technological constraints in the nonlinear-optimization calculus (using the LINEX production function with factor- and time-dependent elasticities) shows that the equilibrium conditions no longer yield the equality of cost shares and output elasticities (Kümmel 2007; Kümmel et al. 2010; Kümmel and Lindenberger 2014). Several studies suggest that energy is a much more important production factor than its small cost share may indicate, and reveal that the output elasticity of energy is much larger than energy's share in total factor cost, while for labour the opposite is true. Energy and its conversion into physical work accounts for most of the growth that mainstream economics attributes to 'technological progress' and related concepts (Lindenberger and Kümmel 2011). The provision of adequate and affordable quantities of sufficient energy is a pre-condition for economic growth, making energy systems the 'technology incubators' for a prosperous twenty-first century (Ayres and Voudouris 2014). This would imply that for considerably higher energy prices due to growing scarcity and reduced extraction of oil and natural gas, and the implementation of an effective climate policy within and outside the framework of the Paris agreement, continued economic growth along the historical trend cannot safely be assumed (Ayres et al. 2013).

1.7.2.2 Empirical Results

Empirical results are presented here regarding the linkages between energy use and economic growth for selected studies with different regions and periods, testing for the correlation, significance and direction of the relationship. Further results for other regions and time periods are condensed in Table 1.3 which demonstrates the relevance as well as the diversity of the relationship between energy and economic growth.

1. *Contribution of production factors:* Economic growth in the USA, Japan, and Germany during the second half of the twentieth century is modelled by production functions that depend on capital, labour, energy, and elasticities, including the recessions during the energy crises of the 1970s and the structural change of German reunification in the 1990s (Kümmel et al. 1997). Determining the relative contributions of the production factors to the value added, there is good agreement between theoretical and empirical research, showing that the production elasticity of energy is approximately the same as the corresponding elasticities of capital and labour, pointing to significant improvements in the efficiency of energy conversion techniques after the first oil price explosion (Kümmel et al. 1997): 'In all three countries the time-averaged elasticities of production of energy exceed the share of energy cost in total factor cost by about an order of magnitude, and those of labour are much less than labour's cost share. Only for capital, elasticities, and shares are roughly in equilibrium' (Kümmel et al. 2002). Time changes of these parameters are driven by technological innovation diffusion and creativity (Kümmel et al. 2010).
2. *Role of energy costs*: The application of flexible semi-parametric statistical techniques to data fitting of macroeconomic models explains past economic growth rates in the US, the UK and Japan since 1900 when output elasticities of capital, labour, and useful energy have been highly variable. Results confirm that growth since the industrial revolution has been driven largely by declining energy costs due to the discovery and exploitation of relatively inexpensive fossil fuel resources (Ayres and Voudouris 2014). A related model by Warr and Ayres (2012) explains US growth from 1900 to 1973–1974 with satisfactory accuracy. A slightly underestimated growth for the period after 1975 is explained by subdividing capital stock into traditional components and those based on Information and Communication Technologies (ICT). Results are also extended to Japan (Warr and Ayres 2012). Due to strong decreases in the costs of renewable energy (on the development of solar cell prices see also Section 1.5), these new technologies have been

Table 1.3 Relationship between energy consumption (EC) and economic growth for various regions and time periods, based on the selected sources.

Location/Period	Method	Result	Source
25 OECD countries (1981–2007)	Principal component analysis; cointegration techniques	International developments dominate long-run relationship between EC and real GDP; bi-directional causal relationship between EC and economic growth	Belke et al. (2011)
Greece (1970–2011)	Multivariate, cointegration techniques and vector error correction model	In the long-run electricity demand appears to be price inelastic and income elastic; in the short-run relevant elasticities are below unity; bi-directional relationship between electricity consumption and economic growth; directed energy conservation policies boost economic growth; renewable energy sources provide significant benefits ensuring sufficient security of supply	Polemis and Dagoumas (2013)
OPEC countries	Granger causality runs	In the short-run, Granger causality runs from income to EC for Iran, Iraq, Qatar, UAR, and Saudi Arabia; for the rest of OPEC Granger causality runs from EC to income; no long-run Granger causality for all; in some countries (Qatar, Saudi Arabia, Nigeria) EC affects economic growth, yet very minimal	Hossein et al. (2012)
15 Asian countries (1980–2011)	Panel cointegration and Granger causality analysis	Positive impact of economic growth and trade openness on EC; bi-directional causality between economic growth and EC, trade openness and EC	Nasreen and Anwar (2014)
India (1971–2014)	Autoregressive distributed lag model for cointegration analyses	Environmental Kuznets curve is validated; EC has a positive relationship with carbon emissions; feedback effect exists between economic growth and carbon emissions; energy-efficient technologies in domestic production mitigate carbon emissions	Ahmad et al. (2016)

Table 1.3 (Continued)

Location/Period	Method	Result	Source
5 Asia-Pacific countries (1965–2010)	Bi-variate exponential autoregression model; two-way Granger causality	Economic uncertainty, real oil price and real exchange rate have significant negative effect on EC and/or economic growth; causality of economic growth and EC for Philippines, one-way effect of economic growth on EC for Singapore, neutrality hypothesis for rest of countries	Chiou-Wei et al. (2016)
East and west China groups, 30 provinces, (1985–2007)	Panel unit root, cointegration, and dynamic ordinary least squares to re-investigate comovement	Positive long-run cointegrated relationship: 1% increase in real GDP per capita increases consumption of energy by about 0.48–0.50% and CO_2 emissions by about 0.41–0.43% in China. Economic growth in east China is energy-dependent to a great extent; income elasticity of EC in east China is over two times that of west China.	Fei et al. (2011)
Bangladesh (1972–2006)	Johansen bi-variate cointe-gration; autoregressive distributed lag; Granger short/long-run causality, vector error correction	Uni-directional causality exists from EC to economic growth in short and long-run; bi-directional long-run causality between electricity consumption and economic growth but no causal relationship in the short-run. Strong causality results indicate bi-directional causality for both cases; uni-directional causality runs from EC to CO_2 emission for the short-run but feedback causality exists in the long-run	Jahangir Alam et al. (2012)
India (1990–2020)	Consumption pattern of six income classes; energy and CO_2 coefficients for various production sectors	CO_2 emissions projected to increase from 0.18 tons carbon (tC) per capita in 1990 to 0.62 tC per capita in 2020 under reference scenario with 5.5% GDP growth per annum; scenarios of technology improvement reducing emissions to 0.47 tC per capita 2020.	Murthy et al. (1997)

(continued)

Table 1.3 (Continued)

Location/Period	Method	Result	Source
Selected ASEAN countries (1971–2009)	Autoregressive distributed lag method and Granger causality based on vector error-correction model	Long-run elasticities of EC with respect to CO_2 emissions higher than short-run elasticities; emissions increasing with respect to EC; non-linear relationship of emissions and economic growth in Singapore and Thailand for long-run supports Environmental Kuznets Curve[a] hypothesis; bi-directional Granger causality between EC and CO_2 emissions	Saboori and Sulaiman (2013)
Gulf Cooperation Council	Granger causality; Gini index estimates	Significant positive association of EC with CO_2 and economic growth, in the short- and long-run; mutual Granger causality of EC and CO_2 emissions, unidirectional causal link from economic growth to EC; decoupling occurred in all Gulf cooperation countries (GCC) except Saudi Arabia; divergences in Gini index contributed towards different emissions inequality, depending on energy carriers and economic sectors	Salahuddin and Gow (2014)
27 China provinces (1978–2008)	Panel data with instrumental regression technique	Cross-province integrated energy market reduces response of equilibrium user costs of energy to local demand and production; reduced transportation costs and improved marketization are two important instruments to enhance energy market integration	Sheng et al. (2014)
16 countries in sub-Sahara Africa (1971–2013)	Panel vector autoregression in method of moments frame	Interaction variable (EC, democracy) is positively and significantly related to economic growth; democracy moderates EC-growth nexus; strong evidence of uni-directional relationship from trade openness to EC	Adams et al. (2016)
21 African countries (1970–2006)	Panel cointegration and causality tests	Long-run equilibrium between EC, real GDP, prices, labour and capital for each group and all countries; in/decreasing EC in/decreases growth and vice versa; applies for energy exporters and importers	Eggoh et al. (2011)
Nigeria (1971–2006)	Cointegration and causality; vector error correction	Decline in GDP or economic recession may have adverse effects on EC	Sa'ad (2010)

a) Environmental Kuznets curve is a visual representation of the hypothesis that in a growing economy market forces first increase and then decrease environmental problems.

EC: energy consumption, GDP: gross domestic product.

becoming competitive to unconventional fossil fuel sources and are expected to be competitive with conventional sources in the future (REN21 and ISEP 2013).

3. *Different income groups:* The relationship between GDP and energy consumption is further studied by using panel data from 1971 to 2008 for 95 countries, classified into four income groups (low, lower-middle, upper-middle, and high income). Panel causality tests reveal that there is a 'long-run Granger causality running from GDP to EC for low and high income countries and bidirectional Granger causality between GDP and EC for the lower-middle and upper-middle income countries' (Farhani and Ben Rejeb 2015). Another study using panel data of EC and GDP for 51 countries from 1971 to 2005 finds a long-run Granger causality[2] from GDP to EC for low income countries and a bidirectional causality between EC and GDP for middle income countries but no strong relation between EC and GDP for all income groups (Ozturk et al. 2010). This demonstrates that there is a complex bidirectional relation between EC and GDP depending on income, where in some cases GDP growth is a driver for EC growth, while in other cases EC growth is the main driver.
4. *Effect on energy demand and efficiency:* Newly developed panel data techniques are used to analyse structural, demographic, technological, and temperature drivers in rising global energy demand in recent decades (Huang 2014). Industry and service value added, urbanization and technical innovations have significant effect on energy demand, taking into account spatial lags, a negative price elasticity and positive but declining income elasticity. This has important implications for 'public policies that aim to encourage energy savings, develop the service sector and promote energy-efficient technologies towards a sustainable energy future' (Huang 2014).
5. *Different energy technologies:* The dynamic relationship between energy consumption, industrial output and GDP growth data over the period 1980–2011 in OECD countries is examined in Salim et al. (2014). The results show a long-term equilibrium relationship between non-renewable and renewable energy sources. There is evidence of a bidirectional short-run relationship between GDP growth and non-renewable EC, and a unidirectional causality between GDP growth and renewable energy consumption. The expansion of renewable energy sources is a viable solution for addressing energy security and climate change issues, and could gradually substitute non-renewable energy sources (Salim et al. 2014). Another study focuses on the short-run dynamics and long-run equilibrium relationships among nuclear energy consumption, oil prices, oil consumption, and economic growth for developed countries over the period 1971–2006 (Lee and Chiu 2011). The panel causality results find evidence of unidirectional causality running from oil prices and economic growth to nuclear energy consumption in the long run, while there is no causality between nuclear energy and economic growth in the short run.

1.7.3 Emission-Related Factors

The energy–economy relationship was transformed by the debate on anthropogenic GHG emissions and projected risks of climate change, analysed in reports of the Intergovernmental Panel on Climate Change (IPCC 2014b). Global CO_2 emissions from fossil fuel combustion have been growing at about the same rate as global population in most of the 1970–2010 period, but emissions growth accelerated toward the end of the period (IPCC 2014a). While the emissions per capita (*G/P*) show some convergence over time, they still significantly vary across regions, due to structural and institutional differences, e.g. differences in renewable energy and nuclear energy policies

2 The Granger causality test determines whether one time series is useful in forecasting another, thus predicting the future values of a time series using prior values of another time series.

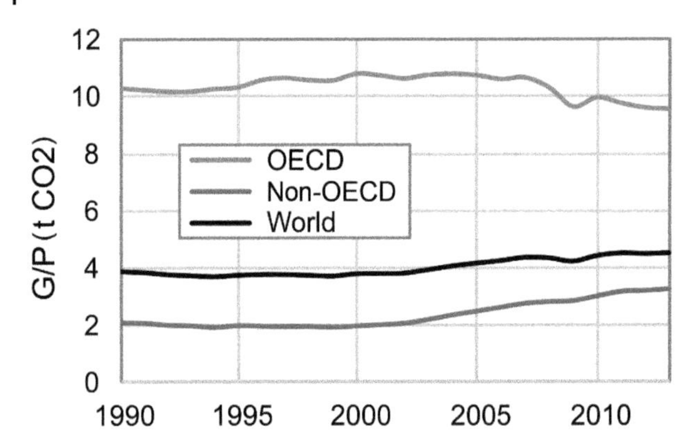

Figure 1.12 CO_2 emissions per capita between 1990 and 2013. Source: The authors with data from IEA 2015a.

etc. (Matisoff 2008). What is striking is that in 1990 per capita emissions of CO_2 in OECD countries (10 tons) start at a level that is five times higher than for non-OECD countries, while in 2013 this ratio shrinks to nearly three, leaving the world at slightly above 4 tons per capita (Figure 1.12). While per capita emissions decline slowly in high-emission regions, there are rapidly increasing per capita emissions combined with relatively fast growth of population and per capita income in Asia, in particular in China (JRC/PBL 2013). Compared to 1990, emissions per capita grew in Asia, were nearly constant in OECD as well as globally, and declined in Economies of Transition (Eastern Europe) following the end of the Cold War, as well as the Middle East and Africa, partly due to a fast growing population (IPCC 2014a).

Each human being contributes to GHG emissions, however individual contributions vary widely depending on socio-economic, demographic, and geographic conditions. A number of empirical econometric studies have used various statistical estimation techniques and data sets (e.g. for different countries, time horizons and other variables included or excluded in the regression model). A review by O'Neill et al. (2012) confirms that GHG emissions increase with population size, whereas the elasticity values (percent increase in emissions per 1% increase in population size) vary widely: from 0.32 to 2.78%. Poumanyvong and Kaneko (2010) estimate elasticities ranging from 1.12 (high-income) to 1.23 (middle-income) to 1.75 (low-income) countries while Jorgenson and Clark (2010) find a value of 1.65 for developed and 1.27 for developing groups of countries. According to Raupach et al. (2007), there is a 91-fold difference in per capita CO_2 emissions from fossil fuels between the highest and lowest emitters across the nine global regions (IPCC 2014a).

The empirical results highlight significant differences between OECD and non-OECD countries regarding the factors that determine GHG emissions. Figure 1.13 shows the evolution of emission intensity of energy (*G/E*) that does not substantially change for the world between 1990 and 2013, staying at nearly 55 Gt CO_2 per EJ, despite some small decline up to 2000. The decreasing emission intensity from around 52 to 47.5 Gt CO_2/EJ is more than compensated for by the rise in non-OECD to around 56 Gt CO_2/EJ (Figure 1.12) (IEA 2015a).

While emission intensity *(G/E)* stayed almost at the same level between 1990 and 2013 globally, emissions per unit of GDP *(G/W)* (Figure 1.14) decreased by more than half in OECD countries and by about two thirds in non-OECD countries, following a pattern of decline similar to that of energy intensity *E/W* (see Figure 1.11), partly due to efficiency gains in power generation and use.

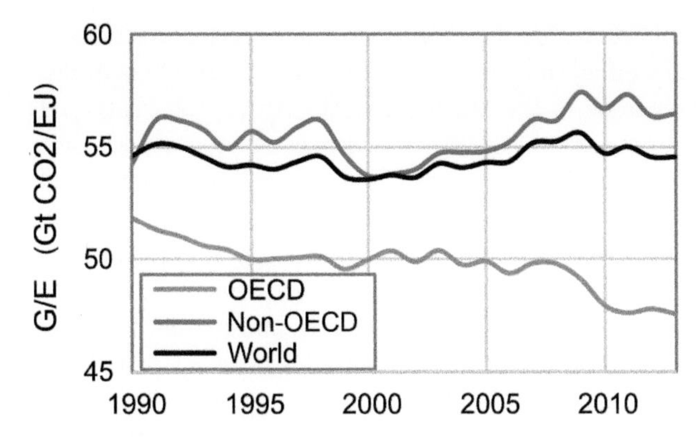

Figure 1.13 CO_2 emissions per energy unit consumed between 1990 and 2013. Source: The authors with data from IEA 2015a.

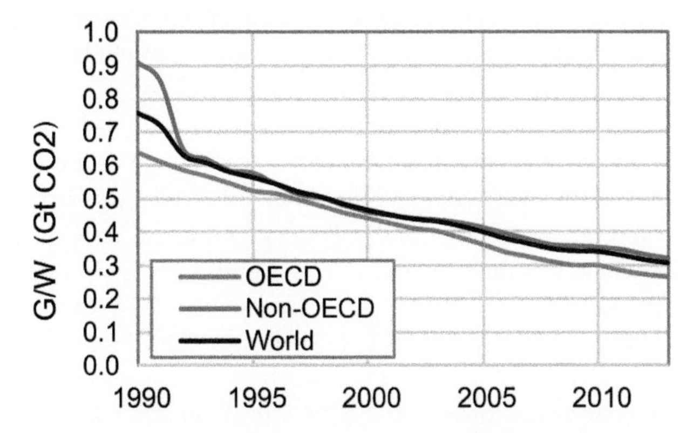

Figure 1.14 CO_2 emissions per unit GDP 1990 and 2013. Source: The authors with data from IEA 2015a.

1.7.4 Energy-Related Impacts

Each energy source changes the environment and has a specific stress and risk profile (Scheffran and Singer 2004; Scheffran and Cannaday 2013). Coal consumption not only causes considerable damage to the landscape and groundwater but also releases photochemical and acid-generating pollutants, as well as greenhouse gases during their combustion, causing climate change with a wide range of consequences for natural and social systems. Similarly, petroleum and natural gas not only pollute waters, soils, and the atmosphere, but also they contribute to global warming. Nuclear power provides the short term gain of reduced CO_2 emissions, but has a potentially catastrophic risk profile (as exemplified by the nuclear disasters of Chernobyl in 1986 and Fukushima in 2011) and burdens natural cycles with long-lasting radioactive emissions. Renewable energy sources have substantially lower CO_2 emissions, but need land and depend on variation of environmental factors. Bioenergy affects natural habitats and food production. The use of water and wind energy changes the landscape.

Both energy reserves and energy consumption are distributed unevenly across the world, which results in a dependence on a few producers. About two thirds of the petroleum reserves are concentrated in the Middle East, while 43% of natural gas reserves are located in the countries of the former Soviet Union (predominantly Russia) and 29% in the Middle East. The debate on the 'limits to growth' was influenced by the oil crisis in the 1970s, when the Organization of the Petroleum

Exporting Countries (OPEC) cartel cut oil production, leading to price increases and economic challenges in oil-dependent countries, as well as counter-measures by industrialized countries. Oil and gas production became a security issue in the Middle East, the Arctic, Caucasus, and Central Asia (Yergin 1991; Klare 2001; Singer 2008). The strategic games entered a new stage with the exploitation of unconventional fossil energy resources, such as oil sands, shales, and fracking.

Growing GHG emissions from fossil fuels have led to an increased anthropogenic radiative forcing and an additional uptake of energy by the climate system, leading to a warming of the atmosphere and the oceans, a decline of snow and ice amounts, sea level rise, and natural feedback responses to a rising global temperature (IPCC 2013, 2014b). Global warming is expected to lead to a degradation of natural resources, human insecurity and societal instability, potentially aggravating environmental conflicts and the 'North–South divide' between industrialized and developing countries (Scheffran et al. 2012). The cost of these degradations cannot be underestimated, and should be taken into account in the economic equation to compare the costs and benefits of the two scenarios that the world economies need to choose from: either continuing business as usual or immediately mitigating global warming. The difficulty here is that most of these costs and risks are still in the future, whereas the expenses to mitigate global warming must be made right away. Climate-induced impacts such as natural disasters and related damages and risks of emissions (R/G) could thus be included by extending the Kaya equation along Figure 1.1, although data on climate damages are still scarce and largely built on model assumptions. Some studies suggest that disasters do not display significant effects on economic growth, and only extremely large disasters have a negative effect on economic output, which was the case only in two events where political revolutions followed those natural disasters (Cavallo et al. 2010). The determinants of the direct effects of disasters have been reviewed in a study that distinguishes between short- and long-run indirect effects (Cavallo and Noy 2011).

There is a complex relationship between economic growth, job creation, peak oil and climate change (Ayres et al. 2013). While 'peak oil' remains a concern in the long run (despite current exploitation of unconventional oil and gas resources), unsolved problems failing to reach an affordable, equitable, environmentally sustainable and future-oriented energy supply bear a considerable risk and conflict potential if a sustainable balance between human requirements and available energy resources cannot be established. Climate change demands a redirection of global economic policy to reverse a 'downward spiral' of the global economy and simultaneously reduce the risk of catastrophic climate change. To this end, major investments in energy efficiency and renewable energy technologies are needed to generate the benefits aimed for (Ayres et al. 2013).

1.7.5 Relative Comparison of the Kaya Factors

The Kaya decomposition analysis reveals and relates the main factors of GHG emissions, including energy use and economic growth. Due to differences in levels of economic, demographic and technological development and growth, emissions evolved at different rates in OECD and non-OECD countries, compared to the baseline year of 1975 (Figures 1.15 and 1.16). Correlating CO_2 emissions with socio-economic indicators shows on the one hand a significant decoupling of energy use (total primary energy supply, or TPES) from economic activity (gross domestic product, or GDP) in OECD countries (TPES/GDP: −45% from 1975 to 2013). On the other hand, total CO_2 emissions of OECD countries in 2013 were 23% higher than in 1975, although they have declined slightly since 2008 (Figure 1.15). Moreover, per-capita economic output (GDP/population: +97%) and population

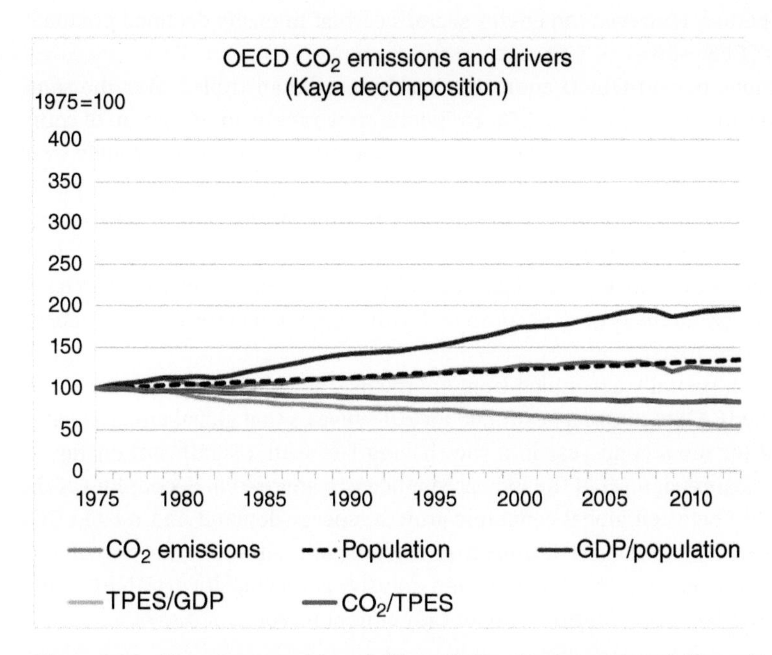

Figure 1.15 Kaya decomposition of CO_2 emission development and drivers between 1975 (reference year) and 2013 in OECD countries. Source: The authors with data from IEA 2015a.

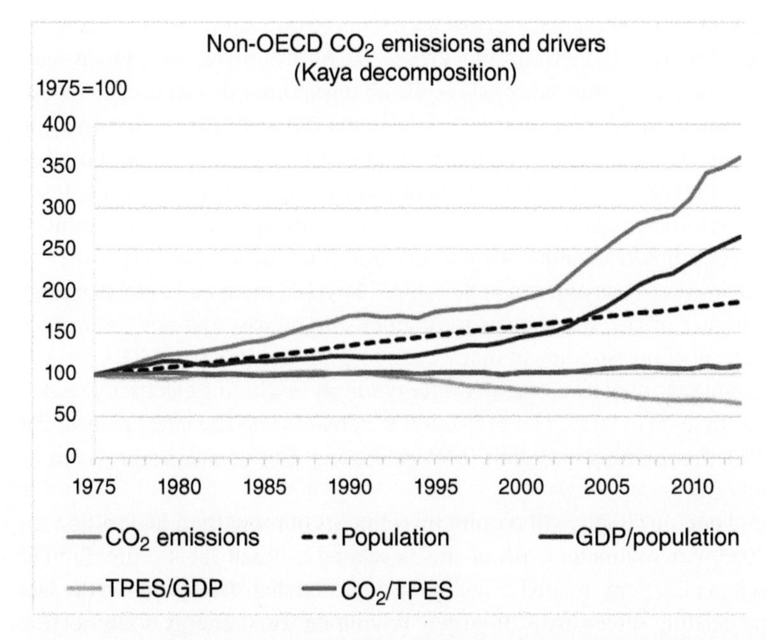

Figure 1.16 Kaya decomposition of CO_2 emission development and drivers between 1975 (reference year) and 2013 in non-OECD countries. Source: The authors with data from IEA 2015a.

(+34%) increased in this period. However, the energy sector's carbon intensity declined gradually in OECD countries (CO_2/TPES: −16%).

In contrast, CO_2 emissions in non-OECD countries clearly more than tripled over the same period (Figure 1.16). Especially in recent years, CO_2 emissions grew rapidly in this group of countries (+171% from 2000 to 2013). Moreover, per-capita economic output increased strongly from 1975 to 2013 (+166%), as did population (+88%), whereas the energy sector's carbon intensity grew only slightly in non-OECD countries (+10%). While the growth of emissions in non-OECD countries was driven by increases in per-capita economic output and population, some trends had a dampening effect on CO_2 emissions since 2008, in particular in OECD countries as a result of a significant reduction in the energy intensity of GDP combined with a slight fall in the carbon intensity of energy generation (by about 10%).

Globally, economic growth partially decoupled from energy use, as energy intensity decreased by 29% over the period. The IEA's preliminary estimate for 2013 reveals that global energy-related CO_2 emissions stayed flat for the second year in a row, having last seen a significant change in 2012. This represents the clearest sign yet of the increased efficiency and partial decoupling of the previously close relationship between global economic growth, energy demand and related CO_2 emissions. Although global GHG emissions continue to grow, indicators show that the growth rate of global carbon dioxide emissions from fossil fuel use and industry is slowing down (UNEP 2016). Energy production generates by far the largest share of CO_2 emissions compared to other sectors (IEA 2015b).

1.8 Development of Energy Investment

Managing the global increase in energy demand and extensive energy transitions requires massive financial and technological resources (Scheffran and Froese 2016). Countries worldwide have recognized this need and have already achieved considerable changes through sustainable investment in the expansion and transformation of their energy systems (Finance 2014). According to the World Energy Investment Outlook 2014, investment in global energy supply amounted to more than $1600 billion in 2013 (IEA 2014b). These estimates ranged from the extraction of fossil fuels, storage and handling facilities, pipelines, tankers, and other transportation equipment to the construction of oil refineries, power stations, wind farms, and solar installations (IEA 2014b). Since 2000, the amount of investment has more than doubled. The rapid increase in global energy demand (see Section 1.5), higher prices, rising costs to produce oil and gas, and new renewable technologies require a high level of investments in many countries all over the world (IEA 2014b).

Figure 1.17 presents the average annual investment in energy supply (excluding electricity related investment) in the period from 2000 to 2035. The projection is derived from the latest runs of the New Policies Scenario (NPS), the main scenario in the World Energy Outlook series provided by the IEA.

The world's projected total energy supply will require investments of more than $40 trillion for the period from 2014 to 2035. Approximately 70% of this is related to fossil fuels. This number increases to almost 100% when electricity-related investments are excluded. This is due to the fact that the transportation and heating sector are still largely relying on fossil energy sources (IEA 2014b). The largest fossil fuel investment requirements are in oil ($13 trillion from 2014 to 2035), followed by investments in natural gas ($8 trillion from 2014 to 2035). According to the IEA, the average annual share of electricity-related investments in the period from 2000 to 2013 amounts to $479 billion and will require more than $16 trillion in the period from 2014 to 2035 (Figure 1.18).

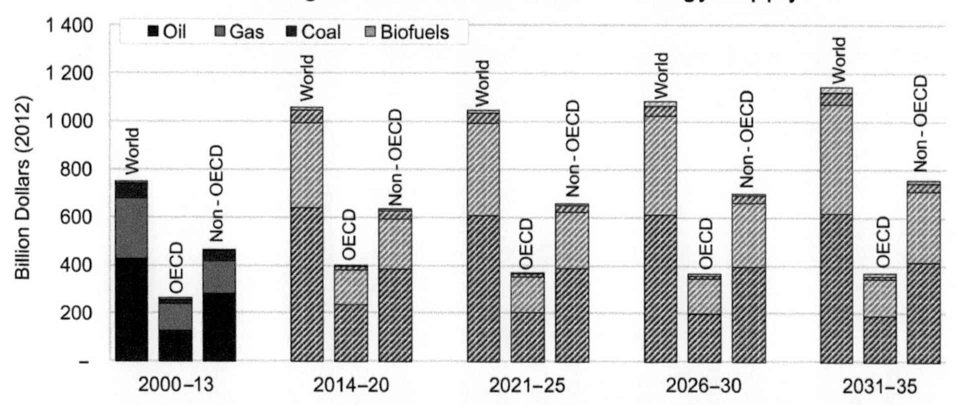

Figure 1.17 Average annual investments in energy supply from 2000 until 2013 and correspondent forecast (NPS) until 2035. Source: The authors with data from IEA 2014b.

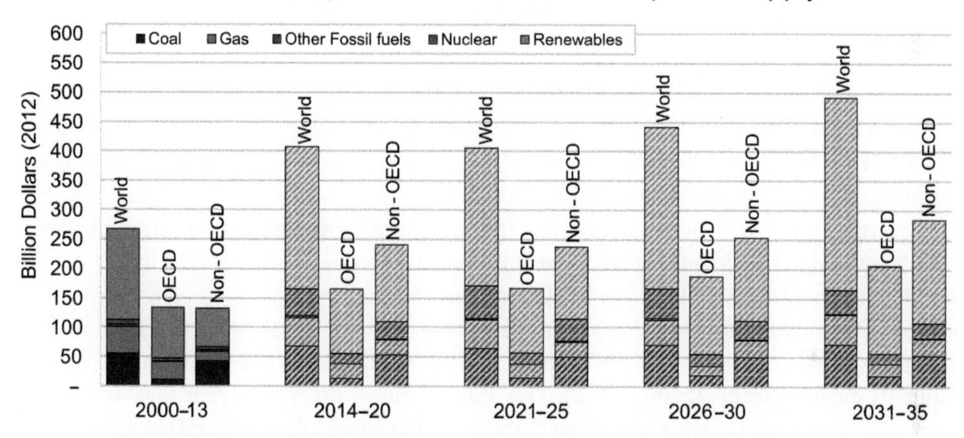

Figure 1.18 Average annual investments in electrical power supply from 2000 until 2013 and correspondent forecast (NPS) until 2035. Source: The authors with data from IEA 2014b.

About two thirds of these are in renewable energies, with an expected increase of an additional 8% to 2035 (IEA 2014b). Moreover, nuclear power is expected to play an important role in future energy finances. The share of nuclear energy investments of total investments in electricity supply is projected to increase from 1.7% in the period from 2000 to 2013 to 5% in the period from 2031 to 2035 (Figure 1.18). However, this does not consider any costs that may arise from possible nuclear disasters. For instance, there have been the significant costs of the Fukushima disaster following the tsunami in Japan on 11 March 2011. The direct costs of the Fukushima disaster are estimated to be about $15 billion in clean-up over the next 20 years, over $60 billion in refugee compensation and $200 billion for replacing nuclear power with fossil fuels which will at least double if the nuclear fleet is mostly restarted by 2020 (Conca 2016; Matsuo and Yamaguchi 2013). The reconstruction costs associated with the earthquake and the tsunami will top $250 billion (Conca 2016). If nuclear disasters occur in the future, the associated costs may be significantly higher.

Since 2000, significant growth in new investment in sustainable power and fuels has been seen globally, with annual growth rates exceeding 50 % (Scheffran and Froese 2016). In 2015, global

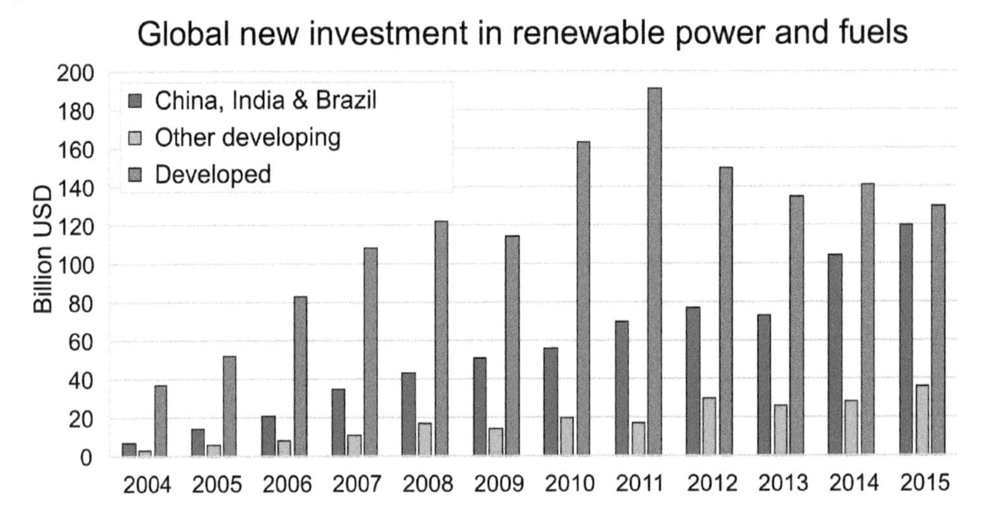

Figure 1.19 Global new investment in renewable power and fuels from 2004 until 2015. Source: The authors with data from Bloomberg 2016.

new investment in renewable power and fuels excluding large hydropower projects (>50 MW) was \$285.9 billion. This represents a rise of 5% compared to the previous year. For the first time in history, investment in renewables excluding large hydro power was higher in developing economies than in developed countries (Bloomberg 2016). This development marks 2015 as a signal year for renewable energy. Figure 1.19 shows global new investment in renewable power and fuels from 2004 to 2015, estimated by Bloomberg New Energy Finance (BNEF). Commitments by the developing world amounted to \$156 billion, up by 19% to a new record.

In contrast, investment in developed countries as a group declined from \$190 to \$130 billion, the lowest amount since 2009. The most significant decrease in investment was seen in Europe (down 21%, despite its record year in offshore wind). This shift in energy investment was to be expected, since developing countries are generally those with fast-rising energy demand and therefore need increased power generation capacity (see Section 1.5).

Figure 1.19 details the increase in capacity investment in the three largest developing economies – China, India, and Brazil. China played a dominant role in this important development and showed the strongest and most consistent increase in dollar commitment, starting with just \$3 billion in 2004, then significantly increasing to a record \$102.9 billion in 2015 (Bloomberg 2016). Renewable energy investment also increased significantly in India (\$10.2 billion in 2015), Brazil (\$7.1 billion in 2015), as well as three 'new markets' in South Africa up by 309% to \$4.5 billion, Mexico doubling to \$4 billion and Chile rising 143% to \$3.4 billion from 2004 to 2015 (Bloomberg 2016).

Looking at the relationship of investments with the Kaya factors, the following trends can be observed:

- As presented in Figure 1.20, there are considerable differences in energy investment between OECD and non-OECD countries. Starting from \$200 in 2000 for the world, per capita investments (*C/P*) have been continuously declining throughout the whole period, both in OECD and non-OECD countries with OECD having nearly three times higher investment in 2000 than non-OECD. In 2015 this ratio is even higher, indicating no convergence.

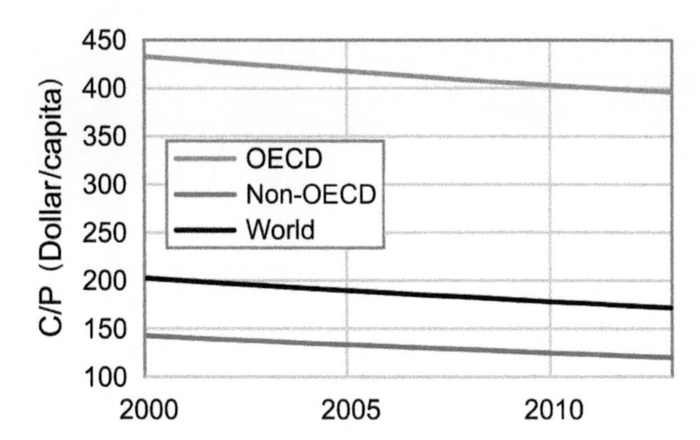

Figure 1.20 Energy investments per capita between 2000 and 2013. Source: The authors with data from IEA 2015a.

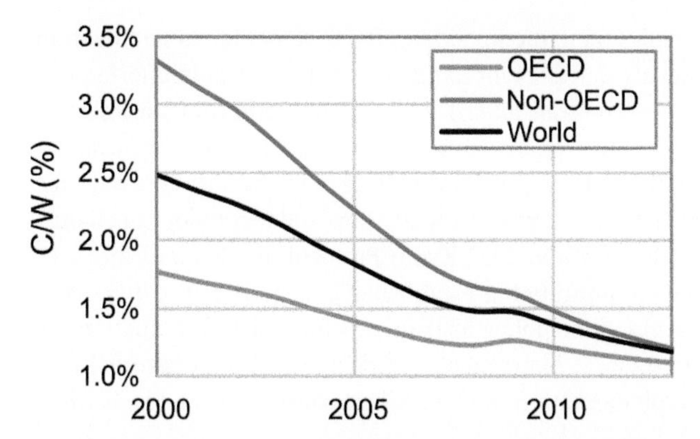

Figure 1.21 Energy investments per unit GDP between 2000 and 2013. Source: The authors with data from IEA 2015a.

- This is different for the fraction of investment per GDP unit (C/W) which converges for the world from 2.5% in 2000 globally to about 1.2% of GDP in 2015, starting from nearly 1.8% for OECD and 3.3% for non-OECD (Figure 1.21).
- Energy generated per investment unit (E/C), which is the inverse cost per energy unit, remains almost constant in OECD in the period 1990 to 2015 (Figure 1.22), with a drop and recovery during the economic crisis of 2008. Non-OECD starts at 50% lower energy output per investment unit and converges to nearly the same value as for OECD and the world. Apparently 1 billion USD is needed to generate 0.5 EJ energy output (Figure 1.22).

1.9 Conditions for Energy Transition and Decarbonization

1.9.1 Targets and Pathways of Climate Policy

In the previous sections, we described expected developments of future emissions as well as emission pathways according to different policy scenarios towards a low-carbon economy. As part of the 'Rio Process', the United Nations Framework Convention on Climate Change (UNFCCC) agreed in 1992 on general principles and rules that were specified by the Kyoto Protocol in 1997 and the Paris

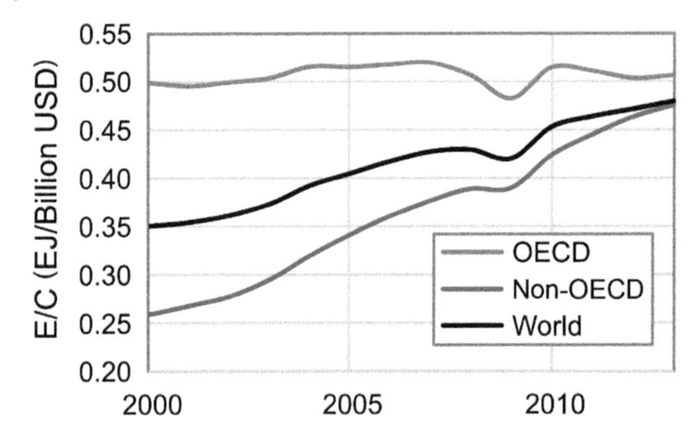

Figure 1.22 Energy generated per investment unit between 2000 and 2013. Source: The authors with data from IEA 2015a.

Agreement of 2015 that set the frame for emission reductions. The UNFCCC demands stabilization of atmospheric GHG concentrations at levels that prevent dangerous anthropogenic interference with the climate system (Baatz and Ott 2017; Scheffran 2008b). As the global community has shown their ability to jointly act on global challenges, though on a different scale, such as the ozone depletion in the 1980s (Benedick 1998), the stabilization of greenhouse gases in the atmosphere requires an unprecedented degree of international action for emission reductions and technological change in the energy sector. With the entry into force of the 1997 Kyoto Protocol in 2005 and the 2015 Paris agreement in 2016 the international community has established cooperative instruments to address the problem of global warming. Among the policy instruments for emission reduction are administrative concepts (emission taxes), market instruments (tradeable emission permits) and cooperative approaches such as Joint Implementation and the Clean Development Mechanism (CDM) (UNFCCC 2015).

While existing measures and their implementation are still insufficient, the question is whether and how fast humanity will be able to develop goals and adaptive strategies to prevent dangerous climate change or at least minimize its negative effects. A slow deceleration of emissions will not be enough to limit global warming to well below 2 °C relative to preindustrial levels as it has been stated in the Paris Agreement, following the UNFCCC's goal of preventing dangerous anthropogenic interference with the climate system (UNFCCC 2015). Figure 1.23 shows five scenarios of emission developments until 2030, the baseline scenario, the current policy trajectory, the scenario of Nationally Determined Contributions (NDC),[3] the 2 °C and the 1.5 °C scenario (UNEP 2016; Rogelj et al. 2016). The collection of scenarios is based on the UNEP Emission Gap Report (UNEP 2016) that draws on different scenarios from the literature. The baseline scenario depicts the emission pathway as it is predicted assuming no additional policy after 2005, whereas the current policy trajectory scenario takes into account currently adopted and implemented policies. The NDC scenario depicts how emissions could develop under full implementation of the NDCs. While the full Emission Gap Report distinguishes between unconditional and conditional NDC cases, only the conditional NDC scenario is displayed here as this is the most ambitious pathway, which requires a mutual tackling of climate change though global action by all parties. The 2 and 1.5 °C scenarios portray global scenarios with a likely chance of limiting global temperatures to below

3 The Nationally Determined Contributions (NDCs) are the national goals of greenhouse gas emission reductions according to the 2015 Paris Treaty. Until the Paris Climate Conference the Intended NDCs (INDCs) were used.

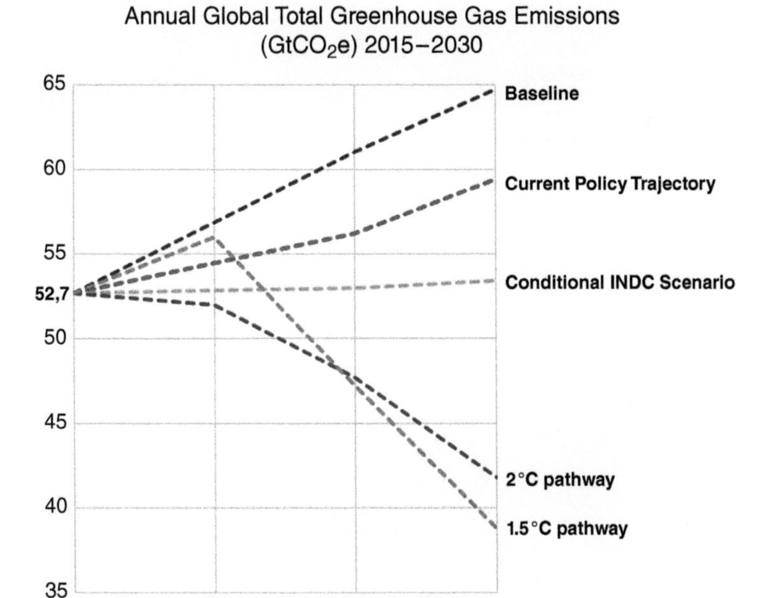

Figure 1.23 Global greenhouse gas emissions under different scenarios and the emission gap in 2030. Source: Adapted from the UNEP Emission Gap Report UNEP 2016.

2 °C and 1.5 °C by 2100 compared to preindustrial levels. Both scenarios represent least-cost global scenarios which include the distribution of emission reductions across regions, gases, and sectors after 2020 in order to minimize the necessary global mitigation costs (Rogelj et al. 2015). Furthermore, the 2 °C scenario is consistent with the full implementation of the Cancun pledges (UNFCCC 2010).

The figure indicates a gap of 12 $GtCO_2e$ between the 2 °C scenario and the conditional NDC scenario in 2030. This gap increases to 15 $GtCO_2e$ comparing the 1.5 °C scenario with the conditional NDC scenario. This gap portrays that current national contributions need to be strengthened since they are not ambitious enough to fulfil the Paris Agreement. Additionally, the current policy trajectory indicates an even stronger increase in emissions showing the fact that stronger actions and additional policies are absolutely necessary to achieve the long-term objectives (UNEP 2016).

1.9.2 Strategies of Implementation

While the Paris Agreement and the NDCs define the overall emission targets, an open question remains regarding how these emission targets are implemented. Different strategies and policies influence the factors in the Kaya equation to diminish emissions and balance human–environment interactions.

In theory, human population (P) is a factor that could be restricted by population control and the demographic transition. However, in practice rather the opposite can be observed. While populations in emerging economies are strongly increasing, China lifted its one child policy and many European countries as well as Japan and South Korea encourage natality due to their aging populations and the need for additional labour force. Changing the labour productivity (W/P) includes technical innovation as well as investment in human and social capital. Further, energy intensity

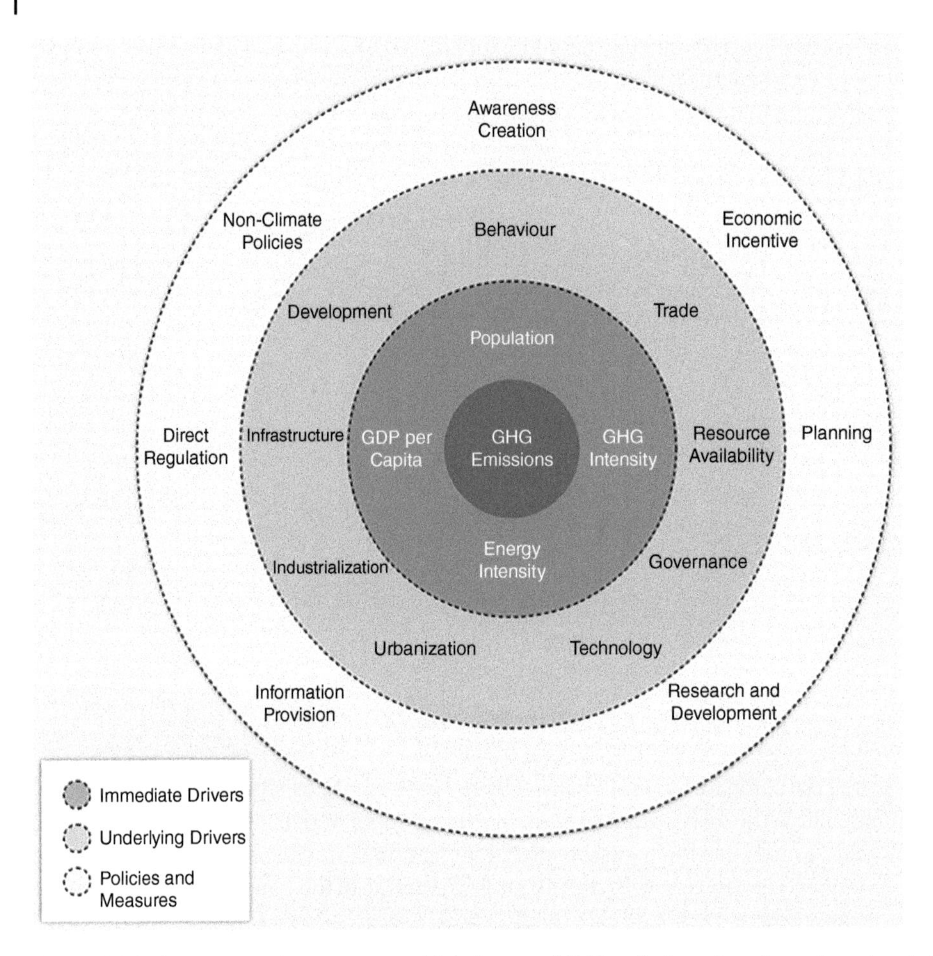

Figure 1.24 Interconnections among multiple layers of GHG emissions, from inner to outer circle: immediate drivers in the composition of emissions in the Kaya identity; underlying drivers (processes, mechanisms, and characteristics) that influence the factors; and policies and measures affecting the underlying drivers (IPCC 2014a).

(E/W) can be reduced by switching to a resource mix with renewable energy, improved energy efficiency and more sustainable lifestyles and consumption patterns during the whole energy life-cycle to diminish the coupling between economic growth and resource use, and satisfy human needs with fewer energy resources. The emission intensity (G/E) can be reduced by a switch to low-carbon energy sources and lifestyles. A broader view of different drivers and policy measures is given in Figure 1.24.

Additional strategies can be applied to other elements of the framework introduced in Figure 1.1 (Scheffran 2008a):

- Risk (R) can be diminished by environmental protection, climate adaptation, hazard prevention and conflict management, e.g. from accidents, disasters, climate change, environmental toxins, or radioactivity.
- Use of investment (C) may be more cost-efficient in achieving economic and climate targets.
- Fair distribution (equity and justice) seeks a balanced resource share around the average energy per capita (E/P) in a democratic participatory process.

- Sufficiency strategies aim for a sufficient value (V^*) of lifestyles and adjust them to the prevailing environmental conditions, combining mitigation and adaptation with improved living quality and happiness at lower resource consumption and environmental stress.
- Nature preservation strategies protect the environment to not exceed the natural carrying capacity, e.g. by regeneration of nature reserves, harvest limits, and conservation of endangered species.

To avoid conflicts and strengthen the positive link between natural and social capital, interdisciplinary approaches from both natural and social sciences are indispensable to jointly contribute to integrated sustainability strategies that implement the fundamental connections between these factors. The challenge is to bring multiple policies and technologies together in a comprehensive integrated framework with lasting impacts on the economy, energy consumption and environment towards sustainable development (Ahmad et al. 2016). Rather than relying on responsive tools, transforming the energy sector over the next decades requires forward looking approaches and capacities to project future technology developments, markets, and energy-related events (IEA 2014a). Powerful action is needed to actively transform energy supply and end use to meet growing energy demand while addressing related concerns for energy security, costs, and energy-related environmental impacts (IEA 2014a).

1.9.3 Integrated Assessment and Decision-Making in Transition Processes

Preventing long-term climate change poses not only a challenge for the decision-making process but also for the decision methods and tools applied in this process. Integrated assessment describes and models the dynamic interaction between natural and socio-economic systems to better understand the implications of decision-making on future climate change (Haurie and Viguier 2005). Economists together with climate scientists have developed integrated assessment models to analyse the interaction between optimal emission and economic pathways, coupling a basic climate model to an economic production function with capital, labour, and technology as production factors. Emission scenarios use computer simulation to project carbon emissions, concentration, and temperature change into the future, based on plausible scenarios and parameter settings, taking into consideration alternative energy pathways, estimated impacts of rising global temperature on society and economy, and the costs of avoiding climate change.

Aggregated optimal growth models such as DICE (Nordhaus 1993; Nordhaus and Boyer 2000), FUND (Anthoff and Tol 2014; Tol 1999) and MIND (Edenhofer et al. 2005; Bauer et al. 2012; Edenhofer et al. 2006) optimize time-discounted economic welfare functions of expected benefits, including costs and climate damages and using energy and emission pathways based on the best available knowledge and investment strategies. Conditions for switching between technology and management paths with different costs and carbon intensities are identified. To take account of the substantial uncertainties intrinsic to these models, the sensitivity of the results to variation of crucial parameters is estimated, in particular time discounting, climate damage, taxes, and time horizon for decision-making (Scheffran 2008a).

While empirical evidence of climate change is mounting, significant uncertainties still remain that affect the energy transition (Murphy et al. 2004; Stainforth et al. 2005). The world is facing deep uncertainty in its understanding of climate thresholds when systems models and probability functions are unknown (Alley et al. 2003; Lempert 2002). Decisions on long-term climate change deal with deep uncertainty and complexity (Lempert et al. 2009). In an unpredictable and changing state of the world, adaptive management approaches define sequential goals for different time-periods, whereupon later goals are made dependent on the achievement of goals in earlier periods (hedging strategy) (Yohe et al. 2004). Rather than maximizing long-term utility, actors adaptively pursue shorter-term targets or seek to stay within admissible domains of their environment according to viability conditions that result from principles, rules, and norms (Scheffran 2008a).

The guardrails approach restrains the set of trajectories to a 'tolerable window' of the climate system, in particular to limitations on the rate of temperature change, to avoid exceeding critical thresholds that can trigger abrupt and dangerous climate change (Bruckner et al. 1999; Aubin et al. 2005). The admissible domains of the climate system and the emission paths within the boundaries are given by changing assessments and judgements, taking into account current vulnerabilities, adaptive capacities, critical thresholds and indicators for qualitative system change. The task is to identify key control variables, learning, and regulation mechanisms for decision-making to maintain the admissible domain and avoid intolerable dangers as envisaged by Article 2 of the UNFCCC. Suggested targets have been 450–550 ppm of atmospheric CO_2 concentration and a temperature change of not more than 1.5–2 °C over its recent pre-industrial average.

Beyond the established optimal control approach, the concepts of adaptive control, inverse modelling, and local optimization to climate change decision-making and management can be combined. An adaptive management framework has been developed to represent the relationship between economic growth, emission reduction, and energy technology outlined in the Kaya framework (Ipsen et al. 2001; Scheffran 2008a, 2000, 2001, 1999). Controls are adjusted towards a moving target under changing conditions while decisions and actions are driven by value functions that include the expected benefit of economic production, diminished by the risk associated with climate change, mitigation, and adaptation costs as well as the payment for taxes and emissions trading. Actors allocate their available investments to competing energy paths that differ in costs, economic output, and carbon emissions per energy unit. Since individually optimal investment strategies depend on those of other actors, the actors can mutually adjust their actions in feedback and learning cycles according to their adaptation rules which result in interdependent solution concepts for a given future time horizon, and translate into admissible emission limits to keep atmospheric carbon concentrations and global mean temperature below a given threshold (Scheffran 2008a).

To make sure that the cumulative emissions of all human beings will not exceed a critical limit, the international community needs to agree on an emission path that avoids dangerous climate change. At the global level, this may lead to a collective action problem for humanity. A way forward might be to focus on the local level, starting local and integrating these efforts to get the global effect – e.g. some cities or federal states implementing concrete climate measures. Assuming that there is an agreed cap on aggregate emissions, it is a challenge to find institutional mechanisms to assign individual limits and ensure compliance to avoid the tragedy of the commons (Scheffran 2008b). Despite diverging interests, some cooperation is essential. To describe, understand, and model the complex micro–macro links and the interaction of conflicting interests across all levels is a challenge for projecting future GHG emissions.

Besides life-style changes, energy savings and technical changes offer an opportunity for emission reductions. Investments into alternative energy sources that emit less carbon per energy unit may induce endogenous technical change and a transition towards more sustainable energy use. Main drivers are represented by value functions, indicating incentives to switch to new technologies for certain actors: producers aiming for higher profits, consumers seeking additional benefits, and countries striving for higher economic welfare. If the expected benefits for new pathways outweigh the costs and risks, then actors tend to accept a transition towards these paths. A major issue is to identify mechanisms, instruments, and enabling environments to invest into new low-carbon energy technologies, including legal regulations, taxes, emissions trading, subsidies, cooperative approaches, and coalition formation (Scheffran and Froese 2016). An energy transition involves the reconfiguration of current patterns and scales of economic and social activities in energy landscapes, which depend on six geographical concepts: location, landscape, territoriality, spatial differentiation, scaling, and spatial embeddedness. More attention to the spaces and places that transition to a low-carbon economy can help better understand what living in a low-carbon economy will be like. It also provides a way to help evaluate the choices and pathways available (Bridge et al. 2013).

Negative incentives are damages and risks associated with climate change that could undermine benefits from economic growth associated with emissions. One problem, however, is that these risks are expected in, and discounted for, a more distant future and hardly affect current investment decisions based on short-term utility calculations. To overcome the initial barriers and to induce early switching, governments can provide incentives to change, e.g. by raising taxes on the old path or providing subsidies to the new path, shifting the composition of technologies (Scheffran 2002). An alternative approach is emissions trading, which defines emission allowances beyond which additional permits have to be paid (Scheffran and Leimbach 2006). In the system of nation states these approaches are difficult to implement beyond the NDCs as there is no established authority that could assign these instruments without the consent of sovereign states (e.g. see Baranzini et al. 2017; Bodnar et al. 2018; Michaelowa et al. 2018; Hof et al. 2017; Avi-Yonah and Uhlmann 2009).

Adaptive approaches will be relevant to find and discuss adequate responses to climate change induced by anthropogenic GHG emissions. Within an adaptive framework of integrated assessment it is possible to study the interaction between key climate variables and the economic system to determine requirements in energy policy for achieving admissible limits of carbon concentration and temperature change that prevent dangerous climate change. Using an established basic climate model, time-dependent functions for concentration and temperature change as well as explicit inverse solutions for emission trajectories that satisfy stabilization limits can be determined. Feeding these into a dynamic model of economic growth with energy as a production factor can be useful to determine boundary conditions for investment policies and optimal alternative energy paths that respect viable climate limits (Scheffran 2008a).

1.10 Perspectives

There is a complex relationship between population, economic growth, global energy demand, and the impacts of climate and environmental change as well as social developments. It is widely admitted that there is a correlation between per capita energy use and per capita GDP over time. As large amounts of energy are required to fuel economic growth, and increase standards of living, a positive relationship between energy use and economic growth can be expected. Global energy intensity has been declining in recent years, which can partly be explained by efficiency improvements. However, the evolution of energy relies on increased specialization of the fuel mix, and on economies which will continue to improve energy intensity of economic output and reduce the reliance on any single energy resource.

Each energy type has a specific stress profile and influences the environment in different ways. Global energy production leads to massive changes in the climate and the environment. Compared to other sectors, fossil energy production generates by far the largest share of CO_2 emissions. Decreasing the reliance of world economies on fossil fuel-derived energy is therefore of urgent relevance for achieving sustainable global development not only on environmental but also on social and economic perspectives.

Global warming is expected to lead to a degradation of natural resources and potential environmental conflicts, affecting human security and societal stability on a global scale. Without addressing the risk and conflict potential of the energy supply chain, it is in practice impossible to achieve an affordable, equitable, environmentally sustainable, and future-oriented energy system that finds a sustainable balance between human requirements and available energy resources. Redirected global economic policy is thus needed to reverse the ongoing development of climate change. It is important to face global warming as fast and as strongly as possible to limit it to well below the 2 °C temperature limit relative to preindustrial levels as it has been stated in the Paris Agreement.

It is still open how humanity will implement the overall emission targets and how quickly it will be able to develop affordable and feasible adaptive strategies to prevent the many dangerous impacts of climate change. The challenge is to bring multiple policies and technologies together in a comprehensive integrated framework with lasting impacts on the economy, energy consumption, and environment towards sustainable development. Powerful action is needed to actively transform energy supply and end use to meet growing energy demand while addressing related concerns for energy security, costs, and environmental impacts. Major investments in energy efficiency and renewable energy technologies are needed, as well as instruments such as legal regulations, taxes, emissions trading, subsidies, cooperative approaches and coalition formation.

Acknowledgments

This research was supported by the German Science Foundation Clusters of Excellence CliSAP and CLICCS.

References

Adams, S., Klobodu, E.K.M., and Opoku, E.E.O. (2016). Energy consumption, political regime and economic growth in sub-Saharan Africa. *Energy Policy* 96: 36–44.

Ahmad, A., Zhao, Y., Shahbaz, M. et al. (2016). Carbon emissions, energy consumption and economic growth: an aggregate and disaggregate analysis of the Indian economy. *Energy Policy* 96: 131–143.

Alley, R.B., Marotzke, J., Nordhaus, W.D. et al. (2003). Abrupt climate change. *Science* 299: 2005.

Anthoff, D. and Tol, R.S.J. (2014). Climate policy under fat-tailed risk: an application of FUND. *Annals of Operations Research* 220 (1): 223–237.

Aubin, J.-P., Bernado, T., and Saint-Pierre, P. (2005). A viability approach to global climate change issues. In: *The Coupling of Climate and Economic Dynamics: Essays on Integrated Assessment* (eds. A. Haurie and L. Viguier). Dordrecht Netherlands: Springer.

Avi-Yonah, R.S. and Uhlmann, D.M. (2009). Combating global climate change: why a carbon tax is a better response to global warming than cap and trade. *Stanford Environmental Law Journal* 28 (1): 3–50.

Ayres, R. and Voudouris, V. (2014). The economic growth enigma: capital, labour and useful energy? *Energy Policy* 64: 16–28.

Ayres, R. U., Van den Bergh, J. C. J. M., Lindenberger, D. & Warr, B. (2013). The Underestimated Contribution of Energy to Economic Growth. INSEAD Working Paper No. 2013/97/TOM/EPS/Social Innovation Centre. Available at SSRN: https://ssrn.com/abstract=2328101 or http://dx.doi.org/10.2139/ssrn.2328101.

Baatz, C. and Ott, K. (2017). In defense of emissions egalitarianism? In: *Climate Justice and Historical Emissions* (eds. L. Meyer and P. Sanklecha). Cambridge, MA: Cambridge University Press.

Baranzini, A., Van den Bergh, J., Carattini, S. et al. (2017). Carbon pricing in climate policy: seven reasons, complementary instruments, and political economy considerations. *Wiley Interdisciplinary Reviews: Climate Change* 8: e462.

Barro, R.J. and Sala-i-Martin, X. (2003). *Economic Growth*. London, England/Cambridge, Massachusetts: The MIT Press.

Bauer, N., Baumstark, L., and Leimbach, M. (2012). The REMIND-R model: the role of renewables in the low-carbon transformation – first-best vs. second-best worlds. *Climatic Change* 114: 145–168.

Belke, A., Dobnik, F., and Dreger, C. (2011). Energy consumption and economic growth: new insights into the cointegration relationship. *Energy Economics* 33: 782–789.

Benedick, R.E. (1998). *Ozone Diplomacy*. Harvard University Press, Cambridge, MA.

Bloomberg (2016). *Global Trends in Renewable Energy Investment*, 2016. Frankfurt am Main: Frankfurt School-UNEP Centre/BNEF.

Bodnar, P., Ott, C., Edwards, E. et al. (2018). Underwriting 1.5°C: competitive approaches to financing accelerated climate change mitigation. *Climate Policy* 18 (3): 368–382.

Bridge, G., Bouzarovski, S., Bradshaw, M., and Eyre, N. (2013). Geographies of energy transition: space, place and the low-carbon economy. *Energy Policy* 53: 331–340.

Brown, J.H., Burnside, W.R., Davidson, A.D. et al. (2011). Energetic limits to economic growth. *BioScience* 61: 19–26.

Bruckner, T., Petschel-Held, G., Tóth, F.L. et al. (1999). Climate change decision-support and the tolerable windows approach. *Environmental Modeling & Assessment* 4: 217–234.

Cavallo, E. and Noy, I. (2011). Natural disasters and the economy: a Survey. *International Review of Environmental and Resource Economics* 5 (1): 63–102.

Cavallo, E., Galiani, S., Noy, I. & Pantano, J. (2010). Catastrophic Natural Disasters and Economic Growth. IDB Working Paper Series, No. IDBWP-183.

CDIAC (2016). *Total Carbon Dioxide Emissions from Fossil Fuel Consumption*. Oak Ridge, TN: Carbon Dioxide Information Analysis Center, Oak Ridge National Laboratory, US Department of Energy.

Chiou-Wei, S.-Z., Zhu, Z., Chen, S.-H., and Hsueh, S.-P. (2016). Controlling for relevant variables: energy consumption and economic growth nexus revisited in an EGARCH-M (exponential GARCH-in-Mean) model. *Energy* 109: 391–399.

Cleland, J. (2013). World population growth; past, present and future. *Environmental and Resource Economics* 55: 543–554.

Conca, J. (2016). After Five Years, What Is The Cost Of Fukushima? Forbes, March 10, 2016. https://www.forbes.com/sites/jamesconca/2016/03/10/after-five-years-what-is-the-cost-of-fukushima/#47bf482f22ed.

Crutzen, P. J. & Stoermer, E. F. (2000). The Anthropocene. Global Change Newsletter, 41, International Geosphere-Biosphere Programme (IGBP).

Edenhofer, O., Lessmann, K., Kemfert, C. et al. (2006). Induced technological change: exploring its implications for the economics of atmospheric stabilization. *The Energy Journal 27, Special Issue 1*: 57–107.

Edenhofer, O., Bauer, N., and Kriegler, E. (2005). The impact of technological change on climate protection and welfare: insights from the model MIND. *Ecological Economics* 54 (2005): 277–292.

Eggoh, J.C., Bangake, C., and Rault, C. (2011). Energy consumption and economic growth revisited in African countries. *Energy Policy* 39: 7408–7421.

Ehrlich, P.R. and Holdren, J.P. (1971). Impact of population growth. *Science* 171: 1212.

EIA (2016). International Energy Outlook 2016, With Projections to 2040. Washington DC, USA.

Eichhorn, W., Henn, R., Neumann, K., and Shephard, R. W. (1992). *Economic Theory of Natural Resources*. Würzburg, Wien: Physica Verlag.

Farhani, S. & Ben Rejeb, J. (2015). Link between Economic Growth and Energy Consumption in Over 90 Countries. Available: https://ideas.repec.org/p/ipg/wpaper/2015-614.html#download.

Fei, L., Dong, S., Xue, L. et al. (2011). Energy consumption-economic growth relationship and carbon dioxide emissions in China. *Energy Policy* 39: 568–574.

Finance (2014). *Global Trends in Renewable Energy Investment 2014*. Frankfurt School of Finance & Management / UNEP Centre / BNEF.

Gilland, B. (1988). Population, economic growth, and energy demand, 1985-2020. *Population and Development Review* 14: 233–244.

Hartshorn, J.E. (1993). *Oil Trade: politics and Prospects*. Cambridge: Cambridge University Press.

Haurie, A. and Viguier, L. (2005). *The Coupling of Climate and Economic Dynamics*. Berlin: Springer.

Hof, A., Den Elzen, M., Admiraal, A. et al. (2017). Global and regional abatement costs of nationally determined contributions (NDCs) and of enhanced action to levels well below 2°C and 1.5°C. *Environmental Science & Policy* 71: 30–40.

Hossein, S.S.M., Yazdan, G.F., and Hasan, S. (2012). Consideration the relationship between energy consumption and economic growth in oil exporting country. *Procedia - Social and Behavioral Sciences* 62: 52–58.

Huang, Y. (2014). Drivers of rising global energy demand: the importance of spatial lag and error dependence. *Energy* 76: 254–263.

Hudson, E.A. and Jorgenson, D.W. (1974). U. S. energy policy and economic growth, 1975-2000. *The Bell Journal of Economics and Management Science* 5: 461–514.

IEA (2014a). Energy Technology Perspectives 2014, Executive Summary. Paris: International Energy Agency/OECD.

IEA (2014b). *World Energy Investment Outlook*. Paris: International Energy Agency/OECD.

IEA (2015a). *CO_2 Emissions from Fuel Combustion*. Paris: International Energy Agency/OECD.

IEA (2015b). *World Energy Outlook 2015*. Paris: International Energy Agency/OECD.

IMF (2016). *World Economic Outlook Database, April 2016*. Washington, DC: International Monetary Fund.

IPCC (2007). *Climate Change 2007: Mitigation of Climate Change. Contribution of Working Group III to the Fourth Assessment Report of the Intergovernmental Panel on Climate Change, 2007* (eds. B. Metz, O.R. Davidson, P.R. Bosch, et al.). Cambridge, United Kingdom/New York, USA: Cambridge University Press.

IPCC (2013). *Climate Change 2013: The Physical Science Basis*. Working Group I Contribution to the IPCC Fifth Assessment Report (eds. T.F. Stocker, D. Qin, G.-K. Plattner, et al.). Cambridge, United Kingdom/New York, USA: Cambridge University Press.

IPCC (2014a). *Climate Change 2014: Mitigation of Climate Change. Contribution of Working Group III to the IPCC Fifth Assessment Report* (eds. O. Edenhofer, R. Pichs-Madruga, Y. Sokona, et al.). Cambridge, United Kingdom/New York, NY: Cambridge University Press.

IPCC (2014b). *Climate Change 2014: Synthesis Report. Contribution of Working Groups I, II and III to the Fifth Assessment Report of the Intergovernmental Panel on Climate Change* (eds. Core Writing Team, R.K. Pachauri and L.A. Meyer). Geneva, Switzerland: IPCC.

Ipsen, D., Rösch, R., and Scheffran, J. (2001). Cooperation in global climate policy: potentialities and limitations. *Energy Policy* 29: 315–326.

Jahangir Alam, M., Ara Begum, I., Buysse, J., and Van Huylenbroeck, G. (2012). Energy consumption, carbon emissions and economic growth nexus in Bangladesh: cointegration and dynamic causality analysis. *Energy Policy* 45: 217–225.

Jorgenson, A.K. and Clark, B. (2010). Assessing the temporal stability of the population/environment relationship in comparative perspective: a cross-national panel study of carbon dioxide emissions, 1960–2005. *Population and Environment* 32: 27–41.

JRC/PBL (2013). *Emission Database for Global Atmospheric Research (EDGAR) Release Version 4.2 FT2010*. European Commission, Joint Research Centre (JRC)/PBL, Netherlands Environmental Assessment Agency.

Kaya, Y. (1990). Impact of Carbon Dioxide Emission Control on GNP Growth: Interpretation of Proposed Scenarios. Paper presented to the IPCC Energy and Industry Subgroup, Response Strategies Working Group, Paris (mimeo).

Klare, M.T. (2001). *Resource Wars – The New Landscape of Global Conflict*. New York: Metropolitan Books.

Kremer, M. (1993). Population growth and technological change: one million B.C. to 1990. *The Quarterly Journal of Economics* 108: 681–716.

Kümmel, R. (1980). *Growth Dynamics of the Energy Dependent Economy*. Königstein/Ts, Cambridge, MA: Hain; Oelgeschlager, Gunn & Hain.

Kümmel, R. (2007). *The Productive Power of Energy and its Taxation*. Talk presented at the 4th European Congress Economics and Management of Energy in Industry. Porto, Portugal, 27-30 November 2007.

Kümmel, R. and Lindenberger, D. (2014). How energy conversion drives economic growth far from the equilibrium of neoclassical economics. *New Journal of Physics* 16: 125008.

Kümmel, R., Lindenberger, D., and Eichhorn, W. (1997). Energie, Wirtschaftswachstum und technischer Fortschritt. *Physik Journal* 53: 869–875.

Kümmel, R., Henn, J., and Lindenberger, D. (2002). Capital, labor, energy and creativity: modeling innovation diffusion. *Structural Change and Economic Dynamics* 13: 415–433.

Kümmel, R., Ayres Robert, U., and Lindenberger, D. (2010). Thermodynamic laws, economic methods and the productive power of energy. *Journal of Non-Equilibrium Thermodynamics* 35 (2): 145–179.

Lee, C.-C. and Chiu, Y.-B. (2011). Oil prices, nuclear energy consumption, and economic growth: new evidence using a heterogeneous panel analysis. *Energy Policy* 39: 2111–2120.

Lempert, R.J. (2002). A new decision sciences for complex systems. *Proceedings of the National Academy of Sciences* 99: 7309–7313.

Lempert, R.J., Scheffran, J., and Sprinz, D.F. (2009). Methods for long-term environmental policy challenges. *Global Environmental Politics*, 08/2009 9 (3): 106–133.

Lindenberger, D. & Kümmel, R. (2011). Energy and the State of Nations. *Energy* 36(10): 6010-6018. https://doi.org/10.1016/j.energy.2011.08.014.

Matisoff, D.C. (2008). The adoption of state climate change policies and renewable portfolio standards: regional diffusion or internal determinants? *Review of Policy Research* 25: 527–546.

Matsuo, Y., Yamaguchi, Y. (2013). The Rise in Cost of Power Generation in Japan after the Fukushima Daiichi Accident and Its Impact on the Finances of the Electric Power Utilities. The Institute of Energy Economics (IEEJ), Japan. http://eneken.ieej.or.jp/data/5252.pdf.

Mattick, C.S., Williams, E., and Allenby, B.R. (2010). Historical trends in global energy consumption. *IEEE Technology and Society Magazine* 29: 22–30.

Meadows, D.H., Meadows, D.L., Randers, J., and Behrens, W.W. III, (1972). *The Limits to Growth - A Report to the Club of Rome's Project on the Predicament of Mankind*. New York: Universe Books.

Michaelowa, A., Allen, M., and Sha, F. (2018). Policy instruments for limiting global temperature rise to 1.5°C – can humanity rise to the challenge? *Climate Policy* 18 (3): 275–286.

Mittelstraß, J. (1984). *Enzyklopädie Philosophie und Wissenschaftstheorie*. Mannheim/Wien/Zürich Bibliographisches Institut AG.

Moreau, V. and Vuille, F. (2018). Decoupling energy use and economic growth: counter evidence from structural effects and embodied energy in trade. *Applied Energy* 215: 54–62.

Murphy, J.M., Sexton, D.M.H., Barnett, D.N. et al. (2004). Quantification of modelling uncertainties in a large ensemble of climate change simulations. *Nature* 430: 768–772.

Murthy, N.S., Panda, M., and Parikh, J. (1997). Economic growth, energy demand and carbon dioxide emissions in India: 1990-2020. *Environment and Development Economics* 2: 173–193.

Nasreen, S. and Anwar, S. (2014). Causal relationship between trade openness, economic growth and energy consumption: a panel data analysis of Asian countries. *Energy Policy* 69: 82–91.

Newman, P. (2017). Decoupling economic growth from fossil fuels. *Modern Economy* 8: 791–805.

Nordhaus, W.D. (1993). Rolling the DICE: an optimal transition path for controlling greenhouse gases. *Resource and Energy Economics* 15: 27–50.

Nordhaus, W.D. and Boyer, J. (2000). *Warming the World: Economics Models of Global Warming*. Cambridge: MIT Press.

OECD (2016). *Fertility. Society at a Glance 2016*. Paris: OECD Social Indicators.

O'neill, B.C., Liddle, B., Jiang, L. et al. (2012). Demographic change and carbon dioxide emissions. *The Lancet* 380: 157–164.

Ozturk, I., Aslan, A., and Kalyoncu, H. (2010). Energy consumption and economic growth relationship: evidence from panel data for low and middle income countries. *Energy Policy* 38: 4422–4428.

Polemis, M.L. and Dagoumas, A.S. (2013). The electricity consumption and economic growth nexus: evidence from Greece. *Energy Policy* 62: 798–808.

Poumanyvong, P. and Kaneko, S. (2010). Does urbanization lead to less energy use and lower CO_2 emissions? A cross-country analysis. *Ecological Economics* 70: 434–444.

Raupach, M.R., Marland, G., Ciais, P. et al. (2007). Global and regional drivers of accelerating CO_2 emissions. *Proceedings of the National Academy of Sciences* 104: 10288–10293.

REN21 (2013). Renewables Global Futures Report. Paris: Renewable Energy Policy Network for the 21st Century and Institute for Sustainable Energy Policies.

REN21 (2016). *Renewables 2016 Global Status Report*. Paris: REN21 Secretariat.

Renshaw, E.F. (1981). Energy efficiency and the slump in labour productivity in the USA. *Energy Economics* 3: 36–42.

Rockstrom, J., Steffen, W., Noone, K. et al. (2009). A safe operating space for humanity. *Nature* 461: 472–475.

Rogelj, J., Luderer, G., Pietzcker, R.C. et al. (2015). Energy system transformations for limiting end-of-century warming to below 1.5 °C . *Nature Climate Change* 5: 519–527.

Rogelj, J., Den Elzen, M., Höhne, N. et al. (2016). Paris agreement climate proposals need a boost to keep warming well below 2 °C. *Nature* 534: 631–639.

Rühl, C., Appleby, P., Fennema, J. et al. (2012). Economic development and the demand for energy: a historical perspective on the next 20 years. *Energy Policy* 50: 109–116.

Sa'ad, S. (2010). Energy consumption and economic growth: causality relationship for Nigeria. *OPEC Energy Review* 34: 15–24.

Saboori, B. and Sulaiman, J. (2013). CO_2 emissions, energy consumption and economic growth in Association of Southeast Asian Nations (ASEAN) countries: a cointegration approach. *Energy* 55: 813–822.

Salahuddin, M. and Gow, J. (2014). Economic growth, energy consumption and CO_2 emissions in gulf cooperation council countries. *Energy* 73: 44–58.

Salim, R.A., Hassan, K., and Shafiei, S. (2014). Renewable and non-renewable energy consumption and economic activities: further evidence from OECD countries. *Energy Economics* 44: 350–360.

Scheffran, J. (1999). Environmental conflicts and sustainable development: a conflict model and its application in climate and energy policy. In: *Environmental Change and Security: a European Perspective* (eds. A. Carius and K.M. Lietzmann). Berlin-Heidelberg: Springer.

Scheffran, J. (2000). The dynamic interaction between economy and ecology: cooperation, stability and sustainability for a dynamic-game model of resource conflicts. *Mathematics and Computers in Simulation* 53: 371–380.

Scheffran, J. (2001). Economic growth, emission reduction and the choice of energy technology in a dynamic-game framework. In: *Operations Research Proceedings 2001: Selected Papers of the International Conference on Operations Research (OR 2001) Duisburg, September 3–5, 2001* (eds. P. Chamoni, R. Leisten, A. Martin, et al.), 329–336. Berlin, Heidelberg: Springer Berlin Heidelberg.

Scheffran, J. (2002). Conflict and cooperation in energy and climate change: the framework of a dynamic game of power-value interaction. In: *Yearbook New Political Economy*, vol. 20 (eds. M. Holler et al.), 229–254. Tubingen, Germany: Mohr Siebeck.

Scheffran, J. (2008a). Adaptive management of energy transitions in long-term climate change. *Computational Management Science* 5: 259–286.

Scheffran, J. (2008b). Preventing dangerous climate change - adaptive decision-making and cooperative management. in long-term climate policy. In: *Global Warming and Climate Change - Ten Years After Kyoto and Still Counting*, vol. 2008 (ed. V.L. Grover), 449–482. Science Publisher.

Scheffran, J. and Cannaday, T. (2013). Resistance against climate change policies: the conflict potential of non-fossil energy paths and climate engineering. In: *Global Environmental Change: New Drivers for Resistance, Crime and Terrorism?* (eds. B. Balasz, C. Burnley, C. Burnley, et al.). Berlin: Springer.

Scheffran, J. and Froese, R. (2016). Enabling environments for sustainable energy transitions: the diffusion of technology, innovation and investment in low-carbon societies. In: *Handbook on Sustainability Transition and Sustainable Peace* (eds. H.G. Brauch, Ú. Oswald Spring, J. Grin and J. Scheffran). Cham: Springer International Publishing.

Scheffran, J. and Leimbach, M. (2006). Policy-business interaction in emissions trading between multiple regions. In: *Emissions Trading and Business* (eds. R. Antes, B. Hansjürgens and P. Letmathe). Heidelberg: Physica-Verlag HD.

Scheffran, J. and Singer, C. (2004). Energy and security – from conflict to cooperation. *INESAP Information Bulletin* 24: 65–70.

Scheffran, J., Brzoska, M., Kominek, J. et al. (2012). Climate change and violent conflict. *Science* 336: 869.

Sheffield, J. (1998). World population growth and the role of annual energy use per capita. *Technological Forecasting and Social Change* 59: 55–87.

Sheng, Y., Shi, X., and Zhang, D. (2014). Economic growth, regional disparities and energy demand in China. *Energy Policy* 71: 31–39.

Sieferle, R.P. (1981). Energie. In: *Besiegte Natur - Geschichte der Umwelt im 19. und 20 Jahrhundert* (eds. F.-J. Brüggemeier and T. Rommelspacher). München: C.H. Beck.

Singer, C. (2008). *Energy and International War: from Babylon to Baghdad and Beyond.* Singapore/Hackensack, NJ: World Scientific.

Smil, V. (2010). *Energy Transitions: History, Requirements, Prospects.* Santa Barbara, CA: Praeger Publishers.

Sorrell, S. and Ockwell, D. (2010). *Can We Decouple Energy Consumption from Economic Growth? Policy Briefing No. 7*. Brighton: Sussex Energy Group.

Stainforth, D.A., Aina, T., Christensen, C. et al. (2005). Uncertainty in predictions of the climate response to rising levels of greenhouse gases. *Nature* 433: 403–406.

Tol, R.S.J. (1999). Time discounting and optimal emission reduction: an application of FUND. *Climatic Change* 41 (3): 351–362.

UN (2015). World Population Prospects: The 2015 Revision, Key Findings and Advance Tables. Working Paper No. ESA/P/WP.241., New York, United Nations Department of Economic and Social Affairs, Population Division.

UNEP (2011). Decoupling natural resource use and environmental impacts from economic growth. A Report of the Working Group on Decoupling to the International Resource Panel, United Nations Environment Programme.

UNEP (2016). *The Emissions Gap Report 2016*. United Nations Environmental Programme (UNEP): Nairobi.

UNFCCC (2010). *Adoption of the Cancun Agreements*. Paris: United Nations Framework Convention on Climate Change.

UNFCCC (2015). Adoption of the Paris Agreement. Paris: UN.

Warr, B. and Ayres, R.U. (2012). Useful work and information as drivers of economic growth. *Ecological Economics* 73: 93–102.

WCED (1987). Our Common Future: Brundtland Report of the World Comission on Environmentand Development. Oxford.

von Weizsäcker, E.U. (1994). *Earth Politics*. London, UK: Zed Books.

von Weizsäcker, E.U., de Larderel, J., Hargroves, K., Hudson, C., Smith, M. & Rodigues, M. (2014). *Decoupling 2: technologies, opportunities and policy options*. A Report of the Working Group on Decoupling to the International Resource Panel.

Wrigley, E.A. (2013). Energy and the English industrial revolution. *Philosophical Transactions of the Royal Society. A: Mathematical, Physical and Engineering Sciences* 371 (1986): 20110568.

Yergin, D. (1991). *The Prize: The Epic Quest for Oil, Money, and Power*. New York: Simon & Schuster.

Yohe, G., Andronova, N., and Schlesinger, M. (2004). To hedge or not against an uncertain climate future? *Science* 306: 416.

2

The Energy Mix in Japan Post-Fukushima

Seiji Nakagame

The Faculty of Applied Bioscience, Kanagawa Institute of Technology 1030 Shimoogino, Atsugi, Kanagawa, 243-0292, Japan

CHAPTER MENU

2.1 Greenhouse Gas (GHG) Emissions by Japan

The average temperatures on Earth have been gradually noticeably increasing for the past three decades; this phenomenon is now unambiguously ascribed to increasing greenhouse gas (GHG) emissions such as emissions of CO_2, methane, nitrous oxide, and other gases produced by human activity (WMO 2016). The level of atmospheric CO_2 was 400 ppm in 2015, which was elevated by 122 ppm compared with pre-industrial times, defined as before 1750 (WMO 2016). It is becoming increasingly clear that this continuing rise of the concentrations of GHGs in the atmosphere is increasing the average temperature and causing negative impacts on human society and natural ecosystems, notably by increasing the frequency and severity of major storm or typhoon events (IPCC 2015). The year 1999 was a landmark year in the mission to limit and mitigate climate change, as the Kyoto protocol became effective that year. The initial commitment of the Japanese government in this global attempt to protect the atmosphere was to achieve the reduction of GHG emission by 6% compared with that of 1990. However, Japan had to pull out of the Kyoto protocol as this goal became unattainable after the shutdown of the nuclear plants in the country following the major nuclear disaster that occurred in Fukushima, a consequence of the black swan event that constituted the earthquake of the millennium and resulting tsunami in that region. A direct

Green Energy to Sustainability: Strategies for Global Industries, First Edition.
Edited by Alain A. Vertès, Nasib Qureshi, Hans P. Blaschek and Hideaki Yukawa.

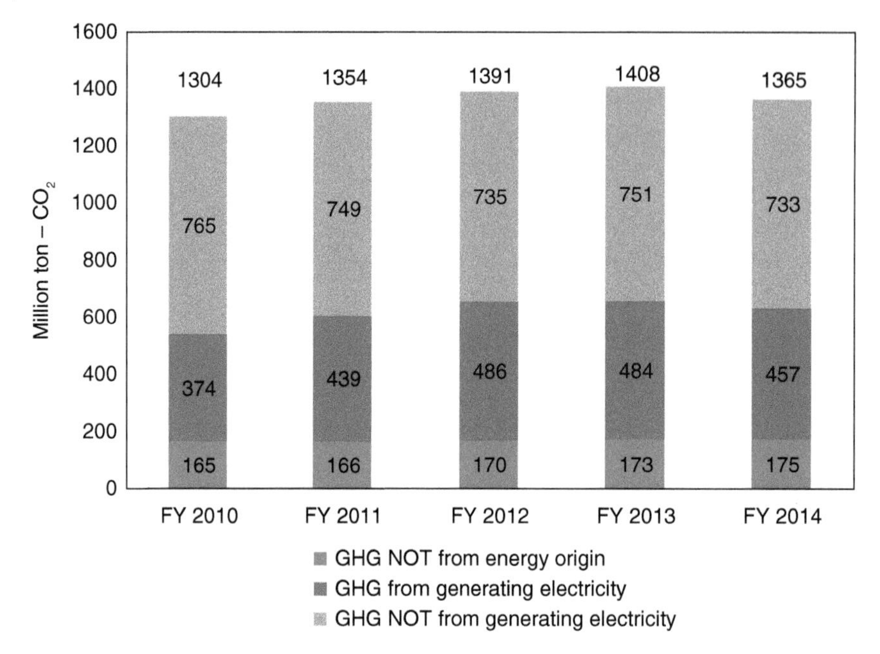

Figure 2.1 Trends in the amounts of greenhouse gas (GHG) emission in Japan (Agency for Natural Resources and Energy 2016).

consequence of this necessary shutdown was the increased dependency on thermal plants that inevitably led to increased CO_2 emissions. In FY 2014, Japan emitted about 1365 million tons of CO_2 (Figure 2.1), a total which represents approximately 3% of the annual GHG emission of the world. Using a parameter that comprises an economic term however, the GHG of Japan in FY 2013 was 0.29 kg CO_2/\$GDP, which was almost identical to that of the European Union, whereas the US emitted 0.47 kg CO_2/\$GDP. In the present configuration, it is extremely challenging for Japan to attain a level of 0.16 kg CO_2/\$GDP by FY 2030, which represents a 40% decrease in the emission of CO_2 per \$GDP and a level lower that the presentCO_2 emissions of either the US or the EU.

2.2 Energy Dependence

As highlighted earlier, although Japan had been highly dependent on fossil fuels such as oil, natural gas, and coal for generating electricity, the Great East Japan Earthquake in 2011 drastically increased that dependence due to the shutdown of its nuclear plants. The big earthquake-triggered tsunami broke the safety systems of the nuclear plant Fukushima Daiichi Nuclear. As a result, 130 PBq of ^{131}I and 15 PBq of ^{137}Cs were discharged from the Fukushima Daiichi Nuclear plant within a few days following the tsunami (NISA 2011a). Although the catastrophe in Fukushima was rated at the top level 7 on the International Nuclear and Radiological Event Scale (NISA 2011b), which is same as the catastrophic Chernobyl accident in 1986, the amounts of discharged radionuclides in Fukushima is about one-seventh of that in the Chernobyl accident (NISA 2011a). The catastrophe deeply contaminated soil and water with ^{131}I, and ^{137}Cs, with half-lives of eight days, and 30.2 years, respectively, which were detected in Tokyo, located only 200 km from the Fukushima Daiichi Nuclear plant. As would be expected, almost all agricultural crops in the Fukushima prefecture just after the accident showed levels of these radioactive nuclides that were much higher than the maximal regulatory limits (Hamada and Ogino 2012). About 85 000 people who lived in the Fukushima prefecture had to be evacuated from the area due to the catastrophe (Fukushima prefecture 2016a). Before the nuclear plant accident, the electricity generated by nuclear plants, coal-fired power plants, natural gas-fired plants, and oil-fired plants accounted for 29%, 25%, 29%, and 8%, respectively of the total generated electricity in Japan (Figure 2.2). As of February 2017, only three nuclear plants are operating, while 40 nuclear plants have been shut down. The ratio of coal-fired plants, natural gas-fired plants, and oil-fired plants reached 30%, 43%, and 15%, respectively of the total generated electricity in Japan in FY 2013 (Figure 2.2). As a result, the energy self-sufficiency rate of Japan dramatically dropped from 20% in FY 2010 to only 6% in FY 2012. What is more, the proportion of electricity generated by fossil fuels reached about 90% in FY 2014. In addition to the high dependency on fossil fuels for generating electricity, more than 99% of the oil and coal that is used in Japan needs to be imported, mainly from Middle-East countries and Australia. Similarly, 97% of the natural gas used in Japan needs to be imported, also mainly from the Middle-East. Consequently, in a list of the energy self-sufficiency rates for countries in the OECD (Figure 2.3), Japan was 33rd, or second last, in comparison the energy self-sufficiency

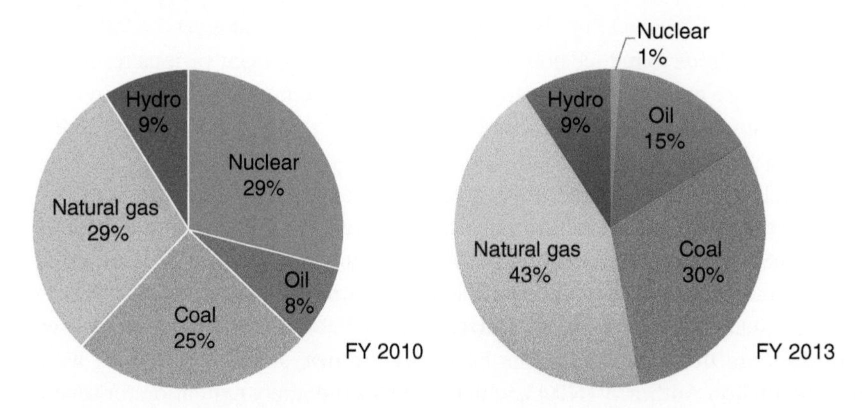

Figure 2.2 Change of the mix of electricity production in Japan before and after the nuclear accident in Fukushima (The Federation of Electric Power Companies of Japan 2015).

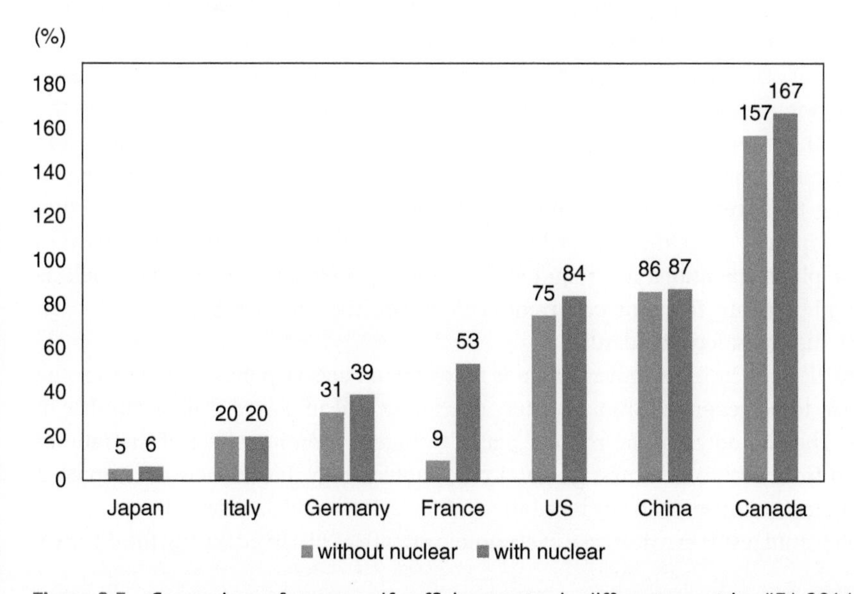

Figure 2.3 Comparison of energy self-sufficiency rates in different countries (IEA 2014a,b).

rate of countries such as Spain (26.7%), Italy (20.1%), and Korea (17.5%) is much higher, in spite of the lack of fossil fuel resources of these countries. Increasing the energy self-sufficiency rate and diversifying the sources of imported fossil fuels are critically important goals for Japan to make its economy less subject to systemic risk. Here, opportunities exist on the global fossil fuels market since it is reported that the fossil fuel demand of China will not increase as much as previously, notably because the growth rate of the Chinese economy has not been increasing as fast as it did in recent years (BP 2016). In addition, the production of shale gas and shale oil, which are mainly produced in North America, are dynamic and depend on oil prices, thus mitigating further the supply and demand tension. Thus, Japan can actively monitor the global energy market and respond with flexibility to global tendencies such as to secure a stable energy supply for key economic sectors such as industry and household in order to ensure its continuous economic growth (IEA 2016)

2.3 The Energy Policy of Japan

The Japanese government has established that the energy used in Japan should meet the following criteria: (i) highest safety standards to reduce pollution risks, (ii) obtained from various resources and various countries to reduce exposure to sourcing risks, (iii) high economic efficiency to optimize GDP, and (iv) environmentally friendly to prevent national pollution and contribute to the worldwide goal of limiting climate change.

The accident at the nuclear plant in Fukushima caused a long-lasting environmental catastrophe. About 640 km^2 in the Fukushima prefecture remains restricted given the high radioactivity that remains (above 20 mSv per year) (Fukushima prefecture 2016b). In addition, it is expected that no less than 30–40 years is required to decommission the damaged nuclear plant (TEPCO 2013). However, three nuclear plants have been restarted as of 2016 after the national shutdown of all plants in 2013, because these nuclear plants have met the new safety criteria that were made by the Nuclear Regulation Authority (NRA), which includes the safety regulation for severe accidents and the adoption of new regulations even for the already licenced facilities (NRA 2013). The remaining 40 nuclear plants in Japan have still not yet resumed, either because they do not meet the revised criteria or because of the fierce opposition of the local residents and local governments. Although the restart of the nuclear plants would very significantly contribute to the reduction of the CO_2 emission of Japan, in the aftermath of the Fukushima nuclear disaster, the Japanese government believes that both careful inspection of the nuclear plants based on the revised criteria, now the most stringent in the world, and the approval of the local residents and the local government are required before each of the nuclear plants can operate again. Beyond strictly implementing these revised criteria, the Japanese government has also established that the continuous development of technology for increasing safety and the education of employees involved in nuclear plant operations are required. Considering that natural disasters such as earthquakes and typhoons are frequent occurrences in Japan, thermal plants also have been required to fulfil the highest safety standards.

As emphasized earlier, the Japanese government is following an energy policy that aims for the energy used in Japan to be generated from a variety of sources and imported from a number of different countries. The shutdown of the nuclear plants initiated a domino effect of the ratio of electricity generated from coal, oil, natural gas, and renewable energy. Fossil fuels are imported mainly from Middle-East countries, and any instability in this region could decrease the availability of fossil fuels, which would result in a decrease in economic growth as observed during the oil crisis in the 1970s.

High economic efficiency for generating electricity and the development of environmentally-friendly energy production processes that emit lower amounts of atmospheric CO_2 are important, as emphasized by the increased in electricity prices and atmospheric CO_2 levels that directly resulted from the increase in the use of oil, coal, and natural gas thermals due to the shutdown of the nuclear plants in Japan post-Fukushima. A direct implication of this is that generating electricity from renewable sources such as solar, wind, and biomass energy has been encouraged by the government of Japan especially after the shutdown of its nuclear plants. The ratio of electricity from renewable resources was 10.4% before 2011, while that level increased to 12.2% in FY 2014 due to the introduction of the policy to buy electricity generated from renewable resources at a fixed price that provide incentives for Japanese businesses and homeowners (Agency for Natural Resources and Energy 2016). The reasons that the electricity price for domestic use and industry increased during that same period by about 20% and 30%, respectively, was that the fixed price from renewable energy and electricity price from thermal plants was higher than the energy generated by nuclear plants (Agency for Natural Resources and Energy 2016).

Generating electricity from lignocellulose is one of the most promising ways to reduce CO_2 emission in Japan. Japan comprises approximately 250 thousand km^2 of forests, representing about 67% of its total area. However, most of these forests are located in mountainous areas, which makes their exploitation difficult thus the price of Japan-grown lignocelluloses tends to be high. As a result, about 70% of the wood used in Japan, even for housing purposes, is imported from other countries (Forestry Agency 2014). The cost of renewable electricity varies with the type of renewable energy, and there are even important variations within a renewable energy class depending notably on the price of raw materials (IEA 2016). Of the electricity generated from lignocellulosic biomass, forest thinning in Japan is bought at the highest price (IEA 2016). The fixed price policy implemented by Japan to incentivize the production of electricity generated from renewable resources will continue to be enforced for 20 years as a means to increase investment in this type of renewable electricity and thus to build the critical mass that is required for Japan to develop economies of scales that will positively and significantly impact Japan's GDP and economic performances. Thermal power is a term used generically here to comprise oil-fired power plants, coal-fired power plants, and natural gas-fired power plants. Japan's current policy is to decrease its dependence on oil-fired power plants, because these power plants are expensive compared with coal-fired power plants or natural gas-fired plants. Although power generated by coal-fired power plants is cheaper and is available as a stable supply, its major drawback is the high amounts of CO_2 that are emitted per unit of energy. Consequently, high energy efficient processes to decrease CO_2 emissions by coal-fired power plants have been developed. Moreover, these new technologies are planned to be used in developing countries, which will contribute to further decrease the amounts of CO_2 emission.

2.4 Paris Agreement

Japan is a signatory of the Paris agreement that stipulates that each committed country should take actions to reduce the increase of the average global temperature to less than 2 °C compared with the pre-industrial era (before 1750)(UNFCCC 2015). Via this agreement, Japan committed to reduce its 2030 CO_2 emissions by 26.0% from that of 2013. To this end, Japan decided in 2016 upon the strategies necessary to attain the target of the Paris agreement, namely, the Innovative Energy Strategy and the National Energy and Environment Strategy for Technological Innovation Towards 2050 (NESTI 2050) (Council for Science Technology and Innovation 2016).

2.5 Prospective Energy Demand

The current population in Japan is 126 million; this number is decreasing due to a declining birth rate. On the other hand, although the declining birth rate combines with a growing proportion of elderly people, with as possible consequence a decrease of the GDP, the Japanese government estimates that the energy demand will in any case increase based on the assumption of an economic growth rate of 1.7% per year. In 2015, the Japanese government forecast that the energy demand in FY 2030 will be 326 million kl, which is a 35 million kl energy saving compared to FY 2013. This assumption is based on the premise that a 50 million kl energy saving will be made possible by the development of energy saving technology.

2.6 Improvement in Energy Efficiency

To achieve the goal of a 50 million kl energy saving, the industrial and business sectors are expected to reduce the energy consumption in these areas by 10 million kl and 12 million kl, respectively. Although the energy efficiency of the industrial sector has dramatically improved by more than 40% since the oil crisis in 1970s, further improvements will be necessary to achieve the ambitious goals of 10 million kl reduction in consumption of the industrial sector. Efforts required from the business sector will be even higher. To improve the energy efficiency of the business sector, the Japanese government is planning to make use of a benchmark system, which can subjectively evaluate how much each company is working on to improve its energy efficiency. This benchmark calculated from the 10–20% of the top companies will set the specific goal that each company needs attain. In a program that has started in FY 2016, about 10 000 companies, that need to submit a periodic report based on the energy efficient law will be classified from S to C based on how hard they are working to improve their energy efficiencies. The companies that fall within the class C category will receive training by the Japanese government to improve their energy efficiencies. Although some companies are already efficiently re-using the heat generated in their plants, there remains a very large amount of in situ-generated heat energy that is remains unrecycled. To promote the efficient use of heat produced in industrial plants, the Japanese government recently implemented a new policy to promote the recycling use of heretofore discarded heat or vapour energy (Agency for Natural Resources and Energy 2016).

However, the industrial sector is not the only sector that is encouraged to improve its overall efficiency. In particular, the household sector is also required to significantly improve its energy efficiency. An obvious start is to significantly increase the deployment of heat insulation for houses, as saving remains the most powerful and immediate way to improve energy efficiency for this sector, notably considering that about 30% of the energy used in a typical household is for air conditioning and heating. In addition, since about 40% of the houses in Japan (that is, an approximate total of 20 million houses) are thought not to be equipped with appropriate heat insulation, the Japanese government is planning to implement a policy that would require that newly constructed private houses and residential buildings contain the heat insulation material necessary to achieve energy saving goals in the household sector. The Japanese government is aiming at reducing the energy in the household sector by 11.6 million kl in FY 2030 (Agency for Natural Resources and Energy 2016). The ambitious objective of the Japanese government is that on average newly constructed private houses and residential buildings should exhibit a zero energy balance, a project called 'net zero energy house (ZEH)', by 2030 (Agency for Natural Resources and Energy 2016). Within this scenario, the Japanese energy self-sufficiency ratio in FY 2030 will be 24.3%.

2.7 Reduction of CO_2 Emission in Electric Generation

Reducing CO_2 emissions resulting from the generation of electricity is a very effective way to reduce GHG, considering that electricity is the primary driving force of economic activities. As a result, the power industry as a whole has tried to reduce its CO_2 emissions, as exemplified in the Kyoto protocol where the power industry set the goal to reduce its CO_2 emissions by 20% compared with that of FY 1990, which is equal to 0.34 kg CO_2/kWh (Kyoto protocol 2005). However, the shutdown of Japanese nuclear plants, indirectly resulting from the black swan earthquake and tsunami of 2011, made the attainment of this goal totally impossible, because thermal power generation had to compensate for this loss in energy generation capacity (Agency for Natural Resources and Energy 2016). In 2015, the power industry revised its target based on the prospect of the energy demand in FY 2030, setting the goal at 0.37 kg CO_2/kWh (Agency for Natural Resources and Energy 2016). Importantly, it was decided by the act on the rational use of energy that the efficiency of electric power generation for coal, natural gas, and oil, which is newly constructed, should at a minimum be 42.0%, 50.5%, and 39.0%, respectively. In addition, the ratio of electricity generated from renewable resources should represent a minimum of 44% of the total electricity sold to the end consumers by electricity retailers in FY 2030 (Agency for Natural Resources and Energy 2016).

2.8 Development of New Technologies for Decreasing GHG Emissions

As lignocellulosic biomass is produced by photosynthesis, it is regarded as carbon neutral when it is burned. Thus, research and development of biofuels from lignocellulosic biomass have been primary objectives in Japan.

2.9 Production and Use of Bioethanol in Japan

Bioethanol has been produced by the fermentation of edible biomass such as corn or sugar cane for several decades already (Guo et al. 2015). Bioethanol is used not only as a fuel for combustion engine powered vehicles, but also as a sustainable chemical feedstock (Morschbacker 2009). As there are not enough available areas for producing crops cheaply in Japan, the crops for producing ethanol for industrial use have to be imported, in addition to the import of bioethanol for motor vehicles and chemicals from Brazil and the US. In 2015, Japan imported about 606 million litres of ethanol for fuel (including the ethanol in imported ethyl tertiary butyl ether [ETBE]) (USDA Foreign Agricultural Service 2016). To produce fuels for motor vehicles, bioethanol is reacted with isobutene to produce ETBE. In FY 2010, 210 000 kl (crude oil equivalent) of ETBE was blended with gasoline and used for motor vehicles, the concentration of ETBE in the final gasoline product ranged from 1% to 8% vol (Petroleum Association of Japan 2015). However, as bioethanol from edible biomass has been perceived by the general public as one of the main causes of increasing food prices, the development of bioethanol from inedible biomass such as wood and corn stover is currently being promoted. In spite of intensive research into the production of bioethanol from inedible biomass being conducted in Japan and many other countries for the past decade, the production cost of bioethanol produced from these feedstocks still impedes the industrialization of this particular source of bioethanol (Tan et al. 2016). One example of a research project in this area is the national project conducted by the New Energy and Industrial Technology Development

Organization (NEDO) by way of various grants and incentives for companies in the pulp and paper industry and the oil refinery industry. Although pilot scale plants were built for defining a path to practicability, it is expected that further reduction in production costs is still required for full scale industrialization to occur (NEDO 2014a).

2.10 Production and Use of Hydrocarbons in Japan

One of the disadvantages of using ethanol as a fuel for motor vehicles is that this alcohol has a lower calorific value than that of gasoline, given its much higher oxygen content (Brinkman 1981). A natural hedge against this chemical constraint is to produce hydrocarbons from lignocellulosic materials rather than ethanol, of course the production process needs also to compete on an economic basis with the production of ethanol by fermentation, which is already a well-developed, simple and industrially robust process (Xue et al. 2015). For example, the pyrolysis of lignocellulose to produce syngas, and particularly carbon monoxide followed by its reaction with water was studied (Alonso et al. 2010). Another exciting area of research is the production of carbohydrates using algae from glucose derived from lignocellulosic biomass; similarly, the direct photosynthesis by algae is a very promising process (IHI 2015). These new biofuels could be integrated into the supply of jet fuels by mixing them with conventional fuels at a concentration of up to 10% of total jet fuel (Shoji 2015).

2.11 Production and Use of Hydrogen in Japan

In addition to electricity and heat, hydrogen is expected to play an important role as a secondary energy source in the near future (Mekhilef et al. 2012). For example, fuel-cell vehicles, which are powered not by a combustion engine but rather by electricity generated from the chemical reaction between oxygen and hydrogen, have been sold since 2002 by Japanese automobile companies such as Toyota and Honda. In addition, more than 150 000 household fuel cells are currently in use in Japan to generate electricity and heat from natural gas (Ren and Gao 2010). Although the price of household cells remains very expensive (about $15 000), the Japanese government is planning to spread their use by subsidizing this segment of the industry. The establishment of a national hydrogen supply from renewable resources will be beneficial for Japan, due to its current high dependency on imported energy. Currently, hydrogen gas can be obtained from the reaction of fossil fuels with steam, co-products of sodium hydroxide, and purification of coke (Agency for Natural Resources and Energy 2016; Dincer and Acar 2015). In the future, the production of hydrogen from lower grade coal and the electrolysis of water using electricity generated from renewable energy such as solar and wind power would be possible (Agency for Natural Resources and Energy 2016; Dincer and Acar 2015). The production of cheaper hydrogen gas in Japan would enable the use of fuel cell vehicles and household fuel cells more cost effectively, which in turn would significantly contribute to reducing CO_2 emission levels in Japan.

2.12 Contributions of the Japanese Government to Fundamental Research and Development

Reducing CO_2 emissions at the global level is critical for reducing global warming and climate change caused by anthropogenic activities. However, to attain this goal meaningfully, further fundamental research is required as highlighted, for example, during the conference COP21 that was held in Paris in 2015 (UNFCCC 2015). Japanese companies tend to tackle industrial projects that can be industrialized within a decade (Innovation100 Committee Secretariat 2016). Technologies that would require a longer term to reach commercial maturity are typically beyond the commercial horizon of most private enterprises. As a result, various Japanese governmental agencies such as the Ministry of Education, Culture, Sports, Science and Technology (MEXT) and the Ministry of Economy, Trade and Industry (METI), which comprises NEDO, have been subsidizing research and development on technologies for reducing CO_2 emissions that have a long time horizon to commercial maturity (Matsumoto et al. 2009; USDA 2009). Notably, the Japanese government not only provides incentives for research but is also encouraging partnerships and collaborative projects among government, industry, and academia (Inui et al. 2010; NEDO 2014b).

2.13 Perspectives

The nuclear plant catastrophe that occurred in 2011 at Fukushima substantially changed the energy policy of Japan. Notably, its dependency on fossil fuels was drastically increased by the shutdown of the nuclear plants. As a result, the amounts of CO_2 released into the atmosphere by Japanese companies dramatically increased compared to those before 2011. Japan is currently challenged to reduce its CO_2 emission, but is implementing specific strategies towards this goal, which include increasing energy efficiency and the use of renewable energy. Technological innovations as well as operational improvements will enable Japan to overcome the difficulty that resulted from the necessary shutdown of its nuclear plants. Achieving this goal will in turn enable the country to decrease its CO_2 emissions as it committed to do by implementing the Paris agreement (UNFCCC 2015).

References

Agency for Natural Resources and Energy, FY2015 Annual Report on Energy (Energy White Paper). 2016.

Alonso, D.M., Bond, J.Q., and Dumesic, J.A. (2010). Catalytic conversion of biomass to biofuels. *Green Chemistry* 12 (9): 1493–1513.

BP, BP Statical Review 2016 China's energy market in 2015. 2016.

Brinkman, N.D., Ethanol Fuel-A Single-Cylinder Engine Study of Efficiency and Exhaust Emissions. 1981, SAE International.

Council for Science Technology and Innovation, National energy and environment strategy for technological innovation towards 2050. 2016.

Dincer, I. and Acar, C. (2015). Review and evaluation of hydrogen production methods for better sustainability. *International Journal of Hydrogen Energy* 40 (34): 11094–11111.

Forestry Agency, Ministry of Agriculture, Forestry and Fisheries, Annual Report on Forest and Forestry in Japan Fiscal Year, 2014.

Fukushima prefecture. Number of evacuees. 2016a [cited 2017 6 March]; Available from: https://www.pref.fukushima.lg.jp/site/portal-english/en03-08.html.

Fukushima prefecture. Transition of evacuation instruction zones. 2016b [cited 2017 6 March]; Available from: https://www.pref.fukushima.lg.jp/site/portal-english/en03-08.html.

Guo, M.X., Song, W.P., and Buhain, J. (2015). Bioenergy and biofuels: history, status, and perspective. *Renewable & Sustainable Energy Reviews* 42: 712–725.

Hamada, N. and Ogino, H. (2012). Food safety regulations: what we learned from the Fukushima nuclear accident. *Journal of Environmental Radioactivity* 111: 83–99.

IEA, Energy Balances of OECD contries 2014. 2014a.

IEA, Energy Balances of Non-OECD countries 2014. 2014b.

IEA, Energy Policies of IEA Countries -Japan Review 2016-. 2016.

IHI. Success in massive scale algae cultivation for Biofuel. 2015 [cited 2017 6 March]; Available from: www.ihi.co.jp/en/all_news/2015/press/2015-5-21/index.html.

Innovation100 Committee Secretariat, Ministry of Economy Trade and Industry, Japan, Innovation Network World Innovation Lab, Who is responsible for creating innovation in companies? Five guidelines for action based on the insights gained from the pioneering work of 17 corporate executives. 2016.

Inui, M., Vertès, A.A., and Yukawa, H. (2010). Advanced fermentation technologies. In: *Biomass to Biofuels: Strategies for Global Industries* (eds. N. Qureshi, A.A. Vertès, H.P. Blaschek and H. Yukawa). Oxford, UK: Blackwell Publishing Ltd.

IPCC, Climate Change 2014-Impacts, Adaptation, and Volunerability. Part A: Global and Sectoral Aspects. 2015.

Kyoto protocol, Kyoto protocol target achievement plan. 2005 [cited 2017 6 March]; Available from: http://japan.kantei.go.jp/policy/kyoto/050428plan_e.pdf.

Matsumoto, N., Sano, D., and Elder, M. (2009). Biofuel initiatives in Japan: strategies, policies, and future potential. *Applied Energy* 86: S69–S76.

Mekhilef, S., Saidur, R., and Safari, A. (2012). Comparative study of different fuel cell technologies. *Renewable & Sustainable Energy Reviews* 16 (1): 981–989.

Morschbacker, A. (2009). Bio-ethanol based ethylene. *Polymer Reviews* 49 (2): 79–84.

NEDO, Development of an Innovative and Comprehensive Production System for Cellulosic Bioethanol/R&D for Comprehensive Bioethanol Production System/Development of a Comprehensive Bioethanol Production System from Fast Growing Trees Using Mechanochemical Pulping (FY2009-FY2013) Final Report 2014a.

NEDO, Development of an integrated system for the low-cost cellulosic bio-ethanol production from energy crops cultivation to conversion process based on environmentally-friendly pre-treatment technology (FY2009-FY2013). 2014b.

NISA. Evaluation of the Status of Reactor Cores in Units 1–3 of the Fukushima Nuclear Power Plant 1. 2011a [cited 2017 6 March]; Available from: http://www.meti.go.jp/earthquake/nuclear/pdf/20110606-1nisa.pdf.

NISA. INES (the International Nuclear and Radiological Event Scale) Rating on the Events in Fukushima Daiichi Nuclear Power Station by the Tohoku District - off the Pacific Ocean Earthquake. 2011b [cited 2017 6 March]; Available from: http://warp.ndl.go.jp/info:ndljp/pid/3514506/www.nisa.meti.go.jp/english/files/en20110412-4.pdf.

NRA. New Regulatory Requirements for Light-Water Nuclear Power Plants. 2013 [cited 2017 6 March]; Available from: https://www.nsr.go.jp/data/000067212.pdf.

Petroleum Association of Japan, Petroleum industry in Japan 2015. 2015: p. 58–60.

Ren, H.B. and Gao, W.J. (2010). Economic and environmental evaluation of micro CHP systems with different operating modes for residential buildings in Japan. *Energy and Buildings* 42 (6): 853–861.

Shoji, Y., Euglena plans Japanese refinery for algae-derived jet fuel, Nikkei Asian Review. 2015.

Tan, H.T., Corbin, K.R., and Fincher, G.B. (2016). Emerging Technologies for the Production of renewable liquid transport fuels from biomass sources enriched in plant cell walls. *Frontiers in Plant Science* 7: 18.

TEPCO. Mid- and-long term roadmap towards the decommissioning of TEPCO's Fukushima Daiichi nuclear power station units 1–4. 2013 [cited 2017 6 March]; Available from: http://www.meti.go.jp/english/press/2013/pdf/0627_01.pdf.

The Federation of Electric Power Companies of Japan, Graphical Flip-chart of Nuclear & Energy Related topics 2015. 2015.

UNFCCC. Paris agreement. 2015 [cited 2017 6 March]; Available from: http://unfccc.int/files/essential_background/convention/application/pdf/english_paris_agreement.pdf.

USDA Foreign Agricultual service, Japan to Focus on Next Generation Biofuels, Japan: Biofuels Annual 2009.

USDA Foreign Agricultural Service, Market for Liquid Biofuels Remains Steady as Japan Remains Focused on Advanced Fuels. Japan: Biofuels Annual, 2016.

WMO, The State of Green Gases in the Atmosphere Based on Global Observations through 2015. WMO Greenhouse Gas Bulletin, 2016 (12).

Xue, Y.P., Jin, M., Orjuela, A. et al. (2015). Microbial lipid production from AFEX (TM) pretreated corn stover. *RSC Advances* 5 (36): 28725–28734.

3

Green Energy in Africa, Asia, and South America

Daniel de Castro Assumpção[1], *Marcelo Hamaguchi*[2], *José Dilcio Rocha*[3] *and Adriano P. Mariano**[,1]

[1] *University of Campinas (UNICAMP), School of Chemical Engineering, Campinas, SP, Brazil*
[2] *Valmet, Araucária, PR, Brazil*
[3] *Embrapa Agroenergy - The Brazilian Agricultural Research Corporation (Embrapa), Brasília, DF, Brazil*

CHAPTER MENU

3.1 Introduction

Green energy has been increasing steadily worldwide, and in 2015 more than half of all added power generation capacity came from renewables. In comparison to 2004, investments grew by more than six times, reaching almost $300 billion. China alone was responsible for about 36% of the rise in investment. On the other hand, while investment in Africa also increased, in Brazil it fell by 10%, mostly due to an economic recession (Frankfurt School, UNEP, BNEF 2016). Whereas these numbers express the widespread consensus among nations that investments in green energy are needed to reduce the risks of climate change, in practice the speed of transformation is dictated by the particular socioeconomic and environmental characteristics of each country or region.

In this chapter, we attempt to elucidate the transformation rates and respective driving forces, as well as contrary forces found in three very distinctive regions: South America, Africa, and Asia (Southeast Asia and China). For example, located in South America, Brazil is one of the largest renewable energy markets globally (Irena 2016). Ethanol and hydro- and biomass-derived power account for approximately 40% of the total energy consumption (World Bank 2014). Nonetheless, the emerging second-generation biofuel industry in that country still relies on imported technologies due to restricted investments in indigenous technological solutions. In Africa, the nascent renewable energy industry is challenged by a profound lack of infrastructure and an intense debate over land-use competition and food security. This debate, or challenge, has also been prominent in Southeast Asia and China. The former is one of the world's major players in the biofuel industry.

Green Energy to Sustainability: Strategies for Global Industries, First Edition.
Edited by Alain A. Vertès, Nasib Qureshi, Hans P. Blaschek and Hideaki Yukawa.

However, the rapid growth of this industry has led to increasing rates of deforestation in recent years. In China, the green energy industry is struggling to find its way in a massive carbon-intensive overpopulated country, with diminishing availability of arable lands and fresh water, as well as with serious environmental pollution problems.

The first aim of this chapter is to inform the status of commercial deployment of green energy technologies – with emphasis on biofuels – in South America, Africa, South East Asia, and China. As such, for each of these regions, the 'Current status' section brings an overview of the current energy mix. This section also investigates the present situation of the renewable energy industry in these locations, including socioeconomic and environmental aspects, government support, and policies in place. The second aim is to collect representative commercial initiatives and identify their major challenges. This topic is covered in the 'Commercial deployment and challenges' section. Finally, the potential of growth of the renewable energy industry in the studied regions is critically analysed in the 'Perspectives' section.

3.2 South America

3.2.1 Current Status

For decades, biomass has been considered as a potential source of renewable energy in South America, and Brazil is naturally at the forefront considering its vast area, its prosperous sugarcane industry, its reserves of planted forest, and its overall resources for producing vegetable oils. In 2015, the area under sugarcane cultivation reached 10.5 million hectares (UNICADATA 2015), concentrated mostly in South–Central Brazil. This resulted in the production of 30 million m^3 of ethanol and the generation of 19 million MWh of surplus electricity through the combustion of sugarcane bagasse (EPE 2015).

The history of ethanol as a fuel in Brazil dates back to the beginning of the twentieth century. However, it was only in the late 1970s, as a government response to the world oil crisis, that the production of bioethanol soared to billions of litres. At that period, the Brazilian federal government put in place an ethanol-blending mandate for gasoline (up to 25%) and incentivized the production of cars powered by hydrous ethanol (93% ethanol and 7% water). Ethanol also received subsidies of about $30 billion, generated from taxes on gasoline, until the early 2000s when production cost was already competitive compared to gasoline (Goldemberg 2007). In 2003, with the advent of flex-fuel cars and the resulting market growth, ethanol started to be projected to the world as a sustainable and promising fuel, which encouraged the ethanol industry to further invest and expand its production capacity. Nevertheless, this prosperous scenario suffered a reversal in the wake of the 2008 financial crisis, reaching a deep downturn when the Brazilian government decided to subsidize imported gasoline as a means of inflation control (Wilkinson 2015). As a consequence, bank credit lines became scarce and the ethanol industry started losing competitiveness. Although signs of recovery have been observed since 2015, these events show how political interventions and the lack of long-term commitment can put the stability of the biofuel market at risk.

In most of the South American countries, biofuels are produced to meet domestic demand driven by local blending mandates (Figure 3.1). For instance, in Brazil there is a mandate for anhydrous ethanol, which is mixed with gasoline (E27 blend), and the demand for hydrous ethanol by flex-fuel cars is dictated by market forces. On the other hand, in Peru some of the ethanol producers find it more attractive to export the biofuels produced locally, meaning that paradoxically this country usually relies on imports to meet its internal demand (Nolte and Beillard 2015). Also, with focus on

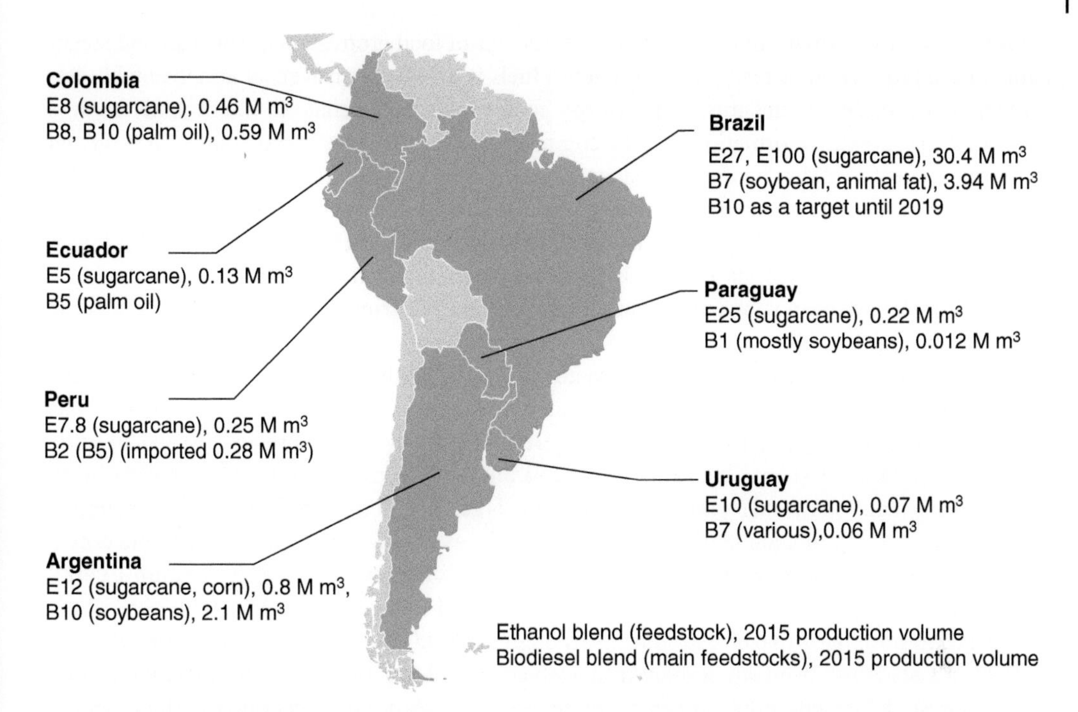

Figure 3.1 Major biofuel (ethanol and biodiesel) producing countries in South America and respective mandates and 2015 production volumes. Source: Data from UNICADATA 2015; Joseph 2015; Nolte and Beillard 2015.

international markets, since 2013 Argentina has been struggling against trade barriers imposed by the European Union (EU). Notably, in 2015, biodiesel exports to the EU declined by 50% (Joseph 2015). This is an alarming situation taking into account that the supply to Europe represents almost 80% of the total Argentinian biodiesel exports. In some parts of the continent, climate conditions and the availability of fertile land can limit the local production of biofuels. In Chile, for example, the area suitable for agriculture is preferably dedicated to grow food crops, which motivates the country to search for alternative solutions such as second-generation biofuels derived from agricultural or forest residues. Despite the limited production of biofuels, Chile is an important player in the forestry industry, the activities of which, including pulp milling and pellet production, are mainly concentrated south of the capital Santiago.

In fact, planted forests are expanding in South America and some pulp companies are signalling that in the near term forest feedstocks will be used not only for the production of biofuels, but also for the production of renewable chemicals and materials (Mariano 2015a). Today, silviculture expansion is mainly driven by the continuous growth of the eucalyptus market pulp industry in Brazil, Chile, and Uruguay. Supporting this market growth, the improved energy efficiency of modern pulp mills has also allowed pulp companies to sell power to the national energy grid. As such, green electricity has been an important source of revenues to these companies although the forestry industry's share of the power market is relatively small. For example, despite the fact that the planted area in Brazil reached 7.7 million hectares in 2014 (IBÁ 2015), pulp mills accounted for only 0.6% of the total electricity commercialized in that country (EPE 2015). Planted forests are also important for the steel sector in Brazil, since wood is used for the production of charcoal as a substitute for coal, providing thermal energy and carbon for the reduction of iron ore.

From a social and environmental perspective, the use of food crops such as soybean and sugarcane for the production of first-generation liquid fuels is usually controversial and inevitably sets off waves of protests around the world. To the 'food versus fuel' debate in South America, it is important to bring the information that the composition of soybeans is approximately 20% oil and 80% protein meal, and that biodiesel producers usually operate in different soybean markets. This allows for the commercialization of a range of products, including biodiesel and cooking oil from the oil portion, and livestock feed and lectin from the protein meal (Dourado et al. 2011; Medic et al. 2014). As for social aspects, biodiesel social seal programs, such as the National Biodiesel Program in Brazil, have been supporting smallholder farmers, securing market access and providing technical assistance (Wilkinson 2016).

The impacts of the sugarcane ethanol expansion have also been subject to intensive debate. Impacts on food production are certainly mitigated by the fact that (i) sugarcane productivity has been increasing, (ii) crop expansion has been occurring mainly over lands previously used for cattle grazing, (iii) rotation systems alternating sugarcane and food crops are used to allow soil recovery (Goldemberg et al. 2008), and (iv) ethanol production is generally integrated with sugar production as a strategy to accommodate market oscillations. As far as environmental aspects are concerned, as mentioned above, more than 90% of the sugarcane plantations are located in South-Central Brazil (mainly in São Paulo State) and the remainder in the Northeast region. Both regions are over 2500 km from the Amazon rainforest. Although the bold expansion of sugarcane cultivation in the last decade has been supposedly responsible for pushing other agricultural activities into the rainforest, deforestation has in fact decreased in the same period. According to the Brazilian National Institute for Space Research (INPE), between August 2014 and July 2015 the deforestation area in the Brazilian Amazon region was 583 000 ha, which is down 79% when compared to 2004 numbers. Whereas indirect land use change impacts may have been overstated, environmental impacts related to, for example, soil degradation, nitrogen dynamics, soil carbon stocks, and an increasing use of pesticides, herbicides, and fertilizers are still a challenge for the implementation of a more sustainable ethanol production in Brazil (Filoso et al. 2015).

3.2.2 Commercial Deployment and Challenges

Oil and natural gas (NG) are still widely produced and consumed in the region. Therefore, the access to financial support is critical to the advancement of the bioenergy industry, especially considering that upfront costs of some renewable energy alternatives are higher when compared to conventional fuel technologies. Whereas investments in the mature bioelectricity sector benefit from contract auctions and fiscal incentives (IRENA 2015), the emerging second-generation (2G) biofuel industry, which still relies on imported technologies, generally faces loan restrictions imposed by local banks. In Brazil, the restrictions (in part due to the intrinsic technical risk of new technologies) are also linked to the fact that government credit lines at attractive rates are rarely approved unless a significant part of the equipment is manufactured locally. On the other hand, since this restriction may slow down the commercial deployment of 2G biofuel plants, investors can benefit from reductions in import tariff rates through programs such as the Brazilian 'Ex-tariff' mechanism (APEX-BRASIL 2017).

Several South American governments support research and development (R&D) activities in renewable energy and fuels with loan and grant programs, including research funding, and graduate student scholarships (IRENA 2015). For more specific applications, dedicated programs have been occasionally formulated. This is the case of the PAISS program, a joint initiative of the Brazilian Development Bank (BNDES) and the Innovation Funding Agency (FINEP). This program was created to provide financial assistance to companies and research centres interested in exploring the potential of sugarcane-based advanced fuels, with emphasis on bringing 2G biofuel technologies to commercial scale (Cortez et al. 2014). However, if in such incentive initiatives only a small share of the funds is available to support research activities, investors will continue to be encouraged to import technology and seek partnership abroad with well-established innovation centres and startup companies to cut off research time and money. A representative example is the two commercial cellulosic ethanol units recently built in Brazil. With financial support from the PAISS program, North American and European technologies were the choice to process sugarcane bagasse and straw into more than 100 million litres of ethanol per year (Damaso et al. 2014).

The emerging 2G biofuel industry is also facing important technical and market challenges (Mariano 2015b). On the technical side, the presence of abrasive impurities carried along with agricultural residual feedstocks has been a major source of equipment failure found in Brazilian mills. With respect to market penetration, with no government program requiring transportation fuel to be sold with a minimum volume of 2G biofuels, competition is direct against lower-cost first-generation biofuels. Market protection during the initial years of this industry could certainly increase the success rate of these companies since production costs are still not competitive.

On the other hand, the urge to reduce costs with improved logistics brings opportunities for the commercialization of densified biomass, especially in the form of bio-oil and pellets. Fibria (Fibria Celulose S/A, São Paulo, SP, Brazil), the major market pulp producer in Brazil, along with its equity investment in a Canadian pyrolysis oil company (Ensyn Corporation, Ottawa, ON, Canada), announced in 2013 its interest in producing bio-oil from eucalyptus forests. Pellets are also a relatively new product for South America and have a great potential in both local and external markets. Nevertheless, many small pellet companies still need quality and sustainability certification in order to penetrate more restrictive markets such as that of the European Union. Moreover, to make exports feasible, large pelletizing plants and suitable infrastructure in ports are needed. Whereas large volumes such as 80 000 t are required to fill up bulk carriers, the total pellet production in Brazil in 2014 was only 50 000 t, which represented less than 1% of the world production (Garcia et al. 2016). Although there are still no large producers in South America, three projects with capacities ranging from 130 000 to 600 000 t/yr have been recently announced in Brazil. Wood will be the feedstock for two of these projects conducted independently by the companies Tanac S/A (Rio Grande, RS, Brazil) and Finagro (Belo Horizonte, MG, Brazil) in the state of Rio Grande do Sul. Most of the production will supply the export market (Europe, South Korea, and Japan) taking advantage of the existing infrastructure (railway and port) for soy exports (Diaz-Chavez et al. 2017). In the third project, led by the joint venture Cosan-Sujimoto (Cosan S/A Industria e Comércio, São Paulo, SP, Brazil; Sumitomo Corporation, Tokyo, Japan) in São Paulo state, pellets will be produced from sugarcane residues and shipped to Japan for power generation (Brazilgovnews 2016).

Finally, a major challenge for the commercial deployment of emerging renewable energy technologies is that South American investors are generally risk averse. This notable risk aversion is due not only to cultural reasons, but also to the lack of consistent government support, partly due to the oil industry lobby; all these factors are to be blamed. As such, cultural and policy factors are still important hurdles to overcome before a more intensive advance of the renewable energy industry can take place in South America.

3.2.3 Perspectives on South America

Brazil is currently leading the advanced biofuel market in South America and expectations are that other countries in the region will take the same road to commercial deployment as soon as technologies reach maturity. Once the production cost of cellulosic liquid fuels becomes competitive, and sustainability criteria are developed and met, South America will be in a favourable position to become a global supplier. For the moment, Brazilian exports of first-generation ethanol to the US are already a common practice. However, shipments are cyclical and dependent on, among other factors, local credit prices perhaps best exemplified by the incentives offered by the Low Carbon Fuel Standard (LCFS) program in California.

The use of biomass for electricity generation remains below its potential in South America but is expected to increase significantly over the coming years. An important driver for expansion is the strong dependence on hydropower, which has been constantly affected by severe droughts. Supply shortages and corresponding growing prices in combination with government incentives have prompted, for example, sugarcane ethanol producers to invest in new and more efficient bagasse boilers. On the one hand, a constant and growing supply of electricity is critical for the economic growth of the region. On the other hand, green electricity and cellulosic liquid biofuels are generally in competition for the same feedstock. As such, energy, agricultural, and forestry policies need to be well balanced to create sustainable conditions for investments in both competing sectors. In the face of this challenge of competing uses for biomass, improvements in feedstock productivity is also constantly needed, with special focus on the development of highly-productive fibrous plants. A good example of such plants is a variety of sugarcane called 'energy cane' that has been introduced in Brazil with the advantage of being able to be cultivated in marginal lands that are not suitable for food crops (Carvalho-Netto et al. 2014).

It is also expected that the emerging South American cellulosic industry will diversify into other feedstock and products.With regard to products, although ethanol and biodiesel are still dominant, applied research on renewable chemicals and advanced fuels is gaining momentum. For example, two aircraft manufactures, Boeing (The Boeing Company, Chicago, IL, USA) and Embraer (Embraer S/A, São José dos Campos, SP, Brazil), have joined forces to build in Brazil an innovation centre exclusively dedicated to developing bio-jet fuel (Boeing 2017). Aircraft manufactures and airline companies have embraced the challenging target of reducing 50% (from 2005 levels) of the aviation CO_2 emissions by 2050 (Boeing, Embraer, FAPESP, UNICAMP 2013). And, to meet this target, these companies have adopted an active role in the also challenging task of bringing bio-jet fuel conversion technologies to commercial scale (Alves et al. 2017; Pereira et al. 2017). As for the feedstocks, the forestry sector will certainly play an important role in the coming years. It is likely that forest biorefineries will focus on lignin-derived materials and new fibre materials such as nanocellulose, avoiding competition against the sugarcane industry.

3.3 Africa

3.3.1 Current Status

Complex social and economic issues make the development of the African biofuel industry even more difficult. In this continent, the biofuel industry is strongly challenged by the lack of infrastructure and a bold debate over land-use competition and food security. Having said that, studies and commercial projects in some African countries demonstrate that renewable energy initiatives can be associated with economic, environmental, and social benefits, including the co-existence of biofuel and food production. This section presents an overview of the main current initiatives and expectations regarding the African renewable energy scenario.

North Africa (Egypt, Tunisia, Algeria, Libya, and Morocco) and the sub-Saharan region differ in several aspects, including energy production and consumption profiles. In the north, the energy

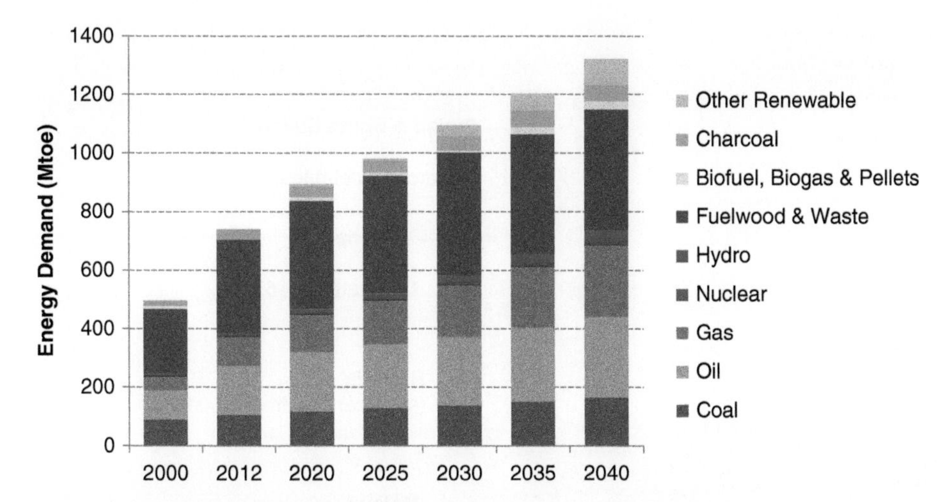

Figure 3.2 Projections of primary energy demand by fuel in the African continent. Other renewables include solar, wind, marine, and geothermal. Source: Data from IEA 2014.

supply relies heavily on fossil fuels, with more than 90% of the energy matrix consisting of local oil and gas resources (Mitchell 2011). Moreover, the dry climate and the Sahara Desert virtually eliminate the possibility of agricultural activities and, therefore, biofuel production. On the other hand, the considerable solar capacity in North Africa (2000–3000 kWh/m^2/year) has been attracting investments in solar energy projects. These projects supply local markets, and in the future North-African countries will export electricity to Europe once transmission lines and a regulatory framework are developed (United Nations 2012). In sub-Saharan Africa, on the other hand, bioenergy dominates and wood and charcoal are the major sources. Even though wood is a renewable resource, demand usually overcomes supply, causing environmental degradation and, in many cases, severe health issues and carbon monoxide poisoning inside houses. South Africa is a distinct case: it is the largest consumer of energy in the continent and 77% of its energy needs are met with coal, and most of the coal consumed in the continent is supplied by this country (IEA 2014). In the continent as a whole, non-renewable hydrocarbons and coal account for at least 50% of the energy supply today. Projections for the next 25 years indicate that this share is not expected to change significantly, and for the bioenergy supply, fuelwood and waste will still be dominant over biofuels (Figure 3.2).

The limited growth of biofuel production can certainly not be associated with a lack of arable lands. As a matter of fact, sub-Saharan Africa has suitable natural conditions for biofuel production, and according to the Food and Agriculture Organization of the United Nations (FAO), more than 1 billion ha of land with potential for rain-fed crop production is available. However, less than 30% is expected to be cultivated by 2030 (FAO 2003). Despite the land underutilization, the lack of infrastructure is in fact the major roadblock for effective industrial development. For example, whereas 99% of the population in Northern Africa has access to electricity, this number drops to only 32% in the 50 countries of sub-Saharan Africa. In countries such as South Sudan and the Central African Republic, only 1% and 3% of the population, respectively, is connected to the power grid (IEA 2015a).

Despite the infrastructure challenges found in the multi-variable equation to be solved for an effective and sustainable growth of the biofuel industry in Africa, in the government policy terms of this equation there are international agreements and internal policies already in place. In the

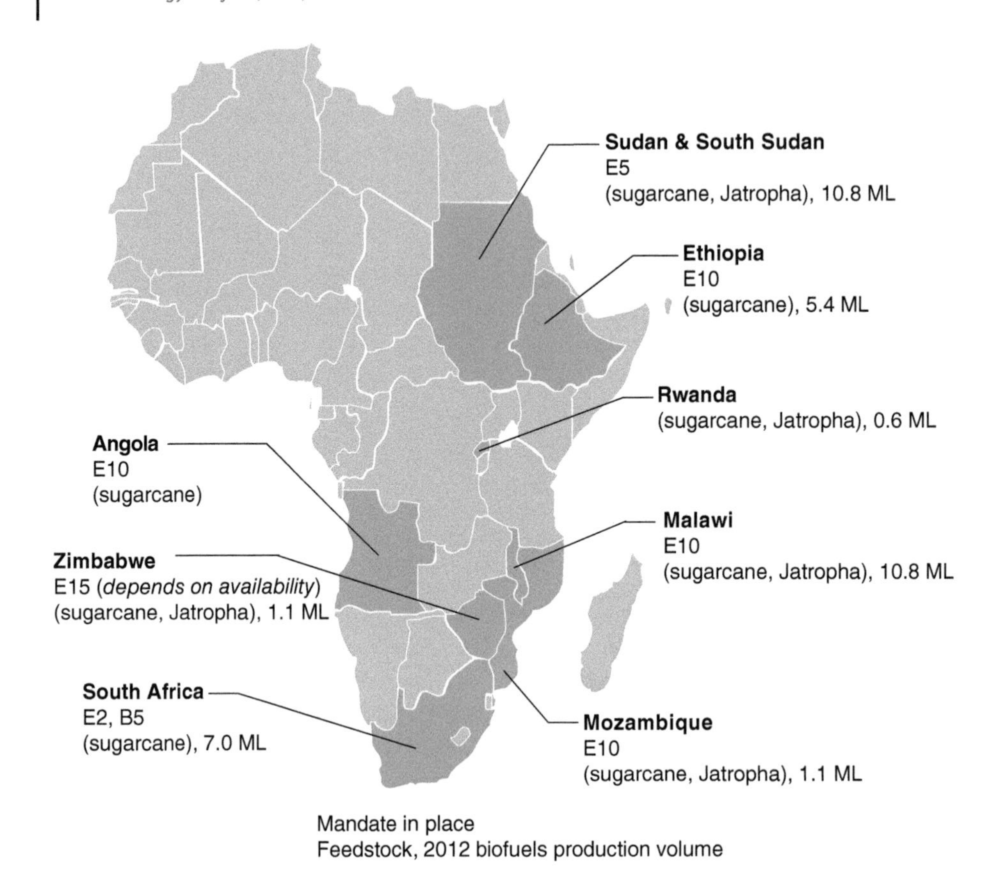

Figure 3.3 Major biofuel (ethanol and biodiesel) producing countries in Africa and respective mandates and 2012 production volumes. Source: Data from UNCTAD 2016.

international context, tariffs on vegetable oils and biodiesel exports to the US and the European Union are lower when compared to the tariffs these regions impose to countries such as Brazil, Indonesia, and South Africa. As for the internal incentives, public and private entities are developing policies and programs and attention has been given mainly to sugarcane molasses to produce ethanol and Jatropha for biodiesel production. For instance, Zimbabwe and Malawi have an ethanol blend mandate of 15% and 10%, respectively (Figure 3.3). This obligation has supported the development of the sugarcane industry in these countries for the last three decades, and mitigated the dependence on oil imports. Other countries, such as Tanzania, Zambia, Nigeria, and Uganda have also provided a stimulus for the private sector. However, social–environmental aspects were barely taken into account in these policies (Jumbe et al. 2009).

3.3.2 Commercial Deployment and Challenges

Renewable energy projects can be found mainly in sub-Saharan Africa. South Africa is the continent's largest producer of renewable energy and in order to reduce its reliance on hydrocarbons, investments are mainly focused on solar and wind projects. In Ethiopia, the construction of a 1000 MW geothermal power plant has been supported by the Iceland and the US governments. Zimbabwe, with a strong dependency on imported fuels, was in 2007 the first country in the continent to have a commercial biodiesel plant. With a production volume of 100 million litres per year, this plant processes mainly Jatropha oil. To meet its mandatory blending policy, the Zimbabwean government has supported investments in new biofuel plants and the expansion of sugarcane and Jatropha crops for ethanol and biodiesel production, respectively.

However, biofuel projects are not always well received by local communities, especially because of food security concerns. This is not surprising considering that, according to the FAO, more than 20% of the African population lives in severe undernourishment conditions (FAO 2016). In some countries, at times of food price crises, governments are pressured to restrict biofuel expansion. For example, in 2009 Tanzania suspended financial incentives for biofuels, and land expansion for biofuel feedstock was halted (Amigun et al. 2011). However, these sorts of restrictions should be carefully analysed because they inhibit income opportunities for farmers, limiting job creation and wages in rural areas, where poverty is often present. Moreover, food prices are actually connected to oil prices, not to the expansion of biofuel production (PANGEA 2012).

In fact, the expansion of biofuel production in Africa should largely be built upon government incentives to smallholder farmers to improve the distribution of wealth. On the other hand, for this strategy to become economically sustainable, the first challenge to overcome is the fact that traditional farming in Africa has low yields and lacks best practices regarding, for example, the use of fertilizers and sugarcane pre-harvest burning. Secondly, if not controlled, biofuel crops may invade areas reserved for indigenous tribes and wildlife. Ultimately, the major challenge for the expansion of the biofuel industry in Africa is to devise a sustainable and economically feasible growth of biofuel crops that leads to fair wealth distribution, environmental and social preservation, food security, and economic development.

3.3.3 Perspectives on Africa

External and internal factors are fostering the development of the bioenergy industry in Africa. Important external drivers are the market demand created by EU and US biofuel mandates, and the tax exemption offered by the US to African ethanol. Ethanol exports to these regions may accelerate the growth of the sugarcane industry in sub-Saharan Africa, advancing from the sole use of molasses for ethanol production to sugarcane juice, and eventually to lignocellulosic material (bagasse and straw). Moreover, several international investors are willing to invest in renewable energy initiatives in Africa, should sustainability standards be met. Internal factors that are pushing the development of the biofuel industry include the heavy dependency on imported fuels, and the need to substitute diminishing forest resources and diesel oil used in residences and stationary power plants, respectively. For example, in Mozambique, surplus cassava is converted to ethanol in a 2-million-litre ethanol plant (CleanStar Mozambique) co-owned by the New York-based venture capital firm CleanStar Ventures and the Danish company Novozymes (Novozymes A/S, Bagsværd, Denmark). The cassava-based ethanol supplies the Mozambican market as a substitute for charcoal and wood used in residential stoves.

The Mozambican cassava ethanol project is an initiative that demonstrates that even a small-scale ethanol plant can bring positive impacts on different aspects: a new income stream for local farmers, a better quality of life for urban households no longer exposed to dirty cooking fuels, and consequently, preservation of forest resources previously used as cooking fuel. On a wider scale, studies suggest that, similar to the Brazilian experience, the emerging bioenergy industry in Africa can potentially be an enabler for economic and social development, conciliating energy production with food security and environmental preservation (Lynd et al. 2015).

3.4 Southeast Asia

3.4.1 Current Status

Southeast Asia is considered here as the area represented by Brunei, Cambodia, Indonesia, Laos, Malaysia, Myanmar, Philippines, Singapore, Thailand, and Vietnam. These are the member countries of the Association of Southeast Asian Nations (ASEAN). In this region, biofuel production at the commercial scale – especially biodiesel from palm oil – only started in 2003. By this time, political and economic conditions were favourable, leading to a quick development of the biofuels sector. Currently, biodiesel and ethanol are mostly produced in five countries of the region and the expected market integration targeted by the ASEAN economic block will certainly foster the local biofuel industry. However, the fast growth of the biofuel industry, especially palm oil plantations, has already led to increased rates of deforestation of the Southeast Asian rainforests in recent years. Thus, conservation of the rainforests and wildlife is the major challenge for a sustainable production of biofuels in the region.

The primary energy demand in the region increased 2.5 times between 1990 and 2013 and it is forecast to almost double in the period 2013–2040. Despite these efforts in producing energy from renewable resources, approximately 75% of the energy consumed in the region is fossil-based. This share is expected to increase even more in the coming years (Figure 3.4). The energy outlook for SE Asia is shaped by two main factors. First, this region has important reserves of coal and natural gas (NG), with Indonesia and Malaysia being major players in the international NG market. Second, the population may grow from 616 to 760 million inhabitants within this period, putting pressure on electricity supplies, which is mainly provided by coal-fired power plants (IEA 2015b).

In spite of the dominance of fossil-based energy supplies, renewable resources have been receiving increasing attention from local governments and the private sector. For instance, the Philippines in terms of geothermal electricity production capacity is only second to the United States, and in 2015 this energy source represented 27% of its energy matrix (Bertani 2015). Vietnam and Laos have great hydropower potential, but most of the renewable energy produced and consumed in these countries is sourced from biomass. Most impressively, upon a short period of time (2006–2013), the biofuels industry in SE Asia jumped from virtually zero capacity to 7% of the total global production of biofuels (biodiesel and ethanol), experiencing an impressive average annual production growth rate of 44% (Chanthawong and Dhakal 2016).

The steep growth of the biofuels industry in the region was prompted by energy security concerns and market pull by the growing demand from European countries, especially for palm oil-based

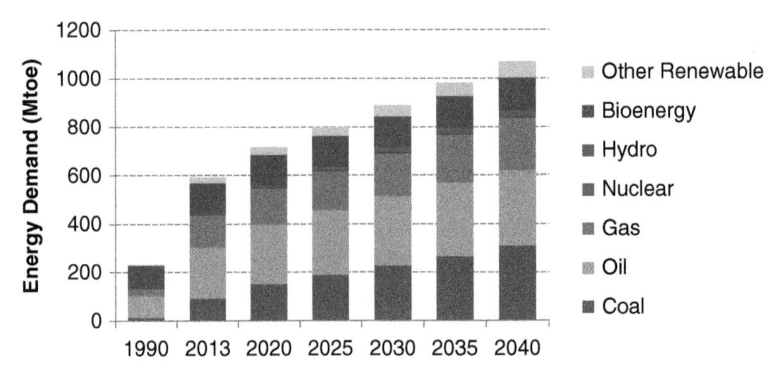

Figure 3.4 Projections of primary energy demand by fuel in Southeast Asia. Source: Data from IEA 2015b.

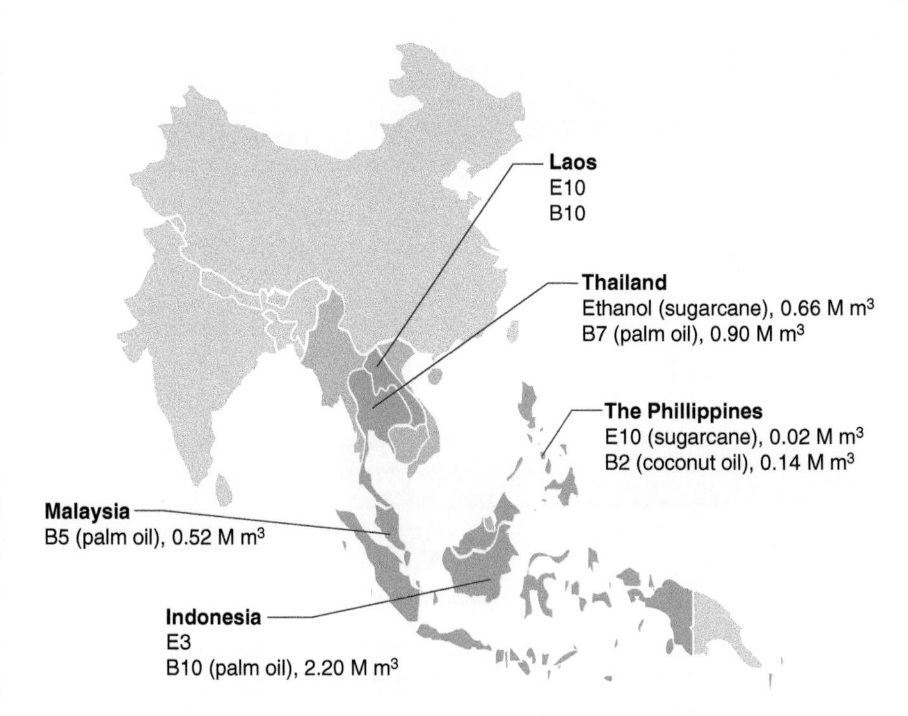

Figure 3.5 Biofuel (ethanol and biodiesel) producing countries in Southeast Asia and respective mandates and 2012 production volumes. Source: Data from Chanthawong and Dhakal 2016.

biodiesel, and local blending mandates (Figure 3.5). As for the latter, agreements have been signed in the context of the ASEAN Economic Community establishing important blending targets (E20 and B10) in long-term development plans for the region (Chanthawong & Dhakal 2016). However, although the biofuels industry in Southeast Asia is supported by favourable political, natural (fertile soil, suitable climate), and economic (accelerated economic growth rates) conditions, these countries have not yet reached their current blending targets because of sustainability concerns, especially deforestation and loss of biodiversity following the expansion of palm oil plantations, and also due to long-standing subsidies for fossil fuels (Dermawan et al. 2012).

3.4.2 Commercial Deployment and Challenges

Southeast Asia is one of the world's major players in biofuels production. In 2013, production volumes were 1325 million liters of ethanol, mainly from sugarcane, and 4028 million liters of biodiesel, mainly from palm oil (Chanthawong and Dhakal 2016). Internal markets are the major consumers (Figure 3.6). As mentioned above, Southeast Asia reached this position in a short period of time. For example, from 2006 to 2012 biofuels production in Indonesia soared from 24 million liters and two mills to 2200 million liters and 26 mills (Slette and Wiyono 2011). Similar growth rates were also observed in Thailand, the Philippines, and Malaysia. However, alongside this exponential growth, social, economic, and environmental issues have appeared and posed significant challenges for the expansion of the local biofuel industry.

Perhaps the most critical of these issues is rainforest deforestation. Southeast Asia is known for its biodiversity. The region has only 3% of the world's land, but 20% of all known species. Indonesia and Malaysia score high in biodiversity and endemism rankings (Mukherjee and Sovacool 2014). It is estimated that this rich and complex ecosystem lost 42 million hectares between 1990 and

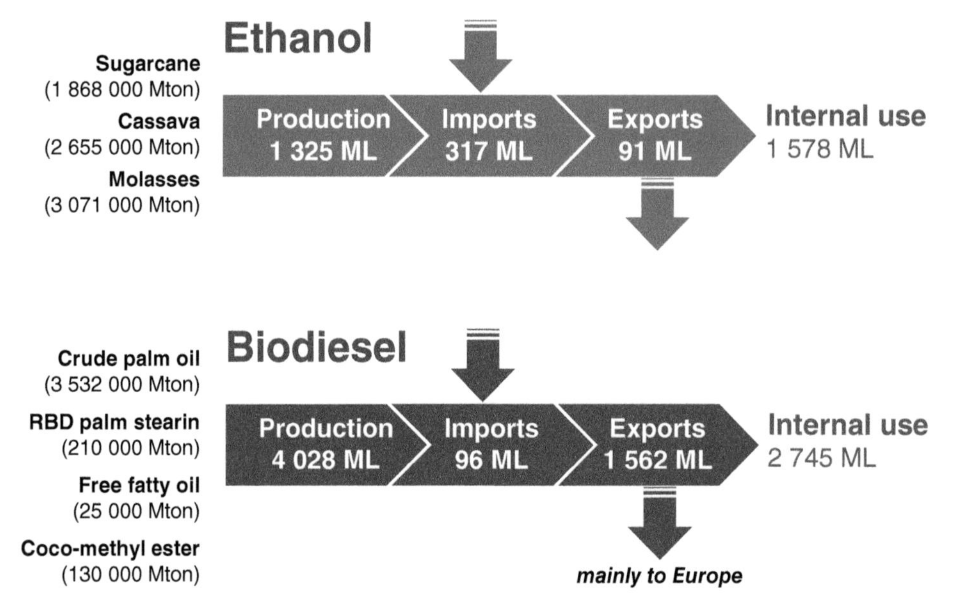

Figure 3.6 Market chain of ethanol (Vietnam and Myanmar not included) and biodiesel (Singapore not included) in Southeast Asia in 2013. Source: Data from Chanthawong and Dhakal 2016.

2010, which is equivalent to approximately 8% of Southeast Asia's land area (FAO 2011). Albeit the deforestation has been caused by several factors, such as illegal logging, it is the expansion of the palm oil plantations (therefore, the biofuel industry) that is often pointed to as the major culprit (Mukherjee and Sovacool 2014). Consequently, major palm oil and derivatives importers, such as the US and the EU have established sustainability standards that Southeast Asia must meet to keep this international trade (Dermawan et al. 2012).

Another challenge for the Southeast Asian biofuel industry is in the food security debate, which is clearly far from a conclusion. For example, the shortage of rapeseed oil in Europe in 2007 – mainly used for biodiesel production – caused the palm oil imports from Southeast Asia to double that year. It was not surprising that the demand growth for palm oil led to local price spikes. It affected harshly the low-income population, since palm oil is present in a wide variety of food and non-food products (Mukherjee and Sovacool 2014). On the other hand, studies indicate that less than 10% of the total crude palm oil has been used for biodiesel production in Indonesia and Malaysia (Dermawan et al. 2012). Moreover, China and India are responsible for approximately half of Malaysia's palm oil exports, and Indonesia supplies about 60% of the edible oil consumed in India (Petrosil 2012) – these numbers suggest that biofuel production may not be the sole cause of the observed price increases.

In addition to these social and environmental challenges, an economic-political barrier limits biofuel production in Southeast Asia: the oil industry subsidies. Local governments heavily subsidize the oil industry and fuel imports during periods of high oil prices. Moreover, it is a common perception among biodiesel and ethanol producers that the oil price must be higher than 100 US dollars per barrel to make biofuels competitive. Even though it has been the case in recent years, feedstock prices followed the oil prices, wiping out the profit margins of biofuel producers (Dermawan et al. 2012).

3.4.3 Perspectives on South East Asia

In the face of various challenges, the Southeast Asian biofuel industry has to make several adjustments to ensure its sustainably. Certainly, the introduction of second-generation technologies will allow production increase without land expansion. Even though production costs of second-generation biofuels are expected to be lower than those of the first generation, there is no commercial project in development in the region currently (JGSEE and STI 2014). Singapore is the only country in Southeast Asia using advanced biofuel technology, producing at commercial scale bio-hydrogenated diesel (BHD) from certified palm oil (Chanthawong and Dhakal 2016).

The introduction of second-generation technologies will certainly support the sustainable replacement of liquid fossil fuels in agreement with the ambitious blending targets (E20 and B10) set by the ASEAN Economic Community, and it will also support producers to meet sustainability standards imposed by both the EU and the United States.

3.5 China

3.5.1 Current Status

Due to its economic and social global reach, China is a major stakeholder whenever international agreements regarding renewable energy strategies and policies are under discussion. The most populous country (1.37 billion inhabitants) and the second economy in the world (with an 11.2 USD trillion GDP in 2016; World Bank 2017a) accounts for 23% of the global energy demand. Thus, the solution for major energy-related global and local issues, such as global warming, energy security, and air pollution, depends heavily on decisions made in Beijing. Moreover, decisions in support of renewable energies mean changes in a massive carbon-intensive energy consumption of 2972 million tons of equivalent oil (in 2014), with fossil fuels (mainly coal) accounting for 89% of this demand (BP 2015).

The high consumption of energy in China has its origins in the economic reforms of the 1970s. Following these reforms, for more than 30 years until 2009, China experienced a GDP annual growth rate of about 10% and the GDP per capita increased from 112 US dollars in 1970 to approximately 8120 in 2016 (World Bank 2017b). In this period, industry became the major economic activity and the largest ever rural exodus took place, with the creation of multiple industrial jobs to serve numerous industries (Huang et al. 2012). Furthermore, many Western companies moved their manufacturing operations to China attracted by lower production costs (especially, labour cost) and less strict environmental regulations. China, thus, has become the nation with the highest energy demand in the world and its energy dependence dramatically increased during that period. Oil imports have been steadily growing and it is expected that imports will surpass domestic production by four times by 2050 (Figure 3.7). Most of the electricity consumed in China (66%) is generated in coal-fired power plants (Figure 3.8). In fact, China is the largest producer and consumer of coal in the world. It has caused serious air pollution problems (Ma et al. 2016), and by 2035 China may account for 30% of the global carbon dioxide emissions per capita (BP 2015). In this manner, ignored by many consumers in Western countries, imports of low-cost goods from China embody large quantities of CO_2 emissions. In 2007, 23% of the emissions in China (1.7 $GtCO_2$) were related to goods exported to Western countries (Liu et al. 2016).

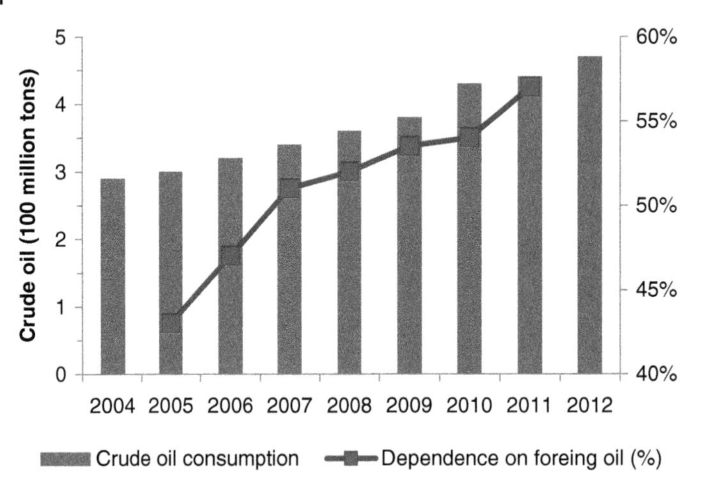

Figure 3.7 Crude oil consumption in China and dependence on foreign oil. Source: Data from the National Bureau of Statistics of China 2016.

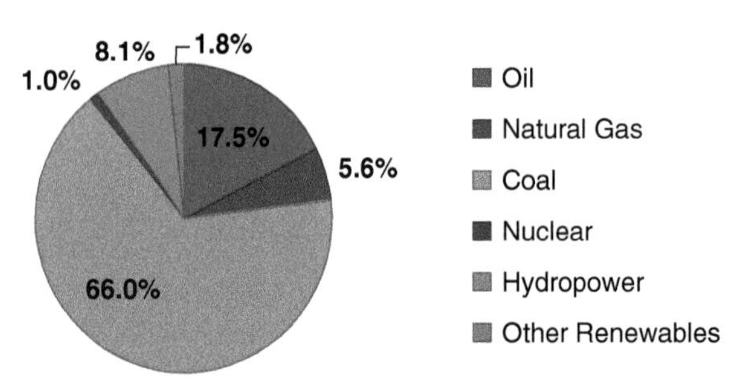

Figure 3.8 China's electricity production breakdown by energy sources. Source: Data from BP 2015.

Not surprisingly, urban air quality deterioration and international pressure have led the Chinese government to 'declare war' on pollution (Reuters 2014). Actions include the shutdown of small coal-fired power plants, and the diversification of the energy matrix with investments in hydroelectric and nuclear power, and in natural gas production. The Chinese government has also put in place policies to support the biofuel industry. In 2005, China promulgated the 'Renewable Energy Law of the People's Republic of China' and, in 2011, general directions for biofuel development were introduced in this law (Zhao et al. 2015). Interestingly, there is no regulation about cellulosic ethanol, biodiesel, biogas, and bio-hydrogen in this law. On the other hand, the government tightly controls grain ethanol production, sales, and distribution. Most of the ethanol must be sold to state-owned oil companies and regulations prevent the use of imported ethanol (Qiu et al. 2012). Although there is no national blending mandate, six provinces have adopted an E10 blending mandate. On the other hand, biodiesel production and sales do not receive significant incentives and is only approved for use in a few cities (Anderson-Sprecher and Ji 2015).

The control over grain ethanol also explains why biofuel production in China is still far behind that of other countries such as the US and Brazil. While ethanol production in these two countries was respectively 15.3 and 7.3 million gallons per year (MGPY) in 2016, China barely reached 0.85 MGPY (RFA 2017). The need to feed 1.3 billion people has led the government to block the expansion of biofuel production from edible sources, avoiding food security issues. Competition for land and water is also a barrier for biofuel expansion. In the face of these challenges, and the limited availability of alternative feedstocks, it is still unclear how China will keep pace with its ambitious targets of substitution of fossil-based liquid fuels (Anderson-Sprecher and Ji 2015).

3.5.2 Commercial Deployment and Challenges

Current biofuel production in China still falls short of its needs and feedstock shortage certainly is the largest constraint. In 2015, the country produced approximately 1 and 3 billion litres of biodiesel and ethanol mainly from waste cooking oil and grains (corn and wheat), respectively. This amount accounts for less than 1% of the Chinese liquid fuel production and its effects on air pollution and oil dependency issues are marginal (Anderson-Sprecher and Ji 2015). Moreover, land and water availability for biofuel crops adds to the challenge. While in southern China, most of the lands are already taken by rice, maize, and sugarcane crops, northern China is suffering from water scarcity due to increasing industrial and domestic consumption (Yang et al. 2009).

With the government restriction on the use of food grains (corn and wheat) for biofuel production and high corn prices, sweet sorghum and cassava have emerged as an alternative solution and are supported by the government. Currently, sweet sorghum and cassava account for 1% and 8% of ethanol production, respectively. However, these crops also compete for land, and production in marginal lands is not sufficient to supply large-scale ethanol plants. In the case of biodiesel, current production corresponds to less than one third of the installed capacity. Biodiesel plants have difficulties in sourcing enough affordable waste cooking oil due to the competition from illegal food use and a lack of large-scale collection infrastructures (Anderson-Sprecher and Ji 2015).

Another challenge for the biofuel industry in China is the competition from imports and methanol fuel blending. Imported ethanol is still a potential challenge since it is only allowed for industrial use (not for fuel purposes). However, given the high domestic production cost and competitive international prices, the economics of importing ethanol is favourable. As for imported biodiesel, the elimination of taxes has resulted in increased purchases of Indonesian biodiesel (Anderson-Sprecher and Ji 2015). In the case of methanol, the abundant coal resources in China have led the country to be the world's largest methanol producer and consumer. The large economies of scale thus gained resulted in low production costs, which combined with government incentives and policies; overall, they have made methanol an affordable alternative fuel that can improve urban air quality (Chen et al. 2014). Nevertheless, as a fossil-based fuel, coal-derived methanol does not contribute to reducing global carbon emissions.

In this challenging scenario for the expansion of the biofuel industry in China, cellulosic biofuels can certainly play a key role. On-going commercial efforts are moving from demonstration- to commercial-scale production (Kim 2017). Nonetheless, the feedstock-constrained Chinese biofuels industry has also been attracted by the abundant natural resources available in Africa. In fact, Chinese companies have acquired land in Africa for biofuel crops (cassava and sugarcane) (Yang 2015). The major investor is the China National Complete Import and Export Corporation Group (COMPLANT; Beijing, China), which is planning to build ethanol plants in several sub-Saharan African countries (GRAIN 2013; ACORD 2016).

3.5.3 Perspectives on China

In spite of several challenges, studies indicate that biofuel production in China will more than double in the next 15 years with increased use of biomass residues, and other non-food feedstocks such as municipal solid waste. In this period, annual ethanol and biodiesel production is expected to increase from 64 to 118 and from 12 to 34 million tons, respectively (Chen et al. 2015). However, given the enormous extension of carbon emissions in China and the urgency of mitigating climate change, the contribution that liquid biofuels can offer in the short- and mid-term remains limited. With an expected share of 6–16% of the total transport energy consumption by 2050, biofuels will cut the oil dependence ratio by only 1–7% and annual CO_2 emissions by 100–370 million tons (Zhao et al. 2015). More effectively in the short term, in the 2015 Paris Climate Conference, China committed to reducing its carbon emissions by modernizing its coal power plants by 2020. If accomplished, this initiative alone could save more than 100 million tons of raw coal per year and cut carbon emissions by 60% (COP 21 2015).

3.6 Global Perspectives

Although most of the countries in South America, Africa, and Asia (Southeast Asia, and China) are committed to advancing investments in green energy, not surprisingly the driving forces and hurdles are significantly different. It is evident that given the sub-Saharan African economic underdevelopment and China's massive carbon footprint, the low-carbon transformation in these countries is even more challenging. Nevertheless, decarbonization of the Chinese coal-based power sector is essential to alleviate the international pressure to add the cost of pollution to goods produced in China. In Southeast Asia, the immediate challenge is to conciliate the production of palm oil biodiesel with the conservation of the rainforest. In South America, the ambition of countries such as Brazil and Argentina to become global biofuel suppliers is to a great extent dependent on the development and fulfilment of sustainability criteria. It is unquestionable that for these three regions the use of non-edible feedstocks such as agricultural and forestry residuals and municipal solid wastes is critical for a sustainable growth of the biofuels industry. However, it is also critical to sustain efforts to promote investment in indigenous technological solutions, the creation of efficient feedstock supply chains, and government policies in support of second-generation biofuels.

References

ACORD (2016) Chinese, Brazilian and Indian investments in African agriculture: impacts, opportunities and concerns. Report. Agency for Cooperation and Research in Development (ACORD).

Alves, C.M., Valk, M., de Jong, S. et al. (2017). Techno-economic assessment of biorefinery technologies for aviation biofuels supply chains in Brazil. *Biofuels, Bioproducts and Biorefining* 11: 67–91.

Amigun, B., Musango, J.K., and Stafford, W. (2011). Biofuels and sustainability in Africa. *Renewable and Sustainable Energy Reviews* 15 (2): 1360–1372.

Anderson-Sprecher, A., Ji, J. (2015) China: Biofuels Annual. Report, USDA Foreign Agricultural Service. Available at: http://www.fas.usda.gov/data/china-biofuels-annual-1 (Downloaded: 05 May 2016).

APEX-BRASIL (2017) The Brazilian Trade and Investment Promotion Agency. Investment guide to Brasil 2017. Available at: www.apexbrasil.com.br/uploads/Investment%20Guide%202017.pdf (Downloaded: 08 November 2017).

Bertani, R. (2015). Geothermal power generation in the world 2010-2014 update report. In: *Proceedings World Geothermal Congress 2015, Melbourne, Australia, 19–25 April 2015*, 2–3.

Boeing (2017). The Boeing Company 2017 Environment report. Available at: http://www.boeing.com/resources/boeingdotcom/principles/environment/pdf/2017_environment_report.pdf (Downloaded: 09 November 2017).

Boeing, Embraer, FAPESP, UNICAMP (2013) Flightpath to aviation biofuels in Brazil: Action plan. Report. São Paulo, Brazil. Available at: http://www.fapesp.br/publicacoes/flightpath-to-aviation-biofuels-in-brazil-action-plan.pdf (Downloaded: 09 November 2017).

BP – British Petroleum (2015) BP Statistical Review of World Energy June 2015. Report. British Petrol. 64th edition.

Brazilgovnews (2016). Japan wants to increase biomass exports from Brazil. Available at: www.brazilgovnews.gov.br/news/2016/12/japan-wants-to-increase-biomass-exports-from-brazil.

Carvalho-Netto, O.V., Bressiani, J.A., Soriano, H.L. et al. (2014). The potential of the energy cane as the main biomass crop for the cellulosic industry. *Chemical and Biological Technologies in Agriculture* 1: 20.

Chanthawong, A. and Dhakal, S. (2016). Liquid biofuels development in southeast Asian countries: an analysis of market, policies and challenges. *Waste and Biomass Valorization* 7 (1): 157–173.

Chen, H., Yang, L., Zhang, P.-H., and Harrison, A. (2014). The controversial fuel methanol strategy in China and its evaluation. *Energy Strategy Reviews* 4: 28–33.

Chen, W., Wu, F., and Zhang, J. (2015). Potential production of non-food biofuels in China. *Renewable Energy* 85: 939–944.

COP 21 (2015) COP 21 records. Available at: http://www.cop21.gouv.fr/en/china-has-promised-to-cut-emissions-from-its-coal-power-plants-by-60-by-2020 (Downloaded: 05 August 2016).

Cortez, L.A.B., Souza, G.M., Brito Cruz, C.H., and Maciel, R. (2014). An assessment of Brazilian Government initiatives and policies for the promotion of biofuels through research, commercialization, and private investment support. In: *Biofuels in Brazil – Fundamental Aspects, Recent Developments, and Future Perspectives* (eds. S.S. Silva and A.K. Chandel), 31–60. Springer.

Damaso, M.C.T., Machado, C.M.M., Rodrigues, D.S. et al. (2014). Bioprocesses for biofuels: an overview of the Brazilian case. *Chemical and Biological Technologies in Agriculture* 1: 6.

Dermawan, A., Obidzinski, K., Komarudin, H. (2012) Withering before full bloom? Bioenergy in Southeast Asia. CIFOR Working Paper no. 94.

Diaz-Chavez, R., Walter, A., Gerber, P. (2017) Socio-economic assessment of forestry production for a developing pellet sector. IEA Bioenergy: Task 40. Available at: http://task40.ieabioenergy.com/wp-content/uploads/2013/09/IEA-Bioenergy-Task_-Brazil-Sept-2017-with-cover-final.pdf. (Downloaded: 09 November 2017).

Dourado, L.R.B., Pascoal, L.A.F., Sakomura, N.K. et al. (2011). Soybeans (*Glycine max*) and soybean products in poultry and swine nutrition. In: *Recent Trends for Enhancing the Diversity and Quality of Soybean Products* (ed. D. Krezhova), 175–190. Rijeka, Croatia: InTech.

EPE – Empresa de Pesquisa Energética (2015) "Brazilian Energy Balance 2015". Report, Ministry of Mines and Energy – MME, Empresa de Pesquisa Energética – EPE, Brazil. Available at: https://ben.epe.gov.br/downloads/Relatorio_Final_BEN_2015.pdf (Downloaded: 12 June 2016).

FAO - Food and Agriculture Organization (2003) World agriculture: towards 2015/2030. Report, Earthscan. London, United Kingdom.

FAO - Food and Agriculture Organization (2011) Southeast Asia subregional Report – Asia Pacific Forestry Sector Outlook Study II. Bangkok, Thailand. Available at: http://www.fao.org/docrep/013/i1964e/i1964e00.htm (Downloaded: 09 May 2016).

FAO - Food and Agriculture Organization (2016). FAO databank. Available at: http://www.fao.org/statistics/en (Downloaded: 27 March 2016).

Filoso, S., Carmo, J.B., Mardegan, S.F. et al. (2015). Reassessing the environmental impacts of sugarcane ethanol production in Brazil to help meet sustainability goals. *Renewable and Sustainable Energy Reviews* 52: 1847–1856.

Frankfurt School – UNEP – Bloomberg New Energy Finance (2016) "Global Trends in Renewable Energy Investment 2016". Report, Frankfurt School of Finance & Management gGmbH.

Garcia, D.P., Caraschi, J.C., Ventorim, G., and Vieira, F.H.A. (2016). Trends and challenges of Brazilian pellets industry originated from agroforestry. *Cerne* 22 (3): 233–240.

Goldemberg, J. (2007). Ethanol for a sustainable energy future. *Science* 315 (5813): 808–810.

Goldemberg, J., Coelho, S.T., and Guardabassi, P. (2008). The sustainability of ethanol production from sugarcane. *Energy Policy* 36: 2086–2097.

GRAIN (2013). Land grabbing for biofuels must stop: EU biofuel policies are displacing communities and starving the planet. Report. Available at: https://www.grain.org/article/entries/4653-land-grabbing-for-biofuels-must-stop.

Huang, J., Yang, J., Msangi, S. et al. (2012). Global biofuel production and poverty in China. *Applied Energy* 98: 246–255.

IBÁ – Brazilian Tree Industry (2015) 2015 Annual Report. Available at: http://iba.org/images/shared/iba_2015.pdf (Downloaded: 16 July 2016).

IEA - International Energy Agency (2014) Africa Energy Outlook. Report, Organization for Economic Co-operation and Development. Paris, France.

IEA - International Energy Agency (2015a) Electricty Database. Organization for Economic Co-operation and Development. Available at: http://www.worldenergyoutlook.org/resources/energydevelopment/energyaccessdatabase (Downloaded: 05 May 2016).

IEA - International Energy Agency (2015b). Southeast Asia Energy Outlook. Report Organization for Economic Co-operation and Development. Paris. France.

IRENA (2015), Renewable Energy in Latin America 2015: An Overview of Policies. Report. The International Renewable Energy Agency, IRENA, Abu Dhabi.

IRENA (2016) Renewable energy market analysis – Latin America. Report. The International Renewable Energy Agency, IRENA, Abu Dhabi.

JGSEE, STI (2014) Asian Bioenergy Technology Status. Report. The Joint Graduate School of Energy and Environment (JGSEE) and National Science Technology and Innovation Policy Office (STI).

Joseph, K. (2015) "Argentina – Biofuels annual". Report, USDA Foreign Agricultural Service, USA.

Jumbe, C.B.L., Msiska, F.B.M., and Madjera, M. (2009). Biofuels development in sub-Saharan Africa: are the policies conducive? *Energy Policy* 37 (11): 4980–4986.

Kim, G. (2017) Biofuels demand expands, supply uncertain. USDA Biofuels Annual Report. Peoples Republic of China. Available at: https://gain.fas.usda.gov/Recent%20GAIN%20Publications/Biofuels%20Annual_Beijing_China%20-%20Peoples%20Republic%20of_1-18-2017.pdf (Downloaded: 09 November 2017).

Liu, Z., Davis, S.J., Feng, K. et al. (2016). Targeted opportunities to address the climate–trade dilemma in China. *Nature Climate Change* 6: 201–206.

Lynd, L.R., Sow, M., Chimphango, A.F.A. et al. (2015). Bioenergy and African transformation. *Biotechnology for Biofuels* 8: 18.

Ma, X., Bi, L., and Wang, Z. (2016). Effect of air pollution on provincial fiscal investment for environmental protection in China. *Nature Environment and Pollution Technology* 15 (1): 27–34.

Mariano, A.P. (2015a). How Brazilian pulp mills will look like in the future? *O Papel* 76: 55–61.

Mariano, A.P. (2015b). Due diligence for sugar platform biorefinery projects: technology risk. *Journal of Science & Technology for Forest Products and Processes* 4 (5): 12–19.

Medic, J., Atkinson, C., and Hurburgh, C.R. (2014). Current knowledge in soybean composition. *Journal of the American Oil Chemists' Society* 91: 363–384.

Mitchell, D. (2011) Biofuels in Africa. Opportunities, Prospects and Challenges. Report, The World Bank. Washington D.C., The United States.

Mukherjee, I. and Sovacool, B.K. (2014). Palm oil-based biofuels and sustainability in Southeast Asia: a review of Indonesia, Malaysia, and Thailand. *Renewable and Sustainable Energy Reviews* 37: 1–12.

National Bureau of Statistics of China (2016). Available at: www.stats.gov.cn/english (Downloaded: 21 April 2016).

Nolte, G.E., Beillard, M.J. (2015) "Peru – Biofuels annual". Report, USDA Foreign Agricultural Service, USA.

PANGEA (2012). Who's fooling whom? The real drivers behind the 2010/2011 food price crisis in Sub-Saharan Africa. PANGEA Exclusive Report. Available at: http://www.pangealink.org/members/wp-content/uploads (Downloaded: 05 March 2016).

Pereira, L.G., MacLean, H.L., and Saville, B.A. (2017). Financial analyses of potential biojet fuel production technologies. *Biofuels, Bioproducts and Biorefining* 11: 665–681.

Petrosil (2012). Edible oil report – April 12, 2012. Petrosil, Mumbai.

Qiu, H., Sun, L., Huang, J., and Rozelle, S. (2012). Liquid biofuels in China: current status, government policies, and future opportunities and challenges. *Renewable and Sustainable Energy Reviews* 16 (5): 3095–3104.

Reuters (2014). China to 'declare war' on pollution, premier says. Available at: https://www.reuters.com/article/us-china-parliament-pollution/china-to-declare-war-on-pollution-premier-says-idUSBREA2405W20140305.

RFA (2017) World Fuel Ethanol Production. Renewable Fuels Association. Available at: http://www.ethanolrfa.org/resources/industry/statistics/#1454099103927-61e598f7-7643.

Slette, J., Wiyono, I.E. (2011) Indonesia biofuels annual 2011. Global Agriculture Information Network (GAIN). Report No. ID1134. USDA Foreign Agricultural Services, Jakarta.

UNCTAD (2016) Second Generation Biofuels Markets: state of play, trade and developing country perspectives. Report, UNCTAD/DITC/TED/2015/8. Geneva.

UNICADATA (2015) "2015 Sugarcane production data" Sugarcane Industry Union, Brazil. Available at: www.unicadata.com.br (Downloaded: 10 July 2016).

United Nations (2012). The renewable energy sector in North Africa: current situation and prospects. Economic Commission for Africa, ECA-NA/PUB/12/01.

Wilkinson, J. (2015) "The Brazilian Sugar Alcohol Sector in the current national and international conjuncture". Report, Action Aid Brazil.

Wilkinson, J. (2016) Agricultural models and best practices from Brazil and the Southern Cone: Lessons for Africa? Policy Brief, 15 Jan. 2016. Available at: http://www.gmfus.org/file/7299/download (Downloaded: 08 November 2017).

World Bank (2014) Renewable energy consumption (% of total final energy consumption) Available at: https://data.worldbank.org/indicator/EG.FEC.RNEW.ZS?name_desc=false&view=map.

World Bank (2017a) World Bank national accounts data and OECD National Accounts data files. Available at: https://data.worldbank.org/indicator/NY.GDP.MKTP.CD?locations=CN.

World Bank (2017b) World Bank national accounts data and OECD National Accounts data files. Available at: https://data.worldbank.org/indicator/NY.GDP.PCAP.CD?locations=CN.

Yang, X. (2015). An analysis of China's biofuels policy and Chinese discourse on land acquisition for biofuels in Africa. Master's Thesis, Michigan Technological University.

Yang, H., Zhou, Y., and Liu, J. (2009). Land and water requirements of biofuel and implications for food supply and the environment in China. *Energy Policy* 37 (5): 1876–1885.

Zhao, L., Chang, S., Wang, H. et al. (2015). Long-term projections of liquid biofuels in China: uncertainties and potential benefits. *Energy* 83: 37–54.

4

The Development of Solar Energy Generation Technologies and Global Production Capabilities

F. John Hay[1] *and N. Ianno*[2]

[1] *Extension Educator – Energy, Department of Biological Systems Engineering, University of Nebraska-Lincoln, Lincoln, NE, 68583, USA*
[2] *Professor, Department of Electrical Engineering, University of Nebraska-Lincoln, Lincoln, NE, 68182, USA*

CHAPTER MENU

4.1 Introduction

Sunlight is the major provider of energy on earth. A majority of our energy use comes from either stored sunlight such as coal, oil, and natural gas, or more recently arriving sunlight in the form of hydroelectricity, wind, waves, biomass, and direct use of sunlight such as solar thermal or sunlight to electricity technologies like photovoltaic cells. This chapter will discuss sunlight energy use by photosynthesis and photovoltaics (PV). In both energy conversions photons from the sun excite electrons, these excited electrons drive the reactions in photosynthesis and create an electric current in photovoltaic cells. Photosynthesis and PV are both impacted by the quantity and quality of the sunlight striking the earth. The atmosphere reflects some radiation back into space while diffusing and scattering some radiation. Approximately 1000 W m^{-2} reaches the earth's surface.

Sunlight quantity varies by time of day and time of year due to the rotation of the earth, and the tilt of the earth on its axis as it orbits the sun. Figure 4.1 represents how the tilt of the earth changes the angle of the sunlight reaching the earth. In the northern hemisphere winter the sunlight reaching the earth has a steeper angle and travels through more atmosphere, thus providing less intense sunlight. Northern hemisphere summer has a less steep angle and sunlight is more directly overhead and intense. The angle of sunlight will determine the optimum angle and direction of placement for solar PV systems. Solar PV Systems in the northern hemisphere will face south and systems in the southern hemisphere will face north.

Human utilization of renewable energy includes hydroelectric, wind, biomass, liquid biofuels, and solar as the top five largest sources of renewable energy respectively in 2016. Total renewable energy generation in the United States in 2016 is just over 10 quadrillion British thermal units

Green Energy to Sustainability: Strategies for Global Industries, First Edition.
Edited by Alain A. Vertès, Nasib Qureshi, Hans P. Blaschek and Hideaki Yukawa.

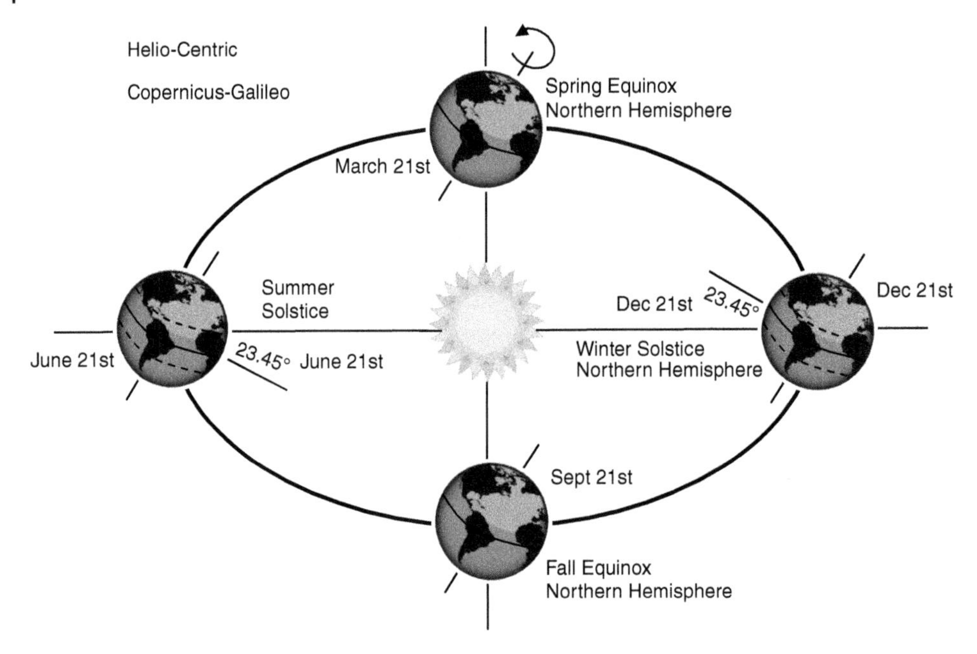

Figure 4.1 Helio-centric representation of the sunlight reaching the earth's surface. The sunlight in northern hemisphere summer is more directly overhead, thus sunlight intensity is greatest in summer when the light has less atmosphere to travel through to reach the earth's surface.

(Btu) with 582 trillion Btu coming from solar electric systems and 3.9 quadrillion Btu coming from biomass in all forms. Total primary energy consumption in the United States is just over 97 quadrillion Btu which puts renewables at over 10% of total energy use (EIA Monthly Energy Review 2017).

Solar photovoltaic system price declines and favourable government programs have driven a rapid increase in installations. Capacity of world solar PV installations has grown from 5.1 GW in 2005 to over 227 GW in 2015. The greatest capacity increases in 2015 have happened in China, Japan, and the United States (REN21 2016).

4.2 Sunlight and Photosynthesis

Photosynthesis is the basis for the food, fuel, and fiber which we rely on for survival. Plants, algae, and some bacteria can use sunlight to fix carbon dioxide into carbohydrates. This chapter will not endeavour to explain the details of photosynthesis, but instead focus on the efficiency of photosynthetic systems. More detailed information on photosynthetic pathways can be found in other chapters of this book as well as numerous other print and on-line resources.

Incident solar radiation on the earth's surface is approximately 1000 W m^{-2} of which only 4.6–6% can be converted into carbohydrates in plants (Xin-Guang et al. 2008). There are three major types of photosynthesis C3, C4, and Crassulacean Acid Metabolism (CAM). The three all have in common the same basic pathways of the Calvin Cycle where the enzyme Ribulose-1,5-bisphosphate carboxylase/oxygenase (RuBisCO) catalyses the reaction fixing carbon dioxide from the air. In C4 photosynthesis an additional step is used to concentrate carbon dioxide inside the cell where it is then exposed to RuBisCO. The CAM photosynthetic pathway goes further by concentrating and storing carbon dioxide only at night to conserve moisture. These differences have a major impact on the efficiency of a plant's use of sunlight.

4.2.1 Photosynthetic Efficiency

Sunlight minus light outside photosynthetic spectrum minus reflected light minus inefficiency of light absorbing pigments minus inefficiency of carbon synthesis minus photorespiration minus respiration equals photosynthetic efficiency (Xin-Guang et al. 2008).

- An estimate of sunlight incident on earth's surface averages about 1000 W m^{-2} at an air mass of 1.5.
- Light outside photosynthetic spectrum (Plants can only use sunlight in the 400–740 nm wavelength range) = loss of 51.3%.
- Reflected light = minimum loss of 4.9% of incident radiation.
- Inefficiency of light absorbing pigments = minimum loss of 6.6% of incident radiation.
- Inefficiency of carbon synthesis = minimum loss of 24.6% of incident radiation for C3 plants and minimum loss of 28.7% of incident radiation for C4 plants.
  - C4 plants have greater losses for carbon synthesis due to the additional step of concentrating carbon dioxide inside the cell.
- Photorespiration = loss of 49% for C3 plants and 0% for C4 plants.
  - Since RuBisCO is exposed to atmospheric oxygen in C3 plants O^2 can be fixed instead of CO^2 causing the cell to complete the Calvin cycle without gaining a fixed carbon molecule.
  - The concentration of CO^2 inside the cells of C4 plants avoids photorespiration, thus increasing overall efficiency of photosynthesis.
- Respiration = minimum loss of 1.9% of incident radiation in C3 plants and minimum loss of 2.5% of incident radiation in C4 plants.
  - Respiration and other photosynthetic processes are impacted by temperature and efficiency falls with increasing temperature (Rowan and Kubien 2007).

$$\textit{Maximum efficiency of C3 plants}: \quad 100 - 51.3 - 4.9 - 6.6 - 24.6 - 49 - 1.9 = 4.6\%$$
$$\textit{Maximum efficiency of C4 plants}: \quad 100 - 51.3 - 4.9 - 6.6 - 28.7 - 0 - 2.5 = 6.0\%$$

4.2.2 Actual Efficiencies

Real world efficiencies are lower than theoretical maximum efficiencies due to differences in sunlight energy, spectrum, efficiency of photosynthetic processes, temperature, and atmospheric gas concentrations. Barstow (Barstow 2015) estimates real world efficiencies are in the range of about 2.4% for C3 plants and about 3.4% for C4 plants when averaged across a full growing season. Global efficiency estimated by Barber suggests photosynthetic efficiency declines to about 0.2% of the sunlight reaching the earth's surface. This low value for global efficiency is due to seasonal changes and areas of land with little or no plant life.

Production of biomass for food, fuel, and fiber to support a growing population will remain a challenge even though enough sunlight reaches the earth's surface in one hour to provide energy for human use for one year (Barber 2007). The low efficiencies of photosynthesis to capture and store solar energy lead to many of the challenges faced by the bioeconomy.

4.3 Photovoltaic Devices

A photovoltaic device, more commonly referred to as a solar cell is a power/energy convertor that converts light energy (electromagnetic energy) to electric energy. In order to develop a general understanding of how this device functions, it is best to begin with the input energy, which is light.

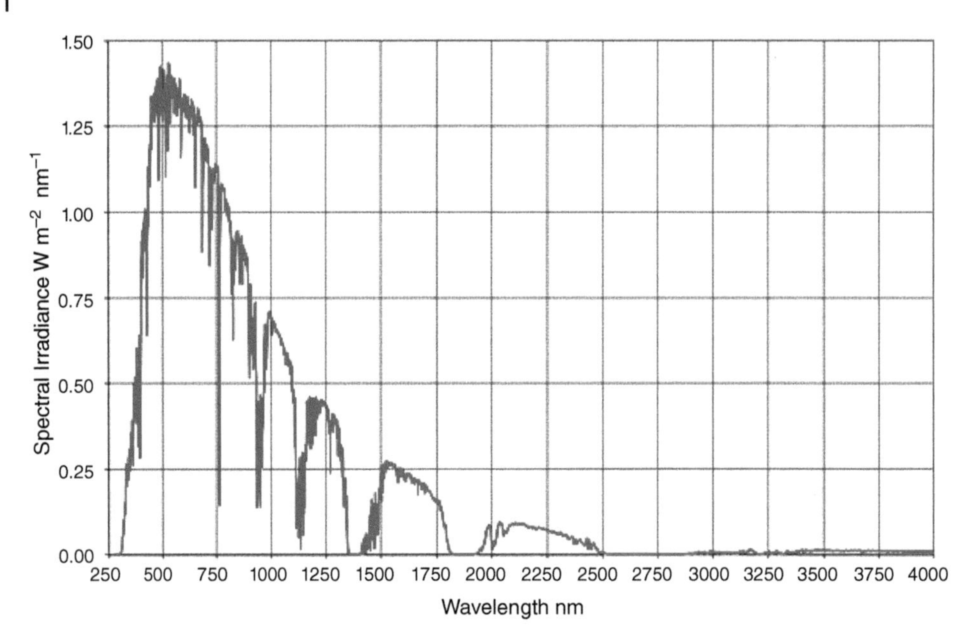

Figure 4.2 Wavelengths of sunlight and their energy.

In this case it is simplest to consider light as an energy particle (photon) whose energy is related to its frequency (color) as seen in Eq. (4.1), where E_{ph} is the energy of individual light particle (photon), h is Planck's constant (6.62×10^{-34} J s^{-1}), c is the speed of light (3×10^{8} m s^{-1}) and the λ is the wavelength (color) of the light (m).

$$E_{ph} = h(c/\lambda) \tag{4.1}$$

The sun produces an enormous amount of light over a range of frequencies (colors), as seen in Figure 4.2, and therefore produces a range of photons of different energies. Based on the value of h and the wavelength of light it is clear that a single photon does not possess much energy. This is compensated for by the fact that the sun produces an enormous number of photons, so many that at noon on a clear day about 1000 W m^{-2} of power strikes a flat plane, which is equivalent to 1 kW per hour of energy. Extending this calculation further shows that in about 20 days the amount of light energy that strikes the earth is equal to all the energy stored in all fossil fuels found in the earth.

In order for the light energy to be converted to electrical energy it must be 'absorbed' by the solar cell. We are all familiar with a simple form of light energy absorption and conversion; that is light energy to heat energy when a surface becomes hot upon exposure to the sun. The mechanism involved in the solar cell is slightly different. In a solar cell the photon must have enough energy to create a free or unbound electron in the device. When this occurs the energy of the electron is increased by the energy of the photon. Photons with less energy than this simply pass through the device and do not contribute to the energy output. If a photon has more than the required minimum energy, the energy above the minimum is wasted with respect to electric power conversion. These two facts (photons that are not absorbed and photons where not all the energy is converted) account for the limited efficiency of solar cells, where the 'semi-empirical' maximum theoretical efficiency (power output/power input) of a simple device is just above 33% (Shockley and Queisser 1961).

After absorption the photon no longer exists (it was pure energy) and its energy has been given to the electron. The solar cell extracts this energy in the form of electric power. In order to understand

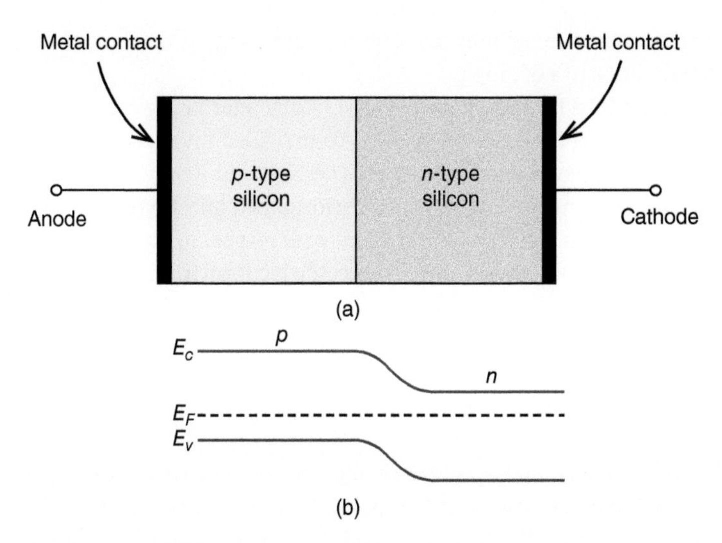

Figure 4.3 (a) Schematic cross section of a *p–n* junction solar cell. (b) Energy band diagram of the junction.

this we must examine the heart of the solar cell device, the *p–n* junction diode. The simplest form of the diode consists of two regions (*p* and *n*) joined together as seen in Figure 4.3a. Although the regions are the same basic material (silicon, the most common semiconductor material for example) they are subtly different. The *n* region has a small amount of impurity added to it to lower its internal energy structure while the *p* region has small amount of material added to it to raise its energy structure. The energy structure or more precisely the band diagram of the diode is seen in Figure 4.3b. The device is purposely designed to have a thick *p* region and thin *n* region. The *p* material has a large number of electrons at the energy level indicated by E_v. The next allowed energy level is energy E_c. You can think of this like rungs on a ladder; you can only place your foot where a rung is not at any arbitrary point in space. The photons with enough energy to raise the electron to level E_c are absorbed by the electrons at E_v imparting their energy into the electron and raising its energy state to level E_c. The electron's potential energy has been increased. This energy is converted to electrical energy when the electron rolls down the hill seen in Figure 4.3b. The total amount depends on the product of the number of electrons and the height of the hill. The number of electrons depends on how many photons are absorbed and the height depends on the material used to make the device. This is the trade-off; if we make the hill bigger, the energy required from a photon increases so fewer electrons are created from the sunlight, but each electron has more energy. If we make the hill smaller, more electrons are created but they have less energy.

Based on the solar spectrum, calculations show that the energy separation between E_c and E_v (also known as the band gap E_g) should be approximately 2.2×10^{-19} J, or 1.38 eV in order to yield a solar cell with the maximum possible efficiency, which is on the order of 44% (Shockley and Queisser 1961). This corresponds to a wavelength of about 9×10^{-7} m, which is in the infrared, outside the visible spectrum. Based on the previous discussion all photons with wavelength less than this value (energy greater than 1.38 eV) will be absorbed and converted to electrons and photons with wavelength greater than this value (energy less than 1.38 eV) will pass through the device.

Each cell is capable of producing an output voltage on the order of 0.7 times the band gap, which for silicon cell is about 0.7 V and a current of about 0.03 A cm^{-2} assuming a cell efficiency of 20%, which is about half the maximum theoretical value. Useful PV power systems consist of panels of individual cells wired together in such a manner that the total output voltage is on the order of 100's of volts and currents on the order of 10's of amps.

Also, the PV output is direct current (dc) while grid power is 60 Hz 120 V alternating current (ac), therefore power electronic systems (included in a complete PV system) called inverters are necessary to convert the PV dc power into grid compatible ac power. The National Renewable Energy Laboratory (NREL) has developed a straightforward software performance and financial model called the System Advisor Model (SAM) that can provide guidance with respect to virtually all the aspects involved in assessing which type of system is right for a specific local environment (System Advisor Model 2019).

4.4 Overview of Solar Photovoltaic Applications

Solar PV cells are wired in series and parallel then encased in clear glass to make a solar PV module. Solar panels are a common term used to describe solar PV modules. Modules are put on racks and wired together to make an array. A solar farm is a large solar array Figure 4.4. Many applications exist for solar PV. These can be simplified to three general categories of grid tied Figure 4.5, grid tied with battery backup Figure 4.6, and off grid Figure 4.7. The specific application may be to provide power to the grid, house, business, pump, light, or other options however each can be placed in one of these three categories. Based on modern 60 cell modules with power of 250–300 W each it would take 20–30 solar modules to generate enough energy for one home for one year.

Figure 4.4 From left to right, solar cell, solar module, solar array, solar farm. Modules commonly have 60–72 cells, arrays are a group of modules wired together, a solar farm is a large area of land used for solar energy production. Source: All images are public use images.

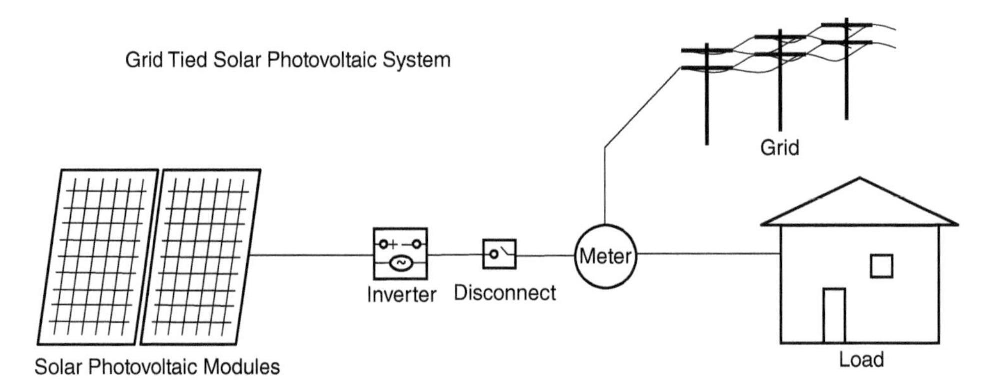

Figure 4.5 Grid tied solar PV has only two major components; the solar modules, inverter to convert direct current electricity from the modules into alternating current for the load and or grid.

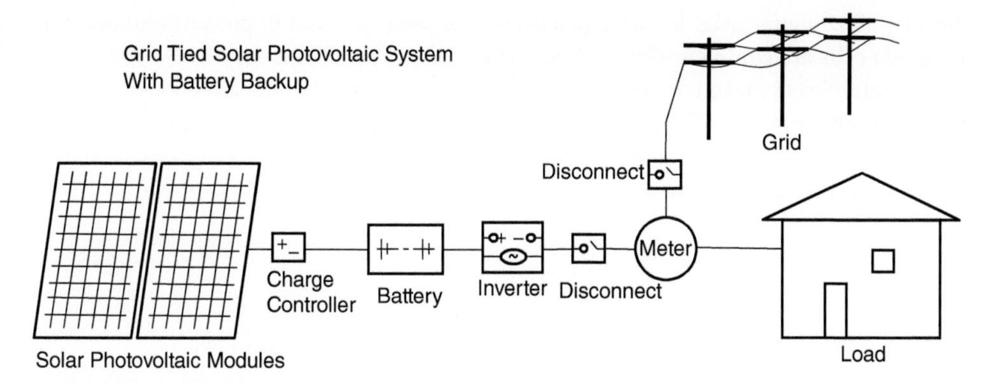

Figure 4.6 Grid tied solar PV with battery backup allows for backup power when the grid is down. The addition of battery bank, charge controller and disconnect to separate the load and system from the grid are needed.

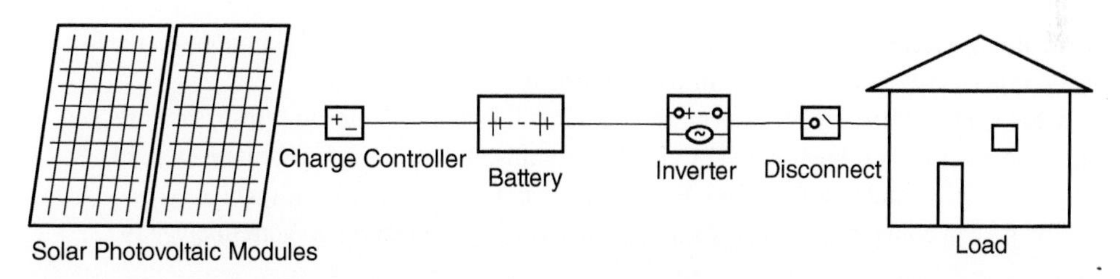

Figure 4.7 Off grid solar PV would be used for applications where electricity is needed and the electric grid is not available or too costly to connect to. Systems might be either direct current and not use an inverter or alternating current and have an inverter. Solar water pumping, off grid cabins, and remote monitoring stations are common applications for off grid systems.

4.5 Perspectives

The efficiency of plant photosynthesis depends on the type of photosynthetic pathway while the efficiency of a crop, in yield per unit area will also depend on numerous agronomic and climatic conditions. Plant breeding and research into artificial photosynthesis are paths to increasing efficient use of sunlight. Our utilization of plants for use in food, fuel, and fiber around the world will continue, yet consider which crops we use today. Of the staple food crops in the world only maize and sorghum are C4 plants with higher photosynthetic efficiency. Other staples such as cassava, potatoes, rice, wheat, and soybeans are all C3. Efficiency gains due to agronomic improvements such as fertilizers and irrigation can be substantial, but crop yields of C3 plants will lag behind the greater efficiency of C4 plants. If farm land becomes limited or world population grows to stress the world food system, gains in efficiency will become more important.

Efficient use of sunlight for electrical energy in the form of photovoltaic cells can improve over the 15–20% range of modern simple solar cells. The semiempirical efficiency of 33% and max theoretical efficiency of 44% from Shockley and Queisser are limits for simple or single junction cells. Layered multijunction cells already exist which can absorb sunlight from a larger range of wavelengths allowing for even higher efficiencies. The price per unit of power produced will influence which cells are commercially available. The future of solar cells may be the utilization of different chemistries to make low cost single junction cells or may be the reduction of price of high efficiency multijunction cells. Additionally, the use of energy storage technology combined with renewables such as solar may allow greater expansion and use of solar PV because its stored energy can be used at times of electricity demand.

References

Barber, J. (2007). Biological solar energy. *Philosophical Transactions of the Royal Society A: Mathematical, Physical and Engineering Sciences* 365 (1853): 1007–1023.

Barstow, B. (2015). Molecular mechanisms for the biological storage of renewable energy. *Advanced Science Engineering and Medicine* 7: 1066–1081.

REN21, Renewable Energy Policy Network for the 21st Century 2016, Renewables 2016 Global Status Report, viewed February 2017, online at https://bioenergy.inl.gov/InternationalReports/Renewables%20Global%20Status%20Report.pdf

Rowan, S. and Kubien, D. (2007). The temperature response of C3 and C4 photosynthesis. *Plant, Cell & Environment* 30 (9): 1086–1106.

Shockley, W. and Queisser, H. (1961). Detailed balance limit of efficiency of p-n junction solar cells. *Journal of Applied Physics* 32: 510–519. https://doi.org/10.1063/1.1736034.

System Advisor Model (2019), National Renewable Energy Laboratory online at https://sam.nrel.gov

U.S. Energy Information Administration (EIA) 2017, Monthly Energy Review February 24, 2017.

Xin-Guang, Z., Long, S., and Ort, D. (2008). What is the maximum efficiency with which photosynthesis can convert solar energy into bioenergy? *Current Opinion in Biotechnology* 19 (2): 153–159.

5

Recent Trends, Opportunities and Challenges of Sustainable Aviation Fuel

Libing Zhang[1], Terri L. Butler[2] and Bin Yang[*,1]*

[1]*Bioproducts, Sciences, and Engineering Laboratory, Department of Biological Systems Engineering, Washington State University, Richland, WA 99354, USA*
[2]*Buerk Center for Entrepreneurship, University of Washington, Seattle, WA 98195, USA*

CHAPTER MENU

5.1 Introduction

Climate change is a global challenge to be addressed by governments and industries around the world. It is estimated by the Intergovernmental Panel on Climate Change (IPCC) that by the year 2100 the world will see a temperature change of 2.5–7.8 °C above the average for the years between 1850 and 1900 (Bernstein et al. 2007). The aviation industry contributes approximately 2% of total carbon emissions across all industrial sectors and air transportation growth is increasing rapidly due to globalization of trade and travel. Aviation has no credible near-term fuel alternative in ethanol or electric power due to the energy density required although aviation fuel can be used in fuel cells to produce electricity locally in dispersed applications. Current jet fuel production processes require huge facilities that are complex to operate while using fossil oil as feedstock. Carbon emission reduction in the aviation industry is essential to combat global warming. The aviation industry has achieved several milestones in emission reduction including improved aircraft fuel efficiency and better air traffic control to promote safe, efficient and sustainable air travel. However, these improvements have led to carbon emission reductions of less than 15% (The global aviation industry 2010). Compared to these non-fuel approaches, the use of alternative jet fuel can achieve a further 50–80% reduction in carbon emission and is, therefore, considered to be the most efficient way to achieve carbon neutral aviation operation. Thus, addressing the source of fuel for the aviation industry is an important part of the answer to achieving a material reduction in greenhouse gas emissions.

Green Energy to Sustainability: Strategies for Global Industries, First Edition.
Edited by Alain A. Vertès, Nasib Qureshi, Hans P. Blaschek and Hideaki Yukawa.

Demands from regulatory agencies for compliance with international carbon emission standards have increased. Domestic interests in biojet fuel powered planes have also been rising. Both of these factors are driving the growth of biojet fuel markets. By 2028, the carbon emission standards for international flights will be regulated, requiring airlines to purchase carbon offsets to compensate for growth in their emissions. This has led to a sense of urgency for airlines and their shareholders, resulting in their encouragement of technologies that are likely to bring new biojet fuel products to market. Although the research, development, and commercialization of biojet fuels are still in the early stages, several activities have been initiated by a number of stakeholders in the past 10 years, including the newly approved by American Society for Testing and Materials (ASTM) biojet standards, biojet fuel development community formation, fuel certification test flights, and off-take fuel agreements between airline companies and early-stage biojet fuel suppliers (Radich 2015).

The US jet fuel market represents 20 billion gallons per year. All emerging biojet fuel technologies that meet ASTM standards are included in this number. However, as of this date, no clear winning sustainable jet fuel technology exists, a situation that encourages continued research and development in innovative biojet fuel technologies. For biomass-based biojet fuels, as in the case of biofuels for ground transportation, one of the most important deciding factors in commercial feasibility is having accessible, low-cost feedstock. This presents a challenge because, unlike crude oil, biomass contains a high level of oxygen (up to 45%) in its major heteropolymers. Biojet technologies have made progress in dealing with these diverse biomass substrates, but more effort is still needed to improve the biomass conversion efficiencies and to reduce the cost of production.

The aim of this chapter is to provide an overview of current opportunities for the development of biojet fuels and highlight the reasons behind the burgeoning efforts. Information gathered through interviews with parties from the private sector and government agencies indicated a growing interest in purchasing and investment for biojet fuels. This chapter also provides a brief summary of the current biojet fuel technologies, and the status of several specific challenges facing the industry is explained, which help the reader to better understand the overall landscape of biojet fuels.

5.2 Overview of the Jet Fuel Market

The current US market for jet fuel is dominated by large petroleum-based refineries that produce a total of approximately 23.6 billion gallons of jet fuel per year (Figure 5.1). Companies such as Exxon Mobil (Irving, TX, USA), Chevron Corporation (San Ramon, CA, USA), BP (London, UK), Valero (San Antonio, TX, USA), and Marathon Petroleum Group (Findlay, Ohio, USA) account for around 50% of that production volume. The global consumption is about four times that much (Indexmundi 2017; US Energy Information Administration 2017a). The domestic jet fuel consumption is expected to increase by more than 40% between 2016 and 2040 as projected by the Energy Information Administration (EIA) (US Energy Information Administration 2017a) because the increased demand for air travel will more than offset projected aircraft efficiencies gained.

5.2.1 Driving Force of Growing Biojet Fuel Opportunities

Airlines are exploring biojet fuel options to diversify their fuel supplies, lower fuel costs in the long run, reduce price volatility, comply with future emission regulations, and minimize the dependence of the overall aviation industry on fossil fuel. Jet fuel cost represents the highest percentage of total cost in aviation operations. By 2040, the jet fuel price paid by airlines is projected to be 40% higher than that price in 2014 (Radich 2015). The rising jet fuel prices are driving commercial and military aviation operators to look for substitutes or alternative drop-in fuels. The availability of

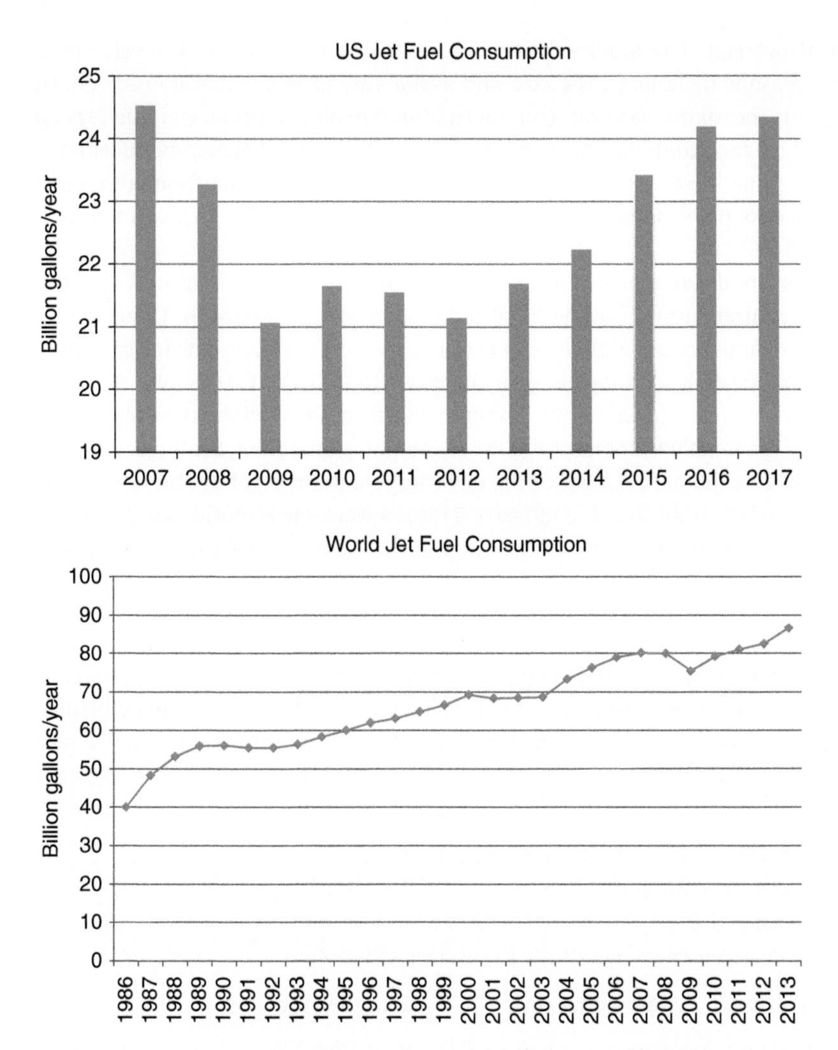

Figure 5.1 Global and US Jet fuel consumption (U.S. Energy Information Administration 2017a).

more drop-in fuels can provide increased options for the airline industry, reducing dependence on conventional jet fuel made from crude oil and reducing overall costs by reducing fuel price volatility (Radich 2015).

The price fluctuation of jet fuel ranged, according to estimations of the US Gulf Coast Kerosene-Type Jet Fuel Spot Price provided by the EIA, from between 80 cents to $4.5/gal over the past 10 years (US Energy Information Administration 2017c). Traditionally, fuel hedging, the contractual tool used by the airlines (Morrell and Swan 2006), can reduce the exposure of airline companies to the volatility of jet fuel cost. However, this financial approach is not always as effective as expected. Fuel contracts usually last only 12–18 months (Amanda Peterka 2013). The pricing of jet fuel is not easy to predict, and this uncertainty represents an important operational risk as it causes uncertainties in finance and regular aviation operations. For example, Delta Airlines (Atlanta, GA, USA) reported a loss of $4 billion due to hedging fuel, a contract strategy that airlines applied to reduce exposure to jet fuel volatility (Ed Hirs and Forbes 2016). Biojet fuel is generally expected to reduce pricing volatility (Winchester et al. 2013), although this is somewhat controversial. The arguments refer back to the reasons why the price of jet fuel is volatile. These reasons include: (i) the cost and availability of raw material crude oil, (ii) competitors' pricing, (iii) transportation cost, (iv) international relations, (v) geographic regions, (vi) supply and demand balance, and (vii) the global political climate, etc. (US Energy Information Administration 2017b). Biojet fuels are also influenced by similar factors even though there are arguments from proponents pointing out the more available and sustainable biomass feedstock for producing biojet fuels.

One of the most important driving forces for the biojet fuel market is the increasingly stringent regulatory compliance requirements of international carbon emission standards. In October 2016, the International Civil Aviation Organization (ICAO) stated the new requirements in the Carbon Offsetting and Reduction Scheme for International Aviation (CORSIA) (ICAO's 39th Assembly 2016) to regulate the industry's carbon emissions. Commercial airlines are facing pressure to pay for renewable credits if they are unable to comply with the regulations. Therefore, the biojet fuel business is not only motivated by the need to reduce the environmental impacts but also has become important to the profit margins of the aviation industry. This is a welcome evolution as such a profitability pressure provides an impetus for the earlier adoption of these fuels and at a large scale.

5.2.2 Biojet Fuel Types and Specifications

Currently there are five main biojet fuel technologies that have been developed and approved by ASTM as meeting the standard specifications ASTM D7566 (ASTM 2014), and so can be blended with Jet A or Jet A-1 fuel certified to Specification D1655 and 16 certifications in preparation (International Air Transport Association 2009):

1. Fischer–Tropsch Synthetic Paraffinic Kerosene (FT-SPK)
2. Fischer–Tropsch Synthetic Kerosene with Aromatics (FT-SKA)
3. Hydrotreated Esters and Fatty Acids (HEFA)
4. Synthesized Iso-Paraffinic (SIP)
5. Alcohol (isobutanol) to Jet Synthetic Paraffinic Kerosene (ATJ-SPK)

Five more biojet fuel technologies are under review as listed in the following:

1. High Freeze Point Hydrotreated Esters and Fatty Acids (HFP-HEFA)
2. Virent BioForm Synthesized Aromatic Kerosene (SAK) Jet Fuel (SAK)
3. LanzaTech ATJ-SPK (Ethanol to Jet)
4. Applied Research Associates Catalytic Hydrothermolysis Jet (ARA-CHJ)
5. BioForm® Synthesized Kerosene (SK) Jet Fuel (Virent SK)

Data is being collected for three additional technologies including ATJ-SKA developed by Byogy (Byogy Renewables, Inc., San Jose, USA), Swed Biofuels (Swedish Biofuels AB, Stockholm, Sweden), and the IH^2 demonstration scale by Shell (Royal Dutch Shell, UK) which acquired the technology from the Gas Technology Institute (GTI) (Des Plaines, USA) in 2009.

Eight more technologies, listed below, are in exploratory discussions. Figure 5.2 shows some of the technologies that are currently being studied.

1. *Vertimas*. One-step catalytic conversion of ethanol to jet, petrol, diesel fuel and chemicals which was originally invented at Oak Ridge National Laboratory.
2. *SBI bioenergy*. Continuous catalytic process that converts fat, oil or grease into renewable gasoline, diesel and jet fuel with proprietary Process intensification and Continuous Flow through Processing technologies (PICFTR).

3. *Joule.* CO_2-derived fuels 'Sunflow-J' jet fuel from specially engineered photosynthetic bacteria, waste carbon dioxide, sunlight and water.
4. *Global bioenergies.* Biological production of isobutene and process to jet.
5. *Eni.* hydrogenated vegetable oil (HVO).
6. *Enerkem.* Municipal waste gasification and catalytic conversion to ethanol followed conversions to biofuels and chemicals.
7. *POET.* Agricultural waste derived alcohol to jet fuel.
8. *Washington State University.* Lignin to Jet Fuel (LJ-D&HDO) through one-step proprietary catalytic upgrading of lignin waste to jet fuel.

As shown in Figure 5.2, biojet fuel technologies use a broad range of feedstocks such as lignocellulosic biomass, municipal solid wastes, vegetable oils, animal fats, lignin, flue gas, and even seawater. Current biojet fuel technology pathways have been summarized by referring to industry publications, academic research and technology reports (Fahim et al. 2009; Wang and Tao 2016; Wang et al. 2016b; Yang and Laskar 2016). Information has also been gleaned from biofuel development companies and institutions (Biochemtax 2017; LanzaTech 2017; Fellet 2016; US Naval Research Laboratory 2016; Virent 2017; WSU Maegan Murray 2016). Compared to current oil refineries, most of the biojet fuel technologies require more than four functional process units, which is likely to be cost intensive, requiring capital, energy, labor, and water to commercialize (Bridgwater 1975; Gerrard 2000). Among the biojet fuels, the lignin-based jet fuel technology is one of the most promising, and potentially most sustainable, due to its one-step process conversion. By 2022, it is projected that 62 million tons of lignin side-product will be generated annually by cellulosic biomass biorefineries (Ragauskas et al. 2014). The pathways of lignin based jet fuel include depolymerization and hydrodeoxygenation (LJ-D&HDO) of lignin to a mixture of long-chain hydrocarbons (C7–C18 cyclic hydrocarbons) that can be made into jet fuel. (Laskar et al. 2014; Laskar and Yang 2012; Laskar et al. 2013; Wang et al. 2015, Wang et al. 2016a; Wang et al. 2017a,b,c; Yang and Laskar 2016) Scaling this process and putting it into production alongside current biorefinery production facilities would significantly improve biomass conversion efficiency and the economics of biofuels and chemicals production.

In discussions of capital cost savings, the co-processing of biomass by co-feeding of biomass feedstock with petroleum streams in conventional oil refinery units has received attention (Al-Sabawi et al. 2012; Huber and Corma 2007; Yanik et al. 2012). Recently, an oil refinery raised great interest by co-processing biomass and crude oil distillates for instance. Chevron Corporation (San Ramon, CA, USA), Phillips 66 (Houston, USA) and BP (London, UK) are researching the feasibility of co-processed fuels up to 5 vol% bio-oils from HEFA and middle crude oil distillates to produce hydrocarbons that could be used in jet fuel (Joseph Sorena and Clark 2015). The compatibility of biojet production united with existing infrastructure motivated the interests in co-feeding of the biojet feedstock/intermediates with crude oil-based distillates. Figure 5.2 shows several potential research opportunities in co-feeding desired by oil refiners.

In 2009, ASTM International Committee on Petroleum Products and Lubricants issued ASTM Standard Specification D7566, titled as Standard Specification for Aviation Turbine Fuel Containing Synthesized Hydrocarbon, to certify drop-in jet fuel from alternative feedstock. It contains five Annexes that define the specifications of the five biojet fuels blended with conversional jet fuel certified by D1655 ASTM standard. In Table 5.1, the specifications of the five approved biojet fuels are presented. The rest of the specifications of biojet fuel pathways in Figure 5.2, such as syngas fermentation and CH, are still under development. The blending limit of each biojet fuel is different. The FT-SPK, HEFA, and SIP biojet fuels were approved for higher blending feasibility testing (ASTM Committee 2017).

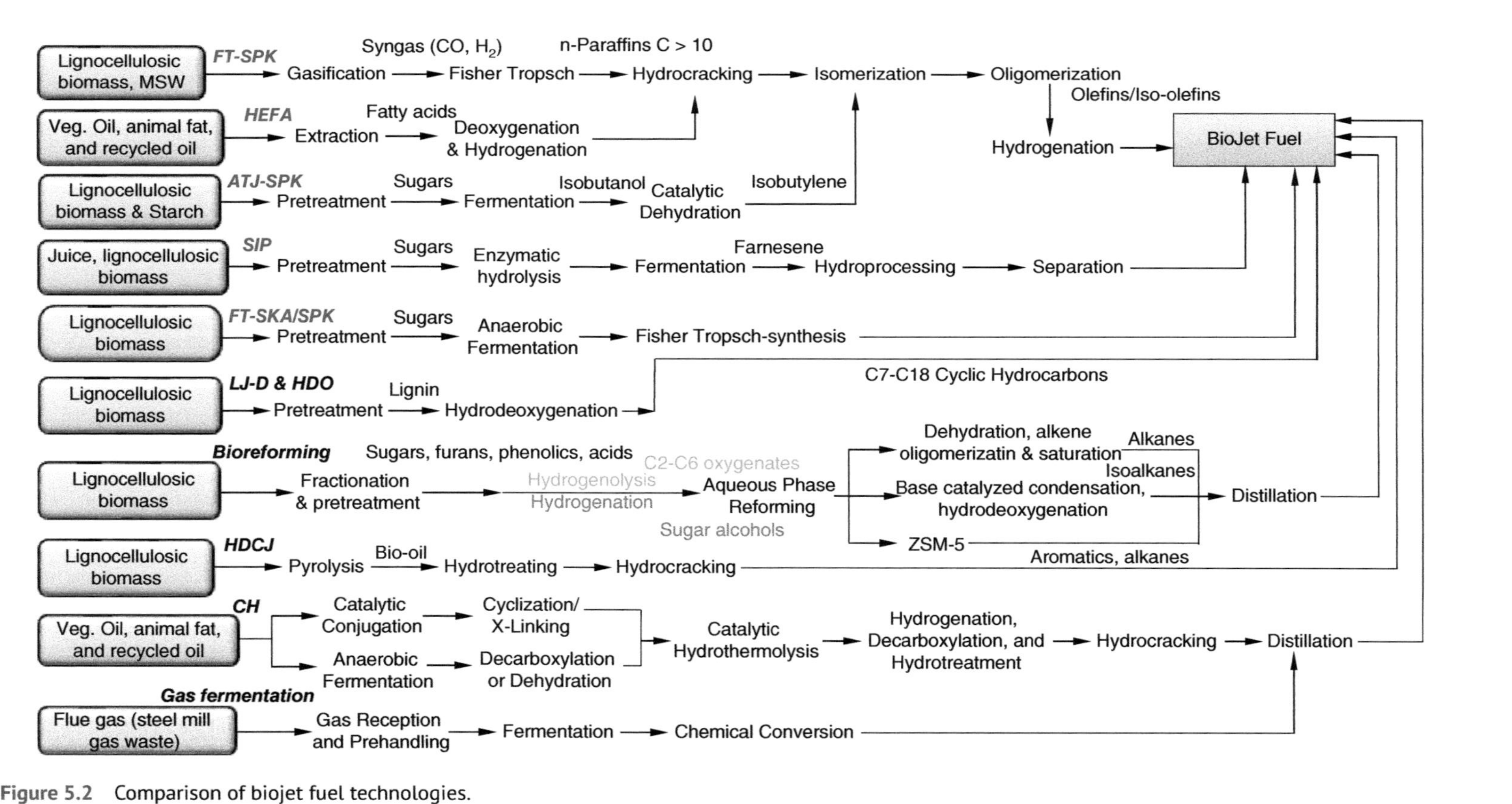

Figure 5.2 Comparison of biojet fuel technologies.

Table 5.1 Approved biojet fuels specifications (ASTM Committee 2017).

		ASTM D7566-16b					
Jet fuel tests		**Jet A/A-1**	**FT-SPK**	**FT-SKA**	**HEFA**	**SIP**	**ATJ-SPK**
Approved blend, %		**100**	**50**	**10**	**50**	**50**	**30**
Composition							
Acidity, total (mg KOH/g)	Max	0.1	0.015	0.015	0.015	0.015	0.015
Aromatics (vol %)	Max	8–25	0.5	20/21.2	0.5	0.5	0.5
Cycloparaffins, mass %	Max		15	15	15		15
Saturated hydrocarbons, mass %	Min					98	
Farnesane, % mass	Min					97	
Carbon and hydrogen, mass %	Min		99.5	99.5	99.5	99.5	99.5
Nitrogen, mg kg^{-1}	Max		2	2	2	2	2
Water, mg kg^{-1}	Max		75	75	75	75	75
Sulfur, mg kg^{-1}	Max	0.3	15	15	15	2	15
Metals[a)], mg kg^{-1}	Max		0.1 per metal	0.1 per metal	0.1 per metal	0.1 per metal	0.1 per metal
Halogens, mg kg^{-1}	Max		1	1	1	1	1
Hexahydrofarnesol	Max					1.5	
Olefins, $mgBr_2$ per 100 g	Max					300	
Contaminants							
Existent gum, mg per 100 ml	Max	7		4	7		
FAME, ppm	Max				5		
MSEP	Min			90			
Volatility							
Distillation temperature, °C							
10% recovered, temperature (T10)	Max	205	205	205	205	250	205
Final boiling point, °C	Max	300	300	300	300	255	300

(Continued)

Table 5.1 (Continued)

		ASTM D7566-16b					
Jet fuel tests		**Jet A/A-1**	**FT-SPK**	**FT-SKA**	**HEFA**	**SIP**	**ATJ-SPK**
Approved blend, %		**100**	**50**	**10**	**50**	**50**	**30**
Distillation (D86) T90-T10, °C	Min		22	22	22	5	21
Flash point, °C	Min	38	38	38	38	100	38
Fluidity							
Freezing point, °C	Max	−47 Jet A-1 −40 Jet A	−40	−40	−40	−60	−40
Density @ 15 °C, kg l^{-1}		775–840	730–770	755–800	730–770	765–780	730–770
Viscosity −20 °C, $mm^2 s^{-1}$	Max	8.0					
Combustion							
Energy density, MJ kg^{-1}	Min	42.8			44.1	43.5	
Thermal stability							
Temperature, °C		325	325	328	325	355	325
Filter pressure drop, mmHg	Max	25	25	25	25	25	25
Additives							
Antioxidants, mg l^{-1}	Max	24	17–24	17–24	17–24	17–24	17–24

a) Metal includes Al, Ca, Co, Cr, Cu, Fe, K, Li, Mg, Mn, Mo, Na, Ni, P, Pb, Pd, Pt, Sn, Sr, Ti, V, Zn.

The US civilian jet fuels are mainly categorized into Jet A-1, Jet A, and Jet B, while the US military sets its own standards, which are JP-1, JP-2 to JP-10, each for specific applications. Jet A-1 is similar to JP-8 and Jet B is similar to JP-4. Biojet fuels imitate Kerosene-type fuels, which have a carbon number between 8 and 16, for example Jet A, Jet A-1, JP-5 and JP-8. While a lot of interest in biojet fuels has been generated, so far there have been little in the way of commercial products available in the market because research and development in biojet fuels is still in the early stages.

Notably, Jet A/A-1 and the five approved biojet fuels differ in composition. The energy content per unit weight increases in the following order; aromatic, naphthenic, paraffinic for hydrocarbons with the same carbon number (Hemighaus et al. 2006). As shown in Table 5.1, although biojet fuel possesses more cycloparaffins than paraffins, it remains hard to compare fuel efficiencies between biojet fuel and conventional jet fuel due to variations in hydrocarbon profiling. Whether the composition of biojet fuel, especially the new contaminants that arise in these processes, such as FAMES and Halogens, could cause problems to the propulsion and energy system of an airplane engine needs to be confirmed by engine manufacturers (FAA Aviation Rulemaking Advisory Committee 1998). Biojet fuels with low sulfur content have been achieved, which reduces aviation impact on air pollution (Hileman et al. 2009). Jet A/A-1 has a minimum requirement of 8% vol of aromatics, so the lack of enough aromatics in biojet poses a theoretical risk of losing proper functions of elastomeric materials in jet engines (Karatzos et al. 2014). However, the performance characteristics of biojet fuels can be adjusted by additives, which is generally required to be compatible with aircraft operations.

Hydrocarbons that can be used in jet fuel can be categorized as: paraffins, olefins, naphthenes, and aromatics (Hemighaus et al. 2006). Hydrocarbons in jet fuel are mixtures of compounds from each of these classes that contain different numbers of carbon atoms. Since biomass and crude oil are both converted to a mixture of hydrocarbon compounds, taking a look at the elemental profiling of various biomass and crude oil compositions is an efficient way to understand the different chemical compositions of jet fuels and conversion efficiency in biomass and crude oil refineries (Table 5.2) (Baedecker et al. 1993; Bilba et al. 2013; Canoira et al. 2008; Valkenburg et al. 2008). For example, the oxygen element in biomass is much more than that of crude oil, which requires more energy input to effectively remove excess oxygen and produce hydrocarbons consisting of only carbon and hydrogen atoms. This is one of the reasons why hydrogen hydrotreating is needed in nearly all biojet fuel conversion pathways, and the cost and availability of these industrial processes are considered a risk in the research and development of biojet fuel (National Academies of Sciences, Engineering, and Medicine 2016). What is more, understanding the different element compositions is extremely useful for estimating the potential abundance of air pollutants that are emitted during fuel combustion.

5.3 Assessment of Environmental Policy and Economic Factors Affecting the Aviation Industry

Aviation should be environmentally sustainable, cause minimal pollution to air and water, and contribute to high quality human life. More and more stringent environmental regulations have been enforced to fight against global warming, affecting the worldwide efforts to reduce carbon footprints. Renewable biojet fuels are significant in such efforts because of the increasing demand for jet fuels.

Table 5.2 Feedstock elemental composition (Baedecker et al. 1993; Bilba et al. 2013; Canoira et al. 2008; Valkenburg et al. 2008).

Raw materials, % dry weight	Crude oil	Coal	Wood waste	Switchgrass	Sugarcane bagasse	MSW	Animal fat /vegetable oil
Carbon	84–87	73.6	48.0	46.7	48.6	33.9–84.8	76.6
Hydrogen	11–14	4.7	5.5	5.9	6.3	1.72–15.16	12.0
Oxygen	0–2	5.1	39.1	39.0	45.1	15.8–43.7	12.4
Nitrogen	0–1	1.3	1.4	0.54	0.48	0.12–2.37	2.2
Sulfur	0–3	1.6	0.1	0.13	0.21	0.06–1.4	0.15
Ash mineral	0.1–0.2	13.7	5.9	7.7	3.7	4.4–44.2	0.1–3.7

5.3.1 Momentum Building of International Carbon Emission Regulations

As stated earlier, the aviation industry is responsible for approximately 2% of global CO_2 emissions (Lister et al. 1999). The ICAO is a UN agency charged with managing all aspects of international civil aviation such as aviation sustainability, air navigation, and air travel safe (ICAO 1994). The ICAO has set a goal of capping net carbon emissions from 2020 to achieve 50% reduction by 2050 from 2005 emission levels (The global aviation industry 2010). The Carbon Offsetting and Reduction Scheme (CORSIA) is a historic move for the global international aviation industry to meet ICAO's goal of carbon neutrality from 2020 by limiting and offsetting emissions from the aviation sector. A total of 70 countries intend to voluntarily participate in the global market-based measure (MBM) scheme implemented in May 2017; this indicates that more than 87.7% of international aviation activities are committed to achieving significant emission reductions (ICAO 2016).

In October 2016, the ICAO announced the CORSIA for International Aviation aiming to offset approximately 80% of the emissions above 2020 levels. The ICAO requires airlines to offset emissions by purchasing eligible emission credits (e.g. renewable energy) equivalent to offsetting requirements from the carbon markets. This emission offset is currently voluntary. The first voluntary transition will happen between 2021 and 2027, but it will become mandatory in 2027. This creates an additional opportunity for biojet fuel to become a beneficial fuel alternative for commercial airlines, helping them meet the emission allowance each year. The regulation is legally binding for countries that are signatories to the Kyoto Protocol. In 2016, at the 39th session of the ICAO Assembly, it was decided that CORSIA will become enforceable in 2021, so at that point, the aviation industry will have the obligation to buy carbon offsets to accommodate any growth in their carbon emissions.

CORSIA is taking a route-based approach where all operators on the same route will have the same offsetting obligations. Before ICAO's CORSIA, the efforts to control greenhouse emissions can be traced back to the1940s, as the International Air Transport Association was founded in 1945 and has since grown to 274 airline members. CORSIA has set up a timeline for execution of the new plan for carbon emission reduction, although ICAO is still working on the implementation plan for quantifying and offsetting regulations. The proposed timeline has three phases, including phase I during years 2021–2023, and phase II during years 2024–2027, as the voluntary participation phases. The carbon offset requirements will become obligatory to airlines from 2028 to 2035 and will be enforced by requiring payment for generating excess carbon emissions above 2020 levels during international flights.

The Paris Agreement consolidates nations aiming to combat climate change within the United Nations Framework Convention on Climate Change (UNFCCC), which was drafted in December 2015, and signed by 195 nations worldwide in 2016, and became effective in 2016. It requires the assessment of carbon emission progress in 2018 and regular reviews every five years. The domestic carbon emission commitments of these countries will be addressed under the Paris Agreement. This is an international agreement that sets out a global action plan to limit global warming, due to be enforced from 2020. As listed in Table 5.3, the commitment to carbon emission reduction receives common consensus internationally. The endorsement of global leaders for reducing carbon emissions and fighting climate change has finally become mainstream.

Emissions Trading Schemes (ETSs) for greenhouse gas emissions are operational in several countries and regions (Talberg and Swoboda 2013). In the United States, the Regional Greenhouse Gas Initiative (RGGI), implemented in 2009, is known as the first mandatory CO_2 cap-and-trade program and it involved nine states. Also, the California Air Resources Board (CARB) has established a Californian cap-and-trade scheme to reduce greenhouse gas emissions to 1990 levels by 2020 (Reuters 2012).The ICAO has endorsed the potential of MBM such as capping, open trading and charging as a means of quantification and reducing greenhouse gas production for international civil aviation.

The European Union Emissions Trading Scheme (EU ETS) has become a well-established practice. In 2005, EU ETS launched Phase I (2005–2007) in January and is one of several options that allow the quantification and promote the reduction of carbon emissions. Airlines receive 85% of their proportionally-allocated allowances based on 2010 emissions for free. The EU suspended full compliance for international flights in and out of the EU until December 31, 2016 for ICAO to set a global scheme. By 2017, EU ETS capped carbon emissions of 1 889 411 334 ton in CO_2 emissions (tCO_2e), a threshold that is projected to evolve to 1 777 105 173 tCO_2e by 2020 (Fund 2013). Whether CORSIA more effectively reduces carbon emissions than does the EU ETS requires more study.

The EU ETS functions using a system of emission allowances, each allowance being equivalent to one ton of CO_2. Each airline is required to submit an Annual Emissions Report with a number or certification to depict the free portion of the emission allowance and the renewable credits that should be purchased to compensate the emission difference between the actual and the allowed emission amounts. The trading system allows aviation companies to sell their excess allowance to other parties. The Environmental Protection Agency (EPA) has also determined that aircraft emissions contribute to climate change and, eventually, is expected to move forward on standards that would be at least as stringent as the ICAO's standards (US Environmental Protection Agency 2016). The uncertainties of the regulatory situation for airlines operating in domestic and international markets is a motivator for the development of alternative jet fuels.

5.3.2 Increasing Activities to Address the Carbon Emission Control

Many efforts have focused on emission reduction including lighter aircraft design, better engine efficiency, rigorous air traffic management, and improvement in ground transportation efficiency. The fuel efficiency of aircraft has been improved by 70% over the years, compared with the early days of the airline industry. Boeing (Boeing Company, Chicago, IL, USA) has claimed that fleet efficiency is improving by an estimated 2.9% per year and emphasized that the lifecycle of biofuel can reduce CO_2 emissions by 50–80% as compared to fossil fuel (Boeing Company 2015). ICAO reported 1.5% fleet efficiency improvements every year (The global aviation industry 2010). US air travel rose 2.6% in 2016 and 6.7% in global air markets compared to 2015 (International Air Transport Association 2017). The growth rates of air travel were estimated to be 2.8% and 3.7%

Table 5.3 Events in carbon emission reduction development.

Timeline	Organization	Agreements	Ratified nations	Legacy	Notes
1947	ICAO		191	Not enforced	ICAO formation
1981	Chicago Convention	Aircraft engine emission	191	Not enforced	
1992	UNFCCC		154	Not enforced	UNFCCC formation
1992	UNFCCC	Kyoto Protocol	US and Canada not ratified	Legally binding	Commit to reduce emissions of greenhouse gases (Berlin mandates)
2005	Directive of the European Parliament and the European Council	EU ETS	Linked to Kyoto Protocol members	Legally binding	Phase I (2005–2007) launched
2007	UNFCCC	Bali Action Plan-Cop13	114	Legally binding	Set emission measurement; 30 billion fast-start financing (in 2010–2012)
2008	Directive of the European Parliament and the European Council	EU ETS	Linked to Kyoto Protocol members	Legally binding	Phase II (2008–2012)
2009	UNFCCC	Copenhagen Accord	114	Not enforced	Set 2 °C limit in global warming
2010	UNFCCC	Cancun-Cop 16	196	Not enforced	Establish approaches to achieve carbon reduction
2012	UNFCCC	Doha Amendment-Cop 18	196	Legally binding	Regulate 2013–2020
2013	Directive of the European Parliament and the European Council	EU ETS	Linked to Kyoto Protocol members	Legally binding	Phase III (2012–2013)
2015	UNFCCC	Paris Agreement Cop 21	196	Layout	Emission contribution submitted
2015	UNFCCC	Paris Agreement Cop 22	196	Layout	Method to evaluate contribution
2016	UNFCCC	Paris Agreement	196	Enforced	
2016	ICAO	CORSIA	191	Enforced from 2028	Carbon offsetting

annually in the US and internationally, respectively (International Air Transport Association 2016). The Federal Aviation Agency (FAA) has a similar estimation in the continuous growth rates to 2037 (Federal Aviation Administration 2017b). Therefore, simply increasing fleet efficiency is not likely to be enough to achieve the carbon emission reduction goals given the rising rate of air travel and increased activity from airport ground transportation.

According to the ICAO environmental report in 2010, several approaches are being implemented to reduce emissions from ground transportation associated with the airlines including: (i) providing public transportation tools for public transport, (ii) regulating emissions from private vehicles, (iii) centralizing shuttles for hotel, rental car or other services, (iv) encouraging the use of alternative fuel or hybrid vehicles, and (v) improving infrastructures to support carbon emission reduction activities (Secretariat 2010). The air traffic management and operational improvements are one of the five pillars in NextGen, a program to ensure sustainable aviation (Federal Aviation Administration 2017a). It was reported that continuous improvement of air traffic and operational management can reduce carbon emissions by 10% (The global aviation industry 2010).

5.3.3 New Technologies and Aviation Operation Improvement

The US accounts for 29% of all greenhouse gas emissions from global commercial aviation as reported by the EPA (US Energy Information Administration 2017b). Three unconventional sources of petroleum including Canadian oil sands, Venezuelan Very Heavy Oils (VHOs), and shale oil can supply jet fuel in addition to fuel produced from conventionally sourced crude oil (Hileman et al. 2009). Life-cycle studies of these three sources revealed that GHG emissions associated with oil sands and VHOs can be 10–20% higher and oil-shale can be 50% higher than conventionally produced Jet A (Hileman et al. 2009). The energy return on investment (EROI) is the ratio of the amount of usable energy obtained from a particular resource to the amount of energy needed to obtain that resource. It is a criterion used to evaluate the energy production efficiency. The EROI ratio of a resource should be at least 3:1 to be viable and feasible as a practical energy source. The EROI ratios of several main energy sources have been reported and indicate the exceptional energy-intensive cost of oil sands compared with the values of coal at 50% and crude oil at 22%, with oil from tar sands at only 3% (Hall et al. 2014).

The prospects for oil shale development remain unclear given the high production cost and environmental risks (Bartis et al. 2005). It was estimated that $20 billion per year of shale oil production is possible with a capacity of 3 million barrels per day, if the production cost proves to be economic (Bartis et al. 2005). The EROI value of oil shale was reported to be <10%, which suggests that the refining of shale oil is an energy-, capital- and water-intensive process (Hall et al. 2014).

Additionally, efforts have been made to investigate non-liquid fuel technologies such as electricity, fuel cells, batteries, and hydrogen energy. However, the current energy density of these technologies does not meet jet fuel requirements and non-liquid fuels are also not compatible with the current liquid fuel distribution infrastructure.

Biodiesel and bioethanol have been widely used as clean energy sources in surface transportation; however, these fuels are not options for jet fuel. Table 5.4 shows a comparison of performance characteristics of jet fuel with ethanol and biodiesel (Renewable Fuels Association 2011; Reynolds et al. 2017). Biodiesel possesses a higher freezing point than what is allowed for jet fuel and its viscosity is too high to meet standards. In addition, it has less energy content per volume compared with regular jet fuel, significantly jeopardizing aircraft range (Appadoo 2009). For example, typical bio-diesel possesses 38 MJ kg^{-1}, which is much lower than 48 MJ kg^{-1} of Jet A/Jet A-1. Assessing the feasibility of ethanol, it has a high volatility and about 40% less energy content than jet fuel, thus posing a real practical problem for aircraft range.

Table 5.4 Comparison of characteristics of ethanol and biodiesel.

Fuel types	Jet A/A-1	Biodiesel (B100)	Ethanol
ASTM standards	ASTM D7566-16b	ASTM D 975	
Viscosity, $mm^2\ s^{-1}$	Max 8.0 at −20 °C	1.9–6 at 40 °C	1.2
Flashpoint, °C	Min 38	Min 130	17
Freezing Point, °C	Max −47 Jet A-1 and −40 Jet A	0	−84
Energy Content, MJ kg^{-1}	Min 42.8	38	25

5.4 Current Activities Around Biojet in the Aviation Industry

Biojet fuels remain the only true alternative for the commercial aviation industry and the military, both facing ambitious near-term greenhouse gas reduction targets. Thus, the aviation industry is more interested in the biojet fuels as many offtake agreements have been made even though biojet fuels are not commercially available yet. These agreements of taking millions of gallons of biojet fuels by airlines shows a strong commitment towards a green aviation industry. Many test flights have been executed by biojet shareholders in order to enter the jet fuel market.

5.4.1 Alternative Jet Fuel Deployment and Use

The major emission constituents of jet fuel combustion include particulates, SO_2, NO_X, carbon monoxide, and carbon dioxide. Biojet fuel can greatly reduce air pollution from sulfur and particulates as discussed before. Calculating the levels of carbon emissions of flights is a complex challenge, since it incorporates several factors including aircraft configuration, fuel burning, and flight distance. The Sabre Holdings calculator was reported to be the most accurate compared to others such as the DEFRA, ICAO (2008), and ClimateCare calculators (Jardine 2009). The EU ETS scheme has set up aircraft operators and guidelines to assist monitoring and reporting annual emissions and tonne km^{-1} data for EU emissions trading (Authority 2009).

Over the past 10 years, the interest in biojet fuel has been growing rapidly and the beginnings of a biojet fuel industry has formed. The Commercial Aviation Alternative Fuels Initiative (CAAFI), a coalition of aviation stakeholders interested in bringing commercially viable, sustainable, alternative jet fuels to the marketplace, was established in 2006 to enable and facilitate the near-term development and commercialization of alternative aviation fuels.

CAAFI is a coalition of airlines, aircraft and engine manufacturers, energy producers, researchers, and federal agencies including FAA, DOT, NASA, DOE's National Renewable Energy Laboratory and Energy Efficiency and Renewable Energy Office and USDA, among others. CAAFI has developed a base technology roadmap and identified milestones along the path to the production, processing, certification, and commercial availability of biofuel for use in commercial aviation. By 2050, the International Air Transportation Association (IATA) aims to reduce the net CO_2 production by 50% compared with 2005 levels. The large-scale production of alternative jet fuels could achieve this goal while improving national energy security, and helping to stabilize fuel costs for the aviation industry. The goal of the Federal Aviation Administration (FAA) is to catalyse the production of 1 billion gallons of 'drop in fuels' by 2018.

Private companies have been leading the way in promoting biojet fuel technologies (Mawhood et al. 2016). However, there is only one currently available commercial biojet fuel, that from AltAir

(CA, USA). United Airlines (Chicago, IL, USA) recently began using AltAir's fuel in routine flights to and from Los Angeles and regular ground operations at the airport (United Airlines 2016). AltAir (CA, USA) has a production capacity of 2500 bbls per day and has made more than 30 military and commercial flights for certification testing purposes.

Airline customers have been investing in biojet fuel companies to help commercialize the aforementioned alternative fuel technologies to keep their future fuel options open. Since, as of this date, there is no clear market leader in the still-emerging biojet fuel market, there remains a strong business opportunity for all emerging biojet fuel technology companies. Airlines and oil companies have created multiple offset agreements with biojet fuel companies to support their development in the past five years, even though the products are not yet available. For example, the airliner JetBlue (Long Island, USA) and SG Preston (Philadelphia, PA, USA), a biojet fuel company, have agreed upon JetBlue purchasing 33 million gallons per year of biojet fuel for 10 years starting in 2019. This agreement and other examples in Table 5.5 demonstrate the industry's interest in alternative jet fuels (BP Press Office 2016; FedEx 2015; IATA 2017; SBI bioenergy and Shell Press

Table 5.5 Interests in biojet fuel by commercial airlines and oil refinery (BP Press Office 2016; FedEx 2015; IATA 2017; SBI bioenergy and Shell Press 2017; United Airlines and Fulcrum BioEnergy Press 2015; Virent and Tesoro Press 2016).

Commercial airlines	Biojet company	Contract length in years	Volume (Million gallons per year)	Year to deliver
JetBlue	SG Preston	10	33	2019
United Airlines	Fulcrum	10 ($1.58 billion)	90	2017
Cathay Pacific	Amyris	2	Toulouse to Hong Kong Flights	2016
Cathay Pacific	Fulcrum	10	37.5	2019
American Airlines	Amyris	N/A	N/A	N/A
Lufthansa	Gevo	5	8	N/A
World Fuel (FBO)	Altair	N/A	N/A	N/A
Alaska	Hawii Bioenergy	5	N/A	2015
FedEx	RedRock	8	3	2017
Southwest	RedRock	N/A	3	2016
South African Airways	Altair	N/A	N/A	N/A
United Airlines	Altair	3	17 000	2016
KLM	SkyNRG-Altair	3	Undisclosed	2016
BA (United Kingdom)	Solena	11	16	2017
Oil companies				
BP (investor as well)	Fulcrum	10	500	2018
Tesoro	Virent		Company Acquirement	2016
Shell	GTI		IH^2 Technology Acquiring	2009
Shell	SBI Bioenergy		Exclusive Licensees	2017
ExxonMobil	Wisconsin Madison University		Continuous 2-years Funding	2017

2017; United Airlines and Fulcrum BioEnergy Press 2015; Virent and Tesoro Press 2016). The demand is much greater than what biojet fuel technologies can supply right now, this tension in demand realization and biojet fuel manufacturing capacity creates a clear opportunity for additional market entrants.

In recent years, more than 20 airlines have flown over 1600 demonstration flights using biojet fuel, especially with conventional fuel blended with biojet fuel (Fellet 2016). Additionally, several test flights were performed using biojet fuel alone, such as the recent flight from Seattle to Washington DC by Alaska Airlines (Seattle, WA, USA) using Gevo's (Englewood, CO, USA) biojet fuel. These examples demonstrate the industry's interest in alternative jet fuel.

5.4.2 Test Flights of Commercial Airlines

As mentioned above, Boeing stated that alternative jet fuel can reduce CO_2 emissions by 50 to 80% compared to fossil fuel through an airplane life cycle. As part of an NSF I-Corps grant received in 2016, the authors' Lignin Biojet team interviewed over 100 jet fuel end users and other stakeholders such as the US Navy, Air Force, Defense Logistics Agency (DLA), Alaska Airlines, JetBlue, United Air, Air Canada (Montreal, Canada), FedEx (Memphis, TN, USA), UPS (Atlanta, GA, USA), Delta Airlines and others. The end users expressed a strong desire to purchase as much cost-efficient biojet fuel as possible. Chevron, Shell, PBF, BP (London, UK) and other refiners have shown great interest in developing biojet fuel technologies. Boeing and Airbus have partnered with Honeywell, International Aero Engines and JetBlue Airways in the pursuit of developing a sustainable second-generation biofuel for commercial jet use, with the hope of reducing the aviation industry's environmental footprint.

In biojet fuel development stakeholders include, but are not limited to, oil refineries, various government agencies, biofuel companies, feedstock suppliers, environmental NGOs and aircraft manufacturers. The success of FT-SPK biojet fuel is mainly attributed to the collaboration of different stakeholders including Boeing, Honeywell/UOP, Air New Zealand (ANZ), Continental Airlines (CAL), Japan Airlines (JAL), General Electric, CFM, Pratt and Whitney, and RollsRoyce (Table 5.6). Table 5.6 shows detail from some test flights (FOCAC 2010; AirportWatch 2012; AirportWatch 2013; Alaska Airlines Press 2011; BBC News 2008; Biofuels International 2015; Boeing Company 2008; ConventryTelegraph 2013; GreenAir Communications 2008, 2009a,b, 2011; LATAM Airlines Group 2013; United Airlines 2016; United Airlines 2011). There has been criticism regarding the emissions of the fuel blends which are, in many cases, the same as emissions from regular jet fuel, given they have similar specifications. Additionally, the land use in the production of biomass feedstock, the necessary hydrogen and conversion inputs for developing biojet fuel, cause carbon emissions, too. The carbon in the fuel goes back to biomass via photosynthesis while the life cycle analysis considers different factors using various methods. In any case, the life cycle analysis of biojet fuel has mostly reported fewer GHG emissions than those of fossil jet fuel (Agusdinata et al. 2011; Budsberg et al. 2016; de Jong et al. 2017).

5.5 Challenges of Future Biojet Fuel Development

As highlighted above, feedstock is a critically important contributor to the cost of biojet fuel and its availability and price are an indicator of the commercialization feasibility of each of the biojet fuel technologies. It remains a question whether or not the currently available biomass is sufficient to meet the jet fuel demand. The Department of Energy Biomass Billion Ton report used a simulated

Table 5.6 Selected flight tests with biojet fuels through different commercial airlines (FOCAC 2010; AirportWatch 2012; AirportWatch 2013; Alaska Airlines Press 2011; BBC News 2008; Biofuels International 2015; Boeing Company 2008; ConventryTelegraph 2013; GreenAir Communications 2008; GreenAir Communications 2011; GreenAir Communications 2009a; GreenAir Communications 2009b; LATAM Airlines Group 2013; United Airlines 2016; United Airlines 2011).

Year	Airlines	Aircraft manufacturers	Other stakeholders	Technologies	Feedstock	Biofuel %	Destination
2008	Virgin Atlantic	Boeing and GE		HEFA	Brazilian babassu nuts and coconuts	20	Heathrow and Amsterdam
2008	United States Air Force	Airbus A380	Shell International; Rolls Royce	FT	Natural gas	50	Filton to Toulouse
2008	Air New Zealand	Boeing	Boeing Rolls-Royce; UOP	HEFA	Jatropha	50	Auckland
2009	Continental	Boeing	Boeing; GE Aviation; CFM; Honeywell UOP	HEFA	2.5% Algae and 47.5% Jatropha	50	Houston
2009	Japan	Boeing	Pratt and Whitney engines	HEFA	Camelina (84%), jatropha (under 16%) and algae (under 1%)	50	Tokyo
2009	Qatar Airways	Airbus	Shell	FT Synthesis	Natural gas	50	London Gatwick to Doha
2010	South African	Boeing	Sasol	Coal to Liquid	Coal	100	Lanseria to Cape Town
2011	United Airlines	Boeing	Solazyme; Honeywell	HEFA	Algae oil	40	Houston to Chicago and Los Angeles to San Francisco
2011	Alaska Airlines	Boeing	Dynamic Fuels; SkyNRG	HEFA	Algae and waste cooking oil	20	Seattle to W ashington, D.C
2011	KLM Royal Dutch Airlines	Boeing	SkyNRG; Dynamic Fuels	HEFA	Waste cooking oil	50	Schiphol bound for Charles de Gaulle

(Continued)

Table 5.6 (Continued)

Year	Airlines	Aircraft manufacturers	Other stakeholders	Technologies	Feedstock	Biofuel %	Destination
2011	Thomson Airways	Boeing	N/A	HEFA	Virgin plant oil from the US and babassu nuts from Brazil	50	Birmingham to Palma
2011	Lufthansa	Airbus	Boeing	FT	Jatropha, camelina and animal fats	50	Hamburg and Frankfurt
2011	United Airlines	N/A	AltAir Paramount	HEFA	Algae and waste cooking oil	15 million over three years 20%	Los Angeles
2012	Air China	Boeing	Honeywell's UOP; PetroChina	HEFA	Jatropha	50	Beijing mainland
2012	Air Canada	Applied Research Associates, Chevron Lummus Global, and Agrisoma Bioscience Inc.		CH	Carinata	100	Ottawa
2013	LAN Colombia	Airbus	Gevo	HEFA	Camelina	30	Bogota and Cali
2013	KLM Royal Dutch Airlines	SkyNRG	Schiphol Group, Delta Air Lines, the Port Authority of New York and New Jersey	HEFA	Cooking oil	25	New York to Amsterdam
2015	Hainan	Boeing	Sinopec	HEFA	Waste cooking oil	50	Shanghai to Beijing

Table 5.7 Feedstock availability, pricing, and potential biojet production (Langholtz et al. 2016; Ragauskas et al. 2014).

Raw Materials	Availability, million dry tons yr^{-1}	Effective year of availability	Price, $/ton	Conversion yield, gallons per dry ton		Biojet, billion gallons yr^{-1}
Animal Fat and Vegetable oil	5	2012–2014	550–1200	HEFA	28–87	0.14–0.44
MSW	51–55	2017–2040	40–60	FT	9–88	0.46–4.8
Forestry and wood wastes	36–56	2017–2040	40–60	ATJ	11–79	0.40–4.4
				HDCJ	19	0.68–1.1
				FT	9–88	0.32–4.9
Energy crops	78–411	2012–2040	40–60	ATJ	11–79	0.89–3.2
				HDCJ	19	1.5–7.8
				FT	9–88	0.70–36.2
Algae	47–132	Present to future	490–2900	HEFA	28–87	1.3–11.5
				CH	8–122	0.38–16.1
Lignin	66	2016	80–250	LJ-D&HDO	30–61	2.0–4.0
Total						4.0–66.5

price of $40–60 per dry ton for some biomass feedstock. The energy crop cost is about $650–890 per acre (Downing et al. 1998). Table 5.7 shows the availabilities and prices of each biomass feedstock that can be converted to biojet fuel through various pathways (Langholtz et al. 2016; Ragauskas et al. 2014). The production yields of each biojet fuel technology varied in different research studies (Wang et al. 2016b). The cost of crude oil is $46.10 per barrel as of June 2017, which is equivalent to $319 per ton, calculated by converting gallons of oil to barrels by dividing the volume by 42 gallons per barrel and the specific gravity of crude oil is assumed to be 915 kg m^{-3}. Compared to the $319 per ton crude oil, animal fat and vegetable oil as well as algae-derived lipids are much more expensive (Table 5.7). As of June 2017, jet fuel is priced at $1.29 per gallon which is equivalent to $375 per ton. Thus, the final product crude-derived jet fuel is much cheaper than the starting raw materials of vegetable oil, animal fat and algae-derived lipids, making the economics of converting these materials into biojet fuel impractical.

An estimation of the US biojet fuel production capacity from biomass is shown in Table 5.7; it reveals that 4–66.5 billion gallons of biojet fuel per year can be produced according to the biomass availability illustrated in DOE's Biomass Billion Ton report. It indicated that the US has the capacity to produce enough biojet fuel to meet the annual demand of jet fuel if all biomass were used in biojet fuel conversions. However, the biojet fuel production has to compete with ethanol and biodiesel production for biomass feedstock. The ethanol mandate is about 15 billion gallons per year since 2010 while the biodiesel mandate is 1–2 billion gallons per year 2011 to 2017 (US Environmental Protection Agency 2017). For example, the mandates for biodiesel and ethanol in 2016 were 1.90 and 18.11 billion gallons per year, respectively. If the conversion efficiency (gallons per ton) of biomass to ethanol and biodiesel are assumed to be 85 gallons per ton and 267 gallons per ton, 220 million tons biomass would be used in ethanol and biodiesel production. The biojet fuel production would lose about 12 billion gallons per year if 55 gallons per dry ton conversion efficiency is assumed. Thus, the feedstock production and distribution system needs to be well established to ensure an appropriate logistics of biomass feedstock.

The cost of biojet fuels remains a challenge, especially when the cost of crude oil is as low as $46.10 per barrel and the cost of jet fuel is about $1.29 per gallon as of June 2017 (US Energy Information Administration 2017c). It has been reported that the minimum selling price of alternative jet fuels can range from $2–24 (Staples et al. 2014). Biojet fuels can achieve cost-competitiveness with crude oil-based jet fuel if crude oil is at least priced at $120 per barrel (National Academies of Sciences, Engineering, and Medicine 2016). The operational cost of biojet fuels is currently high and numerous technical challenges remain. To achieve cost-competitiveness, biojet fuels will need more research and development effort. Also, it is costly and time consuming to scale up biojet fuel technologies, which is reported to range from US$20 to 50 million over ten to fifteen years (Philippe Novelli 2013). Industrial partnerships are therefore crucial in the scale-up and demonstration of the viability of drop-in bio-based fuels.

In addition, the certification and approval process of biojet fuels takes a long time. One of the reasons may be attributed to the involvement of several functional departments specializing in the certification of biojet fuel. The qualification process starts with Tier 1 which tests fuels for specification properties. Tier 2 focuses on the fit-for-purpose properties. Tiers 3 and 4 involve aircraft manufacturers and airlines, including component/rig testing and engine/APU testing. The FAA and ASTM participate heavily in the process (Rumizen 2016). While the first phase of the jet fuel certification is relatively straightforward, which requires 100 gallons of fuel and costs nearly half a million dollars, the requirements for the second phase can vary widely. So far, testing for these phases have required from 30 000 to 200 000 gallons of fuel and have run up costs surpassing $3 million (Rumizen 2017). In addition, there is some level of uncertainty inherent in the certification process.

Private companies are actively involved in biojet fuel commercialization activities (Mawhood et al. 2016). Biojet fuel development is still in the early stages, requiring a lot of effort to overcome challenges. Support from well-established industrial sectors such as oil refineries and engine or aircraft manufacturers is important for the overall success in biojet fuels. Besides industrial efforts, many federal organizations and agencies have collaborated together in developing alternative jet fuels, including the Department of Agriculture (USDA), Environmental Protection Agency (EPA), Department of Transportation (DOT), Department of Commerce (DOC), Department of Defense (DOD), Department of Energy (DOE) and the National Science Foundation (NSF) with projects such as the Defence Production Act (DPA) and Farm to Fleet (Aeronautics Science and Technology Subcommittee 2016). However, there is still a lack of government incentives for the development of renewable aviation fuel. The only credit biojet fuels can claim is the Renewable Fuel Standards (RFS) (US Environmental Protection Agency 2005), but there is no mandate within RFS that applies to jet fuels (Team-CAAFI, Reasearch and Development Team 2014). This is an important consideration as it means that the aviation industry is not obligated to use biojet fuels, albeit biojet fuels can generate renewable fuel credits to meet RFS mandates for ground transportation.

5.6 Perspectives

Sustainable aviation fuel is the only option for the airline industry in the future. The mitigation of greenhouse gas emissions in the aviation industry is crucial to combat global climate change. A broad range of renewable alternative jet fuels, also known as biojet fuels (also referred to as biojet), are in development as drop-in fuels that possess performance characteristics and chemical compositions essentially identical to conventional kerosene jet fuels. Most of the technologies to produce biojet fuels are still in the early stages of research, development, and certification. Notwithstanding this early stage, biojet fuel technologies are the most promising options for alternative energy sources for the airline industry. They present both short- and long-term solutions for replacing crude oil-derived jet fuels. For example, techno-economic analyses have shown that a corn stover ethanol plant with an annual capacity of 57.2 million gallons of ethanol would be able to produce an additional 20 million gallons of lignin-based jet fuel if the catalytic process is applied to upgrade the waste lignin stream (Shen et al. 2019). Results indicated that coproduction of jet fuel from waste lignin can dramatically improve the overall economic viability of an integrated process for corn stover ethanol production. Lignin-derived jet fuel would offer unique advantages compared to the above-mentioned six varieties of biojet fuels: (i) uses low cost raw materials without conflicting with food or other biofuel production, (ii) has higher thermal stability, (iii) has higher energy density, (iv) is produced at a lower cost, and (v) reduces greenhouse emissions (Ruan et al. 2019). However, it is also important to realize that the commercialization of new technology presents difficulties even greater than those that have already been overcome in the past for alternative biojet technology development, as well as tremendous dedication, persistence, and financial strength are required to clear this last remaining hurdle. Furthermore, the aviation industry is facing an increasingly stringent regulatory compliance environment, mainly from international carbon emission regulations associated with quantifying emissions and payment of renewable fuel credits. There are both environmental and economic benefits for aviation and biojet fuel groups to be engaged. In about 10 years, commercial airlines will have mandatory and legally-binding international rules to comply with. It is thus in the interest of all the stakeholders to invest in biojet fuel development now, since biojet fuel technologies will take a long time to be fully developed and be producing at a commercial scale, and to significantly penetrate the conventional jet fuel markets. Many challenges remain, including the availability and price of feedstock, conversion efficiency technical challenges, the tedious and cost-intensive fuel certifications, and government regulatory factors. These uncertainties and challenges, combined with decreasing oil reserves and volatile fuel pricing, justify strategic alliances for biojet fuel stakeholders to collaborate and contribute to the advancement of viable alternatives to conventional jet fuel.

Acknowledgments

This work was supported by the National Science Foundation I-Corps #1655505, Sun Grant-U.S. Department of Transportation (DOT) Award # T0013G-A-Task 8, and the Joint Center for Aerospace Technology Innovation with the Bioproducts, Science & Engineering Laboratory and Department of Biological Systems Engineering at Washington State University.

References

Aeronautics Science and Technology Subcommittee. 2016. Federal alternative jet fuels research and development strategy. *National Science and Technology Council.*

Agusdinata, D.B., Zhao, F., Ileleji, K., and DeLaurentis, D. (2011). Life cycle assessment of potential biojet fuel production in the United States. *Environmental Science & Technology* 45 (21): 9133–9143.

AirportWatch. 2012. Canada claims world's first 100% biofuel-powered civil jet flight.

AirportWatch. 2013. KLM to make one flight per week New York to Amsterdam for 6 months using 25% used cooking oil.

Alaska Airlines Press. 2011. Alaska Airlines Launching Biofuel-Powered Commercial Service in the United States

Al-Sabawi, M., Chen, J., and Ng, S. (2012). Fluid catalytic cracking of biomass-derived oils and their blends with petroleum feedstocks: a review. *Energy & Fuels* 26 (9): 5355–5372.

Amanda Peterka, E.E.N. 2013. Clean, green options lacking as airlines seek alternatives to petroleum, Vol. 2017.

Appadoo, R. (2009). *Insights into Jet Fuel Specifications*. Montreal, Canada: Aviation and Alternative Fuels (ICAO), ICAO.

ASTM. 2014. Revised ASTM Aviation Fuel Standard Paves the Way for International Use of Synthesized Iso-Paraffinic Fuel in Airliners.

ASTM Committee (2017). ASTM Standard, ASTM D7566-16, Standard Specification for Aviation Turbine Fuel Containing Synthesized Hydrocarbons, ASTM International, Est Conshohocken, PA, 2017, www.astm.org.

Authority, D.E. 2009. Monitoring and Reporting Annual Emissions and Tonne km Data for EU Emissions Trading.

Baedecker, M.J., CozzARELLI, I.M., Eganhouse, R.P. et al. (1993). Crude oil in a shallow sand and gravel aquifer—III. Biogeochemical reactions and mass balance modeling in anoxic groundwater. *Applied Geochemistry* 8 (6): 569–586.

Bartis, J.T., LaTourrette, T., Dixon, L. et al. (2005). *Oil Shale Development in the United States: Prospects and Policy Issues*. Rand Corporation.

BBC News. 2008. Airline in first biofuel flight.

Bernstein, L., Bosch, P., Canziani, O. et al. (2007). Climate change 2007: synthesis report. In: *Contribution of Working Groups I, II and III to the Fourth Assessment Report of the Intergovernmental Panel on Climate Change* [Core Writing Team, Pachauri, R.K and Reisinger, A. (eds.)], 104. Geneva, Switzerland: IPCC.

Bilba, K., Savastano Junior, H., and Ghavami, K. (2013). Treatments of non-wood plant fibres used as reinforcement in composite materials. *Materials Research* 16 (4): 903–923.

Biochemtax. 2017. MOGHI project Vol. 2017.

Biofuels International. 2015. First commercial bio-jet flight takes off in China.

Boeing Company (2008). *Boeing, Air New Zealand and Rolls-Royce Announce Biofuel Flight Demo*. Boeing Company.

Boeing Company. 2015. Cut the Carbon: Aviation industry reaffirms pledge to reduce emissions.

BP Press Office. 2016. BP announces investment of $30 million in biojet producer, Fulcrum.

Bridgwater, A. (1975). Operating cost analysis and estimation in the chemical process industries. *Revista Portuguesa de Quimica* 17 (107): 107.

Budsberg, E., Crawford, J.T., Morgan, H. et al. (2016). Hydrocarbon bio-jet fuel from bioconversion of poplar biomass: life cycle assessment. *Biotechnology for Biofuels* 9 (1): 170.

Canoira, L., Rodríguez-Gamero, M., Querol, E. et al. (2008). Biodiesel from low-grade animal fat: production process assessment and biodiesel properties characterization. *Industrial & Engineering Chemistry Research* 47 (21): 7997–8004.

ConventryTelegraph. 2013. From coconuts to cooking oil... flight fuel of the future?

Downing, M., Demeter, C., Braster, M. et al. (1998). Agricultural cooperatives and marketing bioenergy crops: case studies of emerging cooperative development for agriculture and energy. *Proceeding of Bioenergy* 98: 4–8.

Hirs, E., Forbes,. 2016. Delta CEO Admits To $4 Billion Lost In Hedging Fuel Costs.

FAA Aviation Rulemaking Advisory Committee. 1998. Fuel Properties Effect on Aircraft and Infrastructure.

Fahim, M.A., Al-Sahhaf, T.A., and Elkilani, A. (2009). *Fundamentals of Petroleum Refining*. Elsevier.

Federal Aviation Administration. 2017a. The FAA and its NextGen program aim to balance environmental protection with sustained aviation growth, Vol. 2017.

Federal Aviation Administration. 2017b. FAA Forecasts Continued Growth in Air Travel.

FedEx. 2015. Biofuels Take Flight with FedEx.

Fellet, M. (2016). Now boarding: commercial planes take flight with biobased jet fuel. *Chemical & Engineering News* 94: 16–18.

FOCAC. 2010. South Africa launches world first synthetic jet fuel.

Fund, E.D. 2013. The World's Carbon Markets: A case study guide to emissions trading. Retrieved from IETA website: http://www.ieta.org/worldscarbonmarkets.

Gerrard, A. 2000. *Guide to capital cost estimating*. IchemE.

GreenAir Communications. 2008. Airbus completes first commercial aircraft test flight using alternative fuel.

GreenAir Communications. 2009a. Japan Airlines demonstration flight concludes current series of alternative biofuel feedstocks testing.

GreenAir Communications. 2009b. Qatar Airways undertakes first commercial passenger flight powered by a natural gas blended jet fuel.

GreenAir Communications. 2011. First-ever transatlantic aviation biofuel flight sets up week of alternative aviation fuel events at Paris Air Show.

Hall, C.A., Lambert, J.G., and Balogh, S.B. (2014). EROI of different fuels and the implications for society. *Energy Policy* 64: 141–152.

Hemighaus, G., Boval, T., Bacha, J. et al. (2006). *Aviation Fuels Technical Review*. Chevron Products Company.

Hileman, J.I., Ortiz, D.S., Bartis, J.T. et al. (2009). *Near-Term Feasibility of Alternative Jet Fuels*. Santa Monica, CA, USA: RAND Corporation and Massachusetts Institute of Technology.

Huber, G.W. and Corma, A. (2007). Synergies between bio-and oil refineries for the production of fuels from biomass. *Angewandte Chemie International Edition* 46 (38): 7184–7201.

IATA. 2017. Fact Sheet Alternative Fuels

ICAO. 1994. About ICAO

ICAO. 2008. ICAO Carbon Emissions Calculator.

ICAO. 2016. Carbon Offsetting and Reduction Scheme for International Aviation (CORSIA), Vol. 2017.

ICAO's 39th Assembly. 2016. Historic agreement reached to mitigate international aviation emissions Vol. 2017.

Indexmundi. 2017. World Jet Fuel Consumption by Year.

International Air Transport Association. 2009. Fact Sheet: Alternative Fuels.

International Air Transport Association. 2016. IATA Forecasts Passenger Demand to Double Over 20 Years.

International Air Transport Association. 2017. Another Strong Year for Air Travel Demand in 2016.

Jardine, C.N. 2009. Calculating the carbon dioxide emissions of flights. *Final report by the Environmental Change Institute.*

de Jong, S., Antonissen, K., Hoefnagels, R. et al. (2017). Life-cycle analysis of greenhouse gas emissions from renewable jet fuel production. *Biotechnology for Biofuels* 10 (1): 64.

Joseph Sorena, E.L. and Clark, A. (2015). *Co-Processing of HEFA Feedstocks with Petroleum Hydrocarbons for Jet Production*. Commercial Aviation Alternative Fuels Initiative (CAAFI).

Karatzos, S., McMillan, J.D., and Saddler, J.N. (2014). *The Potential and Challenges of Drop-in Biofuels*, vol. 39. IEA Bioenergy Task Force.

Langholtz, M., Stokes, B., Eaton, L. 2016. 2016 Billion-ton report: Advancing domestic resources for a thriving bioeconomy, Volume 1: Economic availability of feedstock.

LanzaTech. 2017. Technical Overview, Vol. 2017.

Laskar, D.D., Yang, B. 2012. Aqueous Phase Depolymerization and Hydrodeoxygenation of Lignin to Jet Fuel. *Abstract of papers at 34th Symposium on Biotechnology for Fuels and Chemicals (SBFC)*, April 30 May 3, 2012, at New Orleans, LA.

Laskar, D.D., Yang, B., Wang, H., and Lee, J. (2013). Pathways for biomass-derived lignin to hydrocarbon fuels. *Biofuels, Bioproducts and Biorefining* 7 (5): 602–626.

Laskar, D.D., Tucker, M.P., Chen, X. et al. (2014). Noble-metal catalyzed hydrodeoxygenation of biomass-derived lignin to aromatic hydrocarbons. *Green Chemistry* 16: 897–910.

LATAM Airlines Group. 2013. LAN Airlines completes biojet flight in Colombia

Lister, D., Penner, J.E., Griggs, D.J. et al. (1999). *Aviation and the Global Atmosphere-Summary for Policymakers*. Cambridge, UK: Cambridge University Press.

Mawhood, R., Gazis, E., de Jong, S. et al. (2016). Production pathways for renewable jet fuel: a review of commercialization status and future prospects. *Biofuels, Bioproducts and Biorefining* 10 (4): 462–484.

Morrell, P. and Swan, W. (2006). Airline jet fuel hedging: theory and practice. *Transport Reviews* 26 (6): 713–730.

National Academies of Sciences, Engineering, and Medicine (2016). *Commercial Aircraft Propulsion and Energy Systems Research: Reducing Global Carbon Emissions*. Washington, DC: The National Academies Press https://www.nap.edu/catalog/23490/commercial-aircraft-propulsion-and-energy-systemsresearch-reducing-global-carbon.

Novelli, P. 2013 The Challenges for the Development and Deployment of Sustainable Alternative Fuels in Aviation - Outcomes of ICAO's SUSTAF Experts Group

Radich, T. (2015). *The Flight Paths for Biojet Fuel*. U.S. Energy Information Administration, Independent Statistics & Analysis.

Ragauskas, A.J., Beckham, G.T., Biddy, M.J. et al. (2014). Lignin valorization: improving lignin processing in the biorefinery. *Science* 344 (6185): 1246843.

Renewable Fuels Association (2011). *Fuel Ethanol Industry Guidelines, Specifications, and Procedures*. Renewable Fuels Association (RFA).

Reuters (2012). *Auction to Kick-Start California Carbon Market*. Reuters.

Reynolds, R.E., Herwick, G., McCormick, R.L. et al. (2017). *Changes in Diesel Fuel - the Service Technician's Guide to Compression Ignition Fuel Quality*. National Biodiesel Board.

Ruan, H., Qin, Y. Heyne, J. Gieleciak, R. Feng, M. and Yang, B. (2019), Chemical Compositions and Properties of Lignin-Based Jet Fuel Range Hydrocarbons, Fuel, 256: 115947.

Rumizen, M. (2016). *Sustainable Alternative Jet Fuel Certification and Qualification*. Washington, DC: CAAFI.

Rumizen, M. 2017. Alternative Jet Fuel Approval Process. *2017 Worldwide Energy Conference*, Workshop 18: Alternative/Renewable Fuels Panel Discussion –Technology Development and Certification.

SBI bioenergy and Shell Press. 2017. Agreement grants Shell exclusive development and licensing rights for SBI Bioenergy patented renewable drop-in biofuels.

Secretariat, I. 2010. Aviation's contribution to Climate Change. *BAN Ki-moon*.

Shen, RC., Tao, L. and Yang, B. (2019). Techno-EconomicAnalysis of Jet Fuel Production from Biorefinery Waste Lignin, BioFPR, 13: 486–501.

Staples, M.D., Malina, R., Olcay, H. et al. (2014). Lifecycle greenhouse gas footprint and minimum selling price of renewable diesel and jet fuel from fermentation and advanced fermentation production technologies. *Energy & Environmental Science* 7 (5): 1545–1554.

Talberg, A., Swoboda, K. 2013. Emissions trading schemes around the world.

Team-CAAFI, Reasearch and Development Team. 2014. Policy Research Needs Relevant to Alternative Jet Fuels *CAAFI*.

The global aviation industry, I. 2010.The right flightpath to reduce aviation emissions. in: *UNFCCC Climiate Talks.*

U.S. Energy Information Administration. 2017a. Annual Energy Outlook 2017.

U.S. Energy Information Administration. 2017b. Sources of Greenhouse Gas Emissions, Vol. 2017.

U.S. Energy Information Administration. 2017c. Spot Prices (Crude Oil in Dollars per Barrel), Products in Dollars per Gallon), Vol. 2017.

U.S. Naval Research Laboratory. 2016. NRL Seawater Carbon Capture Process Receives U.S. Patent, Vol. 2017.

United Airlines. 2011. United enters the biofuel age.

United Airlines. 2016. United Airlines begins using biojet fuel in routine LAX flights.

United Airlines and Fulcrum BioEnergy Press. 2015. United Airlines Purchases Stake in Fulcrum BioEnergy with $30 Million Investment.

US Energy Information Administration. 2017a. Energy Outlook.

US Energy Information Administration. 2017b. Factors Affecting Diesel Prices, Vol. 2017.

US Energy Information Administration. 2017c. U.S. Gulf Coast Kerosene-Type Jet Fuel Spot Price FOB (Dollars per Gallon), Vol. 2017.

US Environmental Protection Agency (2005). *Renewable Fuel Standard Program*. US Environmental Protection Agency.

US Environmental Protection Agency. 2016. EPA Determines that Aircraft Emissions Contribute to Climate Change Endangering Public Health and the Environment.

US Environmental Protection Agency. 2017. Final Renewable Fuel Standards for 2014, 2015 and 2016, and the Biomass-Based Diesel Volume for 2017, Vol. 2017.

Valkenburg, C., Gerber, M., Walton, C., Jones, S., Thompson, B., Stevens, D.J. 2008. Municipal solid waste (MSW) to liquid fuels synthesis, volume 1: Availability of feedstock and technology. *Richland, WA (US): Pacific Northwest National Laboratory, December*. http://www.pnl.gov/main/publications/external/technicalBreports/PNNLY18144.pdf. *Accessed October*, **30**, 2009.

Virent. 2017. Bioreforming Technology, Vol. 2017.

Virent and Tesoro Press. 2016. Tesoro to Acquire Virent in Support of Commercializing Renewable Fuels and Chemicals.

Wang, W.-C. and Tao, L. (2016). Bio-jet fuel conversion technologies. *Renewable and Sustainable Energy Reviews* 53: 801–822.

Wang, H., Ruan, H., Pei, H. et al. (2015). Biomass-derived lignin to jet fuel range hydrocarbons via aqueous phase hydrodeoxygenation. *Green Chemistry* 17 (12): 5131–5135.

Wang, H., Zhang, L., Deng, T. et al. (2016a). ZnCl2 induced catalytic conversion of softwood lignin to aromatics and hydrocarbons. *Green Chemistry* 18 (9): 2802–2810.

Wang, W.-C., Tao, L., Markham, J., Zhang, Y., Tan, E., Batan, L., Warner, E., Biddy, M. 2016b. Review of Biojet Fuel Conversion Technologies. NREL (National Renewable Energy Laboratory (NREL), Golden, CO (United States)).

Wang, H., Feng, M., and Yang, B. (2017a). Catalytic hydrodeoxygenation of anisole: an insight into the role of metals in transalkylation reactions in bio-oil upgrading. *Green Chemistry*: 1668–1673.

Wang, H., Ruan, H., Feng, M. et al. (2017b). One-pot process for hydrodeoxygenation of lignin to alkanes using Ru-based bimetallic and bifunctional catalysts supported on Zeolite Y. *ChemSusChem* 10 (8): 1846–1856.

Wang, H., Wang, H., Kuhn, E. et al. (2017c). Production of jet fuel-range hydrocarbons from hydrodeoxygenation of lignin over super Lewis acid combined with metal catalysts. *ChemSusChem*: 285–291.

Winchester, N., McConnachie, D., Wollersheim, C., Waitz, I.A. 2013. Market cost of renewable jet fuel adoption in the United States. MIT Joint Program on the Science and Policy of Global Change.

WSU Maegan Murray. 2016. WSU Tri-Cities researchers receive NSF grant to test market potential for jet fuel research.

Yang, B., Laskar, D.D. 2016. Apparatus and process for preparing reactive lignin with high yield from plant biomass for production of fuels and chemicals, Google Patents.

Yanik, S., O'connor, P., Bartek, R. 2012. Co-processing solid biomass in a conventional petroleum refining process unit, Google Patents.

6

The Environmental Impact of Pollution Prevention and Other Sustainable Development Strategies Implemented by the Automotive Manufacturing Industry

Sandra D. Gaona[1], *Cheryl Keenan*[2], *Cyril Vallet*[3], *Lawrence Reichle*[3] *and Stephen C. DeVito*[1]

[1] *Toxics Release Inventory Program (mail code 7410M), United States Environmental Protection Agency, Washington, DC, 20460, USA*
[2] *Eastern Research Group Inc., Lexington, MA, 02421, USA*
[3] *Abt Associates Inc., Cambridge, MA, 02138, USA*

CHAPTER MENU

6.1 Introduction

This chapter characterizes chemical release and other waste management quantities as well as pollution prevention activities carried out by the US automotive manufacturing industry over the 2005–2015 time-frame.[1] Analysis of information available from federal databases such as the US Environmental Protection Agency's (EPA's) Toxics Release Inventory (TRI) and industry reports reveals the corresponding environmental impacts, and identifies opportunities for continued progress. Throughout this chapter several terms are used that may not be familiar to the reader. These terms are defined below.

> A **'TRI chemical'** is a chemical that is included on the Toxics Release Inventory (TRI) list of chemicals, as established under Section 313(d)(2) of the *Emergency Planning and Community Right-to-Know Act*. Chemicals included on the TRI list are those that as a result of continuous, or frequently recurring releases are known to cause or can reasonably be anticipated to cause (1) significant adverse acute human health effects at concentration levels reasonably likely to exist beyond facility site boundaries; (2) cancer or teratogenic effects or serious or irreversible reproductive dysfunctions, neurological disorders, heritable genetic mutations, or other chronic health effects; or (3) a significant adverse effect on the

1 At the time of the analysis and writing of this chapter, the 2015 reporting year was the year for which the most recent TRI data were available.

Green Energy to Sustainability: Strategies for Global Industries, First Edition.
Edited by Alain A. Vertès, Nasib Qureshi, Hans P. Blaschek and Hideaki Yukawa.

environment of sufficient seriousness to warrant reporting due to the chemical's toxicity, its toxicity and persistence in the environment, or its toxicity and tendency to bioaccumulate in the environment.

A **'TRI-reported chemical'** refers to chemicals on the TRI list of chemicals for which facilities in the US have submitted reports to the US Environmental Protection Agency (EPA) TRI Program indicating releases to the environment or otherwise managed as waste.

'TRI-reported chemical waste' or **'TRI-reported waste'** refers to the quantity of the TRI chemical(s) contained in waste and reported to EPA by facilities as released to the environment or otherwise managed as waste, such as through recycling, treatment, or combustion for energy recovery.

Beyond TRI-reported chemical waste management, this chapter also reviews data and literature on a range of pollution prevention and sustainability strategies in the industry such as those related to improving energy efficiency and material use. This chapter does not consider, in detail, the environmental impacts from resource extraction or depletion such as water consumption or mining of raw materials.

6.2 Overview of the Automotive Manufacturing Industry

The automotive manufacturing industry, as defined in this chapter, includes three subsectors characterized by the North American Industry Classification System (NAICS) as: Motor Vehicle Manufacturing (NAICS 3361); Motor Vehicle Body and Trailer Manufacturing (NAICS 3362); and Motor Vehicle Parts Manufacturing (NAICS 3363). Facilities in these subsectors do not conduct after-market vehicle repairs and modifications or any other waste-generating processes occurring over a vehicle's lifetime aside from its initial manufacture. However, this chapter considers certain factors that are under manufacturers' control such as fuel economy and recyclability of vehicles at the end of their useful life. Off-road recreational and industrial vehicles are not discussed herein.

6.2.1 History

The US automotive industry was established in the late 1800s with vehicles powered by a variety of energy sources including steam, electricity, gasoline, and even biodiesel, but were too expensive for most households. With pioneers such as Henry Ford and the increased efficiency of new moving assembly lines, production grew rapidly between 1910 and 1920, and vehicles eventually became more accessible and affordable to the general public. Ford Motor Company, General Motors (GM) Company, and Fiat Chrysler Automobiles (previously Chrysler Group) quickly established their dominance and still lead production today. In total, the US is one of the largest auto-producing countries in the world, with about 12 million vehicles produced annually, second only to China with about 25 million vehicles produced annually. By comparison, the countries in the European Union produce about 18 million vehicles per year collectively (OICA 2015).

6.2.2 Production and Economic Trends

In 2015, the automotive manufacturing industry contributed $678 billion (US Bureau of Economic Analysis 2015), or 3.8%, to the US Gross Domestic Product (GDP) and provided approximately 811 000 US jobs (US Census Bureau 2014). From December 2007 through June 2009, production within the US automotive industry slowed due to a global-scale recession. Total US motor vehicle production dropped from 10.7 million vehicles in 2007 to 5.7 million vehicles in 2009. Production

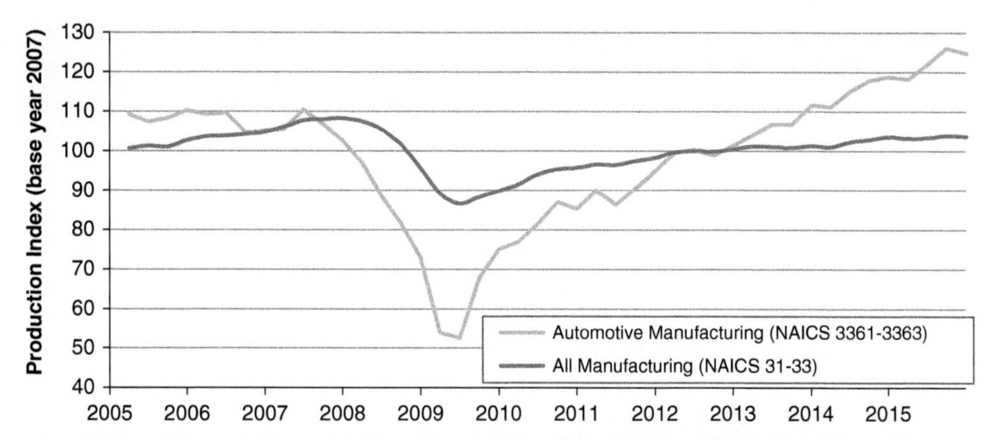

Figure 6.1 Automotive manufacturing production index – seasonally adjusted, 2005–2015. Notes: NAICS, North American Industry Classification System (NAICS). 'All Manufacturing' includes automotive manufacturing. Source: Federal Reserve Board of Governors, 2005–2015 (Federal Reserve Board of Governors 2015).

of cars and light, medium, and heavy trucks declined significantly by 2008 and bottomed out in mid-2009. Between 2007 and 2009, the industry lost over a quarter of a million employees, and two of the three largest automakers (GM and Chrysler) filed for bankruptcy (Katz et al. 2013). Over the course of the next several years, the US Treasury Department invested approximately $80 billion dollars in the industry to help it recover (US Department of the Treasury 2015). GM and Chrysler emerged from bankruptcy in June 2009. The industry gradually created 500 000 new jobs, and total production has largely returned to pre-recession levels (US Department of the Treasury 2015, Automotive News 2015) .

The automotive manufacturing production index indicates that this industry was more affected by the 2007–2009 recession than all manufacturing industries in aggregate, as shown in Figure 6.1. Even before 2008, the industry was experiencing declining market shares as well as significant debt, pension, and health care costs. Additionally, due to the declining availability of loans and rising unemployment, consumers were not able to purchase new vehicles, causing sales to drop rapidly in 2008 (Biesbroeck and Sturgeon 2010).

The financial health of the automotive manufacturing industry is strongly influenced by the cost of vehicle production. In fact, raw materials contribute to about half of the total cost to produce a vehicle. Automobiles are primarily composed of steel (47%), iron (8%), plastic (8%), aluminium (7%), and glass (3%), among other materials. Some automakers are shifting towards higher proportions of aluminium to steel to reduce vehicle weight and increase fuel economy (Kallstrom 2015), and presumably production costs.

6.2.3 Key Players

Despite the 2007–2009 economic recession, the US automobile manufacturers – Ford, GM, and Fiat Chrysler – remained the three largest in terms of US car and light truck production between 2005 and 2015. Foreign automakers with manufacturing operations in the United States did not experience as much of a decline in production during the recession and quickly recovered (Automotive News 2015). Specifically, production of cars and light trucks by facilities in the US that are owned and operated by Honda and Toyota (Japanese automakers) increased significantly after the recession, exceeding pre-recession levels (Young 2014).

6.3 Chemicals and Chemical Waste in Automotive Manufacturing

A variety of chemicals are used throughout the automobile manufacturing process, and many must be managed as waste after their useful life. Additionally, pollutants are emitted during automobile operation. Automotive manufacturing facilities can have a significant impact on the quantity and composition of the TRI chemical wastes they generate and manage through selection of chemicals used, their application, and the design of fuel-efficient vehicles.

6.3.1 Emissions from Fuel Combustion

Fossil fuels are the primary sources of energy used to power motor vehicles and automotive manufacturing facilities. Even if fossil fuels are not used at an automotive manufacturing facility, most power their operations with electricity generated from fossil fuels. The processing and combustion of fuels during facility operation generate pollution in the form of releases of TRI chemicals, smog-forming particles, and greenhouse gases (GHGs). The fuel economy (or fuel efficiency) of motor vehicles also influences the quantity of fuel combusted and, therefore, the amount of pollution emitted.

A comprehensive understanding of the automotive manufacturing industry's environmental impacts includes consideration of the industry's GHG emissions. GHG emissions are typically measured in million metric tons of carbon dioxide equivalent (MMT CO_2-eq) and include direct emissions of carbon dioxide, methane, and nitrous oxide. Direct GHG emissions associated with the industry can be grouped into two sources: automobile manufacturing and automobile operation.

6.3.1.1 Automobile Manufacturing GHG Emissions

The manufacture of automobiles generates GHG emissions from the stationary combustion of fuels to power manufacturing operations. Regulated facilities meeting certain thresholds are required to report their GHG emissions to EPA's Greenhouse Gas Reporting Program (GHGRP).[2] Figure 6.2 presents the annual direct GHG emissions from automotive manufacturing facilities that also reported to TRI from 2010 to 2015. Total direct GHG emissions have fluctuated over the years, but GHG emissions reported for 2010 are essentially equal to GHG emissions reported for 2015. Indirect emissions associated with the use of electricity are not captured here.

6.3.1.2 Automobile Operation GHG Emissions

The operation of motor vehicles generates GHG emissions through fuel combustion. Emissions from this activity can be studied by analysing the EPA Inventory of Greenhouse Gas Emissions and Sinks[3] for the *transportation sector*, which includes 'the movement of people and goods by cars, trucks, trains, ships, aeroplanes, and other vehicles' and does not include emissions from manufacturing (US EPA 2017b). Transportation sector GHG emissions are shown in Table 6.1 by vehicle type (US EPA 2017a). Overall, GHG emissions in 2015 increased by 20% since 1990 but decreased by 12% since 2005, despite a slight increase in vehicle-miles travelled (US Department of

---

2 For more information on EPA's GHGRP, please visit https://www.epa.gov/ghgreporting.
3 For more information on EPA's Inventory of Greenhouse Gas Emissions and Sinks, please visit https://www.epa.gov/ghgemissions/inventory-us-greenhouse-gas-emissions-and-sinks.

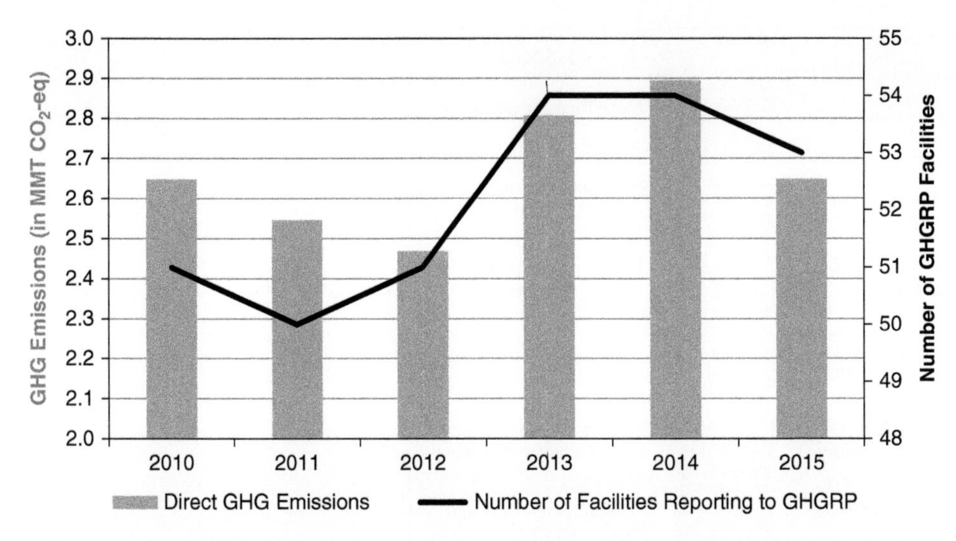

Figure 6.2 Direct GHG emissions from automotive manufacturing facilities, 2010–2015. Source: US EPA Greenhouse Gas Reporting Program 2010–2015 (US EPA 2015b).

Table 6.1 GHG emissions from automobile operation in the United States (MMT CO_2-eq).

Vehicle type	1990[a)]	2005	2010	2011	2012	2013	2014	2015	% change 2005–2015
Passenger cars	657	709	784	774	768	763	763	744	5%
Light-duty trucks	336	552	349	332	326	323	338	302	−45%
Medium- and heavy-duty trucks	231	409	400	398	398	405	415	412	1%
Buses and motorcycles	10	14	20	20	22	22	23	23	64%
Total	1234	1684	1552	1525	1514	1513	1539	1481	−12%

a) Year 1990 is included as a common baseline for assessing trends in GHG emissions. Data were not available for all years.
Source: US EPA Inventory of US Greenhouse Gas Emissions and Sinks: 1990–2015 (US EPA 2017a)

Transportation 2014). GHG emissions from passenger cars increased by 5% from 2005 to 2015 due to a slight increase in market share. However, GHG emissions from light-duty trucks decreased by 45%. This general trend is due to the improved fuel economy and declining market share of new light-duty trucks (US EPA 2017a). Fuel economy is discussed further in Section 6.4.1.9.

6.3.2 TRI-Reported Chemical Waste Management

The automotive manufacturing industry, as defined by facilities classified in NAICS 3361, 3362, or 3363, covers the assembly of automobiles and the manufacture of parts, vehicle bodies, and trailers.

Inputs		Outputs
Metalworking fluids, metals	Casting, Forging, & Stamping	Metal filings, chips, spent fluids
Chemicals, solvents, water	Metal Parts Cleaning & Coating	Spent chemical baths, oils, spills, wastewater
Plastic pellets, cleaning solvents	Plastic Parts Molding	Excess plastic bits, spent solvents
Treatment chemicals, water, facility waste streams	Waste Water Treatment	Treated effluent, sludge
Water, samples, chemicals, rags, solvents, lubricants	Assembly, Testing, & Cleanup	Waste solvent/oils/grease, discarded samples
Paints, organic solvents, water	Body Painting	VOCs, spent paints and solvents
Soldering metals	Vehicle Assembly & Welding	Excess soldering material
Auto bodies and parts	End-of-Life	Auto shredder residue, scrap metals, engine oils, gasoline, batteries, plastics

Figure 6.3 Automobile manufacturing process: material inputs and outputs. Source: Adapted from the International Labour Organization Encyclopaedia of Occupational Health and Safety, 4th Edition (McCann 2012).

Manufacturing practices include 'metal bending, forming, welding, machining, and assembling metal or plastic parts into components and finished products' (US Department of Labor 2012). These operations generate scrap metal that is either recycled or released to air, land, and water (such as in the form of metal dust). Other chemical-intensive practices include the application of surface coatings, paints, and adhesives to vehicle frames, bodies, and other components. For example, painting alone requires pretreatment, a rust prevention layer, sealer, primer-surfacer, and top coats with chemical products containing metals, organic solvents, resins, and pigments (Akufuah et al. 2016, Fettis 1995). Figure 6.3 shows a schematic of common automotive manufacturing practices that generate chemical wastes reportable to the TRI.

6.3.2.1 US EPA Toxics Release Inventory

EPA's TRI Program publicly tracks quantities of chemicals included on the TRI chemical list that are released on-site to air, water, and land, transferred off-site to other facilities, or otherwise managed as waste by facilities throughout the United States, as specified under Section 313 of the Emergency Planning and Community Right-to-Know Act (EPCRA) and Section 6607 of Pollution Prevention Act (PPA). TRI data are reported by facilities subject to the TRI reporting requirements, and reported to EPA's TRI Program, state, and tribal governments (US EPA 2016). The TRI reporting requirements include facilities that (i) are classified by a TRI-covered industry NAICS code, (ii) have the equivalent of 10 or more full-time employees, and (iii) manufacture or process more than 25 000 pounds, or otherwise use more than 10 000 pounds of a TRI chemical within a calendar year. Thresholds for TRI chemicals that are designated as persistent and bioaccumulative, are lower – as low as 0.1 g for dioxin – due to their potentially greater threat to human and environmental health. Currently, the TRI list of chemicals includes almost 600 individually-listed chemicals, and more than 30 chemical categories.

Facilities in a regulated sector, that meet the employment threshold, and exceed any of the activity thresholds discussed above within a given calendar year are required to report quantities of the TRI chemicals that they released on-site to air, land, or water; recycled, combusted for energy recovery, treated on-site, or transferred off-site to other facilities or locations for treatment, recycling, storage, or disposal. Releases to air include stack and fugitive emissions. Releases to land include, for example, disposal in landfills and injection into underground wells. Releases to water include discharges into rivers, streams, or other bodies of water. In addition, for a given chemical for which reporting is required, a facility is also required to disclose any source reduction practices (e.g. process modifications, substitution of raw materials) implemented at a facility for the chemical during the reporting year.

Facilities are required to submit a TRI reporting form by July 1 of the following year for each chemical for which an applicable reporting threshold was exceeded. Each year, EPA's TRI Program receives approximately 80 000 reporting forms from approximately 20 000 facilities. Since not all facilities meet the reporting thresholds, the TRI database does not contain information on all the quantities of TRI chemicals released to the environment or otherwise managed as waste. EPA makes all reported information available to the public through various data tools maintained by EPA.

6.3.2.2 Trends in TRI-Reported Chemical Waste Management

From 2005 to 2015, 1485 unique automotive manufacturing facilities reported to EPA's TRI Program. Automotive facilities reporting to TRI are located primarily in Michigan, Indiana, Ohio, Kentucky, Tennessee, Alabama, and neighbouring states. For 2015, 825 facilities reported, which is a 23% decrease from the 1072 facilities that reported for 2005.[4] The number of TRI forms filed by the automotive manufacturing industry also decreased over the same period from 3959 to 2911 (26% decrease). TRI forms filed for 33 metals and metal compound categories accounted for roughly half of all forms submitted. Other chemicals for which TRI forms were commonly submitted by automotive manufacturing facilities include toluene, xylenes, diisocyanates, and ethylene glycol.

A summary of the automotive manufacturing sector's TRI reporting is included in Table 6.2 and is discussed in more detail throughout this section of the chapter.

For TRI reporting, waste managed includes all chemical waste managed through recycling, energy recovery, treatment, and release except catastrophic or one-time releases. Between 2005 and 2015, total waste managed from automotive manufacturing decreased by 11%. Releases alone decreased by 50% over this time period.

Automotive industry subsectors follow similar trends in quantities of waste managed and released. However, motor vehicle manufacturing has experienced the largest percentage decrease in waste managed with a 28% reduction from 2005 to 2015 compared to a 13% reduction for body and trailer manufacturing and a 1% increase for parts manufacturing. There is also variability in chemicals reported between subsectors. For example, parts manufacturing tends to use more metals while automobile manufacturing tends to use more solvents. Since metals make up much of the final products, it may not be feasible for parts manufacturers to significantly reduce the quantities used.

---

4 The possibility that some of this reduction in facilities reporting may be attributed to outsourcing is discussed in Section 6.3.2.5.

Table 6.2 TRI reporting overview for automotive manufacturing, 2015.

Sector	Number of facilities	Facilities reporting source reduction	Waste managed[a)]		Releases	
			Million pounds	% of sector total	Million pounds	% of sector total
Motor vehicle manufacturing (NAICS 3361)	76	11	61.3	31.7%	11.7	53.0%
Motor vehicle body and trailer manufacturing (NAICS 3362)	169	16	13.8	7.1%	2.3	10.5%
Motor vehicle parts manufacturing (NAICS 3363)	580	83	118.2	61.2%	8.1	36.5%
Total	825	110	193.3	100%	22.1	100%

a) Managed waste quantities include release quantities.
Source: US EPA Toxics Release Inventory – 2015 National Analysis Dataset, Retrieved January 2017

Figure 6.4 presents TRI data on releases and waste managed quantities reported by the automotive manufacturing industry normalized by production volume. As mentioned previously, quantities of chemical waste managed include chemical release quantities. In Figure 6.4, release quantities per vehicle are shown separately from the other waste management quantities. Although production has increased significantly since 2009, releases have remained relatively steady, resulting in significant reductions in TRI-reported releases and waste managed per vehicle. It is important to note that this calculation is only a proxy for the quantities of TRI chemicals used and released because motor vehicle parts manufactured in a given year may not be assembled in that same year. Additionally, parts used in the US automotive industry can be imported into the United States and not all facilities that manufacture parts meet the TRI reporting criteria.

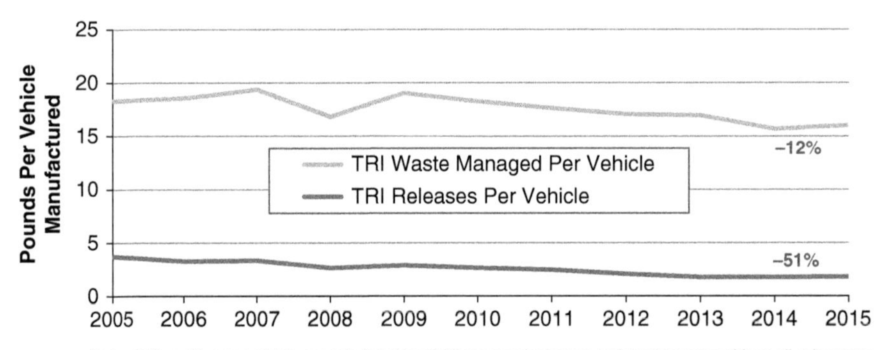

Figure 6.4 Releases and waste managed quantities per motor vehicle produced, 2005–2015. Note: (i) Quantities per vehicle are calculated by dividing annual releases and waste managed from all subsectors by US production. (ii) Managed waste quantities include release quantities. Sources: US EPA Toxics Release Inventory – 2015 National Analysis Dataset, Retrieved January 2017; Federal Reserve Board of Governors, 2005–2015.

6.3.2.3 Waste Management Methods

Recycling is the primary waste management method the automotive manufacturing industry uses to manage chemical waste reported to TRI. The proportion of TRI-reported waste that was recycled increased from 64% to 76% between 2005 and 2015. Recycling is applied mostly to metals and metal compounds, specifically to manganese, copper, chromium, zinc, nickel, and the respective compounds of these metals. Recycling is applied to a lesser extent to organic solvents, specifically xylenes, 1,2,4-trimethylbenzene, n-butyl alcohol, and glycol ethers. For these chemicals, treatment and release to the environment are the more commonly applied waste management practices.

6.3.2.4 Trends in Releases

Air releases make up the largest portion of the automotive manufacturing industry's TRI-reported releases due to the use of volatile compounds in paints and other coatings formulations. Although metals and metal compounds account for about 65% of the quantities managed as waste, organic solvents are released into the environment in greater quantities, as reported to EPA's TRI Program by facilities in the automotive manufacturing industry.

Figure 6.5 compares the 10 chemicals released in the largest quantities in both 2005 and 2015, and shows that the following eight chemicals have remained at the top of the list over this period: xylenes, glycol ethers, n-butyl alcohol, 1,2,4-trimethylbenzene (1,2,4-TMB), methyl isobutyl ketone (MIBK), styrene, zinc, and copper. It is clear from Figure 6.5 that, with the exception of manganese and nickel (and nickel compounds), releases of these chemicals have decreased dramatically from 2005 to 2015.

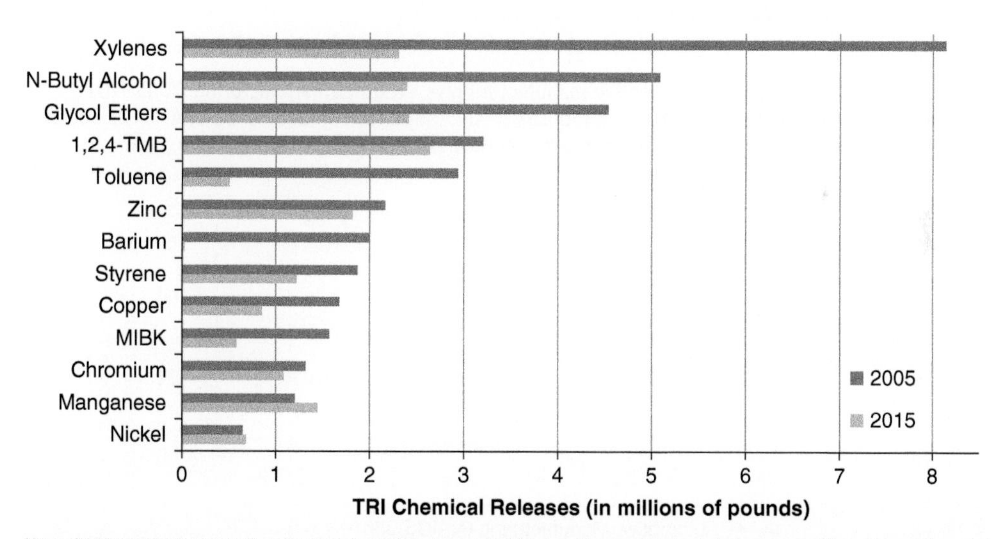

Notes:(1) Metals are grouped with their respective metal compounds. The quantities represent only the mass quantities of the metals, and do not include non-metal portions of metal compounds. (2) Chemicals are included according to the top 10 chemicals released in the largest quantities in 2005 and 2015, with both years having eight of the same chemicals. (3) TMB = Trimethylbenzene, MIBK = Methyl Isobutyl Ketone.

Source: U.S. EPA Toxics Release Inventory-2015 National Analysis Dataset, Retrieved January 2017

Figure 6.5 TRI chemicals released in the largest quantities from automotive manufacturing facilities during 2005 and 2015. Notes: (i) Metals are grouped with their respective metal compounds. The quantities represent only the mass quantities of the metals, and do not include non-metal portions of metal compounds. (ii) Chemicals are included according to the top 10 chemicals released in the largest quantities in 2005 and 2015, with both years having eight of the same chemicals. (iii) TMB, Trimethylbenzene; MIBK, Methyl Isobutyl Ketone. Source: US EPA Toxics Release Inventory – 2015 National Analysis Dataset, Retrieved January 2017.

6.3.2.5 Automotive Manufacturing vs. All Other Manufacturing Sectors

Quantities of TRI chemical waste managed reported by facilities over the 2005–2015 time frame decreased by 11% in the automotive manufacturing industry, but increased by 14% in all other manufacturing industries, as shown in Figure 6.6. Similarly, during the same period, quantities of TRI chemicals released into the environment by the automotive manufacturing industry decreased by 50%, and by 24% in all other manufacturing industries, as shown in Figure 6.7. Although overall waste managed quantities among all manufacturing industries have essentially returned to pre-recession levels, release quantities have remained low. Possible explanations

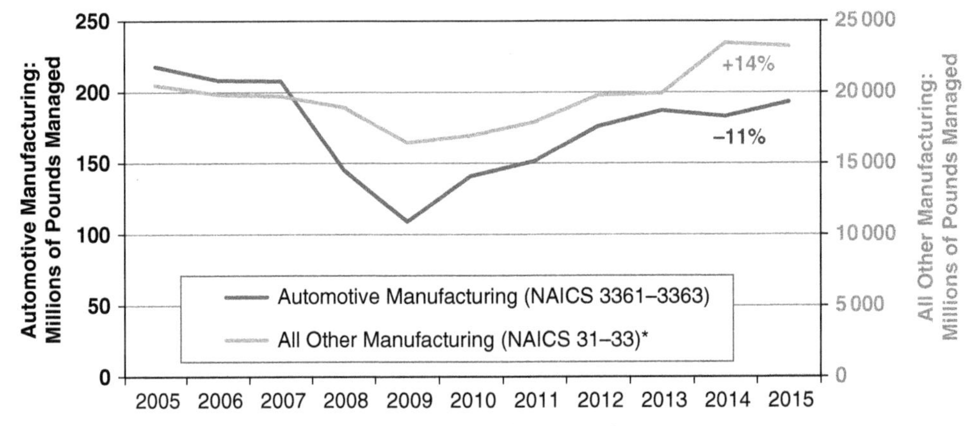

Figure 6.6 TRI waste managed by automotive manufacturing vs. all other manufacturing, 2005–2015. Note: *Excludes the Automotive Manufacturing Industry (NAICS 3361–3363). Source: US EPA Toxics Release Inventory – 2015 National Analysis Dataset, Retrieved January 2017.

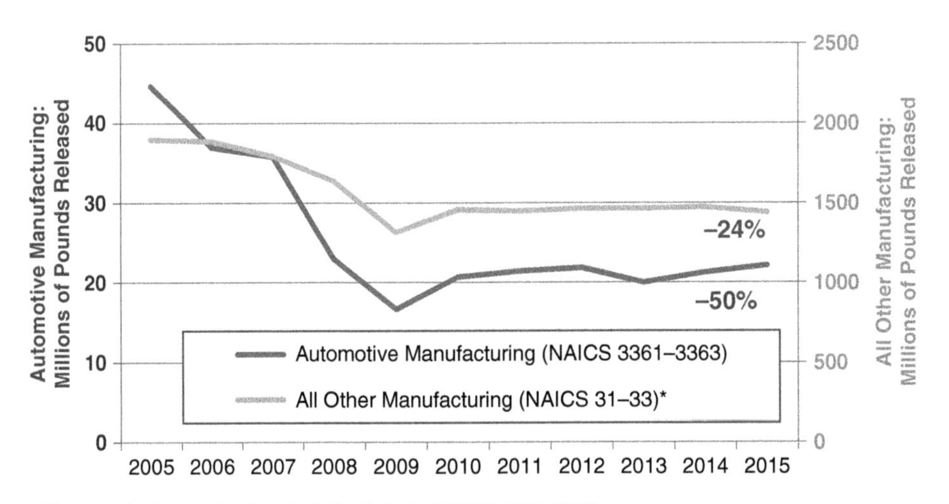

Figure 6.7 TRI releases from automotive manufacturing vs. all other manufacturing, 2005–2015. Note: *Excludes the Automotive Manufacturing Industry (NAICS 3361–3363). Source: US EPA Toxics Release Inventory – 2015 National Analysis Dataset, Retrieved January 2017.

include implementation of preferred waste management practices, outsourcing of processes with significant releases, or changes in chemical composition of the materials used. As noted in Section 6.2.2, total US vehicle production and the automotive manufacturing production index have largely recovered from the 2007 to 2009 recession, indicating that decreased production is not the reason for the reduction in releases. An earlier analysis by the TRI Program also concluded that outsourcing was not the primary driving factor (US EPA 2017c). It therefore appears that the automotive manufacturing industry emerged from the recession with more efficient waste management programs that reduce wastes and releases of TRI chemicals into the environment.

6.4 Pollution Prevention in Automotive Manufacturing

6.4.1 Sustainability Trends in Automotive Manufacturing

This section discusses the activities, innovative research, and developments that have contributed to the automotive manufacturing industry's progress in environmental sustainability. These methods and technologies are compared to source reduction activities reported to TRI to assess their effectiveness and to provide insights as to where additional activities could be implemented.

Pollution prevention is an essential component of sustainable manufacturing practices. In the United States the PPA of 1990 established a national policy that 'pollution should be prevented or reduced at the source whenever feasible; pollution that cannot be prevented should be recycled in an environmentally safe manner, whenever feasible; pollution that cannot be prevented or recycled should be treated in an environmentally safe manner whenever feasible; and disposal or other release into the environment should be employed only as a last resort and should be conducted in an environmentally safe manner'. This hierarchy is illustrated in Figure 6.8. While not specifically mentioned in the PPA of 1990, energy recovery is a preferred practice over treatment and disposal, and hence, is included in the hierarchy illustrated in Figure 6.8.

As established by the PPA, source reduction 'is more desirable than waste management or pollution control. Source reduction refers to practices that reduce hazardous substances from being released into the environment prior to recycling, treatment, or disposal. These practices

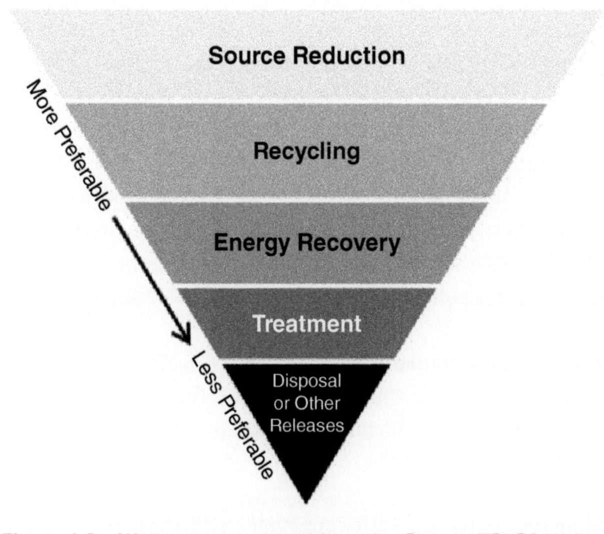

Figure 6.8 Waste management hierarchy. Source: TRI P2 Webpage {US EPA, n.d. #52}.

include equipment or technology modifications, process or procedure modifications, reformulation or redesign of products, substitution of raw materials, and improvements in housekeeping, maintenance, training, or inventory control' (Pollution Prevention Act 1990).

Along with source reduction, further sustainability gains are achieved by implementing preferred waste management practices such as recycling and treatment. Pollution prevention and improved waste management practices have been applied throughout the manufacturing life cycle of automobiles as follows:

- *Process and technology modifications* implemented to reduce energy consumption and the amount of raw materials used;
- *Safer, environmentally friendly alternative materials* substituted throughout the manufacturing process;
- *Recycling of metals and solvents* to reduce environmental impacts after the useful life of a material or product; and
- *Fuel economy improvements* contribute to reducing the industry's environmental impacts by manufacturing vehicles that are more fuel efficient.

6.4.1.1 Corporate Sustainability Reports

Corporate sustainability reports provide an overview of sustainability goals, efforts, and accomplishments for some of the largest automotive manufacturers. Water use, energy use, and waste generated from vehicle manufacturing may also be included in these reports. Companies often report recognition in Leadership in Energy and Environmental Design (LEED), ENERGY STAR building labels, and other sustainability awards. Additionally, they provide specific examples of sustainability progress such as General Motors' 122 landfill-free facilities (General Motor Company 2014); Ford's first automotive application of cellulose-reinforced plastic to replace fibreglass (Ford Motor Company 2014); and the Toyota Prius' ranking for highest fuel efficiency in the compact, midsize, and midsize station wagon classes (Toyota Motor Corporation 2014).

6.4.1.2 Eco-Efficiency

In a study of industry-wide environmental sustainability, manufacturing sectors were ranked by *eco-efficiency score*, a measure of environmental performance against economic output (Egilmez 2013). The score accounted for GHG emissions, energy use, hazardous waste, TRI releases, and water use. Motor vehicle manufacturing (NAICS 3361) ranked among the highest with an *eco-efficiency score* of 100%.[5]

However, when automotive manufacturing and its supply chain (including extraction and processing of raw materials) are compared, the industry's upstream supply chain is responsible for 92.8% of TRI releases and 60.8% of hazardous waste generation. In other words, the production of materials used by the industry generates significantly more pollution than automobile production alone. Some of the highest contributing supply chain sectors include other basic organic chemical manufacturing, iron and steel mills, power generation and supply, and paint and coating manufacturing (Egilmez 2013).

The motor vehicle parts, body, and trailer manufacturing industry (NAICS 3362 and 3363) received an *eco-efficiency score* of 69%. In order to become 100% eco-efficient, the industry would need to 'reduce GHG emissions by 35%, energy use by 31%, hazardous waste by 37%, toxic releases by 38%, and water withdrawals by 32%' (Egilmez 2013, p. 97).

5 Eco-efficiency scores are a comparison of environmental impacts and economic output relative to other manufacturing sectors. A score of 100% does not mean that the industry cannot improve its eco-efficiency.

6.4.1.3 Process and Technology Modifications

Process and technology modifications have the potential to reduce chemical inputs and wastes at all stages of production. The use of metalworking fluids, solvents, lubricants, paints, and other chemicals throughout production result in chemical wastes such as metal filings, spent chemical baths, and emissions of volatile organic compounds (VOCs). Source reduction practices, waste treatment, and recycling can all help to reduce releases of TRI chemicals.

Energy Efficiency Improving energy efficiency in the automotive production process reduces the quantities of fuels burned and thereby reduces the quantities of TRI chemicals formed as byproducts and released to the environment. If energy is generated by combustion of fossil fuels on-site at a facility, and energy efficiency improvements are implemented, the corresponding reductions in the releases of TRI chemicals produced will likely be reflected in the facility's TRI reporting. Most automobile manufacturers have had programs in place for many years to improve the efficiency of material and energy use throughout production. Companies are incentivized by cost savings that they can pass on to consumers (Orsato and Wells 2007).

Government programs also incentivize improvements. For example, the automotive industry collaborated with EPA's ENERGY STAR Program[6] in 2005 to develop an energy performance indicator tool to recognize energy-efficient automobile assembly plants (Boyd 2005). ENERGY STAR certification is awarded to plants ranking in the lowest 25% by energy use per vehicle manufactured. In 2014, 10 facilities received certifications. ENERGY STAR determined that paint booths used the most energy (30–50% of a plant's energy consumption), followed by heating/ventilating/air conditioning, lighting, compressed air, welding, material handling, metal forming, and other processes. Automotive assembly plants reported energy efficiency improvements for all of these processes (US EPA 2015a). ENERGY STAR also produced a report containing recommendations for energy efficiency measures for the automotive assembly industry; providing detailed technical direction, associated cost savings, and payback times. Some automotive-specific recommendations include adopting technologies in infrared and ultraviolet (UV) paint curing, powder-based paints, ultrafiltration/reverse osmosis for wastewater cleaning, energy-efficient welding, and hydroforming (Galitsky and Worrell 2008). Major US automobile manufacturers have also reported advancements in energy efficiency and set future energy consumption goals in their corporate sustainability reports. For example, Ford has reduced the energy consumed per vehicle produced by 9% from 2014 to 2015 and aimed to reduce this amount by 25% from 2011 to 2016 (Ford Motor Company 2016). Similarly, GM has set a goal for a 25% reduction in energy consumption at its US facilities by 2018. In addition to reducing fossil fuel combustion, some companies seek out cleaner sources of energy. GM has invested in more than 100 MW of renewable energy including solar power, landfill gas, hydro-electric power, and waste-to-energy (General Motor Company 2016).

Process Improvements In order to help facilities realize the many possible process improvement options, the US Department of Energy Advanced Manufacturing Office[7] sponsored the development of a process modelling framework to help manufacturers select processes that decrease waste without compromising quality. Some examples of design efficiency can include decreasing cutting fluid, scrap, and overall energy loss (Marusich 2013). This type of research has direct applications in the motor vehicle manufacturing industry, as it helps optimize production time, energy consumption, and use of 'raw materials while assuring quality, performance, and costs' (Marusich 2013).

---

6 For more information on EPA's ENERGY STAR Program, please visit https://www.energystar.gov.

7 For more information on the Department of Energy's Advanced Manufacturing Office, please visit http://energy.gov/eere/amo/advanced-manufacturing-office.

6.4.1.4 Safer, Environmentally Friendly Alternative Materials

The automotive manufacturing industry uses a wide variety of chemicals, many of which are TRI chemicals, throughout the production process including chemicals used in paints, adhesives, coatings, and resins. Research continues to advance the knowledge and availability of safer alternatives for the ingredients used to develop materials including alternatives to metals used in the manufacture of vehicle components.

Alternative Paints and Coatings Before new technologies can be fully developed and deployed by the industry, they must first be recognized as viable and cost-effective alternatives. EPA sponsors the Presidential Green Chemistry Challenge every year to promote the environmental and economic benefits of innovative green chemistry research. Winning technologies developed and used for automobile manufacturing have included (US EPA 2014):

- a UV-curable paint primer by BASF Corporation that significantly decreases VOCs compared to existing primers, eliminates use of diisocyanates, and, therewith, eliminates landfill waste,
- a bio-based oil by the Proctor & Gamble Company and the Cook Composites and Polymers Company to replace petroleum-based solvents in alkyd paints that reduces the amount of solvent required by 50% and reduces VOC and ozone emissions, and
- water-based refinish coatings by PPG Industries, Inc. to reduce metal sludge and VOC emissions.

Despite the health and safety advantages of less toxic materials, other environmental criteria need to be considered. For example, a concern for water-based paints is the requirement for significantly more air conditioning and ventilation needed to dry the paints. Further investigation is needed to ensure that this energy requirement does not undermine the overall environmental benefits of water-based paints (Kanellos 2009). Alternatively, the use of powder-based coatings has experienced rapid growth in the automotive industry in recent years. These coatings completely eliminate the use of solvent because they are applied in powder form and then cured with heat or light (Acmite 2011).

Bio-Based Materials Bio-based materials have long been used to make a variety of vehicle components. Natural fibres from bananas, jute, sisal, and flax are used to reinforce resins and plastics such as those used in door panelling. Cotton, hemp, coir, and ramie have been used to create carpets and insulating mats for vehicle interiors (Center for Automotive Research 2012). These materials are renewable and biodegradable, and can be less toxic than petrochemical-derived plastics (Institute for Local Self-Reliance 1997). Examples include:

- *Interior components.* The plastics research group at Ford is leading research on use of soy-based foams in seat cushions, plant material byproducts to reinforce plastics vehicle storage bins, electrical harnesses, and armrests (Briddell 2015). The Ford team is also investigating the use of other byproducts including shredded currency, tomato skins, and hemp fibres to manufacture vehicle interior components.
- *Mechanical vehicle components.* In terms of core vehicle components such as engines and vehicle bodies, bio-based materials may not provide the durability needed. However, Goodyear has developed bio-based isoprene to replace a petroleum-derived ingredient in synthetic rubber for tyres, and Daimler produces air filter systems made from bio-based polyamide (Andresen et al. 2012). Additionally, the Ford Motor Company is developing brake lines and fuel tubes made from bio-based polyamides (Lee 2013).

Table 6.3 summarizes examples of bio-based materials and other alternatives used in automotive manufacturing.

Table 6.3 Examples of alternatives for vehicle components.

Component	Alternative materials
Paints and coatings	• UV-curable paint primers to reduce energy use and VOC emissions • Bio-based oil to replace petroleum-based solvents in alkyd paints • Water-based coatings to replace petroleum based solvents
Plastics, resins, and adhesives	• Biodegradable plant based resins for vehicle interiors • Plastics reinforced with plant byproducts (corn/rice husks, coconut fibres, and wheat straw) • Tree-based cellulose to replace fibreglass in armrests • Natural fibres from bananas, jute, sisal, and flax are used to reinforce resins and plastics such as those used in door panelling • Use of plant lignin in adhesives and polymer additives
Interior surfaces and cushioning	• Soy-based foams for seat cushions • Use of cotton, hemp, coir, and ramie to create carpets and insulating mats • Recycled clothing used in sound insulation materials
Metals	• Replacement of steel with aluminium or magnesium to produce lighter, more fuel-efficient vehicles

Sources: Institute for Local Self-Reliance 1997; Lee 2013; Briddell 2015; and Ford Motor Company 2014

Greater industry use of these alternative materials has significant potential to reduce TRI-reported chemical waste quantities because the alternatives can completely replace use of certain TRI chemicals or materials that contain TRI chemicals with safer alternatives that serve the same purpose, sometimes more effectively. (Section 6.4.2 discusses how innovations in paints, coatings, and other materials have contributed to reducing the generation of TRI chemical wastes from solvents and metals such as copper, zinc, and lead.)

6.4.1.5 Recycling of Metals and Solvents

While the automotive manufacturing industry has made progress in changing production practices to reduce the quantity of scrap metal generated, recycling of scrap metal continues to play a crucial role in achieving sustainability, both during the manufacturing process and at the end of the vehicle life. The automotive recycling industry reports over $32 billion in annual sales (Automotive Recyclers Association 2015). Approximately 95% of vehicles are recycled, and 86% of materials from those vehicles are recycled, reused, or used for energy recovery (Duranceau and Sawyer-Beaulieu 2011, US EPA 2013).

6.4.1.6 Metal Scrap and Waste

Most vehicle bodies and mechanical components are constructed from metals. Metal stamping, moulding, and machining results in large quantities of scrap metal. In addition, metals are used in coatings, brake pads, catalytic converters, and electronics. As discussed earlier, TRI-reported metal wastes from automotive manufacturing are often recycled at rates above 90%. Some of the most commonly recycled metals (and their metal compounds) include copper, manganese, chromium, zinc, nickel, and lead.

6.4.1.7 Fluids and Solvents

Solvent recycling, which started occurring in automotive manufacturing in the 1980s, applies mainly to paint solvents. Some waste management companies now provide full solvent recycling services to manufacturing facilities. These companies partner with facilities to install recycling equipment free of charge; facilities only pay the discounted costs of the recovered chemicals and save money on disposal costs. Some recycled paint thinners include methyl ethyl ketone, acetone, naphtha, xylene, and toluene (CleanPlanet 2015). As of this analysis, most of the TRI-reported solvent recycling that occurs is done off-site.

6.4.1.8 End-of-Life Vehicles (ELVs)

After a vehicle's useful life, some parts can be reused in other vehicles including engines, transmissions, doors, and bumpers. Starters, alternators, and water pumps can be remanufactured, while batteries, catalytic converters, tyres, and plastics can be recycled into new products via shredding, magnetic separation, metal smelting, and other methods. By avoiding the mining and processing of virgin materials, the current use of recycled copper, aluminium, and steel results in approximately 5600 pounds of GHG reductions per vehicle (Automotive Recyclers Association 2015).

In addition, the Automotive Recyclers Association estimates reuse or recycling of '100.8 million gallons of gasoline and diesel fuel, 24 million gallons of motor oil, 8 million gallons of engine coolant, 4.5 million gallons of windshield washer fluid, 96% of all lead acid batteries' from end-of-life vehicles (ELVs) every year (Automotive Recyclers Association 2015).

6.4.1.9 Fuel Economy

The fuel economy of motor vehicles is a measure of how efficiently fuel is being used and therefore has a direct influence on the quantity of GHG emissions. Fuel economy is measured in miles per gallon (MPG) for gasoline and diesel powered vehicles. A miles per gallon equivalent (MPGe) can also be calculated for electric vehicles by associating the electricity required per vehicle with the quantity of fossil fuel needed for electricity generation. FuelEconomy.gov, the US government's official website on fuel economy information for most gas-powered, electric, and hybrid vehicles sold in the United States, combines highway and urban fuel economy to calculate a final MPGe value. A strong drive for fuel economy improvements has been the Corporate Average Fuel Economy standards, which require manufacturers to maintain a certain average fuel economy across their fleet of new vehicles each year. Automotive manufacturers have improved their vehicles' fuel economy through engine modifications or other features including:

- regenerative braking to charge hybrid electric vehicle batteries;
- hybrid electric and combustion propulsion systems;
- continuous variable transmission to allow engines to run at their most efficient rotations per minute;
- smaller engines that rely on turbo systems for power; and
- use of lighter materials and reduction of air resistance.

Figure 6.9 shows average fuel economy across all models by vehicle type in 2014 (US Department of Energy 2015). Electric vehicles have the highest MPGe, while vehicles running on conventional gasoline or compressed natural gas have the lowest MPGe. Although electric vehicles, plug-in hybrid electric vehicles, and hybrid vehicles have traditionally had very low production volumes, they have become more commercialized in recent years (Automotive News 2015).

6.4.2 Pollution Prevention Activities Reported to TRI

As discussed in Section 6.3.2, the quantities of TRI chemicals reported by the automotive manufacturing sector as released to the environment or otherwise managed as waste decreased significantly between 2005 and 2015. This section identifies the chemicals targeted for pollution prevention and

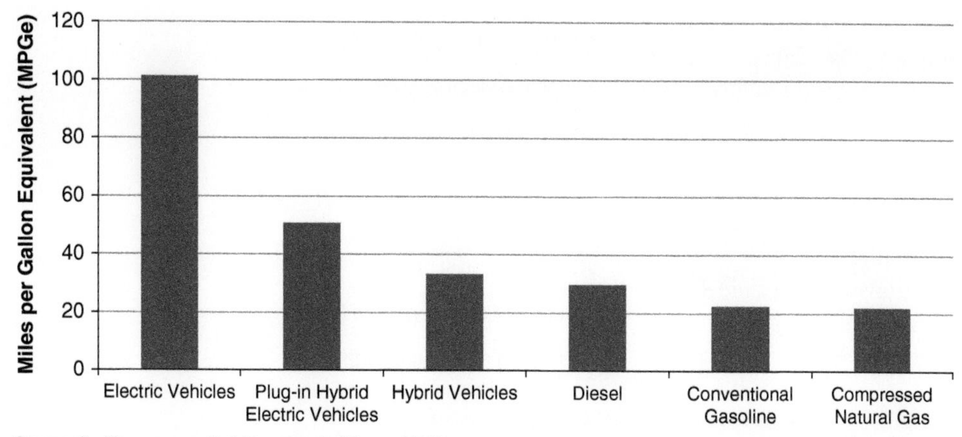

Figure 6.9 Average fuel economy across vehicle models by vehicle type in 2014. Source: FuelEconomy.gov (US Department of Energy 2015).

the types of pollution prevention activities implemented. Additionally, the automotive manufacturing industry's progress towards reducing TRI chemical waste managed and released is described.

Facilities subject to the TRI reporting requirements, such as those within the automotive manufacturing industry, are required to report any source reduction practices (e.g. process modifications, substitution of raw materials) that were implemented at the facility during the year for the chemical for which they are reporting. TRI Reporting Form R contains specific fields for these required data elements. In Section 8.10 of the TRI Form R, for example, facilities select source reduction activities[8] from a list of coded processes or improvements (all source reduction activity codes are grouped into the eight categories listed in Figure 6.10). In addition, facilities have the option to provide additional text to describe their source reduction activities in Section 8.11 of Form R. It is a unique opportunity for manufacturers to share their achievements in pollution prevention with other automotive facilities, users of TRI data, and the public. For reporting year 2015, 23% of automotive manufacturing facilities included text in Section 8.11 of their Form R reports.

For the 2015 reporting year, 110 facilities in the automotive manufacturing industry reported 285 source reduction activity codes for 63 chemicals and chemical categories. This represents approximately 13% of the 825 automotive manufacturing facilities that reported to EPA's TRI Program for the 2015 reporting year. 'Good operating practices' was the most commonly reported source reduction category as shown in Figure 6.10. 'Process modifications' and 'raw material modifications' were also frequently reported.

Within the eight source reduction categories listed in Figure 6.10, there are nearly 50 specific, predefined source reduction activities that facilities can select. Six of these are specific to 'green chemistry' (US EPA 2016). These green chemistry-specific source reduction codes were added to the list of source reduction codes beginning with the 2012 reporting year. From 2012 to 2015 auto manufacturing facilities have reported these codes 105 times (mostly by facilities that manufacture parts, i.e. facilities classified in NAICS 3363). The most frequently reported green chemistry source reduction activities are:

---

8 The terms 'source reduction' and 'pollution prevention' are used interchangeably here. In this chapter, the term 'source reduction' is used in order to be consistent with TRI reporting terminology.

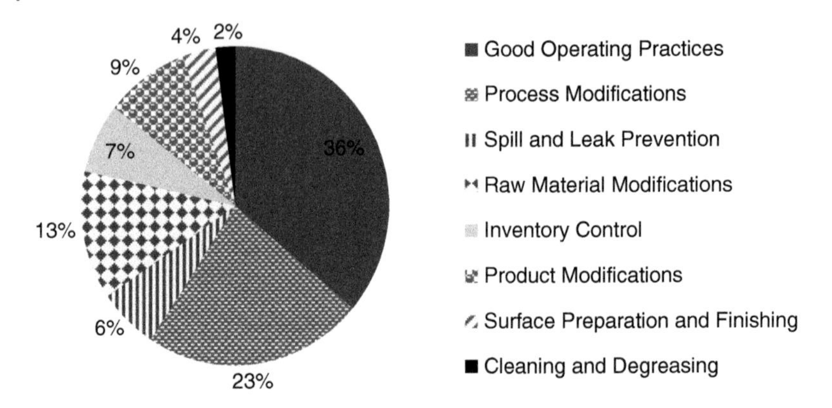

Figure 6.10 Source reduction activities reported by automotive manufacturing facilities for 2015. Source: US EPA Toxics Release Inventory – 2015 National Analysis Dataset, Retrieved January 2017.

- quality monitoring or analysis systems for products or processes involving manganese, nickel, and chromium;
- optimizations of reactions using ammonia and other various chemicals; and
- the reduction of use of solvents such as xylenes and other organic solvents.

Figure 6.11 shows how the total number of reported source reduction activities changed from 2005 through 2015. The decline from 2007 to 2009 corresponds to an overall decrease in TRI

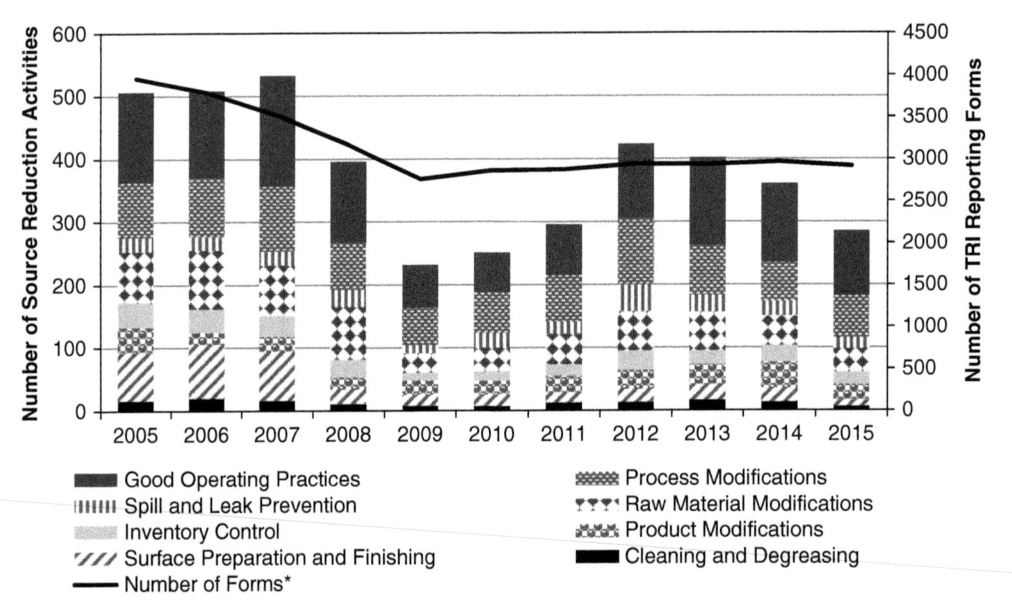

Figure 6.11 Source reduction activities reported by automotive manufacturing facilities, 2005–2015. Note: *Total number of forms reported by the automotive manufacturing industry. Source: US EPA Toxics Release Inventory – 2015 National Analysis Dataset, Retrieved January 2017.

reporting due to reduced production during the economic recession. Since 2005, the percentage of facilities with source reduction and the distribution of source reduction types have remained fairly constant except for 'surface preparation and finishing', which has decreased in recent years. In the mid-2000s, solvent waste had already been significantly minimized by the use of water-based paints and improved spraying techniques. Following this change, facilities may have shifted their efforts towards other areas of source reduction.

Observations from the TRI data regarding source reduction activities for chemicals with the largest decreases in waste managed include:

Metals tend to undergo more 'raw material modifications' because they are primarily used in alloys for vehicle bodies and parts. For example, many 'raw material modifications' were implemented for materials containing lead and lead compounds. This was likely due to reductions in the use of lead solder, which was reported in explanatory text in Section 8.11 of Form R by several facilities. Non-metals and organic solvents tend to undergo more 'surface preparation and finishing' modifications, as they are primarily used in paints and coatings. The use of organic solvent-based paints and coatings has declined significantly over the past decade due to their replacement with powder and water-based paints. This has led to a decline in releases for some of the most common organic solvents, such as xylenes, toluene, and glycol ethers.

6.4.2.1 Examples of Source Reduction Activities Reported to TRI

Facilities subject to the TRI reporting requirements have the option of describing their source reduction activities in explanatory text; Table 6.4 provides examples of this text for reported chemicals showing the greatest decreases in waste managed between 2005 and 2015. Notable activities include the reduction of metals in brake pad friction formulations, paints, and solders as well as leak prevention and solvent substitutions. In many cases source reduction activities can be associated with specific regulatory drivers; for example, state-imposed limits on the copper content of brake pad formulations.

6.4.2.2 TRI Pollution Prevention Analysis – Effectiveness of Source Reduction Activities

A published study estimated how source reduction activities affect quantities of TRI chemical releases reported by facilities that are subject to the TRI reporting requirements (Ranson et al. 2015). The study used a common economics research technique known as a 'differences-in-differences' analysis. This method estimates how releases of a TRI chemical from a facility change in the years before and after implementing a source reduction project by comparing trends in releases of TRI chemicals targeted by source reduction against trends in releases of chemicals not targeted for source reduction. This comparison helps to control for other factors that affect releases, such as changes in production, economic conditions, and environmental regulations.

Applying this technique to the automotive manufacturing industry showed that the *average* source reduction project in the automotive manufacturing industry resulted in a 12–26% decrease in facility-level TRI releases of the targeted chemical. 'Surface preparation and finishing' and 'raw material modifications' were found to be the most effective types of source reduction with average decreases in releases of 26% and 24%, respectively. Between 1991 and 2014, source reduction may have reduced cumulative automotive manufacturing releases reported to EPA's TRI Program by 300–700 million pounds (17–34%), as calculated by the difference between actual releases and simulated releases with no source reduction.

6.4.2.3 Barriers to Source Reduction

Facilities can voluntarily report barriers they face in implementing source reduction activities in Section 8.11 of Form R. These explanatory text data can be used to identify the challenges that

Table 6.4 Example descriptions of source reduction activities reported by automotive manufacturing facilities for TRI chemicals with the greatest decrease in waste managed from 2005 through 2015.

Source reduction type	Example source reduction activity
Good operating practices	*Xylenes*. 'Facility reduced Defects Per Unit (DPU) in paint areas resulting in a paint consumption reduction (i.e. spraying less paint for repair work). This reduces usage, emissions, and waste disposal of chemicals in paint (Xylene)'.
	Zinc and zinc compounds. 'looking into reduced copper formulas for brake pads, which will reduce the amount of brass in some blocks, which will decrease the amount of brass used, thus reducing copper, lead, and zinc'.
	Methyl isobutyl ketone. 'Production to line-up like colour trucks so there is less colour changing resulting in a reduction in flush solvent used (flush solvent and paint contains MIBK). Facility reduced Defects Per Unit (DPU) in paint areas resulting in a paint consumption reduction (i.e. spraying less paint for repair work)'.
Process modifications	*n-butyl alcohol*. 'Using a reformulated cleaning solvent that has replaced the n-Butyl alcohol component with other solvent components that are not TRI listed substances'.
Spill and leak prevention	*Toluene*. 'Improved solvent leak detection (virgin and return lines) through use of a new photoionization detector'.
Raw material modifications	*Lead and lead compounds*. 'Lead solder use is decreasing dramatically. Approximately 25% of all soldering is leaded now. Expect this number to continue to decrease moving forward'.
Inventory control	*Toluene*. 'Vendor substituted less hazardous solvents to reduce VOC and HAPs; also, evaluated existing warehouse and disposed of outdated materials. Implemented new procedures to limit storage of rejected or unused products'.
Product modifications	*Copper and copper compounds*. 'Product formulation has been modified to reduce the amount of copper used. This formulation management method will continue to reduce the amount of copper used in brake pad friction formulations over the next few years'.
Surface preparation and finishing	*Glycol ether*. 'Replaced water-based paints with 2k urethane solvent-based [paints] that are HAP-free'.
	Ethylbenzene. 'Have gone to water based coatings as a standard'.
	Barium and barium compounds. 'Substituted coating material December 2009 to a barium free material'.
Cleaning and degreasing	*Copper and copper compounds*. 'Facility personnel reviewed parts washing processes and installed filtering units that increased the life of wash waters; thus reducing the need for changeovers and reducing the amount of wastewaters generated'.

Source: US EPA Toxics Release Inventory – 2015 National Analysis Dataset, Retrieved January 2017.

facilities face in implementing source reduction. As of reporting year 2014, the TRI reporting form allows facilities to select from a list of barrier codes.

For the 2015 reporting year, 345 barriers were reported by the automotive manufacturing industry. Of these entries, the most commonly reported barrier category was 'no known substitutes or alternative technologies' (abbreviated as 'no substitute'). This barrier may have a variety of causes including a lack of awareness of substitutes, stringent material specifications, or product testing requirements. Barriers in this category were primarily reported for metals and metal compounds, including chromium, copper, lead, and manganese, which are present in raw materials. The second most commonly reported barrier category was 'concern that product quality may decline as a result of source reduction' ('quality'), and the third most commonly reported

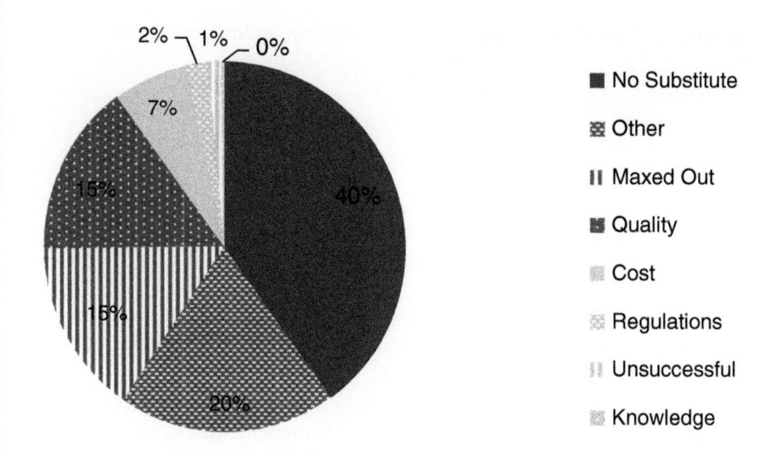

Figure 6.12 Barriers to source reduction reported by automotive manufacturing facilities for 2015. Source: US EPA Toxics Release Inventory – 2015 National Analysis Dataset, Retrieved January 2017.

barrier, 'pollution prevention previously implemented – additional reduction does not appear technically or economically feasible' ('maxed out'), was reported by facilities that have already implemented source reduction activities to the maximum extent possible as shown in Figure 6.12. Specific examples from each category are provided in Table 6.5.

Barrier categories reveal some challenges to source reduction; however, for some chemicals there is no evidence that source reduction was attempted. Table 6.6 lists 10 of the 34 chemicals for which no source reduction activities were reported to EPA's TRI Program between 2005 and 2015, in order of descending quantity of waste managed. Several of these are metals that are present in raw materials and cannot be substituted easily. For example, aluminium oxide is used to give automotive paints their reflective properties. Moving forward, these chemicals should be considered for further pollution prevention efforts.

6.5 Perspectives

Useful perspectives can be inferred from the analysis presented here, and a summary overview of the situation today yields powerful insights for the future of global pollution prevention practice, as follows.

6.5.1 Summary

The US automotive manufacturing industry is a key sector of the American economy with many pollution prevention practices already implemented and with many opportunities for continued progress. Analysis of this sector using data from the TRI shows that:

- The most released chemicals have remained largely the same since 2005. The top chemicals released are organic solvents such as xylenes, n-butyl alcohol, and glycol ethers, which are primarily used in paints, coatings, and cleaners.
- Other chemicals released in large quantities include metals such as zinc, barium, and copper, which are used as components of vehicle parts and in paints and coatings.

Table 6.5 Example descriptions of barriers to source reduction reported to TRI by automotive manufacturing facilities.

Barrier category	Example
No substitute	'The TRI chemicals this facility reports are present as an article component of the steel. The end uses of the products determine the content of the TRI chemicals present in the steel. Therefore, the site cannot opt to reduce or substitute the TRI chemicals present in the steel'.
Quality	'Working with raw material manufacturers to encourage development of materials containing less styrene when possible. Must balance product quality, strength, and other requirements with new materials'.
Maxed out	'Ethylene Glycol based antifreeze are already reused on-site for testing after being put through a filtration unit. Pollution prevention previously implemented as we have switched over as many processes as we could to Propylene Glycol (molding machines). Additional reduction does not appear technically feasible, as we must use Ethylene Glycol for testing radiators and heater cores that use Ethylene Glycol antifreezes'.
Cost	'We have looked at and continue to look at other resins that are classified as more "green," but the costs associated with using them in our process at this point are prohibitive'.
Customer demand	'Customer dictates what camshaft materials are made out of. Manganese compounds are in the lobes we receive for assembling and grinding for the customer'.
Unsuccessful	'Different type of spray nozzles [sic] were tested to reduce over spray from our paint processes. The new type of nozzles [sic] did reduce over spray but the paint thickness was reduced. With the reduced paint thickness our parts would not meet the customer requirements. We were unable to use the nozzles as hoped'.
Regulations	'Ammonia is used in Selective Catalytic Reduction to reduce nitrogen oxide emissions generated during the combustion of natural gas. This emission control equipment/process is required by the local air district'.
Knowledge	'Require technical information on pollution prevention techniques applicable to specific production processes'.

Source: US EPA Toxics Release Inventory – 2015 National Analysis Dataset, Retrieved January 2017

Although US automobile production fell rapidly in the years of the economic recession, the industry has largely recovered, and production has returned to near pre-recession levels. TRI data reflects this fluctuation with a drop in waste managed quantities between 2005 and 2009 and a subsequent increase back to pre-recession levels. However, the quantity of chemicals released to the environment decreased by half due to the many efforts by the automotive industry to use chemicals more efficiently and utilize new technologies and materials. Examples of these sustainable opportunities include:

- eliminating material from the start such as the use of solder that contains lead or reducing the use of copper in brake pads;
- transitioning to UV-curable paints provides manufacturers with low-energy coating options, and water- or bio-based paints greatly reduces the use of organic solvents waste;
- substituting for bio-based materials considered safer, biodegradable alternatives like the use of plant-based materials in vehicle interiors and coatings or bio-based alternatives for more functional vehicle components such as tyres and tubing; and
- recycling of manufacturing materials and end-of-life vehicles at high rates mitigates TRI chemical releases and supplies automakers with recycled materials.

Table 6.6 Top 10 chemicals for which the largest quantities of waste managed and no source reduction activities were reported by automotive manufacturing facilities from 2005 to 2015.

Rank	Chemical	Production-related waste managed (in pounds), 2005–2015
1	Aluminium oxide (fibrous forms)	13 142 042
2	Polychlorinated alkanes	801 478
3	Sodium dimethyldithiocarbamate	519 596
4	N,N-Dimethylformamide	308 856
5	1,1-Dichloro-1-Fluoroethane	282 866
6	Chlorine	195 305
7	3-Iodo-2-Propynyl butylcarbamate	175 374
8	Chlorine dioxide	163 726
9	Vanadium and vanadium compounds	116 868
10	Cyanide compounds	108 177

Note: (i) 24 other chemicals reported by the automotive industry had no source reduction reported. (ii) The top 10 chemicals were chosen based on the highest aggregate TRI chemical waste managed over the period of 2005–2015.
Source: US EPA Toxics Release Inventory – 2015 National Analysis Dataset, Retrieved January 2017

Many of these efforts are reflected in the TRI reporting of source reduction activities, and these actions implemented by the automotive manufacturing industry have reduced releases of a targeted chemical by 12–24% on average, with 'surface preparation and finishing' and 'raw material modifications' categories being the most effective types. For the entire industry, the analysis shows an estimated reduction of 300–700 million pounds of TRI-reported releases from 1990 to 2014 due to source reduction efforts.

6.5.2 Potential Pollution Prevention Opportunities

Further reductions in the automotive manufacturing industry's TRI-reported releases, human health impacts and environmental impacts may require coordinated efforts and research from numerous stakeholders. Since the automotive manufacturing industry's upstream supply chain, including other basic organic chemical manufacturing, iron and steel mills, power generation and supply, and paint and coating manufacturing contribute more to quantities of TRI-reported releases than the industry itself (Egilmez 2013), the supply chain should be further examined to identify where impacts can be reduced. Additional research is also warranted regarding barriers to source reduction options especially those that limit the economic viability of alternatives. Information reported to TRI indicates that the top barriers relate to a lack of known alternatives and the perception that additional reduction is not technically or economically feasible.

TRI can be used as a platform to share the effective pollution prevention methods across the automotive industry that have been successfully employed within a parent company or on a facility level. Communication between EPA and industry will facilitate transitions to alternative materials and processes, and perhaps EPA can inform and incentivize academic research through government grants or by recognizing products that reduce pollution. Quantities of TRI chemical releases, source reduction activities, and barriers to pollution prevention, as reported to TRI, can be communicated back to the industry with insights on what pollution prevention measures have been most effective. EPA's TRI Program has implemented such outreach through the TRI Pollution Prevention

Search Tool,[9] the TRI Pollution Prevention Spotlights,[10] and through this chapter. Moving forward, EPA can continue to expand on these types of outreach efforts to share and promote successful pollution prevention activities throughout the industry.

Disclaimer

The views, statements, opinions, and conclusions expressed in this chapter are entirely those of the authors, and do not necessarily reflect those of the United States EPA, the Eastern Research Group, or Abt Associates Incorporated, nor does mention of any chemical substance, commercial product, or company constitute an endorsement by these organizations.

References

Acmite. (2011). Global Powder Coating Market. Available: http://www.acmite.com/market-reports/chemicals/global-powder-coating-market.html [Accessed October 2015].

Akufuah, N.K., Poozesh, S., Slaimeh, A. et al. (2016). Evolution of the automotive body coating process – a review. *Coatings* 6: 24.

Andresen, C., Demuth, C., Lange, A., Stoick, P. & Rudy, P. (2012). Biobased Automobile Parts Investigation. *Iowa State University* [Online]. Available: http://www.usda.gov/oce/reports/energy/Biobased%20Automobile%20Parts%20Investigation%20Report.pdf [Accessed February 2016].

Automotive News. (2015). North American Production 2005–2017. *Automotive News Data Center* [Online]. Available: http://www.autonews.com/section/datalist [Accessed January 2017].

Automotive Recyclers Association. (2015). Automotive Recycling Industry. Available: http://a-r-a.org/what-we-do/why-use-recycled-parts [Accessed July 2015].

Biesbroeck, J. & Sturgeon, T. J. (2010). Effects of the 2009–09 Crisis on the Automotive Industry in Developing Countries: A Global Value Chain Perspective. *Global Value Chains in a Postcrisis World.* Available: http://feb.kuleuven.be/public/n07057/cv/vbs10WBbook.pdf.

Boyd, G.A. (2005). *Development of a Performance-based Industrial Energy Efficiency Indicator for Automobile Assembly Plants.* Oak Ridge, TN: Argonne National Laboratory Available: https://www.energystar.gov/sites/default/files/buildings/tools/AutoEPIBackground.pdf.

Briddell, C. (2015). Dr. Deborah Mielewski: Greening automobiles from the inside out. *Green Chemistry: The Nexus Blog* [Online]. Available: https://communities.acs.org/community/science/sustainability/green-chemistry-nexus-blog/blog/2015/04/21/dr-deborah-mielewski-greening-automobiles-from-the-inside-out [Accessed April 2016].

Center for Automotive Research. (2012). The Bio-Based Materials Automotive Value Chain. Available: http://www.cargroup.org/?module=Publications&event=View&pubID=29 [Accessed February 2016].

CleanPlanet (2015). Why Recycle? Available: http://cleanplanetchemical.com/solvent-recycling/why-recycle [Accessed July 2015].

Duranceau, C. & Sawyer-Beaulieu, S. (2011). Vehicle Recycling, Reuse, and Recovery: Material Disposition from Current End-of-Life Vehicles. *SAE Technical Paper.*

9 For more information on the TRI Pollution Prevention Search Tool, please visit https://www3.epa.gov/enviro/facts/tri/p2.html.

10 For more information on the TRI Pollution Prevention Spotlights, please visit https://www.epa.gov/toxics-release-inventory-tri-program/tri-p2-spotlight-series-0.

Egilmez, G. (2013). Sustainability assessment of US manufacturing sectors: an economic input output-based frontier approach. *Journal of Cleaner Production* 53: 91–102. Available: http://www.sciencedirect.com/science/article/pii/S0959652613001753.

Federal Reserve Board of Governors (2015). Production Index. *G.17 - Industrial Production and Capacity Utilization.* Available: http://www.federalreserve.gov/datadownload/Build.aspx?rel=G17 [Accessed January 2017].

Fettis, G. (1995). *Automotive Paints and Coatings.* New York: VCH Publishers.

Ford Motor Company. (2014). Sustainability Report 2013/14. Available: http://corporate.ford.com/microsites/sustainability-report-2013-14/doc/sr13.pdf [Accessed July 2015].

Ford Motor Company (2016). Accelerating Today for a Better Tomorrow. Sustainability Report 2015/2016. Available: http://corporate.ford.com/microsites/sustainability-report-2015-16/doc/sr15.pdf.

Galitsky, C. & Worrell, E. (2008). Energy Efficiency Improvement and Cost Saving Opportunities for the Vehicle Assembly Industry. *Ernest Orlando Lawrence Berkeley National Laboratory* [Online]. Available: http://www.energystar.gov/ia/business/industry/LBNL-50939.pdf [Accessed March].

General Motor Company. (2014). 2014 Sustainability Report - Data Center. GM Sustainability [Online]. Available: http://gmsustainability.com/data/environmental.php [Accessed July 2015].

General Motor Company (2016). 2015 Sustainability Report. Available: http://www.gmsustainability.com/impacts/operations.html#renewable-energy [Accessed May 2017].

Institute for Local Self-Reliance (1997). Biochemicals for the Automotive Industry. *Ohio Environmental Protection Agency.* Available: http://www.epa.ohio.gov/portals/41/p2/biochemicalsautomotiveindustry.pdf.

Kallstrom, H. (2015). Investing in the automotive industry – what you need to know. *Market Realist* [Online]. Available: http://marketrealist.com/2015/02/shift-growth-global-automotive-industry [Accessed April 2016].

Kanellos, M. (2009). For Eco-Friendly Paint, Use Solvents, Not Water. *Greentech Media* [Online]. Available: http://www.greentechmedia.com/articles/read/for-eco-friendly-paint-use-solvents-not-water [Accessed October 2015].

Katz, H. C., MacDuffie, John Paul, Pil, Frits K. (2013). Crisis and Recovery in the US Auto Industry: Tumultous Times for a Collective Bargaining Pacesetter. *Collective Bargaining Under Duress: Case Studies of Major North American Industries.* Cornell University Press. Available: https://mgmt.wharton.upenn.edu/files/?whdmsaction=public:main.file&fileID=6562.

Lee, E. C. (2013). Bio-based Materials for Durable Automotive Applications. Available: http://www.lawbc.com/share/bcs2013/Molecules%20to%20Market/lee-presentation.pdf [Accessed February 2016].

Marusich, T. (2013). Sustainable Manufacturing via Multi-Scale, Physics-Based Modeling. Third Wave Systems Inc. Minneapolis: US Department of Energy Advanced Manufacturing Office. Available: http://energy.gov/sites/prod/files/2013/11/f4/sustainable_mfg_process_modeling.pdf.

McCann, M. (2012). Chapter 82 - Metal Processing and Metal Working Industry. *Encyclopaedia of Occupational Health and Safety 4th Edition.* International Labour Organization. Available: http://www.iloencyclopaedia.org/part-xiii-12343/metal-processing-and-metal-working-industry.

OICA (2015). *International Organization of Motor Vehicle Manufacturers. 2015 Production Statistics.* [Online]. Available: http://www.oica.net/category/production-statistics/2015-statistics [Accessed July 2018].

Orsato, R.J. and Wells, P. (2007). The automobile industry & sustainability. *Journal of Cleaner Production* 15 (11–12): 989–993.

Pollution Prevention Act (1990). 42 USC. §13101 et seq.

Ranson, M., Cox, B., Keenan, C. & Teitelbaum, D. (2015). The Impact of Pollution Prevention on Toxic Environmental Releases from US Manufacturing Facilities. Environmental Science & Technology 49 (21): 12951–12957.

Toyota Motor Corporation. (2014). Vehicles. *North American Environmental Report 2013* [Online]. Available: http://www.toyota.com/usa/environmentreport2013/vehicles.html [Accessed July 2015].

US Bureau of Economic Analysis. (2015). Gross-Domestic-Product-(GDP)-by-Industry Data, Gross Output, 1997–2015. Available: http://www.bea.gov/industry/gdpbyind_data.htm [Accessed January 2017].

US Census Bureau (2014). 2014 County Business Patterns (NAICS). Available: https://www.census.gov/programs-surveys/cbp.html [Accessed May 2015].

US Department of Energy (2015). Download Fuel Economy Data. FuelEconomy.gov. Available: https://www.fueleconomy.gov/feg/download.shtml [Accessed October 2015].

US Department of Labor (2012). Bureau of Labor Statistics. Transportation Equipment Manufacturing: NAICS 336. Available: http://www.bls.gov/iag/tgs/iag336.htm.

US Department of the Treasury (2015). TARP Programs - Auto Industry. Available: http://www.treasury.gov/initiatives/financial-stability/TARP-Programs/automotive-programs/Pages/default.aspx [Accessed August 2015].

US Department of Transportation. (2014). Bureau of Transportation Statistics. *Table 1–35: US Vehicle-Miles (Millions)* [Online]. Available: https://www.rita.dot.gov/bts/sites/rita.dot.gov.bts/files/publications/national_transportation_statistics/html/table_01_35.html [Accessed January 2017].

US EPA (2013). Advancing Sustainable Materials Management: Facts and Figures 2013. Available: http://www.epa.gov/osw/nonhaz/municipal/msw99.htm.

US EPA (2014). Office of Pollution Prevention and Toxics - Presidential Green Chemistry Challenge Nominations: Automotive Flagged. Washington DC.

US EPA (2015a). *Automotive Assembly Plants* [Online]. Available: https://www.energystar.gov/sites/default/files/tools/Industry_Insights_Auto_Assembly_2015.pdf [Accessed August 2015].

US EPA (2015b). *GHG Reporting Program Data Sets* [Online]. Available: http://www2.epa.gov/ghgreporting/ghg-reporting-program-data-sets [Accessed January 2017].

US EPA (2016). Toxic Chemical Release Inventory Reporting Forms and Instructions: Revised 2016 Version. Available: https://ofmpub.epa.gov/apex/guideme_ext/guideme_ext/r/files/static/v3321/rfi/RY_2016_RFI.pdf [Accessed May 2017].

US EPA (2017a). *Inventory of US Greenhouse Gas Emissions and Sinks: 1990–2015* [Online]. Available: https://www.epa.gov/sites/production/files/2017-02/documents/2017_complete_report.pdf [Accessed May 2017].

US EPA (2017b). *Sources of Greenhouse Gas Emissions* [Online]. Available: https://www.epa.gov/ghgemissions/sources-greenhouse-gas-emissions#transportation [Accessed June 2017].

US EPA (2017c). Trends in Environmental Releases of Toxic Chemicals Reported Annually by the Automotive Industry to the US EPA's Toxics Release Inventory Program. Washington DC.

Young, A. (2014). Here Are The July 2014 'Big Eight' US Auto Sales Numbers: GM, Ford, Chrysler, Toyota, Honda, Nissan, Volkswagen, Kia/Hyunda. *International Business Times*. Available: http://www.ibtimes.com/here-are-july-2014-big-eight-us-auto-sales-numbers-gm-ford-chrysler-toyota-honda-nissan-1645592 [Accessed August 2015].

7

The Global Demand for Biofuels and Biotechnology-Derived Commodity Chemicals: Technologies, Markets, and Challenges

Stephen R. Hughes[1] *and Marjorie A. Jones*[2]

[1]*Applied DNA Sciences, Stony Brook, NY 11790, USA*
[2]*Department of Chemistry, Illinois State University, Normal, IL 61790, USA*

CHAPTER MENU

7.1 Introduction

The current global demand for biofuels and biotechnology-derived commodity chemicals is presented in the context of total worldwide energy demands, including an itemization of the share of petroleum used for transportation purposes and the share of petroleum used for manufacturing chemicals. This demand is further detailed in each of the major markets including EU, US, Japan, Brazil, China, Russia, and India. The displacement of petroleum-derived chemicals by biomass-derived fuels and chemicals is discussed. Current projections regarding the global demand are provided, based on predicted economic growth rates. Potential impacts on the price volatility of transportation fuels and chemicals are highlighted, taking into account alternative scenarios of the world's economic growth. The projections consider the resulting atmospheric CO_2 levels in the case where no remediation technology is implemented. The worldwide outlook for biofuels and biotechnology-derived chemicals is summarized based on these assessments.

7.2 Overview of Global Energy Demand

Energy consumption worldwide continues to rise, driven by strong demand in countries outside the Organization for Economic Cooperation and Development (OECD). China and India led the

Green Energy to Sustainability: Strategies for Global Industries, First Edition.
Edited by Alain A. Vertès, Nasib Qureshi, Hans P. Blaschek and Hideaki Yukawa.

growth in consumption from 2015 to 2016 with nearly identical increases (3.8% and 3.9%, respectively). However, India's energy consumption grew at a rate similar to the recent past, driven by solid economic growth. In contrast, China's energy consumption reflects a gradual slowing in economic growth from the rates seen over the previous years (British Petroleum Company 2017; United States Department of Energy, United States Energy Information Administration, Office of Energy Analysis 2016). Global energy consumption for 2011–2016 with projected growth for the next two decades is presented in Table. 7.1.

Even though the consumption of non-fossil fuels is expected to grow faster than the consumption of fossil fuels in the next two decades, fossil fuels are still projected to account for 78% of total energy use in 2040 compared to 83% in 2016 (United States Department of Energy, United States Energy Information Administration, Office of Energy Analysis 2016). Global energy consumption by fuel for 2011–2016 and projections for 2020–2040 are presented in Table 7.2.

Natural gas is the fastest-growing fossil fuel with global consumption increasing by an average of about 2% per year (United States Department of Energy, United States Energy Information Administration, Office of Energy Analysis 2016). Although liquid fuels, mostly petroleum-based, remain the largest source of world energy consumption, the liquid fuel's share of the world marketed energy consumption fell from 33.4% in 2011 and 2012 to 32.8% in 2016 and is projected to decline to 30.2% in 2040 (United States Department of Energy, United States Energy Information Administration, Office of Energy Analysis 2016). Coal, the world's slowest-growing energy source, rose 0.56% from 2015 to 2016, and its consumption is projected to be overtaken by natural gas in 2030 as a fraction of worldwide use. Renewable energy consumption, the fastest growing energy source, increased by 6.3% between 2015 and 2016 and is projected to account for 16% of global energy use by 2040, an increase from 12% in 2016. Nuclear power is the world's second fastest-growing energy source, with projected consumption increasing by 6.1% between 2015 and 2016 and is projected to account for 5.6% of global energy use by 2040, an increase from 4.7% in 2016. (British Petroleum Company 2017; United States Department of Energy, United States Energy Information Administration, Office of Energy Analysis 2016).

Renewable energy sources, including biofuels and bio-derived chemicals, are viewed as a response to the major challenges of the rising global energy demand, which are: (i) energy security, (ii) economic development, and (iii) environmental impact (Janda and Stankus 2017). The renewable fuels-led transformation of the power sector has raised new questions about power market design and electricity security, while traditional energy security concerns remain (International Energy Agency 2016). In terms of economic development, the industrial sector accounts for the largest share of global delivered energy consumption to end users in 2016 (39.7%) and is projected to continue to be the largest energy-consuming end-use sector through 2040 (37.9%). However, it is not the fastest-growing end-use sector, with growth in the residential sector (33%), the commercial sector (38%), and the transportation sector (40%) projected to outpace that from the industrial sector (32%) from 2016 to 2040 (United States Department of Energy, United States Energy Information Administration, Office of Energy Analysis 2016). Essentially all growth in transportation energy use occurs in developing, non-OECD economies, where economic growth leads to rising standards of living that result in demand for personal travel and freight transport to meet growing consumer demand for goods (United States Department of Energy, United States Energy Information Administration, Office of Energy Analysis 2016). Global energy consumption by sector for 2011–2016 and projections for 2020–2040 (United States Department of Energy, United States Energy Information Administration, Office of Energy Analysis 2016) are provided in Table 7.3.

Table 7.1 Global energy consumption (quadrillion Btu) by country grouping, 2011–2040.

Region	History						Projections				
	2011	2012	2013	2014	2015	2016	2020	2025	2030	2035	2040
OECD Americas	120.6	118.1	120.7	122.0	121.5	122.7	125.7	128.1	130.7	133.8	138.1
United States	96.8	94.4	97.1	98.5	97.8	98.7	100.8	102.0	102.9	103.8	105.7
Canada	14.5	14.5	14.5	14.6	14.6	14.7	15.1	15.6	16.3	17.1	18.1
Mexico and Chile	9.3	9.2	9.1	8.9	9.1	9.2	9.8	10.5	11.6	12.8	14.3
OECD Europe	82.0	81.4	80.2	80.5	81.2	82.8	84.9	87.5	90.3	93.1	95.5
OECD Asia	39.4	39.0	38.9	39.3	40.2	41.2	43.4	45.0	46.2	47.4	48.5
Japan	21.2	20.8	20.7	20.3	20.6	20.9	21.9	22.3	22.3	22.2	21.5
South Korea	11.3	11.4	11.3	12.2	12.5	13.1	13.9	14.7	15.4	16.1	16.9
Australia and New Zealand	6.9	6.8	6.8	6.8	7.0	7.2	7.6	8.1	8.5	9.2	10.1
Total OECD	242.0	238.4	239.8	241.8	242.9	246.7	253.9	260.6	267.2	274.3	282.1
Non-OECD Europe and Eurasia	49.0	50.7	50.7	51.0	49.4	49.1	51.8	54.8	56.4	57.9	57.6
Russia	30.9	32.1	32.5	32.6	31.7	31.4	33.2	34.7	35.1	35.5	34.5
Other	19.0	18.6	18.2	18.4	17.7	17.7	18.7	20.1	21.3	22.4	23.1
Non-OECD Asia	168.2	175.9	182.8	189.2	194.0	201.1	222.7	246.4	269.9	295.1	322.1
China	109.4	115.0	121.3	126.1	129.2	134.1	147.3	159.4	170.4	180.7	190.1
India	25.0	26.2	26.4	27.2	28.0	29.1	32.8	38.4	44.9	52.8	62.3
Other	33.7	34.7	35.1	35.8	36.8	37.9	42.7	48.6	54.6	61.6	69.6
Middle East	29.9	31.7	32.6	34.0	35.2	36.4	40.8	45.4	50.7	56.6	61.8
Africa	20.1	21.5	21.5	22.3	23.0	23.8	26.1	30.0	33.8	38.4	44.0
Non-OECD Americas	30.5	31.0	31.2	31.2	30.9	31.0	33.5	36.7	39.7	43.3	47.3
Brazil	14.8	15.2	15.3	15.3	15.1	15.1	16.3	18.1	20.0	22.0	24.3
Other	15.7	15.8	16.0	15.9	15.8	15.9	17.2	18.6	19.8	21.2	23.0
Total Non-OECD	298.6	310.8	318.8	327.6	332.5	341.4	375.0	413.3	450.5	491.2	532.8
Total Global	540.5	549.3	558.6	569.4	575.4	588.1	628.9	673.9	717.7	765.6	815.0

OECD, Organization for Economic Cooperation and Development: Australia, Austria, Belgium, Canada, Chile, Czech Republic, Denmark, Estonia, Finland, France, Germany, Greece, Hungary, Iceland, Ireland, Israel, Italy, Japan, Latvia, Luxembourg, Mexico, Netherlands, New Zealand, Norway, Poland, Portugal, Slovakia, Slovenia, South Korea, Spain, Sweden, Switzerland, Turkey, United Kingdom, United States; Btu, British thermal unit.
Source: Adapted from (United States Department of Energy, United States Energy Information Administration, Office of Energy Analysis 2016).

An environmental concern evident from Table 7.3 is that almost one-third of global energy consumption is categorized as electricity-related losses, ranging from 25.8% in 2011 to 26.1% in 2016, and projected to rise to 28.0% by 2040. Losses from electricity transmission and distribution systems result from technical inefficiencies and theft. Technical losses, caused by the resistance of wires and equipment when electricity that passes through them is converted to heat, can be reduced by

Table 7.2 Global delivered energy consumption (quadrillion Btu) by fuel, 2011–2040.

Fuel	History						Projections				
	2011	2012	2013	2014	2015	2016	2020	2025	2030	2035	2040
Liquids (oil)	180.3	183.6	185.6	187.9	189.7	192.7 (33%)	204.2	212.5	221.8	233.2	246.0 (30%)
Natural gas	121.6	124.2	125.4	128.0	128.5	131.5 (22%)	138.3	154.8	173.1	192.5	211.4 (26%)
Coal	152.0	153.3	156.5	160.3	160.8	161.7 (28%)	168.6	173.2	174.4	176.9	180.2 (22%)
Nuclear	26.2	24.5	24.7	25.2	26.2	27.8 (4.7%)	30.9	34.6	40.2	43.4	46.0 (5.6%)
Renewables[a)]	60.4	63.8	66.4	68.0	70.1	74.5 (13%)	87.0	98.8	108.1	119.5	131.4 (16%)
Total	540.5	549.3	558.6	569.4	575.4	588.1 (100%)	628.9	673.9	717.7	765.6	815.0 (100%)

a) Includes biomass, hydro, geothermal, solar, wind, and ocean.
Source: Adapted from (United States Department of Energy, United States Energy Information Administration, Office of Energy Analysis 2016).

Table 7.3 Global delivered energy consumption (quadrillion Btu) by end-use sector, 2011–2040.

Sector	History						Projections				
	2011	2012	2013	2014	2015	2016	2020	2025	2030	2035	2040
Residential	52.4	53.0	55.3	56.9	58.5	58.8	61.1	65.4	69.8	74.1	78.3
Commercial	29.0	29.3	30.4	31.3	32.2	32.8	34.5	37.2	40.0	42.7	45.2
Industrial	217.0	222.3	223.8	227.1	228.8	233.3	245.8	262.6	278.0	294.0	309.1
Transportation	102.8	104.2	106.3	106.3	109.0	110.2	115.8	122.9	131.6	142.2	154.8
Electricity-related losses	139.3	140.4	143.2	147.7	147.1	153.3	172.1	186.1	198.7	212.9	227.9
Total global	540.5	549.3	558.6	569.4	575.4	588.1	628.9	673.9	717.7	765.6	815.0

Source: Adapted from (United States Department of Energy, United States Energy Information Administration, Office of Energy Analysis 2016).

upgrading transmission lines and power transformers and by improving power dispatch planning. Thefts occur when electricity consumed is not accounted for on meters or when customers avoid paying for the electricity they consume. Because they are by far the major energy consumers, China and the US (115.0 and 94.4 quadrillion Btu in 2012, respectively) have the highest electricity-related losses, 32.5 and 25.7 quadrillion Btu in 2012, respectively (United States Department of Energy, United States Energy Information Administration, Office of Energy Analysis 2016). However, the percentages of total energy consumption in 2012, 28.3% in China and 27.2% in US, projected by 2040 to be 29.4% and 27.5%, respectively, are comparable to global values and to those of other countries consuming a total of more than 25 quadrillion Btu (EU, Russia, India) (United States Department of Energy, United States Energy Information Administration, Office of Energy Analysis 2016).

7.3 Petroleum Demand and Petroleum Products for Potential Replacement by Bioproducts

Oil remained the world's leading fuel in 2016, accounting for 32.9% of the global energy consumption (British Petroleum Company 2017). Global consumption of petroleum and other liquid fuels

Table 7.4 World petroleum and other liquids consumption (million barrels per day) by region.

Countries	2013	2014	2015	2016	2017 (estimate)	2018 (forecast)
North America	23.51	23.56	23.95	24.03	24.27	24.59
US (50 states + DC)	18.96	19.11	19.53	19.63	19.95	20.26
Canada	2.45	2.41	2.41	2.43	2.41	2.41
Mexico	2.09	2.04	2.01	1.95	1.90	1.90
Central & South America	7.11	7.27	7.22	7.22	7.20	7.18
Brazil	3.03	3.14	2.99	2.95	2.90	2.83
Europe	14.32	14.24	14.46	14.83	14.93	15.00
Eurasia	4.60	4.85	4.68	4.76	4.87	4.98
Russia	3.49	3.69	3.55	3.60	3.70	3.80
Middle East	8.22	8.51	8.71	8.72	9.04	9.30
Asia and Oceania	30.76	31.19	32.38	33.20	33.85	34.61
China	11.08	11.49	12.02	12.44	12.78	13.12
Japan	4.50	4.27	4.12	3.99	3.84	3.76
India	3.66	3.74	4.14	4.46	4.67	4.93
Africa	3.80	3.98	4.00	4.16	4.29	4.43
Total OECD	46.11	45.81	46.41	46.85	47.14	47.50
Total Non-OECD	46.21	47.80	48.99	50.07	51.32	52.58
World	92.33	93.60	95.40	96.92	98.46	100.08

Other liquids includes lease condensate, natural gas plant liquids, and refinery processing gain.
Source: Adapted from (United States Department of Energy, United States Energy Information Administration, Office of Energy Statistics, Office of Survey Development and Statistical Integration, Integrated Energy Statistics Team 2017; United States Energy Information Administration 2017).

averaged 96.9 million barrels per day (b/d) in 2016, an increase of 1.5 million b/d from 2015 (United States Energy Information Administration 2017). The transportation and industrial sectors account for most of the growth in liquid fuels consumption, with liquid fuels remaining the main source of transportation energy consumption worldwide. The industrial sector is responsible for most of the remaining increase in liquid fuels consumption, with the chemicals industry continuing to demand large quantities of petroleum feedstocks (United States Department of Energy, United States Energy Information Administration, Office of Energy Analysis 2016). World petroleum and other liquids consumption by region is provided in Table 7.4.

Consumption growth from 2015 to 2016 was driven by non-OECD countries, where consumption increased by 1.1 million b/d, while in OECD countries petroleum and other liquid fuels consumption rose by 0.44 million b/d (United States Energy Information Administration 2017). From 2015 to 2016, growth in the US and the EU increased by +0.1 million b/d and + 0.37 million b/d, respectively, while Japan recorded the largest decline in oil consumption (−0.13 million b/d). Outside of the OECD, net oil importing countries recorded significant increases, with China accounting for the largest increase in demand (+0.42 million b/d), while India, with an increase of +0.32 million b/d, surpassed Japan in consumption (British Petroleum Company 2017; United States Energy Information Administration 2017).

The annual growth of OECD consumption of petroleum and other liquid fuels is projected to slow to 0.36 million b/d in 2018 from 0.44 million b/d in 2016 (United States Energy Information Administration 2017). The US consumption is expected to increase by 0.31 million b/d in 2018, while the consumption in Japan is estimated to decline by 0.15 million b/d in 2017 and by 0.08 million b/d in 2018. The largest contributors to the growth of non-OECD petroleum consumption are projected to be China and India. China's consumption is projected to increase by 0.34 million b/d in both 2017 and 2018, driven by increased use of gasoline, jet fuel, and hydrocarbon gas liquids (HGL). In India, a consumption growth greater than 0.2 million b/d in both 2017 and 2018 is projected, primarily from an increased use of transportation fuels (United States Energy Information Administration 2017).

The US consumes more energy from petroleum than from any other energy source (United States Energy Information Administration 2017). In 2016, petroleum accounted for 37% of total the US primary energy consumption. The remaining sources were natural gas accounting for 29% of total consumption, coal 15%, renewable energy (conventional hydroelectric power, geothermal, solar, wind, and biomass) 10%, and nuclear electric power 9% (United States Energy Information Administration 2017), differing slightly from the worldwide (liquids 33%, natural gas 22%, coal 28%, nuclear 4.7%, and renewables 13% [Table 7.2]). Petroleum products include transportation fuels, fuel oils for heating and electricity generation, asphalt and road oil, and feedstocks for making the chemicals, plastics, and synthetic materials. US petroleum and other liquids consumption by product is presented in Table 7.5.

Of the approximately 7 billion barrels of total annual US petroleum consumption in 2016, 48% was motor gasoline, 20% was distillate fuel (heating oil and diesel fuel), 13% was HGL, and 8% was jet fuel (United States Energy Information Administration 2017). Diesel fuel is used in the diesel engines of heavy construction equipment, trucks, buses, tractors, boats, trains, some automobiles, and electricity generators. Heating oil is used in boilers and furnaces for heating homes and buildings, for industrial heating, and for producing electricity in power plants. HGL includes propane, ethane, butane, and other HGL that are produced at natural gas processing plants and oil refineries (United States Energy Information Administration 2017).

In their report on reducing the US dependence on oil (United States Department of Energy, Energy Efficiency & Renewable Energy, Bioenergy Technologies Office 2013), the US Department of Energy, Bioenergy Technologies Office, stated 'A greater emphasis on research, development, and demonstration [is needed] to produce bio-based hydrocarbon fuels and bioproducts that can displace *all* of the products made from a barrel of oil. Since only about 40% of a barrel of crude

Table 7.5 US petroleum and other liquids consumption (million barrels per day) by product.

Product	2016	2017 (estimate)	2018 (forecast)
Motor gasoline	9.33	9.33	9.36
Fuel ethanol blended into gasoline	0.94	0.94	0.94
Distillate fuel oil	3.88	3.96	4.05
Hydrocarbon gas liquids (HGL)	2.49	2.65	2.84
Jet fuel	1.61	1.62	1.61
Other oils	2.00	2.02	2.04
Residual fuel oils	0.36	0.34	0.33
Unfinished oils	−0.03	0.00	0.00
Total consumption	19.63	19.92	20.22

Source: Adapted from (United States Energy Information Administration 2017)

is used to make petroleum gasoline, [novel technologies are needed] that will help [to] convert biomass into affordable diesel, jet fuel, heavy distillates, and other chemicals and products. These bio-based hydrocarbon fuels are nearly identical to the petroleum-based fuels they are designed to replace [making them] compatible with today's engines, pumps, and other infrastructure.... Breakthroughs in bioconversion technologies and successes in scaling up technologies for commercial operations promote U.S. leadership in global clean energy technology. Advances can provide benefits in such related areas as agricultural production and food processing. Investments in bioprocessing will also help to reduce production costs, improve process and product reliability, and increase profitability'. (United States Department of Energy, Energy Efficiency & Renewable Energy, Bioenergy Technologies Office 2013).

7.4 Role of Biofuels and Biobased Chemicals in Renewable Energy Demand

In 2015, the United Nations General Assembly adopted the 2030 Agenda for Sustainable Development with 17 Sustainable Development Goals (SDGs), including SDG-7 to ensure access to affordable, reliable, sustainable and modern energy for all, and at the 21st Conference of the Parties (COP21) in Paris, 195 countries signed an accord committing to low-carbon development by conformity to the SDGs (Renewable Energy Policy Network for the 21st Century (REN21) 2016; United Nations Conference on Trade and Development (UNCTAD) 2016). That year also saw the greatest increase in annual renewable energy consumption in spite of falling fossil fuel prices. The worldwide consumption of renewable energy and that in the United States is currently the highest in history. Renewable energy, including conventional hydroelectric power, geothermal, solar, wind, and biomass (13%), and nuclear electric power (5%) provided about 18% of global final energy consumption in 2016 (British Petroleum Company 2017; United States Department of Energy, United States Energy Information Administration, Office of Energy Analysis 2016). An estimated 5% of global fuel for the transportation sector was supplied by renewable energy in 2016, which was 14% of renewable energy consumption (United States Department of Energy, United States Energy Information Administration, Office of Energy Statistics, Office of Survey Development and Statistical Integration, Integrated Energy Statistics Team 2017). Similarly, for the US in 2016, renewable energy (12%) and nuclear electric power (10%) provided about 22% of total primary energy consumption (United States Energy Information Administration 2017; United States Energy Information Administration 2017), and 14% of renewable energy consumed was used for transportation, which constituted 5% of the energy used for transportation (United States Energy Information Administration 2017; United States Energy Information Administration 2017). Liquid biofuels continued to represent the vast majority of the renewable energy contribution to the transportation sector (United States Department of Energy, United States Energy Information Administration, Office of Energy Statistics, Office of Survey Development and Statistical Integration, Integrated Energy Statistics Team 2017; United States Department of Energy, United States Energy Information Administration, Office of Energy Analysis 2016).

Renewable energy covers a wide range of resources and technologies. Renewable resources are abundant, but they differ in the difficulty involved in converting them into energy (International Energy Agency 2016). Biomass is the largest source of renewable energy in the US, representing 4.0 of 10.2 quadrillion Btu in 2016 (United States Energy Information Administration 2017). Total biomass (39%) includes wood and wood-derived materials, municipal solid waste from biogenic sources, landfill gas, sludge waste, agricultural byproducts, and other biomass; as well as

Table 7.6 US renewable energy consumption (quadrillion Btu) by type.

Energy type	2016	2017 (estimate)	2018 (forecast)
Hydroelectric power[a)]	2.477	2.851	2.604
Wood biomass	1.959	1.935	1.945
Waste biomass	0.522	0.532	0.534
Wind	2.155	2.298	2.459
Geothermal	0.226	0.230	0.236
Solar[b)]	0.587	0.776	0.949
Ethanol	1.186	1.188	1.190
Biomass-based diesel	0.289	0.292	0.312
Biofuel losses and co-products[c)]	0.796	0.817	0.814
Total	10.190	10.877	11.017

a) Conventional hydroelectric power only and does not include hydroelectricity generated by pumped storage.
b) Includes small-scale solar thermal and photovoltaic energy used in the commercial, industrial, and electric power sectors.
c) Losses and co-products from the production of fuel ethanol and biomass-based diesel.
Source: Adapted from (United States Energy Information Administration 2017).

fuel ethanol and biomass-based diesel (United States Energy Information Administration 2017). Hydroelectric energy is the second largest source of renewable energy with 24% of consumption, followed by wind (21%), solar (6%), and geothermal (2%). Biofuel losses account for the remaining 7% (United States Department of Energy 2016; United States Energy Information Administration 2017). US renewable energy consumption by type is presented in Table 7.6.

The process of converting a biomass feedstock into a product that can be used for biofuels or other bioproducts consists of several steps, including biomass production, collection or harvesting, pre-processing, transport, storage before and after transport, conversion of biomass to energy or energy carrier, transport of energy, and energy consumption. These steps depend on the type, location and source of the biomass, the conversion process, and the final energy form (International Energy Agency, Food and Agriculture Organization of the United Nations 2017). Technologies are commercially available for producing first-generation or conventional biofuels such as ethanol from fermentation of sugar and starch-based feedstocks. Ongoing optimization of the conversion process and development of by-products has reduced costs and improved yields and conversion efficiencies. Technologies for production of second-generation or advanced biofuels, which use primarily non-edible feedstocks, are either at the early stages of commercialization, such as hydrogenation of vegetable oils or thermal liquefaction of biomass to produce diesel fuel, hydrolysis of lignocellulosic materials followed by fermentation to ethanol or butanol, gasification of biomass to produce syngas for liquid hydrocarbons or for gas-powered vehicles, or at the research and development or pilot stage, such as the production and processing of algae and microalgae or growing of dedicated energy crops to serve as feedstocks (Bond et al. 2014; International Energy Agency, Food and Agriculture Organization of the United Nations 2017).

Although more sustainable production of first-generation biofuels in developing countries still has considerable potential, efforts worldwide are increasingly being focused on the production of advanced biofuels and biomaterials. The production of such advanced biofuels began at a commercial scale in 2015 (Renewable Energy Policy Network for the 21st Century (REN21) 2016; United

Nations Conference on Trade and Development (UNCTAD) 2016). Advanced biofuels can be classified by process type, estimated emissions reductions, or feedstock type. Process improvements have been a crucial factor in decreasing costs for the industry and increasing market share. Countries vary significantly not only in their technological approaches but also in the feedstocks they use for biofuel production, including the use of corn stover, sugarcane bagasse, municipal solid waste, and forestry residues, among others. Historically, the US has had the largest installed capacity for cellulosic ethanol production, followed by China, Canada, the EU, and Brazil (United Nations Conference on Trade and Development (UNCTAD) 2016).

Advances in renewable energy technologies are continuing along with ongoing energy efficiency improvements, significant progress in hardware and software to support the integration of renewable energy, and progress in energy storage development and commercialization (International Energy Agency 2016; Renewable Energy Policy Network for the 21st Century (REN21) 2016; United Nations Conference on Trade and Development (UNCTAD) 2016). While biomass is currently the most abundant single source of renewable energy in the United States, there are still economic and technological barriers, including crop yields, environmental impacts, logistical operations, and systems integration across production, harvest, and conversion, that must be overcome to exploit biomass resources for more biofuels and bioproducts (United States Department of Energy 2016).

7.5 Achieving Petroleum Replacement with Biobased Fuels and Chemicals

The petroleum refining capacity in the US in 2016 was 18.40 million barrels per day (United States Energy Information Administration 2017), which is equivalent to approximately 6.72 billion barrels per year. Roughly 7% of the total refining capacity in the US is biorefining with about 1% of that used to manufacture biobased products and the rest used for biofuels. This estimate is based on the primary feedstock sources that are used as input for biorefineries, including wet corn milling, soybeans, fats and oils, sugar beets, and sugarcane milling (Golden et al. 2015). In 2016, the Renewable Fuels Association estimated that the production of biorefineries was 15.56 billion gallons per year (Renewable Fuels Association, 2017), which is equivalent to approximately 370 million barrels per year. Fuel from several sources, such as corn, sorghum, wheat, starch, and cellulosic biomass, is included in this amount (Golden et al. 2015). Biobased products can reduce consumption of petroleum in two ways: first, by replacing chemical feedstocks previously derived from crude oil refineries with chemical feedstocks now being derived from biorefineries, currently estimated to be about 150 million gallons per year; and second by using biobased materials as substitutes for synthetic (petroleum-based) materials in widespread use, such as natural fibres for packing and insulating materials in place of synthetic foams, which currently is roughly equal to 150 million gallons per year as direct replacement. Thus, the use of biobased products is currently displacing about 300 million gallons of petroleum annually (Golden et al. 2015). Both biobased products and biofuels can be manufactured from first-generation or second-generation biomass feedstocks. First-generation products are manufactured from edible biomass, such as starch-rich or oily plants. Second-generation products utilize biomass consisting of the crop residues or other non-food sources, such as perennial grasses, and generally have a significantly higher potential for replacing fossil-based products (Golden et al. 2015).

Biomass has an advantage over other renewable energy sources because it can be readily stored until needed and can produce liquid fuel alternatives for petroleum-based transportation fuels. To prevent the exacerbation of global warming and to curtail the strong dependence of the transportation sector on imported oil and reduce crude oil imports, cost-effective technologies are needed to sustainably convert biomass into the full range of fuels and products currently derived from crude oil. Biomass conversion technologies are being developed that will produce intermediate products that can be substituted for high-value chemicals to enhance the profitability of biorefining. Sustainable production of biofuels and biobased chemicals from biomass can be integrated into the extensive existing infrastructure of oil refineries, pipelines, and distribution networks. (United States Department of Energy, Energy Efficiency & Renewable Energy, Bioenergy Technologies Office 2013). The challenge of replacing petroleum-based fuels and products can be met by innovation in developing technologies for using the wide range of biomass available, including residual biomass, as well as cultivating new energy crops, improving carbon and energy efficiency in processing, and increasing agricultural productivity (Kircher 2015).

The EU market for industrial biotechnology-derived products is expected to increase from 28 billion euros in 2013 to 50 billion euros in 2030, representing a compounded growth rate of 7% per year (BIO-TIC 2014). This growth will be largely driven by projected increases in the consumption of bioethanol and biobased plastics. New products such as aviation biofuels are likely to be commercialized in this period and gain market share. However, despite this very large projected market demand, significant barriers to full development of industrial biotechnology production in Europe still remain. The principal barrier to fully exploiting the industrial biotechnology opportunities in Europe relates to product cost-competitiveness, both compared to fossil alternatives and to equivalent products from elsewhere in the world. Cost-competitiveness is affected by many factors including the cost of feedstock, technology maturity, and the level of market support for biobased products. The EU has significant technological strength but is limited by high feedstock costs (BIO-TIC 2014).

In Europe, several product groups have been identified as being particularly promising based on their future market prospects, as well as their potential to introduce widely applicable technology and to respond to consumer needs (BIO-TIC 2014). These are: (i) advanced biofuels (advanced bioethanol and biobased jet fuels), where the EU market could be worth 14.4 billion euro and 1.4 billion euro, respectively, by 2030. For biobased jet fuels, the proportion filled by industrial biotechnology-based processes is unclear given the range of technologies available and their early stage of development. (ii) Chemical building blocks which can be transformed into a wide range of products which are either similar or offer additional functionality compared to petroleum-based products, where the EU market could reach 3.2 billion euro by 2030. (iii) Biobased polymers where the EU market could reach 5.2 billion euro in 2030. (iv) Biosurfactants derived from fermentation typically used in detergents, where the EU market could reach 3.1 million euro in 2030 (BIO-TIC 2014).

Consumer product manufacturers are eager to use renewable chemicals to meet consumer demand for environmentally preferable products. The main challenge that producers have is securing reliable, competitive supplies for large-scale product applications. Some renewable chemicals, such as succinic acid and polylactic acid, are already being produced commercially by multiple, competing companies and could potentially have commodity applications, while butanol and isoprene, are approaching the same status (Biotechnology Innovation Organization 2016). Several other renewable chemicals, such as 1,3-propanediol, propylene glycol and some diacids, are being produced at commercial levels by a single company with production tailored to niche product markets (Biotechnology Innovation Organization 2016). Forming partnerships with consumer product manufacturers or larger mid-market chemical producers is a common strategy for emerging companies commercializing new renewable chemicals. Ensuring that consumers receive the environmental, economic, and performance benefits of renewable chemicals requires an integrated effort across this entire production chain (Biotechnology Innovation Organization 2016).

Currently, biobased compounds are commercially successful either because they are available only from biomass (such as enantiomerically pure amino acids from sugar) or they are cost-competitive with their petroleum-based counterparts. The former product segment primarily consists of high-value but low volume chemicals (pharmaceuticals, cosmetics, high-performance polymers). The latter are biobased versions of existing petrochemicals with established markets. They are referred to as 'drop-ins' because they are chemically identical to existing petroleum-based chemicals. Biobased drop-in chemicals have not yet reached the market on a broad scale due to high cost. However, improvements in fuel-biorefineries in combination with the development of advanced bio-catalysts will lower the cost of sustainable chemicals, and they are expected to grow to a 22% share of sales in 2020. The production volume of biobased chemicals is and will be much smaller than biofuels because biobased chemicals are to this date not promoted as much by governmental incentives (Kircher 2015).

An obvious but critically important factor in the commercialization of biorefinery processes based on microbial fermentation is that the production process should be economically competitive with the petroleum refinery process. Thus, it is fundamental to maximize the performance of the production microorganism with respect to the titre, productivity, and yield of the desired product. In 2004, the US Department of Energy (DoE) developed a list of chemicals (Werpy et al. 2004) by selecting those that can be produced via microbial fermentation or simple chemical processes starting from sugars and can be further converted to other valuable or commodity products to replace those based on fossil resources.

This list was re-analysed in 2010 (Bozell & Petersen 2010), mainly based on literature citations (as a measure for technology development) and multiple product applicability. Among the top 10 chemicals in the 2010 report, ethanol, glycerol, biohydrocarbons, lactic acid, succinic acid, 3-hydroxypropionic acid, levulinic acid, sorbitol, and xylitol have been or are being commercialized. Clearly, the selected molecules, including carboxylic acids, alcohols, ethers, and aldehydes, contain a high degree of chemical functionality, allowing them to serve as entry points for further chemical transformation. Due to the unique oxygen-rich composition of carbohydrates, their conversion into renewable chemicals that preserve the functional groups is an advantage over the current petroleum and natural gas conversion routes. Biomass conversion with high atom efficiency is a key aspect of the competitive synthesis of chemicals and chemical-based products (Dusselier et al. 2014).

More recently, Dusselier et al. (2014) explored a new assessment method to evaluate the competitiveness and sustainability of the use of carbohydrates, preferably from lignocellulosic feedstocks, to produce chemicals based on the preservation of functionality and a high atom economy in the process. This latter method is based on the concept that highly functionalized chemicals are the most versatile and cost-effective derivatives of carbohydrates, while chemical conversions should involve maximum conservation of atoms in the desired products. Based on these criteria, a list of about 25 chemicals was proposed and several preferred reaction types were discussed. High scoring, drop-in chemicals from carbohydrates are formic acid, ethylene glycol, acetic acid, glycolic acid, and acrylic acid, while furfural, furfuryl alcohol, sorbitol, lactic and levulinic acids, and isosorbide are already exclusively produced commercially from carbohydrates in the chemical industry. Ultimately, the usefulness and potential market demand of the products determines their real value (Dusselier et al. 2014).

Choi et al. (2015) reviewed the current technological achievements and progress on the commercialization for the production of the value-added chemicals on the 2010 list and their derivatives.

Table 7.7 Top biobased chemicals in commercial production.

Chemical	Process	Companies in Commercial Production
Succinic acid	Microbial fermentation; Feedstock for other chemicals (now mainly from PET directly) including 1,4-butanediol (1,4-BDO), tetrahydrofuran, gamma-butyrolactone, and poly(butylene succinate)	Reverdia (joint venture of DSM and Roquette), Succinity (joint venture of BASF and Corbion Purac), BioAmber (joint venture of DNP Green Technology and ARD), and Myriant; these and other companies are working on conversion to various derivatives
Ethanol	Microbial fermentation; Dehydration to ethylene; polymerization of ethylene to polyethylene (PE), poly(ethylene terephthalic acid) PET, polyvinylacetate (PVA), polyvinylchloride (PVC), polypropylene, and many others.	Braskem, Dow Chemical–Mitsui Chemicals, Solvay, and India Glycols Ltd. have commercialized biobased ethylene production; first 2 are developing PE process, Solvay developing PVC process; numerous companies working on PET, ethyl lactate, ethyl acetate, acetic acid, acetaldehyde, ethylene glycol, and butadiene processes
Polylactic acid (PLA)	Microbial fermentation of dextrose to lactic acid; conversion to lactide; polymerization to PLA	NatureWorks
1,3-Propanediol	Fermentation	Dupont, Tate & Lyle Bioproducts
Epichlorohydrin	Chlorination of biobased glycerol	Solvay, Dow Chemical
Propylene glycol	Hydrogenolysis of biobased glycerol	ADM, Dow Chemical
Glycerol carbonate	Carbonation of biobased glycerol	Huntsman
2,5 Furandicarboxylic acid (FDCA)	Chemical conversion of biobased carbohydrate	Avantium is currently producing FDCA at a 40 metric ton per year pilot plant in Geleen, Netherlands and plans to start commercial production of FDCA and PEF at 50000 ton per year plant in 2017
Polyethylene furanoate (PEF) to replace PET	From patented biobased FDCA combined with plant-based monoethylene glycol (MEG)	Avantium
Furfuryl alcohol	Hydrogenation of biobased furfural	International Furan Chemicals; Shijiazhuang Worldwide Furfural and Furfuryl Alcohol and Furan Resin Co. Ltd
Sorbitol	Hydrogenation of glucose	Roquette, ADM
Isosorbide	Dehydration of sorbitol	Roquette, ADM
Xylitol	Hydrogenation of xylose	Danisco
Isoprene	Bacterial fermentation	GlycosBio
Levulinic acid	Dehydration and chemical processes from plant derived biomass	Avantium, GFBiochemicals
Itaconic acid	Microbial fermentation	Qingdao Kehai Biochemistry Co., Ltd., Itaconix Corporation
Poly(itaconic acid)	Polymerization of itaconic acid	Itaconix Corporation

Source: Adapted from (Biotechnology Innovation Organization 2016; Choi et al. 2015).

Many companies around the world have been producing and are about to produce several platform chemicals and their derivatives from biomass with increasingly higher efficiencies. Metabolic engineering has been playing the most critical role in improving the performance of microorganisms for the efficient production of a desired chemical. Biobased chemicals in commercial production are listed in Table 7.7.

Although it is likely that many chemicals that are currently produced from petroleum will be produced from biomass in the future, several challenges still remain to accelerate their large-scale commercialization. Notably, the development of a high-performance microbial strain is still the greatest challenge. Furthermore, systems metabolic engineering and other advanced tools and strategies will play significant roles in more rapidly and inexpensively developing industrial strains. High-throughput gene target identification systems such as the sRNA technology, new genome-scale computational tools, and the high-throughput robotics system will further speed up the strain development process. Efforts will be focused on cost-effective preparation of fermentable substrates from non-food biomass, the use of waste biomass, and recycling of process water (Biotechnology Innovation Organization 2016; Choi et al. 2015).

7.6 Projections of Global Demand for Biobased Fuels and Chemicals

The US Energy Information Administration (EIA) International Energy Outlook 2016 (IEO2016) Reference Case projects significant growth in worldwide energy demand over the 28-year period from 2012 to 2040 (United States Department of Energy, United States Energy Information Administration, Office of Energy Analysis 2016). The Reference Case assumes known technologies and technological and demographic trends, generally reflects the effects of current regulations, and does not anticipate new governmental incentives that have not been announced. The IEO2016 Reference Case projections do not include the effects of the recently finalized Clean Power Plan (CPP) regulations in the United States. The US EIA's preliminary analysis of the proposed CPP9 shows potential reductions of 21% (about 4 quadrillion Btu) in US coal consumption in 2020 and 24% (almost 5 quadrillion Btu) in 2040 relative to the IEO2016 Reference Case projection. With the CPP, US renewable energy use in 2020 would be 7% (about 1 quadrillion Btu) higher than in the Reference Case, and in 2040 it would be 37% (4 quadrillion Btu) higher than in the Reference Case. The US consumption of petroleum and other liquid fuels and of natural gas would be slightly lower with the CPP than in the Reference Case (United States Department of Energy, United States Energy Information Administration, Office of Energy Analysis 2016).

Projections in the US EIA Annual Energy Outlook 2017 (AEO2017) are developed using the National Energy Modelling System (NEMS), an integrated model that considers the various interactions of economic changes and energy supply, demand, and prices (United States Energy Information Administration 2017). Energy market projections are subject to much uncertainty, as many of the events that shape energy markets and future developments in technologies, demographics, and resources cannot be predicted with certainty. The Reference Case projection assumes trend improvement in known technologies, along with a view of economic and demographic trends reflecting the current central views of leading economic forecasters and demographers. It generally assumes that current laws and regulations affecting the energy sector, including sunset dates for laws that have them, are unchanged throughout the projection period. The potential impacts of proposed legislation, regulations, or standards are however not reflected in the Reference Case (United States Energy Information Administration 2017).

The overall US energy consumption remains relatively flat in the AEO2017 Reference Case, rising 5% from the 2016 level by 2040 and somewhat close to its previous peak (United States Energy Information Administration 2017). Varying assumptions about economic growth rates or energy prices considered in the AEO 2017 side cases affect projected consumption. Natural gas use increases more than other fuel sources in terms of quantity of energy consumed, led by demand from the industrial and electric power sectors. Petroleum consumption remains relatively flat as increases in energy efficiency offset growth in the transportation and industrial activity measures. Coal consumption decreases as coal loses market share to natural gas and renewable generation in the electric power sector. On a percentage basis, renewable energy grows the fastest because capital costs fall with increased penetration and because current state and federal regulations and incentives favour its use. Liquid biofuels growth is constrained by relatively flat transportation energy use and blending limitations (United States Energy Information Administration 2017).

The EIA forecasts that, in the short term, gasoline consumption will increase by 40 000 b/d (0.5%) in 2017 and by 90 000 b/d (0.9%) in 2018. In the same forecast, gasoline consumption growth is expected to decrease slightly from 2016 levels, as highway travel growth slows to 1.2% and 1.6% in 2017 and 2018, respectively, reflecting slower employment growth and rising gasoline prices (United States Energy Information Administration 2017). In the short term, continued growth in passenger and freight activity is offset by fuel efficiency increases, resulting in roughly unchanged jet fuel consumption through 2018. Consumption of distillate fuel, which includes diesel fuel and heating oil, declined by an estimated 140 000 b/d (3.5%) in 2016 as a result of warmer-than-normal winter temperatures, reduced oil and natural gas drilling, and declining coal production. However, stronger expected economic growth, increasing oil and natural gas drilling activity, and an assumption of normal temperatures should contribute to distillate fuel consumption growth of 110 000 b/d (2.9%) in 2017 and 70 000 b/d (1.9%) in 2018 (United States Energy Information Administration 2017).

Algae are considered a promising renewable fuel feedstock. Currently, the production of algae biomass is not cost-effective due to numerous technological barriers that must be overcome in order to fully use their biomass potential. Notably, the harvesting of algae does not compete with human or animal food crops and they can grow on non-arable land, as well as in freshwater, brackish, salt water or wastewater. Nevertheless, their carbon fixation capacity is much higher than other land grown plants, with harvesting cycles of less than ten days with very high yield-per-area. However, the global market is still in its emerging stage and companies are working on establishing pilot plants (United Nations Conference on Trade and Development (UNCTAD) 2016).

The production of renewable chemicals is expected to grow most rapidly in Asia in response to the region's demand for products, supply of biomass raw material, and favourable policies. Future value is dependent on the price of competing petroleum-based chemicals, the price of oil, and environmental policies that are expected to be stricter as the harmful impacts of global warming become more evident (Biotechnology Innovation Organization 2016). In the US, it has been estimated that industrial biotechnology industry revenues in 2016 reached 105 billion dollars at an average annual growth rate of 12% over the past decade, with the share of renewable chemicals representing 66 billion dollars. Biobased polymers in 2016 represented a 2% share of the total 256 million metric ton market for polymers (up from 1.5% of the 235 million metric ton market in 2011). By 2020, the 17 million metric tons of biobased polymers are expected to represent 4% of a 400 million metric ton market. The strongest growth in market demand for biobased polymers will be in food packaging and utensils. In parallel, the production capacity for biobased polyethylene terephthalate (PET) is projected to grow from 600 000 metric tons in 2013 to 7 million metric tons in 2020 (Biotechnology Innovation Organization 2016).

7.7 Potential Impacts on Price of Transportation Fuels and Chemicals Assuming Various Scenarios of World Economic Growth

Strong economic growth drives increasing demand for energy because growing economies require energy, and the petroleum products made from crude oil and other hydrocarbon liquids account for about 33% of the energy consumed worldwide (United States Department of Energy, United States Energy Information Administration, Office of Energy Analysis 2016). The world's gross domestic product (GDP), which is a measure of economic growth, is projected to rise by 3.3% per year from 2012 to 2040 (United States Department of Energy, United States Energy Information Administration, Office of Energy Analysis 2016). The fastest rates of growth are forecast to occur in emerging, non-OECD countries, where combined GDP increases by 4.2%/year (United States Department of Energy, United States Energy Information Administration, Office of Energy Analysis 2016). In contrast, in OECD countries, the average GDP grows at a much slower rate of 2.0%/year over the projection as a result of their more mature economies and slow or declining population growth trends. The strong projected economic growth rates in the non-OECD countries drive the rapid growth in future energy demand among those nations (United States Department of Energy, United States Energy Information Administration, Office of Energy Analysis 2016).

Crude oil prices are determined by global supply and demand, which are impacted by numerous factors. Among others, seasonal changes significantly influence the supply and demand balance for crude oil and its market price. Crude oil markets are stronger in the fourth quarter of the year, when global demand for heating oil is increased by cold weather and by inventory building, and weaker in late winter when demand for heating oil decreases with warmer weather. Other causes of disruption in world crude oil prices include political events, catastrophic weather events, and buyer/seller transactions making it difficult to forecast oil prices (United States Energy Information Administration 2017).

The Organization of the Petroleum Exporting Countries (OPEC) also can influence world oil supply and prices by setting production targets for its members. OPEC countries control about 73% of the world's total proved oil reserves, and in 2015, they produced 43% of the world's total crude oil. OPEC sets crude oil output targets, or quotas, for each member. Compliance of member countries with OPEC quotas is not certain because production decisions are in the hands of the individual countries. How effectively OPEC can influence oil prices depends on how unwilling or unable consumers are to move away from using oil, how competitive non-OPEC producers become when oil prices change; and how efficiently OPEC producers can supply oil compared with non-OPEC producers (United States Energy Information Administration 2017).

7.8 Projection of Energy-Related CO_2 Emissions With or Without Remediation Technology

In the United States, about 80% of all CO_2 emissions in 2012 were related to energy (United States Department of Energy, United States Energy Information Administration, Office of Energy Analysis 2016).

The transportation sector and the electric power sector are two of the largest sources of US energy-related CO_2 emissions (United States Department of Energy, United States Energy Information Administration, Office of Energy Analysis 2016). For transportation, the main mechanism to reduce emissions is increasing the stringency of fuel economy and greenhouse gas emissions standards for both light-duty vehicles and heavy trucks. For electric power, the US Environmental Protection Agency (EPA) finalized a CPP on August 3, 2015 to reduce carbon emissions from power plants for the first time (United States Energy Information Administration 2017). These standards were developed under the Clean Air Act, an act of Congress that requires the EPA to take steps to reduce air pollution that harms the public's health. The CPP, which is currently stayed pending judicial review, requires states to develop plans to reduce CO_2 emissions from existing generating units that use fossil fuels (United States Energy Information Administration 2017). Combined with lower natural gas prices and the extension of renewable tax credits, the CPP would accelerate a shift towards less carbon-intensive electricity generation. In the electric power sector, coal-fired plants would be replaced primarily with new natural gas, solar, and wind capacity, which reduces electricity-related CO_2 emissions. Transportation CO_2 emissions would remain relatively flat after 2030 as consumption and the carbon intensity of transportation fuels stay relatively constant (United States Energy Information Administration 2017).

Worldwide energy-related CO_2 emissions from the use of liquid fuels, natural gas, and coal increase in the IEO2016 Reference Case, with the relative contributions of the individual fuels shifting over time. CO_2 emissions from the consumption of liquid fuels accounted for the largest portion (43%) of global emissions in 1990. In 2012, 36% of total emissions was associated with liquid fuels, second to coal (43%) and the level is projected to remain the same through 2040 in the IEO2016 Reference Case. Coal, which is the most carbon-intensive fossil fuel, became the leading source of world energy-related CO_2 emissions in 2006, and it is projected to remain the leading source through 2040. However, its share is projected to stabilize and then decline to 38% in 2040, only slightly higher than the liquid fuels share (United States Department of Energy, United States Energy Information Administration, Office of Energy Analysis 2016).

Much of the growth in energy-related CO_2 emissions is attributed to developing non-OECD countries, many of which continue to rely heavily on fossil fuels to meet the rapid growth of energy demand. In the IEO2016 Reference Case, total non-OECD emissions in 2040 are projected to be 29.4 billion metric tons, or about 51% higher than the 2012 level. In comparison, total OECD emissions are projected to be 13.8 billion metric tons in 2040, or about 9% higher than the 2012 level. The IEO2016 Reference Case estimates do not include effects of the recently finalized CPP regulations in the United States, which would reduce projected US energy-related CO_2 emissions in 2040 by 0.5 billion metric tons, based on EIA's analysis of the CPP proposed rule 8, from 5.3 billion metric tons in 2012 to 5.5 billion metric tons in 2040 without the CPP or 5.0 billion metric tons with the CPP (United States Department of Energy, United States Energy Information Administration, Office of Energy Analysis 2016).

7.9 Government Impact on Demand for Biofuels and Biobased Chemicals

Numerous factors simultaneously limit and accelerate the growth of the global market for biofuels and biobased chemicals. Among them are government subsidies, national commitments to mitigate climate change, oil prices, and other political/environmental factors. Countries that have supportive regulations and governmental incentives for research in this area, such as North America and Europe, are gaining the majority market share (United Nations Conference on Trade and Development (UNCTAD) 2016). Biofuel production faces different challenges around the world. Africa suffers from overestimated expectations and agricultural difficulties with some feedstocks, but countries such as Mali, Ghana, and Nigeria have established mandates for the use of biofuels. India has a biofuel target of 10%, and the government has created incentives to production in the form of capital subsidies, tax breaks, and public bidding processes to achieve it. In Latin America, fuel demand is rising and fossil-fuel subsidies are being slowly phased out, while novel biofuel models are being developed, since the smaller countries with less land cannot replicate Brazil's experience (United Nations Conference on Trade and Development (UNCTAD) 2016). Economically, biofuels sustain about 1.7 million jobs around the world, including 845 000 in Brazil alone (about 268 400 jobs in sugar cane and 190 000 in ethanol processing in 2014; also 200 000 indirect jobs in equipment manufacturing and 162 600 jobs in biodiesel in 2015) and 282 000 in the United States (227 562 jobs for ethanol and 49 486 jobs for biodiesel in 2015) (Renewable Energy Policy Network for the 21st Century (REN21) 2016).

In the United States, the Agriculture Act of 2014 (H.R. 2642 2014 Farm Bill) has strengthened the USDA BioPreferred program's mandatory federal purchasing initiative to require biobased-only procurement targets for supplies and services in Federal agencies, reporting of biobased products by procuring agencies, audits to ensure compliance, and a study of the economic impacts of the program. It now includes forest products and will assist landowners in determining whether products are eligible for the 'USDA Certified Biobased Product' label. Mandatory funding of $3 million per year is provided for 2014–2018. The Biorefinery, Renewable Chemical, and Biobased Product Assistance Program adds coverage for the production of renewable chemicals, manufactured biobased products, and other biorefinery byproducts to encourage advanced cellulosic biofuel technologies and renewable chemical/biobased product manufacturing. The Biomass Crop Assistance Program (BCAP) promotes the production of cellulosic biofuels to meet the Renewable Fuel Standard (RFS) and now includes material collected or harvested directly from the National Forest System, Bureau of Land Management land, non-Federal land, and tribal lands in a manner that is consistent with conservation stewardship plans. BCAP provides funding for crop and woody biomass production; and collection, harvest, storage, and transportation, as well as the establishment of crop and forest lands for biomass production (United States Department of Agriculture 2014).

On November 23, 2016, the US EPA finalized a rule setting volume requirements and associated percentage standards that apply under the RFS program in calendar year 2017 for cellulosic biofuel, biomass-based diesel, advanced biofuel, and total renewable fuel. EPA also finalized the volume requirement for biomass-based diesel for 2018. The final volumes represent continued growth over historic levels. The final standards meet or exceed the volume targets specified by Congress for total renewable fuel, biomass-based diesel, and advanced biofuel (United States Environmental Protection Agency 2016). The US renewable fuel volume requirements for 2014–2018 under the RFS program are presented in Table 7.8.

Table 7.8 US renewable fuel volume requirements for 2014–2018 (billion gallons).

	2014	2015	2016	2017	2018
Cellulosic biofuel	0.033	0.123	0.230	0.311	n/a
Biomass-based diesel	1.63	1.73	1.9	2.0	2.1
Advanced biofuel	2.67	2.88	3.61	4.28	n/a
Renewable fuel	16.28	16.93	18.11	19.28	n/a

Source: Adapted from (United States Environmental Protection Agency 2016).

Total renewable fuel volumes are increased by 1.2 billion gallons from 2016 to 2017, a 6% increase. Advanced renewable fuel – which requires 50% life cycle carbon emissions reductions – grows by roughly 0.70 billion gallons between 2016 and 2017. Non-advanced or 'conventional' renewable fuel increases in 2017 meet the 15 billion gallon congressional target for conventional fuels. The standard for biomass-based biodiesel – which must achieve at least 50% lifecycle greenhouse gas emission reductions compared to petroleum-based diesel – grows by 0.10 billion gallons. The required volume of biomass-based diesel for 2017 is twice as high as the minimum congressional target. Cellulosic biofuel – which must achieve at least 60% lifecycle greenhouse gas emissions reductions – grows by 35% over the 2016 standard. The advanced biofuel standard, which is comprised of biomass-based diesel, cellulosic biofuel, and other biofuel that achieves at least 50% lifecycle greenhouse gas emissions reductions, increases by 19% over the 2016 standard. In the Final Rule, the EPA states: 'The final standards are expected to continue driving the market to overcome constraints in renewable fuel distribution infrastructure, which in turn is expected to lead to substantial growth over time in the production and use of renewable fuels'. (United States Environmental Protection Agency 2016).

7.10 Perspectives

The consumption of renewable energy worldwide is the highest in the history of mankindin nominal terms, however not in relative terms. Liquid biofuels continue to represent the vast majority of the renewable energy contribution to the transport sector. Biomass is the largest source of renewable energy, but economic and technological hurdles still limit efforts to exploit biomass resources for more biofuels, biopower, and bioproducts. To effectively reduce crude oil use, biorefineries must be able to convert biomass into the full range of fuels and chemicals currently derived from petroleum. Metabolic engineering has been playing the most critical role in improving microorganisms for the efficient production of a desired fuel or chemical.

Numerous factors simultaneously affect the growth of the global market for biofuels and biobased chemicals. These include potential impacts on food and environment, availability of natural resources, government subsidies, national commitments to mitigate climate change, oil prices, and other political/environmental factors. Countries worldwide are turning towards increased production of biofuel and biobased chemicals to improve energy security, advance economic development, and address environmental concerns.

Because transportation accounts for nearly three-fourths of total US petroleum consumption, using electric vehicles (EVs) can have an impact on production of biofuels and biotechnology-derived commodity chemicals. Today only a small percentage of US oil demand is due to electricity generation, thus charging the battery of an EV will not increase oil consumption even with large-scale deployment of these vehicles. The strategic importance of vehicle electrification derives from the lower cost-per-mile of electricity compared to gasoline or diesel, even when oil prices are relatively low. Additionally, EVs produce no tailpipe emissions when in all-electric mode. In areas using relatively low-polluting energy sources for electricity production, EVs have a life cycle emissions advantage over similar conventional vehicles. In regions depending on fossil fuels for electricity generation, EVs may not demonstrate such a strong benefit. The upfront cost of electrified vehicles is higher, but prices are likely to decrease as production volumes increase and battery technologies mature. Also, tax credits for EVs have been enacted into law as a policy tool aimed at accelerating the move to this technology past early adopter reluctance and into the mass market.

References

Biotechnology Innovation Organization (2016). *Advancing the Biobased Economy: Renewable Chemical Biorefinery, Commercialization, Progress, and Market Opportunities, 2016 and Beyond*. Washington, DC: Biotechnology Innovation Organization (BIO).

BIO-TIC. 2014. The bioeconomy enabled: a roadmap to a thriving industrial biotechnology sector in Europe. http://www.industrialbiotech-europe.eu/new/wp-content/uploads/2015/08/BIO-TIC-roadmap.pdf [accessed 8 July 2017]

Bond, J.Q., Upadhye, A.A., Olcay, H. et al. (2014). Production of renewable jet fuel range alkanes and commodity chemicals from integrated catalytic processing of biomass. *Energy & Environmental Science* 7 (4): 1500–1523. https://doi.org/10.1039/c3ee43846e.

Bozell, J.J. and Petersen, G.R. (2010). Technology development for the production of biobased products from biorefinery carbohydrates – the US Department of Energy's "top 10" revisited. *Green Chemistry* 12: 539–554.

British Petroleum Company (2017). *BP Statistical Review of World Energy, June 2017*, 66e. London: British Petroleum Co.

Choi, S., Song, C.W., Shin, J.H., and Lee, S.Y. (2015). Biorefineries for the production of top building block chemicals and their derivatives. *Metabolic Engineering* 28: 223–239.

Dusselier, M., Mascal, M., and Sels, B.F. (2014). Top chemical opportunities from carbohydrate biomass: a chemist's view of the biorefinery. *Topics in Current Chemistry* https://doi.org/10.1007/128_2014_544.

Golden, J.S., Handfield, R.B., Daystar, J., and McConnell, T.E. (2015). An economic impact analysis of the U.S. biobased products industry: a report to the congress of the United States of America. A joint publication of the Duke Center for Sustainability & Commerce and the supply chain resource cooperative at North Carolina State University. *Industrial Biotechnology* 11 (4): 201–209.

International Energy Agency (2016). *World Energy Outlook 2016, Part B: Special Focus on Renewable Energy*. IEA Publications https://www.iea.org/media/publications/weo/WEO2016SpecialFocusonRenewableEnergy.pdf [accessed 8July2017].

International Energy Agency, Food and Agriculture Organization of the United Nations (2017). *How2Guide for Bioenergy, Roadmap Development and Implementation*. IEA Publications https://www.iea.org/publications/freepublications/publication/How2GuideforBioenergyRoadmapDevelopmentandImplementation.pdf.

Janda K, Stankus E. 2017. Quantification of biofuels potential of post-soviet countries in the context of global biofuels development. Munich Personal Research Papers in Economics (RePEc) Archive (MPRA) Paper No. 76728, posted 12 February 2017. https://mpra.ub.uni-muenchen.de/76728

Kircher, M. (2015). Sustainability of biofuels and renewable chemicals production from biomass. *Current Opinion in Chemical Biology* 29: 26–31.

Renewable Energy Policy Network for the 21st Century (REN21). 2016. Renewables 2016 Global Status Report. Paris: REN21 Secretariat.

Renewable Fuels Association (2017), Biorefinery Locations, http://www.ethanolrfa.org/resources/biorefinery-locations [accessed 31Aug2017]

United Nations Conference on Trade and Development (UNCTAD). 2016. Second Generation Biofuel Markets: State of Play, Trade and Developing Country Perspectives. UNCTAD/DITC/TED/2015/8 UNITED NATIONS PUBLICATION Copyright © United Nations http://unctad.org/en/PublicationsLibrary/ditcted2015d8_en.pdf [accessed 27June2017]

United States Department of Agriculture. 2014. Agricultural Act of 2014: Highlights and Implications. https://www.ers.usda.gov/agricultural-act-of-2014-highlights-and-implications/energy [accessed 27June2017]

United States Department of Energy (2016). 2016 Billion-ton report: advancing domestic resources for a thriving bioeconomy, volume 1: economic availability of feedstocks. In: *ORNL/TM-2016/160* (eds. M.H. Langholtz, B.J. Stokes and L.M. Eaton), 448. Oak Ridge, TN: Oak Ridge National Laboratory https://doi.org/10.2172/1271651, http://energy.gov/eere/bioenergy/2016-billion-ton-report [accessed 8July2017].

United States Department of Energy, Energy Efficiency & Renewable Energy, Bioenergy Technologies Office. 2013. Replacing the Whole Barrel To Reduce U.S. Dependence on Oil, July 2013. DOE/EE-0920

United States Department of Energy, United States Energy Information Administration, Office of Energy Analysis. 2016. International Energy Outlook 2016, With Projections to 2040, May 2016. DOE/EIA-0484(2016). https://www.eia.gov/outlooks/ieo/pdf/0484(2016).pdf [accessed 8July2017] https://www.eia.gov/outlooks/aeo/data/browser/#/?id=15-IEO2016&cases=Reference&sourcekey=0 [interactive table; accessed 27August2017]

United States Department of Energy, United States Energy Information Administration, Office of Energy Statistics, Office of Survey Development and Statistical Integration, Integrated Energy Statistics Team. 2017. Monthly Energy Review June 2017. DOE/EIA-0035(2017/06)

United States Energy Information Administration. January 2017. Annual Energy Outlook 2017 With Projections to 2050 (AEO2017), Independent Statistics and Analysis. https://www.eia.gov/outlooks/aeo/#AEO2017 [accessed 8July2017]

United States Energy Information Administration. January 2017. Energy Explained, Nonrenewable Sources, Oil: Crude and Petroleum Products Explained, Oil Prices and Outlook. https://www.eia.gov/energyexplained/index.cfm?page=oil_prices [accessed 8July2017]

United States Energy Information Administration. January 2017. Energy Explained, Nonrenewable Sources, Oil: Crude and Petroleum Products Explained, Use of Oil. https://www.eia.gov/energyexplained/index.cfm?page=oil_use [accessed 8July2017]

United States Energy Information Administration. May 2017. US Energy Facts Explained, Consumption and Production. https://www.eia.gov/energyexplained/?page=us_energy_home [accessed 8July2017]

United States Energy Information Administration. May 2017. US Short-Term Energy Outlook (STEO), Independent Statistics and Analysis.

United States Environmental Protection Agency. December 12, 2016. Renewable Fuel Standard Program: Standards for 2017 and Biomass-Based Diesel Volume for 2018. Final Rule. Federal Register Vol. 81, No. 238 (40 CFR Part 80), pp 89746–89804 https://www.gpo.gov/fdsys/pkg/FR-2016-12-12/pdf/2016-28879.pdf [accessed 8July2017]

Werpy, T., Petersen, G., Aden, A. et al. (2004). *Top value added chemicals from biomass, Vol 1: Results of screening for potential candidates from sugars and synthesis gas*, 76. Washington, DC: Pacific Northwest National Laboratory, National Renewable Energy Laboratory and Department of Energy.

Part II

Chemicals and Transportation Fuels from Biomass

8

Sustainable Platform Chemicals from Biomass

Ankita Juneja and Vijay Singh

Agricultural and Biological Engineering, University of Illinois Urbana-Champaign, Urbana, IL 61801, USA

CHAPTER MENU

8.1 Introduction

With increasing concerns about environmental safety, rapid economic growth, limited fossil reserves, and volatile oil prices, there is rising interest in the production of fuel and industrial chemicals from renewable resources and biomass. Over the last several decades, significant efforts have been made to develop technologies and approaches for successful commercial scale biofuel production. Currently, the US produces approximately 15.3 billion gallons of ethanol every year, and most of it is derived from corn (RFA 2017). To meet the supply without interfering with food sources, lignocellulosic biomass, which is abundant in nature, can play an important role. Similar to biofuel production, during the past few years, much progress has been made in the production of renewable value chemicals and biomaterials.

Starch- and protein-derived products have long been available in the market, however, current efforts are focused on sustainable production of platform chemicals, which are the building blocks of several other industrial high-value chemicals and biomaterials (Kamm 2007). It is predicted that by 2025, renewable sources would contribute about 25% of all the raw materials used in the chemical industry, compared to only 3–4% in 2010 (Fiorentino et al. 2016; Kamm 2007). Wide varieties of renewable feedstock have been used for sugar production as a feedstock for biochemicals, including corn and lignocellulosic biomass. Despite intense research and development activities in the arena, the production process of sugars from lignocellulosic biomass remains, to this date, more challenging compared to corn, due to the recalcitrant structure of this material. Whereas the enzymes for achieving the breakdown of starch to glucose are commercialized at the industrial scale by several companies including Novozymes (Denmark) and DowDuPont (Wilmington, USA), the enzymes

Green Energy to Sustainability: Strategies for Global Industries, First Edition.
Edited by Alain A. Vertès, Nasib Qureshi, Hans P. Blaschek and Hideaki Yukawa.

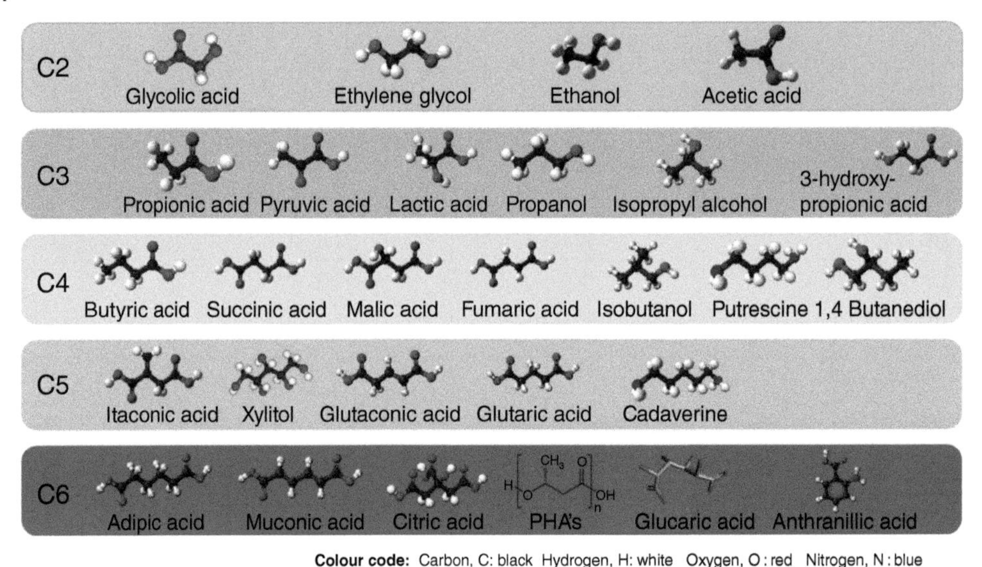

Figure 8.1 Platform chemicals that are shifting from petro-based to bio-based production.

for biomass hydrolysis are still at the research or pre-commercialization phase due to the varying compositions of the different biomass raw materials that are being considered by the industry as potential large scale feedstock. In the case of lignocellulosic biomass, enzymes for the release of sugars represent the key cost centre (Kumar and Murthy 2011). These concerns still make corn a more practical option for the production of sugars and their commercial-scale conversion into industrial biochemical. As a result, from an ever increasing market demand driven mostly by the biofuel industry, the production of corn has increased from 10.75 billion in 2012 to 15.15 billion bushels in 2016, whereas the price of corn has declined in parallel from US$6.89 in 2012 to US$3.40/bu. in 2016 (http://www.worldofcorn.com/). Increase in availability and decline in price has made corn more attractive as a feedstock for high value chemical production.

The manufacture of an increasing number of different platform chemicals originally produced with fossil sources is now being attempted with renewable sugar sources using microorganisms (Figure 8.1). The US Department of Energy has listed 12 high potential bio-based platform chemicals, selected by integrating multiple factors, including biomass precursors, building blocks, process platforms, intermediates, final products, and applications (Werpy and Petersen 2014). These chemicals can be produced either through conventional chemical processes or through the biological conversion of sugar. In spite of the very large market potential of both commodity chemicals and high value products, the current production volumes of bio-based chemicals remain low because of technical challenges in the conversion processes and still unfavourable production costs. Nonetheless, this sluggish uptake of the new manufacturing technology of chemicals is bound to accelerate in the coming years with the development of advanced technologies, the identification of novel feedstocks, the large scale production of tailor-made engineered crops, and the increasing quantities of starch that are available for the bioprocessing industry due to improved corn yields (Dale 2017), the footprint of bio-based chemicals is expected to steadily grow in the coming years. According to a recent report, the global market of renewable chemicals (including alcohols) is expected to reach up to US$84.3 billion by 2020 (ResearchandMarkets 2015). Assuming optimal market conditions, the production of bulk chemicals from renewable feedstocks could reach up to 113 million

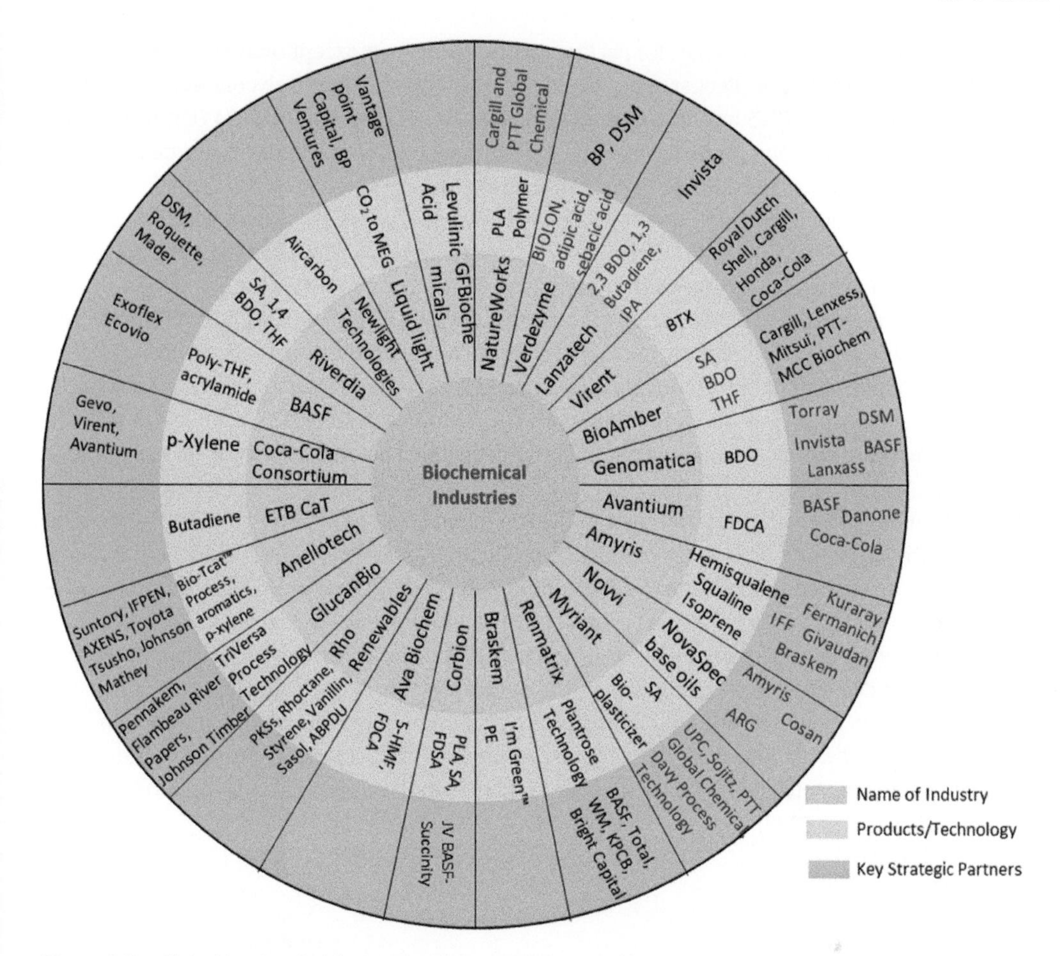

Figure 8.2 Global leaders in biochemical/biomaterial production.

tonnes (38% of total organic chemical production) by 2050 (Fiorentino et al. 2016). Many companies have invested in the production of biobased chemicals, either standalone or in strategic partnership with other companies. Figure 8.2 provides information on different companies that have implemented various initiatives for the commercial production of sustainable platform chemicals. In this chapter, we review current research projects for the production of commercially important platform organic acids and alcohols.

8.2 2-Carbon

This section explains the industrial importance, production, and scope of glycolic acid, one of the largest produced C2 platform chemicals from sugar.

8.2.1 Glycolic Acid

Glycolic acid ($C_2H_4O_3$) is a widely used chemical compound with applications in the cosmetic, pharmaceutical, and biopolymer industries. Glycolic acid can be polymerized to polyglocolic acid

(PGA), which is an excellent polymer for packing material due to its exceptional gas-barrier properties and mechanical strength (Becker et al. 2015). The global market size of glycolic acid was valued at USD 159.6 million in 2015 and is projected to reach USD 415 million by 2024 (GrandViewResearch 2016a). Such a strong projected expansion of market demands a commensurate growth in the global production capacity. Some of the current methods of glycolic acid production involve acid catalysis of carbon monoxide and formaldehyde under conditions of high temperature and pressure (Shattuck 1948), and the chemical hydrolysis of chloroacetic acid or glycolonitrile (Yunhai et al. 2006), or biocatalytic synthesis from ethylene glycol (Gao et al. 2014) or formaldehyde and hydrogen cyanide (Panova et al. 2007). These methods provide yields of greater than 99%; however, hazardous chemicals and effluents are used and produced in the synthesis process. Bio-based routes have thus been investigated for producing glycolic acid from renewable sources, using the yeasts *Saccharomyces cerevisiae* and *Kluyveromyces lactis* (Koivistoinen et al. 2013) of the gram positive bacterium *Corynbecterium glutamicum* and the gram negative bacterium *Escherichia coli* (Zahoor et al. 2014).

These developments were demonstrated by Koivistoinen et al. (Koivistoinen et al. 2013), who engineered the glyoxylate cycle for glycolic acid production in *S. cerevisiae.* Specifically, this was accomplished through the expression of a high affinity glyoxylate reductase (GLYR1) and the deletion of the two malate synthase encoding genes (MLS1) to prevent the conversion of glyoxylate into malate (Figure 8.3). The gene encoding the cytosolic form of the isocitrate dehydrogenase (IDP2) was also deleted in combination with the overexpression of the isocitrate lyase encoding gene (ICL1) for the increased production of glycolic acid. Similar genetic modifications were also performed in *K. lactis.* Commercially, METabolic EXplorer (METEX) (Saint-Beauzire, France), in agreement with Roquette (Nord-Pas-de-Calais, France), produces glycolic acid by fermentation (Lambert and Lapeyre 2009).

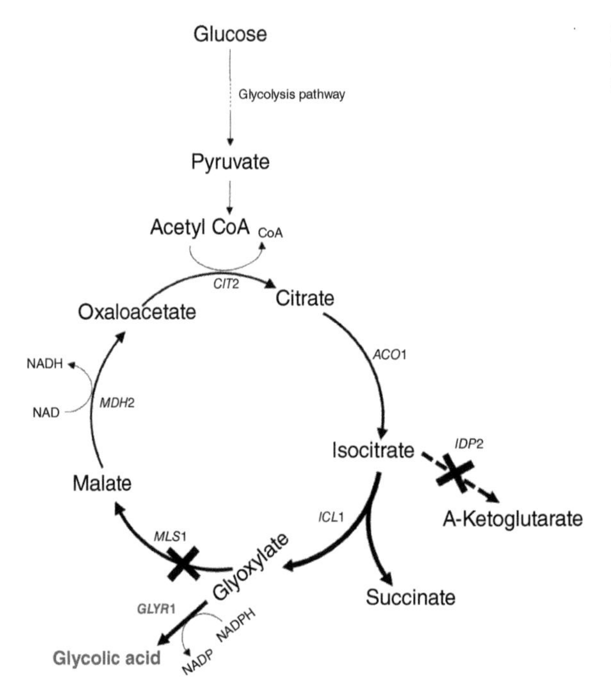

Figure 8.3 Engineering of the glyoxylate cycle in *Saccharomyces cerevisiae* for glycolic acid production (Koivistoinen et al. 2013).

8.3 3-Carbon

This section explains the industrial importance, production, and scope of a few important C3 platform chemicals derived from sugar: propionic acid, pyruvic acid (PA) and lactic acid.

8.3.1 Propionic Acid

Propionic acid (C_2H_5COOH) is a C3 carboxylic acid with widespread industrial applications as a specialty chemical; notably, it is used in food and feed preservatives and used in the production of cellulose fibres, herbicides, perfumes, and pharmaceuticals. The global market for propionic acid was 399.4 kilo tons in 2013 and is expected to exceed 470.0 kilo tons (USD 1.53 billion) by 2020 (GrandViewResearch 2015b). Currently, propionic acid is produced by petrochemical routes via the Reppe process or the Larson process (Boyaval and Corre 1995). On the other hand, propionic acid is naturally produced by a group of bacteria of the genus *Propionibacterium* when they are fermented anaerobically over a variety of substrates including glucose, glycerol, lactate, and sucrose (Wang et al. 2015). The pathway in propionibacteria to produce propionic acid is shown in Figure 8.4.

Other than pure sugars, low cost feedstock such as corn fibre and corn steep liquor (two low-value co-products from the corn wet milling industry), whey (the main co-product of the cheese industry), and crude glycerol (the co-product from biodiesel production) have been utilized (refer to Table 8.1 for details). Production yields obtained with these and some other low value substrates are summarized in Table 8.1. Metabolic enhancements have also been attempted for increasing the industrial yields of propionic acid. Theoretically, the yield of propionic acid from glucose via glycolysis is limited to 0.55 g/g glucose; however, it is possible to exceed this limit by utilizing the pentose phosphate pathway (PPP) and carbon fixation in the carboxylic acid pathway (Wang and Yang 2013). However, the propionic acid yields attained with *Propionibacterium* have been observed to be typically lower than 0.55 g/g, which shows that this group of bacteria has low carbon fixation rates and a tendency to rely primarily on glycolysis for glucose catabolism. As a result, engineering *P. shermanii*, was attempted, for example by over expressing three biotin-dependent carboxylases (pyruvate carboxylase [PYC], methylmalonyl-CoA decarboxylase [MMD], and methylmalonyl-CoA carboxyltransferase [MMC]) in the dicarboxylic acid pathway. The mutants overexpressing these carboxylases showed significantly improved yields in propionic acid compared to the wild type (Wang et al. 2015).

8.3.2 Pyruvic Acid

Pyruvate ($C_3H_4O_3$) is a key intermediate in multiple metabolic pathways in all organisms and a chemical precursor for various compounds, polymers and bioproducts. Industrial applications of

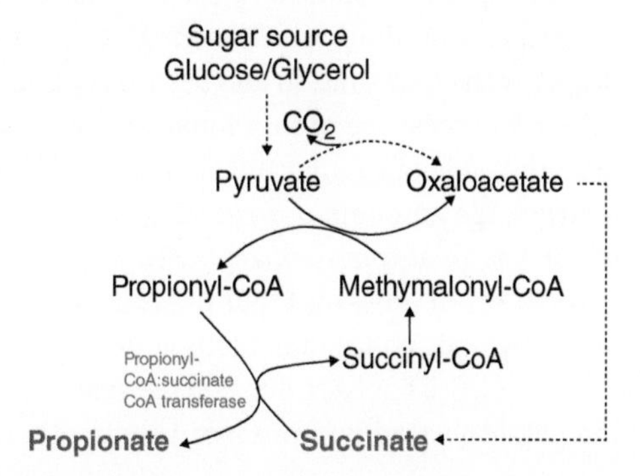

Figure 8.4 The propionic acid metabolic pathway in *Propionibacterium*. Source: Adapted from Wang et al. 2015.

Table 8.1 Production of propionic acid from low cost renewable substrates.

Substrate	Bacterium	Process	Titre (g/l)	Productivity (g/l·h)	Yield (g/g)	References
Cheese whey	*Propionibacterium acidipropionici*	Continuous with cells recycled by ultrafiltration	25	14.3	–	Boyaval and Corre (1987)
Whey permeate	*Propionibacterium acidipropionici*	Recycle-batch with fibrous bed bioreactor	65	0.22–0.68	0.5	Yang et al. (1995)
Cane molasses	*Propionibacterium freudenreichii*	Fed-batch with fibrous bed bioreactor	91.9	0.36	0.46	Feng et al. (2011)
Corn meal	*Propionibacterium acidipropionici*	Batch with fibrous bed bioreactor	40	2.12	0.58	Huang et al. (2002)
Corncob molasses	*Propionibacterium acidipropionici*	Fed-batch	71.8	0.28	–	Liu et al. (2012)
Jerusalem artichoke	*Propionibacterium acidipropionici*	Fed-batch with fibrous bed bioreactor	68.5	1.55	0.43	Liang et al. (2012)
Sugarcane bagasse	*Propionibacterium acidipropionici*	Fed-batch with fibrous bed bioreactor	58.8	0.38	0.37	Zhu et al. (2012)
Glycerol	*Propionibacterium acidipropionici*	Fed-batch with fibrous bed bioreactor	106	0.04	0.56	Zhang and Yang (2009)

PA encompass a wide range, such as the pharmaceutical, food, agrochemical, polymer, and cosmetic industries (Li et al. 2001a). It was also observed that diet supplementation with PA increases fat loss and minimizes the associated loss of body protein (Stanko et al. 1992). Pyruvate is found at the biochemical junction of glycolysis and the tricarboxylic acid cycle. The biochemical formation of pyruvate from glucose via glycolysis generally follows the following stoichiometric equation (Maleki and Eiteman 2017):

$$\text{Glucose} + 2\text{NAD} + 2\text{Pi} + 2\text{ADP} \rightarrow 2\text{pyruvate} + 2\text{NADH} + 2\text{ATP}$$

Under aerobic conditions, PA supplies energy to cells via the citric acid cycle, whereas under anaerobic conditions it is fermented to produce lactate. On the industrial scale, PA is usually produced by the dehydration and decarboxylation of tartaric acid or by the catalytic oxidation of lactic acid in the presence of iron ions, phosphates, and palladium (Li et al. 2001a). Due to its location at a vital node of cellular metabolism, it is difficult to obtain strains that can accumulate extracellularly large amounts of pyruvate, however, several such strains have been successfully derived, which has great commercial potential. Remarkably, the cost of obtaining pyruvate by fermentation is about seven times and four times lower than the cost of pyruvate obtained by chemical method and enzymatic conversion method, respectively (Li et al. 2001a).

A lot of work has been done on the yeast strain *Torulopsis glabrata*, which was modified to weaken the pathways leading from pyruvate to other products (Figure 8.5) (Miyata and Yonehara 1996).

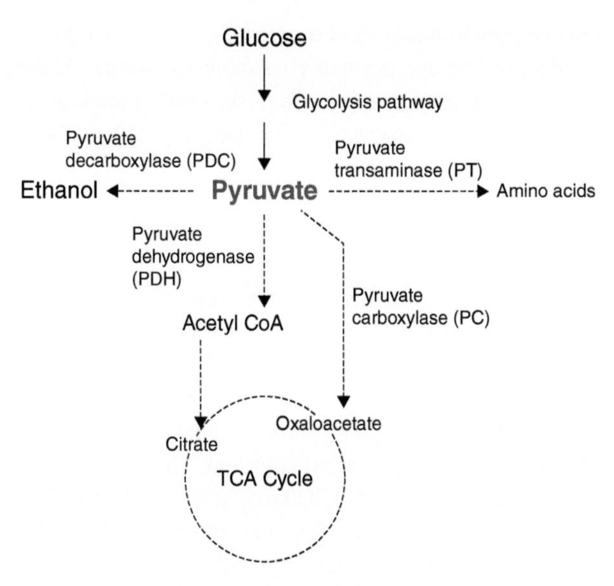

Figure 8.5 Pyruvate production pathway in engineered *Torulopsis glabrata*. Weakened pathways are shown in dotted line. Source: Adapted from Miyata and Yonehara 1996.

Table 8.2 Pyruvic acid production performance of the most relevant microbial production hosts.

Organism	Substrate	Process operation	Max. titre (g/l)	Max. yield (g/g)	References
S. cerevisiae	Glucose	Fed-batch	135	0.54	van Maris et al. (2004)
E. coli	Glucose	Fed-batch	90	0.68	Zhu et al. (2008)
C. glutamicum	Glucose	Fed-batch	44	0.47	Wieschalka et al. (2013)
T. glabrata	Glucose	Fed-batch	69	0.62	Li et al. (2001b)

Other microorganisms modified for PA production include *S. cerevisiae* (van Maris et al. 2004), *E. coli* (Causey et al. 2004; Tomar et al. 2003; Zelić et al. 2004), *Bacillus megaterium* (Hollmann and Deckwer 2004), *Trichosporon cutaneum* (Wang et al. 2002), *Debaryomyces hansenii* (Yanai et al. 1994), and *Yarrowia lipolytica* (Morgunov et al. 2004; Otto et al. 2012; Rywińska et al. 2013; Yin et al. 2012). The yields and concentrations of PA produced using microbes is listed in Table 8.2. Maximum production titres of 90 and 135 g/l have been reported with *E. coli* (Zhu et al. 2008) and *S. cerevisiae* (van Maris et al. 2004) respectively. The yields of pyruvate with the yeast *Blastobotrys adeninivorans* were 0.77 g/g of glycerol as substrate (Rywińska et al. 2013).

Since pyruvate is biochemically located at the end of glycolysis and at the beginning of the tricarboxylic acid pathway, metabolic engineering of strains for improved production focused mainly on restricting or eliminating the further metabolism of pyruvate, preventing by-product formation, and increasing the rate of glycolysis (Maleki and Eiteman 2017). Furthermore, any metabolic modification must account for system constrains such as the need to provide cells with sufficient reducing equivalents such as NADPH (nicotinamide adenine dinucleotide phosphate) and biochemical precursor molecules to satisfy biochemical demand. An engineered *E. coli* with an increased glycolysis rate of generated 31 g/l pyruvate from glucose in 32 h at a yield of 0.64 g/g in batch culture

(Yokota et al. 1997; Yokota et al. 1994). However, the biomass yield decreased from 0.26to 0.14 g/g. Future metabolic engineering of strains could focus on tolerance to high osmolarity, low pH, higher temperature, and multiple stresses. Although a vast amount of research has used glucose as the carbon source, it will be interesting to see the processes and strains, which would be able to utilize mixed substrates for PA production.

8.3.3 Lactic Acid

Lactic acid (LA) ($CH_3CHCOOH$) is a commodity organic acid with a very wide range of applications in the food, pharmaceuticals, textile, and cosmetics industries. The global market of lactic acid is projected to be worth US$9.8 billion by 2025 (GrandViewResearch 2017b). The demand for lactic acid production is increasing continuously due to its extensive application as a precursor of polylactic acid, which is a very promising biodegradable polymer. Archer Daniels Midland Company (Chicago, USA), NatureWorks (Minnetonka, USA), Corbion (Amsterdam, The Netherlands), Galactic (Brussels, Belgium), and other companies (such as CCA [Changzhou] Biochemical Co. Ltd., Henan Jindan Lactic Acid Co. Ltd., and Musashino Chemical Co. Ltd) are major manufacturers of lactic acid. Lactic acid can be produced by chemical synthesis by the hydrolysis of lactonitrile or by microbial fermentation; however, 90% of current LA is produced by the latter (Benninga 1990).

LA is usually produced by fungi and bacteria. Fungi can release extracellular amylases, which is beneficial to avoid a hydrolysis step for starchy materials. Fungi of the *Rhizopus* genus have been extensively employed with starches from corn, wheat, potato, pineapple, and hydrolyzed corn cobs, pine wood, and waste paper (Martinez et al. 2013). Lactic acid bacteria have the natural ability to produce lactic acid by sugar fermentation. For industrial applications, bacteria of the genus *Lactobacillus* are most diversely used due to their high growth rate and productivity, along with special resistance against an acidic environment (Kylä-Nikkilä et al. 2000).

A very high titre of 280 g/l lactic acid was reported when using immobilized *Rhizopus oryzae* on glucose (Yamane and Tanaka 2013). With the overexpression of five glycolytic genes of the glycolytic pathway encoding glucokinase (GLK), glyceraldehyde phosphate dehydrogenase (GAPDH), phosphofructokinase (PFK), trio sephosphate isomerase (TPI), and bisphosphate aldolase (FBA), the bacterium *C. glutamicum* achieved high titres of 195 g/l D-LA under anaerobic conditions (Tsuge et al. 2015). *S. cerevisiae* and *E. coli* have also been used to achieve maximum titres of 138 and 122 g/l respectively (Becker and Wittmann 2015). In an attempt to produce low cost LA, a mixture of sugars, including xylose, arabinose, and glucose, in corncob molasses was utilized by *Bacillus* sp. strain XZL9 for L-lactic acid production of 74.7 g/l in fed-batch fermentation (Wang et al. 2010).

8.4 4-Carbon

This section explains the industrial importance, production, and scope of a few important C4 platform chemicals derived from sugar: butyric acid, succinic acid, malic acid, and putrescine.

8.4.1 Butyric Acid

Butyric acid ($CH_3CH_2CH_2$-COOH) is an important specialty chemical with wide range of applications in the chemicals, food, pharmaceuticals, and plastic industry. The global butyric acid market is expected to reach US$289.3 million by 2020 (MarketsandMarkets n.d.). Commercially, it is synthesized from petrochemical routes by the oxidation of butyraldehyde using propylene, or by sugar fermentation. Butyric acid production occurs from the glycolytic pathway via condensation of acetyl-CoA to acetoacetyl-CoA. Several bacteria of the genera *Clostridium* (Van Andel et al. 1985; Wu and Yang 2003), *Butyrivibrio* (Bryant and Small 1956), and *Enterococcus* (Centeno et al. 1999) have been shown to produce butyric acid, however, specific strains of *Clostridium* such as *C. butyricum*, *C. beijerinckii*, *C. acetobutylicum*, *C. tyobutyricum*, *C. populeti*, and *C. thermobutyricum* are considered to be superior strains and thus are used extensively. These strains have been shown to produce high titres of butyric acid with acetic acid as a byproduct (Jang et al. 2012). The necessity to reduce byproduct formation to enhance the butyric acid yield has led researchers to engineer strains by deleting the acetate producing pathway (Liu et al. 2006).

Bacteria producing butyric acid can utilize a large variety of substrates, however, the most common substrates used for research are glucose (Wu and Yang 2003) and xylose (Zhu and Yang 2004) as pure sugars, and wheat straw (Baroi et al. 2015), sugarcane bagasse (Wei et al. 2013) and corn straw (Li et al. 2011b) as lignocellulosic feedstock. Batch (titre up to 71.6 g/l (Baroi et al. 2015)), fed-batch (titre up to 62.8 g/l (Fayolle et al. 1990)), continuous fermentation (titre up to 57.9 g/l (Zhu and Yang 2004)), and immobilized cell fermentation (titres up to 13.70 g/l (Jiang et al. 2010)) are most frequently used for butyrate production. A high titre of 86.9 g/l with a yield of 0.46 g/g glucose was achieved in a repeated fed-batch fermentation with immobilized *Clostridium tyrobutyricum* in a fibrous bed bioreactor (Jiang et al. 2011). Recently, METEX (Saint-Beauzire, France) signed an agreement with Total Développement Régional (TDR) (France) for the production of butyric acid (total 4 kt butyric acid in two stages) using renewable sources via fermentation technology.

8.4.2 Succinic Acid

Succinic acid (SA) ($C_4H_6O_4$) is an intermediate of the tricarboxylic acid cycle and has numerous applications in several markets comprising the detergent/surfactant market, the ion chelator market, the food market (e.g. acidulants, flavours, or antimicrobials) and the pharmaceutical market. SA has been identified among the top 12 building block chemicals for high value biobased chemicals or materials by the US Department of Energy (Werpy and Petersen 2014). SA can be converted to succinonitrile, which is a precursor for the production of putrescine (discussed later in the chapter) (Tsuge and Kondo 2017). With the growing preference of biobased over fossil fuel-based SA production, application areas are expanding into biodegradable plastic, garbage bags, packaging film, and plastic bottles. The overall global market of SA is projected to reach US$1.1 billion by 2022 (Global Industry Analysts 2017). This large market potential has driven the global biobased SA production capacity of 55 000 metric tons per year, which is similar to the petroleum based SA production. Future worldwide production of SA are projected to reach 400 000 metric tons per year (Biddy et al. 2016). The price of bio-based SA (US$1600–2000 per ton (Villadson 2013)) was reported in recent years to be much lower than petroleum-derived SA (US$6000–9000/ton (Biddy et al. 2016)).

Commercially, biosuccinate is produced by four companies. Succinity (Düsseldorf, Germany), which is a joint venture between BASF (New Jersey, USA) and Corbion (Netherlands), uses *Basfia succiniciproducens*, a natural succinate-producing bacterium to produce succinate from glycerol and sugar as feedstock. Another company, Myrant (Quincy, USA) uses engineered *E. coli* and sorghum as feedstock. BioAmber (Plymouth, USA)/Mitsui (Tokyo, Japan) started the production with *E. coli*, but has recently changed to yeast *Candida krusei* (Jansen and van Gulik 2014). Reverdia (a joint venture of DSM/ Roquette) (The Netherlands) constructed a strain of *S. cerevisiae* with increased SA production, commercialized as ‘Biosuccinium’.

Most of the natural succinate-producing bacteria have been isolated from the rumens of ruminants. Carbon dioxide, methane, and traces of hydrogen produced create unique anaerobic conditions for microbial succinic acid production (Kamra 2005). Some of the succinic acid producing bacteria isolated from the rumen are *Actinobacillus succinogenes*, *Actinobacillus succiniciproducens*, *Mannheimia succiniciproducens*, and *Bacteroides fragilis*. Fungi such as *Fusarium* spp. *Aspergillus* spp. and *Penicillium simplicissimum* are known to excrete succinic acid (Magnuson and Lasure 2004). These rumen bacteria produce 52–106 g/l succinic acid with yields of 0.76–0.88 g/g glucose, using a CO_2-fixing anaplerotic pathway in anaerobic fed-batch fermentations (Guettler et al. 1996; Lee et al. 2006; Meynial-Salles et al. 2008; Okino et al. 2008). The combined fungal-bacteria two-step production of SA showed high yields of 2.2 g/l/h and 0.95 g/g with *Rhizophus* sp. and *Enterococcus faecalis* respectively. *Rhizophus* sp. produces fumarate in the first step, which is then transferred to a second reactor with *E. faecalis*, which converts fumarate to succinic acid (Moon et al. 2004). Other than using these organisms naturally, metabolic engineering of the bacteria is also underway to increase the yield and range of substrates, and alleviate the problem of product inhibition and byproduct formation.

8.4.3 Malic Acid

Malic acid ($C_4H_6O_5$) is a naturally occurring C4 building block that is a key intermediate of the tricarboxylic acid cycle. It is extensively used in the food industry as flavour enhancer and acidulant. It also has applications in the pharmaceutical, textile, and polymer industries. Malic acid can furthermore be converted to other derivatives that are used to produce plastics, polymers, and resin materials. The global market for malic acid is approximately 70 000 tons per year with a worth of US$152 million; this market is expected to reach US$244.0 million by 2024 (GrandViewResearch 2016b). Malic acid can be extracted from plants, chemically synthesized by the hydration of either maleic or fumaric acid, or produced by enzymatic conversion from fumaric acid.

The enzymatic conversion (fumarate hydratase) can be achieved by biological hydration or the fungal fermentation of simple sugars (Battat et al. 1991; Neufeld et al. 1991). The most successful fermentative production of malic acid to this date has been achieved using *Aspergillus flavus*, achieving 113 g/l L-malic acid with a yield of 0.95 g/g glucose and a productivity of 0.59 g/l/h in batch fermentation (Battat et al. 1991). Malic acid production was also attempted with other organisms, including the yeast *S. cerevisiae*. Since wild-type *S. cerevisiae* strains produce only low levels of malate, metabolic engineering is required to achieve efficient malate production with this yeast. However, with the overexpression of the cytosolic isoenzyme of malate dehydrogenase (Mdh2p), *S. cerevisiae* is only able to produce 12 g/l malic acid (Pines et al. 1997). For achieving higher production yields with *S. cerevisiae*, the alcoholic fermentation pathway needs to be eliminated, as this pathway is the preferred one when a high concentration of sugars are present, as is the case in industrial fermentations. A natural isolate of another yeast, *Zygosaccharomyces rouxii*, has shown concentrations of 75 g/l of malic acid with 300 g/l glucose (Taing and Taing 2007). *E. coli* has also been evaluated for malic acid production with metabolic modifications. Malate metabolism-related genes encoding fumaric acid reductase (FrdBC), malic enzyme (SfcA and MaeB), and fumarase (FumB and FumAC) were deleted in succinic acid-producing *E. coli*. This engineered strain showed production of 33.9 g/l L-malic acid and a yield of 1.06 g/g on glucose (Zhang et al. 2011).

8.4.4 Putrescine

Putrescine has gained recent attention as a component of polymers, pharmaceuticals, agrochemicals, surfactants, and other additives. In particular, putrescine is used to produce polyamide Nylon

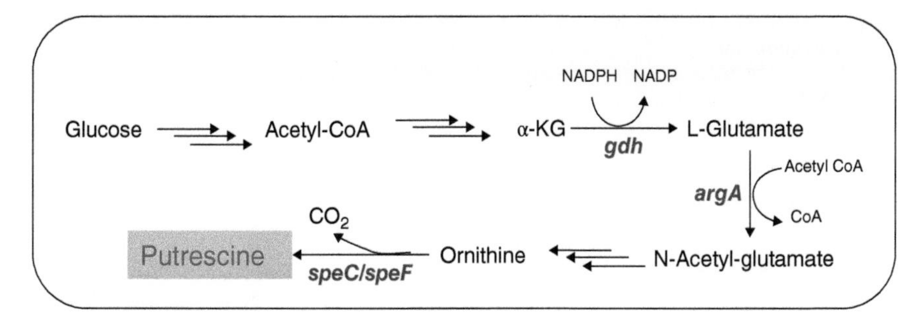

Figure 8.6 Metabolic pathways that produce putrescine from TCA cycle intermediates. Source: Adapted from Jang et al. 2012.

46, by reacting with adipic acid (Scott et al. 2007), the product being marketed by DSM (Heerlen, Netherlands) under the trade name Stanyl. It is chemically produced on an industrial scale by the hydrogenation of succinonitrile, which uses hydrogen cyanide and acrylonitrile (Ott et al. 2012). However, the biobased production of putrescine is a promising alternative to the chemical synthesis process. Putrescine is found in a wide range of organisms (Tabor and Tabor 1984) (pathway shown in Figure 8.6) including *E. coli*. This molecule plays an important role in cell proliferation and normal cell growth (Tabor and Tabor 1985). Attempts have been made to produce putrescine by overexpressing the ornithine decarboxylase (ODC) gene in *E. coli*, which directly coverts ornithine to putrescine (Qian et al. 2009). A metabolically engineered strain of *E. coli* that overexpresses the *speC* gene (encoding ODC), produces about 25 mg/l putrescine (Kashiwagi and Igarashi 1988). Another attempt using *E. coli* LJ110 by overexpressing the *speF* gene (encoding another form of ODC) yields 5.1 g/l putrescine by fed batch culture (Eppelmann et al. 2016). Recently, *C. glutamicum* has also been engineered for putrescine production (Nguyen et al. 2015a,b) with theoretical metabolic yields of 0.627 and 0.653 mol per mol sugar with glucose and glycerol as feedstock respectively (Nguyen et al. 2015a).

8.5 5-Carbon

This section explains the industrial importance, production, and scope of a few important C5 platform chemicals derived from sugar: itaconic acid, glutaconic acid, glutaric acid, and xylitol.

8.5.1 Itaconic Acid

Itaconic acid (2-Methylidenebutanedioic acid or $C_5H_6O_4$) (IA), an unsaturated dibasic acid, has gained attention commercially as a flexible building block for detergents, coatings, and rubber. IA can also be converted to methyl methacrylate (MMA), which is the monomer of poly(methyl methacrylate), commercially known as Plexiglass. Driven by the shift from petrochemicals to biobased materials, the global market of itaconic acid is projected to exceed US$216 million by 2020 (Global Industry Analysts 2016). It is also predicted to replace methacyclic acid, which is currently produced by the petrochemical industry (Lee et al. 2011). It is industrially produced by fermentation with the natural producer, *Aspergillus terreus* via a pathway shown in Figure 8.7 (Bonnarme et al. 1995; Jaklitsch et al. 1991; Klement and Büchs 2013; Li et al. 2011a). For *A. terrerus*, PYC, which is a critical enzyme to replenish tricarboxylic (TCA) intermediates, is exclusively located in the cytosol (Jaklitsch et al. 1991), where it may utilize the CO_2 cleaved from

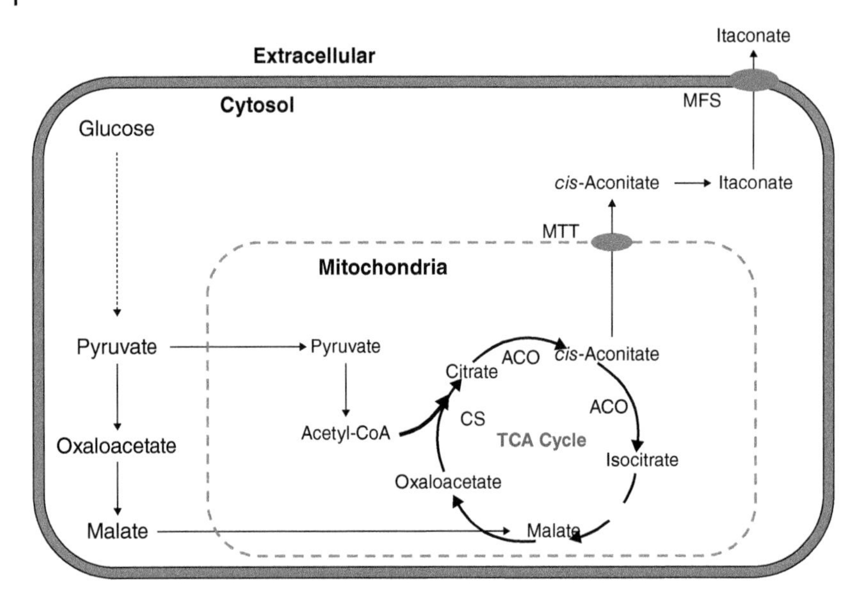

Figure 8.7 Biosynthetic pathway of itaconic acid in *A. terreus*. Source: Adapted from Huang et al. 2014.

pyruvate or cis-aconitate. This recycling of CO_2 greatly increases the efficiency of the biosynthesis of IA.

Nonetheless, the cost of biotechnologically producing itaconic acid is high. The strain, TN-484, used in industrial fermentations was developed by selection using an itaconic acid gradient agar plate technique after N-methyl-N′-nitro-N-nitrosoguanidine(NTG)-treatment (N-Methyl-N-nitro-N-nitrosoguanidine) of *A. terrerus*, mainly to overcome the product inhibition occurring in naturally occurring strain (Yahiro et al. 1995). The yield of this strain is 0.54 g/g glucose in 6-day flask culture. A typical industrial feedstock are molasses, but corn starch and corn steep liquor have also been used as a low cost feedstock with >50% yield (Yahiro et al. 1997; Yahiro et al. 1995). For a successful IA production, *A. terreus* is cultivated at high substrate concentrations and a low pH under phosphate-limited conditions (Kuenz et al. 2012). A patent describes PCI 519, an engineered strain of *E. coli*, which reaches production titres and yields of 4.2 g/l and 0.52 g/g glucose in 72 hours (Liao and Chang 2012). Other hosts such as *Ustilago maydis* (Maassen et al. 2014), *S. cerevisiae* and *Aspergillus. niger* are not cost-effective due to lower yields but provide promising proof of concepts for further research.

8.5.2 Xylitol

Xylitol ($C_5H_{12}O_5$) is a naturally occurring five carbon sugar alcohol that has been used as a sweetener and is of interest to the food, odonatological, and pharmaceutical industries. This compound has a sweetening capacity that is similar to that of glucose, as well as antimicrobial and anti-cariogenic properties (Antunes et al. 2017). It can also be used for synthesizing esters and polymers. The global xylitol market is estimated at 190.9 thousand metric tons, and is valued at US$725.9 million in 2016 and expected to reach 266.5 thousand metric tons valued at just above US$1 billion by 2022 (ResearchandMarkets 2017). The chemical synthesis route of xylitol employs acid hydrolysis of lignocellulosic biomass under high temperature and pressure with metal catalyst, to obtain pure xylose for hydrogenation. However, biobased xylitol production can be carried out by suitable microorganisms under much milder conditions.

Xylitol fermentations have been performed as batch, fed-batch, and continuous processes, as well as with the use of immobilized cells and bioreactors (Antunes et al. 2017). Due to their higher natural xylitol production capacities, yeasts from the genus *Candida* such as *Candida guilliermondii, Candida tropicalis, C. parapsilosis* and *C. boidinii* have been considered to constitute the best industrial options. For example, a natural xylitol producer, *C. tropicalis*, produces 182 g/l xylitol in oxygen limited culture with cell recycling, however, without cell recycling, the productivity is in the range of 0.5–3.9 g/l/h (Granström et al. 2007). Xylitol production in yeasts occurs as a natural intermediate during D-xylose metabolism, which occurs in two steps. The first step involves the reduction of xylose to xylitol by xylitol reductase, which is then oxidized to xylulose by the enzyme xylitol dehydrogenase (Pal et al. 2013). Cofactor imbalances during these steps result in the secretion of xylitol as a by-product of D-xylose fermentation. Filamentous fungi and bacteria have also been used for xylitol production, but the production yields remain much lower than those achieved with yeast. Further metabolic engineering for the biotechnological production of xylitol is explored on strains that can consume xylose as a carbon source. The modifications are focused on optimizing the metabolic pathway, disrupting xylitol uptake and enhancing the secretion of xylitol (Su et al. 2013).

8.5.3 Glutaconic Acid and Glutaric Acid

Glutaconic acid ($HO_2CCH{=}CHCH_2CO_2H$) was first produced in *E. coli* by overexpressing six enzymes, found in glutamate-fermenting bacteria; this engineering design resulted in the creation of a new anaerobic pathway (Djurdjevic et al. 2011). An intracellular concentration of glutaconic acid of 2.1 g/l was achieved, but only 351 mg/l were found in the medium, which indicated extracellular transport limitations. Theoretically, the reduction of glutaconic acid should yield glutaric acid, however this has not been observed experimentally.

Glutarate is a building block for polyesters and polyamides like nylon-4,5 and nylon-5,5 (copolymers of glutarate with putrescine and cadaverine, respectively) (Wang et al. 2017). Glutaric acid is naturally produced during the metabolism of lysine (Revelles et al. 2005; Revelles et al. 2004). L-lysine catabolic pathway has also been utilized in *E. coli* for the production of glutaric acid (Adkins et al. 2013; Park et al. 2013). Glutaric acid was first produced in *E. coli* by the production of 5AVA (5-aminovalerate). The gene encoding 5AVA aminotransferase (gabT) and the gene coding for glutarate semialdehyde dehydrogenase (gabD) were used to subsequently convert 5AVA into glutaric acid. Titres of 1.7 g/l of glutaric acid from 10 g/l of l-lysine with supplementation of α-ketoglutarate are typically achieved with recombinant strain (Park et al. 2013).

8.6 6-Carbon

This section explains the industrial importance, production, and scope of a few important C6 platform chemicals derived from sugar: adipic acid, muconic acid, citric acid, and glucaric acid.

8.6.1 Adipic Acid

Adipic acid (AA) is one of the most important bulk chemicals and is a major building block for the bio-based plastic industry. Its primary use is as a precursor for the synthesis of Nylon-6,6 polyamide although it is also widely used to produce polyester and polyurethane resins. The current market of AA is over 3.5 million metric tons per year, a market that is growing at a rate of approximately 5%

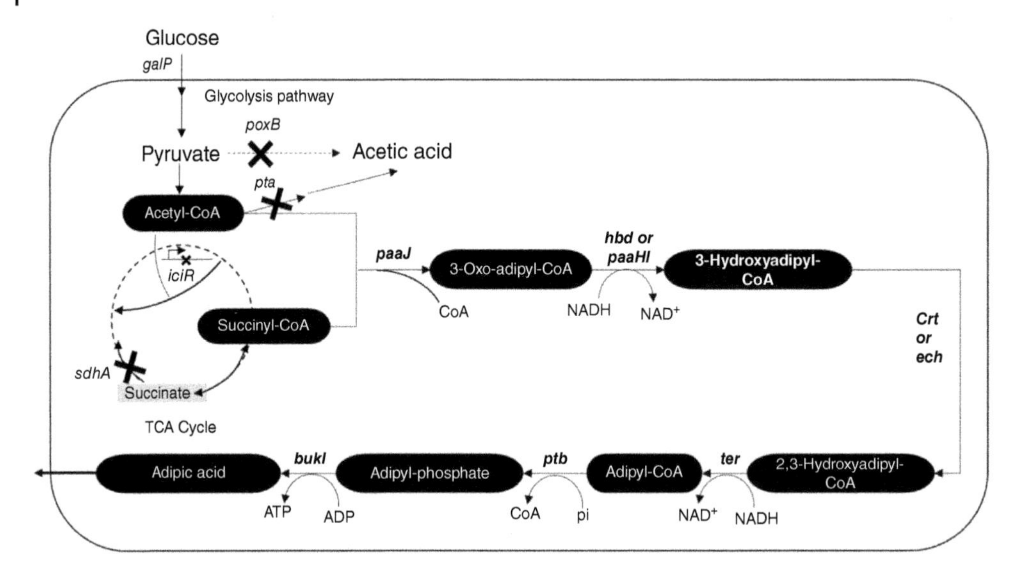

Figure 8.8 Adipic acid production by reversal of adipic acid degradation pathway. Source: Adapted from Yu et al. 2014.

per year (BusinessWire 2016). Commercially available AA is obtained by chemical routes, either by oxidation of a mixture of cyclohexanone and cyclohexanol, which are derived from benzene, or by hydrocyanation of butadiene, followed by hydroisomerization to adiponitrile, which is then hydrolyzed (Polen et al. 2013). Biobased AA production is a sustainable alternative to the manufacturing route via benzene, which is carcinogenic.

AA has been reported to be produced through the ω-oxidation of fatty acids by engineered yeasts, a process that has been patented (Picataggio and Beardslee 2016). In another patent, a alpha-keto acid pathway was proposed for AA biosynthesis in *E. coli*, where alpha-ketopimelic acid is converted into adipic acid via decarboxylation and oxidation (Raemakers-Franken et al. 2010). One of the most interesting biosynthetic routes for AA production employs the reversal of the adipic acid degradation pathway (Figure 8.8) (Yu et al. 2014). Muconic acid (MA) can also serve as an intermediate in the chemo-catalytic conversion to AA (Polen et al. 2013).

8.6.2 Muconic Acid

Muconic acid (MA) is a platform chemical that serves as a precursor to several bio-plastics. It is also a synthetic precursor to terephthalic acid, a chemical used for manufacturing polyethylene terephthalate (PET) and polyester. The market for MA is driven by the demand for AA. The global market of MA in 2015 was US$61.47 million and is projected to reach an estimated value of US$86.54 million by 2020 (TransparencyMarketResearch 2016). Two different production routes have been investigated for muconic acid: the biotransformation of aromatics or de novo synthesis from glucose. Several microorganisms have been reported to have a natural ability to produce *cis,cis*-MA from aromatics (van Duuren et al. 2012), including yeasts such as *Candida* sp., and bacteria such as *Acinetobacter* sp., *Rhodococcus* sp., and *Sphingo* bacterium sp., among others (Neidle et al. 1988; Tsai et al. 2005; Warhurst et al. 1994; Wu et al. 2004). Since aromatics are biologically toxic to microorganisms, highly tolerant and efficient mutants are essential for improving those bioprocesses.

The earliest study on artificial biosynthesis of MA in *E. coli* from glucose was reported by Draths and Frost (1994). The carbon flux was redirected from the native shikimate pathway to MA production in *E. coli* by introducing three heterologous enzymes: 3-dehydroshikimate dehydratase, protocatechuic acid decarboxylase and catechol 1,2-dioxygenase (CDO). The yield achieved with this modification were up to 2.4 g/l of MA via a two stage bioconversion in batch flask (Draths and Frost 1994), and 38.6 g/l via fed batch fermentation (Niu et al. 2002). A similar pathway was also reconstituted in *S. cerevisiae*, with the highest titre of 141 mg/l reported to this date (Curran et al. 2013).

8.6.3 Citric Acid

Citric acid is the most important organic acid produced at industrial scale by fermentation. The global market for this organic acid was USD 2.16 Billion in 2014, and is projected to reach USD 3.66 billion by 2022 (CredenceResearch 2015). ADM (Chicago, USA), Cargill (Wayzata, USA), Tate and Lyle (London, UK), and DSM (Heerlen, Netherlands) are the leading producers of citric acid in North America and Europe. Citric acid has widespread applications in food, pharmaceuticals, detergents, cosmetics, and toiletries (Soccol et al. 2006). Although a large number of microorganisms including fungi and bacteria such as *Arthrobacter paraffinens*, *Bacillus licheniformis,* and *Corynebacterium* sp., *Aspergillus aculeatus*, *A. carbonarius*, *A. awamori*, *A. foetidus*, *A. fonsecaeus*, *A. phoenicis* and *Penicillium janthinellum*; and yeasts such as *C. tropicalis*, *C. oleophila*, *C. guilliermondii*, *C. citroformans*, *Hansenula anamola,* and *Y. lipolytica* have been employed for citric acid production, *A. niger* remains the fermentor of choice for commercial scale production (Grewal and Kalra 1995; Ikeno et al. 1975; Kubicek et al. 1985; Pandey et al. 2001; Vandenberghe et al. 1999; Yokoya 1992). *A. niger* is preferred due to its high yield per unit time and ease of handling, as well as its ability to ferment a variety of cheap raw materials. Under specific environmental conditions, citric acid is produced by *A. niger* as an overflow product due to the faulty operation of the tricarboxylic acid cycle (Ramakrishnan et al. 1955). In addition, yeasts of the genus *Candida* have also been used industrially for citric acid production, but the bottleneck of using yeast to produce citrate is that isocitrate is produced simultaneously (Max et al. 2010). Similarly, *Y. lipolytica* has been developed as a microbial cell factory for citric acid to gain access to n-paraffins and fatty acids, which are not converted by *A. niger* (Papanikolaou et al. 2006). *Y. lipolytica* can yield about 140 g/l citric acid and has proved efficient on glucose and sucrose (Förster et al. 2007).

Solid state and liquid fermentation are two viable formats of production for citric acid; each of these methods can lead to different yields even with the same organism (Soccol et al. 2006). Although a variety of sugar sources have been used as feedstock, molasses are typically preferred for microbial production of citric acid due to their relatively low cost and high sugar content of 40–55% (Soccol et al. 2006). On the other hand, solid state fermentation of cassava bagasse with *A. niger* reached a maximum of 88 g/g dry matter (Vandenberghe et al. 2000). Corn cobs were used as another substrate for *A. niger* fermentation supplemented with Rapidase Pomaliq, which is an enzyme derived from *A. niger* and *Trichoderma reesei,* with a reported maximum yield of 585 g/kg dry matter at 72 hours (Hang and Woodams 2001).

8.6.4 Glucaric Acid

Glucaric acid, also known as saccharic acid, is a natural nontoxic compound produced in small amounts by mammals, including humans. D-glucaric acid can be used as a building block for biopolymers as well as in the formulation of detergents and corrosion inhibitors (Goswami and Mishra 2017). Glucaric acid can be used for the production of a wide range of products with applicability in high-volume markets. The global glucaric acid market was estimated at US$550.4 million in 2016, and is expected to reach USD 1.30 billion by 2025 (GrandViewResearch 2017a). Commercially, Rivertop Renewables (Missoula, USA) in collaboration with DTI (Danchem Technologies and Innovations, Danville, USA), has scaled up glucaric acid production. Rennovia is another firm that has been using glucaric acid for the production of biobased adipic acid and is expecting consumption of about 500 million pounds per year of glucaric acid (http://www.rennovia.com/markets/). Recently, ADM (Chicago, USA) has signed a license agreement with Rennovia (Santa Clara, USA), and Johnson Matthey (Johnson Matthey) to access this catalyst technology and has entered the field of glucaric acid production (Matthey 2017).

The chemical production of glucaric acid includes the oxidation of glucose with nitric acid or facile nitroxide-mediated and photooxidation with titanium photocatalysts (Goswami and Mishra 2017). Notably, an engineered *E. coli* strain expressing several heterogeneous genes including myo-inositol-1-phosphate synthase (Ino1) from *S. cerevisiae*, myo-inositol oxygenase (MIOX) from mice, and uronate dehydrogenase (Udh) expressed from *Pseudomonas syringae*, was reported to produce 1.1 g/l glucaric acid from 10 g/l glucose (Moon et al. 2009). In *E. coli*, the glucaric acid pathway competes with the endogenous metabolism, a phenomenon that may limit carbon flux into the pathway. Following initial demonstration, the MIOX enzyme was further expressed to improve the yields of glucaric acid (Moon et al. 2010). Another research group developed an *E. coli* strain with dynamically controlled Pfk activity, demonstrating improved yields (Reizman et al. 2015).

8.7 Perspectives

Many platform chemicals are currently primarily manufactured using fossil resources. However, due to an increasing demand for sustainable production of bio-based platform chemicals, several large chemical companies have started to adopt renewable resources and biotechnological processes. In spite of this increasing interest, limitations remain that need to be addressed such as the development of natural or engineered microorganisms for efficient production, compatibility with existing industrial infrastructures, optimization of downstream processes, and access to cost-competitive and sustainable resources. One of the major challenges is the development of efficient microorganisms that are capable of producing target compounds at sufficiently high titres, yields, and commercially feasible levels on a variety of low cost substrates. The amount of additional sugar (from low cost substrates) needed to meet the projected demands of some important biochemicals is summarized in Table 8.3. These factors will play an important role in reducing production costs and allowing bio-based processes to compete against the current petrochemical processes. Metabolic engineering approaches have helped demonstrate that the industrial-scale production of numerous commodity chemicals by biotechnology is a cost-efficient proposition that holds enormous promise to transition the chemical industry from a petroleum base to a biotechnology base.

Table 8.3 Calculated amount of additional sugar needed to meet the rising demands of biobased production of platform chemicals.

Biochemical	Yield (g/g sugar)	Current production (×1000 MT/yr)	CAGR (%)	Expected production in 2030[a] (×1000 MT/year)	Additional sugar required (×1000 MT/year)	References
Propionic acid	0.55	399.4	2.7	628.2	416	GrandViewResearch (2015b)
Lactic acid	0.99	1220	16.2	9983	8852	Becker and Wittmann (2015) and GrandViewResearch (2017b)
Succinic acid	1.12	50.1	27.4	3283.8	2887	AlliedMarketResearch (2014) and Nghiem et al. (2017)
Fumaric acid	0.85	233.3	5.9	618.2	453	Cao et al. (1996) and RadiantInsights (2014)
Malic acid	0.95	70	4.2	130	63.2	Battat et al. (1991) and GrandViewResearch (2016b)
Aspartic acid	0.3	39.3	5.6	94	182	GrandViewResearch (2015a)
Adipic acid	0.17	3500	5	7280	22 235	BusinessWire (2016) and Niu et al. (2002)
Itaconic acid	0.54	70	16.8	719	1202	Global Industry Analysts (2016) and GreenChemcalsBlog (2015)
Glycerol	0.51	2470	6.6	6860	8608	GlobalMarketInsights (2016) and Taherzadeh et al. (2002)
Xylitol	0.9	190.9	5.7	415	249	Canilha et al. (2003) and ResearchandMarkets (2017)
Citric acid	0.59	180	5.3	411.3	392	CredenceResearch (2015) and Hang and Woodams (2001)

a) Calculated based on CAGR.

References

Adkins, J., Jordan, J., and Nielsen, D.R. (2013). Engineering *Escherichia coli* for renewable production of the 5-carbon polyamide building-blocks 5-aminovalerate and glutarate. *Biotechnology and Bioengineering* 110 (6): 1726–1734.

AlliedMarketResearch. 2014. Bio Succinic Acid Market by Application (1-Butanediol,4-Butanediol (BDO), Polyester Polyols, PBS, Plasticizers, Solvents & lubricants, Alkyd Resins, Resins, coatings, pigments, De-icer solutions) – Global Opportunity Analysis and Industry Forecast, 2013-2020 Vol. 2018.

Antunes, F.A.F., dos Santos, J.C., da Cunha, M.A.A. et al. (2017). Biotechnological production of xylitol from biomass. In: *Production of Platform Chemicals from Sustainable Resources* (eds. Z. Fang, R.L. Smith Jr., and X. Qi), 311–342. Springer.

Baroi, G., Baumann, I., Westermann, P., and Gavala, H. (2015). Butyric acid fermentation from pretreated and hydrolysed wheat straw by an adapted *Clostridium tyrobutyricum* strain. *Microbial Biotechnology* 8 (5): 874–882.

Battat, E., Peleg, Y., Bercovitz, A. et al. (1991). Optimization of L-malic acid production by *Aspergillus flavus* in a stirred fermentor. *Biotechnology and Bioengineering* 37 (11): 1108–1116.

Becker, J., Lange, A., Fabarius, J., and Wittmann, C. (2015). Top value platform chemicals: bio-based production of organic acids. *Current Opinion in Biotechnology* 36: 168–175.

Becker, J. and Wittmann, C. (2015). Advanced biotechnology: metabolically engineered cells for the bio-based production of chemicals and fuels, materials, and health-care products. *Angewandte Chemie International Edition* 54 (11): 3328–3350.

Benninga, H. (1990). *A History of Lactic Acid Making: A Chapter in the History of Biotechnology*. Springer Science and Business Media.

Biddy, M.J., Scarlata, C., and Kinchin, C. (2016). Chemicals from biomass: a market assessment of bioproducts with near-term potential. In: *National Renewable Energy Laboratory (NREL)*. Golden: CO (United States).

Bonnarme, P., Gillet, B., Sepulchre, A. et al. (1995). Itaconate biosynthesis in *Aspergillus terreus*. *Journal of Bacteriology* 177 (12): 3573–3578.

Boyaval, P. and Corre, C. (1987). Continuous fermentation of sweet whey permeate for propionic acid production in a CSTR with UF recycle. *Biotechnology Letters* 9 (11): 801–806.

Boyaval, P. and Corre, C. (1995). Production of propionic acid. *Le Lait* 75 (4–5): 453–461.

Bryant, M.P. and Small, N. (1956). The anaerobic monotrichous butyric acid-producing curved rod-shaped bacteria of the rumen. *Journal of Bacteriology* 72 (1): 16.

BusinessWire. 2016. Adipic Acid (ADPA): 2016 World Market Outlook and Forecast up to 2020.

Canilha, L., e Silva, J.B.A., Felipe, M.G., and Carvalho, W. (2003). Batch xylitol production from wheat straw hemicellulosic hydrolysate using *Candida guilliermondii* in a stirred tank reactor. *Biotechnology Letters* 25 (21): 1811–1814.

Cao, N., Du, J., Gong, C., and Tsao, G. (1996). Simultaneous production and recovery of fumaric acid from immobilized *Rhizopus oryzae* with a rotary biofilm contactor and an adsorption column. *Applied and Environmental Microbiology* 62 (8): 2926–2931.

Causey, T., Shanmugam, K., Yomano, L., and Ingram, L. (2004). Engineering *Escherichia coli* for efficient conversion of glucose to pyruvate. *Proceedings of the National Academy of Sciences of the United States of America* 101 (8): 2235–2240.

Centeno, J., Menendez, S., Hermida, M., and Rodrıguez-Otero, J. (1999). Effects of the addition of *Enterococcus faecalis* in Cebreiro cheese manufacture. *International Journal of Food Microbiology* 48 (2): 97–111.

CredenceResearch. 2015. Global Citric Acid Market Is Projected To Reach USD 3.66 Billion By 2022.

Curran, K.A., Leavitt, J.M., Karim, A.S., and Alper, H.S. (2013). Metabolic engineering of muconic acid production in *Saccharomyces cerevisiae*. *Metabolic Engineering* 15: 55–66.

Dale, B.E. 2017. Feeding a sustainable chemical industry: do we have the bioproducts cart before the feedstocks horse? Faraday Discussions.

Djurdjevic, I., Zelder, O., and Buckel, W. (2011). Production of glutaconic acid in a recombinant *Escherichia coli* strain. *Applied and Environmental Microbiology* 77 (1): 320–322.

Draths, K.M. and Frost, J.W. (1994). Environmentally compatible synthesis of adipic acid from D-glucose. *Journal of the American Chemical Society* 116 (1): 399–400.

Eppelmann, K., Nossin, P.M., Kremer, S.M., Wubbolts, M.G. 2016. Biochemical Synthesis of 1, 4-butanediamine, Google Patents.

Fayolle, F., Marchal, R., and Ballerini, D. (1990). Effect of controlled substrate feeding on butyric acid production by *Clostridium tyrobutyricum*. *Journal of Industrial Microbiology and Biotechnology* 6 (3): 179–183.

Feng, X., Chen, F., Xu, H. et al. (2011). Green and economical production of propionic acid by *Propionibacterium freudenreichii* CCTCC M207015 in plant fibrous-bed bioreactor. *Bioresource Technology* 102 (10): 6141–6146.

Fiorentino, G., Ripa, M., and Ulgiati, S. (2016). Chemicals from biomass: technological versus environmental feasibility. A review. *Biofuels, Bioproducts and Biorefining* 11: 195–214.

Förster, A., Aurich, A., Mauersberger, S., and Barth, G. (2007). Citric acid production from sucrose using a recombinant strain of the yeast *Yarrowia lipolytica*. *Applied Microbiology and Biotechnology* 75 (6): 1409–1417.

Gao, X., Ma, Z., Yang, L., and Ma, J. (2014). Enhanced bioconversion of ethylene glycol to glycolic acid by a newly isolated Burkholderia sp. EG13. *Applied Biochemistry and Biotechnology* 174 (4): 1572–1580.

Global Industry Analysts, I. 2017. Encouraging demand for bio-succinic acid to drive the market for succinic acid, Vol. 2017.

Global Industry Analysts, I. 2016. Itaconic acid (IA) -A global strategic business report.

GlobalMarketInsights. 2016. Glycerol Market size worth $3.04bn by 2022, Vol. 2018.

Goswami, R. and Mishra, V. (2017). Sugar-derived industrially important C6 platform chemicals. In: *Platform Chemical Biorefinery* (eds. S.K. Brar, S.J. Sarma and K. Pakshirajan), 229–248. Elsevier.

GrandViewResearch. 2015a. Global Aspartic Acid Market By Application (Feed Supplements, Medicine, Polyaspartic Acid, Aspartame, L-Alanine) Expected to Reach USD 101.0 Million by 2022, Vol. 2018.

GrandViewResearch. 2015b. Global Propionic Acid Market By Application (Animal Feed, Calcium & Sodium Propionate, Cellulose Acetate Propionate) Expected To Reach USD 1.53 Billion By 2020: Grand View Research, Inc.

GrandViewResearch. 2017a. Glucaric Acid Market Size Worth $1.30 Billion By 2025 | CAGR: 10.1%, Vol. 2017.

GrandViewResearch. 2016a. Glycolic Acid Market Analysis By Application (Personal Care, Household, Industrial), And Segment Forecasts To 2024.

GrandViewResearch. 2017b. Lactic Acid Market Size Worth $9.8Bn By 2025 & PLA To Reach $6.5Bn, Vol. 2017.

GrandViewResearch. 2016b. Malic Acid Market Projected To Reach $244.0 Million By 2024, Vol. 2017.

Granström, T.B., Izumori, K., and Leisola, M. (2007). A rare sugar xylitol. Part II: biotechnological production and future applications of xylitol. *Applied Microbiology and Biotechnology* 74 (2): 273.

GreenChemcalsBlog. 2015. LEAF Technologies enters itaconic acid market.

Grewal, H. and Kalra, K. (1995). Fungal production of citric acid. *Biotechnology Advances* 13 (2): 209–234.

Guettler, M.V., Jain, M.K., Rumler, D. 1996. Method for making succinic acid, bacterial variants for use in the process, and methods for obtaining variants, Google Patents.

Hang, Y. and Woodams, E. (2001). Enzymatic enhancement of citric acid production by *Aspergillus niger* from corn cobs. *LWT-Food Science and Technology* 34 (7): 484–486.

Hollmann, R. and Deckwer, W.-D. (2004). Pyruvate formation and suppression in recombinant *Bacillus megaterium* cultivation. *Journal of Biotechnology* 111 (1): 89–96.

Huang, X., Lu, X., Li, Y. et al. (2014). Improving itaconic acid production through genetic engineering of an industrial *Aspergillus terreus* strain. *Microbial Cell Factories* 13 (1): 119.

Huang, Y.L., Wu, Z., Zhang, L. et al. (2002). Production of carboxylic acids from hydrolyzed corn meal by immobilized cell fermentation in a fibrous-bed bioreactor. *Bioresource Technology* 82 (1): 51–59.

Ikeno, Y., Masuda, M., Tanno, K. et al. (1975). Citric acid production from various raw materials by yeasts. *Journal of Fermentation Technology*.

Jaklitsch, W.M., Kubicek, C.P., and Scrutton, M.C. (1991). The subcellular organization of itaconate biosynthesis in *Aspergillus terreus. Microbiology* 137 (3): 533–539.

Jang, Y.S., Kim, B., Shin, J.H. et al. (2012). Bio-based production of C2–C6 platform chemicals. *Biotechnology and Bioengineering* 109 (10): 2437–2459.

Jansen, M.L. and van Gulik, W.M. (2014). Towards large scale fermentative production of succinic acid. *Current Ppinion in Biotechnology* 30: 190–197.

Jiang, L., Wang, J., Liang, S. et al. (2011). Enhanced butyric acid tolerance and bioproduction by *Clostridium tyrobutyricum* immobilized in a fibrous bed bioreactor. *Biotechnology and Bioengineering* 108 (1): 31–40.

Jiang, L., Wang, J., Liang, S. et al. (2010). Production of butyric acid from glucose and xylose with immobilized cells of *Clostridium tyrobutyricum* in a fibrous-bed bioreactor. *Applied Biochemistry and Biotechnology* 160 (2): 350–359.

Kamm, B. (2007). Production of platform chemicals and synthesis gas from biomass. *Angewandte Chemie International Edition* 46 (27): 5056–5058.

Kamra, D.N. (2005). Rumen microbial ecosystem. *Current Science* 89: 124–135.

Kashiwagi, K. and Igarashi, K. (1988). Adjustment of polyamine contents in *Escherichia coli. Journal of Bacteriology* 170 (7): 3131–3135.

Klement, T. and Büchs, J. (2013). Itaconic acid – a biotechnological process in change. *Bioresource Technology* 135: 422–431.

Koivistoinen, O.M., Kuivanen, J., Barth, D. et al. (2013). Glycolic acid production in the engineered yeasts *Saccharomyces cerevisiae* and *Kluyveromyces lactis. Microbial Cell Factories* 12 (1): 82.

Kubicek, C.P., Röhr, M., and Rehm, H. (1985). Citric acid fermentation. *Critical Reviews in Biotechnology* 3 (4): 331–373.

Kuenz, A., Gallenmüller, Y., Willke, T., and Vorlop, K.-D. (2012). Microbial production of itaconic acid: developing a stable platform for high product concentrations. *Applied Microbiology and Biotechnology* 96 (5): 1209–1216.

Kumar, D. and Murthy, G.S. (2011). Impact of pretreatment and downstream processing technologies on economics and energy in cellulosic ethanol production. *Biotechnology for Biofuels* 4 (1): 27.

Kylä-Nikkilä, K., Hujanen, M., Leisola, M., and Palva, A. (2000). Metabolic engineering of lactobacillus helveticus CNRZ32 for production of purel-(+)-lactic acid. *Applied and Environmental Microbiology* 66 (9): 3835–3841.

Lambert, E.d., Lapeyre, E. 2009. METABOLIC EXPLORER : Technical advances in the Glycolic Acid programme.

Lee, J.W., Kim, H.U., Choi, S. et al. (2011). Microbial production of building block chemicals and polymers. *Current Opinion in Biotechnology* 22 (6): 758–767.

Lee, S.J., Song, H., and Lee, S.Y. (2006). Genome-based metabolic engineering of *Mannheimia succiniciproducens* for succinic acid production. *Applied and Environmental Microbiology* 72 (3): 1939–1948.

Li, A., van Luijk, N., Ter Beek, M. et al. (2011a). A clone-based transcriptomics approach for the identification of genes relevant for itaconic acid production in *Aspergillus. Fungal Genetics and Biology* 48 (6): 602–611.

Li, W., Han, H.-J., and Zhang, C.-h. (2011b). Continuous butyric acid production by corn stalk immobilized *Clostridium* thermobutyricum cells. *African Journal of Microbiology Research* 5 (6): 661–666.

Li, Y., Chen, J., and Lun, S.-Y. (2001a). Biotechnological production of pyruvic acid. *Applied Microbiology and Biotechnology* 57 (4): 451–459.

Li, Y., Chen, J., Lun, S.-Y., and Rui, X.-S. (2001b). Efficient pyruvate production by a multi-vitamin auxotroph of *Torulopsis glabrata*: key role and optimization of vitamin levels. *Applied Microbiology and Biotechnology* 55 (6): 680–685.

Liang, Z.-X., Li, L., Li, S. et al. (2012). Enhanced propionic acid production from Jerusalem artichoke hydrolysate by immobilized *Propionibacterium acidipropionici* in a fibrous-bed bioreactor. *Bioprocess and Biosystems Engineering* 35 (6): 915–921.

Liao, J.C., Chang, P.-C. 2012. Genetically modified microorganisms for producing itaconic acid with high yields, Google Patents.

Liu, X., Zhu, Y., and Yang, S.T. (2006). Construction and characterization of ack deleted mutant of *Clostridium tyrobutyricum* for enhanced butyric acid and hydrogen production. *Biotechnology Progress* 22 (5): 1265–1275.

Liu, Z., Ma, C., Gao, C., and Xu, P. (2012). Efficient utilization of hemicellulose hydrolysate for propionic acid production using *Propionibacterium acidipropionici. Bioresource Technology* 114: 711–714.

Maassen, N., Panakova, M., Wierckx, N. et al. (2014). Influence of carbon and nitrogen concentration on itaconic acid production by the smut fungus *Ustilago maydis. Engineering in Life Sciences* 14 (2): 129–134.

Magnuson, J.K. and Lasure, L.L. (2004). Organic acid production by filamentous fungi. In: *Advances in Fungal Biotechnology for Industry, Agriculture, and Medicine* (eds. J.S. Tkacz and L. Lange), 307–340. Springer.

Maleki, N. and Eiteman, M.A. (2017). Recent progress in the microbial production of pyruvic acid. *Fermentation* 3 (1): 8.

MarketsandMarkets. n.d. Butyric Acid Market worth 289.3 Million USD by 2020

Martinez, F.A.C., Balciunas, E.M., Salgado, J.M. et al. (2013). Lactic acid properties, applications and production: a review. *Trends in Food Science and Technology* 30 (1): 70–83.

Matthey, J. 2017. Johnson Matthey and Rennovia announce licence agreement with ADM for glucaric acid production technology.

Max, B., Salgado, J.M., Rodríguez, N. et al. (2010). Biotechnological production of citric acid. *Brazilian Journal of Microbiology* 41 (4): 862–875.

Meynial-Salles, I., Dorotyn, S., and Soucaille, P. (2008). A new process for the continuous production of succinic acid from glucose at high yield, titer, and productivity. *Biotechnology and Bioengineering* 99 (1): 129–135.

Miyata, R. and Yonehara, T. (1996). Improvement of fermentative production of pyruvate from glucose by *Torulopsis glabrata* IFO 0005. *Journal of Fermentation and Bioengineering* 82 (5): 475–479.

Moon, S.-K., Wee, Y.-J., Yun, J.-S., and Ryu, H.-W. (2004). Production of fumaric acid using rice bran and subsequent conversion to succinic acid through a two-step process. In: *Proceedings of the*

Twenty-Fifth Symposium on Biotechnology for Fuels and Chemicals Held May 4–7, 2003, in Breckenridge, CO (eds. M. Finkelstein and B.H. Davison), 843–855. Springer.

Moon, T.S., Dueber, J.E., Shiue, E., and Prather, K.L.J. (2010). Use of modular, synthetic scaffolds for improved production of glucaric acid in engineered *E. coli*. *Metabolic Engineering* 12 (3): 298–305.

Moon, T.S., Yoon, S.-H., Lanza, A.M. et al. (2009). Production of glucaric acid from a synthetic pathway in recombinant *Escherichia coli*. *Applied and Environmental Microbiology* 75 (3): 589–595.

Morgunov, I.G., Kamzolova, S.V., Perevoznikova, O.A. et al. (2004). Pyruvic acid production by a thiamine auxotroph of *Yarrowia lipolytica*. *Process Biochemistry* 39 (11): 1469–1474.

Neidle, E., Hartnett, C., Bonitz, S., and Ornston, L. (1988). DNA sequence of the Acinetobacter calcoaceticus catechol 1, 2-dioxygenase I structural gene catA: evidence for evolutionary divergence of intradiol dioxygenases by acquisition of DNA sequence repetitions. *Journal of Bacteriology* 170 (10): 4874–4880.

Neufeld, R., Peleg, Y., Rokem, J. et al. (1991). L-Malic acid formation by immobilized *Saccharomyces cerevisiae* amplified for fumarase. *Enzyme and Microbial Technology* 13 (12): 991–996.

Nghiem, N.P., Kleff, S., and Schwegmann, S. (2017). Succinic acid: technology development and commercialization. *Fermentation* 3 (2): 26.

Nguyen, A.Q., Schneider, J., Reddy, G.K., and Wendisch, V.F. (2015a). Fermentative production of the diamine putrescine: system metabolic engineering of *Corynebacterium glutamicum*. *Metabolites* 5 (2): 211–231.

Nguyen, A.Q., Schneider, J., and Wendisch, V.F. (2015b). Elimination of polyamine N-acetylation and regulatory engineering improved putrescine production by *Corynebacterium glutamicum*. *Journal of Biotechnology* 201: 75–85.

Niu, W., Draths, K., and Frost, J. (2002). Benzene-free synthesis of adipic acid. *Biotechnology Progress* 18 (2): 201–211.

Okino, S., Noburyu, R., Suda, M. et al. (2008). An efficient succinic acid production process in a metabolically engineered *Corynebacterium glutamicum* strain. *Applied Microbiology and Biotechnology* 81 (3): 459–464.

Ott, J., Gronemann, V., Pontzen, F. et al. (2012). *Ullmann's Encyclopedia of Industrial Chemistry*. Germany: Wiley-VCH.

Otto, C., Yovkova, V., Aurich, A. et al. (2012). Variation of the by-product spectrum during α-ketoglutaric acid production from raw glycerol by overexpression of fumarase and pyruvate carboxylase genes in *Yarrowia lipolytica*. *Applied Microbiology and Biotechnology*: 1–13.

Pal, S., Choudhary, V., Kumar, A. et al. (2013). Studies on xylitol production by metabolic pathway engineered *Debaryomyces hansenii*. *Bioresource Technology* 147: 449–455.

Pandey, A., Soccol, C., Rodriguez-Leon, J., and Nigam, P. (2001). Production of organic acids by solid state fermentation. In: *Solid State Fermentation in Biotechnology-Fundamentals and Applications* (ed. A. Pandey), 113–126. New Delhi: Asitech Publishers.

Panova, A., Mersinger, L.J., Liu, Q. et al. (2007). Chemoenzymatic synthesis of glycolic acid. *Advanced Synthesis and Catalysis* 349 (8–9): 1462–1474.

Papanikolaou, S., Galiotou-Panayotou, M., Chevalot, I. et al. (2006). Influence of glucose and saturated free-fatty acid mixtures on citric acid and lipid production by *Yarrowia lipolytica*. *Current Microbiology* 52 (2): 134–142.

Park, S.J., Kim, E.Y., Noh, W. et al. (2013). Metabolic engineering of *Escherichia coli* for the production of 5-aminovalerate and glutarate as C5 platform chemicals. *Metabolic Engineering* 16: 42–47.

Picataggio, S., Beardslee, T. 2016. Biological methods for preparing adipic acid, Google Patents.

Pines, O., Shemesh, S., Battat, E., and Goldberg, I. (1997). Overexpression of cytosolic malate dehydrogenase (MDH2) causes overproduction of specific organic acids in *Saccharomyces cerevisiae*. *Applied Microbiology and Biotechnology* 48 (2): 248–255.

Polen, T., Spelberg, M., and Bott, M. (2013). Toward biotechnological production of adipic acid and precursors from biorenewables. *Journal of Biotechnology* 167 (2): 75–84.

Qian, Z.G., Xia, X.X., and Lee, S.Y. (2009). Metabolic engineering of *Escherichia coli* for the production of putrescine: a four carbon diamine. *Biotechnology and Bioengineering* 104 (4): 651–662.

RadiantInsights. 2014. Global fumaric acid market to exceed $760 Million by 2020, Vol. 2018.

Raemakers-Franken, P.C., Schürmann, M., Trefzer, A.C., De Wildeman, S.M.A. 2010. Preparation of adipic acid, Google Patents.

Ramakrishnan, C., Steel, R., and Lentz, C. (1955). Mechanism of citric acid formation and accumulation in *Aspergillus niger*. *Archives of Biochemistry and Biophysics* 55 (1): 270–273.

Reizman, I.M.B., Stenger, A.R., Reisch, C.R. et al. (2015). Improvement of glucaric acid production in *E. coli* via dynamic control of metabolic fluxes. *Metabolic Engineering Communications* 2: 109–116.

ResearchandMarkets. 2015. Renewable Chemicals Market – Alcohols (Ethanol, Methanol), Biopolymers (Starch Blends, Regenerated Cellulose, PBS, Bio-PET, PLA, PHA, Bio-PE, and Others), Platform Chemicals & Others – Global Trends & Forecast to 2020.

ResearchandMarkets. 2017. Xylitol – A Global Market Overview, Vol. 2017.

Revelles, O., Espinosa-Urgel, M., Fuhrer, T. et al. (2005). Multiple and interconnected pathways for L-lysine catabolism in Pseudomonas putida KT2440. *Journal of Bacteriology* 187 (21): 7500–7510.

Revelles, O., Espinosa-Urgel, M., Molin, S., and Ramos, J.L. (2004). The davDT operon of Pseudomonas putida, involved in lysine catabolism, is induced in response to the pathway intermediate δ-aminovaleric acid. *Journal of Bacteriology* 186 (11): 3439–3446.

RFA (2017). *Industry Statistics, Renewable Fuel Association*. DC: Washington.

Rywińska, A., Juszczyk, P., Wojtatowicz, M. et al. (2013). Glycerol as a promising substrate for *Yarrowia lipolytica* biotechnological applications. *Biomass and Bioenergy* 48: 148–166.

Scott, E., Peter, F., and Sanders, J. (2007). Biomass in the manufacture of industrial products – the use of proteins and amino acids. *Applied Microbiology and Biotechnology* 75 (4): 751–762.

Shattuck, M.T. 1948. Continuous glycolic acid process, Google Patents.

Soccol, C.R., Vandenberghe, L.P., Rodrigues, C., and Pandey, A. (2006). New perspectives for citric acid production and application. *Food Technology and Biotechnology* 44 (2).

Stanko, R.T., Tietze, D.L., and Arch, J.E. (1992). Body composition, energy utilization, and nitrogen metabolism with a severely restricted diet supplemented with dihydroxyacetone and pyruvate. *The American Journal of Clinical Nutrition* 55 (4): 771–776.

Su, B., Wu, M., Lin, J., and Yang, L. (2013). Metabolic engineering strategies for improving xylitol production from hemicellulosic sugars. *Biotechnology Letters* 35 (11): 1781–1789.

Tabor, C.W. and Tabor, H. (1984). Polyamines. *Annual Review of Biochemistry* 53 (1): 749–790.

Tabor, C.W. and Tabor, H. (1985). Polyamines in microorganisms. *Microbiological Reviews* 49 (1): 81.

Taherzadeh, M.J., Adler, L., and Lidén, G. (2002). Strategies for enhancing fermentative production of glycerol – a review. *Enzyme and Microbial Technology* 31 (1–2): 53–66.

Taing, O. and Taing, K. (2007). Production of malic and succinic acids by sugar-tolerant yeast *Zygosaccharomyces rouxii*. *European Food Research and Technology* 224 (3): 343–347.

Tomar, A., Eiteman, M., and Altman, E. (2003). The effect of acetate pathway mutations on the production of pyruvate in *Escherichia coli*. *Applied Microbiology and Biotechnology* 62 (1): 76–82.

TransparencyMarketResearch. 2016. Muconic Acid Market – Global Industry Analysis, Size, Share, Growth, Trends and Forecast, 2014–2020.

Tsai, S.-C., Tsai, L.-D., and Li, Y.-K. (2005). An isolated Candida albicans TL3 capable of degrading phenol at large concentration. *Bioscience, Biotechnology, and Biochemistry* 69 (12): 2358–2367.

Tsuge, Y. and Kondo, A. (2017). Production of amino acids (L-glutamic acid and L-lysine) from biomass. In: *Production of Platform Chemicals from Sustainable Resources* (eds. Z. Fang, R.L. Smith Jr., and X. Qi), 437–455. Springer.

Tsuge, Y., Yamamoto, S., Kato, N. et al. (2015). Overexpression of the phosphofructokinase encoding gene is crucial for achieving high production of D-lactate in *Corynebacterium glutamicum* under oxygen deprivation. *Applied Microbiology and Biotechnology* 99 (11): 4679–4689.

Van Andel, J., Zoutberg, G., Crabbendam, P., and Breure, A. (1985). Glucose fermentation byClostridium butyricum grown under a self generated gas atmosphere in chemostat culture. *Applied Microbiology and Biotechnology* 23 (1): 21–26.

van Duuren, J.B., Wijte, D., Karge, B. et al. (2012). pH-stat fed-batch process to enhance the production of cis, cis-muconate from benzoate by Pseudomonas putida KT2440-JD1. *Biotechnology Progress* 28 (1): 85–92.

van Maris, A.J., Geertman, J.-M.A., Vermeulen, A. et al. (2004). Directed evolution of pyruvate decarboxylase-negative *Saccharomyces cerevisiae*, yielding a C2-independent, glucose-tolerant, and pyruvate-hyperproducing yeast. *Applied and Environmental Microbiology* 70 (1): 159–166.

Vandenberghe, L.P., Soccol, C.R., Pandey, A., and Lebeault, J.-M. (1999). Microbial production of citric acid. *Brazilian Archives of Biology and Technology* 42 (3): 263–276.

Vandenberghe, L.P., Soccol, C.R., Pandey, A., and Lebeault, J.-M. (2000). Solid-state fermentation for the synthesis of citric acid by *Aspergillus niger*. *Bioresource Technology* 74 (2): 175–178.

Villadson, J. 2013. Industrial production of succinic acid – A stepwise approach, Vol. 2017.

Wang, J., Wu, Y., Sun, X. et al. (2017). De novo biosynthesis of glutarate via α-keto acid carbon chain extension and decarboxylation pathway in *Escherichia coli*. *ACS Synthetic Biology* 6 (10): 1922–1930.

Wang, L., Zhao, B., Liu, B. et al. (2010). Efficient production of L-lactic acid from corncob molasses, a waste by-product in xylitol production, by a newly isolated xylose utilizing Bacillus sp. strain. *Bioresource Technology* 101 (20): 7908–7915.

Wang, Q., He, P., Lu, D. et al. (2002). Screening of pyruvate-producing yeast and effect of nutritional conditions on pyruvate production. *Letters in Applied Microbiology* 35 (4): 338–342.

Wang, Z., Lin, M., Wang, L. et al. (2015). Metabolic engineering of *Propionibacterium freudenreichii* subsp. shermanii for enhanced propionic acid fermentation: effects of overexpressing three biotin-dependent carboxylases. *Process Biochemistry* 50 (2): 194–204.

Wang, Z. and Yang, S.-T. (2013). Propionic acid production in glycerol/glucose co-fermentation by *Propionibacterium freudenreichii* subsp. shermanii. *Bioresource Technology* 137: 116–123.

Warhurst, A.M., Clarke, K.F., Hill, R.A. et al. (1994). Production of catechols and muconic acids from various aromatics by the styrene-degraderRhodococcus rhodochrous NCIMB 13259. *Biotechnology Letters* 16 (5): 513–516.

Wei, D., Liu, X., and Yang, S.-T. (2013). Butyric acid production from sugarcane bagasse hydrolysate by *Clostridium tyrobutyricum* immobilized in a fibrous-bed bioreactor. *Bioresource Technology* 129: 553–560.

Werpy, T., Petersen, G. 2014. Top value added chemicals from biomass. Results of screening for potential candidates from sugars and synthesis gas. US. Department of Energy. 2004. *There is no corresponding record for this reference.*

Wieschalka, S., Blombach, B., Bott, M., and Eikmanns, B.J. (2013). Bio-based production of organic acids with *Corynebacterium glutamicum*. *Microbial Biotechnology* 6 (2): 87–102.

Wu, C.-M., Lee, T.-H., Lee, S.-N. et al. (2004). Microbial synthesis of cis, cis-muconic acid by Sphingobacterium sp. GCG generated from effluent of a styrene monomer (SM) production plant. *Enzyme and Microbial Technology* 35 (6): 598–604.

Wu, Z. and Yang, S.T. (2003). Extractive fermentation for butyric acid production from glucose by *Clostridium tyrobutyricum*. *Biotechnology and Bioengineering* 82 (1): 93–102.

Yahiro, K., Shibata, S., Jia, S.-R. et al. (1997). Efficient itaconic acid production from raw corn starch. *Journal of Fermentation and Bioengineering* 84 (4): 375–377.

Yahiro, K., Takahama, T., Park, Y.S., and Okabe, M. (1995). Breeding of *Aspergillus terreus* mutant TN-484 for itaconic acid production with high yield. *Journal of Fermentation and Bioengineering* 79 (5): 506–508.

Yamane, T. and Tanaka, R. (2013). Highly accumulative production of L (+)-lactate from glucose by crystallization fermentation with immobilized *Rhizopus oryzae*. *Journal of Bioscience and Bioengineering* 115 (1): 90–95.

Yanai, T., Tsunekawa, H., Okamura, K., Okamoto, R. 1994. Manufacture of pyruvic acid with Debaryomyces. Japanese Patent, JP 0,600,091.

Yang, S.T., Huang, Y., and Hong, G. (1995). A novel recycle batch immobilized cell bioreactor for propionate production from whey lactose. *Biotechnology and Bioengineering* 45 (5): 379–386.

Yin, X., Madzak, C., Du, G. et al. (2012). Enhanced alpha-ketoglutaric acid production in *Yarrowia lipolytica* WSH-Z06 by regulation of the pyruvate carboxylation pathway. *Applied Microbiology and Biotechnology*: 1–11.

Yokota, A., Henmi, M., Takaoka, N. et al. (1997). Enhancement of glucose metabolism in a pyruvic acid-hyperproducing *Escherichia coli* mutant defective in F1-ATPase activity. *Journal of Fermentation and Bioengineering* 83 (2): 132–138.

Yokota, A., Terasawa, Y., Takaoka, N. et al. (1994). Pyruvic acid production by an F1-ATPase-defective mutant of *Escherichia coli* W1485lip2. *Bioscience, Biotechnology, and Biochemistry* 58 (12): 2164–2167.

Yokoya, F. (1992). Citric acid production. *Industrial Fermentation Series*: 1–82.

Yu, J.L., Xia, X.X., Zhong, J.J., and Qian, Z.G. (2014). Direct biosynthesis of adipic acid from a synthetic pathway in recombinant *Escherichia coli*. *Biotechnology and Bioengineering* 111 (12): 2580–2586.

Yunhai, S., Houyong, S., Deming, L. et al. (2006). Separation of glycolic acid from glycolonitrile hydrolysate by reactive extraction with tri-n-octylamine. *Separation and Purification Technology* 49 (1): 20–26.

Zahoor, A., Otten, A., and Wendisch, V.F. (2014). Metabolic engineering of *Corynebacterium glutamicum* for glycolate production. *Journal of Biotechnology* 192: 366–375.

Zelić, B., Gostović, S., Vuorilehto, K. et al. (2004). Process strategies to enhance pyruvate production with recombinant *Escherichia coli*: from repetitive fed-batch to in situ product recovery with fully integrated electrodialysis. *Biotechnology and Bioengineering* 85 (6): 638–646.

Zhang, A. and Yang, S.-T. (2009). Propionic acid production from glycerol by metabolically engineered *Propionibacterium acidipropionici*. *Process Biochemistry* 44 (12): 1346–1351.

Zhang, X., Wang, X., Shanmugam, K., and Ingram, L. (2011). L-malate production by metabolically engineered *Escherichia coli*. *Applied and Environmental Microbiology* 77 (2): 427–434.

Zhu, L., Wei, P., Cai, J. et al. (2012). Improving the productivity of propionic acid with FBB-immobilized cells of an adapted acid-tolerant *Propionibacterium acidipropionici*. *Bioresource Technology* 112: 248–253.

Zhu, Y., Eiteman, M.A., Altman, R., and Altman, E. (2008). High glycolytic flux improves pyruvate production by a metabolically engineered *Escherichia coli* strain. *Applied and Environmental Microbiology* 74 (21): 6649–6655.

Zhu, Y. and Yang, S.-T. (2004). Effect of pH on metabolic pathway shift in fermentation of xylose by *Clostridium tyrobutyricum*. *Journal of Biotechnology* 110 (2): 143–157.

9

Biofuels from Microalgae and Seaweeds: Potentials of Industrial Scale Production

Licheng Peng[1], Freeman Lan[2] and Christopher Q. Lan[2]

[1]*Department of Environmental Science, Hainan University, Haikou, Hainan province, China, 570228*
[2]*Department of Chemical and Biological Engineering, University of Ottawa, Ottawa, ON, Canada K1N 6N5*

CHAPTER MENU

9.1 Introduction

Fossil fuels, composed mainly of petroleum, coal, and natural gas, account for up to 90% of energy supplies in the world and are predicted to contribute more than three-fourths of the world's energy in 2040 (Sieminski 2016). The demand for energy continues to rise with the fast expansion of industrialization and growth of population. In developing countries such as China and India, increases in economic activity as measured by the Gross Domestic Product (GDP) are correlated with energy use (Figure 9.1) (EIA 2016). The Energy Information Administration (EIA) predicts that the world energy consumption will increase at an average rate of 1.4% per year in the period 2012–2040, corresponding to an increase from 549 quadrillion Btu in 2012 to 629 and 815 quadrillion Btu in 2020 and 2040, respectively. The rapid increase of energy consumption will inevitably impact the energy and environment sustainability of the affected regions.

While fossil fuels are essential in modern society, if for nothing else then for transportation, the negative effects of their combustion have become increasingly threatening. The combustion of fossil fuels contributes to the emission of substances such as carbon monoxide (CO), nitrous oxides (NO_x), or sulfur dioxide (SO_2), which intensify global atmospheric pollution (Türe et al. 1997). According to the World Health Organization (WHO), air pollution caused the deaths of around seven million people worldwide in 2012 (World Health Organization 2012). These deaths are concentrated in developing countries of Asia and correlate to the increase in fossil fuels use in the developing economies (Türe et al. 1997). World Energy Outlook (WEO) Special Report 2016 Energy and Air Pollution (IEA 2016) presented a strategy, in the form of a Clean Air Scenario, on the basis

Green Energy to Sustainability: Strategies for Global Industries, First Edition.
Edited by Alain A. Vertès, Nasib Qureshi, Hans P. Blaschek and Hideaki Yukawa.

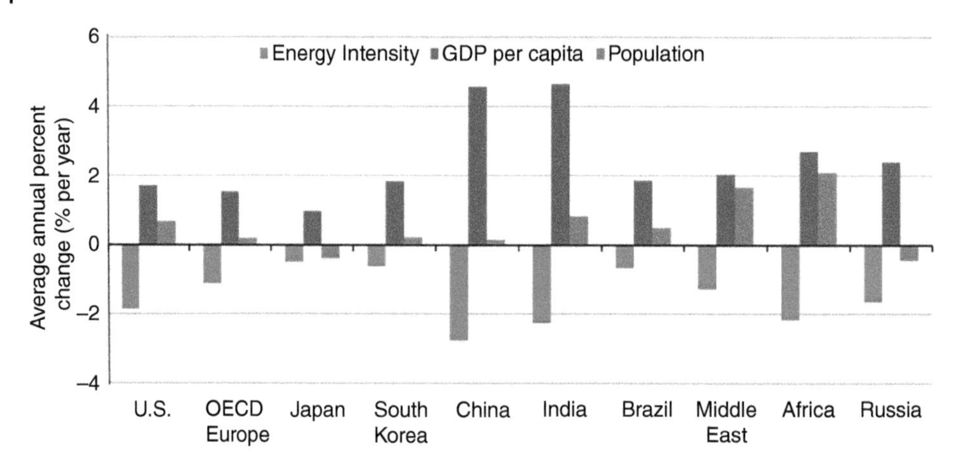

Figure 9.1 Average annual percent change of energy intensity, population, and GDP of selected countries and regions (2012–2040) (EIA 2016).

of existing technologies, tailored to local conditions, and relies on actions as follows: (i) setting an ambitious long-term air quality goal; (ii) executing cost-effective clean air measures for the energy sector; and (iii) ensuring strict monitoring and enforcement, and effective communication. These countermeasures, if executed properly in alignment with the acceleration of transition to clean energy, is expected to help control energy-related pollution levels. The International Energy Agency (IEA) (IEA 2016) estimated that the premature deaths caused by outdoor air pollution would decline by 1.7 million in 2040 under the Clean Air Scenario. Recently, nations have increasingly turned their attention to the development and use of bio-energies (bio-fuels) as a promising method of pollution mitigation (International Energy Agency 2014; International Energy Agency 2016; International Energy Agency 2016).

9.2 Biofuels

Renewable energy sources such as photovoltaic power, hydropower, wind power, geothermal power and tide power, they are produced as electricity or heat, both of which are difficult to store for long-term usage or to transport long distances. For example, some solar projects have aimed to produce electricity in North Africa deserts and transport it to Europe. However, some people remain dubious on the plan, and regarded it as unrealistic, owing to the added transportation costs to get the electricity to the EU, requirements for large road networks and transmission grids (Lisa Friedman 2011). As a comparison, the most important advantages of biofuels lie in the fact that their physical properties and combustion characteristics are very similar to those of fossil fuels and therefore could be utilized without any significant modification of existing infrastructures for storage, transportation, and combustion. In addition, all forms of renewable energy have the common merits of being sustainable, potentially CO_2-neutral, and of low or zero air pollution. As compared with conventional diesel fuels, the biodiesel combustion emission of residual hydrocarbons, carbon monoxide, greenhouse gas (GHG) and particulates are much lowered, and as well as the emission of sulfur dioxide when it burns in an engine because biodiesel is produced from natural sources which contain very small amounts of sulfur (Madiwale and Bhojwani 2016; Carneiro et al. 2017).

9.2.1 Types of Biofuels

Biofuels are produced primarily using biomass as feedstocks. Plants and algae fix carbon dioxide to produce biomasses and, in this process, capture and convert solar energy into chemical potential (Cadenas and Cabezudo 1998). The resulting biomass is then converted into different forms of biofuels through a variety of processes including thermochemical or biochemical processes (Sujith Sukumaran 2013). In other cases, some microalgal species have the capacity to catalyse the photolysis of water utilizing light as an energy source to produce hydrogen. In such a scenario, microalgal cells serve as the catalysts instead of feedstock (Oey et al. 2016). Despite of the enormous technical challenges that have so far impeded their commercial production, algae-based biofuels have been deemed to be the most promising biofuels for the future and transformational technology breakthroughs that would bring the overall process efficiency to a commercially viable level are expected in the coming years (Brennan and Owende 2010; Lin 1985; Desmorieux and Decaen 2005; Liang et al. 2012; Zuorro et al. 2016; Santana et al. 2012).

9.2.1.1 Biodiesel

Biodiesel is the monoalkyl ester of long chain fatty acids derived from renewable feedstock such as algal oil, vegetable oil, or animal fat (Tan et al. 2017; Mankee and Lee 2012). It has attracted the attention of manufacturers and researchers and is named as such because of the similarity between its properties including viscosity, density, energy content, and combustion behaviours and those of conventional diesel (Rawat et al. 2013). Biodiesel can be used in conventional diesel engines with little or no modification while being less eco-toxic. Biodiesel can be blended with any portion of conventional diesel or used by itself (Meher et al. 2006).

Biodiesel is produced through the transesterification of triglycerides with methanol or ethanol. Equations (9.1) and (9.2) illustrate the general equation and methanolysis of triglyceride in transesterification reactions, respectively (Meher et al. 2006).

$$\underset{\text{Ester}}{RCOOR^1} + \underset{\text{Alcohol}}{R^2OH} \xleftrightarrow{\text{Catalyst}} \underset{\text{Ester}}{RCOOR^2} + \underset{\text{Alcohol}}{R^1OH} \tag{9.1}$$

$$\begin{array}{llll} CH_2-OCOR1 & & CH_2OH & R^1COOCH_3 \\ | & & | & \\ CH-OCOR2 + 3CH_3OH & \overset{\text{Catalyst}}{\rightleftharpoons} & CH\ \ OH + R^2COOCH_3 & \\ | & & | & \\ CH_2-OCOR3 & & CH_2OH & R^3COOCH_3 \\ \text{Triglyceride} \quad \text{Methanol} & & \text{Glycerol} & \text{Methyl esters} \end{array} \tag{9.2}$$

Biodiesel can also be produced by the esterification of free fatty acids (Fukuda et al. 2001). The catalysts used for transesterification and esterification could be bases, acids, or lipases. The chemical composition of biodiesel, in particular the chain length of the fatty acids, depends on the source of the triglycerides (Mankee and Lee 2012).

9.2.1.2 Bioethanol

Bioethanol is ethanol produced by sugar fermentation processes. Bioethanol is well established as a transport fuel (Gray et al. 2006) because of its high octane number and the ability to replace lead as an octane enhancer in petrol. Carbohydrates produced from sugar or starch crops such as sugarcane and corn are converted to alcohol (i.e. ethanol) under the action of microorganisms, usually yeasts (Demirbas 2007). The conversion reactions are comprised of two steps. The first step is the hydrolysis of starch or sucrose to produce glucose and the second is the conversion of glucose to ethanol as shown in Eq. (9.3).

$$C_6H_{12}O_6 \rightarrow 2C_2H_5OH + 2CO_2 \tag{9.3}$$

9.2.1.3 Bio-oils and Bio-syngas

Under anoxic conditions at high temperature, biomass can be processed to produce a mixture of vapour, liquid, and solids (Wang and Lan 2009). The liquid product is a complex mixture called bio-oil, the composition of which varies significantly with the choice of biomass and processing conditions. Bio-oils can substitute conventional fuels in internal combustion engines either totally or partially as a blend (Rawat et al. 2013). Biogas, the product (i.e. CH_4, CO_2, N_2, H_2, H_2S) of anaerobic digestion, is an environmentally friendly, clean, cheap, and versatile fuel (Kapdi et al. 2005). It is usually further processed to (i) synthesize products such as methanol, oils, diesel, ammonia, and methane,(ii) be a source of pure hydrogen, and (iii) be fuels to generate electricity and/or heat directly.

9.2.1.4 Bio-hydrogen

Bio-hydrogen is defined as hydrogen produced biologically, commonly by algae and bacteria (Meher Kotay and Das 2008). Biological processes for producing biohydrogen mainly include biophotolysis, photodecomposition, and fermentations (e.g. photo- and dark-fermentation) (Wang and Lan 2009; Levin et al. 2004). Direct biophotolysis can be achieved by photoautotrophic organisms that utilize light energy to split water into hydrogen and oxygen. Indirect biophotolysis is carried out by cyanobacteria in the absence of nitrogen (Benemann 2000). Photodecomposition is the production of hydrogen produced by non-sulfur reducing bacteria in the presence of some organic compound (e.g. CH_3COOH) and light energy, under conditions of nitrogen deficiency. Fermentation approaches include photo-fermentation, which produces hydrogen by algae and photosynthetic bacteria in the presence of light energy and oxygen, and dark fermentation, where hydrogen is produced by hydrogenase in dark and anaerobic fermentation conditions (Benemann 2000).

9.2.2 Feedstock of Biofuel Production Based on Plants

Biofuels are generally divided into three generations according to the source of biomass (Table 9.1) (Brennan and Owende 2010; Chisti 2010; Voet and Luo 2010; Chisti 2008; Zhang et al. 2003; Stephenson et al. 2011). The first-generation biofuels are derived from agricultural edible crops and compete with food production, entailing negative impacts on food security and suffering from the high cost of food crops as well (Rawat et al. 2013). Food wastes such as waste cooking oil and fats could be used to produce biodiesel in a cost-effective manner (Ahmad et al. 2011) and the price is variable and in general is approximately half that of virgin oil Nevertheless, the supplies of these food wastes are too small to be regarded as a meaningful supply of feedstocks for the biofuels intending to substitute for fossil fuels (Rawat et al. 2013).

The second-generation biofuels use cellulosic biomasses (usually terrestrial plants) as feedstock. While the supplies of such biomasses are relatively abundant, the conversion of cellulose to biofuels is of low efficiency and high costs (Balan 2014; Ye and Cheng 2002). The third-generation

Table 9.1 Comparison of some sources of biodiesel (Brennan and Owende 2010; Voet and Luo 2010; Chisti 2008; Zhang et al. 2003; Stephenson et al. 2011; Ahmad et al. 2011).

Biofuel generation	Examples	Products		Land use (M ha)	Percent of existing US cropping area[a]	Land requirement	Oil yield (l/ha)	Potential fuel yield (l biofuel/ha/yr)	Primary product cost (US$/l)		Refs
		Primary	Secondary						Current	Potential	
1st	Corn	Bioethanol	Biomethane	1540	846	High-quality agricultural land	172	200–7500	0.45–0.55	0.40–0.50	Brennan and Owende (2010); Voet and Luo (2010); Chisti (2008); Zhang et al. (2003); Stephenson et al. (2011); Ahmad et al. (2011)
	Soybean	Biodiesel	Distillers grain	594	326		446				
	Canola		Animal feed	223	122		1190				
	Coconut			99	54		2689				
	Oil palm			45	24		5950				
2nd	Jatropha	–	Bioethanol Solid fuel Hydrogen gas	140	77	Marginal Land	1892	5000–5012 000	0.80–1.20	0.55–0.70	Brennan and Owende (2010); Chisti (2008)
	Waste cooking oil	Biodiesel									Stephenson et al. (2011)
3rd	Microalgae[b]	Biodiesel	Bioethanol	2	1.1	Low-quality land	136 900	50 000–140 000	1.50–2.50	0.50–1.00	Zhang et al. (2003); Ahmad et al. (2011)
	Microalgae[c]	Hydrogen gas	Biomethane Glycerol Animal feed Pigments	4.5	2.5		58 700				

a) For meeting 50% of all transport fuel needs of the United States.
b) 70% oil (by wt) in biomass.
c) 30% oil (by wt) in biomass.

biofuels are derived from aquatic plants, which are primarily algae (Chisti 2010; Antoni et al. 2007) including microalgae and macroalgae (seaweeds). Starch and triglycerides produced from algae are easy to process while, at the same time, do not compete for land resources with food production. It is hopeful that, with further advances of the technologies of microalgal farming and ocean farming, biofuels could be produced at a magnitude and cost that make them reliable substitutes for fossil fuels in the future. Thus far, all biofuels have been ultimately derived from plants.

9.2.2.1 Terrestrial Plants

Vegetable oils produced by a range of oil-bearing crops such as rapeseed, sunflower, soybean, oil palm, jatropha, corn, cassava, and sugar beet have been proven to be excellent feedstock for biodiesel production through transesterification (Mankee and Lee 2012). Triacylglycerols, the key molecules for producing biodiesel, are stored in seeds. During seed development, photosynthate from the mother plant is imported in the form of sugars and then converted into precursors of fatty acids through glycolysis (Durrett et al. 2008). As early as the 1930s and 1940s, vegetable oils were already used as fuels (Shay 1993) in many countries including the United States, Brazil, China, Thailand, India, Indonesia, Malaysia, the Philippines, South Africa, and Zimbabwe (Onion and Bodo 1983).

Carbohydrates are typically the feedstock for first-generation bioethanol production. Food crops such as sugar cane (Kim and Day 2011), beets (Razmovski and Vučurović 2012), corn, potatoes, and sweet potatoes (Dewan 2013), and the waste from their processing (e.g. molasses) are excellent sources of sugars, which could be fermented to produce ethanol as biofuel.

As previously mentioned, the production of these first-generation biofuels is in direct competition with food supply, in particular the competition with food production for land use and water resources, thus limiting the sustainability and their potential for large scale production as a meaningful replacement for fossil fuels (Popp et al. 2014). However, some experts do not agree with the perception that expansion of biofuels will seriously compete with food supply (Nogueira et al. 2013), and Food and Agriculture Organization (FAO) reported that the increment in food productivity could fulfil more than 80% of the food and feed global future demand (Devereux 2009).

Cellulosic biomasses comprise a range of materials such as wood, straw, corn stalks, and grass as feedstocks. These biomasses can be converted to bio-oils, bio-chars, and biosyngas through thermochemical processes such as pyrolysis (Guo et al. 2015), while their cellulose contents could be enzymatically converted to sugars (e.g. glucose and xylose) to be further fermented and thereby increase the overall ethanol or butanol yields (AFDC 2012).

9.2.2.2 Aquatic Plants – Algae

Algae are comprised of microalgae and macroalgae (seaweeds). Microalgae are conventionally defined as all unicellular or simple multicellular photosynthetic microorganisms including both eukaryotes (i.e. eukaryotic microalgae) and prokaryotes (i.e. cyanobacteria and other colour bacteria). Algae have been used for the production of a large variety of different products including food, animal feeds, high-value bioactives (Morand et al. 2004), and, most relevantly, biodiesels (Mcdermid and Stuercke 2003). As biofuel feedstock, microalgae are at this stage considered to be more attractive than macroalgae because of their faster growth and higher lipid or starch contents (Abomohra et al. 2016; Abomohra et al. 2016). It is worth noting that the studies on microalgae as biofuel feedstock have increased significantly since 2008 according to the retrieval results of the databas ScienceDirect (Figure 9.2) (Harun et al., 2010; Silva and Bertucco 2016; Sharma and Singh 2017).

9.3 Biofuels from Microalgae and Seaweeds

The potential of microalgae and macroalgae (i.e. seaweeds) as the feedstock for biofuels such as bioethanol, biodiesel, biogas, biohydrogen, and their responding advantages are discussed in this section. The properties of microalgal biodiesel similar to diesels and their advantages as the feedstock of biofuels will be deployed with time, based on the expanding demands of biofuels in the world.

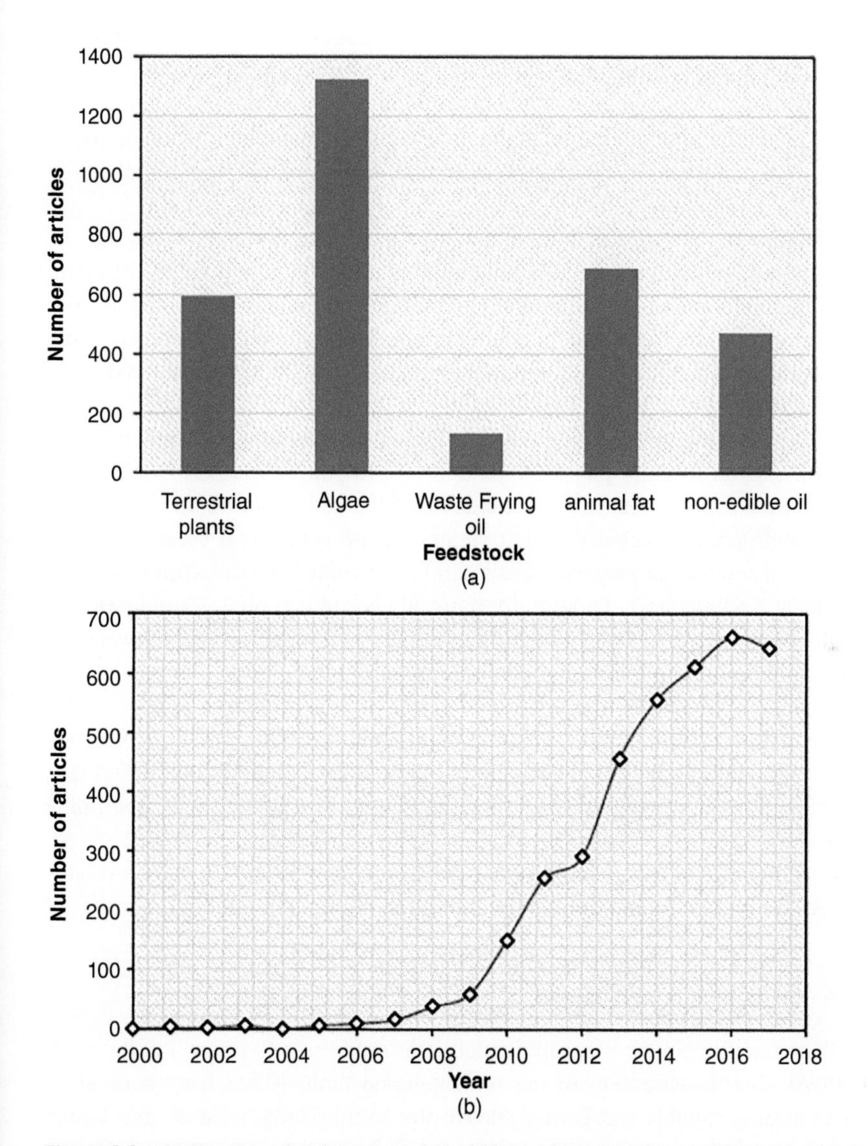

Figure 9.2 (a) Number of articles based on different feedstock in year 2014, and (b) number of articles on algal biodiesel from year 2000–2017 (Harun et al. 2010; Silva and Bertucco 2016).

9.3.1 Biofuels from Microalgae

In this chapter we adopt a conventional definition of microalgae so that they include all unicellular and simple multi-cellular photosynthetic microorganisms including both prokaryotic microalgae (cyanobacteria and other colour bacteria) and eukaryotic microalgae (Peng 2016). Microalgae are widely found in oceans and other habitats such as lakes, rivers, ponds, wetlands, deserts and even the north and south poles (Wang and Lan 2009). There are an estimated one to ten million microalgal species on earth with more than 40 000 species identified to this date (Wang and Lan 2009). Microalgae are generally fast growing and capable of high-efficiency photosynthesis. They contain carbohydrates that can be used as carbon sources for bioethanol production via fermentation (Harun et al. 2010), and some species are potent feedstock for biodiesel production because of high lipid content. Additionally, microalgae are capable of fixing CO_2 from the atmosphere, flue gases (e.g. the gas existing in the atmosphere via a flue in industry), or soluble carbonate while capturing solar energy with efficiency 10–50 times greater than terrestrial plants (Li et al. 2008; Khan et al. 2009).

Microalgae as the feedstock for biofuels have been widely studied for the production of bioethanol, biodiesel, biogas, biohydrogen, and others. Some microalgal species accumulate large quantities of storage materials in the form of carbohydrates (Wang and Lan 2009; Milano et al. 2016), which can in turn be converted into bioethanol. For instance, *Chlorococum* sp., and *Chlamydomonas perigranulata* biomass is reported to be a good feedstock for producing bioethanol (Harun et al. 2010).

The possible routes for producing microalgae-based bioethanol include the traditional process, through metabolic pathways and through photofermentation (Silva and Bertucco 2016). The traditional process involves pretreatment steps (e.g. breakdown of the cell structure and hydrolysis of biomass), enzymatic hydrolysis (to achieve higher yields of conversion), and yeast fermentation. Metabolic pathways of microalgal photosynthesis can be controlled in dark conditions, for producing hydrogen, acids, and alcohols. Though not naturally occurring, photofermentation is made possible through genetic engineering microalgae to redirect the biochemical pathways for photosynthesis into ethanol production (Silva and Bertucco 2016).

The lipid content of microalgae varies from 14% to 30% dry cell weight when under rapidly growing conditions and up to approximately 70% dry cell weight under stressed conditions (Khozin-goldberg and Cohen 2006). Thus, some microalgal species are excellent feedstocks for biodiesel production. Transesterification and esterification are common ways to convert microalgal lipids to biodiesel. In the transesterification process, three reversible steps in series are included, where triglycerides are converted to diglycerides, and diglycerides to monoglycerides, and then monoglycerides are converted to esters (biodiesel) and byproduct glycerol (Mata et al. 2010).

9.3.2 Biofuels from Macroalgae (i.e. Seaweeds)

Macroalgae (seaweed) are multi-cellular plant-like eukaryotic organisms (Annam Renita et al. 2010; Klöser et al. 1996). The development of macroalgae-based biofuels has been studied in numerous countries including notably the United States, the Netherlands, Ireland, and South Korea (Jiang et al. 2016). Macroalgae can be utilized as feedstock for ethanol fermentation (Harun et al. 2010; Adams et al. 2008; Seo et al. 2010; Kraan 2013), with the most efficient macroalgae strains for ethanol production include *Sargassum, Glacilaria, Saccharina latissima, Prymnesium parvum*, and *Euglena gracilis*, owing to their large ratio content of carbohydrates in biomass (Ziolkowska and Simon 2014; Matanjun et al. 2009).

These organisms can also produce lipids and free fatty acids, which can be used as feedstocks for biodiesel production (Chen et al. 2015; Tamilarasan and Sahadevan 2014; Vassilev and Vassileva 2016). The chemical process involved in the conversion of macroalgal oil to biodiesel is transesterification, in which alcohol is added to convert the triglycerides into hydrocarbon chains, with by-product of glycerine and the end product of fatty acid methyl esters (Gallagher 2011).

Macroalgae can also be used for biogas production, owing to their high quantities of polysaccharides, lipids, and low quantities of lignocellulosic materials particularly recalcitrant to degradation (Vergara-Fernández et al. 2008). For example, under the conditions of fermentation investigated by Singh and Gu (2010) and Parmar et al. (2011), the methane production of brown macroalgae *Laminaria digitata* and *Macrocystis* sp. was high and reached 0.500 m^3 CH_4/kg, and 0.390–0.410 m^3 CH_4/kg, respectively. While the methane production of *Gracilaria* sp. and *Laminaria* sp. was 0.28–0.40 m^3 CH_4/kg (Bird et al. 1990) and 0.260–0.280 m^3 CH_4/kg (Chynoweth 2005), respectively.

9.3.3 Advantages of Algae as the Feedstock of Biofuels

As shown in Figure 9.3, first-generation biofuels currently account for more than 95% of the biodiesel on the market (Demirbas 2011). However, the productions of these fuels lead to increases in food cost, making them unattractive for larger-scale use (Rawat et al. 2013; Demirbas 2011). On the other hand, second-generation biofuels, although not competing directly with the food supply, are economically challenging due to the complexity and costs associated with the consumption of cellulosic biomasses. In contrast, microalgae do not compete with food crops and have high oil productivities. Thus, the mass production of microalgal oil potentially provides an alternative and eco-sustainable substitute for the crude fossil petroleum.

Algae, especially microalgae, are fast growing, rich in lipid, and require less land and water compared with conventional terrestrial plants (Table 9.1). The cultivation system of microalgae can be placed on any low-quality and infertile lands to bypass competition with agricultural crops (Stephenson et al. 2011). At the same time, the land use for algae cultivation (based on an assumption of 30% oil by weight in biomass) is roughly 342, 50, and 31 times more efficient than that of corn, canola, and Jatropha, respectively (Chisti 2010).

Besides growing on lower quality land, the oil yield of microalgae is much higher compared to conventional terrestrial plants. For instance, the oil production of some microalgal species could reach up to approximately 136 900 l/ha/yr on a basis of 70% oil by weight in biomass compared to the 5 950 l/ha/yr of oil pam, which is the highest oil producer among investigated terrestrial plants (Table 9.1) (Chisti 2010).

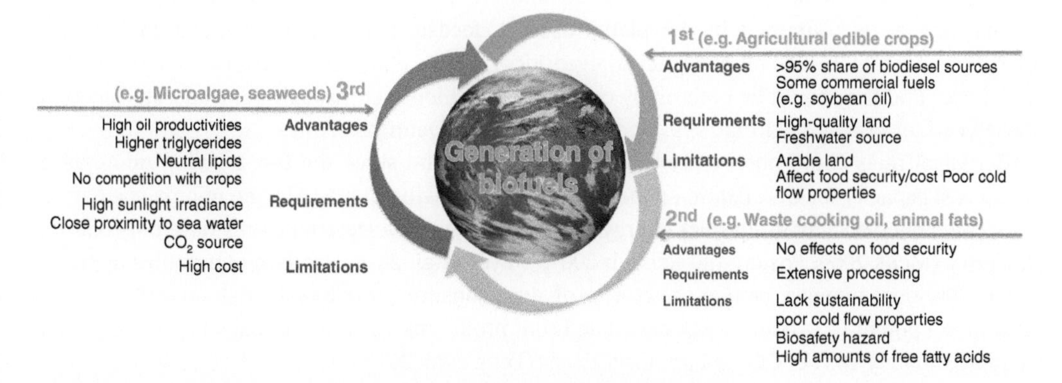

Figure 9.3 Comparisons of first, second, and third generation of biofuels.

Table 9.2 Comparison of properties of biodiesel from microalgal oil and diesel fuel and ASTM biodiesel standard.

Properties	Biodiesel from microalgal oil	Diesel fuel (EN 590:1999)	ASTM D6751 07b	EN 14214:2008
Density (kg/l)	0.864	0.838	–	0.86–0.90
Viscosity (mm^2/s, cSt at 40 °C)	5.2	2.0–4.5	1.9–6	3.5–5
Flash point (°C)	115	>55	>93	>101
Solidifying point (°C)	−12	−50 to 10	–	–
Cold filter plugging point (°C)	−11	*a*	*a*	*a*
Acid value (mg KOH/g)	0.374	Max 0.5	Max 0.5	
Heating value (MJ/kg)	41	40–45	–	–
H/C ratio	1.81	1.81	–	–
Cetane number		51 min	47 min	51 min
Price		£94.00		

a) Country specific
a. BS EN 590 : 2013 European standard for Automotive fuels
b. BS EN 14214–2008 Automotive fuels-Fatty acid methyl esters (FAME) for diesel engines
c. BS EN 14214 : 2012 + A1: 2014, liquid petroleum products. Fatty acid methyl esters (FAME) for use in diesel engines and heating applications. Requirements and test methods
Source: Adapted from (Meher et al. 2006; McKendry 2002).

The properties of microalgal biodiesel are similar to those of petroleum-derived diesel, making it straightforward to replace for it (Rawat et al. 2013). As shown in Table 9.2 (Rawat et al. 2013; Miao and Wu 2006), the density of microalgal biodiesel is 0.864 kg/l, which meets the density range of the European Union biodiesel standard (EN 14214:2008). Other properties such as viscosity, solidifying point, cold filter plugging point, heating value, and H/C ratio are in the ranges of the three selected diesel standards (i.e. EN 590:1999, ASTM D675107b, and EN 14214:2008) in Table 9.2. Flash point and acid value are on the other hand parameters that are slightly higher or lower than the standard, but this could be modified by technologies in the future.

The conversion of lipids into conversion using microalgae biomass is less complex and more cost-effective than that of lignocellulosic biomass (Harun et al. 2010; Seungphill et al. 2010). Similarly, the pyrolysis of microalgae biomass is less complex and more economical than that from lignocellulosic biomass. This is because microalgae contain higher amounts of easily-pyrolysed contents such as cellular lipid, polysaccharide, and protein. Moreover, unlike the high amounts of free fatty acids contained in the plants used as feedstock for second-generation biofuels, microalgae contain 25–250 times higher triglycerides (Ahmad et al. 2011; Harun et al. 2010; Huber et al. 2006) and they could be potentially converted into biodiesel through the transesterification of triglycerides (Ma and Hanna 1999; Demirbas and Fatih Demirbas 2011).

Microalgal fuels could potentially be made carbon-neutral since the fast-growing microalgae have a vast capacity for CO_2 fixation compared to slow-growing plants. Marine microalgae have a 3–15% efficiency of converting solar energy into biomass while terrestrial plants have only 0.2–2% efficiency (Melis 2009; Posten and Schaub 2009; Weyer et al. 2010). Carbon constitutes approximately 50% algae biomass, so that the growth of algae consumes a relatively high quantity of CO_2, making it a potent CO_2 fixer. It was calculated that producing 100 t of microalgal biomass would fix approximately 183 t of CO_2 via photosynthesis (Peng et al. 2015).

9.4 Recent Developments in Algae Processing Technologies

Once the maximum algal biomass is reached in cultivation, the processes such as harvesting and dewatering, extraction of lipids, and conversion to biofuels would follow. Thus, the algae processing technologies directly impact the efficiency of algae-based fuels production.

9.4.1 Harvesting and Dewatering

Harvesting of microalgae can generally be divided into two steps, i.e. bulk harvesting and thickening/dewatering (Chen et al. 2011). Harvesting microalgae is difficult and the cost of this manufacturing operation can represent up to 20–30% of the total biomass production costs (Mata et al. 2010). This is due to a number of factors including the small size and low density of the cells, and the negative charge and excess algogenic organic matters (AOM) that prevent the cells from aggregating (Danquah et al. 2009). Typically, flocculation, flotation, or gravity sedimentation is used to separate microalgal biomass from the bulk suspension yielding 2–7% of the total solid matter (Brennan and Owende 2010). Other major harvesting techniques include precipitation, centrifugation have been applied in some companies (www.oilseedcrops.com; www.oilseed.org) (Figure 9.4), filtration, screening, ion exchange, passage through a charged zone, ultrasonic vibration, as well as electrophoresis techniques (Uduman et al. 2010). The selection of harvesting techniques depends on the cell size, density, and the properties of the algae strain, as well as the economic value of the desired products (Danquah et al. 2009). Filtration and centrifugation are usually applied in the second step of dewatering, which aims to concentrate the slurry. Notably, this step consumes more energy than the first step (Brennan and Owende 2010). Dewatering is one of the most energy intensive processes in microalgal culture, which accounts for 600 kWh/d and drying consumes approximate 3 000 kWh/d of the total energy consumption (Pirwitz et al. 2016).

A variety of different commercial appliances are available for harvesting microalgae. For instance, Model 4 Algae Appliance (© 2012 OriginOil, Inc., www.originoil.com) (Figure 9.5) is an entry-level, low-cost algae harvester that provides a low energy (approximately 0.0002 kWh/l), chemical-free, continuous-flow 'wet harvest' system capable of processing 3 000–6 000 l/d in continuous harvest to efficiently dewater (i.e. removes up to 99% of the water from the algal culture), rupture cell walls, and concentrate the algae harvest. The wet algae can be directly fed from the growth system to the harvester, and the dewatering process uses electromagnetic pulses and a biomass concentrator. This appliance is sold at around $35 000–$50 000 (2017 prices) and is delivered on site.

Compared to microalgae, harvesting seaweed requires more pre-processing steps such as grinding or maceration before the biomass can be pumped into downstream systems (Elliott et al. 2015). Seaweed that is grown on solid substrates needs to be cut before harvesting. For those that are free-floating in water, petroleum-driven rotary cutters are applied. The harvested seaweed is collected by motorboat and then transported to land for further drying in the sun.

Since the chemical composition of macroalgal biomass may vary with the season, the careful timing of harvest (e.g. summer harvests of *L. digitata*) may enable optimal biofuel composition, ensuring the highest proportion of carbohydrate and lowest ash content in the biomass (Adams et al. 2011).

Figure 9.4 (a) Microalgae Centrifuge Extraction (Wang et al. 2008); (b) harvest intake (Yanagi et al. 1995).

9.4.2 Extraction Approaches

Drying of microalgae, the step prior to lipid extraction from microalgae, may be achieved by the approaches of spray-drying, drum-drying, freeze-drying, sun-drying, and oven-drying (Mata et al. 2010; Molina Grima et al. 2003). Then, the dried biomass is continuously extracted to obtain microalgal lipids that are used for producing biodiesel (Rawat et al. 2013). The choice of lipid-extracting approaches usually depends on the type of microalgal cells and the thickness of the cell walls (Cooney et al. 2011). Extraction methods can be classified as mechanical or non-mechanical. Mechanical methods mainly consist of cell homogenizers, bead mills, autoclave,

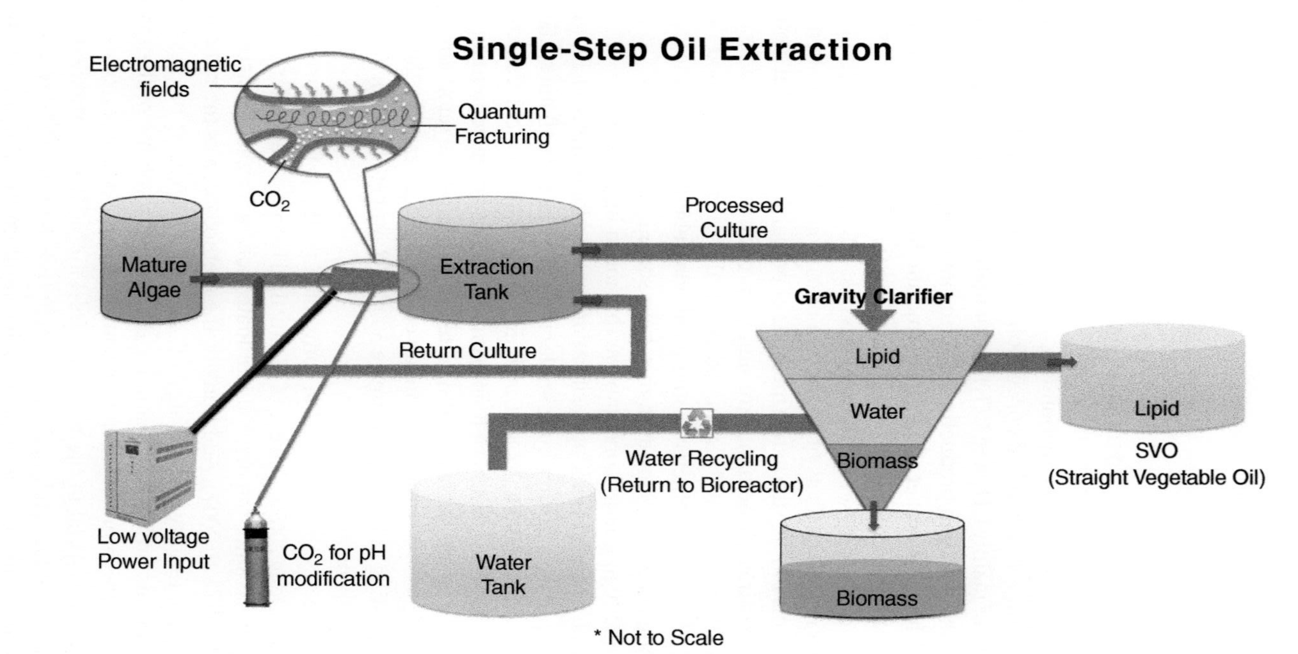

Figure 9.5 Electromagnetic process in single-step microalgal oil extraction designed by Origin Oil (Cheng et al. 2006).

ultrasounds, and spray drying. Non-mechanical methods include freezing, the utilization of organic solvents (e.g. polar and non-polar solvents), osmotic shock, acid, and base as well as enzymatic reactions (Mata et al. 2010), and supercritical carbon dioxide ($SCCO_2$) (Halim et al. 2011). Though many extraction methods for microalgal lipids have been developed at the laboratory scale, the procedures for large-scale extraction are complex and are still at the developmental stage (Mankee and Lee 2012).

9.4.3 Conversion to Biofuels

The technology for biofuel production from microalgae and seaweeds mainly consists of two processes that can be classified as thermochemical and biochemical/biological conversion. Thermochemical conversion processes are combustion (e.g. direct burning), pyrolysis, gasification, and thermochemical liquefaction. Biochemical conversion encompasses anaerobic digestion for producing biogas (a mixture of mainly CH_4 and CO_2), fermentation for producing ethanol (McKendry 2002), and photobiological hydrogen production (Sharma and Singh 2017).

As summarized in Table 9.3 (Rawat et al. 2013; Sharma and Singh 2017; Lin et al. 2011; Patil and Deng 2009; Wei et al. 2013), microalgal oils can be applied for direct use (e.g. direct burning) or blended with petroleum-derived diesel. This method is cost-effective, but it is limited by the high viscosity of oils, and the negative effects of deposition of carbon, coking formation, and thickening lubricating oil (Vellguth 1983). Alternative methods of preparing microalgal biofuels involving micro-emulsions and supercritical methanol are simpler processes that require less reaction time with good adaptability (Tan et al. 2017; Sawangkeaw et al. 2010). The method of supercritical methanol exhibits high conversion efficiency and good adaptability, requires less reaction time, and no alkaline soaps are generated. However, they have some limitations such as high viscosity, bad volatility/stability and high-energy consumption that mean higher costs.

Pyrolysis is an anaerobic thermal cracking gasification method for converting biomass into charcoal, bio-oil, or fuel gas, and then further into simpler molecules such as alkanes, alkenes, aromatics, carboxylic acids and others (Babu 2008). Catalytic pyrolysis could yield a mixture of various hydrocarbon molecules (gasoline) that have high aromatic hydrocarbon content and octane number (Huang et al. 2010). The factors such as heating rate, temperature, and retort atmosphere decide the chemical properties of pyrolysis products (Mohan et al. 2006), for example, the bio-oil yield can be maximized by faster heating rates to approximate 500 °C (Rawat et al. 2013), in consideration of the commercial fast pyrolysis installations.

Transesterification is one of the most suitable conversion methods for industrial-scale production of biodiesel (Patil and Deng 2009). The use of in situ algal biomass transesterification can lower the cost of the overall process since the method eliminates solvent extraction steps and minimizes the reaction time (Michaelj et al. 2007). Additionally, the method enables the reduction of the viscosities while increasing the fluidity of microalgal oil, as well as increasing biodiesel yields (Rawat et al. 2013). However, various problems remain, for example, the volume of the reacting alcohol increases in the process; furthermore, temperature control is required for achieving improved conversion yields. Proper stirring speed in the reaction vessel for biodiesel formation is also needed. Also, microalgae biomass contains high moisture content (greater than 115% w/w on a basis of oil weight), and this has a strong negative influence on the equilibrium—the final product's yield.

Table 9.3 Typical conversion technologies applied in the production of biodiesel (Meher et al. 2006; Silva and Bertucco 2016; Chung 2013).

Variable	Direct use or blending method	Micro-emulsion	Supercritical methanol	Pyrolysis	Transesterification
Theory/Reaction		A colloidal equilibrium dispersion of isotropic fluid microstructures with dimensions in 1–150 nm	Tyiglyceride + methanol – FAMEs + Glycerol	Simple heating in lack of oxygen, breaking chemical bonds to obtain small molecules[a)]	Triglycerides + methanol – fatty acid alkyl ester + glycerol (with catalyst)[b)]
Catalysts	No	No	No	No	Acid, alkali, enzyme[c)]
Advantageous	Availability, portability, heat content, renewability	Simple process	High conversion; less reaction time, good adaptability; no alkaline soaps generated	Simplify, cost effective, no-polluting	Similar fuel properties to diesel, suitable for industrial-scale production, high efficiency of conversion; reduced viscosity; simply
By-product	N/A		Glycerol		Glycerol
Limitations	High viscosity, deposition of carbon, coking formation, thickening of lubricating oil	Carbon deposition; thermal cracking; high temperature; high viscosity, bad volatility/stability	High temperature, pressure, energy consumption	Producing low value products, less purity	Low free fatty acid and water content are required (for alkali catalysis); pollutants produced; side reactions occurred; products separation required.
Cost	Cheap	Expensive	Expensive	Cost effective	Cheap

a) Under the process of pyrolysis, biomass can be converted to: (i) Charcoal (up to 35% yield, via carbonisation, slow pyrolysis); (ii) bio-oil (Up to 80% yield, via flash pyrolysis, low temperature); and (iii) fuel gas (up to 80% yield, via flash pyrolysis, low temperature).
b) Three steps are included, Step 1: Triglycerdises + alcohol (catalyst) – Diglycerides; Step 2: Diglycerides + alcohol (catalyst) – Monoglycerides; Step 3: Monoglycerides + alcohol (catalyst) – Fatty acid alkyl ester.
c) Acid catalysts: sulfuric acid, hydrochloric acid, sulfonic acids; Alkaline catalysts: NaOH, KOH; Enzymatic catalysts: lipase.

9.5 Potential for Industrial Scale Production

The potential of alga-based biofuels for industrial scale production will be considerable, not only in the context of their contribution in energy supply and environmental protection, but also the large marketplace of biofuel demand, the supporting policies for biofuel expansion.

9.5.1 Role of Biofuels in Energy Supply and Environmental Protection

Due to the fast depletion of fossil fuel reserves, renewable fuels including biofuels will play an increasingly important role in sustaining the energy demands of the world. As predicted by the EIA (Figure 9.6) (Sieminski 2016; Demirbas and Fatih Demirbas 2011), though fossil fuels will continue to represent more than three-fourths of the world energy use in 2040, other types of fuels will grow their share of the total energy supplied. Biodiesel is predicted to diversify the global energy sources and play a more crucial role as a potential combustible source among the alternative renewable energy in the near future (Rawat et al. 2013). Since more importantly fuel-bound oxygen in biodiesel which enables more complete combustion than diesel, the direct (tailpipe) emission of particulates can be reduced (Wang et al. 2016), and also decreases in polycyclic aromatic hydrocarbons (PAH) emissions as compared with the combustion of fossil-sourced diesel (Sadiktsis et al. 2014).

The consumption of fossil fuels is not only constrained by their sustainability because of the finite reserves but also by the negative environmental effects clearly perceived today in the form of important climate change. One primary environmental concern of fossil fuel combustion is CO_2 emission, which is considered to be the major contributor to global warming. In comparison, the biodiesel derived from microalgal oil is carbon-neutral and the burning of such biodiesel can therefore theoretically be considered to be zero CO_2-emission, which may be emphasized by the results of life cycle assessment (LCA). Available data of LCA results showed the high potential of algae-based biofuels, in terms of GHG emission reduction and land use occupation, but also showed some limitation in terms of energy efficiency (Carneiro et al. 2017). Importantly, unlike fossil fuels, biodiesel combustion does not emit toxic gases such as NO_x and SO_x into the atmosphere (Hu et al. 2008; Hulatt and Thomas 2011; Williams and Laurens 2010). Moreover, biodiesel fuel is less eco-toxic than fossil fuels to water and soil in the event of a spillage (e.g. biodiesel fuel, and diesel fuel), for example, petroleum-derived fuels were shown to be toxic to soil microorganisms at a concentration of 3% (w/w), whereas biodiesel fuel up to a concentration of 12% was demonstrated to be non-toxic at total soil saturation (Lapinskien et al. 2006).

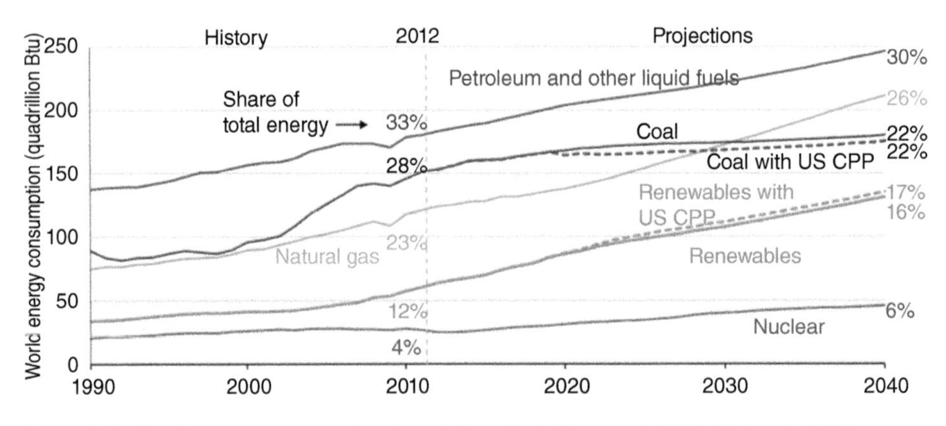

Figure 9.6 World energy consumption (quadrillion Btu) (Sieminski 2016; Wei et al. 2013).

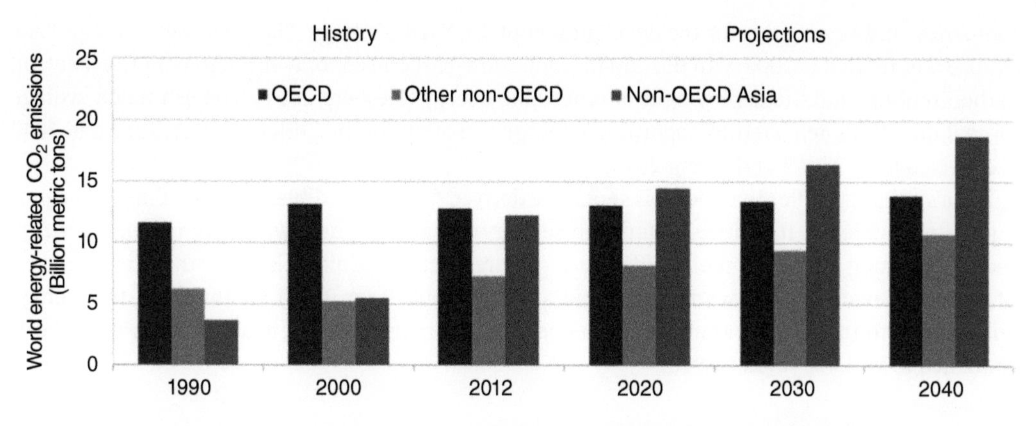

Figure 9.7 World energy-related carbon dioxide emissions (Billion metric tons) (EIA 2016).

9.5.2 Biofuel Demand and Supporting Policies

The correlation between fossil energy use and air pollution is gaining attention in modern society, and many strategies and policies are being adopted to control the worldwide emission of gaseous pollutants. The statistics of the International Energy Outlook 2016 (EIA 2016) projected that those Asian countries not in the OECD (the Organization for Economic Co-operation and Development) will account for about 60% of the world increase in energy-related CO_2 emissions in 2040. The world's energy-related carbon dioxide emissions, which are counted in billions of metric tons are shown in Figure 9.7 (EIA 2016). Thus, with the developing interest in renewable resources and their potential for replacing fossil fuels in the future, more and more governments are turning their attention to renewable fuels, and sustainable development.

The US Environmental Protection Agency announced and implemented the Clean Power Plan (CPP) regulations, which encourages the reduction of carbon pollution from power plants through using cleaner and lower-polluting energy sources, and therefore reducing climate change. More environmental policies and regulations have been enacted by many countries, particularly among the OECD, intended to reduce GHG emissions from electric power plants in terms of decreasing the use of fossil fuels. As a result, while renewables will grow by an average of 2.9%/yr from 2012 to 2040, the coal-fired net generation will increase by 0.8%/yr. This indicates that the dominant role of coal for electric power plants is declining, and that generation from renewable energy sources will equal that from coal on a worldwide basis at the end of the projection period (i.e. 2040) (EIA 2016). Influenced by the 2011 East Japan earthquake and the nuclear accident at Fukushima, Japan has pledged to change its energy policy (IEA 2016) compared to that of before the 2011 earthquake, which was an ambitious 25% cut in GHG emissions from 1990 to 2020 through an increased nuclear power share in electricity supply from 30% to 50% (Siegel et al. 2008). The 2014 Strategic Energy Plan (SEP) implemented in Japan confirmed four important objectives: energy security (stable supply), economic efficiency (cost reduction) and the environment (environmental suitability), and the premise of safety ('3Es plus S'). The SEP particularly highlights renewable energy as promising, multi-characteristic, important, low-carbon, and domestic energy sources. In this scenario, the estimated average annual growth in delivered transportation energy consumption in Japan declines at a rate of 0.7%/yr between 2012 and 2040 (EIA 2016).

China's Action Plan for the Prevention and Control of Air Pollution has been a high priority for the Xi government since 2013, the plan consists of strategies for the power sector, identified industries, and key regions for the development of renewable energy. The Renewable Energy Law of 2005 was revised in 2009 with the aim to reach a low-carbon economy through pollution control, carbon trading, and the deployment of renewable energy (Herberg 2014). The estimated average annual growth in delivered transportation energy consumption in China between 2012 and 2040 is expected to be 2.7%/yr (Herberg 2014).

India's policy specifies that biofuels should be derived from non-edible feedstock that can be produced on wastelands. It is aimed at satisfying energy demands in India while strengthening energy security by using renewable fuels, and ensuring the restrictions of biofuels under the circumference of Declared Goods by the Government of India (Sharma and Singh 2017). On this basis, the average annual growth in delivered transportation energy consumption between 2012 and 2040 has been estimated to be 4.4%/yr (EIA 2016).

9.5.3 Routes to Cost-Effective Alga-Based Biofuels

Exploring more routes to achieve cost-effective alga-based biofuels is one of the main approaches to commercialization. The routes mainly include achieving high biomass concentration of algae, co-producing value-added products to offset overall costs, optimization of cultivation processes, and combining alga-based biofuel production with environment protection.

9.5.3.1 Optimization of Cultivation Processes

Large-scale production of microalgae biofuel involves the cultivation, harvesting and drying, lipid extraction, and algal lipids-to-biodiesel conversion. Steps in cultivation involve strain selection, the generation of a master and working cell bank, seed inoculation, cultivation volumes, selection of cultivation methods (e.g. open or closed system), and system maintenance at the industrial scale. In strain selection, priority is given to the best strain with high growth rate and lipid yield. A long lag phase and culture shock should be avoided in the process, with a seed inoculate at 20–25% of the final culture volume (Grobbelaar 2009), thus ensuring an initial cell density greater than 1×10^7 cells per ml (Rawat et al. 2013). Robust in-process controls ensure optimal microalgal cultivation, notably by adjusting the supply of nutrients, CO_2, or controlling cultural pH, illumination, mixing, and cooling. Some large-scale microalgal cultivations have been implemented, such as the open ponds used by Aurora Algae Inc. (Hayward, CA, www.aurorainc.com) in the USA (Figure 9.8a), and the serpentine photobioreactors designed by Varicon Aqua Solutions Ltd. (Worcester, www.variconaqua.com) in the UK, which furthermore operates a patented self-cleaning mechanism for dramatically reducing production costs (Figure 9.8b).

The first step to improving microalgae as biofuel feedstocks is enhancing algal niomass growth and lipid accumulation. Of particular relevance among the various factors affecting microalgal growth and lipid accumulation are: pH, CO_2, dO_2. The optimal pH range for most algal species is 7–9 with a few exceptions (Wang and Lan 2009). However, relatively few studies have focused on the effects of pH on lipid accumulation (Bartley et al. 2014), or fatty acid composition (Breuer et al. 2013; Rodolfi et al. 2009; Moheimani 2013) in micro-algal species.

CO_2 levels affect the efficiency of photosynthesis and therefore influence cell growth and the biomass productivity of microalgae (Wang and Lan 2009). High levels of dO_2 have strong detrimental effects on microalgal cell growth and may cause culture collapse in the worst case scenario (Peng et al. 2013), or may also lead to lipid peroxidation (Acquisti et al. 2007; Peng et al. 2016).

Photosynthetically active radiation (PAR, at a wavelength of 400–700 nm) influences microalgal growth and biomass production through light intensity, spectrum, and frequency (e.g. light-dark cycle) (Peng et al. 2013). Differing light intensity may cause different microalgal photoadaptive

(a)

(b)

Figure 9.8 (a) Image of large-scale microalgae culture ponds, Aurora Algae Inc. CA, USA (Miyashita et al. 1996); (b) Phyco-Flow™ (≥600 l) serpentine photobioreactor designed by Varicon Aqua Solutions Ltd., UK (Nelis and Leenheer 1991).

responses (photoacclimation) including photolimitation, photoinhibition, and photobleaching (Kim et al. 2015; Takahashi et al. 2013). Light spectrum directly impacts the light utilization efficiency of microalgae and the selection on spectrum utilization depends on the chloroplasts harboured by microalgae species.

9.5.3.2 Achieve High Biomass Concentration of Algae

Achieving high biomass concentration of algae is essential for developing more cost-effective algae-derived biofuels. To achieve high biomass concentration in microalgal cultures, various measures can be implemented, for example enhanced cultivation conditions beyond genetic engineering (Potvin and Zhang 2010), comprising medium composition optimization (Liu et al. 2008), better process control (e.g. protozoan contamination control to ensure the non-sterile cultivation) (Bartley et al. 2014; Peng et al. 2016), and process optimization (e.g. optimizing the light utilization and mitigating oxygen stress) (Li et al. 2008; Pulz 2001; Liu et al. 2007). Whereas obtaining high lipid productivity is the ultimate objective, the optimization of both high biomass concentration and lipid accumulation appears to be conflicting because microalgal cells shift from a biomass generation stage where the cells actively divide to an energy storage stage when cells are subjected to various stress conditions (Bartley et al. 2014), such as nitrogen starvation (Li et al. 2008), unfavourable salinity or pH (Peng et al. 2015; Bartley et al. 2014; Santos et al. 2012), abnormal temperature and unfavourable radiances (Breuer et al. 2013).

9.5.3.3 Co-producing Value-Added Products to Offset Overall Costs

Microalgae can produce a large variety of novel value-added products, which could be exploited to improve the process economics of algae-based biofuel production. For instance, high value natural pigments can be extracted from algal biomass including chlorophyll D (Miyashita et al. 1996), carotenoids (Nelis and Leenheer 1991), β-carotene (Choudhari et al. 2008), astaxanthin (Kobayashi et al. 1991), and phycobilin (Borowitzka 2013); these compounds find numerous uses as human and animal feed supplements (Nelis and Leenheer 1991), or as alternatives to chemical dyes and colouring agents (Babitha 2009). Supplements can also be produced from the various vitamins and other elements (e.g. iron, magnesium, calcium, etc.) in microalgal cells. What is more, several long-chain polyunsaturated fatty acids docosahexaenoic acid (DHA) and eicosapentaenoic acid (EPA) can be extracted as pharmaceutical compounds (Borowitzka 2013). Microalgae have also been applied as additives of cosmetics, biodegradable plastics (Hempel et al. 2011), stabilizing substances, and agricultural fertilizers (Pulz and Gross 2004).

9.5.3.4 Combining Alga-Based Biofuel Production with Environment Protection

Microalgae are capable of utilizing nutrients in wastewater as growth substrates. Thus, the combination of the removal of nutrients from wastewater and the cost-reduction of microalgae production for biofuels is a win–win situation benefiting the environment overall. Remarkably, the integration of microalgae-based bioremediation and biofuel technology can be a route for sustainable production of biofuels (Rawat et al. 2013).

1. *Microalgae in wastewater treatment*
   First of all, microalgae can be applied in the final steps of wastewater treatment (Razzak et al. 2013) to remove contaminants such as nitrogen and phosphate (Mart Nez et al. 2000; Lee and Lee 2002), organic contaminants (Xu and Jiang 2013), heavy metals (Zeraatkar et al. 2016), or pathogens from domestic wastewater (Marques et al. 2006). For example, *Scenedesmus obliquus* was used to remove nitrogen and phosphorus from urban wastewater (Mart Nez et al. 2000).

Likewise, *Neochloris oleoabundans* is capable of depleting nitrogen and phosphorus from secondary municipal wastewaters containing high concentrations of nitrate and phosphate while achieving high biomass productivity of 350 mg DCW/l/d (Wang and Lan 2011). These results clearly indicate that *N. oleoabundans* could potentially be applied for combined wastewater treatment and biofuel production.

2. *CO_2 mitigation*

   Among the methods of CO_2 fixation widely considered, microalgae stand out as an effective approach to biological CO_2 fixation (Wang et al. 2008; Yanagi et al. 1995). This is due to the high carbon composition of microalgal biomass. In general, microalgae are capable of utilizing CO_2 from three different sources including atmospheric CO_2, CO_2 emission from power plants and industrial processes, and CO_2 from soluble carbonate (Wang et al. 2008). The efficiency of CO_2 removal or fixation in closed photobioreactors (PBRs) mainly depends on four factors: microalgal species, inlet CO_2 concentration, design of PBR, and operating cultivation conditions (de Morais and Costa 2007; Cheng et al. 2006). While the CO_2 fixation efficiency for *Spirulina* sp. and *S. obliquus* observed in these latter experiments was respectively 27–38% and 7–13% when aerated with an inlet stream with 6% CO_2 in a three serial tubular PBR, the CO_2 fixation efficiency decreased to 7–17% for *Spirulina* sp. and 4–9% for *S. obliquus* when the inlet air stream increased to a 12% CO_2 (de Morais and Costa 2007). In comparison, the CO_2 removal efficiency by *Chlorella vulgaris* can reach up to 55.3% at an inlet air stream with 0.15% CO_2 in a membrane photobioreactor (Cheng et al. 2006). In practice, flue gas can be used as a carbon source for microalgal cultivation. The microalgal species that can tolerate both high temperature (above 30 °C) and a high concentration of CO_2 (Peng et al. 2015) are appropriate for CO_2 capture directly from flue gas (Wang and Lan 2011). However, the use of CO_2 gas derived from power plants for microalgal growth is still at a research and development stage; this research has a large economic significance since companies that implement such technologies will be awarded carbon credits as a result of mitigating the emissions of harmful gases.

9.6 Progresses in the Commercial Production of Alga-Based Biofuels

There has been increasing attention on the concept of microalgae-based biofuel since it was first formulated in the 1950s (Sheehan et al. 1998; Li et al. 2008). Many nations, including the USA, Japan, the Europe Union, Australia, China, and Brazil have sponsored projects or have implemented policies to incentivize research on microalgae, a timeline of which is provided as follows (Sheehan et al. 1998).

1978–1982: the US Department of Energy (DOE) invested more than US $25 million in the Aquatic Species Program (ASP) to develop renewable transportation fuels (e.g. hydrogen) from microalgae;

1980–1996: ASP switched the emphasis on algae for biodiesel production;

1995: DOE eliminated funding for algae research within the Biofuels Program;

1990–2000: the Japan Ministry of International Trade and Industry (MITI) funded 'the earth update technology research plan', utilizing microalgae for CO_2 mitigation and producing biofuels;

2005: US Energy Policy Act of 2005 (FERC 2005) stated that the oil industry is required to blend 7.5 billion gallons of renewable fuels into gasoline by 2012;

2005–2008: first period of Sustainable Energy Europe Campaign (SEEC) ran, and the European Biomass Industry Association (EUBIA) has been involved in this campaign since 2007;

2006–2011: A five-year project named 'GHG emissions reduction using seaweeds' was funded in 2006, by the Ministry of Maritime Affairs and Fishery and the Ministry of Land, Transport and Maritime Affairs of South Korea (Chung 2013);

2009–2011: the latest phase of SEEC; this program ended on 31 December 2011;

2007–2013: Algae Cluster, an EU seventh framework programme (FP7), started to demonstrate the production of biofuels from algae. The total budget of €31 million (c. US $42 million) was funded on three algae-related projects, including Biofuel from Algae Technologies BioFat Project, All Gas Project, and InteSusAl Project Algae Cluster (www.algaecluster.eu), a EU seventh framework programme (FP7), started to demonstrate the production of biofuels from algae. The total budget of €31 million (c. US $42 million) was funded on three algae-related projects, including Biofuel from Algae Technologies BioFat Project, All Gas Project, and InteSusAl Project.

2010: The American Recovery and Reinvestment Act (ARRA) was activated and $80 million in government funding was sponsored for biofuel research and development. The bulk of the funding was designated to algae research and development (National Institute of Health, https://recovery.nih.gov);

2010–2015: the Australian Government sponsored a $2.3 million research grant to the Algal Fuel Consortium *(www.algaeconsortium.com)* for producing fuels and chemicals from microalgae. The cultivation system at 0.4 ha of raceway ponds was adjacent to a gas-fired power station, simultaneously mitigating CO_2 emission.

2011: National Aeronautics and Space Administration (NASA , http://climate.nasa.gov) inspired the Project of Sustainable Energy for Spaceship Earth, projected to cascade microalgal cultivation for producing 'clean energy' biofuels and sewage treatment;

2011–2015: the European Union sponsored a €14 million (c. US $19 million) initiative, the Energetic Algae project (EnAlgae, *www.enalgae.eu*), which is a four year Strategic Initiative of the INTERREG IVB North West Europe Programme, engaging 19 partners and 14 observers across 7 EU Member States;

2016: The production of liquid biofuels is projected to reach approximately 105 and 173 million gallons per day in 2020 and 2040, respectively, according to the International Energy Outlook 2016 of EIA (Sieminski 2016).

The development of alga-based biodiesel technologies and the strong supporting policies around the world have attracted more and more private companies' attention to microalgae-based biofuels (Table 9.4). These companies are mainly located in the USA, founded between 2003 and 2010, reflecting the heightened interest and supporting governmental policies at the time. Many manufacturers focus on the sole product of biodiesel while others produce a diversity of products such as foods, human nutritional products, and antioxidant diets, with biofuels only a part of the portfolio.

Several process models have been proposed for microalgal biomass production with minimal capital requirement. One example is three phase separation, which is shown in Figure 9.9 (Benemann and Oswald 1994; Scott et al. 2010; Pienkos and Al 2010). This process envisions a continuous open-pond cultivation of algae with a chemical flocculant being used to harvest the biomass (Mankee and Lee 2012; Pienkos and Al 2010). The process integrates the cultivation of microalgae with the harvesting, lipid extraction, and separation. In this process, carbon dioxide, nutrients, and water are provided for the cultivation of algae, while a chemical flocculant produces a biomass slurry, which is subjected to extraction with hot diesel to recover the lipids with the stream being fed to a three-phase centrifuge to separate the oil, water, and cellular debris (Pienkos and Al 2010). The process eliminates the necessity of drying the biomass and therefore leads to savings on energy

Table 9.4 Progress of biofuels based on algae in the world.

S.No.	Region	Country	Company/location	Startup year	Products	Link
1	America	USA	Algae Floating Systems, Inc., South San Francisco, CA	2007	Biodiesel	www.algaefloatingsystems.com
2			Algenol Biofuels, Bonita Springs, FL, Fort Meyers	2006	Ethanol, oil	www.algenolbiofuels.com
3			Aurora Algae Inc., Hayward, CA	2006	Biodiesel, pharmaceutics, food, supplement	www.aurorainc.com
4			Aquatic Energy, LLC, Lake Charles Louisiana	2007	Fuel, food	www.aquaticenergy.com
5			A2BE Carbon Capture, Boulder Colorado	2007	Biofuels, foods, fertilizers	www.algaeatwork.com
6			Bioalgene, Seatle, WA	2009	Biofuels	www.bioalgene.com
7			Blue Marble Energy, Seattle, Washington	2007	Biochemical and bioenergy products	www.bluemarblebio.com
8			Bionavitas, Inc., Redmond, WA.	2006	Biodiesel, nutraceuticals, and environmental remediation	www.bionavitas.com
9			Bodega Algae, LLC, Boston, MA	2006	Biofuel	www.bodegaalgae.com
10			Cellana, Inc., Hawaii	2004	Biofuel, feed, human nutritional products	www.cellana.com
11			Circle Biodiesel and Ethanol Corp., San Marcos, CA	2006	Biodiesel, algae ethanol	www.circlebio.com
12			Community Fuels, Encinitas, CA	2004	Biofuels	www.communityfuels.com
13			Diversified Energy Corporation, Gilbert, AZ	2005	Biofuels, antioxidant diet	http://www.diversified-energy.com
14			Inventure Chemical, Seattle	2007	Synthetic Jet fuel	www.inventurechem.com
15			LiveFuels, Inc., San Carlos, CA	2006	Bio-crude-oil	www.livefuels.com
16			Parabel Inc. formerly PetroAlgae Inc., Melbourne, FL	2007	Biofuels, liquid biocrude	www.parabel.com
17			PhycoBiosciences Inc., Chandler, AZ	2009	Algae biomass	www.phyco.net
18			PetroSun, Scottsdale, AZ	2007	Algal oil, jet fuels	www.petrosuninc.com
19			Sapphire Energy, San Diego, CA	2007	Bio-crude oil	www.sapphireenergy.com
20			Solena Fuels, WA	1995	Syn-gas, jet fuel	www.solenafuels.com

(Continued)

Table 9.4 (Continued)

S.No.	Region	Country	Company/location	Startup year	Products	Link
21			Solix Biofuels, Fort Collins, Colorado	2006	Biofuel	www.solixbiofuels.com
22			Solazyme, Inc., San Francisco, CA	2003	High-performance oil, biofuels, Solodiesel, jet fuel, green chemicals, cosmetics	www .solazymeindurstrials.com
23			XL Renewables, Phoenix, AZ	2007	Integrated biorefineries	www.xlrenewables.com
24		Canada	Centurion Biofuels, Hamilton, ON	2007	Biofuel	http://www.dev .centurionbiofuels.com
25			Pond Biofuels Inc., Toronto, ON	2007	Biodiesel	www.pondbiofuels.com
26			AlgaBloom International, Vancouver, BC		Food	www.algabloom.com
27			Solarvest BioEnergy, Vancouver, BC	2006	Clean energy (H_2), Omega, protein	www.solarvest.ca
28		Argentina	Oil Fox, S.A, Buenos Aires, Argentina		Biodiesel, Biogas	www.oilfox.com.ar
29		Mexico	Recursos Renovables Alternativos, Mérida, Yucatán	2008	Bioethanol	http://rra.mx/en
30	Europe	Spain	AlgaelLink	2007	Food (e.g. nutraceuticals, food additives), Feed (e.g. aquaculture feed) and Fuel industry; Jet Fuel	http://algaelink.com
31			BFS Biopetróleo, San Vicente del Raaspeig	2006	Bio-crude oil	www.biopetroleo.com
32			Biofuel Systems	2005	Biopetroleum, additives	www.biofuelsystems.com
33		Netherlands	Ingrepo B·V, Zutphen		Industrial-scale algae production	www.ingrepro.nl
34			AlgaeLink NV, Roosendaal	2007	Biofuels, food, feed	www.algaelink.com
35		Italy	Teregroup		Biodiesel	www.teregroup.net
36		Portugal	AlgaFuel S.A., Sines		Biofuel	http://www.algafuel.pt
37			Galp Energia, Lisbon		Biodiesel	www.galpenergia.com
38	Asia	Iran	Biofuel Research Team (BRTeam)	2010	Biofuel research	http://www.brteam.ir
39		Israel	Seambiotic, Ashkelon	2003	Biofuels, CO_2 capture	www.seambiotic.com
40		Japan	Euglena Co., Ltd. Tokyo	2005	Functional food and cosmetics, biofuel, jet fuels	www.euglena.jp
41		China	ENN group, Langfang, Hebei	1989	CO_2 mitigation, aviation fuels	http://www.enn.cn/en

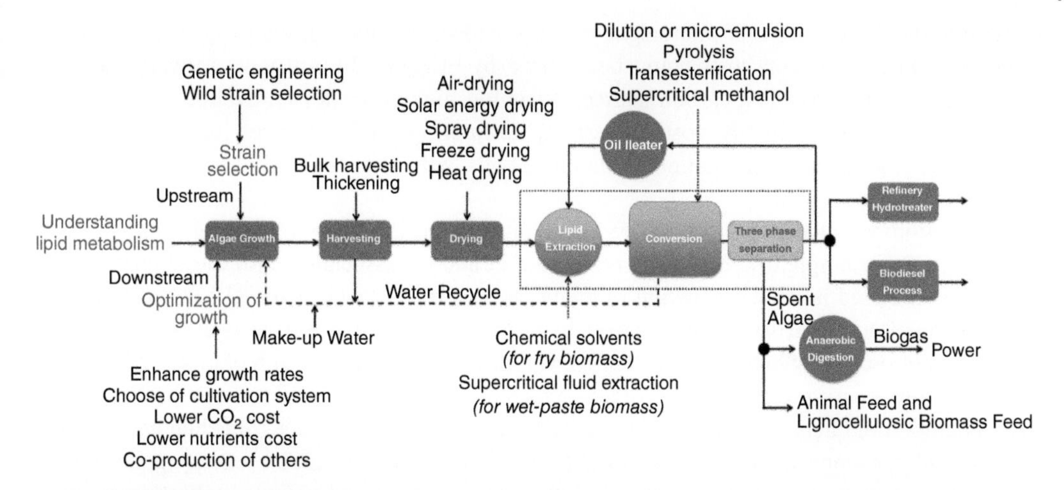

Figure 9.9 Process flow diagram for a model algal lipid production system. Source: Adapted from (Scott et al. 2010; Pienkos and Al 2010).

and operation costs. It also provides a lipid stream that can be sent directly to biofuel processing without the need for recovering organic solvents typically used in an extraction. The residual slurry is delivered to an anaerobic digester for producing biogas (methane) or to be used as animal feed or cellulosic feedstock for biofuels production through other forms of conversion processes (Pienkos and Al 2010).

9.7 Perspectives

The industrialization of algae-derived biofuels still faces major obstacles, the two biggest of which is the high production cost (Borowitzka 2013) and low ratio between the energy obtained from the biofuel product and the energy required to generate it, and the net energy ratio (Mankee and Lee 2012; Quinn et al. 2013). The current cost of algal biomass production is about US \$5 kg^{-1}, but it needs to be reduced to approximately US \$0.25 kg^{-1} to compete with fossil fuels in the current market (Rodolfi et al. 2009).

Among many contributors are the intense energy demands in the cultivation that results from illumination, preventing contamination, circulation, mixing, and cooling (Wijffels and Barbosa 2010; Sousa et al. 2012). These energy demands not only drive up the cost of production, but also result in a net energy ratio as low as two orders of magnitude below that of traditional fuels, and also significantly lower than first-generation biofuels (Quinn et al. 2013).

Downstream processing of microalgae also faces many obstacles. The harvesting and drying of microalgae is more difficult than other cells due to the small cell size and high water content. Moreover, large-scale extraction of microalgal lipid is complex and industrializable methods are still in development. In addition, more efficient methodologies for converting microalgal oils to biodiesels are required. The existing high amounts of free fatty acids increase the energy required for the conversion process while also affecting the overall quality of biodiesel products (Demirbas 2008). The bacterial oxidation of the free fatty acids in low quality biodiesel may induce internal corrosion of the storage tanks of vehicles (Antoni et al. 2007).

While microalgal biofuels are far from being commercially competitive, there are also many ways to improve their process economics through technological improvements. For instance, increasing the lipid content of microalgae through screening lipid-rich microalgae strains could improve yields; cultivation costs could be reduced by utilizing the nutrients in wastewater or digestate, using flue gas as a carbon source, designing high efficiency cultivation systems, and developing techniques for low-cost harvesting, dewatering, and extraction of algal biomass. Meanwhile, producing a combination of microalgal oils with other high value co-products is also one of the available routes.

The environmental benefits are a key factor for the attractiveness of biofuels compared to fossil fuels. For example, the notorious black soot emissions associated with diesel engines are virtually eliminated in the combustion of biodiesel. Emissions from biodiesel combustion have reduced emissions of particulate matter, CO, hydrocarbons, SO_x, and NO_x as well as carcinogens relative to petro-diesel (Sheehan et al. 1998). Overall, the commercialization of biofuels has been making steady progress since the early 2000s, which is demonstrated by the commercialization of first-generation biofuels (i.e. biodiesel made from rapeseed oil) in Europe (Sheehan et al. 1998). A large and sustained demand for renewable fuels as well as favourable energy policies will continue to push forward the progress of biofuel development and commercialization by providing appropriate incentives for private companies and investors to tackle the new technology. Sustaining development of advanced technologies to optimize microalgae cultivation, harvesting, drying, oil extraction, and fuel-conversion processing will be the key to making microalgae-based biofuels commercially viable.

References

Abomohra, E.F., Jin, W., Tu, R. et al. (2016). Microalgal biomass production as a sustainable feedstock for biodiesel: current status and perspectives. *Renewable & Sustainable Energy Reviews* 64 (64): 596–606.

Abomohra, E.F., El-Sheekh, M., and Hanelt, D. (2016). Screening of marine microalgae isolated from the hypersaline Bardawil lagoon for biodiesel feedstock. *Renewable Energy* 101: 1266–1272.

Acquisti, C., Kleffe, J., and Collins, S. (2007). Oxygen content of transmembrane proteins over macroevolutionary time scales. *Nature* 445 (7123): 47–52.

Adams, J.M., Gallagher, J.A., and Donnison, I.S. (2008). Fermentation study on *Saccharina latissima* for bioethanol production considering variable pre-treatments. *Journal of Applied Phycology* 21 (5): 569–574.

Adams, J.M.M., Ross, A.B., Anastasakis, K. et al. (2011). Seasonal variation in the chemical composition of the bioenergy feedstock *Laminaria digitata* for thermochemical conversion. *Bioresource Technology* 102 (1): 226–234.

AFDC (2012). *Drop-in Biofuels*. Washington, DC: Alternative Fuel Data Center, Department of Energy.

Ahmad, A.L., Yasin, N.H.M., Derek, C.J.C., and Lim, J.K. (2011). Microalgae as a sustainable energy source for biodiesel production: a review. *Renewable and Sustainable Energy Reviews* 15 (1): 584–593.

Annam Renita, A., Amarnath, D.J., Padhmanabhan, A., and Dhamodaran, B. (2010). *Production of Bio-Diesel from Marine Macro Algae*, 430–432. Recent Advances in Space Technology Services and Climate Change.

Antoni, D., Zverlov, V.V., and Schwarz, W.H. (2007). Biofuels from microbes. *Applied Microbiology and Biotechnology* 77 (1): 23–35.

Babitha, S. (2009). Microbial pigments. In: *Biotechnology for Agro-Industrial Residues Utilisation* (eds. P.S. nee' Nigam and A. Pandey), 147–162. Springer.

Babu, B.V. (2008). Biomass pyrolysis: a state-of-the-art review. *Biofuels, Bioproducts and Biorefining* 2 (5): 393–414.

Balan, V. (2014). Current challenges in commercially producing biofuels from lignocellulosic biomass. *ISRN Biotechnology* 2014 (3): 1–31.

Bartley, M.L., Boeing, W.J., Dungan, B.N. et al. (2014). pH effects on growth and lipid accumulation of the biofuel microalgae *Nannochloropsis salina* and invading organisms. *Journal of Applied Phycology* 26 (3): 1431–1437.

Bartley, M.L., Boeing, W.J., Corcoran, A.A. et al. (2014). Effects of salinity on growth and lipid accumulation of biofuel microalga *Nannochloropsis salina* and invading organisms. *Journal of Applied Phycology* 54 (3): 83–88.

Benemann, J.R. (2000). Hydrogen production by microalgae. *Journal of Applied Phycology* 12 (3): 291–300.

Benemann, J. R.; Oswald, W. J., Systems and economic analysis of microalgae ponds for conversion of CO_2 to biomass. *Nasa Sti/recon Technical Report N* 1994, 95.

Bird, K.T., Chynoweth, D.P., and Jerger, D.E. (1990). Effects of marine algal proximate composition on methane yields. *Journal of Applied Phycology* 2 (3): 207–213.

Borowitzka, M.A. (2013). High-value products from microalgae-their development and commercialisation. *Journal of Applied Phycology* 25 (3): 743–756.

Brennan, L. and Owende, P. (2010). Biofuels from microalgae a review of technologies for production, processing, and extractions of biofuels and co-products. *Renewable & Sustainable Energy Reviews* 14 (2): 557–577.

Breuer, G., Lamers, P.P., Martens, D.E. et al. (2013). Effect of light intensity, pH, and temperature on triacylglycerol (TAG) accumulation induced by nitrogen starvation in *Scenedesmus obliquus*. *Bioresource Technology* 143 (6): 1–9.

Cadenas, A. and Cabezudo, S. (1998). Biofuels as sustainable technologies: perspectives for less developed countries. *Technological Forecasting and Social Change* 58 (1–2): 83–103.

Carneiro, M.L.N.M., Pradelle, F., Braga, S.L. et al. (2017). Potential of biofuels from algae: comparison with fossil fuels, ethanol and biodiesel in Europe and Brazil through life cycle assessment (LCA). *Renewable & Sustainable Energy Reviews* 73: 632–653.

Chen, C., Yeh, K., Aisyah, R. et al. (2011). Cultivation, photobioreactor design and harvesting of microalgae for biodiesel production: a critical review. *Bioresource Technology* 102 (1): 71–81.

Chen, H.H., Dong, Z., Gang, L. et al. (2015). Macroalgae for biofuels production: progress and perspectives. *Renewable & Sustainable Energy Reviews* 47: 427–437.

Cheng, L., Zhang, L., Chen, H., and Gao, C. (2006). Carbon dioxide removal from air by microalgae cultured in a membrane-photobioreactor. *Separation and Purification Technology* 50 (3): 324–329.

Chisti, Y. (2008). Biodiesel from microalgae beats bioethanol. *Trends in Biotechnology* 26 (3): 126–131.

Chisti, Y. (2010). Biodiesel from microalgae. *Biotechnology Advances* 25 (3): 294–306.

Choudhari, S.M., Ananthanarayan, L., and Singhal, R.S. (2008). Use of metabolic stimulators and inhibitors for enhanced production of β-carotene and lycopene by *Blakeslea trispora* NRRL 2895 and 2896. *Bioresource Technology* 99 (8): 3166–3173.

Chung, I.K. (2013). Installing kelp forests/seaweed beds for mitigation and adaptation against global warming: Korean Project Overview. *ICES Journal of Marine Science* 70 (5): 1038–1044.

Chynoweth, D.P. (2005). Renewable biomethane from land and ocean energy crops and organic wastes. *Hortscience A Publication of the American Society for Horticultural Science* 40 (2): 57–73.

Cooney, M.J., Young, G., and Pate, R. (2011). Bio-oil from photosynthetic microalgae: case study. *Bioresource Technology* 102 (1): 166–177.

Danquah, M.K., Ang, L., Uduman, N. et al. (2009). Dewatering of microalgal culture for biodiesel production: exploring polymer flocculation and tangential flow filtration. *Journal of Chemical Technology and Biotechnology* 84 (7): 1078–1083.

Demirbas, A. (2007). Progress and recent trends in biofuels. *Progress in Energy & Combustion Science* 33 (1): 1–18.

Demirbas, A. (2008). Biofuels sources, biofuel policy, biofuel economy and global biofuel projections. *Energy Conversion and Management* 49 (8): 2106–2116.

Demirbas, M.F. (2011). Biofuels from algae for sustainable development. *Applied Energy* 88 (10): 3473–3480.

Demirbas, A. and Fatih Demirbas, M. (2011). Importance of algae oil as a source of biodiesel. *Energy Conversion and Management* 52 (1): 163–170.

Desmorieux, H. and Decaen, N. (2005). Convective drying of *Spirulina* in thin layer. *Journal of Food Engineering* 66 (4): 497–503.

Devereux, S., High Level Expert Forum on How to Feed the World 2050(2009).

Dewan, A. (2013). *Biofuel from Waste Agricultural Product and Lignocellulosic Biomass*, 178–186. Agricontrol.

Durrett, T.P., Benning, C., and Ohlrogge, J. (2008). Plant triacylglycerols as feedstocks for the production of biofuels. *The Plant Journal* 54 (4): 593–607.

EIA (2016). *International Energy Outlook 2016*. EIA.

Elliott, D.C., Biller, P., Ross, A.B. et al. (2015). Hydrothermal liquefaction of biomass: developments from batch to continuous process. *Bioresource Technology* 178: 147–156.

FERC, Energy Policy Act of 2005. In Commission, F. E. R., Ed 2005.

Fukuda, H., Kondo, A., and Noda, H. (2001). Biodiesel fuel production by transesterification of oils. *Journal of Bioscience and Bioengineering* 92 (5): 405–416.

Gallagher, B.J. (2011). The economics of producing biodiesel from algae. *Renewable Energy* 36 (1): 158–162.

Gray, K.A., Zhao, L., and Emptage, M. (2006). Bioethanol. *Current Opinion in Chemical Biology* 10 (2): 141–146.

Grobbelaar, J.U. (2009). From laboratory to commercial production: a case study of a *Spirulina* (Arthrospira) facility in Musina, South Africa. *Journal of Applied Phycology* 21 (5): 523–527.

Guo, M., Song, W., and Buhain, J. (2015). Bioenergy and biofuels: history, status, and perspective. *Renewable & Sustainable Energy Reviews* 42: 712–725.

Halim, R., Gladman, B., Danquah, M.K., and Webley, P.A. (2011). Oil extraction from microalgae for biodiesel production. *Bioresource Technology* 102 (1): 178.

Harun, R., Singh, M., Forde, G.M., and Danquah, M.K. (2010). Bioprocess engineering of microalgae to produce a variety of consumer products. *Renewable and Sustainable Energy Reviews* 14 (3): 1037–1047.

Harun, R., Danquah, M.K., and Forde, G.M. (2010). Microalgal biomass as a fermentation feedstock for bioethanol production. *Journal of Chemical Technology & Biotechnology* 85 (2): 199–203.

Hempel, F., Bozarth, A.S., Lindenkamp, N. et al. (2011). Microalgae as bioreactors for bioplastic production. *Microbial Cell Factories* 10 (1): 81.

Herberg, M. E., China's Energy Crossroads: Forging a New Energy and Environmental Balance. 2014.

Hu, Q., Sommerfeld, M., Jarvis, E. et al. (2008). Microalgal triacylglycerols as feedstocks for biofuel production: perspectives and advances. *The Plant Journal* 54 (4): 621–639.

Huang, G., Chen, F., Wei, D. et al. (2010). Biodiesel production by microalgal biotechnology. *Applied Energy* 87 (1): 38–46.

Huber, G.W., Iborra, S., and Corma, A. (2006). Synthesis of transportation fuels from biomass: chemistry, catalysts, and engineering. *Chemical Reviews* 106 (9): 4044–4098.

Hulatt, C.J. and Thomas, D.N. (2011). Productivity, carbon dioxide uptake and net energy return of microalgal bubble column photobioreactors. *Bioresource Technology* 102 (10): 5775–5787.

IEA (2016). *Energy and Air Pollution*. IEA.

IEA (2016). *Energy Policies of IEA Countries Japan 2016*. IEA.

International Energy Agency (2014). *Tracking Clean Energy Progress 2014*. International Energy Agency.

International Energy Agency (2016). *Energy Efficiency in China 2016*, 3–25. International Energy Agency.

International Energy Agency (2016). *Energy Policies of IEA Countries Japan 2016*, 9–158. International Energy Agency.

Jiang, R., Ingle, K.N., and Golberg, A. (2016). Macroalgae (seaweed) for liquid transportation biofuel production: what is next? *Algal Research* 14: 48–57.

Kapdi, S.S., Vijay, V.K., Rajesh, S.K., and Prasad, R. (2005). Biogas scrubbing, compression and storage: perspective and prospectus in Indian context. *Renewable Energy* 30 (8): 1195–1202.

Khan, S.A., Rashmi, Hussain, M.Z. et al. (2009). Prospects of biodiesel production from microalgae in India. *Renewable and Sustainable Energy Reviews* 13 (9): 2361–2372.

Khozin-Goldberg, I. and Cohen, Z. (2006). The effect of phosphate starvation on the lipid and fatty acid composition of the fresh water eustigmatophyte *Monodus subterraneus*. *Phytochemistry* 67 (7): 696–701.

Kim, M. and Day, D.F. (2011). Composition of sugar cane, energy cane, and sweet sorghum suitable for ethanol production at Louisiana sugar mills. *Journal of Industrial Microbiology & Biotechnology* 38 (7): 803–807.

Kim, J., Lee, J.Y., and Lu, T. (2015). A model for autotrophic growth of *Chlorella vulgaris* under photolimitation and photoinhibition in cylindrical photobioreactor. *Biochemical Engineering Journal* 99: 55–60.

Klöser, H., Quartino, M.L., and Wiencke, C. (1996). Distribution of macroalgae and macroalgal communities in gradients of physical conditions in Potter Cove, King George Island, Antarctica. *Hydrobiologia* 333 (1): 1–17.

Kobayashi, M., Kakizono, T., and Nagai, S. (1991). Astaxanthin production by a green alga, *Haematococcus pluvialis* accompanied with morphological changes in acetate media. *Journal of Fermentation and Bioengineering* 71 (5): 335–339.

Kraan, S. (2013). Mass-cultivation of carbohydrate rich macroalgae, a possible solution for sustainable biofuel production. *Mitigation and Adaptation Strategies for Global Change* 18 (1): 1–20.

Lapinskien, A., Martinkus, P., and R B Dait, V. (2006). Eco-toxicological studies of diesel and biodiesel fuels in aerated soil. *Environmental Pollution* 142 (3): 432–437.

Lee, K. and Lee, C.G. (2002). Nitrogen removal from wastewaters by microalgae without consuming organic carbon sources. *Journal of Microbiology and Biotechnology* 12 (6): 979–985.

Levin, D.B., Pitt, L., and Love, M. (2004). Biohydrogen production: prospects and limitations to practical application. *International Journal of Hydrogen Energy* 29 (2): 173–185.

Li, Y., Horsman, M., Wang, B. et al. (2008). Effects of nitrogen sources on cell growth and lipid accumulation of green alga *Neochloris oleoabundans*. *Applied Microbiology and Biotechnology* 81 (4): 629–636.

Li, Y., Horsman, M., Wu, N. et al. (2008). Biofuels from microalgae. *Biotechnology Progress* 24 (4): 815–820.

Liang, K., Zhang, Q., and Cong, W. (2012). Enzyme-assisted aqueous extraction of lipid from microalgae. *Journal of Agricultural & Food Chemistry* 60 (47): 11771–11776.

Lin, L.P. (1985). Microstructure of spray-dried and freeze-dried microalgal powders. *Food Microstructure* 4: 341–348.

Lin, L., Cunshan, Z., Vittayapadung, S. et al. (2011). Opportunities and challenges for biodiesel fuel. *Applied Energy* 88 (4): 1020–1031.

Lisa Friedman, C.W. (2011). *Can North Africa Light Up Europe with Solar Power?* Scientific American.

Liu, W., Au, D.W.T., Anderson, D.M. et al. (2007). Effects of nutrients, salinity, pH and light: dark cycle on the production of reactive oxygen species in the alga *Chattonella marina*. *Journal of Experimental Marine Biology & Ecology* 346 (1–2): 76–86.

Liu, Z.Y., Wang, G.C., and Zhou, B.C. (2008). Effect of iron on growth and lipid accumulation in *Chlorella vulgaris*. *Bioresource Technology* 99 (11): 4717.

Ma, F. and Hanna, M.A. (1999). Biodiesel production: a review. *Bioresource Technology* 70 (1): 1–15.

Madiwale, S. and Bhojwani, V. (2016). An overview on production, properties, performance and emission analysis of blends of biodiesel. *Procedia Technology* 25: 963–973.

Mankee, L. and Lee, K.T. (2012). Microalgae biofuels: a critical review of issues, problems and the way forward. *Biotechnology Advances* 30 (3): 673–690.

Marques, A., Thanh, T.H., Sorgeloos, P., and Bossier, P. (2006). Use of microalgae and bacteria to enhance protection of gnotobiotic Artemia against different pathogens. *Aquaculture* 258 (1–4): 116–126.

Mart Nez, M.E., Sánchez, S., Jiménez, J.M. et al. (2000). Nitrogen and phosphorus removal from urban wastewater by the microalga *Scenedesmus obliquus*. *Bioresource Technology* 73 (3): 263–272.

Mata, T.M., Martins, A.A., and Caetano, N.S. (2010). Microalgae for biodiesel production and other applications: a review. *Renewable and Sustainable Energy Reviews* 14 (1): 217–232.

Matanjun, P., Mohamed, S., Mustapha, N.M., and Muhammad, K. (2009). Nutrient content of tropical edible seaweeds, *Eucheuma cottonii*, *Caulerpa lentillifera* and *Sargassum polycystum*. *Journal of Applied Phycology* 21 (1): 75–80.

Mcdermid, K.J. and Stuercke, B. (2003). Nutritional composition of edible Hawaiian seaweeds. *Journal of Applied Phycology* 15 (6): 513–524.

McKendry, P. (2002). Energy production from biomass (part 2): conversion technologies. *Bioresource Technology* 83 (1): 47–54.

Meher Kotay, S. and Das, D. (2008). Biohydrogen as a renewable energy resource—prospects and potentials. *International Journal of Hydrogen Energy* 33 (1): 258–263.

Meher, L.C., Vidya Sagar, D., and Naik, S.N. (2006). Technical aspects of biodiesel production by transesterification—a review. *Renewable and Sustainable Energy Reviews* 10 (3): 248–268.

Melis, A. (2009). Solar energy conversion efficiencies in photosynthesis: minimizing the chlorophyll antennae to maximize efficiency. *Plant Science* 177 (4): 272–280.

Miao, X. and Wu, Q. (2006). Biodiesel production from heterotrophic microalgal oil. *Bioresource Technology* 97 (6): 841–846.

Michaelj, H., Karenm, S., Thomasa, F., and Williamn, M. (2007). The general applicability of in situ transesterification for the production of fatty acid esters from a variety of feedstocks. *Journal of the American Oil Chemists Society* 84 (10): 963–970.

Milano, J., Hwaichyuan, O., Masjuki, H.H. et al. (2016). Microalgae biofuels as an alternative to fossil fuel for power generation. *Renewable & Sustainable Energy Reviews* 58: 180–197.

Miyashita, H., Ikemoto, H., Kurano, N. et al. (1996). Chlorophyll d as a major pigment. *Nature* 383 (6599): 402–402.

Mohan, D., Pittman, C.U., and Steele, P.H. (2006). Pyrolysis of wood/biomass for bio-oil: a critical review. *Energy & Fuels* 20 (3): 848–889.

Moheimani, N.R. (2013). Inorganic carbon and pH effect on growth and lipid productivity of *Tetraselmis suecica* and *Chlorella* sp. (Chlorophyta) grown outdoors in bag photobioreactors. *Journal of Applied Phycology* 25 (2): 387–398.

Molina Grima, E., Belarbi, E.H., Acién Fernández, F.G. et al. (2003). Recovery of microalgal biomass and metabolites: process options and economics. *Biotechnology Advances* 20 (7): 491–515.

de Morais, M.G. and Costa, J.A.V. (2007). Carbon dioxide fixation by *Chlorella kessleri, C. vulgaris, Scenedesmus obliquus* and *Spirulina* sp. cultivated in flasks and vertical tubular photobioreactors. *Biotechnology Letters* 29 (9): 1349–1352.

Morand, P.; Merceron, M.; Pandalai, S. G., Coastal eutrophication and excessive growth of macroalgae. 2004, 395–449.

Nelis, H.J. and Leenheer, A.P.D. (1991). Microbial sources of carotenoid pigments used in foods and feeds. *Journal of Applied Microbiology* 70 (3): 181–191.

Nogueira, L.A.H., Moreira, J.R., Schuchardt, U., and Goldemberg, J. (2013). The rationality of biofuels. *Energy Policy* 61 (61): 595–598.

Oey, M., Sawyer, A.L., Ross, I.L., and Hankamer, B. (2016). Challenges and opportunities for hydrogen production from microalgae. *Plant Biotechnology Journal* 14 (7): 1487–1499.

Onion, G. and Bodo, L.B. (1983). Oxygenate fuels for diesel engines: a survey of world-wide activities. *Biomass* 3 (2): 77–133.

Parmar, A., Singh, N.K., Pandey, A. et al. (2011). Cyanobacteria and microalgae: a positive prospect for biofuels. *Bioresource Technology* 102 (22): 10163–10172.

Patil, P.D. and Deng, S. (2009). Optimization of biodiesel production from edible and non-edible vegetable oils. *Fuel* 88 (7): 1302–1306.

Peng, L. (2016). *Mitigation of Oxygen Stress and Contamination-Free Cultivation in Microalga Cultures.* University of Ottawa.

Peng, L., Lan, C.Q., and Zhang, Z. (2013). Evolution, detrimental effects, and removal of oxygen in microalga cultures: a review. *Environmental Progress & Sustainable Energy* 32 (4): 982–988.

Peng, L., Lan, C.Q., Zhang, Z. et al. (2015). Control of protozoa contamination and lipid accumulation in *Neochloris oleoabundans* culture: effects of pH and dissolved inorganic carbon. *Bioresource Technology* 197: 143–151.

Peng, L., Zhang, Z., Lan, C.Q. et al. (2016). Alleviation of oxygen stress on *Neochloris oleoabundans*: effects of bicarbonate and pH. *Journal of Applied Phycology*: 1–10.

Peng, L., Lan, C.Q., and Zhang, Z. (2016). Cultivation of freshwater green alga *Neochloris oleoabundans* in non-sterile media co-inoculated with protozoa. *The Canadian Journal of Chemical Engineering* 94 (3): 439–445.

Pienkos, P.T. and Al, D. (2010). The promise and challenges of microalgal-derived biofuels. *Biofuels Bioproducts & Biorefining* 3 (4): 431–440.

Pirwitz, K., Rihko-Struckmann, L., and Sundmacher, K. (2016). Valorization of the aqueous phase obtained from hydrothermally treated *Dunaliella salina* remnant biomass. *Bioresource Technology* 219: 64–71.

Popp, J., Lakner, Z., Harangi-Rákos, M., and Fári, M. (2014). The effect of bioenergy expansion: food, energy, and environment. *Renewable & Sustainable Energy Reviews* 32 (32): 559–578.

Posten, C. and Schaub, G. (2009). Microalgae and terrestrial biomass as source for fuels—a process view. *Journal of Biotechnology* 142 (1): 64–69.

Potvin, G. and Zhang, Z. (2010). Strategies for high-level recombinant protein expression in transgenic microalgae: a review. *Biotechnology Advances* 28 (6): 910–918.

Pulz, O. (2001). Photobioreactors: production systems for phototrophic microorganisms. *Applied Microbiology and Biotechnology* 57 (3): 287–293.

Pulz, O. and Gross, W. (2004). Valuable products from biotechnology of microalgae. *Applied Microbiology and Biotechnology* 65 (6): 635–648.

Quinn, J.C., Smith, T.G., Downes, C.M., and Quinn, C. (2013). Microalgae to biofuels lifecycle assessment-multiple pathway evaluation. *Algal Research* 4 (1): 116–122.

Rawat, I., Ranjith Kumar, R., Mutanda, T., and Bux, F. (2013). Biodiesel from microalgae: a critical evaluation from laboratory to large scale production. *Applied Energy* 103: 444–467.

Razmovski, R. and Vučurović, V. (2012). Bioethanol production from sugar beet molasses and thick juice using *Saccharomyces cerevisiae* immobilized on maize stem ground tissue. *Fuel* 92 (1): 1–8.

Razzak, S.A., Hossain, M.M., Lucky, R.A. et al. (2013). Integrated CO_2 capture, wastewater treatment and biofuel production by microalgae culturing-a review. *Renewable and Sustainable Energy Reviews* 27: 622–653.

Rodolfi, L., Zittelli, G., Bassi, N. et al. (2009). Microalgae for oil: strain selection, induction of lipid synthesis and outdoor mass cultivation in a low-cost photobioreactor. *Biotechnology and Bioengineering* 102 (1): 100–112.

Sadiktsis, I., Koegler, J.H., Benham, T. et al. (2014). Particulate associated polycyclic aromatic hydrocarbon exhaust emissions from a portable power generator fueled with three different fuels-A comparison between petroleum diesel and two biodiesels. *Fuel* 115 (1): 573–580.

Santana, A., Jesus, S., Larrayoz, M.A., and Filho, R.M. (2012). Supercritical carbon dioxide extraction of algal lipids for the biodiesel production. *Procedia Engineering* 42 (10): 1755–1761.

Santos, A.M., Janssen, M., Lamers, P.P. et al. (2012). Growth of oil accumulating microalga *Neochloris oleoabundans* under alkaline-saline conditions. *Bioresource Technology* 104 (1): 593–599.

Sawangkeaw, R., Bunyakiat, K., and Ngamprasertsith, S. (2010). A review of laboratory-scale research on lipid conversion to biodiesel with supercritical methanol (2001–2009). *Journal of Supercritical Fluids* 55 (1): 1–13.

Scott, S.A., Davey, M.P., Dennis, J.S. et al. (2010). Biodiesel from algae: challenges and prospects. *Current Opinion in Biotechnology* 21 (3): 277–286.

Seo, Y.B., Lee, Y.W., Lee, C.H., and You, H.C. (2010). Red algae and their use in papermaking. *Bioresource Technology* 101 (7): 2549–2553.

Seungphill, C., Nguyen, M.T., and Sangjun, S. (2010). Enzymatic pretreatment of *Chlamydomonas reinhardtii* biomass for ethanol production. *Bioresource Technology* 101 (14): 5330.

Sharma, Y.C. and Singh, V. (2017). Microalgal biodiesel: a possible solution for India's energy security. *Renewable and Sustainable Energy Reviews* 67: 72–88.

Shay, E.G. (1993). Diesel fuel from vegetable oils: status and opportunities. *Biomass and Bioenergy* 4 (4): 227–242.

Sheehan J, Terri Dunahay, John Benemann, Paul Roessler. *A look back at the U. S. department of energy's aquatic species programe biodiesel from algae*; 1998.

Siegel, J.; Nelder, C.; Hodge, N., Investing in renewable energy. 2008.

Sieminski, A. (2016). *International Energy Outlook 2016*. Washington, DC: U.S. Energy Information Administration.

Silva, C.E.D.F. and Bertucco, A. (2016). Bioethanol from microalgae and cyanobacteria: a review and technological outlook. *Process Biochemistry* 51 (11): 1833–1842.

Singh, J. and Gu, S. (2010). Commercialization potential of microalgae for biofuels production. *Renewable & Sustainable Energy Reviews* 14 (9): 2596–2610.

Sousa, C., de Winter, L., Janssen, M. et al. (2012). Growth of the microalgae *Neochloris oleoabundans* at high partial oxygen pressures and sub-saturating light intensity. *Bioresource Technology* 104: 565–570.

Stephenson, P.G., Moore, C.M., Terry, M.J. et al. (2011). Improving photosynthesis for algal biofuels: toward a green revolution. *Trends in Biotechnology* 29 (12): 615–623.

Sujith Sukumaran, S.K. (2013). Modeling fuel NO_x formation from combustion of biomass-derived producer gas in a large-scale burner. *Combustion and Flame* 160: 2159–2168.

Takahashi, S., Yoshioka-Nishimura, M., Nanba, D., and Badger, M.R. (2013). Thermal acclimation of the symbiotic alga *Symbiodinium* spp. alleviates photobleaching under heat stress. *Plant Physiology* 161 (1): 477–485.

Tamilarasan, S. and Sahadevan, R. (2014). Ultrasonic assisted acid base transesterification of algal oil from marine macroalgae *Caulerpa peltata*: optimization and characterization studies. *Fuel* 128 (1): 347–355.

Tan, X.B., Man, K.L., Uemura, Y. et al. (2017). Cultivation of microalgae for biodiesel production: a review on upstream and downstream processing. *Chinese Journal of Chemical Engineering* 1: 17–30.

Türe, S., Uzun, D., and Türe, I.E. (1997). The potential use of sweet sorghum as a non-polluting source of energy. *Energy* 22 (1): 17–19.

Uduman, N., Qi, Y., Danquah, M.K. et al. (2010). Dewatering of microalgal cultures: a major bottleneck to algae-based fuels. *Journal of Renewable and Sustainable Energy* 2 (1): 23–571.

Vassilev, S.V. and Vassileva, C.G. (2016). Composition, properties and challenges of algae biomass for biofuel application: an overview. *Fuel* 181: 1–33.

Vellguth, G., Performance of vegetable oils and their monoesters as fuels for diesel engines. Sae Technical Paper 1983.

Vergara-Fernández, A., Vargas, G., Alarcón, N., and Velasco, A. (2008). Evaluation of marine algae as a source of biogas in a two-stage anaerobic reactor system. *Biomass & Bioenergy* 32 (32): 338–344.

Voet, E.V.D. and Luo, R.J.L.L. (2010). Life-cycle assessment of biofuels, convergence and divergence. *Biofuels* 1 (3): 435–449.

Wang, B. and Lan, C. (2009). *Microalgae for Biofuel Production and CO_2 Sequestration*, Energy Science, Engineering and Technology Series, 1–168. New York: Nova Science Publishers, Inc.

Wang, B. and Lan, C.Q. (2011). Biomass production and nitrogen and phosphorus removal by the green alga *Neochloris oleoabundans* in simulated wastewater and secondary municipal wastewater effluent. *Bioresource Technology* 102 (10): 5639–5644.

Wang, B., Li, Y., Wu, N., and Lan, C.Q. (2008). CO_2 bio-mitigation using microalgae. *Applied Microbiology and Biotechnology* 79 (5): 707–718.

Wang, Y., Liu, H., and Lee, C.F.F. (2016). Particulate matter emission characteristics of diesel engines with biodiesel or biodiesel blending: a review. *Renewable & Sustainable Energy Reviews* 64: 569–581.

Wei, C.Y., Huang, T.C., and Chen, H.H. (2013). Biodiesel production using supercritical methanol with carbon dioxide and acetic acid. *Journal of Chemistry* 2013: 1–6.

Weyer, K.M., Bush, D.R., Darzins, A., and Willson, B.D. (2010). Theoretical maximum algal oil production. *Bioenergy Research* 3 (2): 204–213.

Wijffels, R.H. and Barbosa, M.J. (2010). An outlook on microalgal biofuels. *Science* 329 (5993): 796–799.

Williams, P.J.L.B. and Laurens, L.M.L. (2010). Microalgae as biodiesel & biomass feedstocks: review & analysis of the biochemistry, energetics & economics. *Energy & Environmental Science* 3 (5): 554–590.

World Health Organization, Burden of disease from household air pollution for 2012 2012.

Xu, H. and Jiang, H. (2013). UV-induced photochemical heterogeneity of dissolved and attached organic matter associated with cyanobacterial blooms in a eutrophic freshwater lake. *Water Research* 47 (17): 6506–6515.

Yanagi, M., Watanabe, Y., and Saiki, H. (1995). CO_2 fixation by *Chlorella* sp. HA-1 and its utilization. *Fuel & Energy Abstracts* 36 (6–9): 713–716.

Ye, S. and Cheng, J. (2002). Hydrolysis of lignocellulosic materials for ethanol production: a review. *Bioresource Technology* 83 (1): 1–11.

Zeraatkar, A.K., Ahmadzadeh, H., Talebi, A.F. et al. (2016). Potential use of algae for heavy metal bioremediation, a critical review. *Journal of Environmental Management* 181: 817–831.

Zhang, Y., Dubé, M.A., McLean, D.D., and Kates, M. (2003). Biodiesel production from waste cooking oil: 1. Process design and technological assessment. *Bioresource Technology* 89 (1): 1–16.

Ziolkowska, J.R. and Simon, L. (2014). Recent developments and prospects for algae-based fuels in the US. *Renewable & Sustainable Energy Reviews* 29 (7): 847–853.

Zuorro, A., Maffei, G., and Lavecchia, R. (2016). Optimization of enzyme-assisted lipid extraction from *Nannochloropsis* microalgae. *Journal of the Taiwan Institute of Chemical Engineers* 67: 106–114.

10

Advanced Fermentation Technologies: Conversion of Biomass to Ethanol by Organisms Other than Yeasts, a Case for *Escherichia coli*

K. T. Shanmugam, Lorraine P. Yomano, Sean W. York and Lonnie O. Ingram

Department of Microbiology and Cell Science, University of Florida, Gainesville, FL 32611, USA

CHAPTER MENU

10.1 Introduction

The fermentation of sugars by *Saccharomyces cerevisiae* to ethanol is described elsewhere, (Chapter 13) in this monograph. The critical enzymes of this pathway are pyruvate decarboxylase (PDC) and alcohol dehydrogenase (ADH) that convert pyruvate produced by glycolysis to ethanol and CO_2 in stoichiometric amounts. Any anaerobe or facultative anaerobe endowed with these two genes is expected to ferment sugars to ethanol at yields that are comparable to that of yeast. However, yeast dominates the ethanol fermentation industry due to its historic significance. Although evidence on yeast fermentation of fruit juices to ethanol goes back over 7000 years (McGovern et al. 2004), the history of fermentation of concentrated sugar solutions such as plant saps by bacteria is more recent. The description of pulque, a fermentation product of agave sap, is reported in ancient stone carvings and archaeological evidence dates it to about 200 CE (common era) (Escalante et al. 2016). Fermented agave sap is another ancient drink of the Aztecs and the Mayans. It has been suggested that one of the dominant microorganisms of agave sap fermentation is *Zymomonas* spp. (Swings and De Ley 1977). This bacterium was first isolated from spoiled apple cider in 1911 by Barker and Hillier (1912) and the name *Zymomonas mobilis* was coined by Lindner for the bacterium isolated in 1931 from fermented Agave sap in Mexico (Swings and De Ley 1977). *Z. mobilis* is a homoethanol fermenting Gram-negative facultative anaerobe. Although

Green Energy to Sustainability: Strategies for Global Industries, First Edition.
Edited by Alain A. Vertès, Nasib Qureshi, Hans P. Blaschek and Hideaki Yukawa.

the pathway for the conversion of glucose to pyruvate is different (Entner–Doudoroff pathway) from that of yeast (Embden–Meyerhof–,Parnes pathway), both microorganisms use PDC and ADH to convert pyruvate to ethanol (Swings and De Ley 1977; Conway 1992; Rogers et al. 2007; Yang et al. 2016). The enzyme PDC is not common among bacteria and is known to be present in only about 15 genera based on DNA sequence databases. Many of these bacteria were not studied in detail, especially regarding their ethanol production characteristics. Among them, *Z. mobilis* and *Zymobacter palmae* are known homoethanol-producing bacteria (Rogers et al. 2007; He et al. 2014; Okamoto et al. 1994). Bacteria, such as *Sarcina ventriculi*, a Gram-positive bacterium, produces acetate, formate and ethanol during growth and fermentation at neutral pH. As the culture pH decreases due to the accumulation of acids, the *pdc* gene is activated and ethanol is produced at a higher yield (Lowe and Zeikus 1991). The PDC has been purified from some of these bacteria and the apparent *K*m for pyruvate for this enzyme range between 0.3 and 5 mM with the PDC from *Z. mobilis* having an apparent *K*m for pyruvate of 0.3 mM (Neale et al. 1987; Raj et al. 2002).

The *pdc* and *adh* genes from *Z. mobilis* have been transferred to various other bacteria in the construction of recombinant homoethanol producing microorganisms. These studies have been extensively reviewed (Dien et al. 2003; Ingram et al. 1999; Jarboe et al. 2007, 2010; Liu and Khosla 2010). In addition, metabolic engineering also yielded a group of non-recombinant ethanol producers (Argyros et al. 2011; Cripps et al. 2009; Kim et al. 2007; Munjal et al. 2012; San Martin et al. 1992; Su et al. 2010; Zhou et al. 2008). These microorganisms utilize pyruvate dehydrogenase, and alcohol/aldehyde dehydrogenase as the enzymes in the conversion of pyruvate to ethanol at high yield. The two alternate ways of converting pyruvate to ethanol are shown in Figure 10.1. Anaerobic bacteria dissimilate pyruvate using the enzyme pyruvate-ferredoxin oxidoreductase (PFOR). The reductant is released as H_2 by hydrogenase. Because of this loss of reductant, the acetyl-CoA generated by PFOR cannot be efficiently reduced to ethanol since this reaction catalysed by alcohol/aldehyde dehydrogenase requires two molecules of NADHs (Figure 10.2). Some of the thermophilic anaerobes appear to have evolved to address this redox imbalance by generating NAD(P)H using reduced ferredoxin and the enzyme ferredoxin-NAD(P) oxidoreductase (FNOR), and producing ethanol as a major fermentation product (Argyros et al. 2011; Shaw et al. 2008). This pathway is discussed later in this chapter.

As stated earlier, the pyruvate produced by glycolysis is decarboxylated to acetaldehyde by PDC and then reduced to ethanol by an ADH (Figure 10.1). This is exemplified by yeast as well as by the few bacteria that produce PDC. None of the microorganisms isolated from nature possess the PDH-based pathway for homoethanol production (Figure 10.1). In this pathway, the pyruvate is oxidatively decarboxylated to acetyl-CoA and NADH. With this second NADH produced at this step, the conversion of glucose to two acetyl-CoAs yields four NADHs. The reduction of acetyl-CoA by a bifunctional oxidoreductase to ethanol with acetaldehyde as an intermediate consumes the two NADHs generated during the conversion of glyceraldehyde-3-phosphate to acetyl-CoA. This pathway also leads to ethanol production as the main product of sugar fermentation with redox balance. However, it is not clear why evolution did not support this pathway for ethanol production and only favoured the PDC/ADH pathway. The pyruvate dehydrogenase, a complex of three major enzymes, PDC, acetyl-transferase, and dihydrolipoamide dehydrogenase, is found in all aerobic organisms and also in facultative anaerobes (de Kok et al. 1998). This large enzyme complex is inhibited by NADH. Due to this inhibition, PDH produced by *Escherichia coli* during anaerobic growth remains inactive under O_2-limitation conditions that normally have a higher $NADH/NAD^+$ ratio (Kim et al. 2008; Snoep et al. 1993; Sun et al. 2012). Mutations that minimize NADH inhibition of the complex as well as the elevated expression of PDH support homoethanol production but only when pathways that are competing for pyruvate and/or acetyl-CoA have been deleted (Kim et al. 2007; Munjal et al. 2012; Zhou et al. 2008; Wang et al. 2010).

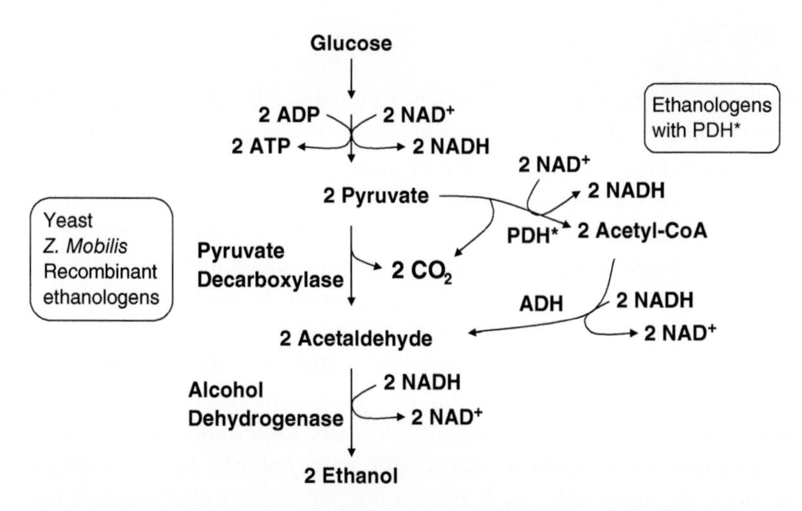

Figure 10.1 Alternate fermentation pathways for production of ethanol from sugars. PDH* represents a PDH complex with a mutation in dihydrolipoamide dehydrogenase subunit of the complex that reduced the sensitivity of the complex to NADH inhibition or the genes encoding the PDH complex is expressed from a non-native promoter. ADH, alcohol dehydrogenase.

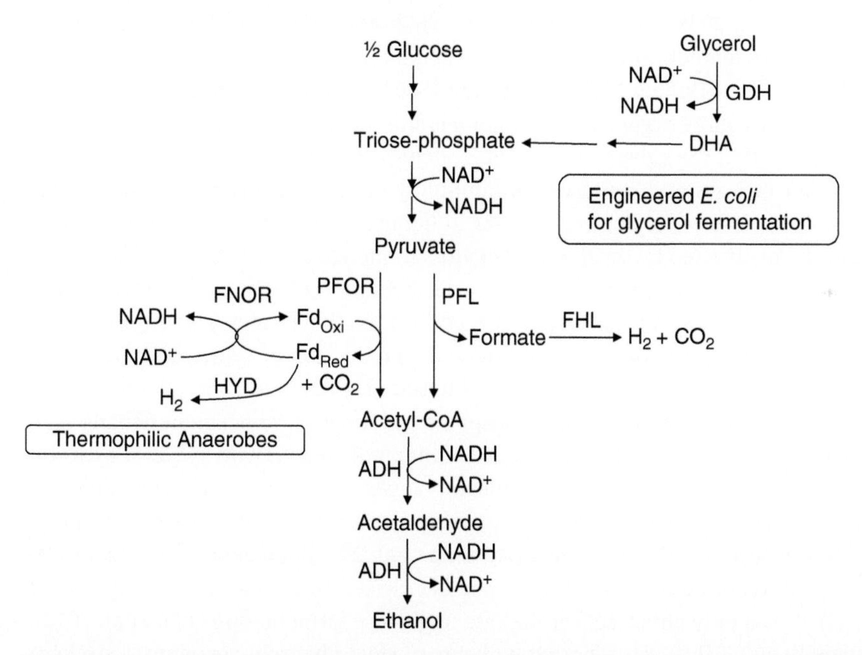

Figure 10.2 Additional metabolic pathways that can support homoethanol production. GDH, glycerol dehydrogenase; DHA, dihydroxyacetone; PFL, pyruvate formate-lyase; FHL, formate hydrogen-lyase; PFOR, pyruvate ferredoxin oxidoreductase; FNOR, ferredoxin $NAD(P)^+$ oxidoreductase; HYD, hydrogenase; ADH, alcohol dehydrogenase. See text for details.

10.2 *Zymomonas mobilis*

Although PDC is uncommon in bacteria, it is still found in a few genera. Among them, *Z. mobilis* is the most studied example of native ethanol-producing non-yeast microorganisms. Because of the interest in this organism several reviews are available describing its physiological, biochemical, and fermentation characteristics (Conway 1992; Rogers et al. 2007; Yang et al. 2016; He et al. 2014). Genome sequences of at least eight *Z. mobilis* strains are available at GenBank and these provide additional input to our understanding of the organism. Ethanol yields and titres attained from *Z. mobilis* grown on glucose are comparable to the values reported for yeast. What is more, the specific productivity of ethanol from glucose by *Z. mobilis* can be higher than 5 g ethanol per gram cell (dry weight) per hour, a value that is significantly higher than the 3.4 g/g cell per hour reported for *S. cerevisiae* (Lee et al. 1981; Rajoka et al. 2005). This high specific productivity is apparently a consequence of the glucose fermentation pathway utilized by the bacterium that yields only one ATP per glucose fermented to two ethanols. The high specific activity of the terminal enzymes PDC and ADH (about 1 μmol/min. mg protein) in *Z. mobilis* cells apparently also accounts for the high glycolytic flux to the product (An et al. 1991; Mejia et al. 1992; Osman et al. 1987). The volumetric productivity of ethanol by the engineered derivative of *Z. mobilis* CP4 can reach about 9 g/(l.h) in batch mode with glucose (Ma et al. 2012; Osman and Ingram 1987). However, the volumetric productivity reported by various investigators for wild type *Z. mobilis* strains is comparable to that of yeast (Rajoka et al. 2005; Osman and Ingram 1987; Morimura et al. 1997; Pinilla et al. 2011). The optimum growth temperature for this organism is 30 °C although fermentation at 35 °C is not uncommon (Swings and De Ley 1977). Increasing O_2 concentration minimally influenced fermentation characteristics of the organism but ethanol is still the main fermentation product.

However, a limitation is that the substrate range of *Z. mobilis* strains is limited to glucose, fructose, and sucrose. Other hexoses and pentoses are minimally utilized by this bacterium (Swings and De Ley 1977). Since xylose is the dominant sugar in hemicellulose, recombinant derivatives of *Z. mobilis* have been constructed by several investigators to increase the substrate range of this unique microbial biocatalyst for achieving an efficient fermentation of the sugars derived from pre-treated lignocellulosic biomass. A recombinant *Z. mobilis* strain that ferments xylose effectively normally contains xylose isomerase and xylulose kinase for production of xylulose-5-phosphate, and transaldolase and transketolase to support the pentose-phosphate pathway for the conversion of xylulose-5-phosphate to glyceraldehyde-3-phosphate and fructose-6-phosphate (Zhang et al. 1995). The calculated volumetric productivity of the recombinant *Z. mobilis* with xylose as a carbon source was about 0.43 g/(l.h), a value that was significantly lower than the rate of fermentation of glucose by the same strain (1.9 g/(l.h)) (Ma et al. 2012). A similar lower rate of xylose fermentation compared to glucose was also reported by others (Agrawal et al. 2012; Lau et al. 2010). In a comparative study, Lau et al. reported that the average rate of xylose fermentation by an engineered *Z. mobilis* strain AX101 was only about 40% of the rate of glucose fermentation (Lau et al. 2010). An ethanol productivity of 2 g/(l.h) was reported with a corn stover hydrolysate adapted strain of *Z. mobilis* (strain SS3), one of the highest values reported to this date (Mohagheghi et al. 2015). The genetic changes that led to this enhanced productivity remains, however, to be defined. These results show that although xylose fermentation can be engineered into *Z. mobilis*, the xylose flux is yet to reach the high values of glucose flux in this bacterium. In a parallel set of experiments, arabinose fermenting recombinant *Z. mobilis* strains also exhibit similar lower rates of fermentation compared to glucose (Deanda et al. 1996; Lawford and Rousseau 2002; Mohagheghi et al. 2002). These recombinant strains still preferentially utilize glucose over xylose or arabinose when these sugars are present together in the medium.

Z. mobilis has been further engineered to tolerate stress conditions from high sugar concentrations (osmotic), temperature, ethanol as well as various inhibitors produced during pretreatment of lignocellulosic biomass to expand its utility. These studies revealed that sorbitol, as a compatible solute, at 10 mM supported growth of the bacterium at glucose concentrations reaching 250 g/l (Loos et al. 1994). Sorbitol appears to play a crucial role in minimizing stress induced by ethanol and higher temperature also (Sootsuwan et al. 2013). In addition to these stress conditions, a fermenting organism encounters several inhibitory compounds generated during pretreatment of biomass (Franden et al. 2013) including various aldehydes, such as furfural, hydroxymethyl furfural, vanillin, etc. Primary means of detoxifying these aldehydes is to reduce them to less toxic alcohols that require reductant. Various oxidoreductases as well as other gene products that contribute to stress tolerance have been identified in *Z. mobilis* in addition to strains that are adapted for growth in plant biomass hydrolysates (An et al. 1991; Agrawal et al. 2012; Mohagheghi et al. 2014; Tan et al. 2016; Yang et al. 2010; Yi et al. 2015). These enhancements of *Z. mobilis*, multiple sugar utilization, stress and inhibitor tolerance, are leading to development of this unique bacterium for industrial level fermentation of sugars derived from lignocellulosic biomass to ethanol.

10.3 *Escherichia coli*

Due to the limitation of *Z. mobilis* as well as *S. cerevisiae* to utilize pentose sugars at rates that are comparable to the glucose flux to ethanol, alternate microbial biocatalysts are sought to ferment all the sugars in lignocellulosic biomass at high rate. Although, there are several yeasts that can ferment pentose sugars to ethanol, they have other disadvantages as described in other sections of this monograph. Anaerobic and facultative anaerobic bacteria produce ethanol from both hexoses and pentoses, but the ethanol yield is not high enough for industrial applications. As a result, metabolic engineering techniques have been applied to bacteria with high substrate diversity for achieving the production of ethanol as a major product of fermentation since such a bacterium is not naturally available. Among the facultative anaerobes, *E. coli* is the most-studied, genetically amenable, physiologically defined bacterium. One of the present authors and his colleagues have introduced the genes encoding PDC and ADH from *Z. mobilis* (portable ethanol operon; [PET] operon) into *E. coli* W (strain TC4(pLOI295)) and altered the native fermentation pathways from that of a mixed acid and ethanol fermentation to homoethanol fermentation (ethanol yield of 0.49 g/g glucose) as seen with *S. cerevisiae* and *Z. mobilis* (Ingram et al. 1987). This was the first demonstration of a switch in fermentation products of a microorganism without genetically altering the native fermentation pathways and thereby demonstrating that metabolic engineering can unlock the potential of microorganisms for industrial production of various fuels and chemicals. The initial ethanologenic construct was further improved by incorporating the *pdc* and *adh* genes into the genome (Ohta et al. 1991) for achieving an increased stability of the new catabolic capabilities and deleting genes encoding competing pathways. However, the ethanol yield (0.13 g/g glucose) of this initial construct (strain KO3) was lower than the ethanol yield of strain TC4(pLOI295) and the theoretical yield of 0.51 g/g sugar, due to co-production of lactate. Increasing the levels of PDC and ADH in the cytoplasm to a level that is comparable to that of *Z. mobilis* (about 1 μmol/(min. mg protein); strain KO4) improved the ethanol yield to close to the theoretical level (Osman et al. 1987; Ohta et al. 1991). In a subsequent step of improvement, removal of fumarate reductase (strain KO11) further increased the volumetric productivity to about 1.7 g/(l.h) with glucose and 1.3 g/(l.h) with xylose in rich medium. Both sugars were fermented to ethanol at close to the theoretical yield. The genome sequence of strain KO11 revealed that the *pdc–adh* module is present in the chromosome at about 15–20 copies accounting for the high level of enzyme activity (Turner et al. 2012). Moreover, Lawford and Rousseau reported that an *E. coli* strain carrying a plasmid encoding *Z. mobilis pdc* and *adh* genes produced ethanol at close to the theoretical yield at a volumetric productivity of 1.6 g/(l.h), a value that is close to that of strain KO11 with chromosomal integration of the two needed genes (Lawford and Rousseau 1991).

There are four native branch points at the pyruvate node in *E. coli* as well as in other facultative anaerobes; the flow of carbon through any of the four branches depends on the growth stage of the culture as well as oxygen availability (Figure 10.3). For example, pyruvate dehydrogenase complex (PDH) is not active in anaerobic cells to support growth due to the inhibition of the enzyme complex by NADH (Kim et al. 2007, 2008; de Graef et al. 1999). Pyruvate formate-lyase (PFL) is the preferred pathway during exponential anaerobic growth since this pathway generates an additional ATP and increases ATP yield per fermented glucose to three. Since anaerobic cultures are limited by ATP, this additional ATP increases the growth rate to improve volumetric productivity. As the energy demand for growth decreases during late exponential to the early stationary phase of growth, lactate dominates as the fermentation product. Succinate is only a minor fermentation product of these organisms. With this background, it is interesting to note that introducing and increasing the intracellular concentration of PDC and ADH shifts the glucose flux away from these alternate routes to ethanol even in the presence of functional, native PFL and lactate dehydrogenase (LDH) (Ingram et al. 1987; Ohta et al. 1991). This could be a result of significant differences in the apparent *K*m for pyruvate for the three enzymes, PDC, PFL, and LDH (Table 10.1). In addition, the presence of high levels of PDC and ADH is also expected to divert the carbon flux from pyruvate to ethanol, even if the apparent *K*m is comparable to that of PFL.

The highest specific productivity of ethanol by an *E. coli* strain with plasmid pLOI297 that carries the *Z. mobilis pdc* and *adh* genes was reported to be 2.2 g/(g cell dry weight. h) (Lawford and Rousseau 1991), a value that is slightly lower than that of yeast (Rajoka et al. 2005). However, the ethanologenic *E. coli* fermented pentose sugars at a rate (1.6 g/g cells. h) that is about 65% of that of the glucose consumption rate (2.4 g/g cells. h) (Gonzalez et al. 2002). Although these initial experiments utilized rich medium, the mineral salts medium with only sugars as the organic nutrients improves the economics of a biorefinery producing ethanol in both the initial cost of nutrients and waste treatment (Gubiczaa et al. 2016). Mineral salts medium (AM1) that only contains 4.2 g/l of inorganic salts including phosphate and ammonium salts notably leads to the production of about 3.3 g cells (dry weight)/l with xylose or glucose as the carbon source. In this medium, xylose fermentation (90 g/l) leads to an ethanol yield of 0.49 g/g xylose with an average volumetric productivity

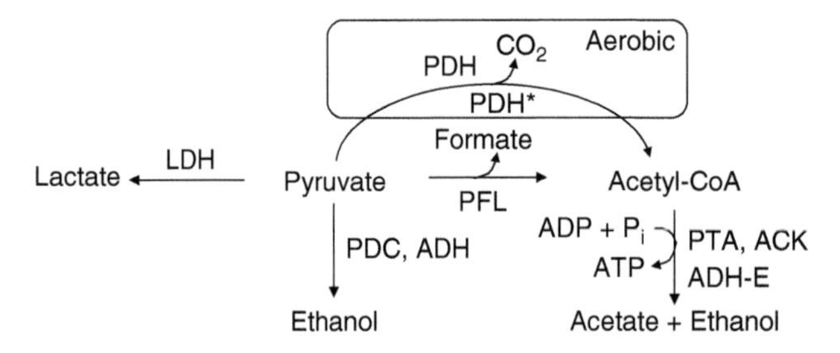

Figure 10.3 Competing pathways at the pyruvate node in native and recombinant ethanologenic bacteria. PDH represents the enzyme complex active under aerobic condition and PDH* indicates the same complex that is active under fermentative condition. LDH, D-lactate dehydrogenase; PFL, pyruvate formate-lyase; PTA, phosphotransacetylase; ACK, acetate kinase; PDC, pyruvate decarboxylase; ADH, alcohol dehydrogenase.

Table 10.1 Apparent *Km* values for pyruvate for various enzymes at the pyruvate node.

Organism	Enzyme	Apparent *Km* (mM)	Reference
E. coli	PFL	2.0	Ingram and Conway (1988)
	D-LDH	7.0	Ingram and Conway (1988)
	PPC	15.0	Izui et al. (1981)
	PPC*	0.25	Izui et al. (1981)
	PDH	0.4	Kim et al. (2008)
Z. mobilis	PDC	0.3	Neale et al. (1987)

PPC*, includes acetyl-CoA and fructose-1,6-bisphosphate as activators. PFL, pyruvate formate-lyase; D-LDH, D-lactate dehydrogenase; PPC, phosphoenolpyruvate carboxylase; PDH, pyruvate dehydrogenase; PDC, pyruvate decarboxylase.

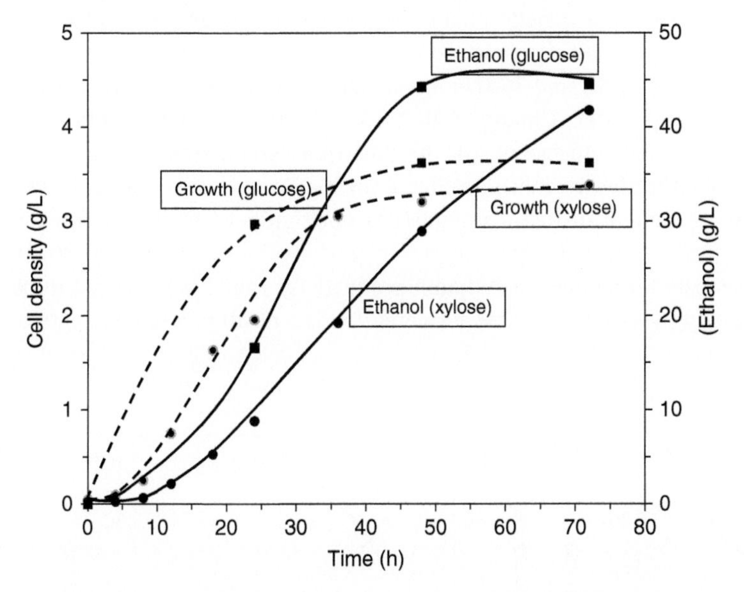

Figure 10.4 Fermentation profile of *E. coli* strain LY180 with glucose and xylose. Fermentations were conducted in mineral salts medium with either glucose or xylose (100 g/l) with pH control at 6.5 and 37 °C.

(over 48 hours) of 0.9 g ethanol/(l. h) (Martinez et al. 2007). With the medium change from rich to mineral salts medium, the PDC and ADH activities could be rate-limiting in fermentations with a decrease in ethanol yield (Huerta-Beristain et al. 2008). This was overcome by integrating the *pdc* and *adh* genes downstream from a strong promoter of *E. coli*, *rrlE*. The resulting strain with chromosomally integrated *pdc* and *adh* genes, LY180, ferments glucose in AM1 medium at a volumetric productivity of about 1.2 g ethanol/(l. h), a value that is only slightly lower than the initial construct strain KO11 that carried multiple copies of the *pdc* and *adh* genes in rich medium (1.6 g/g cells h) (Figure 10.4). The average volumetric productivity of ethanol production from xylose was calculated to be about 0.85 g/(l.h) (Miller et al. 2009). It should be noted that neither *Z. mobilis* nor yeast can grow and ferment in simple mineral salts medium without organic supplements.

Whether the ethanologenic microorganism is native or recombinant for xylose fermentation, both preferentially utilize glucose over xylose. Due to this preference, xylose is not fermented effectively by these microorganisms if glucose is present in the medium (Figure 10.5). Significant interest in fermentation of all the sugars in lignocellulosic biomass without a lag phase led to the need to achieve rapid consumption of pentose sugars in a medium containing a mixture of glucose and xylose. Although *E. coli* strain KO11 preferentially uses glucose, it also co-ferments xylose due to its high glucose flux (Figure 10.5). This co-fermentation property is also present in other ethanologenic *E. coli* strains with a chromosomally integrated PET operon (with a single copy of the *pdc*, *adhA*, and *adhB*), strain LY168, even in mineral salts medium (Yomano et al. 2008). This is apparently due to the higher glycolytic flux that is not met by the glucose fermentation alone. It is interesting to note that a mutation in *mgsA* encoding methylglyoxal synthase improves the co-fermentation of multiple sugars in the presence of glucose (Yomano et al. 2009). Methylglyoxal is a by-product of differential rates of glucose flux between triose-phosphate production and conversion to pyruvate by the lower part of glycolysis. To date, how a mutation in this enzyme improves the co-utilization of sugars is however, still unclear.

The co-fermentation of glucose and xylose has been approached by other investigators using a genetic approach. The most common genetic change in ethanologenic *E. coli* that led to this phenotype is the replacement of the native glucose transport system (PTS system) with a non-specific galactose permease (*galP*) and further metabolic evolution to improve glucose transport (Balderas-Hernandez et al. 2011; Chiang et al. 2013; Lindsay et al. 1995; Ohta et al. 2012). Even in these modified strains, glucose is still the preferred carbon source with xylose fermentation only starting when the glucose concentration in the medium reaches a critical level. The volumetric productivity of ethanol in these mutants has not significantly changed in the presence of both glucose and xylose over that of the values from glucose alone (Chiang et al. 2013). If both sugars are simultaneously fermented, the ethanol productivity would be expected to be higher. This lack of increase suggests that the xylose fermentation is a result of a reduced rate of glucose fermentation.

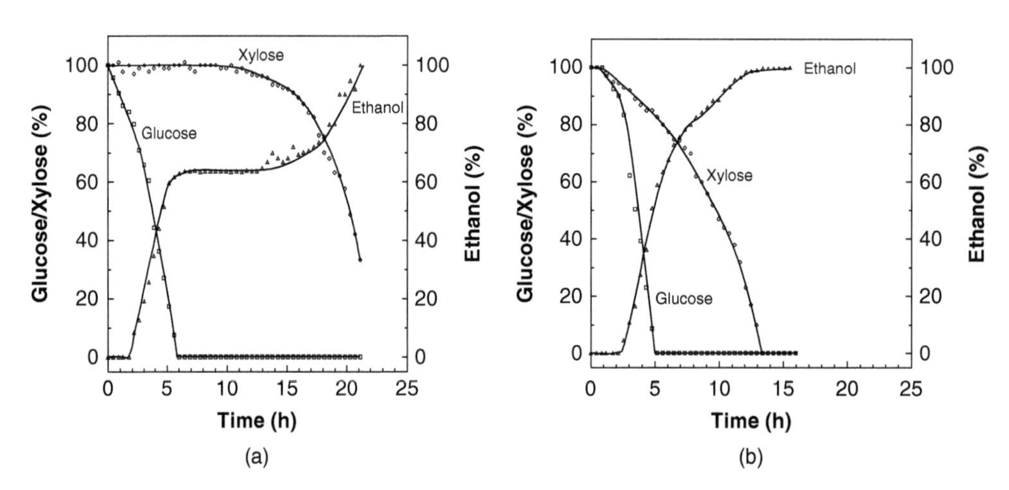

Figure 10.5 Effect of high glucose flux in *E. coli* strain KO11 supporting co-utilization of xylose. (a) *E. coli* with plasmid pLOI555. (b) *E. coli* strain KO11. Cultures were grown with $^{13}C_1$- glucose and $^{13}C_1$- xylose at 37 °C in the NMR without pH control. The amount of glucose or xylose consumed and ethanol produced were continuously monitored by H-NMR. In both fermentations ethanol was the major product accounting for over 95% of the total products.

10.4 Osmotic Stress of High Sugar Concentration

Although ethanologenic *E. coli* ferments xylose readily in mineral salts medium, increasing the pentose concentration to 100 g/l leads to a decline in growth rate, cell yield, and ethanol productivity and yield (Underwood et al. 2002). This limitation, observed even with corn steep liquor supplementation of the mineral salts medium, is not observed when strain KO11 is grown in rich medium. However, supplementing the mineral salts medium with pyruvate, acetaldehyde, acetate, or glutamate overcomes the growth inhibition seen with 100 g/l xylose. This indicates that the high flux rate of pyruvate conversion to ethanol starves the cells of needed carbon skeletons for synthesis of osmoprotectants, such as glutamic acid (Figure 10.6). Increasing the glutamate pool size or pyruvate pool size over the level needed to support ethanol production overcomes the osmotic stress imposed by 100 g/l xylose. This is confirmed by the ability of betaine, a known osmoprotectant, to overcome this limitation in productivity. From a mechanistic view, the citrate synthase appears to be a rate-limiting step in the synthesis of glutamate, the natural osmoprotectant produced by *E. coli*. The *E. coli* citrate synthase is sensitive to inhibition by NADH that is present at a higher concentration during anaerobic growth (de Graef et al. 1999; Weitzman 1966). Because of this inhibition, the citrate synthase activity in the cytoplasm is apparently not high enough to scavenge the small amount of oxaloacetate and acetyl-CoA available in the ethanologenic strain to produce citrate and then ultimately glutamate, due to the high flux of pyruvate to ethanol. This is indeed the case as shown by replacing the native citrate synthase with the ATP-regulated enzyme from *Bacillus subtilis* that is less sensitive to NADH (Jin and Sonenshein 1996; Underwood et al. 2002).

10.5 Inhibitor-Tolerant Ethanologenic *E. coli*

Almost all ethanol produced today comes from food sources such as starch-derived glucose or sucrose. There is increasing interest world-wide to shift the feedstock for industrial production of ethanol from food sources to non-food carbohydrates such as cellulose and hemicellulose in lignocellulosic biomass. The plant biomass has evolved to be resistant to various biological and

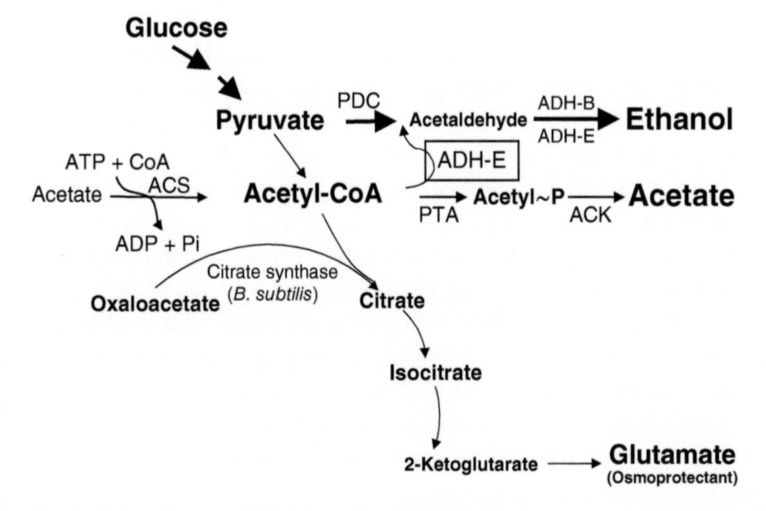

Figure 10.6 Fermentation of xylose (100 g/l) by *E. coli* strain KO11 limits carbon flow to glutamate production as an osmoprotectant. See text for details.

environmental insults and as a result lignocellulosic biomass is very slow to decompose. A chemical treatment at high temperature is needed to enable cellulases to hydrolyse the primary carbohydrate cellulose to glucose for fermentation by ethanologenic microorganisms. During this pretreatment step, several inhibitors that retard fermentation are released (Jonsson and Martin 2016). Pretreatment of biomass with phosphoric acid, a milder acid compared to sulfuric acid, was found to generate less of these inhibitors. Our phosphoric acid based pretreatment process is built upon the steam explosion work of Fontana et al. with sugar cane bagasse (Fontana et al. 1984; Geddes et al. 2013).

An optimal microbial biocatalyst is expected to ferment both glucose and xylose without any lag even in the presence of these inhibitors. To address this need, strain LY180 was evolved in mineral salts medium supplemented with acid-hydrolysate of sugar cane bagasse that contained various growth inhibitors. This metabolic evolution led to the identification of several gene products that contribute to the observed inhibitor tolerance, especially furfural and hydroxymethyl furfural (Miller et al. 2009; Geddes et al. 2011). It was observed that the primary mechanism of furfural and hydroxymethyl furfural tolerance is the reduction of aldehyde to alcohol by the *yqhD* gene product, a non-specific nicotinamide adenine dinucleotide phosphate (NADPH)-dependent aldehyde oxidoreductase (Jarboe et al. 2010; Miller et al. 2009). It is surmised that the enzyme depletes the NADPH pool during reduction of furfural and thus starves the cell of NADPH that is needed for biosynthesis, especially S-containing amino acids, since supplementing the medium with cysteine partially overcomes this inhibitory effect of furans (Miller et al. 2009). A strain that was further evolved for tolerance to inhibitors in hemicellulose hydrolysate, strain SL100, ferments the sugars released during liquefaction followed by simultaneous saccharification and fermentation (L + SScF) to ethanol at a yield of 0.28 g/g of bagasse (sugar cane or sorghum) (Geddes et al. 2013). This translates to a yield of about 72% of the maximum theoretical ethanol (70% carbohydrate content of bagasse) in about 72 hours. Unfortunately, similar results for xylose fermenting *S. cerevisiae* or *Z. mobilis* is not readily available in public databases for enabling a comparison between the two systems.

In an interesting study, strain LY180 was used to ferment crude glycerol from biodiesel production to ethanol resulting in a titre of about 60 g/l. In that same experiment, when a fosmid clone from a metagenomics library was introduced into strain LY180, the ethanol titre increased to about 80 g/l indicating the potential of this organism to tolerate ethanol at this high concentration (Loaces et al. 2016).

Similarly, a plasmid-containing derivative of an *E. coli* strain lacking lactate dehydrogenase (*ldhA*) and pyruvate formate-lyase (*pfl*) (strain FBR5) produces ethanol at a productivity of 0.9 g/(l.h) in batch mode with xylose and at a yield of 0.47–0.5 g ethanol/g sugar, values that are comparable to those attained with *E. coli* strain KO11 with chromosomally integrated PET operon genes (Qureshi et al. 2006). Due to the absence of the two fermentative enzymes, strain FBR5 maintains plasmid pLOI297 containing the PET operon since in that genetic background ethanol production is required for anaerobic growth to occur. In a cell recycle reactor, the specific productivity of this strain with xylose increases to 1.72 g ethanol/(g cells. h) when compared to a free cell continuous reactor (1.06 g ethanol/g cells h) (Qureshi et al. 2012). In a continuous fermentation, strain FBR5 produces ethanol for over 105 days without any loss of productivity with wheat straw hydrolysate as feedstock (Saha and Cotta 2011). This strain also produces over 40 g/l ethanol from dilute acid pre-treated wheat straw (Saha et al. 2011). Strain FBR5 also performs well at the pilot plant scale fermenting pretreated wheat straw to an ethanol yield of 0.44 g/g sugar and a titre of 36 g/l (Saha et al. 2015).

It would suffice to say that ethanologenic *E. coli* has come a long way in the last 30 years from the introduction of the *pdc* and *adh* genes from *Z. mobilis* into *E. coli* to redirect the fermentation to ethanol to compete as a potential industrial microbial biocatalyst for ethanol production using lignocellulosic biomass as a feedstock (Ingram et al. 1987; Gubiczaa et al. 2016; Geddes et al. 2013; Saha et al. 2015).

10.6 Engineering Bacterial Biocatalysts Other than *E. coli* for the Production of Ethanol Using the PDC/ADH Pathway

The seminal studies on the metabolic engineering of *E. coli* for ethanol production opened the critical area of metabolic engineering of microbial biocatalysts for industrial production of several other fuels and chemicals during the last 25 years. As we focus on ethanol in this chapter, a discussion of other bacteria that have been engineered for production of ethanol using the *pdc* and *adh* genes of *Z. mobilis* or the *pdc* gene from *S. ventriculi* is warranted. Many of these bacteria have unique intrinsic physiological characteristics such as growth at high temperature (*Clostridium thermocellum*), the ability to utilize cellulose as a carbon source (*C. thermocellum*; *Clostridium cellulolyticum*), cellobiose utilization (*Klebsiella oxytoca*), xylan utilization (*Enterobacter asburiae*), wide industrial use (lactic acid bacteria, *Corynebacterium glutamicum*, *B. subtilis*), etc. (Bi et al. 2009; Guedon et al. 2002; Jojima et al. 2015; Liu et al. 2006; Nichols et al. 2003; Romero et al. 2007; Talarico et al. 2005; Wood et al. 2005). The highest volumetric productivities of some of these microorganisms reached the values reported for *E. coli* KO11 or LY180.

It is noteworthy that a *K. oxytoca* strain BW21 with *Z. mobilis pdc* and *adh* genes produces about 42 g/l ethanol in 48 hours in mineral salts medium supplemented with 5 g/l corn steep liquor at a yield of 0.47 g/g glucose at pH 5.2 (Wood et al. 2005). The average volumetric productivity of ethanol by this culture reaches about 1.35 g/(l.h), a value that is comparable to that of *E. coli* KO11. These results demonstrate that in addition to ethanologenic *E. coli*, other bacteria with additional unique metabolic capabilities can also be engineered for producing ethanol at rates, titres and yields that are comparable to those reported with ethanologenic *E. coli*. Although the native ethanol producers, yeast and *Z. mobilis*, produce ethanol from glucose at higher volumetric productivity than the recombinant ethanol producers, some of these recombinants, such as *E. coli* and *K. oxytoca*, outshine the native ethanologens in their ability to ferment pentoses as well as cellobiose (derived from cellulose) in mineral salts medium to ethanol.

Another interesting bacterium that is engineered for ethanol production using the *pdc* and *adh* genes of *Z. mobilis* is *C. glutamicum*. This aerobic bacterium is an industrial workhorse for production of various amino acids, nucleotides and vitamins, and is engineered to produce other industrial chemicals (Jojima et al. 2013). The robustness of this bacterium supports ethanol production under growth-arrested O_2-deprivation condition at high cell densities (Inui et al. 2004). Under this condition, the volumetric productivity is directly proportional to the cell density, reaching a value of 30 g ethanol/(l. h) at a cell density of 60 g cell dry weight/l. However, a specific productivity of 0.06 g ethanol/(g cells. h) observed during the growth-arrested fermentation of ethanologenic *C. glutamicum* (Jojima et al. 2015) is significantly lower than the values observed with other ethanologens described above. Increasing the expression level of five genes (*pgi*, *pfkA*, *gapA*, *tpi*, and *pyk*) encoding enzymes in the glycolytic pathway did increase the specific productivity by twofold to 0.12 g ethanol/(g cells. h) suggesting a limitation of the glycolytic flux in the O_2-deprived high cell density fermentation. The fermentation characteristics of metabolically engineered ethanol-producing *C. glutamicum* has been recently reviewed and the reader is referred to these for further discussion (Jojima et al. 2013, 2014).

Several other studies with a large diversity of organisms are focused on demonstrating the feasibility of ethanol production by these organisms despite minimal potential industrial application due to low titre (less than 4 g/l) to low volumetric productivity. Many of these utilize the *Z. mobilis pdc* and *adh* genes. With some of the Gram-positive bacteria such as *B. subtilis* and *Lactobacillus plantarum*, the *pdc* from *S. ventriculi* was chosen to overcome limitations in expression of the *pdc* from a Gram-negative bacterium (*Z. mobilis*) in a Gram-positive host (Liu et al. 2006; Talarico et al. 2005). However, a paucity of long-term studies of these recombinant ethanologenic bacteria especially towards improving the ethanol titre, productivity or yield is hampering a critical evaluation of these engineered microbial biocatalysts. Additional work is needed to bring the potential of these recombinant ethanologens to industrial use.

10.7 Ethanol Production by Non-PDC Pathways

Although the main fermentation pathway for homoethanol production is based on PDC and ADH, there are alternate pathways that can lead to ethanol as a fermentation product (Figures 10.1 and 10.2). However, in these organisms, ethanol is one of several products of fermentation. These alternate pathways also generate acetaldehyde as an intermediate, but from acetyl-CoA for reduction to ethanol by ADH. The main characteristic that differentiates the acetyl-CoA pathway (PDH* or PFL) from the PDC pathway is in the need for reducing equivalent. The PDC enzyme only removes the CO_2 from pyruvate during its conversion to acetaldehyde. PDH, PFL, and PFOR catalyse oxidative decarboxylation of pyruvate to acetyl-CoA and due to this oxidation, the removed reductant needs to be put back on acetyl-CoA to produce acetaldehyde. During decarboxylation of pyruvate by PFL, the reductant conserved as formate is further metabolized to H_2 and CO_2, and is subsequently lost. During the decarboxylation of pyruvate to acetyl-CoA by PFOR, the reductant is conserved as reduced ferredoxin and only part of this is recovered as NADH while the other part is lost again as H_2. In contrast to these two enzymes, the oxidative decarboxylation of pyruvate by PDH retains the reductant as NADH that can be readily used to reduce acetyl-CoA to acetaldehyde. Physiologically, this coupled reaction of PDH/ADH is comparable to that of PDC/ADH. However, the main limiting factor with the PDH/ADH pathway is the activity and level of PDH under fermentative condition. Either the PDH complex is not produced during anaerobic growth or the produced enzyme is inhibited by the higher NADH/NAD^+ ratio of the anaerobic cell (Kim et al. 2007, 2008; de Graef et al. 1999).

This limitation in the PDH/ADH pathway for ethanol production is addressed in *E. coli* by introducing mutations in the LPD (lipoamide dehydrogenase) that renders the PDH complex less sensitive to NADH inhibition (PDH*) (Kim et al. 2007, 2008). With a single mutation, E354K or H322Y, in the LPD, *E. coli* strain SE2378 produces ethanol as the major fermentation product at a yield of 0.41 and 0.42 g/g of glucose and xylose, respectively. The specific productivity of ethanol by this strain is 1.34 g/(g cells h) with glucose and 2.24 g/(g cells. h) with xylose. These specific productivity values are comparable to that of the PDC-based ethanol producing *E. coli* KO11 (1.6 g/g cells h) (see above). In an alternate approach to the *lpd* mutation, the promoter of the genes encoding the PDH complex were replaced in *E. coli* B in contrast to *E. coli* K-12 used by Kim et al. (Munjal et al. 2012; Zhou et al. 2008; Chen et al. 2010). Strain KC01 reported by Zhou and his coworkers has about three-times higher PDH activity under anaerobic condition compared to the enzyme level in cells grown under aerobic condition when the PDH is expressed from a *pfl* promoter (Chen et al. 2010). The specific productivity of this strain with xylose is 1.3 g ethanol/(g cells. h) in a fermentation without pH control. However, with pH control at 6.0, a sub-optimal pH for growth of the organism, the specific productivity decreases to 0.2 g ethanol/(g cells. h). The ethanologenic *E. coli* strain reported by Yazdani and his coworkers expresses the *aceEF, lpd* from *gapA* promoter and the specific productivity of ethanol by this mutant is 0.92 g/(g cells. h) at pH 6.8, with glucose as a carbon source (Munjal et al. 2012). With both promoter alterations, the specific productivity of ethanol is lower than that of strain SE2378 carrying an LPD mutation. It is likely that higher level of PDH in the cytoplasm titrates the NADH level and minimizes the inhibitory effect of NADH on PDH activity in these promoter-altered ethanologenic *E. coli*. A combination of promoter exchange and *lpd* mutation could support a higher specific productivity of *E. coli* that can match that of yeast.

10.8 Partition of Carbon at the Pyruvate Node

In the ethanologenic *E. coli* strains KO11 and LY180, the higher level of PDC and ADH shifts the carbon at the pyruvate node to ethanol even in the presence of active LDH and PFL. However, with the PDH/ADH system with the mutated PDH complex discussed above (PDH*), the level of one or both enzymes are not high enough to support homoethanol production if active LDH and/or PFL is also present in the cytoplasm. In a study designed to evaluate the partitioning of pyruvate among the LDH, PFL, and PDH*, *E. coli* mutants carrying different combinations of these enzymes were used. In the presence of all three pathways at the pyruvate node (LDH, PDH*, and PFL) less than 1% of the pyruvate passes through PDH* (Wang et al. 2010), although this enzyme complex has the lowest apparent *K*m for pyruvate (0.4 mM) compared to the other two enzymes (2 and 7 mM for PFL and LDH, respectively; Table 10.1). Under similar conditions, in strain KO11, PDC accounts for over 90% of the pyruvate flux. Upon removal of the LDH and PFL activities, the pyruvate flux through PDH* reaches 100%. The reason for this disparity is unclear. Even if the PDH* concentration in the cell is lower than the other two enzymes (LDH and PFL), the relatively lower apparent *K*m should trigger a certain level of pyruvate flux through PDH. It is possible that the higher NADH/NAD^+ ratio may favour LDH that requires NADH for activity compared to PDH* that needs NAD^+ as a co-factor. Similarly, the PFL/ADH pathway also utilizes NADH. This differential flux favouring NADH utilization reactions is apparently a physiological adaptation to lower the NADH/NAD^+ ratio of an anaerobic cell. If this is indeed the case, bacterial physiology is on the side of production of reduced organic compounds and this evolutionary advantage needs to be fully exploited for the production of reduced fuel molecules.

In addition to engineering PDH activity and ethanol production with *E. coli*, PDH/ADH based ethanol production has been demonstrated in other bacteria. For example, in *Bacillus coagulans*, the deletion of *ldh* encoding L-LDH of the native fermentation pathway elevates the level of PDH activity during anaerobic growth and the PDH/ADH pathway contributes to ethanol production at a yield of 0.39 g/g glucose fermented (Su et al. 2010). Similar results were also reported for mutants of *Geobacillus stearothermophilus* (0.34 g ethanol/g sugar) and *Geobacillus thermoglucosidasius* (0.42 g ethanol/g glucose) (Cripps et al. 2009; San Martin et al. 1992). These studies show that the PDH/ADH pathway is a viable option for the production of ethanol by bacteria, provided however that the needed appropriate genetic background is established.

10.9 Other Metabolic Pathways that Contribute to Ethanol Production

Some of the anaerobic bacteria utilize another enzyme at the pyruvate node for the oxidative decarboxylation of pyruvate; PFOR (Figure 10.2). The reductant generated in this reaction is captured as reduced ferredoxin. Reduced ferredoxin serves as a substrate for hydrogenase that releases the reductant as H_2 or as a substrate to FNOR that in turn generates NAD(P)H. A combination of PFOR and FNOR converts pyruvate to acetyl-CoA, CO_2, and NAD(P)H, a reaction that is analogous to that catalysed by PDH. The acetyl-CoA can be reduced to ethanol using the two-step reduction process with acetaldehyde as an intermediate (Argyros et al. 2011; Shaw et al. 2008; Lovitt et al. 1988; Olson et al. 2015). Some of these thermophilic anaerobes reach an ethanol volumetric productivity of 2.2 g/(l.h) (Shaw et al. 2008) indicating their potential for industrial use, especially at higher fermentation temperatures.

Another enzyme that can produce acetyl-CoA is PFL that can serve as the starting point for ethanol production (Figure 10.2). As stated earlier, the reductant removed during this reaction is lost as H_2. This paucity of reductant for the two-step reduction of acetyl-CoA to ethanol can be mitigated by using a more reduced starting material, such as glycerol. Gonzalez and his coworkers metabolically engineered *E. coli* to ferment glycerol to ethanol using PFL to produce the needed acetyl-CoA for further reduction to ethanol (Wong et al. 2014; Yazdani and Gonzalez 2008). In these studies, when the fermentation was conducted under anaerobic condition, the specific ethanol productivity was about 1.25 g/(g cells. h). However, the highest reported volumetric productivity of ethanol from glycerol was only about 0.22 g/(l.h), a significantly low value apparently due to low cell density of the culture. Introducing small amounts of O_2 to the fermentation mixture increases the volumetric productivity by about threefold due to the higher cell density that is achieved (Durnin et al. 2009). Under these fermentation conditions, 60 g/l of glycerol was converted to ethanol at a theoretical yield of about 90% of glycerol fermented, although, from a volumetric productivity point of view, the PFL pathway is not as efficient as the PDC one. This demonstrates the utility of this pathway for ethanol production when a feedstock that is more reduced than glucose, such as glycerol, a by-product generated in significant amounts during biodiesel production, is used as the feedstock.

10.10 Perspectives

The results from various studies clearly show that bacteria can be engineered to produce ethanol as the major fermentation product. Among these engineered microbial biocatalysts, *E. coli* constructs carrying the *pdc* and *adh* genes from *Z. mobilis*, either in the chromosome (SL100) or in a plasmid (FBR5) and their derivatives appear to be the most efficient in converting sugars derived from lignocellulosic biomass to ethanol. The utility of the ethanologenic *E. coli* is most evident in fermentation of pentose sugars. The ability of this ethanologen to grow in mineral salts medium is an added advantage in industrial fermentations of lignocellulosic sugars.

Acknowledgements

The authors wish to thank all the students and research associates who have contributed to the work performed in the authors' laboratories over the years. This work was supported by grants from the US Department of Agriculture (USDA) (2012-67009-19596), Biomass Research and Development Initiative Competitive Grant (2011-10006-30358) from the USDA National Institute of Food and Agriculture, the US Department of Energy's Office of International Affairs (DE-PI0000031) and the Florida Department of Agriculture and Consumer Services.

References

Agrawal, M., Wang, Y., and Chen, R.R. (2012). Engineering efficient xylose metabolism into an acetic acid-tolerant *Zymomonas mobilis* strain by introducing adaptation-induced mutations. *Biotechnology Letters* 34 (10): 1825–1832.

An, H.J., Scopes, R.K., Rodriguez, M. et al. (1991). Gel electrophoretic analysis of *Zymomonas mobilis* glycolytic and fermentative enzymes - identification of alcohol dehydrogenase-II as a stress protein. *Journal of Bacteriology* 173 (19): 5975–5982.

Argyros, D.A., Tripathi, S.A., Barrett, T.F. et al. (2011). High ethanol titers from cellulose by using metabolically engineered thermophilic, anaerobic microbes. *Applied and Environmental Microbiology* 77: 8288–8294.

Balderas-Hernandez, V.E., Hernandez-Montalvo, V., Bolivar, F. et al. (2011). Adaptive evolution of *Escherichia coli* inactivated in the phosphotransferase system operon improves co-utilization of xylose and glucose under anaerobic conditions. *Applied Biochemistry and Biotechnology* 163 (4): 485–496.

Barker, B.T.P. and Hillier, V.F. (1912). Cider sickness. *The Journal of Agricultural Science* 5: 67–85.

Bi, C., Zhang, X., Rice, J.D. et al. (2009). Genetic engineering of *Enterobacter asburiae* strain JDR-1 for efficient D(−) lactic acid production from hemicellulose hydrolysate. *Biotechnology Letters* 31 (10): 1551–1557.

Chen, K., Iverson, A.G., Garza, E.A. et al. (2010). Metabolic evolution of non-transgenic *Escherichia coli* SZ420 for enhanced homoethanol fermentation from xylose. *Biotechnology Letters* 32 (1): 87–96.

Chiang, C.J., Lee, H.M., Guo, H.J. et al. (2013). Systematic approach to engineer *Escherichia coli* pathways for co-utilization of a glucose-xylose mixture. *Journal of Agricultural and Food Chemistry* 61 (31): 7583–7590.

Conway, T. (1992). The Entner-Doudoroff pathway: history, physiology and molecular biology. *FEMS Microbiology Reviews* 9 (1): 1–27.

Cripps, R.E., Eley, K., Leak, D.J. et al. (2009). Metabolic engineering of *Geobacillus thermoglucosidasius* for high yield ethanol production. *Metabolic Engineering* 11 (6): 398–408.

Deanda, K., Zhang, M., Eddy, C., and Picataggio, S. (1996). Development of an arabinose-fermenting *Zymomonas mobilis* strain by metabolic pathway engineering. *Applied and Environmental Microbiology* 62 (12): 4465–4470.

Dien, B.S., Cotta, M.A., and Jeffries, T.W. (2003). Bacteria engineered for fuel ethanol production: current status. *Applied Microbiology and Biotechnology* 108: 237–261.

Durnin, G., Clomburg, J., Yeates, Z. et al. (2009). Understanding and harnessing the microaerobic metabolism of glycerol in *Escherichia coli. Biotechnology and Bioengineering* 103 (1): 148–161.

Escalante, A., Soto, D.R.L., Gutiérrez, J.E.V. et al. (2016). Pulque, a traditional Mexican alcoholic fermented beverage: historical, microbiological, and technical aspects. *Frontiers in Microbiology* 7 1026.

Fontana, J.D., Correa, J.B.C., and Duarte, J.H. (1984). Aqueous phosphoric acid hydrolysis of hemicelluloses from sugarcane and sorghum bagasses. *Biotechnology and Bioengineering Symposium* 14: 175–186.

Franden, M.A., Pilath, H.M., Mohagheghi, A. et al. (2013). Inhibition of growth of *Zymomonas mobilis* by model compounds found in lignocellulosic hydrolysates. *Biotechnology for Biofuels* 6: 99.

Geddes, C.C., Mullinnix, M.T., Nieves, I.U. et al. (2011). Simplified process for ethanol production from sugarcane bagasse using hydrolysate-resistant *Escherichia coli* strain MM160. *Bioresource Technology* 102: 2702–2711.

Geddes, C.C., Mullinnix, M.T., Nieves, I.U. et al. (2013). Seed train development for the fermentation of bagasse from sweet sorghum and sugarcane using a simplified fermentation process. *Bioresource Technology* 128: 716–724.

Gonzalez, R., Tao, H., Shanmugam, K.T. et al. (2002). Global gene expression differences associated with changes in glycolytic flux and growth rate in *Escherichia coli* during fermentation of glucose and xylose. *Biotechnology Progress* 18: 6–20.

de Graef, M.R., Alexeeva, S., Snoep, J.L., and Teixeira de Mattos, M.J. (1999). The steady-state internal redox state (NADH/NAD) reflects the external redox state and is correlated with catabolic adaptation in *Escherichia coli*. *Journal of Bacteriology* 181: 2351–2357.

Gubiczaa, K., Nieves, I.U., Sagues, W.J. et al. (2016). Techno-economic analysis of ethanol production from sugarcane bagasse using a liquefaction plus simultaneous saccharification and co-fermentation process. *Bioresource Technology* 208: 42–48.

Guedon, E., Desvaux, M., and Petitdemange, H. (2002). Improvement of cellulolytic properties of *Clostridium cellulolyticum* by metabolic engineering. *Applied and Environmental Microbiology* 68 (1): 53–58.

He, M.X., Wu, B., Qin, H. et al. (2014). *Zymomonas mobilis:* a novel platform for future biorefineries. *Biotechnology for Biofuels* 7: 101.

Huerta-Beristain, G., Utrilla, J., Hernandez-Chavez, G. et al. (2008). Specific ethanol production rate in ethanologenic *Escherichia coli* strain KO11 is limited by pyruvate decarboxylase. *Journal of Molecular Microbiology and Biotechnology* 15 (1): 55–64.

Ingram, L.O. and Conway, T. (1988). Expression of different levels of ethanologenic enzymes from *Zymomonas mobilis* in recombinant strains of *Escherichia coli*. *Applied and Environmental Microbiology* 54 (2): 397–404.

Ingram, L.O., Conway, T., Clark, D.P. et al. (1987). Genetic engineering of ethanol production in *Escherichia coli*. *Applied and Environmental Microbiology* 53 (10): 2420–2425.

Ingram, L.O., Aldrich, H.C., Borges, A.C. et al. (1999). Enteric bacterial catalysts for fuel ethanol production. *Biotechnology Progress* 15 (5): 855–866.

Inui, M., Kawaguchi, H., Murakami, S. et al. (2004). Metabolic engineering of *Corynebacterium glutamicum* for fuel ethanol production under oxygen-deprivation conditions. *Journal of Molecular Microbiology and Biotechnology* 8: 243–254.

Izui, K., Taguchi, M., Morikawa, M., and Katsuki, H. (1981). Regulation of *Escherichia coli* phosphoenolpyruvate carboxylase by multiple effectors in vivo. II. Kinetic studies with a reaction system containing physiological concentrations of ligands. *Journal of Biochemistry* 90 (5): 1321–1331.

Jarboe, L.R., Grabar, T.B., Yomanao, L.P. et al. (2007). Development of ethanologenic bacteria. *Advances in Biochemical Engineering/Biotechnology* 108: 237–261.

Jarboe, L.R., Zhang, X., Wang, X. et al. (2010). Metabolic engineering for production of biorenewable fuels and chemicals: contributions of synthetic biology. *Journal of Biomedicine & Biotechnology* 761042.

Jin, S.F. and Sonenshein, A.L. (1996). Characterization of the major citrate synthase of *Bacillus subtilis*. *Journal of Bacteriology* 178 (12): 3658–3660.

Jojima, T., Inui, M., and Yukawa, H. (2013). Biorefinery applications of *Corynebacterium glutamicum*. In: *Corynebacterium glutamicum Biology and Biotechnology* (eds. H. Yukawa and M. Inui), 149–172. New York: Springer-Verlag.

Jojima, T., Vertes, A.A., Inui, M., and Yukawa, H. (2014). Development of growth-arrested bioprocesses with *Corynebacterium glutamicum* for cellulosic ethanol production from complex sugar mixtures. In: *Biorefineries: Integrated Biochemical Processes for Liquid Biofuels* (eds. N. Qureshi, D. Hodge and A.A. Vertes), 121–139. New York: Elsevier.

Jojima, T., Noburyu, R., Sasaki, M. et al. (2015). Metabolic engineering for improved production of ethanol by *Corynebacterium glutamicum*. *Applied Microbiology and Biotechnology* 99 (3): 1165–1172.

Jonsson, L.J. and Martin, C. (2016). Pretreatment of lignocellulose: formation of inhibitory by-products and strategies for minimizing their effects. *Bioresource Technology* 199: 103–112.

Kim, Y., Ingram, L.O., and Shanmugam, K.T. (2007). Construction of an *Escherichia coli* K-12 mutant for homoethanologenic fermentation of glucose or xylose without foreign genes. *Applied and Environmental Microbiology* 73 (6): 1766–1771.

Kim, Y., Ingram, L.O., and Shanmugam, K.T. (2008). Dihydrolipoamide dehydrogenase mutation alters the NADH sensitivity of pyruvate dehydrogenase complex of *Escherichia coli* K-12. *Journal of Bacteriology* 190 (11): 3851–3858.

de Kok, A., Hengeveld, A.F., Martin, A., and Westphal, A.H. (1998). The pyruvate dehydrogensae multi-enzyme complex from Gram-negative bacteria. *Biochimica et Biophysica Acta* 1385: 353–366.

Lau, M.W., Gunawan, C., Balan, V., and Dale, B.E. (2010). Comparing the fermentation performance of *Escherichia coli* KO11, *Saccharomyces cerevisiae* 424A(LNH-ST) and *Zymomonas mobilis* AX101 for cellulosic ethanol production. *Biotechnology for Biofuels* 3: 11.

Lawford, H.G. and Rousseau, J.D. (1991). Ethanol production by recombinant *Escherichia coli* carrying genes from *Zymomonas mobilis*. *Applied Biochemistry and Biotechnology* 28–29: 221–236.

Lawford, H.G. and Rousseau, J.D. (2002). Performance testing of *Zymomonas mobilis* metabolically engineered for cofermentation of glucose, xylose, and arabinose. *Applied Biochemistry and Biotechnology* 98: 429–448.

Lee, K.J., Skotnicki, M.L., Tribe, D.E., and Rogers, P.L. (1981). The kinetics of ethanol production by *Zymomonas mobilis* on fructose and sucrose media. *Biotechnology Letters* 3 (5): 207–212.

Lindsay, S.E., Bothast, R.J., and Ingram, L.O. (1995). Improved strains of recombinant *Escherichia coli* for ethanol production from sugar mixtures. *Applied Microbiology and Biotechnology* 43 (1): 70–75.

Liu, T. and Khosla, C. (2010). Genetic engineering of *Escherichia coli* for biofuel production. *Annual Review of Genetics* 44: 53–69.

Liu, S.Q., Nichols, N.N., Dien, B.S., and Cotta, M.A. (2006). Metabolic engineering of a *Lactobacillus plantarum* double *ldh* knockout strain for enhanced ethanol production. *Journal of Industrial Microbiology & Biotechnology* 33 (1): 1–7.

Loaces, I., Rodriguez, C., Amarelle, V. et al. (2016). Improved glycerol to ethanol conversion by *E. coli* using a metagenomic fragment isolated from an anaerobic reactor. *Journal of Industrial Microbiology & Biotechnology* 43 (10): 1405–1416.

Loos, H., Kramer, R., Sahm, H., and Sprenger, G.A. (1994). Sorbitol promotes growth of *Zymomonas mobilis* in environments with high concentrations of sugar: evidence for a physiological function of glucose fructose oxidoreductase in osmoprotection. *Journal of Bacteriology* 176 (24): 7688–7693.

Lovitt, R.W., Shen, G.J., and Zeikus, J.G. (1988). Ethanol production by thermophilic bacteria: biochemical basis for ethanol and hydrogen tolerance in *Clostridium thermohydrosulfuricum*. *Journal of Bacteriology* 170 (6): 2809–2815.

Lowe, S.E. and Zeikus, J.G. (1991). Metabolic regulation of carbon and electron flow as a function of pH during growth of *Sarcina ventriculi*. *Archives of Microbiology* 155: 325–329.

Ma, Y.Y., Dong, H.N., Zou, S.L. et al. (2012). Comparison of glucose/xylose co-fermentation by recombinant *Zymomonas mobilis* under different genetic and environmental conditions. *Biotechnology Letters* 34 (7): 1297–1304.

Martinez, A., Grabar, T.B., Shanmugam, K.T. et al. (2007). Low salt medium for lactate and ethanol production by recombinant *Escherichia coli* B. *Biotechnology Letters* 29 (3): 397–404.

McGovern, P.E., Zhang, J.H., Tang, J.G. et al. (2004). Fermented beverages of pre- and proto-historic China. *Proceedings of the National Academy of Sciences of the United States of America* 101 (51): 17593–17598.

Mejia, J.P., Burnett, M.E., An, H. et al. (1992). Coordination of expression of *Zymomonas mobilis* glycolytic and fermentative enzymes: a simple hypothesis based on mRNA stability. *Journal of Bacteriology* 174 (20): 6438–6443.

Miller, E.N., Jarboe, L.R., Yomano, L.P. et al. (2009). Silencing of NADPH-dependent oxidoreductase genes (*yqhD* and *dkgA*) in furfural-resistant ethanologenic *Escherichia coli*. *Applied and Environmental Microbiology* 75 (13): 4315–4323.

Miller, E.N., Jarboe, L.R., Turner, P.C. et al. (2009). Furfural inhibits growth by limiting sulfur assimilation in ethanologenic *Escherichia coli* strain LY180. *Applied and Environmental Microbiology* 75 (19): 6132–6141.

Mohagheghi, A., Evans, K., Chou, Y.C., and Zhang, M. (2002). Cofermentation of glucose, xylose, and arabinose by genomic DNA-integrated xylose/arabinose fermenting strain of *Zymomonas mobilis* AX101. *Applied Biochemistry and Biotechnology* 98: 885–898.

Mohagheghi, A., Linger, J., Smith, H. et al. (2014). Improving xylose utilization by recombinant *Zymomonas mobilis* strain 8b through adaptation using 2-deoxyglucose. *Biotechnology for Biofuels* 7: 19.

Mohagheghi, A., Linger, J.G., Yang, S.H. et al. (2015). Improving a recombinant *Zymomonas mobilis* strain 8b through continuous adaptation on dilute acid pretreated corn stover hydrolysate. *Biotechnology for Biofuels* 8: 55.

Morimura, S., Ling, Z.Y., and Kida, K. (1997). Ethanol production by repeated-batch fermentation at high temperature in a molasses medium containing a high concentration of total sugar by a thermotolerant flocculating yeast with improved salt-tolerance. *Journal of Fermentation and Bioengineering* 83 (3): 271–274.

Munjal, N., Mattam, A.J., Pramanik, D. et al. (2012). Modulation of endogenous pathways enhances bioethanol yield and productivity in *Escherichia coli*. *Microbial Cell Factories* 11: 145.

Neale, A.D., Scopes, R.K., Wettenhall, R.E., and Hoogenraad, N.J. (1987). Pyruvate decarboxylase of *Zymomonas mobilis:* isolation, properties, and genetic expression in *Escherichia coli*. *Journal of Bacteriology* 169 (3): 1024–1028.

Nichols, N.N., Dien, B.S., and Bothast, R.J. (2003). Engineering lactic acid bacteria with pyruvate decarboxylase and alcohol dehydrogenase genes for ethanol production from *Zymomonas mobilis*. *Journal of Industrial Microbiology & Biotechnology* 30 (5): 315–321.

Ohta, K., Beall, D.S., Mejia, J.P. et al. (1991). Genetic improvement of *Escherichia coli* for ethanol production: chromosomal integration of *Zymomonas mobilis* genes encoding pyruvate decarboxylase and alcohol dehydrogenase II. *Applied and Environmental Microbiology* 57 (4): 893–900.

Ohta, K., Hamasuna, H., Tsukamoto, J. et al. (2012). Disruption of *ptsG* gene and *manXYZ* operon of ethanol-producing *Escherichia coli* KO11: effects on glucose and xylose utilization and ethanol production. *Journal of Bioscience and Bioengineering* 113 (5): 608–610.

Okamoto, T., Taguchi, H., Nakamura, K., and Ikenaga, H. (1994). Production of ethanol from maltose by *Zymobacter palmae* fermentation. *Bioscience, Biotechnology, and Biochemistry* 58: 1328–1329.

Olson, D.G., Sparling, R., and Lynd, L.R. (2015). Ethanol production by engineered thermophiles. *Current Opinion in Biotechnology* 33: 130–141.

Osman, Y.A. and Ingram, L.O. (1987). *Zymomonas mobilis* mutants with an increased rate of alcohol production. *Applied and Environmental Microbiology* 53 (7): 1425–1432.

Osman, Y.A., Conway, T., Bonetti, S.J., and Ingram, L.O. (1987). Glycolytic flux in *Zymomonas mobilis:* enzyme and metabolite levels during batch fermentation. *Journal of Bacteriology* 169 (8): 3726–3736.

Pinilla, L., Torres, R., and Ortiz, C. (2011). Bioethanol production in batch mode by a native strain of *Zymomonas mobilis*. *World Journal of Microbiology and Biotechnology* 27: 2521–2528.

Qureshi, N., Dien, B.S., Nichols, N.N. et al. (2006). Genetically engineered *Escherichia coli* for ethanol production from xylose - substrate and product inhibition and kinetic parameters. *Food and Bioproducts Processing* 84 (C2): 114–122.

Qureshi, N., Dien, B.S., Liu, S. et al. (2012). Genetically engineered *Escherichia coli* FBR5: part I. Comparison of high cell density bioreactors for enhanced ethanol production from xylose. *Biotechnology Progress* 28 (5): 1167–1178.

Raj, K.C., Talarico, L.A., Ingram, L.O., and Maupin-Furlow, J.A. (2002). Cloning and characterization of the *Zymobacter palmae* pyruvate decarboxylase gene (*pdc*) and comparison to bacterial homologues. *Applied and Environmental Microbiology* 68 (6): 2869–2876.

Rajoka, M.I., Ferhan, M., and Khalid, A.M. (2005). Kinetics and thermodynamics of ethanol production by a thermotolerant mutant of *Saccharomyces cerevisiae* in a microprocessor-controlled bioreactor. *Letters in Applied Microbiology* 40: 316–321.

Rogers, P.L., Jeon, Y.J., Lee, K.J., and Lawford, H.G. (2007). *Zymomonas mobilis* for fuel ethanol and higher value products. *Advances in Biochemical Engineering/Biotechnology* 108: 263–288.

Romero, S., Merino, E., Bolivar, F. et al. (2007). Metabolic engineering of *Bacillus subtilis* for ethanol production: lactate dehydrogenase plays a key role in fermentative metabolism. *Applied and Environmental Microbiology* 73 (16): 5190–5198.

Saha, B.C. and Cotta, M.A. (2011). Continuous ethanol production from wheat straw hydrolysate by recombinant ethanologenic *Escherichia coli* strain FBR5. *Applied Microbiology and Biotechnology* 90 (2): 477–487.

Saha, B.C., Nichols, N.N., and Cotta, M.A. (2011). Ethanol production from wheat straw by recombinant *Escherichia coli* strain FBR5 at high solid loading. *Bioresource Technology* 102 (23): 10892–10897.

Saha, B.C., Nichols, N.N., Qureshi, N. et al. (2015). Pilot scale conversion of wheat straw to ethanol via simultaneous saccharification and fermentation. *Bioresource Technology* 175: 17–22.

San Martin, R., Bushell, D., Leak, D.J., and Hartley, B.S. (1992). Development of a synthetic medium for continuous anaerobic growth and ethanol production with a lactate dehydrogenase mutant of *Bacillus stearothermophilus*. *Journal of General Microbiology* 138 (5): 987–996.

Shaw, A.J., Podkaminer, K.K., Desai, S.G. et al. (2008). Metabolic engineering of a thermophilic bacterium to produce ethanol at high yield. *Proceedings of the National Academy of Sciences of the United States of America* 105 (37): 13769–13774.

Snoep, J.L., de Graef, M.R., Westphal, A.H. et al. (1993). Differences in sensitivity to NADH of purified pyruvate dehydrogenase complexes of *Enterococcus faecalis*, *Lactococcus lactis*, *Azotobacter vinelandii* and *Escherichia coli*: implications for their activity in vivo. *FEMS Microbiology Letters* 114 (3): 279–283.

Sootsuwan, K., Thanonkeo, P., Keeratirakha, N. et al. (2013). Sorbitol required for cell growth and ethanol production by *Zymomonas mobilis* under heat, ethanol, and osmotic stresses. *Biotechnology for Biofuels* 6: 180.

Su, Y., Rhee, M.S., Ingram, L.O., and Shanmugam, K.T. (2010). Physiological and fermentation properties of *Bacillus coagulans* and a mutant lacking fermentative lactate dehydrogenase activity. *Journal of Industrial Microbiology & Biotechnology* 38: 441–450.

Sun, Z.T., Do, P.M., Rhee, M.S. et al. (2012). Amino acid substitutions at glutamate-354 in dihydrolipoamide dehydrogenase of *Escherichia coli* lower the sensitivity of pyruvate dehydrogenase to NADH. *Microbiology* 158: 1350–1358.

Swings, J. and De Ley, J. (1977). The biology of *Zymomonas*. *Bacteriological Reviews* 41: 1–46.

Talarico, L.A., Gil, M.A., Yomano, L.P. et al. (2005). Construction and expression of an ethanol production operon in Gram-positive bacteria. *Microbiology* 151 (12): 4023–4031.

Tan, F.R., Wu, B., Dai, L.C. et al. (2016). Using global transcription machinery engineering (gTME) to improve ethanol tolerance of *Zymomonas mobilis*. *Microbial Cell Factories* 15: 4.

Turner, P.C., Yomano, L.P., Jarboe, L.R. et al. (2012). Optical mapping and sequencing of the *Escherichia coli* KO11 genome reveal extensive chromosomal rearrangements, and multiple tandem copies of the *Zymomonas mobilis pdc* and *adhB* genes. *Journal of Industrial Microbiology & Biotechnology* 39 (4): 629–639.

Underwood, S.A., Zhou, S., Causey, T.B. et al. (2002). Genetic changes to optimize carbon partitioning between ethanol and biosynthesis in ethanologenic *Escherichia coli*. *Applied and Environmental Microbiology* 68: 6263–6272.

Underwood, S.A., Buszko, M.L., Shanmugam, K.T., and Ingram, L.O. (2002). Flux through citrate synthase limits the growth of ethanologenic *Escherichia coli* KO11 during xylose fermentation. *Applied and Environmental Microbiology* 68 (3): 1071–1081.

Wang, Q., Ou, M.S., Kim, Y. et al. (2010). Metabolic flux control at the pyruvate node in an anaerobic *Escherichia coli* strain with an active pyruvate dehydrogenase. *Applied and Environmental Microbiology* 76 (7): 2107–2114.

Weitzman, P.D. (1966). Regulation of citrate synthase activity in *Escherichia coli*. *Biochimica et Biophysica Acta* 128 (1): 213–215.

Wong, M.S., Li, M., Black, R.W. et al. (2014). Microaerobic conversion of glycerol to ethanol in *Escherichia coli*. *Applied and Environmental Microbiology* 80 (10): 3276–3282.

Wood, B.E., Yomanao, L.P., York, S.W., and Ingram, L.O. (2005). Development of industrial medium required elimination of the 2,3-butanediol fermentation pathway to maintain ethanol yield in an ethanologenic strain of *Klebsiella oxytoca*. *Biotechnology Progress* 21: 1366–1372.

Yang, S.H., Pelletier, D.A., Lu, T.Y.S., and Brown, S.D. (2010). The *Zymomonas mobilis* regulator *hfq* contributes to tolerance against multiple lignocellulosic pretreatment inhibitors. *BMC Microbiology* 10: 135.

Yang, S., Fei, G., Zhang, Y. et al. (2016). *Zymomonas mobilis* as a model system for production of biofuels and biochemicals. *Microbial Biotechnology* 9: 699–717.

Yazdani, S.S. and Gonzalez, R. (2008). Engineering *Escherichia coli* for the efficient conversion of glycerol to ethanol and co-products. *Metabolic Engineering* 10 (6): 340–351.

Yi, X., Gu, H.Q., Gao, Q.Q. et al. (2015). Transcriptome analysis of *Zymomonas mobilis* ZM4 reveals mechanisms of tolerance and detoxification of phenolic aldehyde inhibitors from lignocellulose pretreatment. *Biotechnology for Biofuels* 8: 153.

Yomano, L.P., York, S.W., Zhou, S. et al. (2008). Re-engineering *Escherichia coli* for ethanol production. *Biotechnology Letters* 30 (12): 2097–2103.

Yomano, L.P., York, S.W., Shanmugam, K.T., and Ingram, L.O. (2009). Deletion of methylglyoxal synthase gene (*mgsA*) increased sugar co-metabolism in ethanol-producing *Escherichia coli*. *Biotechnology Letters* 31 (9): 1389–1398.

Zhang, M., Eddy, C., Deanda, K. et al. (1995). Metabolic engineering of *Zymomonas mobilis* for ethanol production from renewable feedstocks. *Abstracts of Papers of theAmerican Chemical Society* 209: 115.

Zhou, S., Iverson, A.G., and Grayburn, W.S. (2008). Engineering a native homoethanol pathway in *Escherichia coli* B for ethanol production. *Biotechnology Letters* 30 (2): 335–342.

11

Clostridia and Process Engineering for Energy Generation*

Adriano P. Mariano[1], *Danilo S. Braz*[1], *Henrique C. A. Venturelli*[1] *and Nasib Qureshi*[2]

[1] *University of Campinas (UNICAMP), School of Chemical Engineering, Campinas, SP 13083-852, Brazil*
[2] *United States Department of Agriculture, Agricultural Research Service, National Center for Agricultural Utilization Research, Bioenergy Research Unit, Peoria, IL 61604, USA*

CHAPTER MENU

11.1 Introduction

From a genus of gram-positive anaerobic bacteria, several non-pathogenic Clostridium species are able to produce n-butanol (hereafter referred to as butanol). This four-carbon alcohol has been drawing attention due to its superior fuel properties and its share of the chemicals market of approximately 4 million tons a year or approximately 6 billion dollars. As an automotive fuel, blends of butanol with gasoline offer better engine performance and reduction of emissions when compared to ethanol-gasoline blends (Trindade and Santos 2017). Its drop-in characteristics, or similar chemical structure to alkane hydrocarbons, make butanol more compatible with existing fuel distribution infrastructure (pipelines and gas stations) and car engines. In countries such as the US, where the existing infrastructure is limiting the growth of the domestic ethanol market (the so-called 'blending wall'), the introduction of drop-in biofuels is crucial to increasing the share of biofuels in the transport energy mix. Moreover, the higher energy content in the butanol molecule gives better mileage than ethanol (~30%) and it can also be added to diesel and even catalytically upgraded to jet fuel. Refiners can also have an economic advantage by blending butanol to gasoline. Since butanol has a lower vapour pressure than ethanol, butanol could partially substitute for expensive less evaporative gasoline components, which is advantageous especially during the summer season.

*Mention of trade names or commercial products in this article is solely for the purpose of providing scientific information and does not imply recommendation or endorsement by the United States Department of Agriculture. USDA is an equal opportunity provider and employer.

Green Energy to Sustainability: Strategies for Global Industries, First Edition.
Edited by Alain A. Vertès, Nasib Qureshi, Hans P. Blaschek and Hideaki Yukawa.

In the chemicals market, butanol is a feedstock for manufacturing a wide range of chemicals, mostly butyl acrylate, butyl acetate, and glycol ethers. It is also used as a direct solvent and in various coating applications (Wise Guy Reports 2016). Additionally, with the production expansion of low-cost natural gas from shale deposits in the US, the substitution of naphtha by natural gas in petrochemical manufacturing results in less production of propylene and butane co-products. This trend opens up new markets for butanol as a C4-building block as a substitution for butane. Current butanol production relies almost exclusively on the oxo synthesis that uses propylene as a feedstock. Major global manufacturers include BASF SE (Germany), The Dow Chemical Company (US), BASF-YPC Ltd. (China), OXO Corporation (US), Sasol Ltd. (South Africa), Formosa Plastics Corporation (Taiwan), Eastman Chemical Company (US), Oxichimie SAS (France), KH Neochem Co. Ltd. (Japan) and CNPC (China). The Asia-Pacific region has the largest share (~50%) of the world's market in terms of volume, and China is the biggest consumer and import market for butanol (Markets and Markets 2015). Not surprisingly, recent attempts to resume the bio-butanol industry have been led by Chinese companies. However, several of the corn starch-based butanol plants installed in China in the past 10 years are currently mothballed due to high corn prices and government restrictions on the use of food grains for fuels and chemicals production (the reader is referred to Chapter 3 for more information on the status of the renewable energy industry in China).

Bio-based butanol is produced via the acetone–butanol–ethanol (ABE) fermentation route. In this fermentation, naturally-occurring solventogenic Clostridium species convert sugars into ABE with yields (g product/g sugar) in a range between 32% and 35% (theoretical yield is 40%). The co-product acetone accounts for approximately one third of the converted sugars. It means that the conversion of 1 t of sugars yields roughly 200 kg of butanol and 100 kg of acetone. Ethanol makes up the remainder of ABE (320–350 kg total). These numbers indicate two challenges for the economical production of bio-based butanol for energy generation: (i) low fermentation yield and the increased share of the feedstock cost component; and (ii) the risk of acetone oversupply in case butanol production reaches the projected production scale of billions of gallons per year in the fuels market. For comparison's sake, whereas the US corn ethanol industry has been delivering about 2.7 gal ethanol per bushel of corn, the conversion of ethanol plants to butanol plants would result in a throughput of about 1.9 gal ABE per bushel of corn (1.1 B + 0.6 A + 0.2 E gal/bu). In a scenario in which the butanol industry reaches a production volume of 15 billion gallons a year, equivalent to the current US ethanol industry, there would be an annual production of about 7.7 billion gallons of acetone, or 23 million metric tons. Assuming 100% substitution of oil-based acetone, the current acetone world market would be able to absorb only 30% of that amount, or approximately 6 million tons. Thus, even taking into account the higher energy density of butanol, the economic competitiveness of the ABE process against the ethanol option is strongly dependent on the valorization of the co-product acetone. Furthermore, since acetone, as a chemical, may not benefit from government incentives to biofuels, such as the case of the renewable identification numbers (RINs) system[1] in the US, biofuels policy factors may also play against the competitiveness of the ABE process. Recently, efforts have been made to use CRISPR/Cas9 technology to redirect the ABE pathway to desired end products such as butanol (Wang et al. 2015).

There are also process engineering challenges with respect to the ABE route. The conventional process design, found in most of the commercial plants in the twentieth century, is based on batch fermentors followed by product recovery by distillation. The characteristic low productivity and

1 Renewable Identification Numbers (RIN's) are credits used for compliance under the Renewable Fuel Standard Program.

energy inefficiency of such a design is aggravated by the fact that solventogenic Clostridium species have relatively low tolerance to butanol. Butanol concentration in the end of the batch is rarely above 13 g/l, and together with the low butanol yield (~0.20 g/g), restricts the initial sugars load to approximately 60 g/l. Moreover, the ABE fermentation is slow, and productivity is usually under 0.50 g/l/h. The combination of these factors leads to an increase in capital investment (especially regarding the number of fermentors), and the processing of dilute streams. The latter determines the energy duty of the separation unit and wastewater footprint, i.e. the ratio of the fermentation stillage volume to product volume (m^3 stillage/m^3 butanol). To illustrate, steam consumption in the distillation and the wastewater footprint can be as high as 30 GJ/t butanol and 80 m^3 stillage/m^3 butanol, respectively (Mariano and Maciel Filho 2012). These values are substantial when compared to the energy butanol can deliver as a fuel (36 GJ/t) and, for example, the wastewater footprint of sugarcane ethanol plants in Brazil (~10 m^3 stillage/m^3 ethanol).

An additional challenge for the ABE process, which is also an opportunity, is found in the upstream part of the process. As in the ethanol case, market concerns such as feedstock availability and price created the need to incorporate lignocellulosic feedstocks into the process. Whereas the challenge of producing affordable and fermentable (with low levels of cell inhibitors) lignocellulosic sugars is common to all second-generation fuels and chemicals, the opportunity resides in the fact that naturally-occurring Clostridium species can ferment both hexose and pentose sugars. Lignocellulose typically has significant pentose content. The latter is not metabolized by current industrial *Saccharomyces cerevisiae* strains found in ethanol plants.

To make the production of bio-butanol economically competitive, the several technological and process engineering challenges listed above need to be overcome. As such, scientific and engineering efforts have been focused on developing (i) superior microbial strains with improved butanol yield and product tolerance, (ii) efficient pretreatment and conversion of lignocellulosic feedstocks and novel substrates such as food waste (FW), (iii) bioreactors and in-situ product removal techniques to increase productivity and to circumvent the butanol toxicity issue, and (iv) process intensification with simultaneous saccharification and fermentation (SSF) and product recovery aimed at reduction of investment cost. Recent advances towards these efforts are presented in the first part of this chapter. In the second part, recent representative economic modelling studies on the ABE process are reviewed with focus on the effects of process engineering hurdles on the economics. Lastly, this chapter discusses technical and market aspects in the context of a butanol plant annexed to a Brazilian eucalyptus (EU) kraft pulp mill. In this case study, the competitiveness of butanol is assessed against ethanol taking into account new fermentation technologies, valorization of acetone and the by-product lignin, plant scale (PS), and design restrictions related to the integration of the ABE plant to the host pulp mill.

11.2 Recent Technological Advances

11.2.1 Micro-organisms

There are several microbial cultures that can produce butanol from various substrates including corn, molasses, and hydrolysates of cellulosic residues (Jones and Wood 1986; Qureshi 2010). Although some of the butanol/ABE producing cultures have been reported to produce up to 32.6 g/l ABE (20.1–20.9 g/l butanol) (Chen and Blaschek 1999), on a routine basis the maximum ABE concentration does not exceed 20–22 g/l. These product concentrations are low as the fermentation broth contains 970–980 g/l water from which ABE/butanol has to be recovered, which is energy intensive. In addition to the low butanol titre in the fermentation broth, there are other problems that this fermentation faces including low ABE/butanol productivity in the batch reactors, and production of some of the undesired products such as acetic and butyric acids. Production of these acids complicates their recovery from the fermentation broth, and reduces ABE yield as some of the carbon is directed towards acid production. For these reasons, development of novel strains that can tolerate and produce high concentrations of ABE/butanol is still desired.

To accumulate high concentrations of butanol in the fermentation broth, two factors are of prime importance that include tolerance of high concentration by the culture, and its continual production with the final aim to use these micro-organisms in real fermentation processes. There are a number of studies where butanol tolerant strains capable of tolerating 3.0–6.0% butanol (v/v; or 24.3–48.6 g/l; Table 11.1) have been either isolated or developed (Pomaranski and Tiquia-Arashiro 2016; Zaki et al. 2014; Ravinder et al. 2014; Rühl et al. 2009). The purpose of development or isolation of such strains is to transfer butanol producing genes from less tolerant strains to tolerant strains. However, to the authors' knowledge such genetic transfers or manipulations have not been successful.

Clostridium beijerinckii DSM 6423 is a strain that produces isopropanol, butanol, ethanol (IBE) rather than acetone, butanol, ethanol (ABE). Production of isopropanol is considered to be superior to acetone as it has better fuel characteristics. In a recent report, random mutagenesis and genome shuffling was applied to improve IBE production by mutant strains (Gérando et al. 2016). In this approach, three highly tolerant mutant strains were isolated that were capable of withstanding up to 50 g/l isopropanol on agar plates. Isolation or development of butanol and/or isopropanol tolerant strains is an indication that in the future novel biofuel producing strains could potentially be developed.

Additional requirement of butanol producing and tolerant strains is that they should be able to grow and utilize cellulosic sugars such as glucose, xylose, arabinose, galactose, and mannose since

Table 11.1 Development or isolation of butanol tolerant strains.

Strain	Butanol tolerance (%, v/v)	Butanol tolerance (g/l)	Butanol production	Reference
Carboxydotrophic strains[a]	3	24.3	Produces low conc. of buOH from CO gas	Pomaranski and Tiquia-Arashiro (2016)
Saccharomyces cerevisiae DBVPG1788	4	32.4	None	Zaki et al. (2014)
Saccharomyces cerevisiae DBVPG6044	4	32.4	None	Zaki et al. (2014)
Saccharomyces cerevisiae YPS128	4	32.4	None	Zaki et al. (2014)
Bacillus megaterium	5	40.5	Not mentioned	Ravinder et al. (2014)
Bacillus aryabhattai	5	40.5	Not mentioned	Ravinder et al. (2014)
Bacillus tequilensis	5	40.5	Not mentioned	Ravinder et al. (2014)
Bacillus circulans	5	40.5	Not mentioned	Ravinder et al. (2014)
Pseudomonas putida	6	48.6	None	Rühl et al. (2009)

a) Strains able to convert carbon monoxide into butanol.

the final aim would be to use agricultural residues. At this point the authors are not aware if the highest butanol tolerant strain (*Pseudomonas putida*, butanol tolerance 6.0% [v/v]; Rühl et al. 2009) can utilize cellulosic sugars. Furthermore, they should produce low concentrations of reaction intermediates such as acids (acetic and butyric). These acids are produced in butanol fermentation, before they (acids) are converted to butanol and acetone.

11.2.2 Novel Substrates

Although in the past, substrates such as corn and molasses have been used, in the future they are expected not to be used because they are costly and would force food, feed versus biofuel feedstock competition. For this reason, the use of cellulosic substrates such as wheat straw, corn stover, barley straw, switchgrass, miscanthus, and dried distiller's grains and soluble (DDGS) is preferred. Details of some of these have been published in recent literature (Qureshi 2010; Ezeji and Blaschek 2008). In addition to these substrates, other novel feedstocks or substrates include sweet sorghum bagasse (SSB), FW, domestic waste (DW), and industrial wastes. SSB is a high energy biomass crop that yields 30–45 t of biomass per hectare and is drought tolerant with high photosynthetic activity. As a result of the United States Department of Agriculture's efforts to develop this crop, we investigated its use for butanol production.

Recently, we loaded 86 g/l cellulosic biomass in our reactors. Upon hydrolysis, this releases 50–52 g/l total sugar which produces approximately 19–20 g/l ABE with a yield of 0.36–0.38. As mentioned earlier, such a concentration of ABE is low from an economic point of view. To make it economically attractive, a higher concentration of cellulosic biomass should be used. However, it is not possible until tolerance and production capacity of microbial strain is increased or product is simultaneously removed from the reactor. Hence, we used a simultaneous product recovery technique in combination with a high concentration of cellulosic residues. In our batch reactors, we were successful increasing SSB concentration to 200 g/l followed by hydrolysis to sugars and fermentation to butanol (Qureshi et al. 2016). Table 11.2 shows the use of various feedstocks and process technologies for this fermentation.

Another source of carbohydrates is FW that can be converted to biofuels. In the United States, approximately 60 million tons of food is wasted each year (Food Waste website 2016). FW includes unconsumed food that is discarded by food processing industries, retailers, restaurants, and consumers. It is noteworthy to mention that biological conversion of FW to biofuels or chemicals does not require supplementation with nutrients, because it contains them. We made a successful attempt to convert FW to butanol (Huang et al. 2015). While using FW for butanol production, fermentation was faster than glucose with ABE productivity of 0.46 g/l/h as compared to when using glucose 0.22 g/l/h. Also, the ABE yield, with the use of FW, was higher (0.38) than when using glucose (0.35). With this yield value, 60 million tons of FW can be converted to approximately 16 million tons (5.1 billion gallons) of ABE per year which is 3.8% of gasoline consumption in the United States each year. This conversion is based on the assumption that FW contains 70% carbohydrates. It is also noteworthy that bioconversion of FW to butanol does not require supplementation with amylolytic enzymes for hydrolysis of starch to glucose which would reduce cost of butanol production from this substrate. It is likely that starch content (or composition) of FW may vary from time to time. To combat this variation robust fermentation strains such as newly developed *C. beijerinckii* would be needed.

Sugarcane bagasse (SB) and eucalyptus (EU) are other potential substrates for butanol production. In Brazil, SB is currently used for energy generation in sugarcane ethanol mills and predominantly studied for its use in the production of second-generation ethanol. Recent studies have also

Table 11.2 Examples of butanol production from agricultural residues and process technologies.

Feedstock/product recovery technique	Feedstock conc. (g/l)	Pretreatment	SHFR	SSFR	ABE produced (g/l)	Reference
Without product recovery						
Wheat straw	86	Dil H_2SO_4	No	No	25.0	Qureshi et al. (2007)
Barley straw	86	Dil H_2SO_4	No	No	26.6	Qureshi et al. (2010a)
Corn stover	86	Dil H_2SO_4	No	No	26.3	Qureshi et al. (2010b)
Switchgrass	86	Dil H_2SO_4	No	No	14.6	Qureshi et al. (2010b)
Sweet sorghum bagasse	200	Hot water	No	No	15.4	Qureshi et al. (2016)
Food waste	129	Not required	No	No	18.9	Huang et al. (2015)
With product recovery						
Barley straw (gas stripping)	86[a)]	Dil H_2SO_4	Yes	No	47.2	Qureshi et al. (2014a)
Wheat straw (gas stripping)	86[a)]	Dil H_2SO_4	Yes	No	47.6	Qureshi et al. (2007)
Corn stover (gas stripping)	86[a)]	Dil H_2SO_4	Yes	No	50.1	Qureshi et al. (2014a)
Corn stover (vacuum)	86	Dil H_2SO_4	No	Yes	20.8	Qureshi et al. (2014b)
Food waste (vacuum)	129	Not required	No	Yes	44.4[b)]	Huang et al. (2015)

Dil H_2SO_4 – Dilute H_2SO_4; SHFR – Separate hydrolysis fermentation and recovery (hydrolysis was performed prior to fermentation; fermentation was combined with recovery); SSFR- Simultaneous saccharification fermentation and recovery (hydrolysis, fermentation, and recovery were combined).

a) Supplemented with cellulosic sugars.

b) Calculated value.

demonstrated that SB undergoing either alkali or dilute sulfuric acid pretreatment followed by enzymatic hydrolysis is also suitable for butanol production (Pang et al. 2016; Jonglertjunya et al. 2014). EU is obtained from planted forests (silviculture) and supplies the production of pulp in Brazil. Since pulp companies are interested in expanding their product portfolio and penetrating emerging markets of the bio-economy, including the biofuels market, the potential of using eucalyptus as a feedstock has been studied. Whereas studies for ethanol production from eucalyptus started in the 1980s, the first study on the use of eucalyptus for ABE production was only published recently in 2015 (Zheng et al. 2015). In this study, steam explosion of EU chips followed by enzymatic hydrolysis yielded sugars that were converted to 13.1 g/l ABE. Notably, this production was obtained without nutrients supplementation of the eucalyptus hydrolysate. Techno-economics aspects of using EU chips for butanol production are presented in the second part of this chapter.

11.2.3 Biomass Pretreatment

To convert agricultural residues/biomass to butanol, it is essential that the biomass be pretreated prior to enzymatic hydrolysis. The most commonly used pretreatment techniques include dilute sulfuric acid, dilute NaOH, and ammonia fibre expansion. In a recent study on conversion of wheat straw to butanol, we used dilute (1% v/v) sulfuric acid to pretreat this substrate to produce butanol. However, economic modelling suggested that the use of dilute sulfuric acid is neither economically feasible nor environmentally favourable. Additionally, use of dilute acid to pretreat barley straw and corn stover produced toxic components that inhibited cell growth and butanol production from these substrates. When dilute acid pretreated and enzymatically hydrolysed corn stover was used it completely arrested cell growth. To remove those pretreatment inhibitors, a technique called overliming (Qureshi et al. 2010a,b) was used. The detoxified corn stover hydrolysate was able to support cell growth and fermentation. However, the detoxification process is costly and can impact the economics of butanol production adversely. Hence, we developed a liquid hot water (LHW) treatment method for pretreating SSB at 190 °C for minimum holding time (<1 minute) at this temperature. The pretreated SSB was successfully hydrolysed to monomeric sugars followed by successful fermentation to butanol. It demonstrated that the LHW pretreatment technique at 190 °C did not generate fermentation inhibitors. Although an economic evaluation of this process has not been performed, it is viewed that it would be more comparatively economical, in addition to being environmentally favourable than when using dilute sulfuric acid.

11.2.4 Novel Product Recovery Techniques

We have successfully developed several in situ product removal techniques including adsorption, gas stripping (GS), liquid–liquid extraction, perstraction, and pervaporation. Details of these techniques have been published earlier (Qureshi 2014). These techniques have limitations including low rate of removal (gas stripping), or being toxic to the culture (liquid–liquid extraction). Keeping these disadvantages in view we have developed two other techniques which can be used to remove butanol. They include vacuum fermentation, and liquid CO_2 extraction. Using vacuum fermentation, the rate of ABE removal can be as high as 13 g/l/h as compared to gas stripping of <0.5 g/l/h. Since, rate of removal by vacuum fermentation is high (Mariano et al. 2011, 2012), they can be used for product removal from highly productive reactor systems such as high cell density reactors. In such reactors, productivities range from 6.5 to 16 g/l/h.

Liquid CO_2 extraction requires no other chemical for butanol separation (Solana et al. 2016). Butanol extraction by liquid CO_2 has a number of advantages including its inert nature, non-toxicity to the microbial culture, and recyclability. Using this technique, we successfully recovered butanol from aqueous solutions. This separation resulted in butanol selectivities up to 2500 at a feed butanol concentration of 5 g/l, which are superior to any other previously reported recovery system.

11.2.5 Bioreactors

Bioreactors play an important role in the production of fuels and chemicals such as acetone, isopropanol, ethanol, or butanol. Use of an appropriate reactor system could result in increased productivity and hence reduction of capital and operational costs. The reactor systems that can be used for butanol production include batch, fed-batch, free cell continuous, immobilized cell continuous (including fluidized bed reactors), and cell recycle membrane reactors. Their application to butanol production and other details have been published elsewhere (Mariano et al. 2015). In batch reactors, reactor productivity of the order of 0.50 g/l/h is obtained which can be increased to 15–16 g/l/h in immobilized cell continuous series reactors. Membrane cell recycle continuous reactors also offer great potential and can be used in this fermentation (Cheryan 1986; Afschar et al. 1985).

11.2.6 Combining Unit Operations and Use of By-products

Combining unit operations or process integration results in the reduction of the number of units for a particular process. Butanol fermentation from cellulosic residues requires pretreatment, enzymatic saccharification, fermentation, and product recovery. To reduce the overall cost of butanol production from agricultural biomass, the last three of the four process steps were combined for the production of butanol from corn stover (Qureshi et al. 2014b). Such a process is expected to be beneficial to the microbial culture as hydrolysed sugars can be used simultaneously (by the culture) to produce butanol. At the same time, the produced butanol can be recovered from the system. In this way neither sugars nor toxic butanol accumulate in the system thus benefiting the process. Such a process not only reduces the number of units, it also reduces the reactor size and process streams.

Production of butanol from biomass is a process where several by-products including lignin, cell mass, acetone, ethanol, acids (acetic and butyric; though in small concentrations), CO_2, and H_2, vitamin B12, and polysaccharides are generated. To make butanol production economically feasible, some of these products will need to be marketed. Produced CO_2 and H_2 can be a source or substrate for another fermentation where they would be converted either into acids or butanol by the microbial strains. Additionally, process water recycling needs to be considered.

11.3 Economic Modelling and Case Study

11.3.1 Techno-economic Studies

Given the various technical challenges found in the ABE process, many solutions have been proposed, such as those listed above. Winner solutions certainly will be those that fulfil some basic process engineering prerequisites, including scalability, operational stability, and certainly most important, cost competitiveness. In this manner, several techno-economic studies have assessed the impact of important factors such as feedstock type, up-, mid-, and down-stream technologies, and process integration on the economics of ABE plants. As usual in the early stage design context, this assessment is based on mass and energy balances conducted in process simulators and/or spreadsheets considering conversion and yield factors obtained from the combination of experimental data (generally available in the literature) and assumptions of hypothetical process performances. This approach can not only predict the economics of the technology at its current status (e.g. yield is 80% of the theoretical value), but also set performance targets for researchers and engineers. In other words, performance metrics that should be attained to make the technology economically feasible.

In this section, 12 representative and recent economic modelling studies on the ABE process are reviewed with focus on the effects of process engineering hurdles on the economics. Table 11.3 presents key process design parameters considered in each of these studies, including feedstock, plant scale, technology, and economic criteria. Most shared common factors are: (i) The ABE plants are greenfield projects. Whereas a site-agnostic design is general and not subjected to constraints found when annexing a plant to an existing mill, the ‘bolt-on’ option, on the other hand, offers site-specific integration opportunities (feedstock supply, energy, equipment, waste water treatment (WWT) and others) that can considerably reduce costs. For instance, in studies 1 and 2, for the same plant capacity, investment cost could decrease from 27 to 4 USD/gal if an ethanol plant is converted to an ABE plant rather than building a new one. For the greenfield design, the net present value (NPV) of the project was negative and, for this reason, not economically feasible. (ii) Feedstock is agricultural biomass. Indeed, only one research group (studies 1 and 2) evaluated the potential of a woody biomass. (iii) Upstream technology is pretreatment with dilute sulfuric acid followed by enzymatic hydrolysis. On the other hand, biomass fractionation (studies 1 and 2) is recommended to valorize the by-product lignin. This aspect is investigated in the case study presented

Table 11.3 Recent techno-economic studies on ABE fermentation.

Study	Feedstock	Upstream technology	Fermentation technology	Downstream separation technology	Plant capacity (ton but/yr)	Investment cost (USD/but gallon capacity)	Economic criterion (EC)	Value of EC	Reference
1	Wood (spruce)	Fractionation (ethanol + SO_2)/enzymatic hydrolysis	SSF + ISPR (gas stripping/ perstraction/ pervaporation/ adsorption/ liquid–liquid extraction) *C. saccharoperbuty lacctonicum*	Vacuum distillation	~30 000	27 (greenfield plant in Norway)	NPV	Negative for all scenarios	Dahlbom et al. (2011)
2	Wood (spruce)	Fractionation (ethanol + SO_2)/enzymatic hydrolysis	Conventional batch fermentors considering regular and improved *Clostridium* strains (with increased butanol yield and sugar conversion) *C. beijerinckii* BA 101	Conventional distillation	~30 000	4 (conversion of an ethanol plant)	Production cost	Varied between 0.89 and 1.24 USD/l (or 1.10 and 1.53 USD/kg)	Larsson et al. (2008)
3	Corn starch/ switchgrass	Not informed	Conventional batch fermentors considering a regular strain *C. acetobutylicum*	Conventional distillation	—	Not informed (greenfield plant in the US)	NPV	Negative for all scenarios	Pfromm et al. (2010)
4	Corn starch/ corn stover/ wheat straw/ glycerol	Dilute acid and enzymatic hydrolysis	Conventional batch fermentor and fermentor with ISPR (vacuum/ adsorption) *C. acetobutylicum*/ *C. beijerinckii*	Conventional distillation	69 000	4 (greenfield plant in the US)	Minimum selling price	0.23–1.41 USD/ kg (glycerol was the best case)	Qureshi and Singh (2014)

(Continued)

Table 11.3 (Continued)

Study	Feedstock	Upstream technology	Fermentation technology	Downstream separation technology	Plant capacity (ton but/yr)	Investment cost (USD/but gallon capacity)	Economic criterion (EC)	Value of EC	Reference
5	Corn starch/ corn stover	Dilute acid and enzymatic hydrolysis	Batch fermentors with ISPR (vacuum stripping) *C. acetobutylicum*	Conventional distillation	~76 000 (corn stover)	17.7 (greenfield plant in the US)	Minimum selling price	1.09 USD/kg (4.02 USD/GGE)	Tao et al. (2014a)
6	Corn stover	Dilute acid and enzymatic hydrolysis	Batch fermentors with ISPR (vacuum stripping) *C. acetobutylicum* with improved butanol yield	Conventional distillation	~113 000	11.7 (greenfield plant in the US)	Minimum selling price	0.99 USD/kg (3.66 USD/GGE)	Tao et al. (2014b)
7	Sugarcane juice (sucrose)	—	Conventional batch fermentors considering regular and mutant strains *C. saccharoper-butylacetonicum DSM 2152*/ *C. beijerinckii* BA101	Conventional distillation	~20 000 (mutant strain with improved yield)	3.7 (greenfield annexed to a first-generation sugarcane ethanol plant)	IRR	14.8% (butanol priced as a chemical/mutant strain with improved yield)	Mariano et al. (2013a)
8	Sugarcane bagasse hydrolysate (C5 stream)	Steam explosion	Conventional batch fermentors considering regular and mutant strains	Conventional distillation	~12 000 (mutant strain with improved yield)	5.1 (greenfield annexed to a second-generation sugarcane ethanol plant)	IRR	15.2% (butanol priced as a chemical/mutant strain with improved yield)	Mariano et al. (2013b)

9	Molasses/wheat straw/corn stover/cassava bagasse	Dilute acid and enzymatic hydrolysis	(i) Batch SSF integrated with gas stripping (ii) Continuous SHF (iii) Batch SHF	(i) Double effect distillation (ii) Liquid–liquid extraction and distillation	~120 000	14.9 (greenfield plant)	NPV and IRR	47–150 million USD (NPV) 11–16% (IRR)	Naleli (2016)
10	Corn starch	—	Continuous fermentor with ISPR (vacuum flash fermentation) and cell recycle (microfiltration membrane) *C. acetobutylicum*	Conventional distillation	80 000	1.5 (greenfield plant in Canada)	NPV	87 million USD	Abdi et al. (2016)
11	Corn stover	Dilute acid and enzymatic hydrolysis	SSF + ISPR (vacuum fermentation) *C. beijerinckii* 8052	Conventional distillation	92 000	13.7 (greenfield plant in the US)	Production cost	2.22 USD/kg	Baral and Shah (2016)
12	Wheat straw	Dilute acid and enzymatic hydrolysis	Batch fermentors/batch with ISPR (membranes) *C. beijerinckii* P260	Conventional distillation	150 000	3.9 (greenfield plant in the US)	Production cost	1.30 USD/kg (conventional tech.) 1.00 USD/kg (advanced tech. Annexed to existing ethanol plant)	Qureshi et al. (2013)

SSF: simultaneous saccharification and fermentation; SHF: separate hydrolysis and fermentation; ISPR: in situ product recovery; NPV: net present value; IRR: internal rate of return; bdmt: bone-dry metric ton; GGE: gasoline gallon equivalent.

in the next section. (iv) Downstream separation technology is conventional distillation. As such, attempts to improve energy efficiency concentrated mostly in the use of in situ product recovery (ISPR). The ISPR technologies allow for the processing of more concentrated sugar streams, targeting product titres that could be feasible, from an energy efficiency point of view, the use of distillation. Distillation, as the preferred separation technology of the chemical and bio-ethanol industries, is an off-the-shelf solution with good scalability and controllability, which provides high product recovery (>99%) (Vane 2008). Furthermore, considering the technology risk of having a non-conventional fermentor, the use of alternative downstream separation technologies certainly increases both the chances of failure of the new process at commercial scale and the cost of capital (Mariano 2015a).

The need for increasing butanol yield (and butanol tolerance by the micro-organism) and energy efficiency were the process engineering hurdles pointed out as most important for the economics of ABE plants. Major conclusions from the techno-economic studies (Table 11.3) are presented as follows. In study 1, the membrane-assisted ISPR technique of perstraction provided superior energy consumption reduction, even more than vacuum distillation. As the mill would produce pellets for meeting its energy needs and sell eventual surplus, the lower steam consumption resulting from using ISPR led to pellets surplus and additional revenues. Study 2 assessed the conversion of an ethanol plant to an ABE plant. Given the more dilute nature of the ABE process, the retrofit would need additional fermentation tanks, and both the existing distillation and evaporation steps would need additional equipment. Evaporation was used to concentrate the distillation stillage to 60% solids, and the concentrated residue was burned to generate steam. In fact, a new parallel evaporation line would be needed and this was the largest cost contributor. Study 3 compared the carbon balance of the ABE process against the ethanol process. Given the low fermentation yield in the ABE process, this study concluded that the ethanol process is less capital-intensive and delivers more energy (liquid fuel yield) from the same amount of biomass. Study 4 explored the potential of different feedstocks for butanol production, namely corn grain, corn stover, wheat straw, and glycerol. The low cost of glycerol, which is a by-product of the biodiesel industry, in combination with energy efficiency given by advanced fermentation technologies resulted in a very competitive minimum butanol selling price (MBSP) of 0.23 USD/kg.

The National Renewable Energy Laboratory (NREL) assessed the competitiveness of butanol against ethanol in study 5 and isobutanol in study 6. In both ABE and isobutanol plants the vacuum (flash) fermentation technology was adopted. These studies concluded that both butanol costs are slightly higher than that of cellulosic ethanol when compared by minimum fuel selling prices on the gasoline gallon equivalent basis (GGE) (USD/GGE: 3.27 [ethanol] 3.62 [isobutanol] 3.66 [butanol]). In studies 7 and 8, the Brazilian Bioethanol Science and Technology Laboratory (CTBE) evaluated the production of butanol in sugarcane biorefineries considering either first-generation sugars (sucrose from sugarcane juice) or pentoses from bagasse hydrolysate. The latter is of interest for sugarcane biorefineries considering mixing lignocellulosic glucose with sucrose for ethanol production. As mentioned above, industrial yeasts in ethanol plants are not able to convert pentoses. In the case of the first-generation butanol, investment in a biorefinery with butanol production was more attractive (IRR = 14.8%) than the conventional ethanol–sugar mill (IRR = 13.3%) only if a micro-organism with improved butanol yield was used and butanol was sold in the chemical market. The economics of butanol from SB hemicellulose hydrolysate was more favourable (IRR = 13.1%) than the competing biogas production alternative (IRR = 11.3%) even considering the lower margins of the automotive fuels market and the use of non-engineered micro-organisms.

Study 9 assessed the impact of ISPR, process intensification, and different downstream separation technologies on the economics of the ABE process. Among six technological scenarios, only the combination of simultaneous batch saccharification and fermentation (SSF) with ISPR (gas stripping – GS) followed by either double effect distillation (DD) or liquid–liquid extraction was economically feasible. This study also compared the energy efficiency of lignocellulosic butanol against butanol production from molasses. The energy consumption in the former (58 GJ/t butanol), calculated for the most attractive scenario (SSF-GS/DD), more than doubled the energy duty of the molasses option (23 GJ/t butanol).

As in the NREL's investigation, in studies 10 and 11 the vacuum-assisted ISPR technique was also the technological solution chosen to improve energy efficiency. In the case of study 10, the use of the ISPR resulted in a positive NPV (USD 87 million), which was otherwise negative for the conventional fermentation process. Whereas study 10 was more focused on the fermentation step, study 11 identified operational targets for improvements in the whole biomass-to-butanol process. The percentage of butanol recovered as the final product (between 90 and 98 wt %) was identified to be the most sensitive parameter followed by sugar utilization in the fermentation, feedstock cost, biomass to sugars conversion rate, and heat recovery. Finally, study 12 concluded that the economics of butanol production from wheat straw is greatly affected by the cost of utilities (steam, cooling and chilling water, electricity), which represented the major operating cost (49%) followed by raw materials (27%). This study also shows the economic benefits of integrating the ABE plant into an existing mill. While the production cost was estimated to be 1.30 USD/kg for a greenfield plant, the use of a membrane-assisted ISPR technique in an ABE plant annexed to an existing ethanol mill could reduce the production cost to 1.00 USD/kg.

11.3.2 Case Study: Production of Butanol from Eucalyptus

11.3.2.1 Pulp Mill Case Study

The global pulp and paper industry, including Brazilian eucalyptus pulp producers, is under transformation searching for business opportunities in the bio-economy (Mariano 2015b). Among a myriad of products that may be derived from woody biomass, especially speciality pulp and lignin (and derivatives), chemicals and liquid fuels obtained from the fermentation of lignocellulosic sugars are also of interest. With feedstock supply chains already in place and gigantic wood processing units (notably in South America where pulp mills process more than 5000 bone dry metric ton per day, bdmt/d), the forest industry can certainly be competitive. As such, we present in this section the techno-economics of an ABE plant annexed to a Brazilian kraft pulp mill. Emphasis is given on the impact of new fermentation technologies and plant scale on the MBSP.

The mill case study is an existing kraft pulp facility installed in Brazil, which currently processes 6786 dry metric tons of eucalyptus per day and produces 3531 air dried ton (adt) of bleached pulp per day or 1.25 million adt/yr. Mass and energy flows of the mill, with focus on the black liquor recovery boiler, cogeneration unit, and the WWT plant, are presented in Figure 11.1. As usual in the case of modern kraft pulp mills, the facility has steam surplus (113 t/h run through a condensing turbine) and sells power to the grid (19.4 MW). Upon installation of the ABE plant, thermal energy will be provided by the mill. In this way, costs related to steam consumption are accounted for by adjusting the amount of power sold to the grid.

11.3.2.2 ABE Plant

The ABE plant to be installed annexed to the pulp mill is fed with additional eucalyptus wood (1000–3000 bdmt/d). Water content in the feedstock is 50% and the composition of *Eucalyptus urograndis* chips in this study is (dry matter) 45.5% cellulose, 20.3% hemicellulose, and 29.7% lignin. Wood chips are fractionated in an organosolv pretreatment unit followed by enzymatic hydrolysis

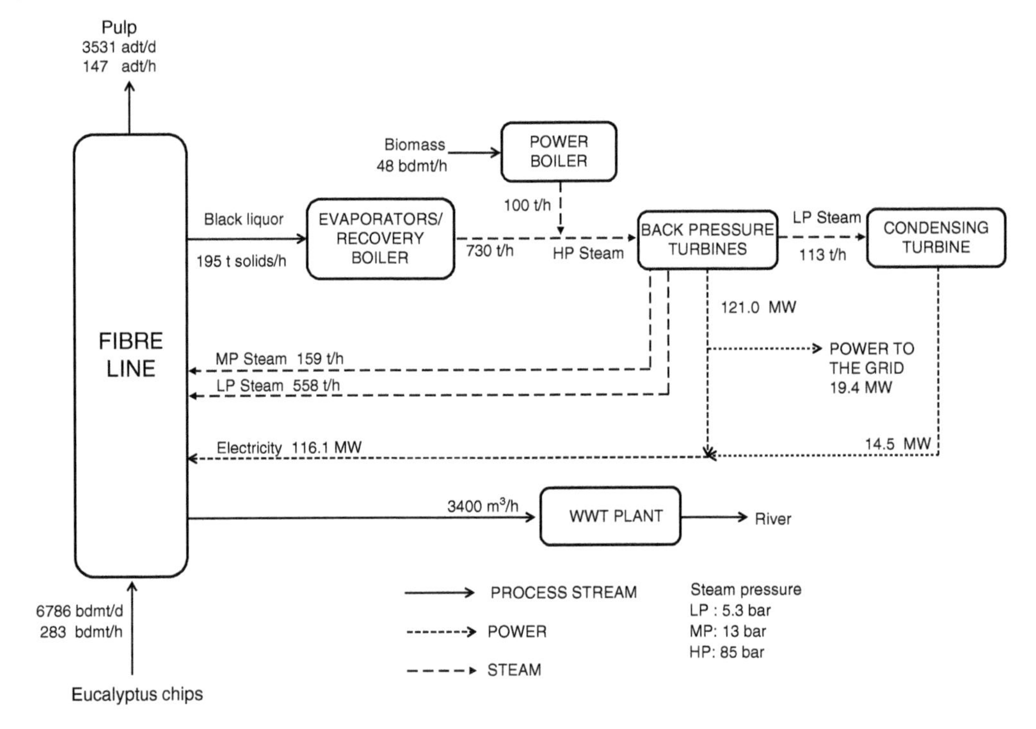

Figure 11.1 Brazilian eucalyptus kraft pulp mill with mass and energy flows. Production capacity is 1 250 000 adt/yr (1 air dry metric ton has 10% moisture content).

of the cellulose. Energy and mass balances for the pretreatment and saccharification units were obtained from Nitzsche et al. (2016), assuming the same conversion factors for eucalyptus. Instead of using the organosolv lignin for heat and power production, this high-quality lignin was assumed to be sold for a minimum price of 100 USD/t (the electricity opportunity cost is 90 USD/t). Fibres from the pretreatment (mainly cellulose) are enzymatically hydrolysed (solids loading of 20 wt.%) to hexoses (C6 sugars, mainly glucose) with molar yield of 0.774, and a minor portion of hemicellulose is hydrolysed to pentoses (C5 sugars, mainly xylose) with molar yield of 0.264. In the organosolv pretreatment, a large portion of the C5 sugars (and xylooligomers) is recovered in a dilute aqueous stream (~16 g/l). This stream could be either biodigested to produce biogas (as in Nitzsche et al. 2016) or fermented after evaporation. In this study, to avoid additional investment cost (biodigestion) or the expensive energy cost of evaporation, the C5 stream is sent to the WWT plant. The residual stream from the hydrolysis (non-converted cellulose and hydrolysis lignin) with 50% moisture content is sent to the black liquor evaporation train of the mill prior to combustion in the recovery boiler. The same pretreatment and saccharification units were considered for both the base-case scenario consisted of an ABE plant (Figure 11.2) and a competing scenario with an ethanol plant (Figure 11.3). Energy and mass balances of the ethanol plant (fermentation + distillation) were calculated based on values reported by Nitzsche et al. (2016). As assumed in the referred study, both C6 and C5 sugars are converted to ethanol by a genetically-modified *S. cerevisiae* strain and the process achieves 92% of the theoretical yield.

In the ABE plant, a naturally-occurring Clostridium species converts 95% and 80% of the C6 and C5 sugars, respectively, in fermentors with working volume of 750 m^3 (each tank has a total volume of 975 m^3). ABE yield is 0.32 (0.10 A, 0.20 B, 0.02 E) and taking into account the sugar

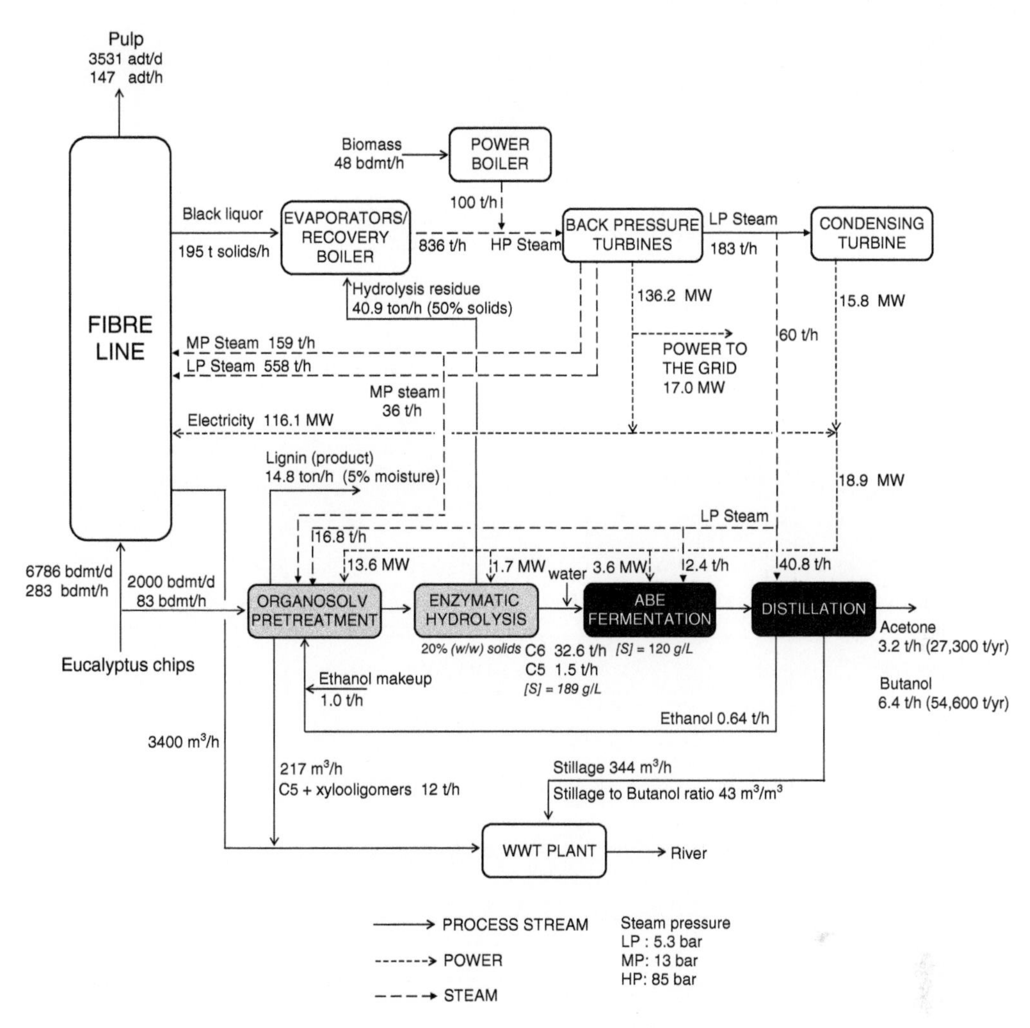

Figure 11.2 Energy and mass balance of an ABE plant integrated to a Brazilian kraft pulp mill.

conversion factors, the process runs at 74% of the theoretical yield. Initial sugar concentration in the fermentation is 60 g/l (diluted from 189 g/l, which is the sugar concentration in the outlet of the enzymatic hydrolysis step), and productivity is 0.50 g/l/h. As described below, a generic approach was adopted to assess the effects of advanced fermentation technologies with ISPR on the techno-economic analysis. In these types of technologies micro-organisms are capable of processing more concentrated sugar solutions (>60 g/l), and the higher the concentration, the better the performance of the technology (the ISPR technology is not specified in this analysis). Investment costs and energy consumption were calculated for specific values of sugars concentration between 80 and 200 g/l. To account for performance, productivity is assumed to start at 2.0 g/l/h (for the 80 g/l case) and increases linearly with sugar concentration. For instance, for a sugar concentration of 200 g/l, productivity is 5 g/l/h. Such productivity values can be considered to be conservative since continuous bioreactors with ISPR using different recovery technologies and feedstocks have been demonstrated to provide productivities higher than 10 g/l/h (Mariano et al. 2015). It is also important to mention that according to Van Dien (2013), fermentation technologies for production

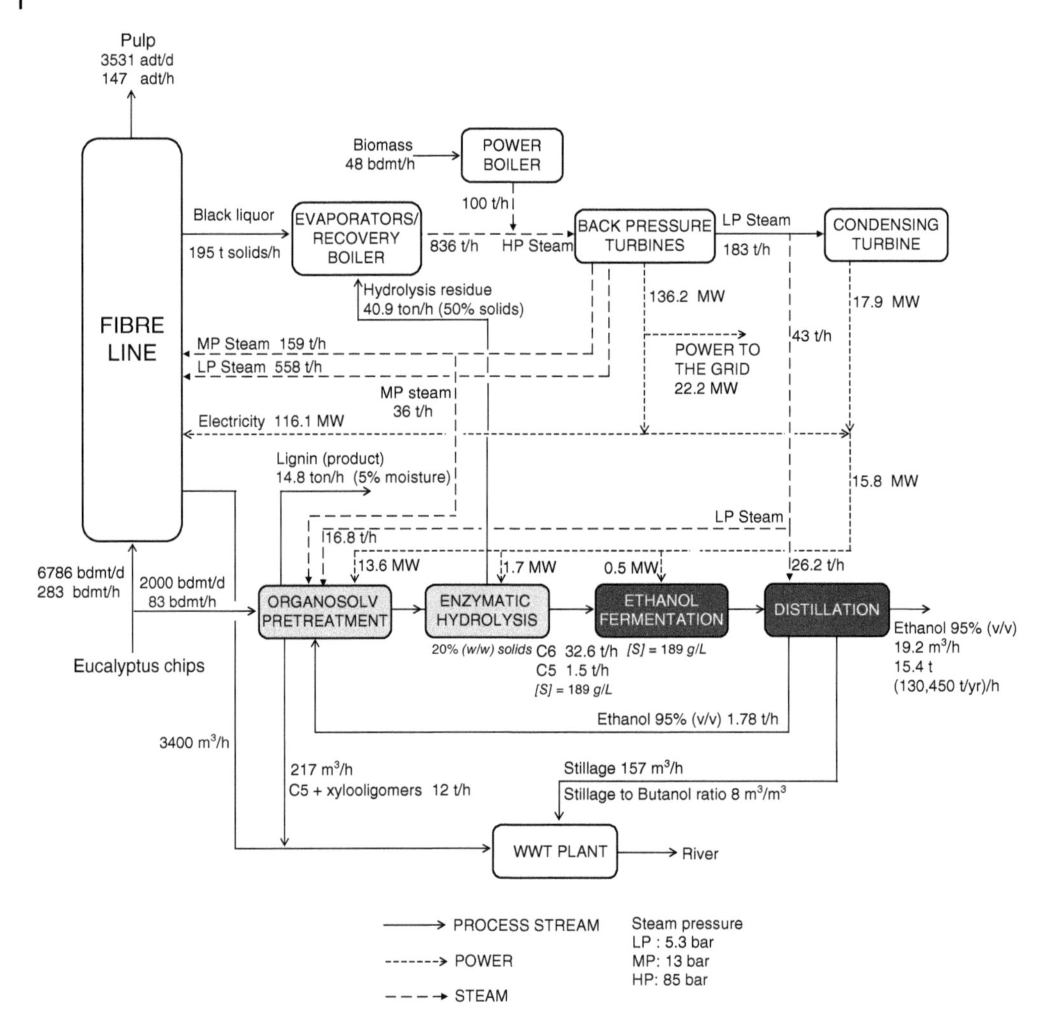

Figure 11.3 Energy and mass balance of an ethanol plant integrated to a Brazilian kraft pulp mill.

of bulk chemicals with productivities below 2.0 g/l/h are generally considered uneconomical due to their high capital cost. Conversions and yields are the same for all cases (as in the 60 g/l case). However, given the higher productivity of these technologies, the fermentor's working volume is 500 m^3 (each tank has a total volume of 750 m^3) for all cases. Since fermentors in advanced technologies are accompanied by pieces of equipment that provide the product recovery (e.g. membranes, vacuum system, condensation system, external tanks, solvent recovery distillation, compressors), the capital investment in the fermentation unit was assumed to be twice the cost of the fermentors (based on the number of fermentors calculated for a given inlet sugars concentration). For similar reasons, energy consumption of the fermentation unit was assumed to be 1 and 2 GJ/t butanol of low pressure (LP) steam and power, respectively. Separation of the fermentation products (ABE) is conducted in a series of five distillation columns and the energy consumption (LP steam) was calculated based on the correlation between butanol concentration in the beer and energy duty of the reboilers presented in Mariano and Maciel Filho (2012). Butanol concentration in the beer was calculated by multiplying the sugar's inlet concentration by butanol yield. Ethanol produced in the

ABE fermentation is completely consumed in the organosolv pretreatment. The flow rate of stillage was calculated based on the butanol concentration in the beer, and this wastewater is sent to the WWT plant of the mill.

11.3.2.3 Approach and Assumptions for the Economic Analysis

The energy and mass balances of the process options annexed to the kraft pulp mill were implemented and integrated with the economic model in Microsoft Excel spreadsheets (version 2010). Prefeasibility grade capital costs were calculated by factoring values reported in the literature according to the capacity power law expression (scaling factor equal to 0.6) and expressed as 2016 US dollars (USDs) using the Chemical Engineering Plant Cost Index. Total capital investment of the pretreatment and saccharification unit (plus fermentation and distillation in the case of the ethanol plant) was calculated considering the purchased equipment cost reported by Nitzsche et al. (2016) and a Lang factor of 3. As for the fermentation unit, total capital cost of fermentors was calculated based on Humbird et al. (2011). The total capital investment of the distillation unit was calculated based on Abdi et al. (2016), and it was assumed that only the cost of the beer column (and associated heat exchangers) would vary as a result of changes of stillage volume (note that stillage is removed from the bottom of the beer column, which is the first column in the distillation train). Total capital investment of evaporators (needed to concentrate the sugars stream in the 200 g/l case) and storage tanks were calculated based on Kazi et al. (2010) and Tao et al. (2014b), respectively. Potential capital costs associated with adjustment of the WWT plant of the mill resulting from additional effluent from the ABE (or ethanol) plant were estimated based on Kazi et al. (2010).

Operating costs included variable costs (e.g. raw materials costs) and fixed operating costs (labour and maintenance defined as 1% and 2% of total capital investment, respectively). The price of eucalyptus wood for the current operations of the pulp mill was estimated to be 62 USD/bdmt based on the average cash cost of pulp in Brazil in 2016 (Source: DEPEC – BRADESCO) and an US Dollar to Brazilian Real exchange rate of USD 1.00 to R$ 3.50. To account for additional wood supply for the ABE (or ethanol) plant, wood was surcharged assuming (i) additional transportation cost (0.14 USD/bdmt/km) resulting from the increase of the average transportation radius, taking into account a eucalyptus productivity of 19 bdmt/h/yr and that 10% of the land would be available for eucalyptus production; and (ii) that wood price would increase by 1.50 USD/bdmt for each additional 500 bdmt/d, since increase in demand tends to push prices up. These assumptions yielded a linear function that correlates wood price (WP in USD/bdmt) with plant scale (PS in bdmt/d): $WP = 62 + 0.0036\,PS$. In this manner, for the maximum scale evaluated (3000 bdmt/d), wood price increases to 72.20 USD/bdmt. The purchase price of enzymes is 4.90 USD/kg (Tao et al. 2014a), and enzymes loading is 5.9 mg/g cellulose (Nitzsche et al. 2016). Electricity price is 58.22 USD/MWh and this value corresponds to the average price obtained in renewable energy auctions in Brazil in the period 2003–2013 (Pereira et al. 2015).

A discounted cash flow analysis was used to calculate the minimum selling price of either butanol (MBSP) or ethanol expressed as USD/GGE. The minimum selling price is defined as the price required to obtain an NPV of zero for a 10% discount rate (or internal rate of return, IRR) after taxes. This economic criterion indicates the cost-competitiveness and market penetration potential of a product (Humbird et al. 2011). The following assumptions were adopted for the discounted cash flow analysis: project starting point in 2017; construction in 1 year; production length of 25 years with an operating factor of 8496 working hours/year; no subsidies on capital costs; 100% of nominal capacity in the first year of production; no debt and 100% equity; 34% income tax; linear depreciation; no scrap value; no premium on green products; and working capital as 5% of total capital investment.

11.3.2.4 Investment Cost and Energy Efficiency of the ABE Plant

The ABE plant with processing capacity of 2000 bdmt/d produces 29 million gallons of ABE annually (87 400 t/yr) or 9.2 (A), 18 (B), and 1.8 (E) million gal/yr (27 300 (A), 54 600 (B), 5500 (E) ton/yr, and yield of 25 gal butanol/bdmt) (Figure 11.2). For this plant size, we first analysed the effects of sugar concentrations (in the fermentation) on investment cost and energy efficiency. Results were benchmarked against an ethanol plant with same biomass processing capacity (Figure 11.3), whose production capacity is 43 million gallons of hydrous ethanol (95% v/v), or 130 450 t/yr (yield of 61 gal/bdmt). The inlet sugar concentration in the fermentation had an important effect on the investment cost of the ABE plant. In comparison to the conventional 60 g/l case, the capital cost of the fermentation and distillation units together decreased by 65% for the 189 g/l case (Figure 11.4). Consequently, the investment cost of the ABE plant decreased from 7.42 to 5.75 USD/gal of ABE capacity (or a decrease of 22%). In terms of absolute values, for sugar concentrations of 60 and 189 g/l, investment costs are 214 and 166 million USD, respectively. In this way, with an advanced technology the capital investment of the ABE plant was even lower than the cost of the ethanol plant (173 million USD). However, as the capacity in gallons of the latter is greater, the cost of the ethanol plant per gallon capacity (4.01 USD/gal) is almost half of the ABE plant. For a sugar concentration of 200 g/l, the capital cost of evaporators needed to concentrate the sugar stream from 189 g/l to the required value surpass further cost gains obtained with smaller fermentation and distillation units. For this reason, we recommend that this trade-off should be taken into account during the development of the fermentation technology. The relationship between solids loading in the hydrolysis step and energy efficiency of the process is explored below.

It is worthwhile noting that in the current design with an organosolv pretreatment technology, the major portion of C5 sugars are not converted to products. From this fact, three observations should be taken into consideration. First, in the competition butanol versus ethanol, the former would be one step ahead in cases where the following two conditions are found: the pretreatment

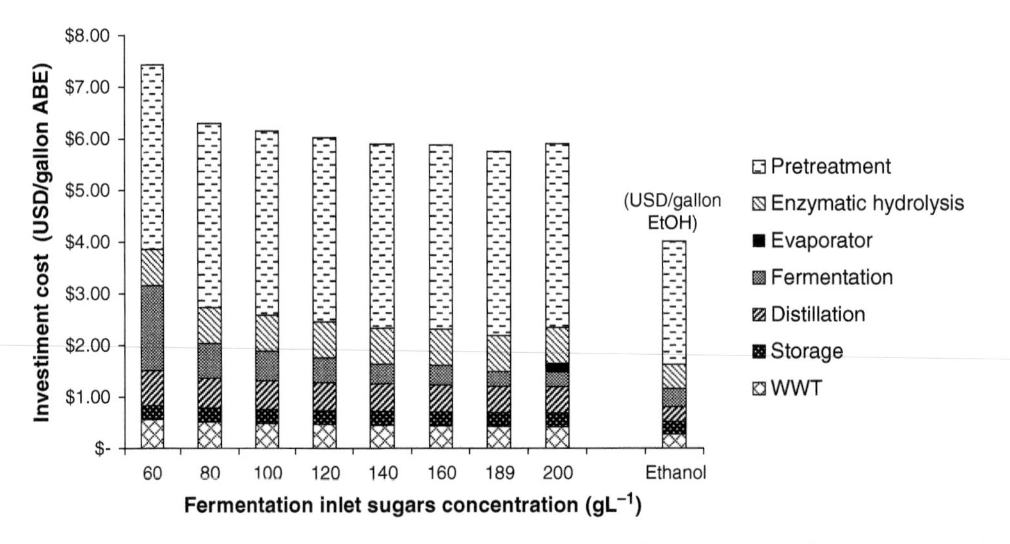

Figure 11.4 Effect of fermentation inlet sugar concentration on investment cost of a second-generation ABE plant with processing capacity of 2000 bdmt/d annexed to a kraft pulp mill.

technology generates a C5 sugars stream (mixed or not with C6 sugars) at concentration values close to the requirement of the fermentation unit (thus, no need for expensive evaporation) and if the micro-organism in the ethanol plant is not able to convert C5 sugars, which is the case of current industrial yeasts. Second, the production capacities of the butanol and ethanol plants are considerably lower than those found in standard designs of biomass-to-ABE or ethanol plants considering other pretreatment technologies such as dilute sulfuric acid. For instance, the conversion of corn stover to ABE or ethanol in plants with the same biomass processing capacity is expected to yield 42 and 61 million gallons of ABE and anhydrous ethanol annually, respectively (Tao et al. 2014a). Nevertheless, whereas the investment costs in the cited study were 10.5 (ABE) and 6.9 (ethanol) USD/gal (or in absolute values, 443 and 423 million USD, respectively), the investment costs are considerably lower in the present case study due to cost reductions offered by the integration with the pulp mill (e.g. the cogeneration unit is already in place, and the use of lower installation cost factors). Third, the organosolv pretreatment produces 126 000 t of high-quality lignin per year (for a 2000 bdmt/d plant scale). Revenues from selling lignin or derivatives to the market can compensate for the C5 sugars not converted (102 000 t/yr) under certain price conditions. The conversion of these C5 sugars to ABE could generate (if the pretreatment was dilute acid, for instance), gross revenues of approximately 33 million USD (85% sugar conversion, 0.32 ABE yield and ABE priced at 1200 USD/t). To match this revenue, lignin should be priced at a conservative value of 260 USD/t. In the case of the organosolv pretreatment, the potential revenue from the conversion of C5 sugars is offset by energy costs. The concentration of the C5 stream to 100 g/l would demand about 120 t LP steam/h. The installation of a new power boiler would be needed to provide the additional steam demand. Assuming an approximate steam price of 30 USD/t, the energy cost would be 31 million USD/yr. Furthermore, 25 GJ/t ABE would be added to the energy consumption of the ABE plant. The biodigestion of the C5 stream, as considered by Nitzsche et al. (2016), is certainly more cost and energy effective.

Sugar concentration in the fermentation also had an important effect on the energy efficiency of the ABE plant. In the conventional 60 g/l case, total energy consumption is 37 GJ/t ABE (Figure 11.5). With increasing sugar concentration, the energy consumption in the distillation

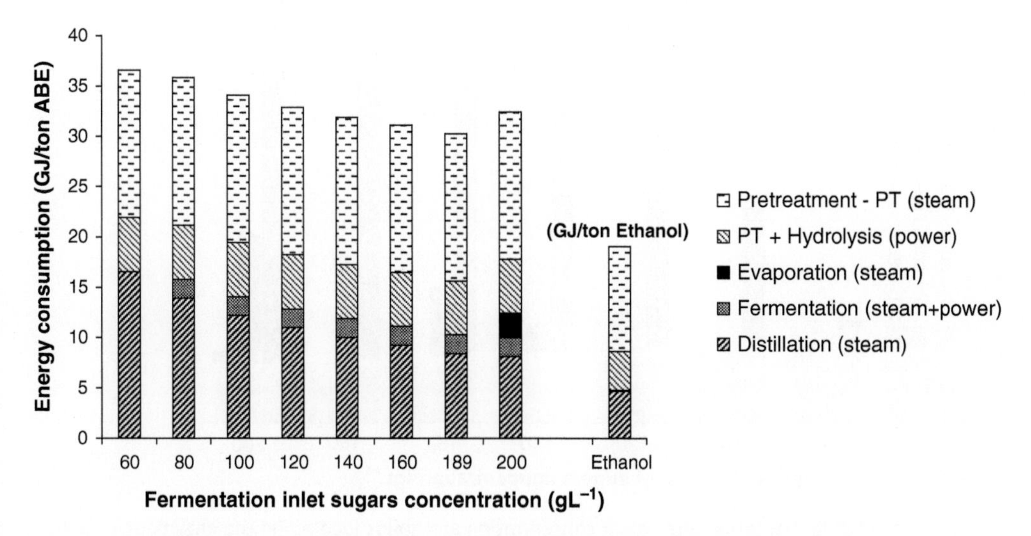

Figure 11.5 Effect of fermentation inlet sugar concentration on energy consumption of a second-generation ABE plant with processing capacity of 2000 bdmt/d annexed to a kraft pulp mill.

decreased by up to 50% in the 189 g/l case, resulting in a total energy consumption in this case of 30 GJ/t ABE (or a total decrease of 19%, which is similar to the cut in investment cost of the plant). Since the pretreatment and saccharification units are responsible for approximately 60% of the energy consumption of the plant, further attempts to improve energy efficiency should target this section of the process. Moreover, for sugar concentration higher than that delivered by the saccharification unit (189 g/l), the energy duty of the evaporation (LP steam) surpassed the additional gain that could be obtained in the distillation. Not surprisingly and in agreement with previous studies (Vane 2008), energy efficiency of the ethanol plant was superior (19 GJ/t ethanol), with the upstream section (pretreatment + saccharification) of the plant accounting for 75% of the energy consumption. By increasing the sugar concentration in the ABE plant, the wastewater footprint decreased from 79 to 29 m^3 stillage/m^3 butanol. In terms of costs, capital necessary to adjust the WWT plant decreased by 25% (from 0.56 to 0.42 USD/gal ABE), and operating cost of the WWT was cut by 37% (from 1.80 to 1.13 million USD/yr).

As mentioned above, the solids loading in the hydrolysis tank determines the sugar concentrations to be sent to the fermentation. As the advantage of using ISPR technologies comes from the fact that these technologies make possible the processing of more concentrated sugars solutions, the use of ISPR technologies is only justified if solids loading is high. To illustrate this point, we compared, for solids loading of 10%, 15%, and 20%, the energy gains in the distillation unit against the energy necessary to concentrate the sugar solution to the desired set point (Figure 11.6). With these values of solids loading, the sugar stream delivered to the fermentation unit has concentrations of 78; 129; and 189 g/l, respectively. The energy balance in this analysis demonstrates that operation with increased sugar concentration (i.e. greater than the concentration in the outlet of the saccharification) only makes sense, from an energy efficiency point of view, if the sugar stream is not evaporated.

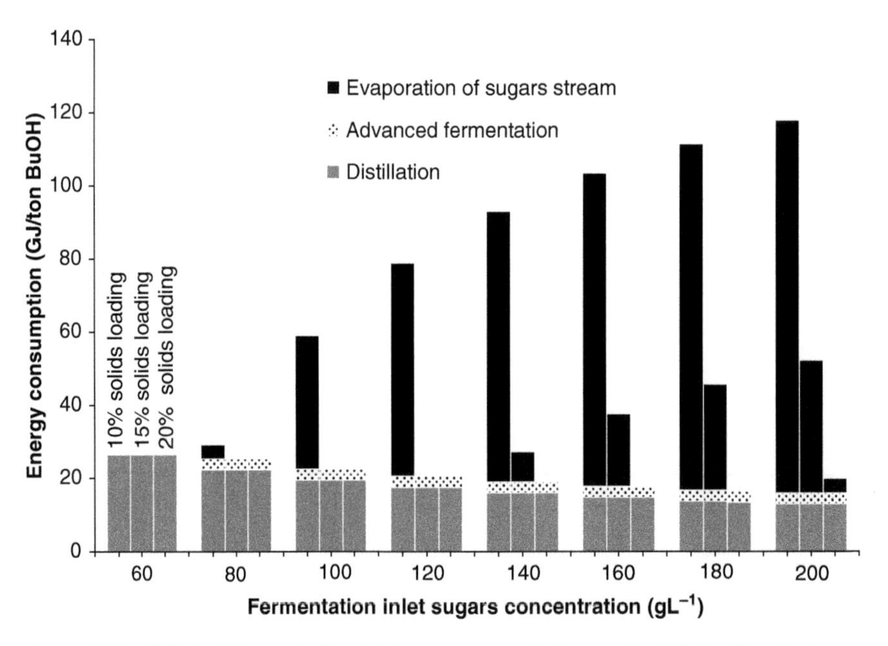

Figure 11.6 Effect of fermentation inlet sugar concentration and solids loading (in the enzymatic hydrolysis step) on the energy consumption of the ABE plant. BuOH is for butanol and plant scale is 2000 bdmt/d.

11.3.2.5 Minimum Butanol Selling Price

The effect of the plant scale and sugar concentration on the MBSP is shown in Figure 11.7a. In this analysis, the price of acetone is set to 1050 USD/t, which is the average price of Brazilian imports during the period 2013–2015 (MDIC 2016). By increasing the plant scale from 1000 to 1500, 2000, and 3000 bdmt/d, the MBSP decreases by 7%, 11%, and 14%, for all sugar concentrations, respectively. Although the MBSP criterion points out that the best option is the largest scale (despite the low gain by moving from 2000 to 3000 bdmt/d), other factors have to be considered in this decision. As the selling of lignin has an important share of the revenues (further discussed below), the risk of not finding offtakers for the whole lignin throughput is not negligible, especially because the lignin market is still emerging. For a plant scale of 3000 bdmt/d, lignin production is 189 000 t/yr, which is approximately at least four times larger than the production in current commercial scale kraft lignin plants in the US and Finland (25 000 and 50 000 t/yr, respectively). Besides the market risk, a technical aspect related to the integration of the ABE plant to the host mill also indicates that the 3000 bdmt/d plant scale is not the best option. In the current design, the residual stream from the hydrolysis (non-converted cellulose and hydrolysis lignin) is sent to the recovery boiler of the mill. The total solids load to the recovery boiler increases by 10% and 16% for the 2000 and 3000 bdmt/d plant scales, respectively. Whereas the chances of the recovery boiler having a spare capacity of 10% are real in recently built pulp mills in South America, the 16% level is a bold assumption. Nevertheless, in mills where the recovery boiler is the process bottleneck, additional capital cost related to the installation of a power boiler has to be considered.

For a processing capacity of 2000 bdmt/d, the use of an advanced fermentation technology can reduce the MBSP from 4.31 to 3.67 USD/GGE (reduction of 15%), or 1.18 to 1.00 USD/kg, according to the assumptions adopted in this study. Despite this gain, the ABE plant would not deliver a biofuel (eucalyptus-based butanol) with price advantage against ethanol, whose minimum selling price for this plant scale is 2.85 USD/GGE (0.59 USD/kg). A summary of operating costs of both ABE and ethanol plants is presented in Table 11.4. Interestingly, the minimum ethanol selling price is lower than the average wholesale price of sugarcane hydrous ethanol (0.49 USD/l or 2.95 USD/GGE) traded in Brazil in the period 2013–2015 (CEPEA 2016). As mentioned in the introduction of this chapter, competitiveness of butanol produced in the ABE process is strongly affected by the price of acetone. As such, we investigated the impact of acetone price on the MBSP for two conditions. In Figure 11.7a, the effect is measured considering the regular Clostridium species (RS) adopted in this study. Alternatively, in Figure 11.7b, the ABE plant runs with a genetically-modified Clostridium species (MS) with improved butanol yield that achieves 90% of the theoretical yield (ABE yield is 0.39 (0.08 A, 0.30 B, 0.01E) and 95% and 80% of the C6 and C5 sugars are converted, respectively). With this analysis, a second point also raised in the introduction of this chapter becomes evident: the need to improve the fermentation yield of the ABE process. While for the RS micro-organism (189 g/l), acetone has to be priced at approximately 1500 USD/t in order for butanol to match the ethanol price, with the MS micro-organism this condition is met with acetone priced at 1150 USD/t. The average price of acetone considered in the 12 studies reviewed in this chapter is 1057 USD/t, which is very close to the base value assumed in this study (1050 USD/t). According to Pfromm et al. (2010), the average acetone price over the period 1990–2010 was about 0.92 ± 0.07 USD/kg. These numbers indicate that, for the present case study, only an improved Clostridium species could deliver butanol at competitive prices against ethanol. Furthermore, if an eventual acetone oversupply pushes the acetone price to the 500 USD/t level, the competitiveness of eucalyptus-based butanol as a liquid fuel would be very low.

Extending the analysis of the impact of co-products prices on the MBSP, lignin, on the other hand, is expected to have a strong positive impact on the competitiveness of butanol. However, since both

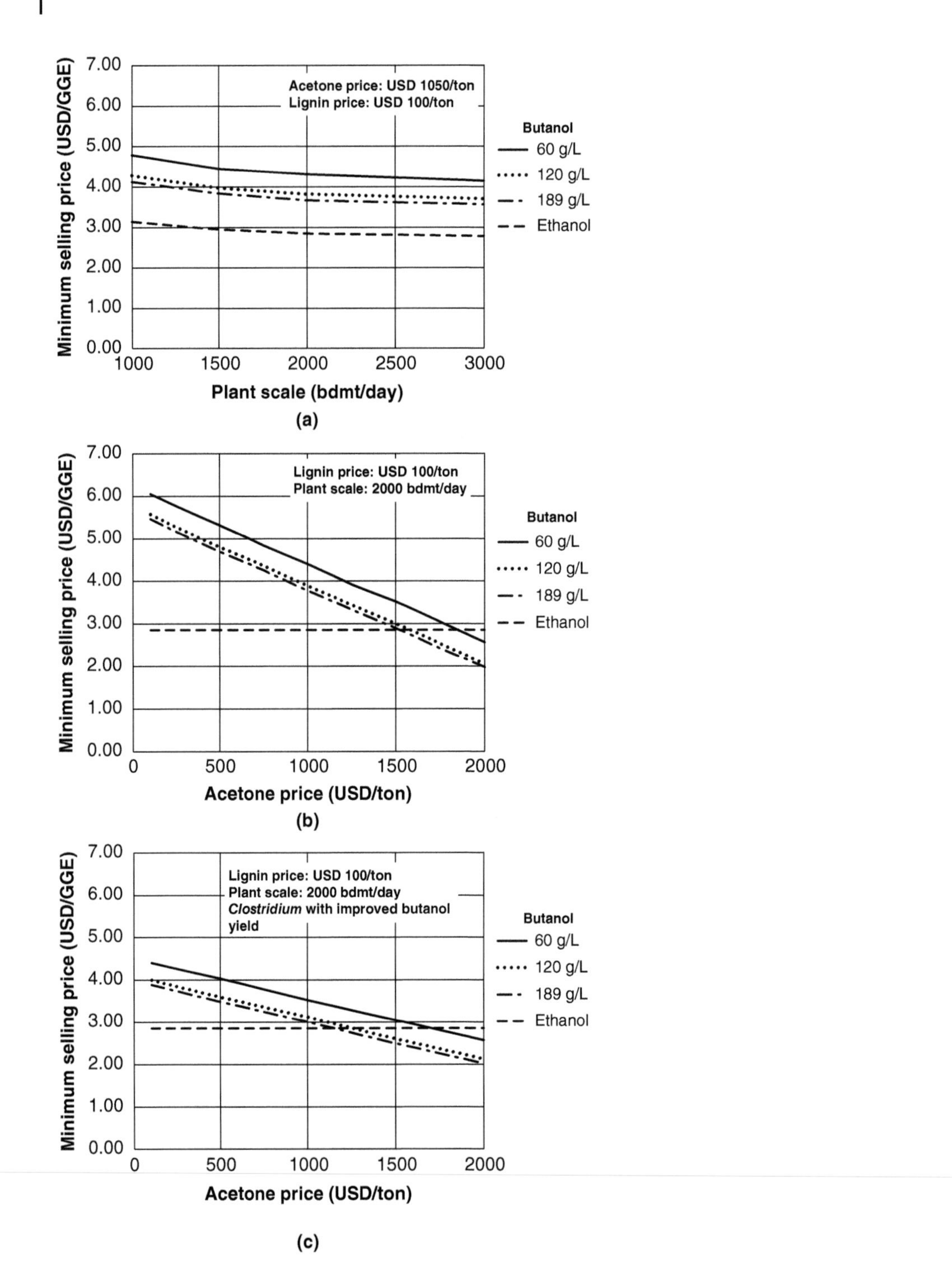

Figure 11.7 Minimum butanol selling price (MBSP) as a function of (a) plant scale, (b) acetone price considering a regular Clostridium species, and (c) acetone price considering a Clostridium species with improved butanol yield. MBSP is expressed in the gasoline equivalent basis (GGE) (1 USD/kg butanol = 0.273 × USD/GGE; 1 USD/kg ethanol = 0.209 × USD/GGE).

Table 11.4 Operating costs of eucalyptus-based butanol and ethanol, 2000 bdmt eucalyptus per day.

	Operating costs (USD/GGE)[a]	
	Butanol[b]	Ethanol
Feedstock	3.29 (50%)	1.82 (54%)
Enzymes	0.62 (9%)	0.35 (10%)
Power	0.08 (1%)	−0.04 (−)
Ethanol (organosolv make-up)	0.36 (5%)	0.00 (0%)
WWT	0.09 (1%)	0.04 (1%)
Fixed costs	0.35 (5%)	0.17 (5%)
Capital depreciation	0.46 (7%)	0.26 (8%)
Average income tax	0.45 (7%)	0.25 (7%)
Average return on investment	0.87 (13%)	0.48 (14%)
Co-products credits[c]	−2.77 (−)	−0.47 (−)
Total (minimum selling price)	**3.81** (100%)	**2.86** (100%)

a) GGE: gasoline gallon equivalent.
b) Sugars inlet concentration of 120 g/l.
c) Lignin price: 100 USD/t; acetone price: 1050 USD/t.

butanol and ethanol can equally benefit from the selling of lignin, competitiveness was examined against gasoline (Figure 11.8a). In the US, the 2016 average wholesale price of regular gasoline was 2.25 USD/gal (Source: EIA). In order for the MBSP to reach that value, lignin should be priced at 268 USD/t. This price is not unrealistic and the outlook for organosolv lignin is promising. For instance, in the study (Nitzsche et al. 2016) that provided the design of the organosolv pretreatment unit used in the present case study, lignin is priced at 660 USD/t (630 €/t). This estimate was determined by the 'German Lignocellulose Feedstock Biorefinery Project' led by the Society for Chemical Engineering and Biotechnology (DECHEMA) (Michels 2014), which identified the processing of beech wood through the organosolv pretreatment as a promising platform to produce chemicals. The importance of selling lignin for the economics of the ABE plant can also be quantified by examining the annual revenues of the project. The revenue breakdown shows that at 100 USD/t, lignin accounts for 12% of total revenues (107 million USD/yr). By increasing the price to 300 USD/t, the share is 28% of 132 million USD/yr.

Since new micro-organisms with improved butanol yield and improved robustness in industrial operation are not readily available, a reasonable alternative is the production of butanol for the chemicals market. In fact, this has been the strategy of technology developers such as the UK-based company Green Biologics, which in 2016 converted an ethanol plant in Little Falls, MN (US) to ABE production. In the period 2013–2015, the average price of butanol imports in Brazil was 1210 USD/t (MDIC 2016). Assuming this price, the IRR of the ABE plant is more attractive than that of the ethanol plant (Figure 11.8b). Moreover, at a pessimistic lignin price of 100 USD/t, the IRR can still achieve 15%, which is a value above the hurdle (10–12% IRR) for acceptable projects in many chemical companies.

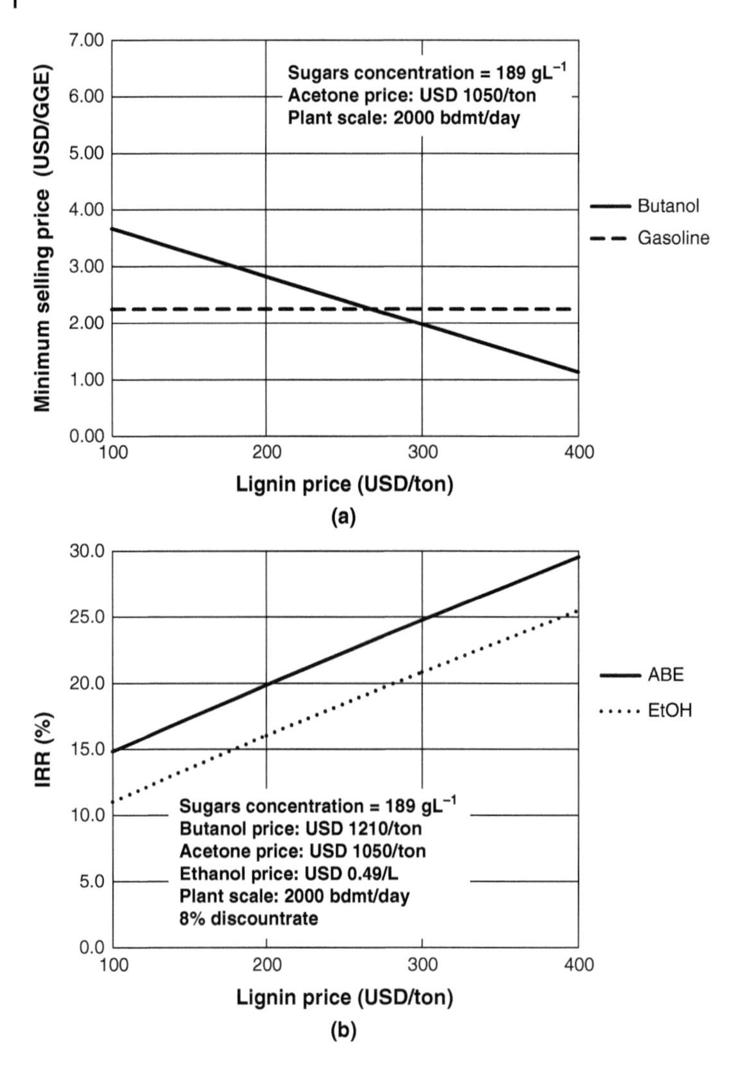

Figure 11.8 Effect of lignin selling price on (a) minimum butanol selling price and (b) IRR (1 USD/kg butanol = 0.273 × USD/GGE).

11.3.2.6 Value Creation to Pulp Mills

Finally, from the point of view of eucalyptus pulp producers, these companies are searching for business opportunities that can bring more value for their planted forests. This target can be assessed by comparing the alternatives according to the value generated by each dry ton of biomass processed. Value creation can be expressed as the ratio EBIDTA/bdmt, in which EBIDTA stands for Earnings Before Interest, Depreciation, Tax, and Amortization. Assuming (i) the 2016 pulp price in the Asian market of 540 USD/adt (Source: Exame), (ii) the 2016 average cash cost of pulp in Brazil of 235 USD/adt (Source: DEPEC – BRADESCO), (iii) revenues with power sold to the grid for 9.6 million USD, and (iv) yield of 0.52 adt pulp per dry metric ton of wood, the case study pulp mill generates 163 USD for each dry ton of eucalyptus converted to pulp. The ABE plant can reach this value (having butanol and acetone prices fixed at market prices) only if lignin is sold for 724 USD/t (which yields an IRR of 45%). At this price, lignin accounts for 49% of the revenues of the ABE plant. Given the relatively low yields (compared to the pulping yield) of fuels and chemicals obtained from biochemical conversion of biomass, these numbers indicate that pulp companies interested in using eucalyptus wood to produce fuels and chemicals should give preference to pretreatment technologies that produce lignin with high quality (i.e. low level of impurities), and the company should seek markets for lignin with price ranges starting above the 500 USD/t level. However, one caveat must be noted: applications for lignin in commercial scale are still under development, so difficulties in finding offtakers in the short term should be considered.

11.4 Perspectives

The supply of lignocellulosic butanol at competitive prices in the biofuels market is dependent on new technologies and valorization of co-products. On the technology side, new fermentation equipment, such as fermentors with ISPR, can offer substantial cost reductions in the fermentation and distillation steps of the process, as demonstrated in the case study. Energy efficiency is also improved. However, in the biochemical conversion of biomass to liquid fuels and chemicals, the pretreatment and saccharification steps are still responsible for at least 50% of the investment cost and energy consumption. With process experience the cost contribution of the upstream part of the process will certainly decrease, giving more weight for new mid- and down-stream technologies in the economics of ABE plants. On the other hand, new micro-organisms with improved butanol yield are linked to increase in revenues and a broader impact on the economic feasibility. In this manner, new fermentation equipment also has the important role of providing the environment for the best performance of superior micro-organisms. Improved butanol yield also means less production of acetone, which is an important market risk in the competition butanol versus ethanol, especially in countries where the growth of the ethanol market is not capped by restrictions in the fuel distribution infrastructure. Additional revenues from other co-products such as lignin and the use of low-cost feedstocks bring benefits for butanol and ethanol projects equally, and for this reason, these factors have more importance on the competitiveness against oil-based fuels. Assuming a rapid advance in terms of costs and efficiency (mass yields and energy) of upstream technologies, it is clear, from a process engineering perspective, that the competitiveness of lignocellulosic butanol in the fuels market is strongly dependent on the design of a process with high solids concentration in the hydrolysis step. As a result of high toxicity of butanol to the microbial culture, the conversion of high concentration sugar streams by improved Clostridium species can only be achieved in tanks with ISPR and productivities higher than 2 g/l/h. Although significant advances have been made, it is viewed that further improvements in culture development, fermentation, hydrolysis, and product recovery are needed to reduce prices of ABE production. These developments would make this process technologically advanced.

Acknowledgements

N. Qureshi would like to thank Bruce Dien (United States Department of Agriculture, Agricultural Research Service, National Center for Agricultural Utilization Research, Bioenergy Research Unit, Peoria, IL, USA) for reading it and providing valuable comments that improved the quality of the chapter.

Nomenclature

ABE	Acetone - butanol - ethanol
adt	Air dried ton
bdmt	Bone dry metric ton per day
buOH	Butanol
C5	Pentose sugars
C6	Hexose sugars
CTBE	Brazilian Bioethanol Science and Technology Laboratory
DD	Double effect distillation
DDGS	Dried distiller's grains and soluble
DECHEMA	Society for Chemical Engineering and Biotechnology
DW	Domestic waste
EBIDTA	Earnings before interest, depreciation, tax, and amortization
EIA	Energy Information Administration
EtOH	Ethanol
EU	Eucalyptus
FW	Food waste
GGE	Gasoline gallon equivalent
GS	Gas stripping
IBE	Isopropanol - butanol - ethanol
IRR	Internal rate of return
ISPR	In situ product recovery
LHW	Liquid hot water
LP	Low pressure
MBSP	Minimum butanol selling price
MS	Genetically-modified Clostridium species
NPV	Net present value
NREL	National Renewable Energy Laboratory
PS	Plant scale
R$	Brazilian Real
RIN's	Renewable identification numbers
RS	Regular Clostridium species
SB	Sugarcane bagasse
SSB	Sweet sorghum bagasse
SSF	Simultaneous saccharification and fermentation
USD	United States dollar
WP	Wood price
WWT	Waste water treatment plant

References

Abdi, H.K., Alanazi, K.F., Rohani, A.S. et al. (2016). Economic comparison of a continuous ABE fermentation with and without the integration of an in situ vacuum separation unit. *Canadian Journal of Chemical Engineering* 94: 833–843.

Afschar, A.S., Biebl, H., Schaller, K., and Schugerl, K. (1985). Production of acetone and butanol by *Clostridium acetobutylicum* in continuous culture with cell recycle. *Applied Microbiology and Biotechnology* 22: 394–398.

Baral, N.R. and Shah, A. (2016). Techno-economic analysis of cellulosic butanol production from corn stover through acetone–butanol–ethanol fermentation. *Energy & Fuels* 30: 5779–5790.

CEPEA (2016) Center for Advanced Studies on Applied Economics. Available from http://www.cepea.esalq.usp.br (retrieved 06.30.16).

Chen, C.K. and Blaschek, H.P. (1999). Acetate enhances solvent production and prevents degeneration in *Clostridium beijerinckii* BA101. *Applied Microbiology and Biotechnology* 52: 170–173.

Cheryan, M. (1986). *Ultrafiltration Handbook*. Lancaster, PA: Technomic Press.

Dahlbom S, Landgren H, Fransson P (2011) Alternatives for Bio-Butanol Production. "Feasibility study presented to Statoil", Technical Report. Lund University

Ezeji, T.C. and Blaschek, H.P. (2008). Fermentation of dried distiller's grains and soluble (DDGS) hydrolysates to solvents and value-added products by solventogenic clostridia. *Bioresource Technology* 99: 5232–5242.

Food Waste website: https://www.midwestdairy.com/food-waste/?gclid=CIeDqIGc-88CFYtahgodGAsIeA. Accessed October 27, 2016.

Gérando, H.M., Fayolle-Guichard, F., Rudant, L. et al. (2016). Improving isopropanol tolerance and production of *Clostridium beijerinckii* DSM 6423 by random mutagenesis and genome shuffling. *Applied Microbiology and Biotechnology* 100: 5427–5436.

Huang, H., Singh, V., and Qureshi, N. (2015). Butanol production from food waste: A novel process for producing sustainable energy and reducing environmental pollution. *Biotechnology for Biofuels* 8: 147.

Humbird D, Davis R, Tao L, Kinchin C, Hsu D, Aden A (2011) Process design and economics for biochemical conversion of lignocellulosic biomass to ethanol – dilute-acid pretreatment and enzymatic hydrolysis of corn stover. Report No. NREL/TP-5100-47764

Jones, D.T. and Wood, D.R. (1986). Acetone-butanol fermentation revisited. *Microbiological Reviews* 50: 484–524.

Jonglertjunya, W., Chinwatpaiboon, P., Thambaramee, H., and Prayoonyong, P. (2014). Butanol, ethanol and acetone production from sugarcane bagasses by acid hydrolysis and fermentation using *Clostridium* sp. *Advanced Materials Research* 931-932: 1602–1607.

Kazi FK, Fortman J, Anex R, Kothandaraman G, Hsu D, Aden A, Dutta A (2010) Techno-economic analysis of biochemical scenarios for production of cellulosic ethanol. Report No. NREL/TP-6A2-46588

Larsson E, Max-Hansen M, Pålsson A, Studeny R (2008) A feasibility study on conversion of an ethanol plant to a butanol plant. "Feasibility study presented to Statoil", Technical Report. Lund University

Mariano, A.P. (2015a). Due diligence for sugar platform biorefinery projects: technology risk. *The Journal of Science and Technology for Forest Products and Processes* 4 (5): 12–19.

Mariano, A.P. (2015b). How Brazilian pulp mills will look like in the future? *O Papel* 76: 55–61.

Mariano, A.P. and Maciel Filho, R. (2012). Improvements in biobutanol fermentation and their impacts on distillation energy consumption and wastewater generation. *Bioenergy Research* 5 (2): 504–514.

Mariano, A.P., Qureshi, N., Filho, R.M., and Ezeji, T.C. (2011). Bioproduction of butanol in bioreactors: new insights from simultaneous *in situ* butanol recovery to eliminate product toxicity. *Biotechnology and Bioengineering* 108: 1757–1765.

Mariano, A.P., Qureshi, N., Filho, R.M., and Ezeji, T.C. (2012). Assessment of in situ butanol recovery by vacuum during acetone butanol ethanol (ABE) fermentation. *Journal of Chemical Technology and Biotechnology* 87: 334–340.

Mariano, A.P., Dias, M.O.S., Junqueira, T.L. et al. (2013a). Butanol production in a first-generation Brazilian sugarcane biorefinery: technical aspects and economics of greenfield projects. *Bioresource Technology* 135: 316–323.

Mariano, A.P., Dias, M.O.S., Junqueira, T.L. et al. (2013b). Utilization of pentoses from sugarcane biomass: techno-economics of biogas vs. butanol production. *Bioresource Technology* 142: 390–399.

Mariano, A.P., Ezeji, T.C., and Qureshi, N. (2015). Butanol production by fermentation: efficient bioreactors. In: *Commercializing Biobased Products: Opportunities, Challenges, Benefits, and Risks* (ed. A.W. Snyder), 48–70. The Royal Sociaty of Chemistry.

Markets and Markets (2015) N-Butanol market by application (butyl acrylate, butyl acetate, glycol ethers, direct solvent, plasticizers) and by region (North America, Asia-Pacific, Europe, RoW) - Global Trends & Forecasts to 2020, Technical Report, Code CH 1543

MDIC (2016) System Information Analysis of Foreign Trade – Butanol and Acetone prices. Available from http://aliceweb.desenvolvimento.gov.br (retrieved 06.29.16).

Michels J (2014) Lignocellulose Biorefinery – Phase 2 – final scientific and technical report of all project partners. DECHEMA Gesellschaft für Chemische Technik und Biotechnologie. Available from: http://edok01.tib.uni-hannover.de/edoks/e01fb15/837304261.pdf (retrieved 12.29.16).

Naleli K (2016) Process Modelling In Production of Biobutanol from Lignocellulosic Biomass via ABE Fermentation. MSc Thesis. Faculty of Engineering at Stellenbosch University

Nitzsche, R., Budzinski, M., and Gröngröft, A. (2016). Techno-economic assessment of a wood-based biorefinery concept for the production of polymer-grade ethylene, organosolv lignin and fuel. *Bioresource Technology* 200: 928–939.

Pang, Z.-W., Lu, W., Zhang, H. et al. (2016). Butanol production employing fed-batch fermentation by *Clostridium acetobutylicum* GX01 using alkali-pretreated sugarcane bagasse hydrolysed by enzymes from *Thermoascus aurantiacus* QS 7-2-4. *Bioresource Technology* 212: 82–91.

Pereira, L.G., Dias, M.O.S., Mariano, A.P. et al. (2015). Economic and environmental assessment of n-butanol production in an integrated first and second generation sugarcane biorefinery: fermentative versus catalytic routes. *Applied Energy* 160: 120–131.

Pfromm, P.H., Amanor-Boadu, V., Nelson, R. et al. (2010). Bio-butanol vs. bio-ethanol: a technical and economic assessment for corn and switchgrass fermented by yeast or *Clostridium acetobutylicum. Biomass & Bioenergy* 34: 515–524.

Pomaranski, E. and Tiquia-Arashiro, S.M. (2016). Butanol tolerance of carboxydotrophic bacteria isolated from manure composts. *Environmental Technology* 37 (15): 1970–1982.

Qureshi, N. (2010). Agricultural residues and energy crops as potentially economical and novel substrates for microbial production of butanol (a biofuel). *CAB Reviews* 5 (59): 1–8.

Qureshi, N. (2014). Integrated processes for product recovery. In: *Biorefineries: Integrated Biochemical Processes for Liquid Biofuels* (eds. N. Qureshi, D. Hodge and A. Vertès), 101–118. Amsterdam, Netherlands: Elsevier BV.

Qureshi, N. and Singh, V. (2014). Process economics of renewable biorefineries: butanol and ethanol production in integrated bioprocesses from lignocellulosics and other industrial by-products. In: *Biorefineries: Integrated Biochemical Processes for Liquid Biofuels* (eds. N. Qureshi, D. Hodge and A. Vertes), 237–254. Elsevier.

Qureshi, N., Saha, B.C., and Cotta, M.A. (2007). Butanol production from wheat straw hydrolysate using *Clostridium beijerinckii. Bioprocess and Biosystems Engineering* 30: 419–427.

Qureshi, N., Saha, B.C., Dien, B. et al. (2010a). Production of butanol (a biofuel) from agricultural residues: Part I – use of barley straw hydrolysate. *Biomass and Bioenergy* 34: 559–565.

Qureshi, N., Saha, B.C., Hector, R.E. et al. (2010b). Production of butanol (a biofuel) from agricultural residues: Part II- use of corn stover and switchgrass hydrolysates. *Biomass and Bioenergy* 34: 566–571.

Qureshi, N., Saha, B.C., Cotta, M.A., and Singh, V. (2013). An economic evaluation of biological conversion of wheat straw to butanol: a biofuel. *Energy Conversion and Management* 65: 456–462.

Qureshi, N., Cotta, M.A., and Saha, B.C. (2014a). Bioconversion of barley straw and corn stover to butanol (a biofuel) in integrated fermentation and simultaneous product recover bioreactors. *Food Bioproducts Processing* 92: 298–308.

Qureshi, N., Singh, V., Liu, S. et al. (2014b). Process integration for simultaneous saccharification, fermentation, and recovery (SSFR): production of butanol from corn stover suing *Clostridium beijerinckii* P260. *Bioresource Technology* 154: 222–228.

Qureshi, N., Liu, S., Hughes, S. et al. (2016). Cellulosic butanol (ABE) biofuel production from sweet sorghum bagasse (SSB): impact of hot water pretreatment and solid loadings on fermentation employing *Clostridium beijerinckii* P260. *Bioenergy Research* 9: 1167–1179.

Ravinder, J., Arulselvi, I., and Elangovan, N. (2014). Isolation and molecular characterization of butanol tolerant bacterial strains for improved biobutanol production. *Journal of Environmental Biology* 35 (6): 1131–1136.

Rühl, J., Schmid, A., and Blank, L.M. (2009). Selected *Pseudomonas putida* strains able to grow in the presence of high butanol concentrations. *Applied and Environmental Microbiology* 75: 4653–4656.

Solana, M., Qureshi, N., Bertucco, A., and Eller, F. (2016). Recovery of butanol by counter-current carbon dioxide fractionation with its potential application to butanol fermentation. Special issue on "Optimisation and Scale-up of Supercritical Fluid Extraction Processes". *Materials* 9: 530. (1-11).

Tao, L., He, X., Tan, E.C.D. et al. (2014a). *Biofuels, Bioproducts and Biorefining* 8 (3): 342–361.

Tao, L., Tan, E.C.D., McCormick, R. et al. (2014b). Techno-economic analysis and life-cycle assessment of cellulosic isobutanol and comparison with cellulosic ethanol and n-butanol. *Biofuels, Bioproducts and Biorefining* 8: 30–48.

Trindade, W.R.S. and Santos, R.G. (2017). Review on the characteristics of butanol, its production and use as fuel in internal combustion engines. *Renewable and Sustainable Energy Reviews* 69: 642–651.

Van Dien, S. (2013). From the first drop to the first truckload: commercialization of microbial processes for renewable chemicals. *Current Opinion in Biotechnology* 24: 1061–1068.

Vane, L.M. (2008). Separation technologies for the recovery and dehydration of alcohols from fermentation broths. *Biofuels, Bioproducts and Biorefining* 2: 553–588.

Wang, Y., Zhang, Z.T., Seo, S.O. et al. (2015). Markerless chromosomal gene deletion in *Clostridium beijerinckii* using CRISPR/Cas9 system. *Journal of Biotechnology* 200: 1–5.

Wise Guy Reports (2016) Global n-butanol industry market research 2016. Technical Report, Code WGR693762

Zaki, A.M., Wimalasena, T.T., and Greetham, D. (2014). Phenotypic characterization of *Saccharomyces spp*. for tolerance to 1-butanol. *Journal of Industrial Microbiology & Biotechnology* 41 (11): 1627–1636.

Zheng, J., Tashiro, Y., Wang, Q. et al. (2015). Feasibility of acetone–butanol–ethanol fermentation from eucalyptus hydrolysate without nutrients supplementation. *Applied Energy* 140: 113–119.

12

Fuel Ethanol Production from Lignocellulosic Materials Using Recombinant Yeasts

Stephen R. Hughes[1] *and Marjorie A. Jones*[2]

[1] *Applied DNA Sciences, Inc., Stony Brook, NY 11790, USA*
[2] *Department of Chemistry, Illinois State University, Normal, IL 61790, USA*

CHAPTER MENU

12.1 Review of Current Fuel Ethanol Production

The current global energy system is characterized by the rapid deployment and falling cost of clean energy technologies. Cellulosic ethanol has theoretical advantages over ethanol derived from starch and sugar such as lower cost feedstock and significant reduction in carbon emissions; however, numerous technical challenges must be overcome to make production of cellulosic ethanol cost-competitive with petroleum-derived transportation fuel or with fuel ethanol from starch or sugar. These challenges are being met with the development of new technologies that are not only energy efficient, but also effectively utilize a greater percentage of carbon into higher value products and coproducts as well as cellulosic ethanol. It will take committed investors to translate technological breakthroughs into commercial success in the cellulosic ethanol industry. Recently, the commercial development of the cellulosic ethanol industry has made progress on several fronts.

12.1.1 Technological Aspects

Energy, environmental, and economic concerns have fuelled the drive to replace petroleum with biomass-based alternatives. One of the four large-scale shifts currently underway in the global energy system, described by the International Energy Agency in its World Energy Outlook 2017, is the rapid deployment and falling cost of clean energy technologies (International Energy Agency 2017). Although global energy demands are rising more slowly than in the past, they are still projected to expand by 30% between today and 2040. The largest contribution to energy demand growth will come from India, for which the share of global energy use is estimated to rise to 11% by 2040, while the demand in Southeast Asia is growing twice as rapidly as that in China (International

Green Energy to Sustainability: Strategies for Global Industries, First Edition.
Edited by Alain A. Vertès, Nasib Qureshi, Hans P. Blaschek and Hideaki Yukawa.

Energy Agency 2017). Overall, developing countries in Asia account for two-thirds of the growth in global energy demand, with the rest originating mainly from the Middle East, Africa, and Latin America (International Energy Agency 2017). Although the industrial sector accounted for the largest share of global delivered energy consumption to end users in 2016 (39.7% of total) and is projected to continue to be the largest energy-consuming end-use sector through 2040 (37.9%), the growth of energy consumption in the transportation sector (40% increase) is projected to outpace that of the industrial sector (32%) from 2016 to 2040 (United States Department of Energy et al. 2016; EIA 2017).

Currently, bioethanol produced from either starch or sugar cane is the leading alternative transportation biofuel produced worldwide, and the microbes, process technology, and yields are well established. In 2015, the United States was the world's leading producer of fuel ethanol, providing 57% of global production, followed by Brazil with 28% and the European Union with 5% (Renewable Fuels Association 2016). In 2016, more than 200 ethanol biorefineries in 29 states in the US produced a record 15.3 billion gallons of bioethanol. A fraction of this volume was from seven facilities that produced ethanol from cellulosic materials, including crop residues, waste streams, and biomass crops. Of the total capacity of 15 730 million gallons, cellulosic ethanol facilities contributed 53 million gallons (Renewable Fuels Association 2016; US Ethanol Plants 2017).

Cellulosic biomass has tremendous potential for contributing significantly to long-term energy sustainability in which biomass-derived fuels displace much of the petroleum used for transportation fuels (United States Department of Energy 2016). Among other things, it does not compete for food and it is widely available. In addition, cellulosic ethanol has theoretical advantages over ethanol derived from starch and sugar such as lower cost feedstock and a significant reduction in carbon emissions. The key question has been whether developers of fuel ethanol produced from lignocellulosic materials, especially those using microbial platforms, can overcome the technical challenges that make it difficult to render the production of cellulosic ethanol cost-competitive with petroleum-derived transportation fuel or fuel ethanol from starch or sugar. Industrial microbial workhorses, with all the necessary traits for utilizing to the theoretical maximum lignocellulosic materials in bulk production, must be developed. Although significant progress has been made in the genetic engineering of microbes designed for advanced biofuels production, titre and yield of these strains are currently too low for these products to compete with their petroleum-derived equivalents. Furthermore, recycling of microbial cells is difficult in a lignocellulosic system, and genetically engineered strains seem highly susceptible to contamination, which also increases operating costs (Koukoulas 2016). Other challenges include the conversion of multiple sugars in lignocellulosic hydrolysates, the toxicity to microbes and enzymes of compounds or inhibitors both found naturally in the lignocellulosic substrates and formed in the conversion process, recovery of sugars at high concentration, and recycling of the enzymes. Toxicity to the microbial cell membrane from over-expression of the products is one of the most severe limitations in achieving high yields (Koukoulas 2016; Veettil et al. 2016). Historical approaches have also included inefficient intermediate conversion step(s) or required fossil energy inputs to drive the processing plants, making them economically unfeasible and unsustainable (Ahmed 2016).

These challenges are being met with the development of new technologies that are not only energy efficient, thus generating higher renewable energy output to fossil energy input ratios, but also designed to effectively utilize a greater percentage of carbon into higher value products and coproducts as well as cellulosic ethanol. Integrating production of cellulosic ethanol with that of higher-value chemicals is a critical strategy for controlling and minimizing financial risks inherent in such large-scale industrial projects. Supporting a higher-value chain through the production of higher value chemicals and coproducts in conjunction with cellulosic ethanol enables the manufacturer to significantly improve profit margins. Continuous processing techniques allow smaller facility sizes, thus reducing capital cost of the cellulosic ethanol biorefineries by a factor of two to five (Ahmed 2016). However, it will take farsighted and committed investors to translate technological breakthroughs into commercial success in the cellulosic ethanol industry (Ahmed 2016).

12.1.2 Commercialization of Cellulosic Ethanol

In 2017, the commercial development of the cellulosic ethanol industry made news on several fronts, reflecting the progress of a maturing technology. On November 2, DuPont (E. I. du Pont de Nemours), Wilmington, Delaware, USA, announced plans to sell its cellulosic ethanol facility in Nevada, Iowa as a result of consolidation after its merger with Dow Chemical Company, Midland, Michigan, USA (McCoy 2017). The facility, which opened in October 2015 and is intended to operate using DuPont enzymes and yeast to convert corn cobs, stems, and leaves into ethanol, is the largest in the US (30 million gallons). The company stated, 'While we still believe in the future of cellulosic biofuels we have concluded it is in our long-term interest to find a strategic buyer for our technology, including the Nevada, Iowa biorefinery', and added it will continue to participate in the biofuels market with enzymes and engineered yeast, and will work with local, state, and federal partners to assure a smooth transition (McCoy 2017). Around the same time in Europe, Clariant Specialty Chemicals Company, Muttenz, Switzerland, announced plans to build a cellulosic ethanol plant in Romania based on its Sunliquid process to make ethanol from agricultural waste. The facility is intended to demonstrate the competitiveness of the company's technology and support a licensing strategy (Lane 2017). Also in November 2017, the second largest US cellulosic ethanol producer (20 million gallons/year), the Poet-DSM Advanced Biofuels joint venture (POET, LLC, Sioux Falls, South Dakota, USA and Royal DSM, Heerlen, NL) in Emmetsburg, Iowa, announced it had achieved a breakthrough in overcoming one of the major challenges in cellulosic biofuels production: pretreating cellulosic feedstock so that enzymes and yeast can access the cellulosic sugars and ferment them into biofuel (POET-DSM 2017; Robertson et al. 2017). In South America, Brazil's largest producer of sugar cane ethanol, Raízen Energia S.A. (São Paulo, Brazil, South America), is planning to increase cellulosic ethanol production more than fivefold within two years, making the new technology competitive with traditional ethanol and harnessing potentially millions of tons of plant material that currently go to waste. ST1, a Nordic energy group headquartered in Helsinki, Finland, has begun producing cellulosic ethanol from sawdust in Kajaani, Finland. In India, Praj Industries Ltd., a process and project engineering company, headquartered in Pune, Maharashtra, India, has demonstrated its technology and is working in close collaboration with the Indian oil marketing companies to build several cellulosic ethanol plants (Think Bioenergy and Cellulosic Ethanol 2017).

On November 7, 2017, Enerkem, headquartered in Montréal, QC, Canada, announced it had received approval from the United States Environmental Protection Agency (EPA) to sell cellulosic ethanol produced at its Edmonton, AB, Canada facility under the US Renewable Fuels Standard (RFS) (Ethanol Producer Magazine 2017). The company uses its proprietary technology to convert non-recyclable, non-compostable municipal solid waste into methanol, ethanol, and other widely-used chemicals. It operates a full-scale commercial facility in Alberta as well as an innovation centre in Quebec. Earlier this year, Enerkem expanded its Edmonton biofuel facility to produce some 13 million gallons of cellulosic ethanol annually following the commissioning of its methanol-to-ethanol conversion unit (Ethanol Producer Magazine 2017).

A future cellulosic ethanol project was announced in November 2017 by Aemetis Inc., Cupertino, California, USA, which operates a 60 million gallon starch/sugar ethanol production plant in Keyes, California. The company plans a LanzaTech (Chicago, Illinois, USA) cellulosic ethanol upgrade to the plant. Aemetis has also signed exclusive licenses for key rights to LanzaTech microbial fermentation of gases and the InEnTec (Richland, Washington, USA) gasification technology. The three companies built an integrated demonstration unit (IDU) at InEnTec's Technology Center in Richland, Washington, earlier in 2017 for production of cellulosic ethanol. The unit has completed about four months of operation, and Aemetis reported that the yield of cellulosic ethanol significantly exceeded their original expectations. Its Aemetis Advanced Products Keyes subsidiary has signed a long-term lease at a former US Army munitions facility located in Riverbank, California that provides for the construction of a cellulosic ethanol biorefinery using the process technologies developed by Aemetis, LanzaTech, and InEnTec to convert waste orchard wood and nutshells into cellulosic ethanol. The project will lower greenhouse gas (GHG) emissions in the Central Valley by significantly reducing the practice of burning orchard waste (Ethanol Producer Magazine 2017). Aemetis Advanced Products Keyes has received third party validation of ethanol and protein yields produced from operation of the IDU in Richland, Washington, enabling Aemetis to move forward towards finalizing its USDA 9003 Biorefinery Assistance Program guaranteed loan and related financings.

12.2 Evolution of Cost of Cellulosic Ethanol Production

In 2000 the US Department of Agriculture, Agricultural Research Service (USDA, ARS) and the US Department of Energy (DOE), with the National Renewable Energy Laboratory (NREL) published a study of the corn ethanol and the cellulosic ethanol industries, looking at both of them on a similar process design and engineering basis and making the substantial assumption that both were mature technologies (McAloon et al. 2000). The costs of producing 25 million gallons per year of fuel ethanol from each process were determined.

In 2007, the US DOE Bioenergy Technologies Office (BTO) in collaboration with NREL established a goal \$1.41 per gallon ethanol for conversion only (exclusive of feedstock) for cost-competiveness of cellulosic ethanol technology by 2012. In 2012 NREL published a report on the achievement of this cost target for biochemical ethanol fermentation via dilute-acid pretreatment and enzymatic hydrolysis of corn stover (Tao et al. 2012). State of Technology (SOT) analyses for cellulosic ethanol production in 2007, 2009, 2011, and 2012 showed that to make the production of cellulosic ethanol more competitive with petroleum-derived transportation fuel, the pretreatment and enzymatic steps required further research to enhance sugar yield and concentration with minimum formation of inhibitors and minimum use of enzymes and chemicals. The report also provided a cost sensitivity analysis to help identify the factors that have the most significant impact on production costs (Tao et al. 2012).

12.2.1 Analysis Comparing Costs of Producing Ethanol from Corn Starch and Lignocellulosic Feedstocks

In a joint project sponsored by the US Department of Agriculture and Agricultural Research Service (USDA, ARS) and the US Department of Energy (DOE), with the National Renewable Energy Laboratory (NREL) studied the corn ethanol and the cellulosic ethanol industries, looking at both of them on a similar process design and engineering basis (McAloon et al. 2000). Both were treated as mature technologies, making the substantial assumptions that the technology improvements needed to make the cellulosic process commercially viable were achieved, and enough cellulosic ethanol plants were built to make the design well-understood. The costs of producing 25 million annual gallons of fuel ethanol from each process for the year 2000 were determined. The collection of stover was a new industry, therefore limited data were available on the collection costs. Because stover is considered a residue, it is expected that its price might not fluctuate as much as a commodity crop like corn. For this analysis, $35/dry ton was used. The authors point out that because there is no operating plant for processing lignocellulose to ethanol, the process design and costing were based on laboratory and pilot scale data, cost estimations of similar industries, and vendor knowledge of equipment design, which increases the margin for cost uncertainty compared to the established corn ethanol industry (McAloon et al. 2000). Each process has the same general flow, from feedstock handling through fermentation to product and coproduct recovery; however, there are some major differences in the processing of corn starch versus stover. Stover requires more feed handling than corn; it must be washed, shredded, and then milled to achieve a particle size that can be conveyed to the process. Corn requires only milling. The steps to reduce the carbohydrate polymers in stover to simple sugar monomers take considerably longer and are more energy intensive than for the starch in corn (McAloon et al. 2000). The cellulose requires heat and dilute acid pretreatment, after which the cellulase enzyme and fermentation organism require about seven days for conversion to ethanol, compared to two days for starch. The longer residence time increases not only the production and capital costs but also the chance for contamination. Utilities such as steam, chilled water, and cooling tower water were treated as equipment in the cellulosic ethanol model, contributing to the capital cost of the plant rather than the variable operating cost as in the starch model (McAloon et al. 2000). Production costs for both processes are summarized in Table 12.1.

Although significant assumptions were made regarding the cellulosic ethanol production process, two results are instructive. The cost of the corn was the single greatest cost in the production of ethanol from corn, and the cost with the greatest variability. Corn prices vary from year to year and will also vary in different locations due to shipping distance from the field to the plant. The largest cost contributor for the lignocellulosic process was the capital cost, which is represented

Table 12.1 Production costs (US dollars/gal) for starch and cellulosic ethanol processes (NREL 2000 Analysis; 1999$).

Item	Starch ethanol process	Cellulosic ethanol process
Feedstock (shelled corn or corn stover)	$0.68	$0.49
Corn steep liquor		$0.05
Other raw materials	$0.06	$0.11
Denaturant	$0.03	$0.03
Waste disposal		$0.03
Utilities	$0.16	
Labour, supplies, and overhead Expenses	$0.13	$0.36
Depreciation of capital	$0.11	$0.54
DDG credit	−$0.29	
Electricity credit		−$0.11
Total production cost	$0.88	$1.50

Source: Adapted from (McAloon et al. 2000).

Table 12.2 Capital costs (US dollars) for starch and cellulosic ethanol processes (NREL 2000 Analysis; 1999$).

Item	Starch ethanol process	Cellulosic ethanol process
Feedstock handling	$2 600 000	$5 400 000
Pretreatment/detoxification		$29 800 000
Saccharification (starch)	$2 300 000	$14 400 000
Simultaneous Saccharification/co-fermentation (Cellulosic)		
Fermentation	$4 600 000	
Cellulase production		$18 100 000
Distillation	$5 300 000	$5 100 000
Solid/syrup separation	$10 500 000	$9 200 000
Drying (starch)		
Storage/load out	$1 500 000	$2 100 000
Wastewater treatment (WWT)	$1 000 000	$9 000 000
Boiler/turbogenerator		$37 500 000
Air compressor	$100 000	
Utilities		$5 500 000
Total capital investment	$27 900 000	$136,100 000

Source: Adapted from (McAloon et al. 2000).

by depreciation cost on an annual basis (McAloon et al. 2000). Capital costs for both processes are listed in Table 12.2.

The report pointed out that some of this cost was due to the higher complexity of the lignocellulose conversion process and suggested that a more accurate comparison is with the early corn ethanol industry. The cost of corn ethanol plants has dropped since the industry's inception, because the technology used in the ethanol dry milling process has evolved and plants have become more efficient. In 1978 ethanol was estimated to cost $2.47/gal to produce (in year 2000 US dollars) and in the 2000 analysis fuel ethanol production costs are estimated to be about $0.88/gal for dry mill operations. Several factors were responsible, among them that ethanol production became less energy intensive due to new techniques in energy integration and the use of molecular sieves for ethanol dehydration, the amount of pure ethanol produced from a bushel of corn has increased from 2.5 gal to more than 2.7 gal, and the capital costs of dry mill ethanol plants have also decreased. In 1978, costs (in $2000) for a 50 million gallon per year plant were reported to be about $2.07/gal. In the 2000 report, these costs were estimated to be between $1.25 and $1.50/gal. It is likely that lignocellulosic ethanol plant costs would also be reduced as more plants are built and economies of learning are gained, and as the technology further improves (McAloon et al. 2000).

12.2.2 Cost Analysis by Tao et al. for Cellulosic Ethanol

In 2007, the US DOE Bioenergy Technologies Office in collaboration with NREL established the goal of demonstrating cost-competiveness of cellulosic ethanol technology by 2012. Assuming feedstock cost for corn stover of $58.50/ton, this goal was set at $1.41/gal ethanol for conversion only (exclusive of feedstock) and at $2.15/gal for total production cost or minimum ethanol selling

Table 12.3 State of technology (SOT) analyses for cellulosic ethanol production in 2007, 2009, 2011, and 2012 (NREL 2012 Analysis).

Item	Year			
	2007 SOT	2009 SOT	2011 SOT	2012 SOT
Minimum ethanol selling price	$3.64	$3.18	$2.56	$2.15
Feedstock contribution ($/gal)	$1.12	$0.95	$0.76	$0.83
Conversion contribution ($/gal)	$2.52	$2.24	$1.80	$1.32
Yield (gal/dry ton)	69	73	78	71
Feedstock cost ($/dry ton)	$77.20	$69.65	$59.60	$58.50
Pretreatment				
Xylan to xylose (including enzymatic)	75%	84%	88%	81%
Xylan to degradation products	13%	6%	5%	5%
Hydrolysate solid-liquid separation	Yes	Yes	Yes	No
Xylose sugar loss	2%	2%	1%	0%
Glucose sugar loss	1%	1%	1%	0%
Enzymatic hydrolysis and fermentation				
Enzyme contribution ($/gal ethanol)	$0.39	$0.36	$0.34	$0.36
Combined saccharification and fermentation time (d)	7	7	5	5
Corn steep liquor loading (wt%)	1%	1%	0.25%	0.25%
Overall cellulose to ethanol	86%	84%	89%	74%
Xylose to ethanol	76%	82%	85%	93%
Arabinose to ethanol	0%	51%	47%	54%
Operating parameters				
Pretreatment solids loading (wt%)	30%	30%	30%	30%
Pretreatment temperature (°C)	190	158	152	160
Acid loading (mg/g dry biomass)	38.0	24.5	15.0	9.0
Secondary oligomer hold step	No	Yes	Yes	No
Ammonia loading (g/l of hydrolysate)	12.9	9.8	3.8	1.6
Conditioning mode	Liquor	Liquor	Liquor	Whole slurry
Saccharification mode	Washed solids	Washed solids	Washed solids	Whole slurry
Enzymatic hydrolysis solids loading (wt%)	20%	20%	17.5%	20%

Source: Adapted from (Tao et al. 2012).

price (MESP) (Tao et al. 2012). The SOT analyses for 2007, 2009, 2011, and 2012 are presented in Table 12.3.

It can be seen that although significant progress was made towards the final cost and technical targets through 2011, further improvements were needed. Instrumental in improving ethanol yield and thus reducing costs are the xylan-to-xylose, xylose-to-ethanol, and arabinose-to-ethanol yields. The 2011 SOT established that these yield targets must be demonstrated in a whole-slurry mode, because the cost associated with separating and washing the solids was too high to overcome (Tao et al. 2012). In 2012 a number of technical strategies were introduced to improve these yields. One strategy was incorporating a deacetylation step upstream of the pretreatment step, following the rationale that removing a significant amount of acetic acid from the process initially would result in reduced inhibition of both the enzyme package (improving cellulose-to-glucose and xylan-to-xylose yields) and the fermentation organism (improving total sugar to ethanol yields and raising the final ethanol titre). The deacetylation step also enables less sulfuric acid usage in pretreatment, resulting in lower furfural formation and less salt formation by requiring less ammonia for neutralization (Tao et al. 2012). With regard to enzyme loading, as expected, cellulose-to-glucose yield decreased at lower enzyme loadings; as a result, an economic evaluation is necessary to balance the costs. Other technical accomplishments in 2012 that allowed cost reductions to be realized between the 2011 and 2012 SOTs included: whole-slurry enzymatic hydrolysis (at 20% solids loadings) with an industrial enzyme package capable of converting cellulose to glucose at ~80% with an enzyme loading of <20 mg/g and fermentation with an industrial organism capable of co-fermenting 5- and 6-carbon sugars at yields of >90% while achieving a final ethanol titre of >70 g/l (Tao et al. 2012). The economic impact of the slight cellulose-to-glucose yield reduction in 2012 compared to 2011 was offset by the lower enzyme loading (19 mg/g versus 40 mg/g). Ethanol yield (gal/ton) was roughly 10% lower than the target case; however, this is offset by a similar 10% cost reduction in combined capital and operating costs, as a result of process modifications that enabled several downstream process improvements that reduced costs (Tao et al. 2012).

It was observed during the 2012 studies that the loss of sugars to lactic acid production, usually from a *Lactobacillus* bacterium (a common contaminant in ethanol fermentations), reduces the amount of ethanol produced. The authors suggested that the health of the cells used to inoculate the main bioreactor is an important factor in the success of the fermentation. If a healthy inoculum of seed culture is used, given their much higher cell density and rapid growth rate, these cells quickly convert most of the sugars to ethanol before the contaminating bacteria can utilize them, but if the cells are not growing rapidly, contaminating bacteria would be able to compete (Tao et al. 2012).

Reduction in capital costs from 2011 to 2012 was noted in pretreatment and wastewater treatment (WWT) as a result of modifications made in the analyses directly tied to process improvements demonstrated in the pilot plant (for example, deacetylation, lower severity pretreatment, and associated WWT updates). Typical capital costs (US dollars) for cellulosic ethanol production from the NREL 2000, 2011, and 2012 analyses (McAloon et al. 2000; Tao et al. 2012) are provided in Table 12.4.

In addition, to help identify the factors that have the most significant impact on production costs, a cost sensitivity analysis was performed (Tao et al. 2012). Enzyme loading (mg/g) is the best-known factor impacting production cost. The baseline case for the 2012 cost analysis used 19 mg/g. Assuming 10 mg/g as a reasonable improvement in the near term for developing commercial enzyme cocktails, this could decrease the MESP by $0.27/gal; on the other hand, increasing the enzyme loading to 30 mg/g would increase the price by $0.21/gal. Sugar and ethanol yield sensitivities have considerable effect on cost as well. Because biomass has a high cellulose content, cellulose-to-glucose yield has the largest impact. If this ratio was 12% higher than the 2012 baseline of 78%, the cost would decrease by $0.15/gal, and if 8% lower, the ethanol price would increase by $0.12/gal. The 2012 baseline run achieved 95% glucose-to-ethanol yield. A lower yield of 85% would result in a $0.12/gal cost increase. Similarly, if xylose-to-ethanol yield decreased from 93% to 80%, the price would increase by $0.09/gal. Baseline xylan-to-xylose yield was 81%. Assuming 90% xylose yield in pretreatment would reduce cost by $0.07/gal, while assuming 70% yield would result in a price increase of more than $0.09/gal. For bulk production, contamination loss also must be addressed. Going from 0% to 10% contamination loss would increase the price by $0.20. In the pretreatment step, 2–6% xylan is lost after the deacetylation step depending on the feedstock and

Table 12.4 Comparison of capital costs for cellulosic ethanol production from 2000, 2011, and 2012 Analyses.

Item	2000 ($1999)	2011 ($2007)	2012 ($2007)
Feedstock handling	$5 400 000		
Neutralization/conditioning		3 000 000	4 000 000
Pretreatment/detoxification	$29 800 000	$30 000 000	$25 000 000
Simultaneous saccharification/Co-fermentation	$14 400 000	$31 000 000	$25 000 000
Cellulase production/On-site enzyme production	$18 100 000	$18 000 000	$17 000 000
Distillation	$5 100 000	$11 000 000	$10 000 000
Solid/syrup separation	$9 200 000	$11 000 000	$10 000 000
Storage	$2 100 000	$5 000 000	$4 000 000
Wastewater treatment (WWT)	$9 000 000	$49 000 000	$41 000 000
Boiler/turbogenerator	$37 500 000	$66 000 000	$68 000 000
Utilities	$5 500 000	$7 000 000	$7 000 000
Total capital investment	$136,100 000	$232 000 000	$210 000 000

Source: Adapted from (McAloon et al. 2000; Tao et al. 2012).

the strength of the alkaline treatment. It was determined that increasing the loss from 2% to 6%, increased the production cost (or MESP) by $0.03/gal (Tao et al. 2012).

12.3 Technological Opportunities to Reduce Cellulosic Ethanol Production Costs

Most commercial ventures in cellulosic ethanol production are not based on completely optimized processes. In general, the lignocellulosic bioconversion scheme has three process steps: (i) pretreatment of biomass, (ii) enzymatic hydrolysis, and (iii) pentose/hexose co-fermentation using a recombinant fermentative organism. Ethanol is separated from the fermentation broth, water is recycled back to the process, and lignin and other residues are burned to produce steam and electricity (Tao et al. 2012). To make the production of cellulosic ethanol more competitive with petroleum-derived transportation fuel, the pretreatment and enzymatic steps require further research to enhance sugar yield and concentration with minimum formation of inhibitors and minimum use of enzymes and chemicals. Although numerous attempts have been made to develop a single microbial strain to carry out hydrolysis and fermentation, it is an objective that is proving difficult to reach because it requires the organism to perform numerous, sometimes competing, functions without interfering with its basic physiological characteristics, growth rate, and tolerance. For only the fermentation step, the strains require extensive optimization to achieve rapid utilization of multiple sugars as well as tolerance to inhibitors and to high levels of product (Koukoulas 2016).

Optimization of the deacetylation step incorporated prior to pretreatment will be required to maximize overall sugar and ethanol yields. In the 2012 study, the deacetylation process was not optimized, thus it may be possible to reduce sodium hydroxide use and decrease xylan losses (although the cost is not highly sensitive to xylan loss) and still obtain acceptable fermentation. The deacetylation step also enables less sulfuric acid usage in pretreatment and less ammonia for neutralization, in addition to dramatically lowering acetic acid concentration, which enables the fermentation organism to more completely utilize available sugars and achieve high ethanol titres. Additionally, strategies to recover solubilized xylan, acetic acid, or lignin in the black liquor represent an opportunity to capture additional value from this stream and to further reduce WWT costs (Tao et al. 2012).

For cellulosic ethanol production to be economically viable, an ethanol titre of at least 100 g/l (10%) with a productivity of 2 g/l/h and a theoretical yield of >90% are required (Koukoulas 2016). Although first-generation bioethanol production has reached these performance levels, advanced biofuel systems lag far behind in these properties (Koukoulas 2016). When using lignocellulosic feedstocks, in addition to exhibiting product tolerance, the microbial strains used to produce bioethanol should be tolerant to a range of inhibitors derived from the feedstock. Developing a microbial strain with increased substrate and product tolerance will be crucial to minimize any risk of contamination (Koukoulas 2016). In addition, the strain must efficiently metabolize both pentoses and hexoses (C5 and C6 sugars) while maintaining all of the required traits such as high sugar and ethanol tolerance and inhibitor tolerance for industrial-scale ethanol production. Extensive research is still required to convert both C5 and C6 sugars to bioethanol in reasonable titre and yield (Koukoulas 2016).

A novel approach to improve yeast tolerance to toxic inhibitory compounds in lignocellulosic hydrolysates and to achieve high fermentation efficiency with minimum detoxification steps after biomass pretreatment was investigated by Liu et al. (2016). The strategy described by these researchers involves in-situ detoxification by polyethylene glycol (PEG) exo-protection of an industrial dry *Saccharomyces cerevisiae* yeast (starch-base). They studied the toxicities of various inhibitors on ethanol production in the fermentation of glucose and found that phenol showed the strongest inhibition on ethanol production. Phenol drastically inhibited ethanol productivity and glucose consumption during the 96 hour fermentation period, producing 111 g/l of ethanol compared to 158 g/l of ethanol in the control. However, addition of 0.25 g/ml of PEG-1000 to the same fermentation broth improved the ethanol productivity and glucose consumption so they became comparable to the fermentation without phenol inhibitor (control) after 48 hours. Similarly, in the presence of mixed inhibitors (1.0 g/l each), addition of 0.25 g/ml of PEG-1000 to a glucose fermentation, increased the ethanol concentration from 93 to 147 g/l. The authors also investigated the conversion of poplar pretreated by hydrolysis combined with steam explosion using the starch-base industrial yeast in an enzymatic saccharification and simultaneous co-fermentation (SSCF). It was observed that although the glucose concentration in the poplar hydrolysate from the enzymatic hydrolysis was sufficiently high (>30.5 g/l), only 7.0 g/l of ethanol was obtained without PEGs. Addition of increasing amounts of PEG-1000 to the SSCF system resulted in conversion of an increasing amount of glucose and formation of increasing amounts of ethanol. The fermentation was apparently inhibited by the presence of lignin. Addition of PEG-1000 evidently alleviated the lignin toxicity in fermentation. A maximum ethanol concentration of 24.0 g/l was achieved with only 0.4 g/l glucose remaining. In the presence of PEG-1000, ethanol productivity was enhanced threefold as compared to that in the absence of PEGs in the fermentation broth (Liu et al. 2016).

Considerable resources are still being applied in the effort to develop a microbial strain that combines the required characteristics for production of bioethanol from lignocellulosic materials. For example, the Australian Renewable Energy Agency recently announced $4.03 million in funding for Sydney-based yeast developer Microbiogen to develop commercially viable production of biofuels from non-food waste along with high value feed coproduct by optimizing microbial strains to more efficiently convert sugars to ethanol. The goal of the project is to accelerate the process of fermentation as well as lower residual sugars, reduce production of unwanted by-products and allow higher solid loadings (Ethanol Producer Magazine 2017).

Enzyme loading is the best-known factor impacting the cost of industrial cellulosic ethanol production. The cost of cellulase alone is estimated to be as high as 25–50% of the total ethanol production costs (Lee et al. 2017). To improve the level of cellulase secretion in the yeast *S. cerevisiae* and thus achieve cost-competitive production of ethanol and other bio-based chemicals from lignocellulosic biomass by consolidated bioprocessing (CBP), various cellulases from different sources were tested by screening an optimal translational fusion partner (TFP) as both a secretion signal and fusion partner. Four cellulases indispensable for cellulose hydrolysis, *Chaetomium thermophilum* cellobiohydrolase (CtCBH1), *Chrysosporium lucknowense* cellobiohydrolase (ClCBH2), *Trichoderma reesei* endoglucanase (TrEGL2), and *Saccharomycopsis fibuligera* β-glucosidase (SfBGL1), were identified to be highly secreted in active form in yeast. Each recombinant yeast could secrete approximately 0.6–2.0 g/l of cellulases into the fermentation broth. Co-fermentation of these yeast strains produced approximately 14 g/l ethanol from pre-treated rice straw containing 35 g/l glucan with threefold higher productivity than that of wild type yeast using a reduced amount of commercial cellulases (Lee et al. 2017). The authors point out that although the secretion level of cellulases was apparently improved, it was still far from being perfect. Further monitoring of each cell behaviour during co-fermentation under different seed ratios and culture conditions will be required for the practical application of this system (Lee et al. 2017).

12.4 Perspectives: Approaches to Optimize the Use of Lignocellulosic and Waste Materials as Feedstocks

Abundant lignocellulosic biomass in the form of agricultural and woody residues or energy crops presents a near-term solution to supporting energy demands that can alleviate global climate change caused by the consumption of fossil resources. Biomass-derived fuels are projected to displace a sizable portion of petroleum-derived transportation fuels. Future production of renewable transportation fuels such as ethanol must rely on lignocellulosic biomass. New approaches will be needed to utilize the abundant lignocellulosic biomass materials for microbial fermentation (Robertson et al. 2017; Nguyen et al. 2017).

Delays associated with the pretreatment phase are one of the biggest technical hurdles faced by commercial cellulosic ethanol projects. The biomass feedstock must be preprocessed to release the cellulosic sugars that enzymes and yeast convert into ethanol. Challenges have been associated with getting the feedstock material to move consistently and fluidly through the pretreatment process. A new technique for pretreatment of biomass to enhance ethanol yield in the simultaneous saccharification and fermentation (SSF) of solid biomass has been developed by Nguyen et al. (2017). SSF of solid biomass can reduce the complexity and improve the economics of lignocellulosic ethanol production by consolidating process steps and reducing end-product inhibition of enzymes. A limitation with SSF has been the low ethanol yields at the high-solids loading of biomass needed during fermentation for more economical ethanol recovery. This study demonstrated that this limitation can be overcome by integrating co-solvent-enhanced lignocellulosic fractionation (CELF) pretreatment with SSF. CELF augments dilute acid pretreatment with tetrahydrofuran (THF) as a water co-solvent. The THF promotes delignification of biomass and hydrolysis of recalcitrant cellulose in water. SSF of CELF-pretreated corn stover, using a yeast strain isolated from cheese whey, *S. cerevisiae* D5A, achieved ethanol titres of 85.6 g/l at 23 wt% solids loading for an enzyme loading of 15 mg protein/g glucan, matching the highest ethanol titre from glucose fermentation for that yeast. Further increasing solids loadings or reducing enzyme loadings decreased ethanol yields due to ethanol toxicity or carbon starvation, demonstrating that the process was limited by the metabolic capabilities and tolerance to ethanol of the yeast strain. These findings suggest that cell viability is the primary factor limiting high ethanol yields in this process, emphasizing the need for strain improvement to address fermentation stresses such as ethanol toxicity (Nguyen et al. 2017).

A consideration of the technical and economic aspects of the biotechnological ethanol industry highlights the need for radical innovation to complement the current manufacturing model (Vertès et al. 2008). One such innovative technological option is the use of non-yeast-based production methods. Sasaki et al. (2019) regarded the choice of a microbial host as a means to bypass product and process reagent toxicities that would be encountered in the industrial process. *Corynebacterium glutamicum* is a rapidly growing, generally recognized as safe (GRAS), Gram-positive, aerobic bacterium used in industrial biotechnology to produce several million tons of amino acids annually, in particular the flavour enhancer l-glutamate and the feed additive L-lysine. Furthermore, *C. glutamicum* is an efficient host for the production of heterologous proteins. Recently it has been metabolically engineered for the production of bulk chemicals such as succinic acid, isobutanol, and ethanol (Witthoff et al. 2013; Shen et al. 2019). For ethanol, this was accomplished by using recombinant *C. glutamicum* engineered with genes for pyruvate decarboxylase (*pdc*) and alcohol dehydrogenase (*adhB*) that can produce ethanol under oxygen deprivation (Jojima et al. 2015). Jojima et al. (2015) investigated the effects of elevating the expression levels of glycolytic genes, as well as *pdc* and *adhB*, on ethanol production. Overexpression of four glycolytic genes (*pgi*, *pfkA*, *gapA*, and *pyk*) in *C. glutamicum* significantly increased the rate of ethanol production. Overexpression of *tpi*, encoding triosephosphate isomerase, further enhanced productivity. Elevated expression of *pdc* and *adhB* increased ethanol yield, but not the rate of production. Fed-batch fermentation using an optimized strain resulted in ethanol production of 119 g/l from 245 g/l glucose with a yield of 95% of the theoretical maximum. Further metabolic engineering, including integration of the genes for xylose and arabinose metabolism, enabled consumption of glucose, xylose, and arabinose, and ethanol production (83 g/l) at a yield of 90% (Jojima et al. 2015). This study suggests potential application of *C. glutamicum* for the production of cellulosic ethanol.

Fuel ethanol production from lignocellulosic materials using recombinant yeasts is an important part of the bioenergy picture. Bioenergy has a significant mitigation potential, provided that resources are developed sustainably and that bioenergy systems are efficient. Bioenergy production can be integrated with food production in developing countries, for example through perennial cropping systems, use of biomass residues and wastes, and advanced conversion systems. Biomass from cellulosic bioenergy crops figure substantially in future energy systems, especially in the framework of global climate policy. Efficient biomass production for bioenergy requires a range of sustainability requirements to safeguard food production, biodiversity, and terrestrial carbon storage (Intergovernmental Panel on Climate Change, Climate Change 2014). Land demand for bioenergy depends on (i) the amount derived from wastes and residues; (ii) the extent to which bioenergy production can be integrated with food production to minimize land-use competition; (iii) the extent to which bioenergy can be grown on marginal lands; and (iv) the quantity of dedicated energy crops and their yields. Considerations of tradeoffs with water, land, and biodiversity are crucial to avoid adverse effects. The impacts on livelihood and large-scale consequences depend on global market factors, income-related food security, and site-specific factors such as land ownership and culture (Intergovernmental Panel on Climate Change, Climate Change 2014). A successful, sustainable cellulosic bioenergy initiative requires integration across the entire value chain from field to product, and overall success requires sufficient understanding of the system to identify key factors that affect environmental sustainability and their potential management (Robertson et al. 2017).

References

Ahmed I. 2016. Cellulosic ethanol, what happened, what's happening? Biofuels Digest. August 1, 2016. http://www.biofuelsdigest.com/bdigest/2016/08/01/cellulosic-ethanol-what-happened-whats-happening [accessed 28 November 2017]

EIA https://www.eia.gov/outlooks/aeo/data/browser/#/?id=15-IEO2016&cases=Reference&sourcekey=0 [interactive table; accessed 28 November 2017]

EPA registers Enerkem to sell cellulosic ethanol under the RFS. 2017. Ethanol Producer Magazine. November 7, 2017. http://www.ethanolproducer.com/articles/14795/epa-registers-enerkem-to-sell-cellulosic-ethanol-under-the-rfs [accessed 28 November 2017]

Aemetis produces cellulosic ethanol from orchard waste. 2017. Ethanol Producer Magazine August 9, 2017. http://www.ethanolproducer.com/articles/14548/aemetis-produces-cellulosic-ethanol-from-orchard-wasteAemetis [accessed 28 November 2017]

Cellulosic. 2017. Ethanol Producer Magazine. November 2017 (various articles). http://www.ethanolproducer.com/tag/cellulosic [accessed 28 November 2017]

Intergovernmental Panel on Climate Change, Climate Change 2014: Mitigation of Climate Change, Contribution of Working Group III to the Fifth Assessment Report of the Intergovernmental Panel on Climate Change (Cambridge Univ. Press, 2014).

International Energy Agency. 2017. World Energy Outlook 2017, Part A: Global Energy Trends. https://www.iea.org/weo2017 [accessed 28 November 2017]

Jojima, T., Noburyu, R., Sasaki, M. et al. (2015). Metabolic engineering for improved production of ethanol by *Corynebacterium glutamicum*. *Applied and Environmental Microbiology* 99 (3): 1165–1172. http://doi.org/10.1007/s00253-014-6223-4.

Koukoulas AA. 2016. A critical look at cellulosic ethanol and other advanced biofuels. Biofuels Digest. December 12, 2016. http://www.biofuelsdigest.com/bdigest/2016/12/12/a-critical-look-at-cellulosic-ethanol-and-other-advanced-biofuels [accessed 28 November 2017]

Lane J. 2017. Clariant to build flagship cellulosic ethanol plant in Romania: 8-figure sales potential envisioned. Biofuels Digest. October 31, 2017. http://www.biofuelsdigest.com/bdigest/2017/10/31/clariant-to-build-flagship-cellulosic-ethanol-plant-in-romania-8-figure-sales-potential-envisioned [accessed 28 November 2017]

Lee C-R, Sung BH, Lim K-M, Kim M-J, Sohn MJ, Bae J-H Jung-Hoon Sohn J-H. 2017. Co-fermentation using recombinant *Saccharomyces cerevisiae* yeast strains hypersecreting different cellulases for the production of cellulosic bioethanol. DOI:10.1038/s41598-017-04815-1. http://www.nature.com/scientificreports [accessed 15 Dec 2017]

Liu X, Xu W, Mao L, Zhang C, Yan P, Xu Z, Zhang ZC. 2016. Lignocellulosic ethanol production by starch-base industrial yeast under PEG detoxification. DOI: 10.1038/srep20361. http://www.nature.com/scientificreports/[accessed 15 Dec 2017]

McAloon A, Taylor F, Yee W, Ibsen K, Wooley R. 2000. Determining the Cost of Producing Ethanol from Corn Starch and Lignocellulosic Feedstock. A joint study by the United States Department of Agriculture, Agricultural Research Service (USDAARS) and the U.S. Department of Energy (DOE) with the National Renewable Energy Laboratory (NREL). October 2000. NREL/TP-580-28893.

McCoy M. 2017. DuPont seeks to sell cellulosic ethanol plant. Chem Eng News. Nov 7, 2017. https://cen.acs.org/articles/95/web/2017/11/DuPont-seeks-sell-cellulosic-ethanol.html [accessed 28 November 2017]

Nguyen, T.Y., Cai, C.M., Kumar, R., and Wyman, C.E. (2017). Overcoming factors limiting high-solids fermentation of lignocellulosic biomass to ethanol. *Proceedings of the National Academy of Sciences of the United States of America* 114 (44): 11673–11678. https://doi.org/10.1073/pnas.1704652114.

POET-DSM achieves cellulosic biofuel breakthrough, new pretreatment clears path for increased cellulosic production. 2017. Press Release. November 2, 2017. https://poet.com/pr/poet-dsm-achieves-cellulosic-biofuel-breakthrough [accessed 28 November 2017]

Renewable Fuels Association. 2016. Fueling a high octane future. https://www.ethanolrfa.org/wp-content/uploads/2016/02/Ethanol-Industry-Outlook-2016.pdf [accessed 28 November 2017]

Renewable Fuels Association. 2016. Industry Statistics. 2016 World Annual Fuel Ethanol Production. http://www.ethanolrfa.org/resources/industry/statistics/#1454099103927-61e598f7-7643 [accessed 28 November 2017]

Robertson, G.P., Hamilton, S.K., Barham, B.L. et al. (2017). Cellulosic biofuel contributions to a sustainable energy future: choices and outcomes. *Science* 356: eaal2324. https://doi.org/10.1126/science.aal2324.

Sasaki, Y., Eng, T., Herbert, R.A. et al. (2019). Engineering *Corynebacterium glutamicum* to produce the biogasoline isopentenol from plant biomass hydrolysates. *Biotechnology for Biofuels* 12: 41–56.

Shen, J., Chen, J., Jensen, P.R., and Solem, C. (2019). Development of a novel, robust and cost-efficient process for valorizing dairy waste exemplified by ethanol production. *Microbial Cell Factories* 18: 51–61. https://doi.org/10.1186/s12934-019-1091-3.

Tao L, Schell D, Davis R, Tan E, Elander R, Bratis A. NREL 2012 Achievement of Ethanol Cost Targets: Biochemical Ethanol Fermentation via Dilute-Acid Pretreatment and Enzymatic Hydrolysis of Corn Stover. National Renewable Energy Laboratory (NREL). April 2014. NREL/TP-5100-61563.

Think Bioenergy, Cellulosic Ethanol. 2017. Cellulosic biofuels: an important part of the future energy mix. Conversations led by Novozymes. November 17, 2017. http://thinkbioenergy.com/cellulosic-biofuels-an-important-part-of-the-future-energy-mix [accessed 28 November 2017]

U.S. Ethanol Plants. 2017. Ethanol Producer Magazine. September 23, 2017. http://www.ethanolproducer.com/plants/listplants/US/Operational/Cellulosic [accessed 28 November 2017]

United States Department of Energy. 2016. 2016 Billion-Ton Report: Advancing Domestic Resources for a Thriving Bioeconomy, Volume 1: Economic Availability of Feedstocks. M. H. Langholtz, B. J. Stokes, and L. M. Eaton (Leads), ORNL/TM-2016/160. Oak Ridge National Laboratory, Oak Ridge, TN. 448p. doi: 10.2172/1271651. https://energy.gov/sites/prod/files/2016/12/f34/2016_billion_ton_report_12.2.16_0.pdf [accessed 28 November 2017]

United States Department of Energy, United States Energy Information Administration, Office of Energy Analysis. 2016. International Energy Outlook 2016, With Projections to 2040, May 2016. DOE/EIA-0484(2016). https://www.eia.gov/outlooks/ieo/pdf/0484(2016).pdf [accessed 28 November 2017]

Veettil, S.I., Kumar, L., and Koukoulas, A.A. (2016). Can microbially derived advanced biofuels ever compete with conventional bioethanol? A critical review. *BioResources* 11 (4): 10711–10755.

Vertès, A.A., Inui, M., and Yukawa, H. (2008). Technological options for biological fuel ethanol. *Journal of Molecular Microbiology and Biotechnology* 15: 16–30. https://doi.org/10.1159/000111989.

Witthoff, S., Mühlroth, A., Marienhagen, J., and Bott, M. (2013). C1 metabolism in *Corynebacterium glutamicum*: an endogenous pathway for oxidation of methanol to carbon dioxide. *Applied and Environmental Microbiology* 79 (22): 6974–6983.

13

Enzymes for Cellulosic Biomass Hydrolysis and Saccharification

Elmar M. Villota[1,3], *Ziyu Dai*[2], *Yanpin Lu*[1] *and Bin Yang**[1]

[1] *Bioproducts, Sciences, and Engineering Laboratory, Department of Biological Systems Engineering, Washington State University, Richland, WA 99354, USA*
[2] *Chemical & Biological Process Development Group, Pacific Northwest National Laboratory, Richland, WA 99352, USA*
[3] *Department of Agricultural and Biosystems Engineering, Central Luzon State University, Nueva Ecija 3120, Philippines*

CHAPTER MENU

13.1 Introduction

Lignocellulosic biomass is primarily composed of 20–50% cellulose, 15–35% hemicellulose, and 5–30% lignin (Lynd et al. 1991; Yang and Wyman 2007). Due to this complex composition, its hydrolysis necessitates a variety of enzymes, which can be in free or in complexed forms, to attain the complete degradation of its polysaccharides into simple sugars (Yang et al. 2011; Yang and Wyman 2008). These enzymes are typically derived from fungi and bacteria, and have been used in many industries such as the food industry and the emerging and highly anticipated industry of second-generation bioethanol. The appeal of the enzymatic processing of pretreated biomass in the second-generation biofuel production lies in its relatively lower energy and chemical requirement, mild operating conditions, and higher yield of fermentable sugars (Jovanovic et al. 2009; Yang et al. 2011).

The most expensive operations in bioconversions of cellulosic biomass to fuels and chemicals are for releasing sugars from this naturally recalcitrant material. Unfortunately, such bioprocessing technologies have not yet been commercialized, at least in part because high enzyme doses are currently required to hydrolyze cellulose to glucose at high yields that are vital to economic success (Wooley et al. 1999) and to compensate for the rapid fall-off in hydrolysis rate as conversion progresses (Desai and Converse 1997; Eriksson et al. 2002; Ragauskas et al. 2006; Yang et al. 2006a). For example, typical cellulase loadings of about 15 FPU/g cellulose used to achieve high

Green Energy to Sustainability: Strategies for Global Industries, First Edition.
Edited by Alain A. Vertès, Nasib Qureshi, Hans P. Blaschek and Hideaki Yukawa.

yields of sugars from pretreated biomass could be translated into about 30 g of enzyme per litre of ethanol made, an extremely high and expensive dose (Yang et al. 2011). Thus, enzyme costs must either be reduced below $2/kg protein or novel strategies developed to substantially reduce the required enzyme dose (Himmel et al. 1999; Wingren et al. 2005; Wyman 2007; Yang et al. 2011). The mechanisms of the hydrolysis reaction and the factors that limit the effectiveness of hydrolysis are however less defined. Consequently, this limits many promising applications in the real world (Ding et al. 2012; Falls et al. 2011; Himmel et al. 2007; Igarashi et al. 2011; Pallapolu Venkata et al. 2011; Shi et al. 2010; Shi et al. 2017). Although plant cell walls are complex, recent advances in analytical chemistry and genomics have substantially enhanced our understanding but have also highlighted how much of a knowledge-gap remains (Marita et al. 2003; Somerville et al. 2004). Notably, there are several challenges in the study of enzymatic hydrolysis of cellulosic biomass. First, the nature of the crystalline substrate, cellulose, is known only in theoretical terms or very small experimentally-based models. Cellulose can exhibit several different supra-molecular structures, including amorphous, para-crystalline, and crystalline (Marita et al. 2003). It was proposed by some researchers that the hydrolysis rate depends on cellulose crystallinity (Fan et al. 1980; Ghose and Bisaria 1979; Lee and Fan 1983; Wood et al. 1989). However, cellulose crystallinity index (CrI) was found to be fairly consistent although the hydrolysis rate declines rapidly as more material is hydrolyzed (Fan et al. 1981; Sasaki et al. 1979; Sinitsyn et al. 1991) but others found the opposite effect (Grethlein 1985; Puri 1984). Further studies are needed to confirm and elucidate the role of crystallinity on processivity of processive or pseudo-processive enzymes originated from various micro-organisms, and on dynamic cellulase's interaction with cellulose as enzymatic hydrolysis of cellulose progresses. Because enzymatic hydrolysis of cellulose is a surface reaction, the cellulose surface area available to cellulases is a potential determinant of hydrolysis rate and effectiveness (Cowling 1975; Cowling and Kirk 1976; King 1966; Mandels et al. 1971; Peitersen et al. 1977; Yang and Wyman 2006). The cellulose surface is thought to contain both distinct and common adsorption sites for cellulases, with the extent to which they are occupied depending on the enzyme components involved and the order in which they are added (Medve et al. 1997). Studies on cellulase adsorption and reversibility of cellulase components indicated that adsorption and desorption of cellulases on cellulose is a dynamic process that requires free diffusion of enzyme across the surface (Bothwell et al. 1997; Creagh et al. 1996; Jervist et al. 1997; Shi et al. 2016; Yang et al. 2006a). The accessible surface of cellulose to cellulase, limited by the structure of cellulose fibres in plant cell walls, is related to many substrate factors, including cellulose crystallinity, degree of polymerization (DP), accessibility of enzyme, available pore volume, and specific surface area (Converse 1993). In addition to crystallinity, several studies and literature reviews discuss the fate of DP of insoluble and soluble cellulose during and after hydrolysis with complete cellulase or its purified component (Eremeeva et al. 2001; Hilden et al. 2005; Kanda et al. 1976; Kleman-Leyer et al. 1996; Mansfield and Meder 2003; Pala et al. 2007; Yang et al. 2011; Zhang and Lynd 2004 2005). However, the understanding of the impact of the cellulose chain length on its hydrolysis, including on cellulase processivity (Valjamae et al. 1999) and adsorption, is still limited. Although the cellulase may adsorb on the surface effectively, its catalytic efficacy may further be dictated by physical parameters such as pH, temperature, ionic strength, and the presence of inhibitors (Andreaus et al. 1999; Chundawat et al. 2006; Panagiotou and Olsson 2007; Reinikainen et al. 1995; Tengborg et al. 2001) in addition to factors related to the substrate and enzyme. It was reported that cellulases are inhibited by xylooligomers more strongly than by glucose or cellobiose (Qing et al. 2010). It is plausible that oligomeric cellulose derivatives are crucial factors to enzymatic hydrolysis of cellulose. Furthermore, it was demonstrated by computer modelling that a high-density water layer is formed on the surface of cellulose thus possibly inhibiting the release of some cellulose derivatives into bulk water (Himmel et al. 2007). Through interrupting hydrolysis then introducing fresh enzyme to enzyme-free and partially converted cellulose to initiate hydrolysis reactions once again, such 'restart' experiments provide a valuable tool to compare interrupted hydrolysis with continual hydrolysis in order to assess how substrate reactivity and enzymes change with conversion at the time of enzymatic hydrolysis and their effects on hydrolysis (Desai and Converse 1997; Ooshima et al. 1991; Zhang et al. 1999). Overall hydrolysis rates and rates per amount of adsorbed enzyme decline dramatically as enzymatic hydrolysis of cellulose progresses, leading to decreased yields and long processing times (Nutor and Converse 1991; Wang and Converse 1991). This has become a bottleneck of commercialization with current cellulosic technologies. Several factors have been proposed to explain such phenomena, including a potential thermal instability of cellulases (Caminal et al. 1985; Converse et al. 1990; Eriksson et al. 2002; Gonzalez et al. 1989), hydrolysis product inhibition (Eriksson et al. 2002; Gan et al. 2003; Gusakov and Sinitsyn 1992; Holtzapple et al. 1990; Kadam et al. 2004; Todorovic et al. 1987; Yang and Wyman 2006), inactivation of cellulases (Converse et al. 1988; Gusakov and Sinitsyn 1992; Gusakov et al. 1987; Mukataka et al. 1983; Ooshima et al. 1990; Reese 1982; Sinitsyn et al. 1986; Sutcliffe and Saddler 1986), enzyme activity decreasing or stopping (Desai and Converse 1997), substrate transformation into a less digestible form (Zhang et al. 1999), and the heterogeneous structure of the substrate (Nidetzky and Steiner 1993; Zhang et al. 1999). The drop in rate was ascribed to declining substrate reactivity as the more easily reacted material was thought to be consumed preferentially (Zhang et al. 1999) although other reports concluded that substrate reactivity was not the principal cause of the long residence time required for good cellulose conversion (Desai and Converse 1997). Since continual hydrolysis results in most studies provide limited data on the dynamic changes of substrate and enzymes, 'interrupt' and 'restart' experiments, in which cellulose hydrolysis is interrupted then partially hydrolyzed cellulose is further hydrolyzed by fresh enzymes, have been used to study the dynamic characteristic of the enzymatic process in order to identify factors that control the rate of cellulose hydrolysis (Desai and Converse 1997; Gusakov et al. 1985; Nidetzky and Steiner 1993; Ooshima et al. 1991; Valjamae et al. 1998; Zhang et al. 1999). Recent reports suggest that kinetic models of cellulose hydrolysis should include correlating cellulose reactivity, oligosaccharide distribution, the effective enzyme binding capacity, and equilibration of depolymerization in order to fully understand the enzymatic hydrolysis of cellulose (Shi et al. 2016; Yang et al. 2006a).

Aerobic and anaerobic bacteria or fungi that produce glycosyl hydrolases have been found in multiple environmental niches. Their glycosyl hydrolases have evolved tolerance to heat, acid, and alkaline due to the unusual growing environment of some of these microorganisms. According to their protein structures, glycosyl hydrolases are classified into four groups: cellulosome, cellulases, hemicellulases, and ligninases. Glycosyl hydrolases are produced by all the microbes cited above, including those with cellulosomal systems. Several filamentous fungi and their glycosyl hydrolases have been extensively researched. Notably, the glycosyl hydrolase of *Trichoderma* and *Aspergillus* strains are commercially produced under extensively optimized conditions (Himmel et al. 1999; Shetty and Marshall 1986). *Trichoderma reesei* typically generates two cellobiohydrolases (CBHI and CBHII), five to six endoglucanases (EGI, EGII, EGIII, EGIV, EGV, and EGVI), β-glucosidases (BGL I and II), two xylanases, and various accessory hemicellulases (Vinzant et al. 2001). The proportion of these components is vital to the efficiency of cellulosic biomass hydrolysis, especially that of crystalline cellulose, due to various synergies that exist among different components (Henrissat et al. 1985; Kanda et al. 1980). Several studies indicated competitive or synergistic binding (Jeoh et al. 2002; Jeoh et al. 2006; Nidetzky et al. 1996), common or separate binding sites, and different binding capacity and affinity among various cellulase components (Ding and Xu 2004) (D.Y. Ryu et al. 1984). However, few studies have been reported on the effects of the molar ratio of these components on binding or bound enzyme ratio on synergism. In addition, it was reported that optimal cellulase's ratio was also dependent on pretreatment conditions and feedstock, although commercial *T. reesei* cellulases typically contain 4~5: 1 of CBHI: EGI, (Banerjee et al. 2010).

The cellulosome is a hydrolytic system that is produced only by anaerobic microbes. In the early 1980s, Bayer, Lamed and their colleagues first discovered the cellulosome system in the anaerobic thermophilic bacterium *Clostridium thermocellum* (Bayer et al. 1983; Lamed et al. 1983a). Since then, more microorganisms equipped with a similar cellusome system have been identified. These microbes and their cellulosome systems have been extensively studied in terms of their protein complex structures, characteristics, cellulosome formation genes, diversities, and interactions with plant cell walls. It was found that cellulosomes are typically integrated with cellulases and hemicellulases and thus constitute efficient machineries to degrade plant cell wall polysaccharides through attaching to plant cell walls by a scaffoldin-borne carbohydrate-binding module (CBM) (Fontes and Gilbert 2010).

The mechanism of enzymatic hydrolysis is complicated by its strong relation with the intrinsic properties of the substrate, distinct functions of various enzyme components, and substrate-enzyme interactions. Therefore, the commercial use of this process has met with limited success to date. To promote the development of novel strategies towards commercializing enzymatic hydrolysis for biomass conversion to fuels and chemicals, it is important first to highlight the current level of knowledge relative to the structures of glycosyl hydrolases and their mechanisms, as well as the current understanding of complex and non-complex cellulase systems, hemicellulases, and enzyme producing microorganisms.

13.2 Glycosyl Hydrolases: General Structure and Mechanism

Enzymes that have the capability to hydrolyze glycosidic bonds between carbohydrates or a carbohydrate and a non-carbohydrate group are generally classified as glycosyl hydrolases (GH) (Lombard et al. 2014). Typically, these are modular enzymes that can include both cellulases and non-cellulosic structural proteins. Their modules usually consist of diverse arrangements of catalytic domains with accessory units that may include carbohydrate-binding modules, dockerins, sortase motifs or cell-adherent modules (Bayer et al. 1998a; Bayer et al. 2006; Miller et al. 2010). Carbohydrate-binding modules, which may contain up to about 200 amino acids, and are either linked to the amino- or carboxy-termini of the catalytic domains and can occur in double or triple domains in a cellulase unit (Shoseyov et al. 2006). The cleavage of the glycosidic bonds through GH can be catalyzed by two amino acid constituent of the enzyme – a general acid, the proton donor, and a basic nucleophile (Davies and Henrissat 1995), and occurs via overall retention or inversion of the anomeric carbon configuration depending on their spatial position. In either the retaining or the inverting mechanisms, the position of the proton donor is identical while that of the nucleophilic catalytic base varies. In retaining catalysts, the nucleophilic catalytic base is near the sugar anomeric carbon while the said catalytic base is more distant in inverting mechanisms to accommodate water molecules between the base and the sugar (McCarter and Withers 1994).

A widely accepted mode of action of proteins with hydrolytic capabilities, as with other enzymes, lies in their unique structure. Cellulases generally consist of distinct integrated modules organized in a complex molecular architecture: a core catalytic domain (CD); one or more carbohydrate-binding modules (CBM); and an unstructured inter-domain linker peptide that connects and holds them together. The CD generally contains the large, globular catalytic domain that expresses the active site, which is basically a lattice of the protein chain that forms either crater, groove, or tunnel-like shaped configurations (Davies and Henrissat 1995; Nimlos et al. 2007). CBM, on the other hand is the module that is attached or directly bound onto the substrate surface, thereby accountable for catalyst and substrate contact and also for maintaining their affinity (Sukumaran 2009). Most importantly, it facilitates hydrolysis by keeping the catalytic domain attached to the substrate, therefore it aids hydrolysis initiation and processivity (Quiroz-Castañeda and Folch-Mallol 2013; Teeri et al. 1998). A higher binding affinity of CBM has been shown to be responsible for higher cellulose conversion (Biswas et al. 2014; Takashima et al. 2007). In addition, the functions of both CD and CBM were found to be also highly influenced by protein glycosylation (Beckham et al. 2012; Dai et al. 2013) What is more, the two modules are connected by linker peptides that can contain from 6 to 59 amino acids. These linker peptides are believed to be highly flexible and may also act as a fulcrum to promote the independent motion and function of catalytic and binding modules (Wilson and Irwin 1999). More recently, these linkers were hypothesized to have the capacity to store kinetic energy akin to a mechanical spring to facilitate a caterpillar-like motion that promotes processivity (McCabe et al. 2010).

13.2.1 Classification of GH

Classifying and grouping the vast pool of enzymes that can degrade carbohydrates is quite a challenging task because similarities and differences of these enzymes can be drawn in many aspects and in a variety of perspectives (Figure 13.1). A classification system established and maintained by the International Union for Biochemistry and Molecular Biology (IUBMB; http://www.chem.qmul.ac.uk/iubmb) assigns cellulases through their functions and is designated by Enzyme Classification (EC) numbers. Under this system, three major groups were identified as

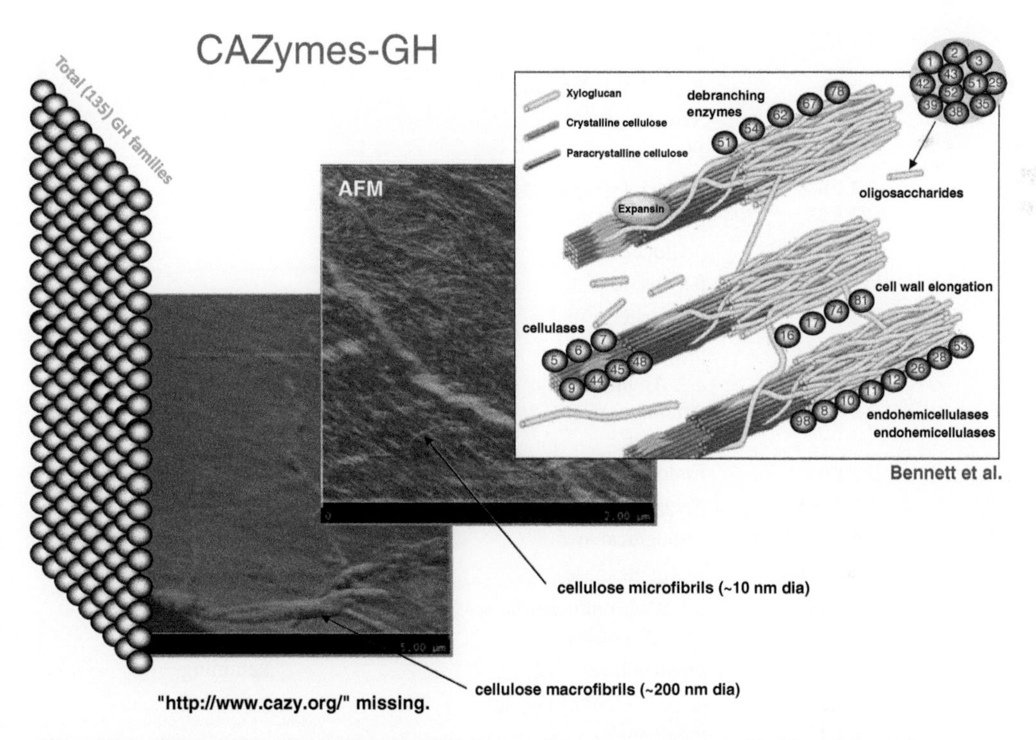

Figure 13.1 Nature diversity of CAZymes-GH. Source: Based on a slide provided by Michael Himmel, National Renewable Energy Laboratory.

Table 13.1 Important glycosyl hydrolase families for biomass hydrolysis.

GH family	EC number	Dominant enzyme groups	Mechanism
1	3.2.1.21	β-Glucosidases	Retaining
3	3.2.1.21	β-Glucosidases	Retaining
5	3.2.1.25 3.2.1.4 3.2.1.78 3.2.1.8 3.2.1.91	β-Mannosidases Cellulases Mannan endo-1,4-β mannosidases Endo-1,4-β-xylanases Cellulose 1,4-β-cellobiosidases	Retaining
6	3.2.1.4 3.2.1.91	Endoglucanases Cellobiohydrolases	Inverting
7	3.2.1.4 3.2.1.-	Endoglucanases Reducing end-acting cellobiohydrolases	Retaining
8	3.2.1.4 3.2.1.8	Cellulase Endo-1,4-β-xylanases	Inverting
9	3.2.1.4 3.2.1.91 3.2.1.21	Endoglucanases Cellobiohydrolases β-Glucosidases	Inverting
10	3.2.1.8 3.2.1.32	Xylanases Endo-1,3- β-xylanases	Retaining
11	3.2.1.8	Xylanases	Retaining
12	3.2.1.4 3.2.1.151 3.2.1.73	Endoglucanases Xyloglucan hydrolases β-1,3-1,4-Glucanases	Retaining
16	3.2.1.6 3.2.1.151	Endo-1,3(4)-β-glucanases Xyloglucanases	Retaining
26	3.2.1.78	Mannanases	Retaining
30	3.2.1.8 3.2.1.21 3.2.1.31 3.2.1.37	Endo-1,4- β-xylanases β-Glucosidases β-Glucuronidase β-xylosidase	Retaining
39	3.2.1.37	β-xylosidase	Retaining
43	3.2.1.37 3.2.1.55 3.2.1.99 3.2.1.8	β-Xylosidase α-L-Arabinofuranosidase Arabinanase Xylanase	Inverting
44	3.2.1.4 3.2.1.151	Endoglucanases Xyloglucanases	Inverting
45	3.2.1.4	Endoglucanases	Inverting
48	3.2.1.4	Endoglucanase	Inverting
51	3.2.1.55 3.2.1.4	α-L-Arabinofuranosidase Endoglucanase	Retaining
52	3.2.1.37	β-Xylosidase	Retaining
54	3.2.1.55 3.2.1.37	α-L-Arabinofuranosidase β-Xylosidase	Retaining

(Continued)

Table 13.1 (Continued)

GH family	EC number	Dominant enzyme groups	Mechanism
61	3.2.1.4	Endoglucanase	Not known
62	3.2.1.55	α-L-Arabinofuranosidase	Not known
64	3.2.1.39	β-1,3-Glucanases	Inverting
67	3.2.1.139 3.2.1.131	α-Glucuronosidases Xylan α-1,2-Glucuronosidases	Inverting
74	3.2.1.4 3.2.1.150 3.2.1.151	Endoglucaases Oligoxyloglucan reducing end-specific cellobiohydrolases Xyloglucanase	Inverting
113	3.2.1.78	β-1,3-mannanases	Retaining

Source: Carbohydrate-Active Enymes (CAZy) database. For full list of most recently identified enzymes please go to http://www.cazy.org.

critical to cellulose hydrolysis, such as EC 3.2.1.4, EC 3.2.1.91, and 3.2.1.21. Separated by dots, the classification number is interpreted as follows: number 3 depicts hydrolase type; number 2 specifically means glycosylases; number 1 further specifies enzymes that act on O-glycosyl compounds. Furthermore, the fourth number represents more specific active sites; for example 4 is specified for endocellulolytic activity or the internal (1 → 4)-β-D-glucosidic linkages in cellulose; 91 for exocellulolytic activity or the ends of the cellulose chains; and 21 for the glycosidic bond of cellobiose, releasing β-D-glucose units (Cunha et al. 2013). Though very useful, this classification falls short when interpreting sequence structures and origins as current groupings may include protein families with entirely different structures and evolutionary origins (Cunha et al. 2013; Henrissat and Davies 1997). In recent years, the ever expanding genetic and protein structural knowledge has enabled a strategic sequence- or structure-based enzyme classification (Xu and Ding 2007). Perhaps most referenced by researchers in cellulose bioconversion technologies is the Carbohydrate-Active Enymes (CAZy) database (http://www.cazy.org) developed by a team led by Coutinho and Henrissat in 1998 (Lombard et al. 2014). The challenging functional annotation in genomes of carbohydrate-active enzymes due to varying modularity and substrate specificity specially for non-specialist has been the motivation for generating this public database (Lombard et al. 2014). This database has been very useful in the sense that it is highly relevant, continuously updated, and is easily accessible. Table 13.1 lists critical cellulases and hemicellulases and their EC number and corresponding GH families under the CAZy classification system.

13.3 The Cellulase Enzyme System

Cellulases and other glycosyl hydrolases are key enzymes in the depolymerization of cellulosic biomass for cellulosic sugar production (Himmel et al. 2007). Cellulose is a major constituent of plant cell walls and is the main target substrate for cellulases. It consists of unbranched, parallel D-glucopyranose units linked by glycosidic (β-1, 4) bonds and forms long (24–36 units) chains of highly organized microfibrils (Endler and Persson 2011; Fernandes et al. 2011). Glucose molecules are positioned in lateral chains within crystalline cellulose in an alternate 180° angle fashion with repeating cellobiose units (Festucci-Buselli et al. 2007). Cleavage of the β(1 → 4) glucosidic bond in cellulose, which results in the formation of glucose and short cellodextrins, can be carried out mainly by a group of enzymes comprising endo-1,4-β-D-glucanase (EG), exoglucanasses or cellobiohydrolase (CBH), and β-glucosidase (BG). The structural makeup of these enzymes reflects their specificity and mode of action. Endo- and exo- modifier refers to the specific spatial position where the enzyme acts on bonds which can be within or at the extremities of the polysaccharide chain. The endo-acting enzymes, also known as endoglucanases (EG), have a distinct cleft- or groove-shaped active site, which allows them to anchor to the substrate anywhere along the mid-section of the polysaccharide. On the other hand, the active site of exo-acting enzyme exoglucanases (e.g. CBH), resembles a drum or a tunnel shape such that the polysaccharide chains are threaded inside and through the active site (Bayer et al. 1998a 2006; Béguin and Aubert 1994; Gilbert and Hazlewood 1993; Miller et al. 2010).

13.3.1 Endoglucanases

One of the best known cellulases, which is of particularly high biotechnological significance in lignocellulosic depolymerization, is the endo-1,4-β-glucanase, more generally referred to as endoglucanase (EG). Endoglucanases are enzymes that are mainly responsible for the random breakdown of the amorphous (non-crystalline) region that occurs within or at the mid-section of a typical cellulose polymer. This cleavage results in more end chains, which will be later sites for the degradative attack of other cellulase systems and particularly the CBH. Size-exclusion with multi-angle laser light scattering (SEC-MALLS) studies revealed that EG action involves rapid and random incision of the cellulose polymer resulting in a decreased degree of polymerization (Zhang et al. 2011b).

EGs have a catalytic core structure classified to more than 10 GH families from the CAZy system, of which GH 6, 7, 9, 12, 44, 45, 48, 61, and 74 are considered as critical enzymes for lignocellulose degradation (Jovanovic et al. 2009). Typical cellulolytic fungi secrete EGs at approximately 20% by weight level in their secretomes (Chundawat et al. 2011; Herpoël-Gimbert et al. 2008; Sipos et al. 2010). Cellulolytic fungi and bacteria produce a variety of EGs and the most commonly reported EGs include GH under families 7, 5, 12, and 45 of the CAZy system that are then tagged as EG I, II, III, and V, respectively. Despite their common specificity, notable differences lie in their activities. GH 45 along with other EG families such as 6, 9, and 48 work via inversion, while families 5, 7, and 12 work via retention (Table 1.1). It was proposed that such EG deviation may relate to either EGs' side-activities on hemicellulose constituents in the substrate (Sweeney and Xu 2012; Vlasenko et al. 2010), or synergism between processive and conventional EGs (Sweeney and Xu 2012; Wilson 2008).

In their study, Murphy et al. documented that EG I, II, and III show a distinct initial burst with a peak rate that is fivefold higher than the rate five minutes after initiation of the hydrolysis process. They further proposed that the EG I action on hydrolysis dominates mostly in the earlier stage, while only swelling occurs in the later stage (Murphy et al. 2012). Swelling promoted by EG exposes individual microfibrils and bundles of microfibrils, resulting in the loosening of the fibre structure and exposure of cellulose. Another proof of the declining hydrolysis rates by EG was concluded by Shu et al. in their work where they proposed that the said phenomenon is not merely affected by adsorption but also by the EGs initial activity (Shu et al. 2014). That is, higher initial EG activities are more likely to deplete the most reactive substrates at faster rates than the adsorption rate that plays no apparent role in the reaction rate decline (Shu et al. 2014).

13.3.2 Cellobiohydrolases

Cellobiohydrolase (CBH) is a widely popular cellulase in bioenergy research because it can hydrolyse cellulose crystalline regions better as compared to other cellulases (Liu et al. 2011; Teeri 1997;

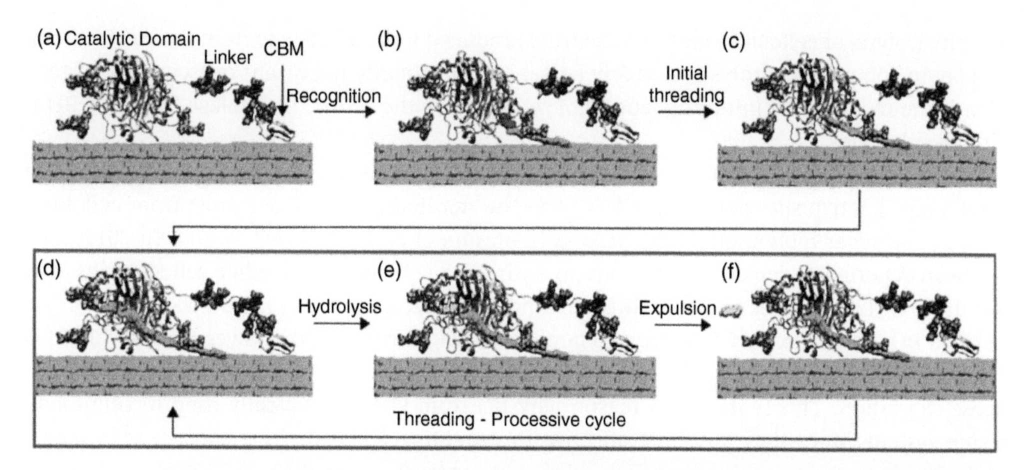

Figure 13.2 Processive catalytic action of CBH on crystalline cellulose. (a) CBH binding to cellulose, (b) recognition of a reducing end of a cellulose chain, (c) initial threading of cellulose chain into the catalytic tunnel, (d) threading and formation of a catalytic active complex, (e) hydrolysis in a processive cycle and (f) product expulsion and threading of another cellobiose, (shown in yellow in e and f). Source: Reprinted from Current Opinion in Biotechnology, Vol. 22 Issue 2, Greg T. Beckham, Yannick J. Bomble. Edward A. Bayer, Michael E. Himmel, Michael F. Crowley, Applications of computational science for understanding enzymatic deconstruction of cellulose, with permission from Elsevier.

Wilson and Kostylev 2012). Crystalline cellulose, being physically well laid out, highly ordered, and packed in a specific lattice, is understood to be most recalcitrant to enzymatic degradation. Given that crystalline regions constitute up to 20% of the cell wall microfibril makes CBH an irreplaceable and indispensable element in any cellulase cocktail.

CBH progressively hydrolyzes from the reducing (CBH I) and non-reducing (CBHII) ends of the cellulose, releasing cellobiose in the process (Quiroz-Castañeda and Folch-Mallol 2013). This organized and processive action is due to the structural architecture that uniquely forms a drum or tunnel pattern where active sites are strategically positioned on the periphery of the geometry (Liu et al. 2011; Vocadlo and Davies 2008). With this configuration, CBHs are able to cleave off cellobiose units from cellulose chains in continuum as they proceed through the active sites. An informative cartoon representation and description of the action of a CBH is this catalytic process is presented in Figure 13.2 (Beckham et al. 2011).

In the SEC-MALLS study done by Zhang et al. on CBH, no significant physical alteration in crystalline cellulose structure was found during the whole hydrolysis period suggesting a very organized layer-by-layer action of CBH on the process (Zhang et al. 2011b). Even if such processive and organized CBH action is expected, the randomness and insolubility of the cellulose substrate itself makes CBH kinetics highly deviant from the typical Michaelis–Menten enzymatic model. In fact, many researchers found significant fractal and localized jamming effects in the process (Igarashi et al. 2011; Warden et al. 2011; Xu and Ding 2007). Also, it has been proposed that processive CBH movement can easily be obstructed by surface irregularities on the cellulose surface and such impact may be a major factor in CBH hydrolytic action kinetics (Kurašin and Väljamäe 2011; Praestgaard et al. 2011). These factors make mathematical modelling of CBH kinetics among the most complex and challenging subjects in the bioengineering and related fields.

13.3.3 Beta-Glucosidases

Beta-glucosidases (BG) are cellulases that are mainly responsible for the last step in converting highly complex cellulose polymer to readily fermentable glucose. These enzymes specifically act on the hydrolysis of cellobiose and cellodextrins produced by CBHs. Aside from this key role that BGs play in complete hydrolysis, a widely perceived and equally important consequence of their action is the reduction of inhibitory effects of cellobiose to the action of cellulase EG and CBH to cellulose substrates.

BGs in general are not modular as they are lacking a distinct CBM component. They have pocket-shaped active sites acting specifically on the non-reducing glucose units from cellobiose and releasing sugar monomers in the process (Langston et al. 2006; Sweeney and Xu 2012). BGs have been reported to have the capability to hydrolyze cellobiose and other cellodextrins with a DP up to 6 (Langston et al. 2006; Sweeney and Xu 2012; Zhang and Lynd 2004). Most of the identified BGs belong to the GH 1, 3, and 9 families (Eyzaguirre et al. 2005; Sweeney and Xu 2012), but GH 1 tends to be more active on different di- or oligosaccharides and is more resilient to glucose inhibition. This is the main reason why BG from GH 1 is typically used in cellulolytic enzyme systems since they are more effective in degrading lignocellulosic substrates (Sweeney and Xu 2012).

The limited capabilities of BGs to hydrolyze soluble rather than insoluble substrates make it possible to observe their kinetics under traditional models (Jeoh et al. 2005). A number of researchers have observed that the kinetics of a typical BG is in very good agreement with a typical Michaelis–Menten type catalysis, with a competitive inhibition effect indicating that BGs are either free, linked with cellobiose, or linked with glucose (Chauve et al. 2013; Shu et al. 2014; Wang et al. 2013).

13.3.4 Polysaccharide Monooxygenases and Cellobiose Dehydrogenases

Recently, a whole new perspective in plant biomass deconstruction was developed because of the findings that shed light to oxidoreductive mechanism lytic polysaccharide monooxygenases (PMO) and cellobiose dehydrogenase (CDH) (Dimarogona et al. 2012; Morgenstern et al. 2014). These newly discovered enzymes are of interest because they show potential in increasing enzymatic conversion rates of biomass which has a direct impact on cost and process economics (Horn et al. 2012).

Previously identified as GH due to traces EG activity, proteins under GH61 have been investigated due to their enigmatic nature and their positive influence on traditional EGs on cellulose hydrolysis (Brown et al. 2010; Harris et al. 2010). Associated CBMs, classified as CBM33 under the CAZy classification (Lombard et al. 2014), have been positively identified to be responsible for increases in accessibility of crystalline cellulose and thereby for promoting better deconstruction efficiency of cellulases in the same analogy as the chitin-binding protein CBP21, which is an auxiliary protein that aids chitin bacterial breakdown (Horn et al. 2012; Vaaje-Kolstad et al. 2005). It was later discovered that the protein CBP21 oxidatively cleaves glycosidic bonds of chitin and leads to the series of works testing the theory that CBM33s have a similar mechanism on cellulose due to the parallelism of chitin and crystalline cellulose structural makeup (Dimarogona et al. 2012; Horn et al. 2012). Some of the evidence of oxidative activity of GH61 that were reviewed by Dimarogona et al. (2012) include works by Quinlan et al. who documented that *Thermoascus aurantiacus* GH61 enzyme shows oxidative activity on cellooligomers and cellulose non-reducing ends (Quinlan et al. 2011); Westereng et al. where the *Phanerochaete chrysosporium* GH61D enzyme acts upon cellulose oxidative cleavage (Westereng et al. 2011); and Phillips et al. (2011), who observed the release of oxidized products from *Neurospora crassa* GH61 in its action on phosphoric acid swollen cellulose. These observations outlined a novel mechanism of action and activity of the then known GH61 family, and subsequently led to the coining of a new class or group of enzymes in the CAZy database (http://www.cazy.org) as family AA9 with a known activity as copper-dependent lytic polysaccharide monooxygenases (LPMOs) (Lombard et al. 2014). In a more recent work, the oxidative action on xylan along with cellulose of a lytic PMO from *Myceliophthora thermophile* has also been documented (Frommhagen et al. 2015).

In conjunction with the oxidative action of lytic PMO, another class of enzymes that can also be found in the genome of most wood-rotting fungi was considered to have complementary roles to the mechanism of fungal PMO. CDHs have been proposed to have a variety of roles in the breakdown of lignocellulosic biomass but the most accounted of these roles is its participation in Fenton-type reactions (Dimarogona et al. 2012; Henriksson et al. 2000). Current works specifically outline its cooperative participation with PMOs by being an electron donor as its promotes the oxidation of cellobiose and in turn aids in the reductive mechanism of PMO (Dimarogona et al. 2012; Wilson 2012). This synergism was documented by Langston et al. in 2011 who tested individual and combined effects of PMO and CDH when used in conjunction with non-complexed enzymes CBH, EG, and BG (Langston et al. 2011). These researchers found that both PMO and CDH are necessary and that the absence of either does not promote any increase in degradation rates of non-complexed enzymes (Langston et al. 2011). Further, Phillips et al. found such a synergistic action in *N. crassa* and suggested that this mechanism of action may be widespread throughout the fungal kingdom since in their study CDH and PMO were found in both ascomycetes and basidiomycetes (Phillips et al. 2011).

13.3.5 Cellulases Synergy and Kinetics

In addition to its appeal in manufacturing as mentioned by Jovanovic et al. (2009), such as mild operating conditions, one of the major advantages of working with free enzymes is their substrate specificities and overall synergy. Clearly, the three elements EG, CBH, and BG in conjunction with the newly discovered PMO and CDH are essential in any industrial enzyme mixture for achieving the complete hydrolysis of cellulose. However, the main challenge currently posed is to optimize such combinations in terms of the quantities of these individual components given the obvious direct impact on the process economy since enzymes are still generally considered expensive. Thus, it is vital to apply innovative techniques to unravel the most critical features of individual cellulases and to define how their synergism dynamically changes while enzymatic hydrolysis progresses to come up with strategies to achieve high yield at a low cost. Numerous researches have focused on this matter because the entire process economics of an industrial-scale second-generation fuels plant still to this date depends strongly on the cost effectiveness of the hydrolysis operation.

It is widely believed that cooperation among non-complexed enzymes is necessary to efficiently depolymerize plant cell walls (Figure 13.3). This cooperative action of different enzymes is one of the most substantial observations in cellulose depolymerization (Andersen et al. 2007; Kumar and Murthy 2013; Lynd et al. 2002; Wang et al. 2012; Zhang and Lynd 2006). The most accepted mechanistic interpretation of this synergism between random site acting EG and chain end-specific processive CBH is that EGs, when splitting long polysaccharide chains, produce new ends that in turn become active sites for CBH. BG, on the other hand, is working synergistically with CBH, since cellobiose is produced as a product from CBH hydrolysis (Jalak et al. 2012; Teeri 1997). In an in situ observational study, Wang et al. (2013), aided by high resolution images from Atomic Force Microscopy (AFM), revealed that the most efficient hydrolysis is attained with the synergistic action of three enzymes EG 1 (TrCel7B), CBH II (TrCel6A), and CBHI (TrCel7A). In this mixture, it is believed that EG 1 softens the fibres while CBH I and CBH II processes it from the reducing and non-reducing ends of the chain (Wang et al. 2013).

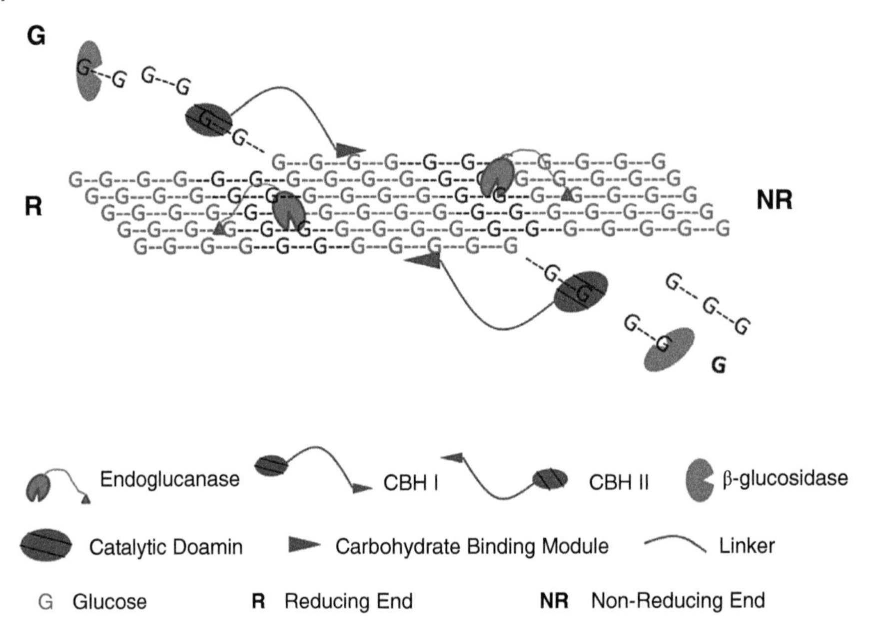

Figure 13.3 Cellulases acting on the surface of cellulose (Kumar and Murthy 2013).

The kinetics of biomass hydrolysis have been generally described by an initial burst or fast hydrolysis rate, then the process rapidly slows down steadily leading to a hydrolysis reaction that takes prolonged periods of time to complete (Jalak et al. 2012; Jalak and Väljamäe 2010; Lebaz et al. 2016; Murphy et al. 2012; Song et al. 2016). Synergistic actions among different enzymes such as endo–exo synergism, exo–exo synergism, and exo–β-glucosidase synergism relevant to this kinetics have been widely documented but it is the endo–exo cooperation that results in a highly effective synergism (Andersen et al. 2007; Kumar and Murthy 2013; Medve et al. 1998; Väljamäe et al. 1999; Yang et al. 2011; Zhang and Lynd 2004). Furthermore, it has been proposed that the degradation of amorphous regions of cellulose by EG avoids the stalling of CBH II and leads to its accelerated recruitment while at the optimal enzyme to substrate ratio, the rate of CBH synergistic hydrolysis is only limited by the rate of processive movement of CBH on the substrate (Fox et al. 2011b; Jalak et al. 2012).

More recent modelling efforts employ a variety of new techniques to investigate the kinetics of synergistic enzymatic activities and important substrate factors to result in meaningful mathematical relationships, which can be very useful in predicting reaction rates and ultimately in process design. For example, in their work, Lebaz et al. used a population balance approach solved by direct quadrature method of moments for mathematical modelling of the enzymatic hydrolysis of cellulose using chain length distributions as a response parameter (Lebaz et al. 2016). Results from this study promoted the view that there is a strong synergy of EG and CBH in cellulose substrate degradation. Another comprehensive work was recently completed by Huron et al. wherein a mechanistic modelling incorporating not only the cocktail composition but also the influence of the morphology of the cellulosic substrate and its evolution with time (Huron et al. 2015). In this experiment, a distinct change in kinetics at the initial and the later stages of the hydrolysis was observed. After an hour of hydrolysis, more glucose was obtained from a cocktail comprising 34% EGI, 30% CBHI, and 36% CBHII; however after 48 hours, the optimal ratio shifted towards higher CBHII concentration, namely 32% EGI, 28% CBHI, and 40% CBHII (Huron et al. 2015). The model predictions suggest that it could be enhanced by lowering the CBHI concentration for the benefit of EGI and CBHII (Huron et al. 2015). This result brings another perspective compared to a previous work that suggested that the best saccharification of crystalline cellulose is achieved with the enzyme blend 20 : 60 : 20 (EGI : CBHI : CBHII) with a saturated level of BG to address inhibition (Baker et al. 1998). Other salient results of the said study include general trends on the impact of specific cellulose characteristics on hydrolysis rate which can help design effective cellulase mixtures, and confirm that the enzyme efficiency depends on substrate characteristics (Huron et al. 2015). Kumar and Murthy also successfully developed models that employed stochastic molecular modelling in kinetics and synergy studies. The action of various cellulases on the cellulose structure was notably modelled using the Monte Carlo simulation technique. A key advantage of the latter proposed model on the mode of action of enzymes and their synergy not only involves cellulase mixture effects but also structural features of cellulose as well as, more importantly, the dynamic morphological changes in the substrate that directly impacts enzyme–substrate interactions during hydrolysis (Kumar and Murthy 2013). Finally, recent discoveries of the oxidoreductive cellulolytic enzyme system has dramatically altered the view on the enzymatic degradation of lignocellulosic biomass(Dimarogona et al. 2012). These recent studies on both synergy of cellulases in combination with oxidoreductive cellulolytic enzymes and LPMOs, acting with electron transfer partners to oxidatively cleave the polysaccharide backbone will lead to further reduction in production costs by decreasing enzyme costs and hydrolysis time, ultimately resulting in dramatically improved efficiencies of lignocellulosic biomass conversions (Agger et al. 2014; Hu et al. 2016; Johansen 2016; Kracher et al. 2016).

13.4 The Hemicellulase Enzyme System

The enzyme system for the hydrolysis of hemicellulose is relatively more diverse than that of cellulose since the substrate itself, the hemicellulose, includes an extensive variety of polysaccharide units. Notably, hemicellulose is comprised of fundamental elements xylan, glucomannan, and glucan backbones depending on the plant source (Pauly et al. 2013). Common among those units are pentoses or hexoses configured with a pyranose ring and linked viaβ-1, 4 bonds. By having C-4 hydroxyl in the equatorial orientation, the hemicellulose backbone appears linear, like that of cellulose, but in contrast hemicellulose is a highly branched polymer with a variety of sugar decorating motifs making it amorphous and more accessible to enzymatic actions (Jovanovic et al. 2009). Conversely, having different backbones and side-chain compositions makes hemicellulose structures complex, thus the degradation of hemicellulose, as expected, necessitates the integrated action of a variety of enzymes with corresponding specificities (Sukumaran 2009). Enzymes degrading hemicellulose, in general, can be divided into depolymerizing enzymes that can cleave the backbone and enzymes that remove substituent side chains (Van Dyk and Pletschke 2012).

Considering recent feedstocks of interest, there are basically two broad classes of hemicellulose that are of great significance to industrial-scale biomass conversion to fuels. The first class is comprised of the xylose-based hemicelluloses, which is a group of polysaccharides with β $(1 \rightarrow 4)$ carbon linked with D-xylopyranosyl and varied O-substitutions by acetyl, arabinosyl, glucuronoyl, and other similar units. The second class is comprised of the mannose-based hemicelluloses, which is a group of β $(1 \rightarrow 4)$-D-mannosyl or glucopyranosyl polymers with variable α $(1 \rightarrow 6)$ D-Galactosyl side chains and O_2 or O_3 acetylation (Sukumaran 2009). The presence of either xylose or mannose in the lignocellulosic substrate depends on its source. The main examples of primary plant-based polysaccharides are listed in Table 13.2 and their backbones and side-chain sugar constituents are

Table 13.2 Primary plant polysaccharides and their side-chain and auxiliary sugar constituents.

Polysaccharide	Plant groups	Backbone	Side chain	Auxiliary groups
Cellulose	All	Glucose	None	None
β-Glucans	Grasses	Glucose	None	None
Arabinoxylan	Grasses	Xylose	Arabinose	Acetate, ferulate, cinnamate
Glucuronoarabinoxylan	Grasses, softwoods	Xylose	4-*O*-Methyl-glucuronic acid, arabinose	Acetate, ferulate, cinnamate
Glucuronoxylan	Hardwoods	Xylose	4-*O*-Methyl-glucuronic acid, arabinose	Acetate
Glucomannan	Softwoods, hardwoods	Glucose, mannose	None	Acetate
Galactoglucomannan	Softwoods, hardwoods	Glucose, mannose	Galactose	Acetate
Xyloglucan	Dicot plants	Glucose	Xylose	Acetate
Galacturonan	All	Galacturonic acid	None	Methyl esters
Rhamnogalacturonan	All	Rhamnose, galacturonic acid	Arabinose, fucose, xylose, apiose	Methyl esters, acetate

Source: Adapted from Jovanovic et al. (2009).

detailed. Furthermore, agricultural wastes such as corn stover, despite their seasonality, are of interest to researchers and practitioners in the biofuels industry, mainly because of the sustainability of this raw material and shorter production cycle.

Paralleling the enzyme system used for celluloses, the hydrolysis of hemicellulose requires an arsenal of enzymes. In grasses and hardwoods, the hydrolysis of xylans requires a synergistic two-step action of mainly endo-xylanases (EX), which can cleave linkages of the xylan backbone thereby releasing xylobiose and xylooligosaccharides, and β-xylosidases (BX), which cleave the xylobiose and xylooligosaccharides to release xylose. On the other hand, the hydrolysis of mannans in softwoods requires both endo-mannanases that cleave internal bonds in mannans and β-mannosidases (BM) that can split mannosyl units from the non-reducing ends of manno-oligosaccharides. In addition to these main enzymes that act upon specific hemicellulose backbone structures, other accessory enzymes such as arabinase, pectinase, galactosidase, feruroyl esterase, and acetyl xylan esterase (AXE) are often recommended because these enzymes cleave side chain constituents. These side chains vary widely with the specific type of biomass considered.

13.4.1 Endo-Xylanases and Beta-Xylosidases

EXs include endo-1,4-β-xylanase that cleave internal β $(1 \rightarrow 4)$ bonds and endo- 1,3-β-xylanases that cleave internal β$(1 \rightarrow 3)$ branch bonds in xylan structures. Generally, EX possesses one or more catalytic modules with or without CBM and other domains (Tao and Kazlauskas 2011; Verjans et al. 2010). The active sites are cleft-shaped with four to seven sub-sites, and EXs bind and clip a segment of a xylan chain in an 'on–off' manner (Tao and Kazlauskas 2011). BXs on the other hand are specifically active in cleaving the ends of xylose polymer backbones and short xylose oligomers (xylobiose) to convert them into the sugar monomer xylose. BXs generally have a pocket-like active site, and xylose residues are cleaved off from the non-reducing end of xylo-oligosaccharides (Jeoh et al. 2005; Sweeney and Xu 2012; Tao and Kazlauskas 2011). This activity can be considered as the second and last step of xylan hydrolysis because substrates are primarily provided by the initial action of EX.

EX has catalytic cores belonging to the GH 8, 10, 11, 30, and 43 families, with GH 10 and GH 11 EX being archetypical (Pollet et al. 2010; Sweeney and Xu 2012). BX has catalytic cores that belong to the GH3, 30, 39, 43, 52, and 54 families. Many BXs also have α-L-arabinofuranosidase activity (Sweeney and Xu 2012). GH 10 and 11 EX differ in substrate specificity. The former has a smaller active site and is able to cleave main chains closer to the substituent while the latter has a larger active site and favours splitting main chains in unsubstituted regions (Van Dyk and Pletschke 2012; Vardakou et al. 2004). Furthermore, GH 10 EXs produce shorter oligosaccharides and have greater activity on the substituted xylan since they may also be active in cleaving the $\beta(1 \rightarrow 3)$ branch bonds on xylan structures and have wider side activity on other polysaccharides such as cellulose (Javier et al. 2007; Tao and Kazlauskas 2011; Ustinov et al. 2008). These characteristics of xylanases are believed to assist cellulases in the hydrolysis of biomass. Because of the presence of xylan with cellulose in many biomass feedstocks such as grasses and hardwoods, xylanases are absolutely required for an effective enzyme cocktail (Gupta et al. 2008; Selig et al. 2008; Tabka et al. 2006; Tao and Kazlauskas 2011).

13.4.2 Endo-Mannanases and Beta-Mannosidase

Mannans can be degraded by EM (EC 3.2.1.78) that is particularly active in cleaving the internal $\beta(1 \rightarrow 4)$ bonds in mannans to produce manno-oligosaccharides (Moreira 2008). Similar to EX, EM may also exhibit CBM or other binding domains and one or more catalytic domains of up to six sub-sites (Tao and Kazlauskas 2011). Manno-oligosaccharides produced by mannanases can be stepwise degraded by BM (EC 3.2.1.25) through cleavage of mannosyl units from its non-reducing ends.

EM is a group of widely dispersed, hydrolytic enzymes with catalytic cores in GH5, 26, and 113 families (Moreira 2008; Sweeney and Xu 2012). BM catalytic cores, on the other hand, belong to GH1, 2, and 5 families. A large number of cellulolytic microorganisms co-secrete mannanases with cellulases and other enzymes (Herpoël-Gimbert et al. 2008; Sweeney and Xu 2012).

13.4.3 Other Hemicellulases and Accessory Enzymes

In more popular feedstocks such as hardwoods and grasses, the typical hemicellulase enzyme cocktail for achieving a complete hemicellulose degradation also contains accessory enzymes that are mainly responsible for cleaving xylan branched components and constituents such as acetyl, feruloyl and other hydroxycinnamoyl groups esterified to the backbone or side-chain sugars. Enzymes such as arabinofuranosidase (AF) and arabinase (AS) remove arabinose constituents from the xylose backbone, and the esterases AXE and ferulic acid esterase (FAE) hydrolyse from the xylan the ester-bonded constituents acetic acid and ferulic acid (Sukumaran 2009).

Side chains anchored by the α-L-arabinofuranosyl α-glycosidic bonds can be removed by AF (EC 3.2.1.55) or by AS (EC 3.2.1.99). In addition, AS can also cleave internal $\alpha\ (1 \rightarrow 5)$ glycosidic bonds in arabinan (Sweeney and Xu 2012). The catalytic modules of AFs belong to the families GH 3, 43, 51, 54, and 64, while those from AS to belong to the family GH 43 (Saha 2000; Sweeney and Xu 2012). Different AFs have different specificities, depending on whether the arabinofuranosyl side chain is linked to the O_2 or O_3 site of the xylan unit (Numan and Bhosle 2006; Shallom and

Shoham 2003). Unlike their GH51 counterparts, GH54 AF acts on both polymeric and oligomeric arabinoxylans (Hashimoto 2006; Tao and Kazlauskas 2011). Also, the exo–endo pair of AF and AS synergistically degrades arabinofuranosyl side chains (Tao and Kazlauskas 2011; Yang et al. 2006b), thus facilitating the action of other enzymes on arabinofuranosyl-substituted hemicelluloses such as on xylan units and on pectin constituents (Sørensen et al. 2007; Tao and Kazlauskas 2011).

A common acetyl ester constituent in xylan or other hemicelluloses can be removed by AXEs (EC 3.1.1.72) by deacetylation of substituted O_2 or O_3 sites of the backbone glysosyl units (Biely et al. 2011; Sweeney and Xu 2012). Another constituent of feruloyl oligosaccharides can be removed by the action of FAE (EC 3.1.1.73), which hydrolyzes feruloyl esters at α-L-arabinofuranosyl residues with the linkage at O_2 or O_5 sites of arabinoxylan (Koseki et al. 2009; Sweeney and Xu 2012). AXEs may belong to the CE 1, 2, 3, 4, 5, 6, 7 or 12 families while FAEs are characterized by the CE 1 family. Due to having a number of CBMs together in their catalytic core, FAEs may also display different specificities towards different hydroxyl-cinnamoyl ester bonds, which are found in hemicellulose to lignin attachment (Benoit et al. 2008; Sweeney and Xu 2012).

A valid argument on the importance of these accessory enzymes depends on the intrinsic characteristic of the primary chain cleaving enzyme, which will have a better activity if side chains are initially removed by debranching enzymes on the basis that those generate a steric hindrance to the main chain depolymerizing enzymes (Van Dyk and Pletschke 2012; Vardakou et al. 2004). Based on this idea, several studies have explored the use of debranching enzymes as a form of pretreatment before introducing the main chain cleaving enzymes. However, no definitive conclusion can yet be drawn as some researchers observed that ultimately the positive effect of debranching xylans prior to hydrolysis depends on the specificity of the enzymes in the reaction, as some main backbone cleaving enzymes exhibit preferential activity in the presence of these units (Sørensen et al. 2007; Van Dyk and Pletschke 2012; Vardakou et al. 2004).

13.4.4 Hemicellulases and Complete Hydrolysis

Enzymes action on cellulose is highly site-specific and can be hindered by any obstruction, notably those resulting from cellulose being an integral part of the plant cell wall. Both lignin and hemicellulose have been found to be physical barriers to the hydrolysis of the cellulose (Öhgren et al. 2007; Yang and Wyman 2004; Zhang et al. 2012). Hemicelluloses may obstruct the access to cellulose just by their presence. While it has been reported that even low amounts of residual xylan can limit the extent and the rate of cellulose hydrolysis (Zhang et al. 2012), it has been claimed that xylanase supplementation significantly increases the rate of cellulose hydrolysis in xylan-containing lignocellulosic materials (Kumar and Wyman 2009; Selig et al. 2008; Zhang et al. 2011a). Thus, enzymatic hydrolysis has been found to be affected by the amount of residual xylans in lignocellulosic materials after pretreatment but ultimately the impact of xylan presence can be positive when considering the process as a whole (Kabel et al. 2007; Palonen et al. 2004).

Zhang et al. further proved that hemicelluloses strongly inhibit cellulase activity. In their work, these researchers demonstrated that xylans inhibit the overall hydrolysis efficiencies of both endoglucanase and cellobiohydrolases (Zhang et al. 2012). In another study, they noted a strong inhibition of cellobiohydrolases, especially of CBHII, by xylooligosaccharides (XOS), while a very limited inhibitory effect of XOS on EGII was observed (Zhang and Viikari 2012). In addition, a strong inhibitory effect of birchwood xylan was observed on the hydrolysis by CBHI and CBHII, as shown by a dramatically decreased formation of cellobiose (Béguin and Aubert 1994), which suggested the total hydrolysis of XOS into a less inhibitory product, xylose, to attain more efficient hydrolysis of cellulose (Zhang and Viikari 2012). Kumar and Wyman also found xylobiose and other XOS with higher degrees of polymerization inhibited the enzymatic hydrolysis of various polysaccharides (Kumar and Wyman 2009; Zhang et al. 2012). In a similar work, Qing et al. found that xylose, xylan, and XOS reduce initial hydrolysis rates by 9.7%, 34.5% and 23.8%, respectively (Qing et al. 2010). A more recent work reported by Baumann et al. suggested that XOS binds to CBHI and that the affinity increases correspondingly with XOS length and results in a negative effect, which suggests that a competitive mechanism of XOS to CBH is at play (Baumann et al. 2011).

13.5 Microorganisms for Biomass Hydrolysis

13.5.1 Diversity of Cellulolytic Microorganisms and Lifestyles

Cellulolytic microorganisms have been isolated from various habitats where cellulose is widely available, such as soil, aquatic environments, and the digestive tracts of animals and insects. In nature, cellulolytic microorganisms usually coexist with other cellulolytic species and non-cellulolytic microorganisms to form a consortium that efficiently utilizes cellulose. Many cellulolytic microorganisms are able to degrade primary plant cell wall components such as cellulose, hemicellulose, and pectin (Wei et al. 2009; Wilson 2008). The degradation of peripheral hemicellulose and pectin results in an increase in the surface area of cellulose that is exposed to microbes; moreover, products of the degradation of hemicelluloses and pectin supply nutrients for non-cellulolytic microorganisms that in return provide growth factors. Aerobic microorganisms, including fungi *Trichoderma reesei, Trichoderma longibrachiatum* and bacteria including *Thermobifida fusca* and *Cellulomonas fimi*, produce non-complexed, cell-free single cellulase components. The cellulases from *T. reesei* have been the most extensively studied. Fungal cellulases usually consist of exoglucanases and endoglucases. These cellulase components act synergistically to hydrolyze cellulose. Commercial cellulases produced from *T. reesei* are readily available but great efforts have continually been devoted to increasing the specific activity of fungal cellulases (Schulein 2000).

Anaerobic bacteria, which are commonly found in the digestive system of ruminants, have a special yet lesser understood mechanism of cell wall degradation using its multi-enzyme complex cellulosome. Cellulosomes are high-activity multienzyme complexes that have the ability to hydrolyze crystalline cellulose and other polysaccharides in plant cell walls (Hyeon et al. 2013; Schwarz 2001). Cellulosome systems have been mostly observed and studied among cellulolytic *Clostridium* species but the systems have been also identified in certain anaerobes under *Acetivibrio* and *Bacteriodes* (Blumer-Schuette et al. 2008). Cellulosome-related signature sequences have also been described in many other microorganisms, such as anaerobic fungi and aerobic bacteria (Shoham et al. 1999). Species from the family of the *Syntrophomonodaceae, Lachnospiraceae, Eubacteriaceae*, and *Clostridiacesae* isolated from hot springs, rumen, or soil and sewage, were also found to exhibit cellulosome-like enzymes to degrade cellulose (Schwarz 2001).

Perhaps the most interesting model to study diversity and synergy of cellulolytic organisms is the digestive system of ruminants. Rumen microorganisms, including 30 predominant bacterial species, approximately 40 species of protozoa, and five species of fungi, degrade forage cell walls in a diverse and complex ecologic system. Synergistic effects among species have been observed to influence fibre digestion (Miura et al. 1983). The major prominent and active fibre-digesting microbial groups are bacteria comprise *Ruminococcus flavefaciens, Ruminococcus albus,* and *Fibrobacter succinogenes.* Microbial populations associated with feed particles are pivotal for feed digestion in the rumen. Mixed cultures of these bacteria species closely adhere to the fibre and can penetrate entirely the intact plant cell wall. Relative proportions of attached bacteria (e.g. *R. flavefaciens, R. albus, and Bacteriodes succinogenes*) were observed to be in the ranges 27–48%, 21–57%, and 35–59%, respectively, depending on the specific strains of bacteria (Minato and Suto 1985). Adhesion of 100% rumen cellulolytic bacteria cells on cellulose has, however, not been observed. Firm adhesion of cellulolytic rumen bacteria was demonstrated to be essential for cellulose degradation (Miron et al. 2001). Some non-cellulolytic rumen bacteria are also able to attach to cellulose, which were presumed to benefit from the efficient utilization of soluble sugars from cellulolytic bacteria and provide essential growth factors for cellulolytic bacteria (Minato and Suto 1985).

It has been shown that *R. albus* and *R. flavefaciens* produce a cellulosome (Ding et al. 2001; Miron et al. 2001; Morrison and Miron 2000; White and Morrison 2001). The cellulosome from *R. albus* is comparable in size to that of *C. thermocellum* (Kim et al. 2001).

13.5.2 Fungi and Their Arsenal for Biomass Hydrolysis

Fungi, mostly of the aerobic kind, typically produce extracellular non-complexed or free enzymes that can act in synergy for plant cell wall degradation. Fungal enzymes are a promising resource in second-generation biofuel processing at an industrial scale because of their capability to secrete large amounts of free enzymes. Sweeney et al. reported that secreted proteins from a typical cellulolytic fungi are nearly 70% CBH (w/w) and 20% (w/w) EG, but only less than 1% (w/w) hemicellulases (Sweeney and Xu 2012).

Table 13.3 lists well-known fungal species used in cellulosic deconstruction. Cellulolytic fungi, belonging to the group of Ascomycetes, *Trichoderma* (*Trichoderma viride*, *Trichoderma longibrachiatum*, *T. reesei*) have long been considered to be the most productive and powerful degraders of crystalline cellulose (Gusakov 2011; Kubicek et al. 2009; Margeot et al. 2009; Merino and Cherry 2007). The typical complete set of enzymes for biomass cell wall degradation from *T. reesei* includes cellobiohydrolases, endoglucanases, cellobiase, xylanases, and several other accessory hemicellulases (Vinzant et al. 2001). Cellulase components synergistically act on insoluble substrates, especially crystalline cellulose (Cruys-Bagger et al. 2012; Fox et al. 2011a; Henrissat et al. 1985; Igarashi et al. 2011; Kanda et al. 1980; Ye and Berson 2011). Commercial cellulase enzymes, primarily CBHs and EGs, which are prepared from mutant strains of *T. reesei* (also known as its anamorph *Hypocrea jecorina*), are produced on an industrial scale by many companies worldwide (Gusakov 2011; Merino and Cherry 2007; Nieves et al. 1997). Among cellulase-producing strains of *T. reesei*, the RUT C30 strain is one of the most powerful and best characterized strains, and has become a standard strain among *T. reesei* high cellulase producers (Le Crom et al. 2009). The primary natural cellulase system produced by *T. reesei* was observed to be 12% EG I, 60% CBH I and 20% CBH II by weight (Zhang and Lynd 2004, 2006; Zhou et al. 2010). The most known non-complexed cellulase system of *T. reesei* consists of two exoglucanases (i.e. CBHI and CBHII), about eight endoglucanases (i.e. EGI to EGVIII), and seven β-glucosidases (i.e. BGI to BGVII) (Aro et al. 2005; Sukumaran 2009).

Another ascomycete of equal industrial significance is *Aspergillus niger* which is indispensable because of its high-level secretion of hemicellulases, which are generally lacking in *Trichoderma* strains. Most enzyme research related to *Aspergilli* has been directed on the degrading hemicellulose and pectin, whereas, cellulases asidc, these have been studied in lesser extents (de Vries and Visser 2001). However, *Aspergillus* species might be useful as a source of β-glucosidase and other accessory enzymes to cellulases, such as xylanases, xyloglucanases, and α-L-arabinofuranosidases (Berlin et al. 2005; de Vries and Visser 2001). Van den Brink et al. list hemicellulases from

Table 13.3 Lists of fungal species known for cellulose hydrolysis.

Microorganism	Enzyme type[a)]	Tolerance[a)]	References
Filamentous fungi (Aerobic)			
Acrophialophora nainiana	NC/HC	M	Ximenes et al. (1999)
Aspergillus acculeatus	NC/HC	M	Adisa and Fajola (1983)
A. fumigatus	NC/HC	M/T	Reese et al. (1950), Vandamme et al. (1982)
A. niger	NC/HC	M	Li and King (1963)
A. oryzae	NC/HC	M	Jermyn (1952)
Fusarium solani	NC/HC	M	Wood and Phillips (1969)
Humicola grisea var. *thermoidea*	NC/HC	T	Fergus (1969)
Irpex lacteus	NC/HC/LN	M	Nisizawa (1955)
Laetiporus sulphureus	NC/HC	M	Valadares et al. (2016)
Myceliophthora thermophile	NC/HC/LN	T	Karnaouri et al. (2014)
Penicillium funmiculosum	NC/HC	M	Wood and McCrae (1982)
P. echinulatum	NC/HC	M	Domsch and Gams (1969)
P. citrinum	NC/HC	T	Olutiola (1976)
Phanerochaete chrysosporium	NC/HC/LN	M	Saddler (1982)
Pleurotus ostreatus	NC/HC	M	Valadares et al. (2016)
Schizophyllum commune	NC/HC	M	Jurasek et al. (1968)
Sclerotium rolfsii	NC/HC	M	Bateman (1969), (1972)
Sporotrichum cellulophilum	NC/HC	T	Komura et al. (1978)
Talaromyces cellulolyticus	NC/HC	T	Inoue et al. (2014)
T. emersonii	NC/HC	T	Folan and Coughlan (1978)
Thermoascus aurantiacus	NC/HC	T	McCleandon et al. (2012)
Thielavia terrestris	NC/HC	T	Skinner and Tokuyama (1978)
Trichoderma koningii	NC/HC	M	Wood and Phillips (1969)
T. reesei	NC/HC	M	Gong et al. (1979)
T. viride	NC/HC	M	Reese and Levinson (1952)
Anaerobic fungi			
Anaeromyces elegans	NC/HC	M	Ho et al. (1990)
A. mucronatus	NC/HC	M	Fliegerova et al. (2002)
A. robustus	NC/HC	M	Solomon et al. (2016)
Caecomyces CR4	NC/HC	M	Matsui and Ban-Tokuda (2008)
Neocallimastic californiae	Cellulosome/NC/HC	M	Solomon et al. (2016)
N. frontalis	Cellulosome	M	Wilson and Wood (1992)
N. hurleyensis	NC/HC	M	Fanutti et al. (1995)
N. Patriciarum	Cellulosome/HC	M	Pai et al. (2010)
Orpinomyces joyonii	Cellulosome	M	Qiu et al. (2000)/Steenbakkers et al. (2001)
O. PC-2	Cellulosome	M	Li et al. (1997)
Piromyces communis	Cellulosome	M	Fanutti et al. (1995)/Li et al. (1997)
P. equi	Cellulosome	M	Raghothama et al. (2001)
P. E2	Cellulosome	M	Dijkerman et al. (1996)
P. finnis	Cellulosome/NC/HC	M	Solomon et al. (2016)

a) AT, Alkali tolerant; HC, Hemicellulase; LN, Ligninase; M, Mesophilic; NC, Noncomplexed cellulase; T, Thermophilic. Adapted from Biofuels, Vol.2 Issue 4, Bin Yang, Ziyu Dai, Shi-You Ding, and Charles E. Wyman, Enzymatic hydrolysis of cellulosic biomass, with permission from Taylor and Francis Online (Journal website: http://www.tandfonline.com/doi.org/10.4155/bfs.11.116).

A. niger to include two endoxylanases (EX), one β-xylosidase (BX), one endomannanase (EM), one β-mannosidase [BM], two α-galactosidases, one β-galactosidase, one β-glucuronidase, one AXE, and two FAE (van den Brink and de Vries 2011).

Basidiomycetes include the wood-rotting fungus *P. chrysosporium*. This strain is well-documented for its property of efficiently degrading lignocelluloses; what is more, it secretes an array of enzymes that hydrolyse cellulose and hemicellulose (Broda et al. 1996; Igarashi et al. 2008). The culture of *P. chrysosporium* grown in the presence of cellulose contains CBHs belonging to GH families 6 and 7 (Uzcategui et al. 1991a), and four EGs in GH families 5 and 12 (Henriksson et al. 1999; Uzcategui et al. 1991b), indicating that basidiomycetes might have cellulolytic systems very similar to that of popular ascomycetes. Igarashi et al. (2008) further characterized the GH 45 protein from *P. chrysosporium* as an EG, showing synergistic capacities with other cellulases.

Less popular, but equally important, candidates for fungal non-complexed enzymes secretion that have been gaining equal attention include the *Penicillium*, *Fusarium*, and *Humicola* species. High β-glucosidase activity in cellulase systems secreted by the *Penicillium* species has been pointed out to have a strong advantage over the *T. reesei* system by a number of researchers (de Castro et al. 2010; Skomarovsky et al. 2006). Similarly, the functional properties of a number of xylanases under the GH 10 and GH 11 families from the wild type fungus *Fusarium oxysporum* have been described, indicating that this fungus is a potential xylanase producer of industrial importance (Gómez-Gómez et al. 2001; Huang et al. 2015). Huang et al. published the genomes from three *Fusarium* species that have a broad range lignocellulolytic activity (Huang et al. 2015). In their comparative analyses, the cellulose and hemicellulose degrading enzymes armament from *Fusarium* species show considerably more β-glucosidases, β-xylosidases, endo-xylanases, endoglucanases, xyloglucanases, and exomannanases than the other test fungi including *A. niger* and *T. reesei*. Lastly, *Humicola insolens*, was found to be homologous to *T. reesei* and contains at least seven cellulases (Sukumaran 2009). Furthermore, *H. insolens* cellulases was shown to efficiently degrade crystalline cellulose (by more than 50%) (Lynd et al. 2002). Current commercial cellulase preparations based on *H. insolens*,however, are mainly for applications such as textile finishing, cleaning cellulase in detergent but not for bioethanol production (Gusakov 2011; Schülein 1997).

In anaerobic fungi, strains of *Anaeromyces* and *Caecomyces* were also found to produce both non-complexed cellulases and hemicellulases. Generally, however, anaerobic fungi are considered to be cellulosome producers. Notable species include the newly described *Neocallimastic*, *Orpinomyces*, and *Piromyces* species. Some of the *Chytridiomycetes* also have multi-enzymatic complexes similar to the cellulosomes of bacteria (Dashtban et al. 2009; Eberhardt et al. 2000; Lynd et al. 2002; Quiroz-Castañeda and Folch-Mallol 2013; Sánchez 2009). Cellulosomes are mainly an attributed feature of anaerobic bacteria and its mechanism is explained in detail in the next section.

13.5.3 Bacteria and Their Cellulolytic Machinery

In aerobic bacteria, the species *Bacillus*, *Cellulomonas*, and *Pseudomonas* are among the most widely used for cellulose hydrolysis. In addition, various aerobic bacteria belonging to the genera *Geobacillus*, *Erwinia*, *Streptomyces*, *Fibrobacter*, as well as *Paenibacillus* are also known to produce diverse kinds of cellulases (Maki et al. 2009; Sethi and Scharf 2013). Similar to aerobic fungi, this group of bacteria externally secretes substantial quantities of non-complexed enzymes that can be used as part of a cell wall-degrading cocktail in a very similar manner to fungal free enzymes. In fact, the mixing of fungal and bacterial free enzymes is one of the possibilities being looked at by researchers. For example, Gao et al. reported the use of a combined mixture of fungal cellulases and bacterial hemicellulases that have effectively maximized the saccharification of AFEX-treated corn (Gao et al. 2010). In a follow-up work, Gao et al. found that the synergism between bacterial hemicellulases and core fungal cellulases results in high glucose (80%) and xylose (70%) and results in a relatively better yield compared to commercial enzymes (Gao et al. 2011).

The most common bacterial strains that are capable of biomass deconstruction are listed in Table 13.4. In these bacteria, different types of cellulose-degrading enzymes and hemicellulolytic enzymes are assembled on the structural scaffolding subunits through strong non-covalent protein–protein interactions between the docking modules called dockerins and the complementary modules termed cohesins (Dashtban et al. 2009). Other thermophilic bacteria, such as *Caldicellulosiruptor* sp., use an additional strategy as they can secrete many free cellulases that contain numerous CDs. In particular, *Caldicellulosiruptor bescii* CelA structure shows GH 9 and a GH 48 CDs, along with three type III carbohydrate-binding modules, drives cellulose hydrolysis, not only through the well-known surface ablation mechanism, but also through excavation into the surface resulting in the formation of cavities in the substrates (Brunecky et al. 2013).

13.5.3.1 The *C. thermocellum* Cellulosome

Evidence of cellulosome activity was first reported in *C. thermocellum* (Lamed et al. 1983b). *C. thermocellum* is an anaerobic, gram-positive, thermophilic and cellulolytic bacterium. Optimal growth is observed at 60–64 °C. In natural environments, *C. thermocellum* commonly associates with other bacteria to form microbial communities, including some anaerobic, thermophilic hemicellulolytic and saccharolytic species. Arguably, the best characterized cellulosome is that produced by *C. thermocellum*. Another well studied cellulosome is that from *Clostridium cellulolyticum* and *Clostridium cellulovorans*. *C. cellulolyticum* structure and the arrangement of its genes are notably different from those of *C. thermocellum* (Belaich et al. 1997) while that of *C. cellulovorans* consists of about 10 catalytic proteins that degrade not only cellulose but also xylan, mannan, lichenan, and pectin (Doi et al. 1994; Tamaru and Doi 1999). The *C. thermocellum* cellulosome has the largest molecular weight and the greatest complexity (Figure 13.4) relative to those described for other species. It contains a non-catalytic scaffolding protein, which readily forms a complex with a large number of glycosyl hydrolases, including cellulases, hemicellulases (e.g. xylanases, mannanases), and carbohydrate esterases. The scaffolding, also known as the cellulosome-integrating protein CipA, has multiple functions besides organizing the enzyme subunits, such as cellulose-binding through CBM and cell-anchor. Schwarz did an intensive review on anaerobic bacteria cellulose degradation and features great deal of information about cellulome *C. thermocellum and other Clostridia species* (Schwarz 2001).

C. thermocellum ATCC 27405 was observed to exhibit substantially the highest first-order cellulose hydrolysis rate constant and high specific growth rate when comparing kinetic parameters for microbial cellulose utilization by various microorganisms, including bacteria and fungi (Lynd et al. 2002). Therefore, *C. thermocellum* is a potential starting point for developing microorganisms for the conversion of cellulosic biomass to ethanol. Numerous studies on batch and continuous cultures of *C. thermocellum* on pure cellulose (Zhang and Lynd 2003; Zhang and Lynd 2005b), soluble sugars (e.g. cellobiose, glucose, and cellodextrin) (Strobel 1995; Zhang and Lynd 2005a) and pretreated cellulosic biomass (Lynd et al. 1989) have been reported. *C. thermocellum* shows a preference for cellobiose over glucose (Strobel 1995). It was found that *C. thermocellum* consumes cellodextrin at an average degree of polymerization of four while it utilizes cellulose, which provides bioenergetic benefits for anaerobic growth on cellulose (Zhang and Lynd 2005a). These results indicate that the mechanism of cellulose hydrolysis by microorganisms is markedly different from that of fungal cellulases.

Table 13.4 List of bacterial strains known for cellulose hydrolysis.

Microorganism	Enzyme type[a]	Tolerance[a]	References
Bacteria (Aerobic)			
Alicyclobacillus cellulosilyticus	NC/HC	T	Kusube et al. (2014)
Acidothermus cellulolyticus	NC/HC	T	Mohagheghi et al. (1986)
Bacillus sp.	NC/HC	M/AT	Fukumori et al. (1986)
Bacillus pumilus	NC/HC	M	Panbangred et al. (1983)
B. sp. SMIA-2	NC/HC	T	Ladeira et al. (2015)
B. substilis	NC/HC	M/T	Murphy et al. (1984)
B. agaradhaerens JAM-KU023	NC/HC	T/AT	Hirasawa et al. (2006)
Brevibacillus sp. strain JXL	NC/HC	T	Liang et al. (2009)
Cellulomonas flavigena	NC/HC	T/AT	Perez et al. (2008)
Geobacillus thermoleovorans	NC/HC	T/AT	Tai et al. (2004), Sharma and Satyanarayana (2006)
Paenibacillus campinasensis BL11	NC/HC	T	Ko et al. (2007)
P. strain B39	NC	T	Wang et al. (2008)
Pseudomonas sp	NC	M	Khatiwada et al. (2016)
Serrati asp.	NC/HC	M	Khatiwada et al. (2016)
Streptomyces sp.	NC/HC	M/T	Hankin and Anagnostakis (1977)
Thermoactinomyces sp.	NC/HC	T	Su and Paulavicius (1975)
Thermomonospora curvata	NC/HC	T	Stutzenberger (1979)
T. fusca	NC/HC	T	Crawford and McCoy (1972)
Bacteria (Anaerobic)			
Acetivibrio cellulolyticus	Cellulosome	M	Ding et al. (1999)
Bacteroides cellulosolvens	Cellulosome	M	Lamed et al. (1991)
Caldicellulosiruptor bescii	NC/HC	T	Yang et al. (2010), Brunecky et al. (2013)
Clostridium acetobutylicum	Cellulosome	M	Sabathe et al. (2002)
C. cellobioparum	Cellulosome	M	Lamed et al. (1987)
C. cellulolyticum	Cellulosome	M	Bagnara-Tardif et al. (1992)
C. cellulovorans	Cellulosome	M	Shoseyov et al. (1992)
C. josui	Cellulosome	M	Kakiuchi et al. (1998)
C. papyrosolvens	Cellulosome	M	Garcia et al. (1989)
C. thermocellum	Cellulosome	T	Bayer et al. (1983), Lamed et al. (1983a)
Ruminococcus albus	Cellulosome	M	Ohara et al. (2000)
R. flavefaciens	Cellulosome	M	Kirby et al. (1997)

a) AT, Alkali tolerant; HC, Hemicellulase; LN, Ligninase; M, Mesophilic; NC, Noncomplexed cellulase; T, Thermophilic. Adapted from Biofuels, Vol.2 Issue 4, Bin Yang, Ziyu Dai, Shi-You Ding, and Charles E. Wyman, Enzymatic hydrolysis of cellulosic biomass, with permission from Taylor and Francis Online (Journal website: http://www.tandfonline.com/doi.org/10.4155/bfs.11.116).

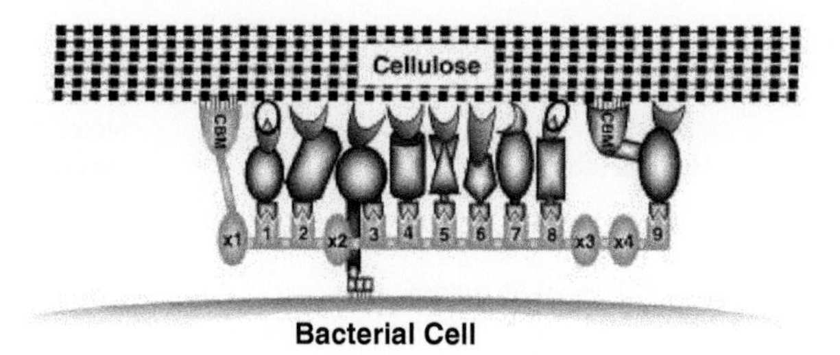

Note: Scaffoldins is drawn in light gray and the enzyme components in dark gray. Catalytic sites are drawn as open bowls for EGs and ellipsoids for CBHs. Image not on scale). Reprinted from Applied Microbiology and Biotechnology, Vol.56 Issue 5, W. Schwarz, The cellulosome and cellulose degradation by anaerobic bacteria, with permission of Springer.

Figure 13.4 Schematic presentation of theoretical cellulosome from *Clostridium*. Note: Scaffoldins is drawn in light grey and the enzyme components in dark grey. Catalytic sites are drawn as open bowls for EGs and ellipsoids for CBHs. (Image not to scale.) Source: Reprinted from Applied Microbiology and Biotechnology, Vol. 56 Issue 5, W. Schwarz, The cellulosome and cellulose degradation by anaerobic bacteria, with permission of Springer.

The cellulosome from *C. thermocellum* is bound to the cell surface via a type-II dockerin domain in the scaffolding. The cellulosome is tightly associated with cells except in the later stages of batch culture. Cellulosomes on the cell surface of *C. thermocellum* are typically arranged as polycellulosomal protuberance-like organelles, which comprise multiple copies of cellulosome and an interior matrix of fibrous material (Bayer et al. 1998b). When binding to cellulose, elongated fibres reformed between cellulose and the cell surface. Lamed et al. 1991 speculated that these fibres might play a role in capturing hydrolysis products (Bayer and Lamed 1986).

Key components of cellulosome are shown in Table 13.5. Cellulase components in the cellulosome include two cellobiohydrolases with different directions of processivity (GHF 48 and 9), and various endoglucanases. CelA accounts for 70% of the endoglucanase activity of the *C. thermocellum* (Schwarz and Staudenbauer 1986). The CelS subunit, which displays exoglucanase and endoglucanase activity, is the main catalytic component of the cellulosome (Morag et al. 1991). The cellulosome has been reported to be particularly effective in solubilizing crystalline cellulose (Boisset et al. 1999).

13.5.3.2 Enzyme-Microbe Synergy

Enzyme-microbe synergy has been proposed by several authors as a driving concept towards superior efficacy. It has been found that the hydrolysis rates of cellulose are higher in growing cultures of *T. reesei* than in cell-free enzyme preparations, suggesting that the association of hyphae with cellulose results in the higher hydrolysis rate in the growing culture (Reese and Mandels 1971). Rolz further concluded that the structure of cell-associated enzyme complex promotes enzyme stability and protection against catabolite repression and provides a natural shield to avoid excessive loss of enzymes via diffusion (Rolz 1986). In anaerobic culture, the cellulosome mediates a close contact between cell and substrate thus minimizing diffusion losses of hydrolytic products, which ultimately benefits the cell (Lamed et al. 1987). In contrast, Shoham et al. (1999) hypothesized that cellobiose (i.e. hydrolysis products) would be maintained at appropriate concentrations for the most efficient use by the cell, thus avoiding cellobiose accumulation and dissipation away from the cell. Some studies may indicate the existence of enzyme-microbe synergism although there has been no direct evidence of such a synergy. It was reported that the mean normalized rate

Table 13.5 Known components of the cellulosome of *Clostridium thermocellum*[a].

Cellulosome components[b]	Description	Modular structure	Molecular mass (kDa)
CipA[c]	Scaffoldin	2(Coh1)-CBM3a-7(Coh1)-X2-DocII	197
CelJ	Cellulase	X-Ig-GH9-GH44-DocI-X	178
CbhA	Cellobiohydrolase	CBM4-Ig-GH9-2(Fn3)-CBM3b-DocI	138
XynY	Xylanase	CBM22-GH10-CBM22-DocI-CE1	120
CelH	Endoglucanase	GH26-GH5-CBM9-DocI	102
CelK	Cellobiohydrolase	CBM4-Ig-GH9-DocI	101
XynZ	Xylanase	CE1-CBM6-DocI-GH10	92
CelE	Endoglucanase	GH5-DocI-CE2	90
CelS[c]	Exoglucanase	GH48-DocI	83
CelF	Endoglucanase	GH9-CBM3c-DocI	82
CelN	Endoglucanase	GH9-CBM3c-DocI	82
CelQ	Endoglucanase	GH9-CBM3c-DocI	82
CelO	Cellobiohydrolase	CBM3a-GH5- DocI	75
XynA, XynU	Xylanase	GH11-CBM4-DocI-NodB	74
CelD	Endoglucanase	Ig - GH9 – DocI	72
XynC	Xylanase	CBM22-GH10-DocI	70
XynD	Xylanase	CBM22-GH10-Doc1	70
ManA	Mannanase	CBM-GH26-DocI	67
CelT	Endoglucanase	GH9 – DocI	65
CelB	Endoglucanase	GH5 – DocI	64
CelG	Endoglucanase	GH5 – DocI	63
CseP	Unknown	UN – DocI	62
ChiA	Chitinase	GH18 – DocI	55
CelA	Endoglucanase	GH8 – DocI	53
XynB, XynV	Xylanase	GH11-CBM4-DocI	50
LicB	Lichenase	GH16 – DocI	38

Components are sorted according to their molecular mass.
a) Modified and updated from Schwarz (2001).
b) Components in bold letters: Localization in the cellulosome was experimentally shown.
c) Main components. Abbreviations: CE, carbohydrate esterase; CBM, substrate binding module; Coh1, typ-I cohesin module; DocI, typ-I dockerin module; FAE, ferulic acid esterase; GH, glycosyl hydrolase (family); Ig, immunoglobulin-like module; UN, unknown module; X, hydrophobic module with unknown function.

constant of cellulose hydrolysis at pH 5.75 is lower than 30%, while it is about 60% at pH 6 in the *in vitro* fermentation of pure cellulose by mixed ruminal microorganisms (Mourino et al. 2001). At pH 6 or higher, ruminal cellulolytic bacteria (RCB) cells grow and cellulose is hydrolyzed by cellulose-enzyme-microbe (CEM) complex. On the other hand, at pH below 6, RCB cannot grow. The adsorbed enzyme continues to hydrolyze cellulose and is even more effective at pH below 6, meanwhile, the acid-tolerant non-cellulolytic cells uptake hydrolytic products of enzymatic cellulose hydrolysis under this condition. Thus, the fact that the rate constant is higher when the CEM complex is functioning suggests that the cellulose utilization by the cell-enzyme complex is more effective than enzyme alone. However, cellulases produced by RCB are covered with glycocalyx and are held tenaciously onto cells so that the direct measurement of cellulase concentration is fairly difficult. Although the evidence is quite clear, there has been no direct quantitative evidence reported to demonstrate the proposed microbe-enzyme synergy. Although a good methodological foundation at the molecular level has been established for microbial physiology studies involving organisms growing on soluble substrates, such methodology for microbial cellulose utilization studies have not been available since the ELISA method to determine the concentration of *C. thermocellum* cells and cellulosome in the presence of cellulose was recently developed (Lynd and Zhang 2002; Zhang and Lynd 2003). In 2006, an enzyme-microbe synergy was also noted by Lu et al. who found that the hydrolysis rates of *C. thermocellum* cultures are up to 4.7 times higher than in pure cellulase under a simultaneous saccharification and fermentation setup (Lu et al. 2006). These researchers further suggested that the presence of cellulose adherent cellulolytic microbe promotes faster hydrolysis by significantly reducing local concentration of inhibitory hydrolysis products – a mechanism coherent with many reported studies.

13.5.3.3 Mechanisms of Cell Adhesion

Several factors (related to bacteria, substrate, and the environment) affect bacterial adhesion on cellulose, including cell age, envelope condition and microbial competition. Pretreatment, such as fine grinding, was reported to improve cell adhesion (Weimer and Schmidt 1989). In addition, the temperature has a profound effect on bacteria adhesion to cellulose. Strains of *B. succinogenes*, *F. flavefaciens*, and *R. albus* adhere to cellulose at 38 °C but not at 4 °C., whereas *Eubacterium cellulosolvens*, *Megasphaera elsdenii*, *Veillonella alcalescens* and *Veillonella parvula* adhere well to cellulose at both 38 °C and 4 °C (Minato and Suto 1985). The effects of pH on bacterial adhesion on cellulose depend on the species and it was reported that cell adhesion is almost unaffected over a range of pH values ranging from 6 to 8 (Miron et al. 2001). The presence of carbohydrates was also found to influence cell adhesion on cellulose. Methylcellulose can almost completely detach attached cells from cellulose, while amylopectin partially inhibits cell adhesion. Cellobiose inhibits the attachment of *B. succinogenes* cells to cellulose (Minato and Suto 1985) while the presence of cations, such as Ca^{2+}, was reported to enhance the cell adhesion of *C. thermocellum* (Shoham et al. 1999).

The adhesion process of the predominant cellulolytic rumen bacteria (i.e. *F. succinogenes, R. flavefaciens,* and *R. albus*), which is similar to the general mode for microbial adhesion to solid substrates that ultimately results in the formation of biofilms, was proposed to occur in four phases: (i) transport of bacteria to the substrate; (ii) initial non-specific adhesion of bacteria to accessible substrate sites mediated by constitutive elements of bacterial glycocalyx; (iii) specific adhesion via adhesin or ligand formation with the substrate; (iv) proliferation of the attached bacteria on potentially digestible tissues of the substrate. This process may be applicable to other cellulolytic bacteria, such as *C. thermocellum*. However, the mechanism of cell adhesion on cellulose might be different for different species (Miron et al. 2001).

The mechanisms of adhesion are not thoroughly understood. Adhesins have been proposed as playing a key role in the adhesion of cellulolytic bacteria on cellulose, such as polycellulosome complexes, fimbriae or pili, glycocalyx capsule, cellulosic fibrils, cellulose binding proteins, and enzyme binding domains. It is believed that cells adhere to cellulose by the cellulose binding domain on their enzymes. For *C. thermocellum,* the adhesion of bacterial cellulosome to cellulose is mediated by the CBM of the scaffolding plus CBD of enzymes connected to the scaffolding. Distinct cellulose binding domains of endoglucanases have been identified in *F. succinogenes* (Gong et al. 1996) and *R. albus* F-40 (Karita et al. 1997). A novel form of cellulose-binding protein (cbpC, 17.7 kDa), which belongs to the Pil protein family, was identified in *R. albus 8* (Miron et al. 2001; Morrison and Miron 2000).

The adhesion of cells on solid substrates mediated by glycocalyx surrounding an individual cell or a colony of cells have been commonly observed in many bacterial species, such as rumen bacteria (Weimer and Odt 1995), or cavity-forming bacteria. The term 'glycocalyx' is used to refer to the bacterial polysaccharide-containing structures lying outside the integral elements of the outer membrane of Gram-negative cells and the peptidoglycan of Gram-positive cells. Glycocalyces can be divided into two types: (i) layers composed of a regular array of glycoprotein subunits; (ii) capsules composed of a fibrous matrix at the cell surface which can be rigid, flexible, integral or peripheral (Rolz 1986). Although the generation and maintenance of a glycocalyx might be energy consuming for bacteria, the glycocalyx can provide essential factors for bacterial growth, as follows: (i) adhere to its food and determine particular locations of bacteria in most natural environments; (ii) conserve and concentrate catalytic enzymes to constantly renew supply of organic nutrients; (iii) resist removal of bacteria by environmental forces; (iv) serve as a food reservoir and mediate nutrient transportation from the solid substrate to cells; (v) provide a micro-environment for bacteria and protect bacteria from their predators and toxic stress. The glycocalyx was thought to be involved in initial non-specific adhesion and sequential specific adhesion of cellulolytic bacteria on cellulose (Miron et al. 2001).

13.6 Perspectives

The mechanism and limiting factors of enzymatic hydrolysis of lignocellulosic biomass have been extensively studied for decades and substantial progress has been made in improving the efficacy of this still expensive step in biomass conversion processes. Non-complexed cellulases produced by a mutant strain of *T. reesei* are commercially available while increasing its specific activity is the subject of great R&D effort. Synergies have been found among components of such non-complexed cellulases system, including EG, CBH, ß-glucosidase, and newly found PMO and CDH. Although modelling studies brought perspectives of optimum enzyme blend and the enzyme–substrate interaction during hydrolysis, it remains a challenge to reveal the optimized combination of EG, CBH, ß-glucosidase, PMO, and CDH. In addition to enzyme-related factors, substrate-related factors (e.g. CrI, DP, adsorption, accessible surface area, etc.) and environmental conditions (e.g. pH, temperature, ionic strength, inhibitors, xylan derivatives) were found to be critical to cellulose hydrolysis. However, because of a lack of analytical tools to monitor the dynamic enzyme–substrate interactions during hydrolysis, the exact mechanism behind cellulose hydrolysis kinetics of rapid slow-down and prolonged completion is still unclear although some theories, including limited processive movement of CBH on cellulose, have been proposed. Unlike pure cellulose, the complex intertwining structure of hemicellulose and cellulose in lignocellulosic biomass requires that the degrading microorganism produce both cellulases and hemicellulases while the proportion of these enzymes varies among distinct species. In addition to the synergy found among components of cellulases and hemicellulases, an enzyme-microbe synergy is observed, especially for microorganisms that produce a complexed enzyme system called cellulosome. Cells of many such microorganisms adhere on lignocellulosic substrates and some are reportedly able to penetrate the biomass cell wall although the adhesion process is not thoroughly understood. Commercial application of such a biomass hydrolysis system has not yet emerged.

In fact, the knowledge from the current set of commonly used enzymes is insufficient to plan a general route with flexibility among feedstock inputs. In addition, many genes, currently annotated as 'other' or 'hypothetical', could provide highly active novel enzymes, either alone, in sequence, or in combination. Overall research targets of biological deconstruction cellulosic biomass are to continuously explore diverse enzymes and microbes to discover new and more effective biocatalysts and to better understand substrate–biocatalyst interactions. Some key approaches to develop innovative technologies to overcome the recalcitrance of cellulosic biomass at high efficiency and low cost are listed below:

- Advanced technologies for the discovery, characterization, and over-expression of glycosyl hydrolases.
- DNA sequence technology for whole genome sequencing of biomass feedstocks, biocatalystic microbes, metagenomics, functional genomics and metagenomics, can build a foundation for advances in biomass production, enzyme categorization and applications.
- Mass spectrometry technology has been applied to examine secretomes and subcellular organelle proteomics of biocatalystic microbes and protein glycosylation.
- Gene transfer technologies improve glycosyl hydrolase production in both homologous and heterologous organisms.
- Cellulase engineering through directed evolution, rational design, post-translational modifications, and their combination may greatly increase cellulase performance and dramatically decrease enzyme use.

Acknowledgement

This work is supported by the USDA/NIFA through Hatch/Multi State Project # WNP00820, U.S. Department of Energy (DOE), the Office of Energy Efficiency & Renewable Energy (EERE) Awards (DE-EE0007104, DE-EE0006112, and DE-EE0008250), and the Bioproducts, Scienceand Engineering Laboratory, Department of Biological Systems Engineering at Washington State University. NREL-authored documents are sponsored by the US Department of Energy under Contract DE-AC36-08GO28308. Accordingly, with respect to such documents, the US Government and others acting on its behalf retain a paid-up nonexclusive, irrevocable world-wide license to reproduce, prepare derivative works, distribute copies to the public, and perform publicly and display publicly, by or on behalf of the Government. Use of documents available from or referenced by this server may be subject to US and foreign Copyright Laws.

References

Adisa, V.A. and Fajola, A.O. (1983). Cellulolytic enzymes associated with the fruit rots of Citrus sinensis caused by Aspergillus aculeatus and *Botryodiplodia theobromae*. *Zeitschrift für allgemeine Mikrobiologie* 23 (5): 283–288.

Agger, J.W., Isaksen, T., Varnai, A. et al. (2014). Discovery of LPMO activity on hemicelluloses shows the importance of oxidative processes in plant cell wall degradation. *Proceedings of the National Academy of Sciences of the United States of America* 111 (17): 6287–6292.

Andersen, N., Stenby, E.H., Michelsen, M.L. 2007. Enzymatic Hydrolysis of Cellulose: Experimental and Modeling Studies, Technical University of DenmarkDanmarks Tekniske Universitet, CenterCenters, Center for Energy Resources Engineering Center for Energy Resources Engineering.

Andreaus, J., Azevedo, H., and Cavaco-Paulo, A. (1999). Effects of temperature on the cellulose binding ability of cellulase enzymes. *Journal of Molecular Catalysis B: Enzymatic* 7: 233–239.

de Aquino Ximenes, F., de Sousa, M.V., Puls, J. et al. (1999). Purification and characterization of a low-molecular-weight xylanase produced by *Acrophialophora nainiana*. *Current microbiology* 38 (1): 18–21.

Aro, N., Pakula, T., and Penttilä, M. (2005). Transcriptional regulation of plant cell wall degradation by filamentous fungi. *FEMS Microbiology Reviews* 29 (4): 719–739.

Bagnara-Tardif, C., Gaudin, C., Belaich, A. et al. (1992). Sequence analysis of a gene cluster encoding cellulases from *Clostridium cellulolyticum*. *Gene* 119 (1): 17–28.

Baker, J.O., Ehrman, C.I., Adney, W.S. et al. (1998). Hydrolysis of cellulose using ternary mixtures of purified cellulases. In: *Biotechnology for Fuels and Chemicals*, 395–403. Springer.

Banerjee, G., Car, S., Scott-Craig, J.S. et al. (2010). Rapid optimization of enzyme mixtures for deconstruction of diverse pretreatment/biomass feedstock combinations. *Biotechnology for Biofuels* 3: 22.

Bateman, D.F. (1969). Some characteristics of the cellulase system produced by *Sclerotium rolfsii* Sacc. *Phytopathology* 59 (1): 37–42.

Baumann, M.J., Borch, K., and Westh, P. (2011). Xylan oligosaccharides and cellobiohydrolase I (Tr Cel7A) interaction and effect on activity. *Biotechnology for Biofuels* 4 (1): 1.

Bayer, E.A. and Lamed, R. (1986). Ultrastructure of the cell surface cellulosome of *Clostridium thermocellum* and its interaction with cellulose. *Journal of Bacteriology* 167 (3): 828–836.

Bayer, E., Kenig, R., and Lamed, R. (1983). Adherence of *Clostridium thermocellum* to cellulose. *Journal of Bacteriology* 156 (2): 818.

Bayer, E.A., Chanzy, H., Lamed, R., and Shoham, Y. (1998a). Cellulose, cellulases and cellulosomes. *Current Opinion in Structural Biology* 8 (5): 548–557.

Bayer, E.A., Shimon, L.J.W., Shoham, Y., and Lamed, R. (1998b). Cellulosomes – Structure and ultrastructure. *Journal of Structural Biology* 124 (2–3): 221–234.

Bayer, E.A., Shoham, Y., and Lamed, R. (2006). Cellulose-decomposing bacteria and their enzyme systems. In: *The Prokaryotes*, 578–617. Springer.

Beckham, G.T., Bomble, Y.J., Bayer, E.A. et al. (2011). Applications of computational science for understanding enzymatic deconstruction of cellulose. *Current Opinion in Biotechnology* 22 (2): 231–238.

Beckham, G., Dai, Z., Matthews, J. et al. (2012). Harnessing glycosylation to improve cellulase activity. *Current Opinion in Biotechnology* 23: 338–345.

Béguin, P. and Aubert, J.-P. (1994). The biological degradation of cellulose. *FEMS Microbiology Reviews* 13 (1): 25–58.

Belaich, J.P., Tardif, C., Belaich, A., and Gaudin, C. (1997). The cellulolytic system of *Clostridium cellulolyticum*. *Journal of Biotechnology* 57 (1–3): 3–14.

Benoit, I., Danchin, E.G., Bleichrodt, R.-J., and de Vries, R.P. (2008). Biotechnological applications and potential of fungal feruloyl esterases based on prevalence, classification and biochemical diversity. *Biotechnology Letters* 30 (3): 387–396.

Berlin, A., Gilkes, N., Kilburn, D. et al. (2005). Evaluation of novel fungal cellulase preparations for ability to hydrolyze softwood substrates–evidence for the role of accessory enzymes. *Enzyme and Microbial Technology* 37 (2): 175–184.

Biely, P., Mastihubová, M., Tenkanen, M. et al. (2011). Action of xylan deacetylating enzymes on monoacetyl derivatives of 4-nitrophenyl glycosides of β-D-xylopyranose and α-L-arabinofuranose. *Journal of Biotechnology* 151 (1): 137–142.

Biswas, R., Persad, A., and Bisaria, V.S. (2014). Production of cellulolytic enzymes. In: *Bioprocessing of Renewable Resources to Commodity Bioproducts*, 105–132. Wiley.

Blumer-Schuette, S.E., Kataeva, I., Westpheling, J. et al. (2008). Extremely thermophilic microorganisms for biomass conversion: status and prospects. *Current Opinion in Biotechnology* 19 (3): 210–217.

Boisset, C., Chanzy, H., Henrissat, B. et al. (1999). Digestion of crystalline cellulose substrates by the *Clostridium thermocellum* cellulosome: structural and morphological aspects. *Biochemical Journal* 340: 829–835.

Bothwell, M.K., Wilson, D.B., Irwin, D.C., and Walker, L.P. (1997). Binding reversibility and surface exchange of *Thermomonospora fusca* E-3 and E-5 and Trichoderma reesei CBHI. *Enzyme Microbiol Technology* 20 (6): 411–417.

van den Brink, J. and de Vries, R.P. (2011). Fungal enzyme sets for plant polysaccharide degradation. *Applied Microbiology and Biotechnology* 91 (6): 1477–1492.

Broda, P., Birch, P.R., Brooks, P.R., and Sims, P.F. (1996). Lignocellulose degradation by *Phanerochaete chrysosporium*: gene families and gene expression for a complex process. *Molecular Microbiology* 19 (5): 923–932.

Brown, K., Harris, P., Zaretsky, E., Re, E., Vlasenko, E., McFarland, K., de Leon, A.L. 2010. Polypeptides having cellulolytic enhancing activity and nucleic acids encoding same, Google Patents.

Brunecky, R., Alahuhta, M., Xu, Q. et al. (2013). Revealing nature's cellulase diversity: the digestion mechanism of *Caldicellulosiruptor bescii* CelA. *Science* 342 (6165): 1513–1516.

Caminal, G., Lopez-Santin, J., and Sola, C. (1985). Kinetic modeling of the enzymatic hydrolysis of pretreated cellulose. *Biotechnology and Bioengineering* 27: 1282–1290.

de Castro, A.M., de Carvalho, M.L.d.A., Leite, S.G.F., and Pereira, N. Jr., (2010). Cellulases from *Penicillium funiculosum*: production, properties and application to cellulose hydrolysis. *Journal of Industrial Microbiology and Biotechnology* 37 (2): 151–158.

Chauve, M., Huron, M., Hudebine, D. et al. (2013). Kinetic modeling of β-glucosidases and cellobiohydrolases involved in enzymatic hydrolysis of cellulose. *Industrial Biotechnology* 9 (6): 345–351.

Chundawat, S.P., Venkatesh, B., and Dale, B.E. (2006). Effect of particle size based separation of milled corn Stover on AFEX pretreatment and enzymatic digestibility. *Biotechnology and Bioengineering.*

Chundawat, S.P., Lipton, M.S., Purvine, S.O. et al. (2011). Proteomics-based compositional analysis of complex cellulase–hemicellulase mixtures. *Journal of Proteome Research* 10 (10): 4365–4372.

Converse, A.O. (1993). Substrate factors limiting enzymatic hydrolysis. In: *Bioconversion of Forest and Agricultural Plant Residues* (ed. J.N. Saddler), 93–106. Wallingford: C.A.B. International.

Converse, A.O., Matsuno, R., Tanaka, M., and Taniguchi, M. (1988). A model of enzyme adsorption and hydrolysis of microcrystalline cellulose with slow deactivation of the adsorbed enzyme. *Biotechnology and Bioengineering* 32: 38–45.

Converse, A.O., Ooshima, H., and Burns, D.S. (1990). Kinetics of enzymatic hydrolysis of lignocellulosic materials based on surface area of cellulose accessible to enzyme and enzyme adsorption on lignin and cellulose. *Applied Biochemistry and Biotechnology* 24-25: 67–73.

Cowling, E.B. (1975). Physical and chemical constraints in hydrolysis of cellulose and Lignocellulosic materials. *Biotechnology and Bioengineering* 5: 163–181.

Cowling, E.B. and Kirk, T.K. (1976). Properties of cellulose and Lignocellulosic materials as substrates for enzymatic conversion processes. *Biotechnology and Bioengineering* 6: 95–123.

Crawford, D.L. and McCoy, E. (1972). Cellulases of Thermomonospora fusca and *Streptomyces thermodiastaticus*. *Applied and Environmental Microbiology* 24 (1): 150–152.

Creagh, A.L., Ong, E., and Jervis, E.J. (1996). Binding of the cellulose-binding domain of exoglucanase Cex from Cellulomonas fimi to insoluble microcrystalline cellulose is entropically driven. *Proceedings of the National Academy of Sciences of the United States of America* 93: 12229–12234.

Cruys-Bagger, N., Elmerdahl, J., Praestgaard, E. et al. (2012). Pre-steady-state kinetics for hydrolysis of insoluble cellulose by cellobiohydrolase Cel7A. *Journal of Biological Chemistry* 287 (22): 18451–18458.

Cunha, E., Hatem, C.L., and Barrick, D. (2013). Natural and designed enzymes for cellulose degradation. In: *Advanced Biofuels and Bioproducts*, 339–368. Springer.

Dai, Z., Uma, K., Qian, W. et al. (2013). Impact of alg3 gene deletion on growth, development, pigment production, protein secretion, and functions of recombinant *Trichoderma reesei* cellobiohydrolases in *Aspergillus niger*. *Fungal Genetics and Biology* 61: 120–132.

Dashtban, M., Schraft, H., and Qin, W. (2009). Fungal bioconversion of lignocellulosic residues; opportunities & perspectives. *International Journal of Biological Sciences* 5 (6): 578.

Davies, G. and Henrissat, B. (1995). Structures and mechanisms of glycosyl hydrolases. *Structure* 3 (9): 853–859.

Desai, S.G. and Converse, A.O. (1997). Substrate reactivity as a function of the extent of reaction in the enzymatic hydrolysis of lignocellulose. *Biotechnology and Bioengineering* 56 (6): 650–655.

Dijkerman, R., den Camp, H.O., and Van der Drift, C. (1996). Cultivation of anaerobic fungi in a 10-l fermenter system for the production of (hemi-) cellulolytic enzymes. *Applied microbiology and biotechnology* 46 (1): 85–91.

Dimarogona, M., Topakas, E., and Christakopoulos, P. (2012). Cellulose degradation by oxidative enzymes. *Computational and Structural Biotechnology Journal* 2 (3): e201209015.

Ding, H. and Xu, F. (2004). Productive cellulase adsorption on cellulose. *ACS Symposium Series* 889 (Lignocellulose Biodegradation): 154–169.

Ding, S.Y., Bayer, E.A., Steiner, D. et al. (1999). A novel cellulosomal scaffoldin from acetivibrio cellulolyticus that contains a family 9 glycosyl hydrolase. *Journal of bacteriology* 181 (21): 6720–6729.

Ding, S.Y., Rincon, M.T., Lamed, R. et al. (2001). Cellulosomal scaffoldin-like proteins from *Ruminococcus flavefaciens*. *Journal of Bacteriology* 183 (6): 1945–1953.

Ding, S.Y., Liu, Y.S., Zeng, Y.N. et al. (2012). How does plant cell wall nanoscale architecture correlate with enzymatic digestibility? *Science* 338 (6110): 1055–1060.

Doi, R.H., Goldstein, M., Hashida, S. et al. (1994). The *Clostridium cellulovorans* cellulosome. *Critical Reviews in Microbiology* 20 (2): 87–93.

Domsch, K.H. and Gams, W. (1969). Variability and potential of a soil fungus population to decompose pectin, xylan and carboxymethyl-cellulose. *Soil Biology and Biochemistry* 1 (1): 29–36.

Eberhardt, R.Y., Gilbert, H.J., and Hazlewood, G.P. (2000). Primary sequence and enzymic properties of two modular endoglucanases, Cel5A and Cel45A, from the anaerobic fungus Piromyces equi. *Microbiology* 146 (8): 1999–2008.

Endler, A. and Persson, S. (2011). Cellulose synthases and synthesis in *Arabidopsis*. *Molecular Plant* 4 (2): 199–211.

Eremeeva, T., Bikova, T., Eisimonte, M. et al. (2001). Fractionation and molecular characteristics of cellulose during enzymatic hydrolysis. *Cellulose (Dordrecht, Netherlands)* 8 (1): 69–79.

Eriksson, T., Karlsson, J., and Tjerneld, F. (2002). A model explaining declining rate in hydrolysis of lignocellulose substrates with cellobiohydrolase I (Cel7A) and endoglucanase I (Cel7B) of Trichoderma reesei. *Applied Biochemistry and Biotechnology* 101: 41–60.

Eyzaguirre, J., Hidalgo, M., Leschot, A. 2005. 23 β-Glucosidases from Filamentous Fungi: Properties, Structure, and Applications.

Falls, M., Shi, J., Ebrik, M.A. et al. (2011). Investigation of enzyme formulation on pretreated switchgrass. *Bioresource Technology* 102 (24): 11072–11079.

Fan, L.T., Lee, Y.-H., and Beardmore, D.H. (1980). Major chemical and physical features of cellulosic materials as substrates for enzymic hydrolysis. *Advances in Biochemical Engineering* 14: 101–117.

Fan, L.T., Gharpuray, M.M., and Lee, Y. (1981). Evaluation of pretreatments for enzymatic conversion of agricultural residues. *Biotechnology and Bioengineering Symposium* 11: 29–45.

Fanutti, C., Ponyi, T., Black, G.W. et al. (1995). The conserved noncatalytic 40-residue sequence in cellulases and hemicellulases from anaerobic fungi functions as a protein docking domain. *Journal of Biological Chemistry* 270 (49): 29314–29322.

Fergus, C.L. (1969). The cellulolytic activity of thermophilic fungi and actinomycetes. *Mycologia* 61 (1): 120–129.

Fernandes, A.N., Thomas, L.H., Altaner, C.M. et al. (2011). Nanostructure of cellulose microfibrils in spruce wood. *Proceedings of the National Academy of Sciences* 108 (47): E1195–E1203.

Festucci-Buselli, R.A., Otoni, W.C., and Joshi, C.P. (2007). Structure, organization, and functions of cellulose synthase complexes in higher plants. *Brazilian Journal of Plant Physiology* 19 (1): 1–13.

Fliegerova, K., Pažoutová, S., Mrazek, J., and Kopečný, J. (2002). Special properties of polycentric anaerobic fungus *Anaeromyces mucronatus*. *Acta Veterinaria Brno* 71 (4): 441–444.

Folan, M.A. and Coughlan, M.P. (1978). The cellulase complex in the culture filtrate of the thermophyllic fungus, *Talaromyces emersonii*. *International Journal of Biochemistry* 9 (10): 717–722.

Fontes, C. and Gilbert, H. (2010). Cellulosomes: highly efficient nanomachines designed to deconstruct plant cell wall complex carbohydrates. *Annual Review of Biochemistry* 79: 655–681.

Fox, J.M., Levine, S.E., Clark, D.S., and Blanch, H.W. (2011a). Initial- and Processive-cut products reveal Cellobiohydrolase rate limitations and the role of companion enzymes. *Biochemistry* 51 (1): 442–452.

Fox, J.M., Levine, S.E., Clark, D.S., and Blanch, H.W. (2011b). Initial-and processive-cut products reveal cellobiohydrolase rate limitations and the role of companion enzymes. *Biochemistry* 51 (1): 442–452.

Frommhagen, M., Sforza, S., Westphal, A.H. et al. (2015). Discovery of the combined oxidative cleavage of plant xylan and cellulose by a new fungal polysaccharide monooxygenase. *Biotechnology for Biofuels* 8 (1): 1.

Fukumori, F., Kudo, T., Narahashi, Y., and Horikoshi, K. (1986). Molecular cloning and nucleotide sequence of the alkaline cellulase gene from the alkalophilic *Bacillus sp.* strain 1139. *Microbiology* 132 (8): 2329–2335.

Gan, Q., Allen, S.J., and Taylor, G. (2003). Kinetic dynamics in heterogeneous enzymatic hydrolysis of cellulose: an overview, an experimental study and mathematical modelling. *Process Biochemistry* 38 (7): 1003–1018.

Gao, D., Chundawat, S.P., Krishnan, C. et al. (2010). Mixture optimization of six core glycosyl hydrolases for maximizing saccharification of ammonia fiber expansion (AFEX) pretreated corn Stover. *Bioresource Technology* 101 (8): 2770–2781.

Gao, D., Uppugundla, N., Chundawat, S.P. et al. (2011). Hemicellulases and auxiliary enzymes for improved conversion of lignocellulosic biomass to monosaccharides. *Biotechnology for Biofuels* 4 (1): 1.

Garcia, V., Madarro, A., Peña, J.L. et al. (1989). Purification and characterization of cellulases from *Clostridium papyrosolvens*. *Journal of Chemical Technology and Biotechnology* 46 (1): 49–60.

Ghose, T.K. and Bisaria, V.S. (1979). Studies on the mechanism of enzymatic hydrolysis of cellulosic substances. *Biotechnology and Bioengineering* 21: 131–146.

Gilbert, H.J. and Hazlewood, G.P. (1993). Bacterial cellulases and xylanases. *Microbiology* 139 (2): 187–194.

Gómez-Gómez, E., Roncero, I.M., Di Pietro, A., and Hera, C. (2001). Molecular characterization of a novel endo-β-1, 4-xylanase gene from the vascular wilt fungus *Fusarium oxysporum*. *Current Genetics* 40 (4): 268–275.

Gong, C.S., Ladisch, M.R., and Tsao, G.T. (1979). Biosynthesis, purification and mode of action of cellulases of *Trichoderma reesei*. *Advances in Chemistry Series* 181: 261–288.

Gong, J., Egosimba, E.E., and Forsberg, C.W. (1996). Cellulose binding proteins of *Fibrobacter succinogenes* and the possible role of a180-kDa cellulose binding glycoprotein in adhesion to cellulose. *Canadian Journal of Microbiology* 42: 453–460.

Gonzalez, G., Caminal, G., De Mas, C., and Lopez-Santin, J. (1989). A kinetic model for pretreated wheat straw saccharification by cellulase. *Journal Chemical Technology and Biotechnology* 44: 275–288.

Grethlein, H.E. (1985). The effect of pore size distribution on the rate of enzymatic hydrolysis of cellulosic substrates. *Bio/Technology* 3: 155–160.

Gupta, R., Kim, T.H., and Lee, Y.Y. (2008). Substrate dependency and effect of xylanase supplementation on enzymatic hydrolysis of ammonia-treated biomass. *Applied Biochemistry and Biotechnology* 148 (1–3): 59–70.

Gusakov, A.V. (2011). Alternatives to *Trichoderma reesei* in biofuel production. *Trends in Biotechnology* 29 (9): 419–425.

Gusakov, A.V. and Sinitsyn, A.P. (1992). A theoretical analysis of cellulase product inhibition: effect of cellulase binding constant, enzyme/substrate ratio, and beta-glucosidase activity on the inhibition pattern. *Biotechnology and Bioengineering* 40 (6): 663–671.

Gusakov, A.V., Sinitsyn, A.P., and Klesov, A.A. (1985). Kinetic model of the enzymic hydrolysis of cellulose in a column type reactor. *Biotekhnologiya* 3: 112–122.

Gusakov, A.V., Sinitsyn, A.P., and Klesov, A.A. (1987). Factors affecting the enzymic hydrolysis of cellulose in batch and continuous reactors: computer simulation and experiment. *Biotechnology and Bioengineering* 29: 906–910.

Hankin, L. and Anagnostakis, S.L. (1977). Solid media containing carboxymethylcellulose to detect Cx cellulase activity of micro-organisms. *Microbiology* 98 (1): 109–115.

Harris, P.V., Welner, D., McFarland, K.C. et al. (2010). Stimulation of lignocellulosic biomass hydrolysis by proteins of glycoside hydrolase family 61: structure and function of a large, enigmatic family. *Biochemistry* 49 (15): 3305–3316.

Hashimoto, H. (2006). Recent structural studies of carbohydrate-binding modules. *Cellular and Molecular Life Sciences* 63 (24): 2954–2967.

Henriksson, G., Nutt, A., Henriksson, H. et al. (1999). Endoglucanase 28 (Cel12A), a new *Phanerochaete chrysosporium* cellulase. *European Journal of Biochemistry* 259 (1–2): 88–95.

Henriksson, G., Johansson, G., and Pettersson, G. (2000). A critical review of cellobiose dehydrogenases. *Journal of Biotechnology* 78 (2): 93–113.

Henrissat, B. and Davies, G. (1997). Structural and sequence-based classification of glycoside hydrolases. *Current Opinion in Structural Biology* 7 (5): 637–644.

Henrissat, B., Driguez, H., Viet, C., and Schuelein, M. (1985). Synergism of cellulases from *Trichoderma reesei* in the degradation of cellulose. *Bio/Technology* 3 (8): 722–726.

Herpoël-Gimbert, I., Margeot, A., Dolla, A. et al. (2008). Comparative secretome analyses of two *Trichoderma reesei* RUT-C30 and CL847 hypersecretory strains. *Biotechnology for Biofuels* 1 (1): 1.

Hilden, L., Valjamae, P., and Johansson, G. (2005). Surface character of pulp fibres studied using endoglucanases. *Journal of Biotechnology* 118 (4): 386–397.

Himmel, M.E., Ruth, M.F., and Wyman, C.E. (1999). Cellulase for commodity products from cellulosic biomass. *Current Opinion in Biotechnology* 10 (4): 358–364.

Himmel, M.E., Ding, S.-Y., Johnson, D.K. et al. (2007). Biomass recalcitrance: engineering plants and enzymes for biofuels production. *Science (Washington, DC, United States)* 315 (5813): 804–807.

Hirasawa, K., Uchimura, K., Kashiwa, M. et al. (2006). Salt-activated endoglucanase of a strain of alkaliphilic *Bacillus agaradhaerens*. *Antonie Van Leeuwenhoek* 89 (2): 211–219.

Ho, Y.W., Bauchop, T., Abdullah, N., and Jalaludin, S. (1990). Ruminomyces elegans gen. et sp. nov., a polycentric anaerobic rumen fungus from cattle. *Mycotaxon* 38: 397–405.

Holtzapple, M., Cognata, M., Shu, Y., and Hendrickson, C. (1990). Inhibition of *Trichoderma reesei* cellulase by sugars and solvents. *Biotechnology and Bioengineering* 36: 275–287.

Horn, S.J., Vaaje-Kolstad, G., Westereng, B., and Eijsink, V. (2012). Novel enzymes for the degradation of cellulose. *Biotechnology for Biofuels* 5 (1): 1.

Hu, J., Pribowo, A., and Saddler, J.N. (2016). Oxidative cleavage of some cellulosic substrates by auxiliary activity (AA) family 9 enzymes influences the adsorption/desorption of hydrolytic cellulase enzymes. *Green Chemistry* 18 (23): 6329–6336.

Huang, Y., Busk, P.K., and Lange, L. (2015). Cellulose and hemicellulose-degrading enzymes in Fusarium commune transcriptome and functional characterization of three identified xylanases. *Enzyme and Microbial Technology* 73: 9–19.

Huron, M., Hudebine, D., Lopes Ferreira, N., and Lachenal, D. (2015). Mechanistic modeling of enzymatic hydrolysis of cellulose integrating substrate morphology and cocktail composition. *Biotechnology and Bioengineering* 113 (5): 1011–1023.

Hyeon, J.E., Jeon, S.D., and Han, S.O. (2013). Cellulosome-based, Clostridium-derived multi-functional enzyme complexes for advanced biotechnology tool development: advances and applications. *Biotechnology Advances* 31 (6): 936–944.

Igarashi, K., Ishida, T., Hori, C., and Samejima, M. (2008). Characterization of an endoglucanase belonging to a new subfamily of glycoside hydrolase family 45 of the basidiomycete *Phanerochaete chrysosporium*. *Applied and Environmental Microbiology* 74 (18): 5628–5634.

Igarashi, K., Uchihashi, T., Koivula, A. et al. (2011). Traffic jams reduce hydrolytic efficiency of cellulase on cellulose surface. *Science* 333 (6047): 1279–1282.

Inoue, H., Decker, S.R., Taylor, L.E. et al. (2014). Identification and characterization of core cellulolytic enzymes from Talaromyces cellulolyticus (formerly Acremonium cellulolyticus) critical for hydrolysis of lignocellulosic biomass. *Biotechnology for biofuels* 7 (1): 151.

Jalak, J. and Väljamäe, P. (2010). Mechanism of initial rapid rate retardation in cellobiohydrolase catalyzed cellulose hydrolysis. *Biotechnology and Bioengineering* 106 (6): 871–883.

Jalak, J., Kurašin, M., Teugjas, H., and Väljamäe, P. (2012). Endo-exo synergism in cellulose hydrolysis revisited. *Journal of Biological Chemistry* 287 (34): 28802–28815.

Javier, P.F., Óscar, G., Sanz-Aparicio, J., and Díaz, P. (2007). Xylanases: molecular properties and applications. In: *Industrial Enzymes*, 65–82. Springer.

Jeoh, T., Wilson, D.B., and Walker, L.P. (2002). Cooperative and competitive binding in synergistic mixtures of *Thermobifida fusca* cellulases Cel5A, Cel6B, and Cel9A. *Biotechnology Progress* 18 (4): 760–769.

Jeoh, T., Baker, J.O., Ali, M.K. et al. (2005). β-D-Glucosidase reaction kinetics from isothermal titration microcalorimetry. *Analytical Biochemistry* 347 (2): 244–253.

Jeoh, T., Wilson, D.B., and Walker, L.P. (2006). Effect of cellulase mole fraction and cellulose recalcitrance on synergism in cellulose hydrolysis and binding. *Biotechnology Progress* 22 (1): 270–277.

Jermyn, M.A. (1952). Fungal Cellulases I. General properties of unpurified enzyme preparations from *Aspergillus oryz.* A'e. *Australian Journal of Biological Sciences* 5 (4): 409–432.

Jervist, E.J., Haynes, C.A., and Kilburn, D.G. (1997). Surface diffusion of cellulases and their isolated binding domains on cellulose. *The Journal of Biological Chemistry* 272 (38): 24016–24023.

Johansen, K.S. (2016). Discovery and industrial applications of lytic polysaccharide mono-oxygenases. *Biochemical Society Transactions* 44 (1): 143–149.

Jovanovic, I., Magnuson, J.K., Collart, F. et al. (2009). Fungal glycoside hydrolases for saccharification of lignocellulose: outlook for new discoveries fueled by genomics and functional studies. *Cellulose* 16 (4): 687–697.

Jurásek, L., Colvin, J., and Whitaker, D. (1968). Microbiological aspects of the formation and degradation of cellulosic fibers. *Advances in Applied Microbiology* 9: 131–170.

Kabel, M.A., van den Borne, H., Vincken, J.-P. et al. (2007). Structural differences of xylans affect their interaction with cellulose. *Carbohydrate Polymers* 69 (1): 94–105.

Kadam, K.L., Rydholm, E.C., and McMillan, J.D. (2004). Development and validation of a kinetic model for enzymatic saccharification of lignocellulosic biomass. *Biotechnology Progress* 20 (3): 698–705.

Kakiuchi, M., Isui, A., Suzuki, K. et al. (1998). Cloning and DNA sequencing of the genes encoding *Clostridium josui* scaffolding protein CipA and Cellulase CelD and identification of their gene products as major components of the Cellulosome. *Journal of bacteriology* 180 (16): 4303–4308.

Kanda, T., Wakabayashi, K., and Nisizawa, K. (1976). Synergistic action of two different types of endo-cellulase components from Irpex lacteus (Polyporus tulipiferae) in the hydrolysis of some insoluble celluloses. *The Journal of Biochemistry (Tokyo)* 79 (5): 997–1005.

Kanda, T., Wakabayashi, K., and Nisizawa, K. (1980). Modes of action of exo- and endo-cellulases in the degradation of celluloses I and II. *The Journal of Biochemistry (Tokyo)* 87 (6): 1635–1639.

Karita, S., Sakka, K., and Ohmiya, K. (1997). Cellulosomes and cellulase complexes of anaerobic microbes: their structure, models and function. In: *Rumen Microbes and Digestive Physiology in Ruminants* (eds. H.I.R. Onodera, K. Ushida, H. Yano and Y. Sasaki), 47–57. Tokyo: Japan Scientific Societies Press.

Karnaouri, A., Topakas, E., Antonopoulou, I., and Christakopoulos, P. (2014). Genomic insights into the fungal lignocellulolytic system of *Myceliophthora thermophila*. *Frontiers in microbiology* 5: 281.

Khatiwada, P., Ahmed, J., Sohag, M.H. et al. (2016). Isolation, screening and characterization of cellulase producing bacterial isolates from municipal solid wastes and rice straw wastes. *Journal of Bioprocessing & Biotechniques* 6: 280–285.

Kim, Y.S., Singh, A.P., Wi, S.G. et al. (2001). Cellulosome-like structures in ruminal cellulolytic bacterium *Ruminococcus albus* F-40 as revealed by electron microscopy. *Asian-Australasian Journal of Animal Sciences* 14 (10): 1429–1433.

King, K.W. (1966). Enzymic degradation of crystalline hydrocellulose. *Biochemical and Biophysical Research Communications* 24 (3): 295.

Kirby, J., Martin, J.C., Daniel, A.S., and Flint, H.J. (1997). Dockerin-like sequences in cellulases and xylanases from the rumen cellulolytic bacterium *Ruminococcus flavefaciens*. *FEMS microbiology letters* 149 (2): 213–219.

Kleman-Leyer, K.M., Siika-Aho, M., Teeri, T.T., and Kirk, T.K. (1996). The Cellulases Endoglucanase I and Cellobiohydrolase II of Trichoderma reesei act synergistically to solubilize native cotton cellulose but not to decrease its molecular size. *Applied and Environmental Microbiology* 62 (8): 2883–2887.

Ko, C.H., Chen, W.L., Tsai, C.H. et al. (2007). Paenibacillus campinasensis BL11: a wood material-utilizing bacterial strain isolated from black liquor. *Bioresource technology* 98 (14): 2727–2733.

(Patent) Komura I, Awao T, Yamada K: US4106989 (1978)

Koseki, T., Fushinobu, S., Shirakawa, H., and Komai, M. (2009). Occurrence, properties, and applications of feruloyl esterases. *Applied Microbiology and Biotechnology* 84 (5): 803–810.

Kracher, D., Scheiblbrandner, S., Felice, A.K.G. et al. (2016). Extracellular electron transfer systems fuel cellulose oxidative degradation. *Science* 352 (6289): 1098–1101.

Kubicek, C.P., Mikus, M., Schuster, A. et al. (2009). Metabolic engineering strategies for the improvement of cellulase production by *Hypocrea jecorina*. *Biotechnology for Biofuels* 2 (1): 1.

Kumar, D. and Murthy, G.S. (2013). Stochastic molecular model of enzymatic hydrolysis of cellulose for ethanol production. *Biotechnology for Biofuels* 6 (1): 1–20.

Kumar, R. and Wyman, C.E. (2009). Effect of xylanase supplementation of cellulase on digestion of corn Stover solids prepared by leading pretreatment technologies. *Bioresource Technology* 100 (18): 4203–4213.

Kurašin, M. and Väljamäe, P. (2011). Processivity of cellobiohydrolases is limited by the substrate. *Journal of Biological Chemistry* 286 (1): 169–177.

Kusube, M., Sugihara, A., Moriwaki, Y. et al. (2014). Alicyclobacillus cellulosilyticus sp. nov., a thermophilic, cellulolytic bacterium isolated from steamed Japanese cedar chips from a lumbermill. *International journal of systematic and evolutionary microbiology* 64 (Pt 7): 2257.

Ladeira, S.A., Cruz, E., Delatorre, A.B. et al. (2015). Cellulase production by thermophilic *Bacillus sp*: SMIA-2 and its detergent compatibility. *Electronic journal of biotechnology* 18 (2): 110–115.

Lamed, R., Setter, E., and Bayer, E.A. (1983a). Characterization of a cellulose-binding, cellulase-containing complex in *Clostridium thermocellum*. *Journal of Bacteriology* 156 (2): 828–836.

Lamed, R., Setter, E., Kenig, R., and Bayer, E.A. (1983b). The Cellulosome – a discrete cell-surface organelle of *Clostridium thermocellum* which exhibits separate antigenic, cellulose-binding and various cellulolytic activities. *Biotechnology and Bioengineering* 13: 163–181.

Lamed, R., Naimark, J.L., Morgenstern, E., and Bayer, E.A. (1987). Specialized surface structure in cellulolytic bacteria. *Journal of Bacteriology* 169: 3792–3800.

Lamed, R., Morag, E., Mor-Yosef, O., and Bayer, E.A. (1991). Cellulosome-like entities inbacteroides cellulosolvens. *Current microbiology* 22 (1): 27–33.

Langston, J., Sheehy, N., and Xu, F. (2006). Substrate specificity of *Aspergillus oryzae* family 3 β-glucosidase. *Biochimica et Biophysica Acta (BBA)-Proteins and Proteomics* 1764 (5): 972–978.

Langston, J.A., Shaghasi, T., Abbate, E. et al. (2011). Oxidoreductive cellulose depolymerization by the enzymes cellobiose dehydrogenase and glycoside hydrolase 61. *Applied and Environmental Microbiology* 77 (19): 7007–7015.

Le Crom, S., Schackwitz, W., Pennacchio, L. et al. (2009). Tracking the roots of cellulase hyperproduction by the fungus Trichoderma reesei using massively parallel DNA sequencing. *Proceedings of the National Academy of Sciences* 106 (38): 16151–16156.

Lebaz, N., Cockx, A., Spérandio, M. et al. (2016). Application of the direct quadrature method of moments for the modelling of the enzymatic hydrolysis of cellulose: I. Case of soluble substrate. *Chemical Engineering Science* 149: 306–321.

Lee, Y.-H. and Fan, L.T. (1983). Kinetic studies of enzymatic hydrolysis of insoluble cellulose: (II). Analysis of extended hydrolysis times. *Biotechnology and Bioengineering* 25: 939–966.

Li, L.H. and King, K.W. (1963). Fractionation of β-glucosidases and related extracellular enzymes from *Aspergillus niger*. *Applied and Environmental Microbiology* 11 (4): 320–325.

Li, X.L., Chen, H., and Ljungdahl, L.G. (1997). Two cellulases, CelA and CelC, from the polycentric anaerobic fungus *Orpinomyces strain* PC-2 contain N-terminal docking domains for a cellulase-hemicellulase complex. *Applied and Environmental Microbiology* 63 (12): 4721–4728.

Liang, Y., Yesuf, J., Schmitt, S. et al. (2009). Study of cellulases from a newly isolated thermophilic and cellulolytic *Brevibacillus sp*. strain JXL. *Journal of industrial microbiology & biotechnology* 36 (7): 961–970.

Liu, Y.-S., Baker, J.O., Zeng, Y. et al. (2011). Cellobiohydrolase hydrolyzes crystalline cellulose on hydrophobic faces. *Journal of Biological Chemistry* 286 (13): 11195–11201.

Lombard, V., Ramulu, H.G., Drula, E. et al. (2014). The carbohydrate-active enzymes database (CAZy) in 2013. *Nucleic Acids Research* 42 (D1): D490–D495.

Lu, Y., Zhang, Y.-H.P., and Lynd, L.R. (2006). Enzyme–microbe synergy during cellulose hydrolysis by *Clostridium thermocellum*. *Proceedings of the National Academy of Sciences* 103 (44): 16165–16169.

Lynd, L.R. and Zhang, Y.H. (2002). Quantitative determination of cellulase concentration as distinct from cell concentration in studies of microbial cellulose utilization: analytical framework and methodological approach. *Biotechnology and Bioengineering* 77 (4): 467–475.

Lynd, L.R., Grethlern, H.E., and Wolkin, R.H. (1989). Fermentation of cellulosic substrates in batch and continuous cultures by *Clostridium thermocellum*. *Applied and Environmental Microbiology* 55: 3131–3139.

Lynd, L.R., Cushman, J.H., Nichols, R.J., and Wyman, C.E. (1991). Fuel ethanol from cellulosic biomass. *Science* 251 (4999): 1318–1323.

Lynd, L.R., Weimer, P.J., van Zyl, W.H., and Pretorius, I.S. (2002). Microbial cellulose utilization: fundamentals and biotechnology. *Microbiology and Molecular Biology Reviews* 66: 506–577.

Maki, M., Leung, K.T., and Qin, W. (2009). The prospects of cellulase-producing bacteria for the bioconversion of lignocellulosic biomass. *International Journal of Biological Sciences* 5 (5): 500–516.

Mandels, M., Kostick, J., and Parizek, R. (1971). Use of adsorbed cellulase in the continuous conversion of cellulose to glucose. *Journal of Polymer Science, Polymer Symposia* 36: 445–459.

Mansfield, S.D. and Meder, R. (2003). Cellulose hydrolysis – the role of monocomponent cellulases in crystalline cellulose degradation. *Cellulose* 10 (2): 159–169.

Margeot, A., Hahn-Hagerdal, B., Edlund, M. et al. (2009). New improvements for lignocellulosic ethanol. *Current Opinion in Biotechnology* 20 (3): 372–380.

Marita, J.M., Ralph, J., Hatfield, R.D. et al. (2003). Structural and compositional modifications in lignin of transgenic alfalfa down-regulated in caffeic acid 3-O-methyltransferase and caffeoyl coenzyme A 3-O-methyltransferase. *Phytochemistry* 62 (1): 53–65.

Matsui, H. and Ban-Tokuda, T. (2008). Studies on carboxymethyl cellulase and xylanase activities of anaerobic fungal isolate CR4 from the bovine rumen. *Current microbiology* 57 (6): 615–619.

McCabe, C., Zhao, X., Adney, W., Himmel, M. 2010. Energy Storage in Cellulase Linker Peptides? National Renewable Energy Laboratory (NREL), Golden, CO.

McCarter, J.D. and Withers, G.S. (1994). Mechanisms of enzymatic glycoside hydrolysis. *Current Opinion in Structural Biology* 4 (6): 885–892.

McClendon, S.D., Batth, T., Petzold, C.J. et al. (2012). Thermoascus aurantiacus is a promising source of enzymes for biomass deconstruction under thermophilic conditions. *Biotechnology for biofuels* 5: 1, 54.

Medve, J., Stahlberg, J., and Tjerneld, F. (1997). Isotherms for adsorption of cellobiohydrolase I and II from *Trichoderma reesei* on microcrystalline cellulose. *Applied Biochemistry and Biotechnology* 66 (1): 39–56.

Medve, J., Karlsson, J., Lee, D., and Tjerneld, F. (1998). Hydrolysis of microcrystalline cellulose by cellobiohydrolase I and endoglucanase II from *Trichoderma reesei*: adsorption, sugar production pattern, and synergism of the enzymes. *Biotechnology and Bioengineering* 59 (5): 621–634.

Merino, S.T. and Cherry, J. (2007). Progress and challenges in enzyme development for biomass utilization. In: *Biofuels*, 95–120. Springer.

Miller, M.E.B., Brulc, J.M., Bayer, E.A. et al. (2010). Advanced technologies for biomass hydrolysis and saccharification using novel enzymes. In: *Biomass to Biofuels: Strategies for Global Industries*, 199–212. Chippenham, Wiltshire, Great Britain: Wiley.

Minato, H. and Suto, T. (1985). Technique for fractionation of bacteria in rumen microbial ecosystem. II. Attachment of bacteria isolated from bovine rumen t cellulose powder in vitro and elution of bacteria attached therefrom. *The Journal of General and Applied Microbiology* 24: 1–16.

Miron, J., Ben-Ghedalla, D., and Morrison, M. (2001). Invited review: adhesion mechanisms of rumen cellulolytic bacteria. *Journal of Dairy Science* 84 (6): 1294–1309.

Miura, H., Horiguchi, M., Oginoto, K., and Matsumoto, T. (1983). Nutritional interdependence among rumen bacteria during cellulose digestion in vitro. *Applied and Environmental Microbiology* 45: 726–729.

Mohagheghi, A., Grohmann, K.M.M.H., Himmel, M. et al. (1986). Isolation and characterization of Acidothermus cellulolyticus gen. nov., sp. nov., a new genus of thermophilic, acidophilic, cellulolytic bacteria. *International Journal of Systematic and Evolutionary Microbiology* 36 (3): 435–443.

Morag, E., Halevy, I., Bayer, E.A., and Lamed, R. (1991). Isolation and properties of a major cellobiohydrolase from the cellulosome of *Clostridium thermocellum*. *Journal of Bacteriology* 173 (13): 4155–4162.

Moreira, L. (2008). An overview of mannan structure and mannan-degrading enzyme systems. *Applied Microbiology and Biotechnology* 79 (2): 165–178.

Morgenstern, I., Powlowski, J., and Tsang, A. (2014). Fungal cellulose degradation by oxidative enzymes: from dysfunctional GH61 family to powerful lytic polysaccharide monooxygenase family. *Briefings in Functional Genomics* 13 (6): 471–481.

Morrison, M. and Miron, J. (2000). Adhesion to cellulose by *Ruminococcus albus*: a combination of cellulosomes and Pil-proteins? *FEMS Microbiology Letters* 185 (2): 109–115.

Mourino, F., Akkarawongsa, R., and Weimer, P.J. (2001). Initial pH as a determinant of cellulose digestion rate by mixed ruminal microorganisms in vitro. *Journal of Dairy Science* 84 (4): 848–859.

Mukataka, S., Tada, M., and Takahashi, J. (1983). Effects of agitation on enzymic hydrolysis of cellulose in a stirred-tank reactor. *Journal of Fermentation Technology* 61 (6): 615–621.

Murphy, N., McConnell, D.J., and Cantwell, B.A. (1984). The DNA sequence of the gene and genetic control sites for the excreted B. subtilis enzyme βglucanase. *Nucleic acids research* 12 (13): 5355–5367.

Murphy, L., Cruys-Bagger, N., Damgaard, H.D. et al. (2012). Origin of initial burst in activity for *Trichoderma reesei* endo-glucanases hydrolyzing insoluble cellulose. *Journal of Biological Chemistry* 287 (2): 1252–1260.

Nidetzky, B. and Steiner, W. (1993). A new approach for modeling cellulase cellulose adsorption and the kinetics of the enzymatic-hydrolysis of microcrystalline cellulose. *Biotechnology and Bioengineering* 42 (4): 469–479.

Nidetzky, B., Claeyssens, M., and Steiner, W. (1996). Cellulose degradation by the major cellulases from *Trichoderma reesei*: synergistic interaction and competition for binding sites on cellulose. In: *Biotechnology in the Pulp and Paper Industry: Recent Advances in Applied and Fundamental Research, Proceedings of the International Conference on Biotechnology in the Pulp and Paper Industry, 6th, June 11–15, 1995*, 537–542. Vienna.

Nieves, R., Ehrman, C., Adney, W. et al. (1997). Survey and analysis of commercial cellulase preparations suitable for biomass conversion to ethanol. *World Journal of Microbiology and Biotechnology* 14 (2): 301–304.

Nimlos, M.R., Matthews, J.F., Crowley, M.F. et al. (2007). Molecular modeling suggests induced fit of family I carbohydrate-binding modules with a broken-chain cellulose surface. *Protein Engineering Design and Selection* 20 (4): 179–187.

Nisizawa, K. (1955). Cellulose-splitting enzymes. *The Journal of Biochemistry* 42 (6): 825–835.

Numan, M.T. and Bhosle, N.B. (2006). α-L-Arabinofuranosidases: the potential applications in biotechnology. *Journal of Industrial Microbiology and Biotechnology* 33 (4): 247–260.

Nutor, J.R.K. and Converse, A.O. (1991). The effect of enzyme and substrate levels on the specific hydrolysis rate of pretreated poplar wood. *Applied Biochemistry and Biotechnology* 28-29: 757–772.

Ohara, H., Karita, S., Kimura, T. et al. (2000). Characterization of the cellulolytic complex (cellulosome) from *Ruminococcus albus*. *Bioscience, biotechnology, and biochemistry* 64 (2): 254–260.

Öhgren, K., Bura, R., Saddler, J., and Zacchi, G. (2007). Effect of hemicellulose and lignin removal on enzymatic hydrolysis of steam pretreated corn Stover. *Bioresource Technology* 98 (13): 2503–2510.

Olutiola, P.O. (1976). A cellulase complex in culture filtrates of *Penicillium citrinum*. *Canadian journal of microbiology* 22 (8): 1153–1159.

Ooshima, H., Burns, D.S., and Converse, A.O. (1990). Adsorption of cellulase from Trichoderma reesei on cellulose and lignaceous residue in wood pretreated by dilute sulfuric acid with explosive decompression. *Biotechnology and Bioengineering Symposium* 36: 446–452.

Ooshima, H., Kurakake, M., Kato, J., and Harano, Y. (1991). Enzymatic activity of cellulase adsorbed on cellulose and its change during hydrolysis. *Applied Biochemistry and Biotechnology* 31 (3): 253–266.

Pai, C.K., Wu, Z.Y., Chen, M.J. et al. (2010). Molecular cloning and characterization of a bifunctional xylanolytic enzyme from *Neocallimastix patriciarum*. *Applied microbiology and biotechnology* 85 (5): 1451–1462.

Pala, H., Mota, M., and Gama, F.M. (2007). Enzymatic depolymerisation of cellulose. *Carbohydrate Polymers* 68 (1): 101–108.

Pallapolu Venkata, R., Lee, Y.Y., Garlock Rebecca, J. et al. (2011). Effects of enzyme loading and Î²-glucosidase supplementation on enzymatic hydrolysis of switchgrass processed by leading pretreatment technologies. *Bioresource Technology* 102 (24): 11115–11120.

Palonen, H., Tjerneld, F., Zacchi, G., and Tenkanen, M. (2004). Adsorption of *Trichoderma reesei* CBH I and EG II and their catalytic domains on steam pretreated softwood and isolated lignin. *Journal of Biotechnology* 107 (1): 65–72.

Panagiotou, G. and Olsson, L. (2007). Effect of compounds released during pretreatment of wheat straw on microbial growth and enzymatic hydrolysis rates. *Biotechnology and Bioengineering* 96 (2): 250–258.

Panbangred, W., Kondo, T., Negoro, S. et al. (1983). Molecular cloning of the genes for xylan degradation of Bacillus pumilus and their expression in *Escherichia coli*. *Molecular and General Genetics MGG* 192 (3): 335–341.

Pauly, M., Gille, S., Liu, L., Mansoori, N., de Souza, A., Schultink, A., Xiong, G. 2013. *Hemicellulose biosynthesis.*

Peitersen, N., Medeiros, J., and Mandels, M. (1977). Adsorption of Trichoderma cellulase on cellulose. *Biotechnology and Bioengineering* 19 (7): 1091–1094.

Pérez-Avalos, O., Sánchez-Herrera, L.M., Salgado, L.M., and Ponce-Noyola, T. (2008). A bifunctional endoglucanase/endoxylanase from *Cellulomonas flavigena* with potential use in industrial processes at different pH. *Current microbiology* 57 (1): 39–44.

Phillips, C.M., Beeson Iv, W.T., Cate, J.H., and Marletta, M.A. (2011). Cellobiose dehydrogenase and a copper-dependent polysaccharide monooxygenase potentiate cellulose degradation by *Neurospora crassa*. *ACS Chemical Biology* 6 (12): 1399–1406.

Pollet, A., Delcour, J.A., and Courtin, C.M. (2010). Structural determinants of the substrate specificities of xylanases from different glycoside hydrolase families. *Critical Reviews in Biotechnology* 30 (3): 176–191.

Praestgaard, E., Elmerdahl, J., Murphy, L. et al. (2011). A kinetic model for the burst phase of processive cellulases. *FEBS Journal* 278 (9): 1547–1560.

Puri, V.P. (1984). Effect of crystallinity and degree of polymerization of cellulose on enzymatic saccharification. *Biotechnology and Bioengineering* 26: 1219–1222.

Qing, Q., Yang, B., and Wyman, C.E. (2010). Xylooligomers are strong inhibitors of cellulose hydrolysis by enzymes. *Bioresource Technology* 101 (24): 9624–9630.

Qiu, X., Selinger, B., Yanke, L.J., and Cheng, K.J. (2000). Isolation and analysis of two cellulase cDNAs from *Orpinomyces joyonii*. *Gene* 245 (1): 119–126.

Quinlan, R.J., Sweeney, M.D., Leggio, L.L. et al. (2011). Insights into the oxidative degradation of cellulose by a copper metalloenzyme that exploits biomass components. *Proceedings of the National Academy of Sciences* 108 (37): 15079–15084.

Quiroz-Castañeda, R.E. and Folch-Mallol, J.L. (2013). Hydrolysis of biomass mediated by cellulases for the production of sugars. In: *Sustainable Degradation of Lignocellulosic Biomass Techniques, Applications and Commercialization*, 119–155. InTech.

Ragauskas, A.J., Williams, C.K., Davison, B.H. et al. (2006). The path forward for biofuels and biomaterials. *Science* 311 (5760): 484–489.

Raghothama, S., Eberhardt, R.Y., Simpson, P. et al. (2001). Characterization of a cellulosome dockerin domain from the anaerobic fungus *Piromyces equi*. *Nature Structural & Molecular Biology* 8 (9): 775.

Reese, E. (1982). Protection of Trichoderma reesei cellulase from inactivation due to shaking. In: *International Symposium on Solution Behavior of Surfactants: Theoretical Application Aspects*, 1487–1504. Boston, MA.: Springer.

Reese, E.T. and Levinson, H.S. (1952). A comparative study of the breakdown of cellulose by microorganisms. *Physiologia Plantarum* 5 (3): 345–366.

Reese, E.T. and Mandels, M. (1971). Enzymatic degradation. In: *Cellulose and Cellulose Derivatives* (eds. N.M. Bikales and L. Segal), 1079–1094. New York, NY: Wiley Interscience.

Reese, E.T., Siu, R.G., and Levinson, H.S. (1950). The biological degradation of soluble cellulose derivatives and its relationship to the mechanism of cellulose hydrolysis. *Journal of bacteriology* 59 (4): 485.

Reinikainen, T., Teleman, O., and Teeri, T.T. (1995). Effects of pH and high ionic strength on the adsorption and activity of native and mutated cellobiohydrolase I from *Trichoderma reesei*. *Proteins* 22 (4): 392–403.

Rolz, C. (1986). Regulation of adsorption-desorption mechanisms during holocellulosis of biomass. *Trends in Biotechnology* 4 (6): 135–136.

Ryu, D.D.Y., Kim, C., and Mandels, M. (1984). Competitive adsoprtion of cellulase components and its significance in a synergistic mechanism. *Biotechnology and Bioengineering* XXVI: 488–496.

Sabathé, F., Bélaïch, A., and Soucaille, P. (2002). Characterization of the cellulolytic complex (cellulosome) of *Clostridium acetobutylicum*. *FEMS microbiology letters* 217 (1): 15–22.

Saddler, J.N. (1982). Screening of highly cellulolytic fungi and the action of their cellulase enzyme systems. *Enzyme and Microbial Technology* 4 (6): 414–418.

Saha, B.C. (2000). α-L-Arabinofuranosidases: biochemistry, molecular biology and application in biotechnology. *Biotechnology Advances* 18 (5): 403–423.

Sánchez, C. (2009). Lignocellulosic residues: biodegradation and bioconversion by fungi. *Biotechnology Advances* 27 (2): 185–194.

Sasaki, T., Tanaka, T., Nanbu, N. et al. (1979). Correlation between X-ray diffraction measurements of cellulose crystalline structure and the susceptibility of microbial cellulase. *Biotechnology and Bioengineering* 21: 1031–1042.

Schülein, M. (1997). Enzymatic properties of cellulases from *Humicola insolens*. *Journal of Biotechnology* 57 (1): 71–81.

Schulein, M. (2000). Protein engineering of cellulases. *Biochimica et Biophysica Acta-Protein Structure and Molecular Enzymology* 1543: 239–252.

Schwarz, W.H. (2001). The cellulosome and cellulose degradation by anaerobic bacteria. *Applied Microbiology and Biotechnology* 56 (5–6): 634–649.

Schwarz, W.H. and Staudenbauer, W.L. (1986). Properties of a *Clostridium thermocellum* endoglucanase produced in Escherichia coli. *Applied and Environmental Microbiology* 51: 1293–1299.

Selig, M.J., Knoshaug, E.P., Adney, W.S. et al. (2008). Synergistic enhancement of cellobiohydrolase performance on pretreated corn Stover by addition of xylanase and esterase activities. *Bioresource Technology* 99 (11): 4997–5005.

Sethi, A., Scharf, M.E. 2013. Biofuels: fungal, bacterial and insect degraders of lignocellulose. eLS.

Shallom, D. and Shoham, Y. (2003). Microbial hemicellulases. *Current Opinion in Microbiology* 6 (3): 219–228.

Sharma, D.C. and Satyanarayana, T. (2006). A marked enhancement in the production of a highly alkaline and thermostable pectinase by Bacillus pumilus dcsr1 in submerged fermentation by using statistical methods. *Bioresource Technology* 97 (5): 727–733.

Shetty, J.K., Marshall, J.J. 1986. Method for determination of transglucosidase, Google Patents.

Shi, J., Mirvat, A.E., Redmond, T., Yang, B., Wyman, C.E., Garlock, R., Balan, V., Dale, B.E., Pallapolu, V.R., Lee, Y. 2010. Interactions of cellulase and non-cellulase enzymes with ideal substrates and switchgrass processed by leading pretreatment technologies. *The 32nd Symposium on Biotechnology for Fuels and Chemicals.*

Shi, J., Wu, D., Zhang, L. et al. (2016). Dynamics changes of substrate reactivity and enzyme adsorption on partially hydrolyzed cellulose. *Biotechnology and Bioengineering* 114 (3): 503–515.

Shi, J., Wu, D., Zhang, L. et al. (2017). Dynamic changes of substrate reactivity and enzyme adsorption on partially hydrolyzed cellulose. *Biotechnology and Bioengineering* 114 (3): 503–515.

Shoham, Y., Lamed, R., and Bayer, E.A. (1999). The cellulosome concept as an efficient microbial strategy for the degradation of insoluble polysaccharides. *Trends in Microbiology* 7 (7): 275–281.

Shoseyov, O., Takagi, M., Goldstein, M.A., and Doi, R.H. (1992). Primary sequence analysis of Clostridium cellulovorans cellulose binding protein A. *Proceedings of the National Academy of Sciences* 89 (8): 3483–3487.

Shoseyov, O., Shani, Z., and Levy, I. (2006). Carbohydrate binding modules: biochemical properties and novel applications. *Microbiology and Molecular Biology Reviews* 70 (2): 283–295.

Shu, Z., Wang, Y., An, L., and Yao, L. (2014). The slowdown of the endoglucanase Trichoderma reesei Cel5A-catalyzed cellulose hydrolysis is related to its initial activity. *Biochemistry* 53 (48): 7650–7658.

Sinitsyn, A., Mitkevich, O., and Klesov, A. (1986). Inactivation of cellulolytic enzymes by stirring and their stabilization by cellulose. *Prikladnaya Biokhimiya i Mikrobiologiya* 22 (6): 759–765.

Sinitsyn, A.P., Gusakov, A.V., and Vlasenko, E.Y. (1991). Effect of structural and physico-chemical features of cellulosic substrates on the efficiency of enzymatic hydrolysis. *Applied Biochemistry and Biotechnology* 30: 43–59.

Sipos, B., Benkő, Z., Dienes, D. et al. (2010). Characterisation of specific activities and hydrolytic properties of cell-wall-degrading enzymes produced by Trichoderma reesei rut C30 on different carbon sources. *Applied Biochemistry and Biotechnology* 161 (1–8): 347–364.

(PATENT) Skinner W, Tokuyama F: US4081328 (1978)

Skomarovsky, A., Markov, A., Gusakov, A. et al. (2006). New cellulases efficiently hydrolyzing lignocellulose pulp. *Applied Biochemistry and Microbiology* 42 (6): 592–597.

Solomon, K.V., Haitjema, C.H., Henske, J.K. et al. (2016). Early-branching gut fungi possess a large, comprehensive array of biomass-degrading enzymes. *Science* 351 (6278): 1192–1195.

Somerville, C., Bauer, S., Brininstool, G. et al. (2004). Toward a systems approach to understanding plant-cell walls. *Science* 306 (5705): 2206–2211.

Song, X., Zhang, S., Wang, Y. et al. (2016). A kinetic study of Trichoderma reesei Cel7B catalyzed cellulose hydrolysis. *Enzyme and Microbial Technology* 87: 9–16.

Sørensen, H.R., Pedersen, S., Jørgensen, C.T., and Meyer, A.S. (2007). Enzymatic hydrolysis of wheat Arabinoxylan by a recombinant "minimal" enzyme cocktail containing β-Xylosidase and novel endo-1, 4-β-Xylanase and α-l-Arabinofuranosidase activities. *Biotechnology Progress* 23 (1): 100–107.

Steenbakkers, P.J., Li, X.L., Ximenes, E.A. et al. (2001). Noncatalytic docking domains of cellulosomes of anaerobic fungi. *Journal of bacteriology* 183 (18): 5325–5333.

Strobel, H.J. (1995). Growth of the thermophilic bacterium *Clostridium thermocellum* in continuous culture. *Current Microbiology* 31: 210–214.

Stutzenberger, F.J. (1979). Degradation of cellulosic substances by *Thermomonospora curvata*. *Biotechnology and Bioengineering* 21 (5): 909–913.

Su, T.M. and Paulavicius, I.R.E.N.E. (1975). Enzymatic saccharification of cellulose by thermophilic actinomyces. *Polymer Composites Symposium* 28: 221–236.

Sukumaran, R.K. (2009). Bioethanol from lignocellulosic biomass part II: production of cellulases and hemicellulases. In: *Handbook of Plant-Based Biofuels* (ed. A. Pandey), 141–158. Boca Raton, FL: CRC Press.

Sutcliffe, R. and Saddler, J.N. (1986). The role of lignin in the adsorption of cellulases during enzymatic treatment of lignocellulosic material. *Biotechnology and Bioengineering Symposium* 17: 749–762.

Sweeney, M.D. and Xu, F. (2012). Biomass converting enzymes as industrial biocatalysts for fuels and chemicals: recent developments. *Catalysts* 2 (2): 244–263.

Tabka, M., Herpoël-Gimbert, I., Monod, F. et al. (2006). Enzymatic saccharification of wheat straw for bioethanol production by a combined cellulase xylanase and feruloyl esterase treatment. *Enzyme and Microbial Technology* 39 (4): 897–902.

Tai, S.K., Lin, H.P.P., Kuo, J., and Liu, J.K. (2004). Isolation and characterization of a cellulolytic Geobacillus thermoleovorans T4 strain from sugar refinery wastewater. *Extremophiles* 8 (5): 345–349.

Takashima, S., Ohno, M., Hidaka, M. et al. (2007). Correlation between cellulose binding and activity of cellulose-binding domain mutants of Humicola grisea cellobiohydrolase 1. *FEBS Letters* 581 (30): 5891–5896.

Tamaru, W. and Doi, R.H. (1999). Three surface layer homology domains at the N terminus of the Clostridium cellulovarans major cellulosomal subunit EngE. *Journal of Bacteriology* 181 (10): 3270–3276.

Tao, J. and Kazlauskas, R.J. (2011). *Biocatalysis for Green Chemistry and Chemical Process Development*. Wiley Online Library.

Teeri, T.T. (1997). Crystalline cellulose degradation: new insight into the function of cellobiohydrolases. *Trends in Biotechnology* 15 (5): 160–167.

Teeri, T., Koivula, A., Linder, M. et al. (1998). *Trichoderma reesei* cellobiohydrolases: why so efficient on crystalline cellulose? *Biochemical Society Transactions* 26 (2): 173–177.

Tengborg, C., Galbe, M., and Zacchi, G. (2001). Influence of enzyme loading and physical parameters on the enzymatic hydrolysis of steam-pretreated softwood. *Biotechnology Progress* 17 (1): 110–117.

Todorovic, R., Grujic, S., and Matavulj, M. (1987). Effect of reaction end-products on the activity of cellulolytic enzymes and xylanase of *Trichoderma harzianum*. *Microbiology Letters* 36 (143–144): 113–119.

Ustinov, B.B., Gusakov, A.V., Antonov, A.I., and Sinitsyn, A.P. (2008). Comparison of properties and mode of action of six secreted xylanases from Chrysosporium lucknowense. *Enzyme and Microbial Technology* 43 (1): 56–65.

Uzcategui, E., Johansson, G., Ek, B., and Pettersson, G. (1991a). The 1, 4-β-d-glucan glucanohydrolases from *Phanerochaete chrysosporium*. Re-assessment of their significance in cellulose degradation mechanisms. *Journal of Biotechnology* 21 (1): 143–159.

Uzcategui, E., Ruiz, A., Montesino, R. et al. (1991b). The 1, 4-β-D-glucan cellobiohydrolases from *Phanerochaete chrysosporium*. I. A system of synergistically acting enzymes homologous to Trichoderma reesei. *Journal of Biotechnology* 19 (2–3): 271–285.

Vaaje-Kolstad, G., Horn, S.J., van Aalten, D.M.F. et al. (2005). The non-catalytic chitin-binding protein CBP21 from Serratia marcescens is essential for chitin degradation. *Journal of Biological Chemistry* 280 (31): 28492–28497.

Valadares, F., Gonçalves, T.A., Gonçalves, D.S. et al. (2016). Exploring glycoside hydrolases and accessory proteins from wood decay fungi to enhance sugarcane bagasse saccharification. *Biotechnology for biofuels* 9 (1): 110.

Valjamae, P., Sild, V., Pettersson, G., and Johansson, G. (1998). The initial kinetics of hydrolysis by cellobiohydrolases I and II is consistent with a cellulose surface erosion model. *European Journal of Biochemistry* 253 (2): 469–475.

Väljamäe, P., Sild, V., Nutt, A. et al. (1999). Acid hydrolysis of bacterial cellulose reveals different modes of synergistic action between cellobiohydrolase I and endoglucanase I. *European Journal of Biochemistry* 266 (2): 327–334.

Van Dyk, J. and Pletschke, B. (2012). A review of lignocellulose bioconversion using enzymatic hydrolysis and synergistic cooperation between enzymes—factors affecting enzymes, conversion and synergy. *Biotechnology Advances* 30 (6): 1458–1480.

Vandamme, E.J., Logghe, J.M., and Geeraerts, H.A. (1982). Cellulase activity of a thermophilic *Aspergillus fumigatus* (fresenius) strain. *Journal of Chemical Technology and Biotechnology* 32 (7-12): 968–974.

Vardakou, M., Katapodis, P., Topakas, E. et al. (2004). Synergy between enzymes involved in the degradation of insoluble wheat flour arabinoxylan. *Innovative Food Science & Emerging Technologies* 5 (1): 107–112.

Verjans, P., Dornez, E., Segers, M. et al. (2010). Truncated derivatives of a multidomain thermophilic glycosyl hydrolase family 10 xylanase from Thermotoga maritima reveal structure related activity profiles and substrate hydrolysis patterns. *Journal of Biotechnology* 145 (2): 160–167.

Vinzant, T.B., Adney, W.S., Decker, S.R. et al. (2001). Fingerprinting *Trichoderma reesei* hydrolases in a commercial cellulase preparation. *Applied Biochemistry and Biotechnology* 91-93: 99–107.

Vlasenko, E., Schülein, M., Cherry, J., and Xu, F. (2010). Substrate specificity of family 5, 6, 7, 9, 12, and 45 endoglucanases. *Bioresource Technology* 101 (7): 2405–2411.

Vocadlo, D.J. and Davies, G.J. (2008). Mechanistic insights into glycosidase chemistry. *Current Opinion in Chemical Biology* 12 (5): 539–555.

de Vries, R.P. and Visser, J. (2001). Aspergillus enzymes involved in degradation of plant cell wall polysaccharides. *Microbiology and Molecular Biology Reviews* 65 (4): 497–522.

Wang, S.S. and Converse, A.O. (1991). On the use of enzyme adsorption and specific hydrolysis rate to characterize thermal-chemical pretreatment. *Applied Biochemistry and Biotechnology* 34-35: 61–74.

Wang, C.M., Shyu, C.L., Ho, S.P., and Chiou, S.H. (2008). Characterization of a novel thermophilic, cellulose-degrading bacterium *Paenibacillus sp.* strain B39. *Letters in applied microbiology* 47 (1): 46–53.

Wang, M., Li, Z., Fang, X. et al. (2012). Cellulolytic enzyme production and enzymatic hydrolysis for second-generation bioethanol production. In: *Biotechnology in China III: Biofuels and Bioenergy*, 1–24. Springer.

Wang, J., Quirk, A., Lipkowski, J. et al. (2013). Direct in situ observation of synergism between cellulolytic enzymes during the biodegradation of crystalline cellulose fibers. *Langmuir* 29 (48): 14997–15005.

Warden, A.C., Little, B.A., and Haritos, V.S. (2011). A cellular automaton model of crystalline cellulose hydrolysis by cellulases. *Biotechnology for Biofuels* 4 (1): 1.

Wei, H., Xu, Q., Taylor, L.E. et al. (2009). Natural paradigms of plant cell wall degradation. *Current Opinion in Biotechnology* 20 (3): 330–338.

Weimer, P. and Odt, C.L. (1995). Cellulose degradation by ruminal microbes: physiological and hydrolytic diversity among ruminal cellulolytic bacteria. *ACS Symposium Series* 618: 291–304.

Weimer, P.J. and Schmidt, J.K. (1989). *Attachment of Fibrobacter succinogenes to Cellulose and Cellulose Derivatives.* U.S. Dairy Forage Research Center, USDA.

Westereng, B., Ishida, T., Vaaje-Kolstad, G. et al. (2011). The putative endoglucanase PcGH61D from *Phanerochaete chrysosporium* is a metal-dependent oxidative enzyme that cleaves cellulose. *PLoS One* 6 (11): e27807.

White, B.A. and Morrison, M. (2001). Genomic and proteomic analysis of microbial function in the gastrointestinal tract of ruminants – review. *Asian-Australasian Journal of Animal Sciences* 14 (6): 880–884.

Wilson, D.B. (2008). Three microbial strategies for plant cell wall degradation. *Annals of the New York Academy of Sciences* 1125 (1): 289–297.

Wilson, D.B. (2012). Processive and nonprocessive cellulases for biofuel production—lessons from bacterial genomes and structural analysis. *Applied Microbiology and Biotechnology* 93 (2): 497–502.

Wilson, D.B. and Irwin, D.C. (1999). Genetics and properties of cellulases. In: *Recent Progress in Bioconversion of Lignocellulosics*, 1–21. Springer.

Wilson, D.B. and Kostylev, M. (2012). Cellulase processivity. *Methods in Molecular Biology (Clifton, NJ)* 908: 93–99.

Wilson, C.A. and Wood, T.M. (1992). The anaerobic fungus Neocallimastix frontalis: isolation and properties of a cellulosome-type enzyme fraction with the capacity to solubilize hydrogen-bond-ordered cellulose. *Applied microbiology and biotechnology* 37 (1): 125–129.

Wingren, A., Galbe, M., Roslander, C. et al. (2005). Effect of reduction in yeast and enzyme concentrations in a simultaneous-saccharification-and-fermentation-based bioethanol process. Technical and economic evaluation. *Applied Biochemistry and Biotechnology* 121–124: 485–499.

Wood, T.M. and McCrae, S.I. (1982). Purification and some properties of a (1→ 4)-β-d-glucan glucohydrolase associated with the cellulase from the fungus *Penicillium funiculosum*. *Carbohydrate Research* 110 (2): 291–303.

Wood, T.M. and Phillips, D.R. (1969). Another source of cellulase. *Nature* 222 (5197): 986–987.

Wood, T.M., McCrae, S.I., and Bhat, K.M. (1989). The mechanism of fungal cellulase action. Synergism between enzyme components of Penicillium pinophilum cellulase in solubilizing hydrogen-bond ordered cellulose. *The Biochemical Journal* 260: 37–43.

Wooley, R., Ruth, M., Sheehan, J. et al. (1999). *Lignocellulosic Biomass to Ethanol Process Design and Economics Utilizing co-Current Dilute Acid Prehydrolysis and Enzymatic Hydrolysis: Current and Futuristic Scenarios*. National Renewable Energy Laboratory.

Wyman, C.E. (2007). What is (and is not) vital to advancing cellulosic ethanol. *Trends in Biotechnology* 25 (4): 153–157.

Xu, F. and Ding, H. (2007). A new kinetic model for heterogeneous (or spatially confined) enzymatic catalysis: contributions from the fractal and jamming (overcrowding) effects. *Applied Catalysis A: General* 317 (1): 70–81.

Yang, B. and Wyman, C.E. (2004). Effect of xylan and lignin removal by batch and flowthrough pretreatment on the enzymatic digestibility of corn Stover cellulose. *Biotechnology and Bioengineering* 86 (1): 88–98.

Yang, B. and Wyman, C.E. (2006). BSA treatment to enhance enzymatic hydrolysis of cellulose in lignin containing substrates. *Biotechnology and Bioengineering* 94 (4): 611–617.

Yang, B. and Wyman, C.E. (2007). Biotechnology for cellulosic ethanol. *Asia Pacific Biotech News* 11 (9): 555–563.

Yang, B. and Wyman, C.E. (2008). Pretreatment: the key to unlocking low-cost cellulosic ethanol. *Biofuels, Bioproducts and Biorefining* 2 (1): 26–40.

Yang, B., Willies Deidre, M., and Wyman Charles, E. (2006a). Changes in the enzymatic hydrolysis rate of Avicel cellulose with conversion. *Biotechnology and Bioengineering* 94 (6): 1122–1128.

Yang, H., Ichinose, H., Nakajima, M. et al. (2006b). Synergy between an. ALPHA.-L-Arabinofuranosidase from *Aspergillus oryzae* and an Endo-Arabinanase from Streptomyces coelicolor for degradation of Arabinan. *Food Science and Technology Research* 12 (1): 43–49.

Yang, S.J., Kataeva, I., Wiegel, J. et al. (2010). Classification of '*Anaerocellum thermophilum*' strain DSM 6725 as *Caldicellulosiruptor bescii sp.* nov. *International journal of systematic and evolutionary microbiology* 60 (9): 2011–2015.

Yang, B., Dai, Z., Ding, S.-Y., and Wyman, C.E. (2011). Enzymatic hydrolysis of cellulosic biomass. *Biofuels* 2 (4): 421–450.

Ye, Z., Berson, R.E. 2011. *Kinetic modeling of cellulose hydrolysis with first order inactivation of adsorbed cellulase.*

Zhang, Y.H. and Lynd, L.R. (2003). Quantification of cell and cellulase mass concentrations during anaerobic cellulose fermentation: development of an enzyme-linked immunosorbent assay-based method with application to *Clostridium thermocellum* batch cultures. *Analytical Chemistry* 75 (2): 219–227.

Zhang, Y.H. and Lynd, L.R. (2004). Toward an aggregated understanding of enzymatic hydrolysis of cellulose: noncomplexed cellulase systems. *Biotechnology and Bioengineering* 88 (7): 797–824.

Zhang, Y.H. and Lynd, L.R. (2005). Determination of the number-average degree of polymerization of cellodextrins and cellulose with application to enzymatic hydrolysis. *Biomacromolecules* 6 (3): 1510–1515.

Zhang, Y.H.P. and Lynd, L.R. (2005a). Cellulose utilization by *Clostridium thermocellum*: bioenergetics and hydrolysis product assimilation. *PNAS* 102: 7321–7325.

Zhang, Y.H.P. and Lynd, L.R. (2005b). Regulation of cellulase synthesis in batch and continuous cultures of *Clostridium thermocellum*. *Journal of Bacteriology* 187 (1): 99–106.

Zhang, Y.H.P. and Lynd, L.R. (2006). A functionally based model for hydrolysis of cellulose by fungal cellulase. *Biotechnology and Bioengineering* 94 (5): 888–898.

Zhang, J. and Viikari, L. (2012). Xylo-oligosaccharides are competitive inhibitors of cellobiohydrolase I from *Thermoascus aurantiacus*. *Bioresource Technology* 117: 286–291.

Zhang, S., Wolfgang, D.E., and Wilson, D.B. (1999). Substrate heterogeneity causes the nonlinear kinetics of insoluble cellulose hydrolysis. *Biotechnology and Bioengineering* 66: 35–41.

Zhang, J., Tuomainen, P., Siika-Aho, M., and Viikari, L. (2011a). Comparison of the synergistic action of two thermostable xylanases from GH families 10 and 11 with thermostable cellulases in lignocellulose hydrolysis. *Bioresource Technology* 102 (19): 9090–9095.

Zhang, M., Rongxin, S., Wei, Q. et al. (2011b). Enzymatic hydrolysis of cellulose with different crystallinities studied by means of SEC-MALLS. *Chinese Journal of Chemical Engineering* 19 (5): 773–778.

Zhang, J., Tang, M., and Viikari, L. (2012). Xylans inhibit enzymatic hydrolysis of lignocellulosic materials by cellulases. *Bioresource Technology* 121: 8–12.

Zhou, W., Xu, Y., and Schüttler, H.B. (2010). Cellulose hydrolysis in evolving substrate morphologies III: time-scale analysis. *Biotechnology and Bioengineering* 107 (2): 224–234.

14

Life Cycle Assessment of Biofuels and Green Commodity Chemicals

Mairi J. Black[1], *Onesmus Mwabonje*[2], *Aiduan Li Borrion*[3] *and Aurelia Karina Hillary*[2]

[1] *Department of Science, Technology, Engineering & Public Policy (STEaPP), University College London, London W1T 6EY, UK*
[2] *Centre for Environmental Policy, Imperial College London, London SW7 2AZ, UK*
[3] *Department of Civil, Environmental and Geomatic Engineering, University College London, London WC1E 6BT, UK*

CHAPTER MENU

14.1 Introduction

The concepts and technologies for making and using biofuels have been variously promoted for nearly 200 years, since the invention of the internal combustion engine in 1826 (Songstad et al. 2009). However, it has been the early twenty-first century that has seen biofuels become a global commercial reality. The reason for their promotion and development at this time has been the result of concerns about energy security; economic development (to promote flagging rural economies and provide alternatives for energy access); concerns about the depletion of fossil fuel reserves; and anthropogenic climate change (Nuffield Council on Bioethics 2011, p. 8). In 2015, it was reported that 64 countries have biofuel policy mandates or obligations in place to support the development and use of biofuels (REN21. 2016, p. 182).

Green Energy to Sustainability: Strategies for Global Industries, First Edition.
Edited by Alain A. Vertès, Nasib Qureshi, Hans P. Blaschek and Hideaki Yukawa.

Biofuels have been generally classified as first generation (1G),[1] second generation (2G),[2] and third generation (3G),[3] based on the feedstock used to produce them, whilst other definitions refer to conventional[4] and advanced biofuels,[5] reflecting the technologies used to convert the feedstock (Murphy et al. 2011). Multiple conversion pathways including biomass to energy (heat and power), biomass to biofuels (liquid fuels) and biomass to green chemicals and bioproducts, are often referred to under the concept of a biorefinery,[6] where one or several conversion processes may occur.

Biofuel policies have often been based on the 'local' availability of feedstocks for first generation biofuels, with countries which grow sugar or starch crops favouring bioethanol, and countries which grow vegetable oil crops favouring biodiesel, however, both bioethanol and biodiesel are now globally traded to meet the demands for renewable fuels (OECD/FAO 2016, p. 116). Many biofuel policy mandates have set aspirational targets, with high expectations for second generation biofuels becoming commercially viable at scale (REN21 2016, p. 39).

For some policies, one of the key reasons and priorities for promoting biofuels has been to reduce Greenhouse Gas (GHG) emissions from fuel use in the transport sector (e.g. UK's Renewable Transport Fuels Obligation 2007 (The Renewable Transport Fuels Obligation 2017); US Renewable Fuel Standard 2007 (US EPA 2017)). The development of these policies has necessitated a means of accounting for GHG emissions associated with the production and use of biofuels, compared to the fossil fuels they are intended to replace. Life Cycle Assessment (LCA) has been widely adopted for this purpose and, as both biofuels policies and biofuels technologies continue to develop, LCA has also been required to develop alongside. In this chapter we review the use of LCA for feedstocks, conversion technologies and end product opportunities for bioethanol (and green commodity chemicals), focussing on variables which arise as the result of choices made in LCA methodology.

14.2 Life Cycle Assessment (LCA)

LCA has evolved into a major decision support tool for sustainability-related issues, including assessing the GHG emissions reductions of biofuels and biochemicals in comparison to fossil sources (e.g. oil or coal as fuels and feedstocks for biochemicals). The quality of the decision support LCA provides, is determined in terms of its relevance to the type of questions to be answered. Originally, the starting point for LCA was with its application to relatively simple choices, for instance, in making technical changes to a product or choosing one material over another e.g. in relation to packaging. The outcomes of the LCA have then been used in influencing consumer choices. Over time, there has been a shift in LCA thinking to more encompassing

1 First Generation (1G) biofuels are derived from commodity crops with easily accessible sugars, starches, and oils which are processed into bioethanol and biodiesel, through relatively simple conversion technologies.
2 Second Generation (2G) biofuels are derived from cellulose, hemicelluloses, and lignocelluloses of plant biomass. Feedstocks can be crop residues, forestry feedstock (residues or dedicated crops) or purpose grown non-food crops. 2G feedstocks require more complex routes to access sugars for further conversion to biofuels.
3 Third Generation (3G) biofuels is a term which has been developed to allow separate consideration of algal biomass as a feedstock source for biofuel production.
4 Conventional biofuels are those biofuels produced using processes which are technologically and market mature (i.e. sugar/starch fermentation or methyl esterification of fatty acids).
5 Advanced biofuels are biofuels produced using pre-commercial technologies. Also defined as those which provide greater GHG reductions than conventional biofuels.
6 Biorefinery has been defined as the sustainable processing of biomass into a spectrum of marketable products and energy (de Jong and Jungmeier 2015, p. 5).

questions, such as the benefits of biofuels and biochemicals versus the fossil-based materials they are replacing (Curran 2012).

LCA is the most reliable method used to verify environmental impacts and support claims. It provides designers, regulators, and engineers with valuable information for exploring decisions in each life cycle stage of materials, buildings, services, and infrastructure. LCA identifies environmental hot spots in products and materials and establishes a benchmark against which improvements can be measured. Many companies use LCA to demonstrate transparency and corporate credibility to stakeholders and customers. LCA is also used in new product research and development, when the environmental footprint of a new product is important to its future marketing or cost structure. LCA is recognized in business rationales, as consumer and regulatory environmental expectations are increasing in demand and sophistication. The benefit of LCA is simple: reliable, transparent data for both manufacturers and consumers, enabling better decisions. The International Organization for Standardization (ISO) defines LCA as a 'Compilation and evaluation of the inputs, outputs and the potential environmental impacts of a product system throughout its life cycle' (ISO 14044: 2006).

14.3 The Origin and Principles of Life Cycle Assessment

LCA owes its roots to the energy crisis dating from the early 1970s (Boustead and Hancock 1979; Curran 1996). At that time, the primary concern was with constraints on the supply of fossil fuel and the impacts of these constraints on energy security. This has since extended into the science of industrial ecology (Socolow et al. 1994), involving a more general study of the interactions between the environment, humans, and activities. A range of techniques have been developed within industrial ecology, for analysing environmental burdens ranging from the specific burdens due to the manufacture of an individual product, to general burdens of a set of manufacturing processes or a set of human activities. A number of methods derived from the first law of thermodynamics form basic analytical tools for industrial ecology. These tools include the approaches of material and systems flow analysis (Matthews et al. 2014). LCA extends these approaches at the product or process level by attempting to quantify the environmental burdens of the use of materials and services, particularly, the burdens of specific production and processing systems (Rebitzer et al. 2004). The framework for accounting for the environmental burdens associated with a specified product, process, or service is one of the primary considerations when conducting a LCA (ISO 14044: 2006). The recognised methodological framework is considered in Figure 14.1.

LCA is therefore the only existing logical methodology for quantifying the effects of human activities on the environment, from extraction of raw materials to production of a product, to end of life disposal. LCA can be applied to agricultural systems and practices, industrial processes and waste management among others, and can be used to assess a number of impact categories (Figure 14.2).

14.4 Developing a Life Cycle Assessment

The first steps in a LCA study involve defining the goal and scope of the analysis. This is followed by the development of a Life Cycle Inventory (LCI), that forms the basis for a Life Cycle Impact Assessment (LCIA) and results interpretation. The definition of the system being assessed, and the function of the system are crucial elements of any LCA study. This involves describing the system that is being analysed, such as an individual product, a production process, the provision of a service

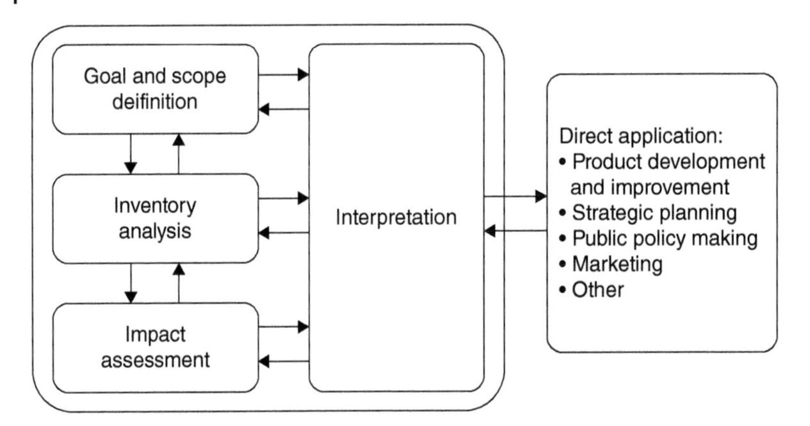

Figure 14.1 The general methodological framework for LCA (European Commission, Joint Research Centre, Institute for Environment and Sustainability, 2010a). The figure describes the Life Cycle Assessment framework for accounting for environmental impacts associated with products and services.

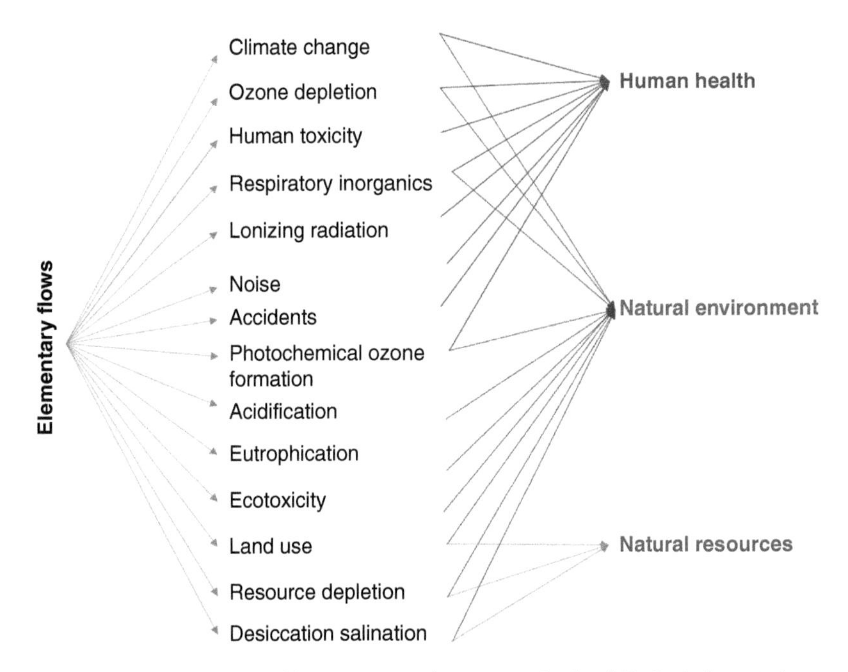

Figure 14.2 Framework of impact categories assessed using Life Cycle Impact Assessment (LCIA) (European Commission, Joint Research Centre, Institute for Environment and Sustainability, 2010b). The figure describes specific impact categories (on health, the environment or on natural resources) which can be selected for assessment singly, or in a multi-criteria approach using LCA methodology and software applications.

or some other human activity, both quantitatively and qualitatively. The extent of the system being assessed can be specified quite flexibly, enabling the use of the LCA framework for research questions related to single products or more widely to a company, an activity or a country or region, the latter sometimes being referred to as scenario or system-level studies. The description of an appropriate system boundary is closely associated with the definition of the system and its functions. The system boundary can be referred to as an imaginary line drawn around all the activities that are

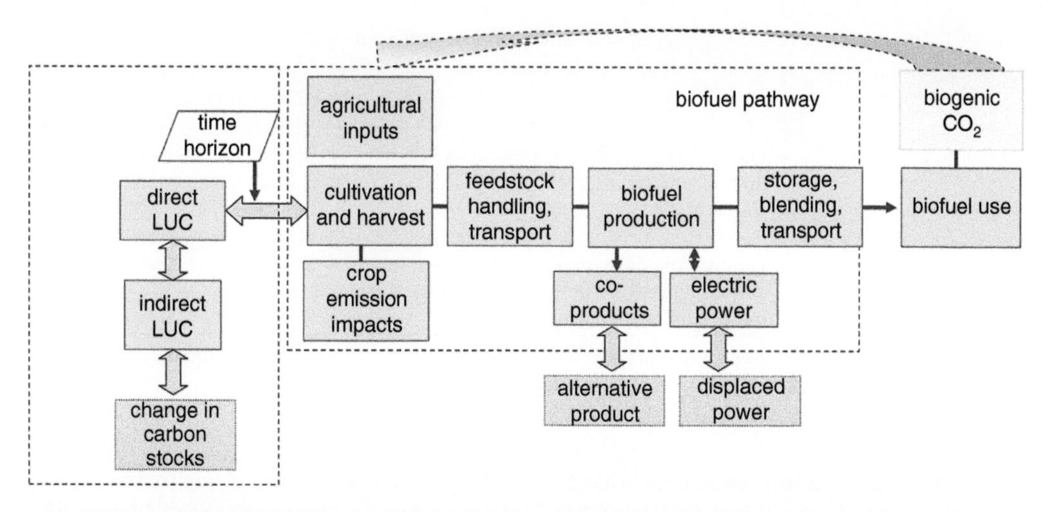

Figure 14.3 System boundary for biofuel production and use, including indirect effects (Sanchez et al. 2012). The figure represents a flowchart for biofuel production and use, and provides an example of how systems boundaries may be drawn.

relevant to the LCA analysis being conducted. Although it is common for the system boundary to be considered as being a spatial concept, it has to be emphasized that it has a time-based dimension that is of equal importance (e.g. the annual or perennial nature of biofuel feedstock). The specific spatial and time-based dimension of a system boundary is significant because it defines what is included and what is excluded from the system and its analysis. All LCAs, including those for biofuels and biochemicals, limit the boundary of the production pathways to a manageable system. Figure 14.3 demonstrates a general system boundary for the biofuel production pathway (Sanchez et al. 2012).

14.5 Scope of the Life Cycle Assessment: Attributional verses Consequential

A critical step in LCA involves clearly stating the goal and scope, which requires a clear understanding of the question that the LCA is intended to address. Similarly, the detailed LCA methods applied must be appropriate for addressing the stated question, to allow the implementation of the LCA. This becomes a very important issue when considering the outcomes of LCA studies on biofuels and biochemicals, especially when comparing results from different studies, policy reporting frameworks and claims made for particular products.

Guidance offered to identify appropriate methodologies whilst undertaking LCA for different aims and objectives, distinguishes two different types of LCA. These are referred to as 'attributional LCA' (aLCA) and 'consequential LCA' (cLCA) (Brander et al. 2009; Curran et al. 2005). The two approaches share common features, fundamentally being variations of the same basic methodology and the distinctions between them can be quite subtle. The key differences between the two approaches have been summarized in Table 14.1 (Brander et al. 2009).

At present, the application and interpretation of aLCA and cLCA creates much debate (Zamagni et al. 2012; Matthews et al. 2014) and the lack of clear definition of these methodologies and associated terms is often the basis of disagreements on the GHG emissions associated with biofuels. The Joint Research Centre's (JRC's) International Reference Life Cycle Data System (ILCD)

Table 14.1 Application differences between aLCA and cLCA (Brander et al. 2009).

	aCLA	cLCA
Appropriate situation	• Quantifying and understanding emissions direct from the life cycle of a product; • Consumption-based emissions.	• Informing consumers and policy-makers on the change in total emissions from purchasing or policy decisions.
Inappropriate situation	• Quantifying the change in total emissions consequence from policies that change the output of a certain product.	• Consumption-based emissions.

The table describes the different scenarios and target audiences to inform the selection of attributional or consequential LCA methodologies.

Handbook is one notable attempt to provide guidance, by expanding and clarifying these detailed options (European Commission, Joint Research Centre, Institute for Environment and Sustainability, 2010b). One vital aspect of consequential modelling is that it is not depicting the actual process as an attributional model does, but it is modelling the forecasted consequences of decisions (European Commission, Joint Research Centre, Institute for Environment and Sustainability, 2010b). In view of this description, cLCA is an approach that has been adopted to answer policy questions and to anticipate potential future outcomes for biofuels and biochemical environmental impacts.

14.6 Biofuels and Green Commodity Chemicals

A large number of feedstock options, conversion technologies and end product outcomes have been anticipated since the promotion of biofuels in policy, and since the development of the biorefinery concept (de Jong and Jungmeier, 2015). Figure 14.4 gives a simple overview of some of the options available for feedstock-conversion routes.

Many specialized technologies have already been developed and are moving from the laboratory to the commercialization stage, with over 100 companies regularly showcased in the biofuel, biorefinery, or bioeconomy press (BiofuelsDigest). A recent review of biomass-fed chemical processes and projects reported in the press[7] from 2013 to 2015, found 28 specific chemicals or homologous series and 20 biomass sources from a total of 243 references (Cooker, 2016).

14.7 Feedstocks for Biofuels

The type of feedstock used for biofuel production impacts the LCA as the result of the inputs required to cultivate, harvest, and process the material pre-conversion. These impacts may be somewhat alleviated if co-products are allocated a 'share' of impacts. One of the benefits of second generation bioethanol is that it can utilize the residues and by-products of a primary growing system and, depending on the LCA convention adopted, share the input burden with the main product.

7 Chemical Engineering Progress and Chemical and Engineering News publications.

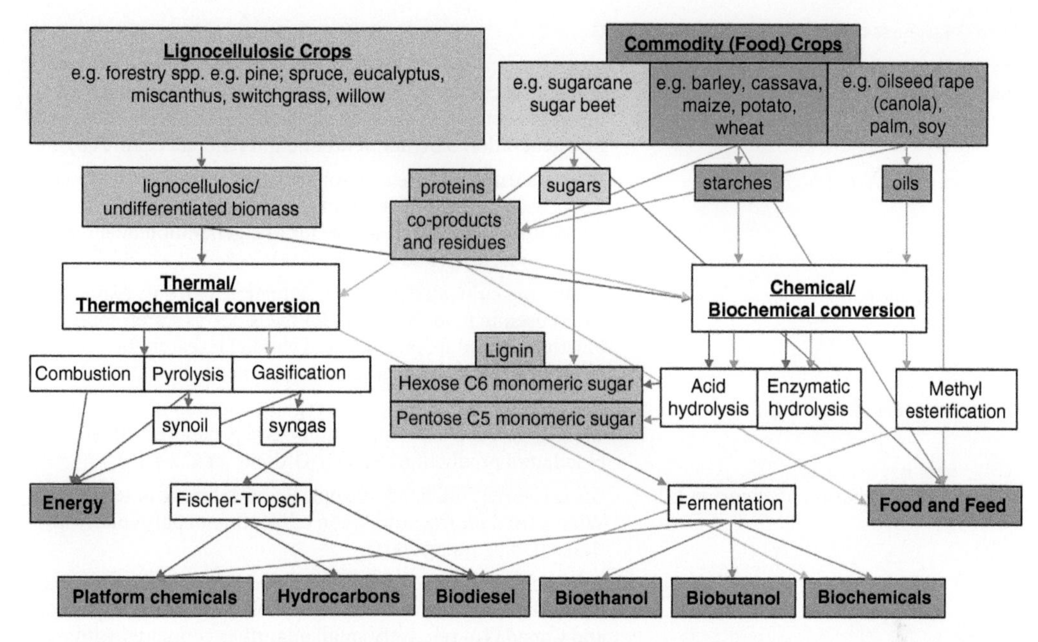

Figure 14.4 Conversion routes for biomass feedstocks to biofuels and green chemicals. Source: Modified from Black et al. (2011). The figure describes possible routes to biofuels and green chemicals and considers feedstock types; conversion technologies and product outcomes which are the focus of research (and increasingly LCA), to provide alternatives to fossil derived fuels and materials.

14.7.1 First Generation Biofuels

Crops which yield sugars and starches have been developed over hundreds of years to become universally traded commodities, supplying the world with food. These crops also have the potential to yield easily accessible feedstock for conversion to biofuels. A list of crops which have been considered for the production of bioethanol using conventional conversion technologies is given in Table 14.2. LCA of first generation biodiesel has also not been included in this chapter, although LCA studies have also been developed for the main commodity and novel biodiesel feedstocks (e.g. oilseed rape [canola] (Stephenson et al. 2008); oil palm (Soraya et al. 2014); soy (Panichelli et al. 2009); and jatropha (Kumar et al. 2012) and camelina (Dangol et al. 2015)).

14.7.2 Second Generation Biofuel Feedstock

Feedstocks for second generation biofuels are the celluloses, hemicelluloses, and lignocelluloses of biomass material. They can be sourced from agricultural residues left after commodity food crops are harvested and processed; forestry residues and sawmill residues from the processing of wood biomass to primary timber products (hardwood and softwood), dedicated energy crops grown on short rotations or biodegradable municipal solid wastes (MSW). A list of lignocellulosic materials which are being considered for second generation bioethanol production is given in Table 14.3.

14.8 Conversion of Feedstock

Biomass can be converted into energy and products via a number of different conversion routes (Figure 14.4) that require significantly different inputs and processes, all of which have a profound

Table 14.2 First generation biofuel feedstocks.

Barley (*Hordeum vulgare* L.)	Grain crop cultivated in temperate regions. Grains used for brewing beer and in food. Shows potential as starch source for first generation bioethanol production (Gibreel et al., 2009).
Cassava (*Manihot esculenta*)	Woody shrub cultivated in tropical and sub-tropical regions. Roots used as carbohydrate source in food (tapioca). Opportunities as starch source for first generation bioethanol production (Marx and Nquma, 2013; Pervez et al., 2014).
Maize (*Zea mays* L.)	Grain crop cultivated widely in temperate to tropic regions. Grain used in food. Now a major feedstock as starch source for bioethanol production subject to varietal research for bioethanol production efficiency (Gumienna et al., 2016).
Sugar beet (*Beta vulgaris* L.)	Root crop grown in temperate regions. Used for the production of sugar for food use. Opportunity as sugar source for bioethanol production (Salazar-Ordóñez et al., 2013).
Sugarcane (*Saccharum officinarum* L.)	Grass species cultivated in tropical and subtropical regions. Widely used for the production of bioethanol particularly in Brazil (Goldemberg 2007).
Wheat (*Triticum* spp.)	Grain crop cultivated worldwide. Used as a staple food in many countries. Used to produce bioethanol in the European Union and Canada (in relatively small quantities compared with sugarcane ethanol and corn ethanol) (Saunders et al. 2011).

The table describes commodity sugar and starch crops used or researched for the production of bioethanol. (This table is not intended to be exhaustive and does not include the many new or 'novel' crops which have been explored as feedstock for biofuels (e.g. sweet sorghum) (Almodares and Hadi, 2009)).

impact on the LCA of the end-product. The main routes for biomass conversion to biofuels and biochemicals have been described by Demirbaş (2001) and McKendry (2002) and are summarized briefly below.

14.8.1 Biochemical Processes

The LCA of converting sugars in biomass to useful products by fermentation is impacted during the conversion phase, as the result of the pre-treatments required to release the sugars prior to fermentation, as well as the fermentation process itself. This is the benefit of using first generation feedstock where the sugars are more easily accessible. Releasing sugars from cellulose, hemicellulose and lignocellulose complexes requires more significant pre-processing to give additional access to acid or enzymes to hydrolyse polysaccharides into sugars.

14.8.2 Thermochemical Processes

Thermochemical processes require heat to convert biomass into a range of end products that depend on the operating conditions of the process. Carbonization occurs at temperatures below 520 K resulting in the production of gases and biochar; liquefaction occurs at 525–600 K (5–20 MPa) producing gases, biochar, and 'biocrude'; pyrolysis at 650–800 K (0.1–0.5 MPa) producing more gases and less biochar and biocrude and gasification at temperatures above 800 K, which produces mainly gases. Products of thermochemical processes may subsequently require further refining or additional conversion to transform them into useful biofuels and biochemicals.

Table 14.3 Second generation biofuel feedstocks.

Agricultural residues	
Cereal straws	Straw has intrinsic value to farmers (in the maintenance of soil quality and as animal bedding) but a proportion is considered surplus and has potential as lignocellulosic feedstock (Talebnia et al., 2010).
Maize (*Zea mays* L.) stover	Corn stover is the field residue of corn production which has intrinsic value in the maintenance of soil quality but is also considered as a lignocellulosic feedstock option (Aden 2008).
Sugar beet (*Beta vulgaris* L.) pulp	Sugar beet pulp is the fibrous residue which remains after processing to extract sugar. Attractive as an option for bioethanol production as does not require drying before conversion (Zheng et al. 2013).
Sugarcane (*Saccharum officinarum* L.) bagasse	Sugarcane bagasse is the process residue remaining after sugar is extracted from sugarcane. It is widely used for energy production in sugarcane processing but is also considered to have great potential as a lignocellulosic feedstock (Cardona et al. 2010).
Forestry and sawmill residues	Hardwood and softwood species from managed forests and wood processing industries have been widely considered as lignocellulosic bioethanol feedstock (Gonzalez et al., 2011).
Energy crops	
Willow (*Salix* spp.)	Deciduous, temperate shrub with options for growth on short rotation for bioenergy/biofuel uses (Ray et al. 2012).
Miscanthus (*Miscanthus* spp.)	Widely distributed genus of grass species, rapidly growing biomass with opportunities as bioenergy or biofuel feedstock (Lee and Kuan 2015).
Switchgrass (*Panicum virgatum* L.)	Native North American perennial prairie grass, highly productive and widely considered for use as biofuel/biochemical feedstock (Keshwani and Cheng, 2009).

The table describes examples of cellulose, hemicellulose, and lignocellulose sources from agricultural, forestry, and energy crops. (This table is not intended to be exhaustive and does not include the new or 'novel' crops which are being researched as feedstock for biofuels/bioenergy or MSW).

14.9 Supply Chain and Logistics

As part of the LCA of biofuels, inputs associated with the transport of biomass from the field to the processing plant, and for the biofuel to the point of distribution, may also be considered within the LCA systems boundary. The logistical component of the biomass-biofuel/bioenergy production chain may have a significant impact, depending on the distances travelled, the mode of transport, and the density of the biomass being transported (Shen et al. 2015).

14.10 Using LCA as a Tool to Assess GHG Emissions and Other Impacts Associated with Bioethanol Production and Supply

As discussed in Section 14.4, developing a LCA should follow strict guidelines that have been developed to ensure both consistency of approach and comparability of LCA studies (e.g. ISO 2006:14040; ISO 2006:14044; JRC 2010a). Even following these guidelines however, there is no single correct methodology for the LCA assessment of biofuel production and supply chains. Where GHG LCA calculations are required for reporting into policy mechanisms, e.g. the UK Renewable Transport Fuels Obligation[8] (2007) and subsequent amendments (2009, 2011, 2013, and 2015), a specific set of guidelines and a dedicated calculator have been developed for use in reporting, to allow consistency and comparability of calculations. Tables 14.4 reviews some recent LCA studies of bioethanol, to highlight where studies which have been developed using different functional units, systems boundaries, data sources and software tools lead to different outcomes which are not necessarily directly comparable. The table also highlights the different conversion technologies that lead to different outcomes from the LCA.

14.11 Discussion on the Suitability of LCA

The complexity and variability of LCA systems can be seen from the sample of papers reviewed in Table 14.4. It is clear that the first consideration when looking at LCA as a methodology, is the question the study is intended to answer. How the question is framed directly influences the boundaries of the LCA and the impact categories that the study is intended to address. For biofuels, that have been promoted in policies to replace fossil fuels and reduce GHG emissions, having transparent and comparable LCA systems is critical to allow biofuel supply chains to be compared and assessed against GHG emissions reductions targets. Gnansounou et al. (2009) have reviewed some of the key environmental impact assessment methodologies that have been used to inform biofuels policies and highlight again the need to understand the system definition and boundaries of the methodology (including the approach taken, whether land-use is included or not, the biofuel blend and the use of the biofuel), the functional unit and allocation methods used. With the development of second generation biofuels and conversion technologies, an increasing number of studies using different LCA approaches have been appearing in the academic press. Singh et al. (2010) identifies the key parameters that influence the outcome of LCAs for lignocellulosic biomass conversion to ethanol as: country; biomass source; system adopted; functional unit; system boundary; land-use change; impact category analysed; sensitivity analysis and reference system. A study by Morales et al. (2015) has reviewed more than 100 case studies of LCAs of lignocellulosic biofuels and reports geographic location, feedstock, functional unit, impact categories, reference system, method of analysis (software used) and allocation. Likewise, Gerbrandt et al. (2016) have also extensively reviewed the LCA literature and have focussed on the impact of LCA methodological choices for corn stover bioethanol.

Apart from the methodological issues associated with using LCAs, data availability and data sourcing are also critical parts of the LCA, with some data available within LCA software tools. However, more specific outcomes might be achieved for studies using actual data that address a specific feedstock, geographic area, supply chain and conversion plant design. For second generation biofuels, studies have had to rely on simulation studies for conversion plants at scale as, until recently, there has been no commercial scale production. In spite of this, the suitability of LCA as a technique for assessing GHG emissions in accounting for climate change of biofuels and biochemicals has been challenged on the basis of the suitability of the data used (Delucchi 2011, 2013). Normally, this does not necessarily translate into substantial criticisms of the fundamental principles and approaches of LCA itself. Rather, these specific challenges are associated with aspects of cLCA of biofuel production, where modelling of indirect land-use change and vegetation carbon stock changes becomes the subject of debate and influence on the outcome of the cLCA. It

8 The UK Renewable Transport Fuels Obligation requires 'obligated parties' to report GHG emissions associated with the biofuels which are blended into road transport fuels.

Table 14.4 Summary of parameters and outcomes of Life Cycle Assessment studies in selected references.

Reference	**Life cycle assessment of bio-based ethanol produced from different agricultural feedstocks (Munoz et al. 2014).**
Type of assessment	Attributional LCA (inventory analysis, economic allocation assuming 100% allocation to ethanol); sensitivity; and uncertainty analysis.
Aims	1. To identify the main activities in the life cycle which drive various environmental impacts 2. To explore the differences between feedstock 3. To benchmark bio-based ethanol with fossil-based alternative from a cradle to-grave perspective
Conversion technology	First generation fermentation
Feedstock	Sugarcane (Brazil south central and north east), maize grain (US) and stover (US); sugar beet (France) and wheat (France)
Functional unit	1 kg of ethanol (5% water content)
Systems boundary	Cradle-to-gate and cradle-to-grave *Includes:* Cultivation; Land use; Crop rotation time; Ethanol production Emissions associated with ethanol disposal (use) to air *Not included:* Distribution
Data sources	Ecoinvent
LCA software used	Simapro v.7.3.2
Impact categories	Biological and Ecosystems Services (BES); Biodiversity Damage Potential (BDP); Climate Regulation Potential (CRP); Biotic Production Potential (BPP); Freshwater Regulation Potential (FWRP); Erosion Regulation Potential (ERP); Water Purification Potential through Physicochemical Filtration (WPP-PCF); and Water Purification Potential through Mechanical Filtration (WPP-MF)
Outcomes	The cradle-to-gate boundary showed that sugar beet production in France had the lowest GHG emissions value, while wheat had the highest (depending of the system for assessing Land Use Change [LUC]). Water footprint results showed a much lower water demand from fossil-based ethanol. BES impact categories identified fossil based ethanol as the preferred routes to produce ethanol. The best environmental performance was shown by sugar beet. The results from bio-based ethanol are very much dependent on the feedstocks, the net product yield per hectare, year, and modelling of emissions caused by LUC.
Reference	**Environmental sustainability analysis of UK whole-wheat bioethanol and CHP system (Martinez-Hernandez et al. 2013).**
Type of assessment	Attributional LCA (inventory analysis, economic allocation) and sensitivity analysis
Aims	1. To assess the environmental impact of the UK wheat bioethanol plant as a stand-alone system or as a whole-wheat system integrated with wheat straw Combined Heat and Power (CHP) plant. 2. To establish the marginal benefits in terms of GWP100 and primary energy savings, compared to the fossil resources to be replaced (natural gas or gasoline) 3. To study the relative LCI of Distillers Dried Grain with Solubles (DDGS) as a commodity to the production of heat and CHP, compared to its use as animal feed

(Continued)

Table 14.4 (Continued)

Conversion technology	First generation fermentation
Feedstock	Wheat grain (UK)
Functional unit	Different common functional units were used; unit mass of dry matter of product, unit energy, year, and ha.
Systems boundary	Cradle-to-grave. *Includes:* Cultivation; land use; ethanol production (including construction materials for each step); transportation
Data sources	Aspen Plus model
LCA software used	Aspen Plus
Impact categories	Cumulative Primary Fossil Energy (CPE); Global Warming Potential over 100 yr (GWP_{100}), Acidification Potential (AP), and Eutrophication Potential (EP), and Abiotic Resource Use (ARU).
Outcomes	85% emissions reduction can be achieved using wheat straw for CHP in an integrated wheat bioethanol plant, compared to fossil fuel. Using DDGS does not reduce emissions but improves the energy use of the system and saves CPE. Hot spots identified and compared in impact categories between bioethanol production system and wheat cultivation.
Reference	**Life cycle analysis for bioethanol production from sugar beet crops in Greece (Foteinis et al. 2011)**
Type of assessment	Attributional LCA (inventory and impact assessment).
Aims	1. To evaluate whether the potential transformation of existing sugar plants of Northern Greece to modern bioethanol plants, under the current state-of-art cultivations of sugar beets would be an environmentally sustainable decision. 2. Use a LCA approach to evaluate bioethanol production from sugar beet and compare with existing sugar production as well as against replacement of gasoline and other types of biodiesel from Greek cultivations.
Conversion technology	First generation fermentation
Feedstock	Sugar beet (Greece)
Functional unit	35 Gcal or 146.46 GJ (based on energy content) of 6800 l/5400 kg bioethanol which is the bioethanol yield from 1 ha of land (based on yield of 65 t sugar beet/ha)
Systems boundary	Cradle-to-gate *Includes*: Cultivation; ethanol production; transportation
Data sources	Ecoinvent, CML2 baseline method; field surveys and interviews; bibliographic data for plant processing
LCA software used	Simapro v 7.14, Eco-Indicator 99
Impact categories	Global Warming Potential (GWP_{100})
Outcomes	The conversion of existing sugar processing plants to bioethanol production plants provides a viable alternative with less environmental impact than the traditional processing facilities.
Reference	**Life cycle assessment of bioethanol from wheat and sugar beet discussing environmental impacts of multiple concepts of co-product processing in the context of the European Renewable Energy Directive (Buchspies and Kaltschmitt, 2016).**

(Continued)

Table 14.4 (Continued)

Type of assessment	Attributional LCA (sensitivity analysis and uncertainty analysis; co-product allocation and substitution by energy content).
Aims	1. To assess potential environmental impacts related to the production of bioethanol and to discuss implications of co-product processing concepts under legislative conditions. 2. Co-product options based on use for energy (or as animal feed)
Conversion technology	First generation fermentation
Feedstock	Sugar beet and wheat grain (Germany)
Functional unit	1 MJ of anhydrous ethanol (0.03% H_2O)
Systems boundary	Cradle-to-gate *Includes:* Cultivation; ethanol production; transport; wastewater treatment; material and energy provision
Data sources	Renewable Energy Directive
LCA software used	Umberto NXT LCA v7.1
Impact categories	Global Warming Potential for 100 year time horizon (GWP_{100}), Cumulative Fossil Energy Demand (CED_{fossil}), Terrestrial Acidification Potential (TAP), Photochemical Oxidant Formation Potential (POFP), Marine Eutrophication (MEP), and Freshwater Eutrophication Potential (FEP)
Outcomes	The outcomes show that assessing the allocation of several co-product options can have a significant impact on the outcomes of the bioethanol LCA, depending on the co-product end use. Accounting for co-products which have not previously been considered in the RED e.g. biogas production from waste water in beet processing can further lower the GWP_{100} of bioethanol.
Reference	**Sustainability assessment of large-scale ethanol production from sugarcane (Pereira and Ortega 2010).**
Type of assessment	Attributional LCA and Embodied Energy Analysis Method (EEA)
Aims	1. To assess the sustainability of ethanol produced in large scale from sugarcane 2. To examine its environmental feasibility through the use of fossil fuel embodies energy and energy assessment
Conversion technology	First generation fermentation
Feedstock	Sugarcane (Brazil)
Functional Unit	1 kg of sugarcane, 1 l of ethanol and 1 J of ethanol
Systems boundary	Cradle-to-gate *Includes:* Cultivation; ethanol production including co-product and by-product use; transport
Data sources	Literature review, official database, actual farm data and interview with sugarcane industry producers and experts, as well as equipment suppliers
LCA software used	None
Impact categories	Transformity, renewability, Energy Yield Ratio (EYR), Environmental Loading Ratio (ELR), Energy Sustainability Index (ESI), and Energy Exchange Ratio (EER)

Table 14.4 (Continued)

Outcomes	In the production chain, the agricultural stage accounted for 83% of the energy flows, while industrial steps accounted for 15% and sugarcane transport for 2% energy needs. The transformity assessment showed that the sugarcane system is more efficient in comparison to corn ethanol. Renewability decreases from 35.4% for sugarcane to 31% for ethanol, which shows that ethanol production depends on the fossil fuels. The EYR results showed that ethanol production has the ability to use the local natural renewable resources' potential even though sugarcane production has a higher EYR. The environmental impact due to industrial and transport for ethanol fell in the 'moderate category', however when the extensive area of sugarcane cultivation is also considered, the impacts become significant, thus a higher ELR results. ESI has calculated that both sugarcane and ethanol have the ability to provide net energy to economy but at the expense of the environment.
Reference	**Life cycle assessment of selected future energy crops for Europe (Rettenmaier et al. 2010).**
Type of assessment	Attributional LCA (sensitivity analysis, allocation according to value, energy content or mass)
Aims	To identify parameters that are crucial for the establishment of successful non-food cropping systems in the EU 27, with the following research questions: 1. What are the environmental advantages of biofuels and bioenergy made from selected dedicated crops in comparison to their fossil equivalents? 2. Which energy crop(s) or which crop group should be chosen from an environmental point of view? 3. Which type of biofuel or bioenergy should be produced from an environmental point of view? 4. What are the effects of the choices made regarding methods and data on the results?
Conversion technology	First and second generation biochemical and thermochemical pathways (fermentation, transesterification, hydrogenation, gasification, hydrolysis as well as direct combustion)
Feedstock	Sugar crops (sugar beet and sweet sorghum) oil crops (rapeseed, sunflower, Ethiopian mustard), lignocellulosic (poplar, willow, eucalyptus, reed canary grass, mischanthus, switchgrass, giant reed and cardoon)
Functional unit	Useful output per hectare in an average year
Systems boundary	Cradle-to-grave *Includes:* Cultivation; land use; biomass conversion; use (as an energy carrier) and transport
Data sources	Literature review and current databases
LCA software used	None
Impact categories	Primary non-renewable energy use, GHG effect, acidification, eutrophication, summer smog, ozone depletion and human toxicity

(Continued)

Table 14.4 (Continued)

Outcomes	The conversion stage has the largest influence on energy and GHG balances, while cultivation stage is most important in terms of acidification, eutrophication, and ozone depletion. The utilization stage has a considerable impact on acidification and eutrophication. Other energy crops show environmental advantages in terms of energy and GHG saving, however ambiguous results regarding acidification, eutrophication, ozone depletion, summer smog and human toxicity which suggest that the overall environmental performance of biofuels and bioenergy from the crops has to be based on the subjective value of choices instead. In terms of the best energy crops, lignocellulosic crops have the lowest energy and GHG savings. In the identification of the best biofuel or bioenergy chain, lignocellulosic crops show the highest potential in saving energy and GHG emissions, and the best pathway is the use of biomass for Combined Heat and Power (CHP), followed by heat generation and second generation bioethanol. With regard to the agricultural LUC, direct LUC only has a significant effect in terms of GHG emissions if grassland on organic soil is converted into cropland. For first generation bioethanol from sugar beet, the indirect LUC supersedes soy production in Brazil.
Reference	**Hybrid input-output life cycle assessment of first and second generation ethanol production technologies in Brazil (Watanabe et al., 2016).**
Type of assessment	Hybrid approach of LCA and I-O Analysis (IOA) and Sensitivity analysis
Aims	To assess current and novel sugarcane ethanol production technologies in Brazil using hybrid LCA approach throughout the Brazilian economy and to assess three autonomous distilleries configurations
Conversion technology	Biochemical pathway (fermentation, hydrolysis)
Feedstock	Sugarcane
Functional unit	None
Systems boundary	Cradle-to-gate Scenarios of three autonomous distilleries configurations: 1. 1G base: basic first generation distillery, represents an average Brazilian autonomous distillery in 2009 2. 1G-optimized: modern first generation distillery, allows selling surplus electricity to the grid using sugarcane straw and bagasse. 3. 1G-base and 1G-optimized: integrated modern first and second generation
Data sources	Journal references
LCA software used	None
Impact categories	Economic (direct and indirect economic effects) and environmental benefits
Outcomes	The direct and indirect economic effects of increased demand analysis showed that the highest economic effect is associated with the 1G optimized scenario. On the other hand, if direct and indirect effects of thermoelectricity displacement are taken into account, the overall effect is reduced to the similar impact of the 1G base scenario. The direct and indirect GHG emissions of increased demand showed that 1G base bioethanol production technology has the highest emissions compared to other technologies. 1G optimized and 1G2G using sugarcane reduced the direct emissions of ethanol production dramatically. Major contributions from indirect GHG emissions were found to be 1G-optimized and 1G2G ethanol production from diesel, gasoline, inorganic chemicals, cements, and other refined petroleum products. The industrial emissions are higher in 1G-optimized scenario because of the bagasse and straw burning to produce more electricity surplus, compare to 1G2G.

(Continued)

Table 14.4 (Continued)

Reference	**Environmental sustainability of bioethanol production from wheat straw in the UK (Wang et al. 2013).**
Type of assessment	Attributional life cycle and sensitivity analysis
Aims	1. To assess the environmental profile of bioethanol produced from wheat straw using different pre-treatment technologies and through well-to-wheel life cycle 2. To compare bioethanol (E100) pathways with conventional petrol
Conversion technology	Second generation biochemical pathway (saccharification and fermentation)
Feedstock	Wheat straw (UK)
Functional unit	To drive 1 km in a Flexible Fuel Vehicle (FFV)
Systems boundary	Cradle-to-grave *Includes:* Scenarios of pretreatment pathways: Wet Oxidation (WO), Liquid Hot Water (LHW), Steam Explosion (SE), Steam Explosion with Acid Catalyst (SEAC), and Dilute Acid (DA)
Data sources	Questionnaires, AspenPlus model, Ecoinvent database, IPCC, 2009 EMEP-EEA Guidebook and literature reviews
LCA software used	Simapro v7.3
Impact categories	Abiotic Depletion Potential (ADP), Acidification Potential (AP), Eutrophication Potential (EP), Global Warming Potential (GWP), Human Toxicity Potential (HTP), Ozone Depletion Potential (ODP), Freshwater Aquatic Ecotoxicity Potential (FAETP), Terrestrial Ecotoxicity Potential (TEP), and Photochemical Oxidants Creation Potential (POCP)
Outcomes	The production of bioethanol from wheat straw is overall environmentally favourable over petrol. However, the results also indicated that wheat straw cultivation stage contributes the highest impacts especially in GWP, ODP, and EP from fertilizer usage, followed by the majority impacts from enzyme production in most of the impact categories. When wheat straw removal process was included in the boundary, the emissions in GWP and EP were affected due to the need of additional fertilizer in replacement of wheat straw. In comparison to petrol, it is shown that wheat straw ethanol produced with SE, LHW, and WO pretreatment methods are overall favourable over petrol particularly in ADP, GWP, ODP, ecotoxicity, and POCP impacts. For AP and EP impact, higher environmental burdens come from bioethanol pathways because of the combustion in the Combined Heat and Power process and the use of fertilizer in the agricultural process.
Reference	**Energetic-environmental assessment of a scenario for Brazilian cellulosic ethanol (Agostinho and Ortega, 2013).**
Type of assessment	Four main methodologies are used: Embodied Energy Analysis, Ecological Rucksack, Energy Accounting, and Gas Emission Inventory. The results of a biorefinery scenario were then compared with conventional ethanol plant (CEP) and the Integrated Systems of Food, Energy and Environmental Services (IFEES) production. Sensitivity analysis was not carried out.
Aims	The objective of this work is to assess, through a multi-criteria approach, the energetic-environmental performance of a Biorefinery scenario in Brazil.

(Continued)

Table 14.4 (Continued)

Conversion technology	A plausible scenario – an adjacent plant to current ethanol plant in Brazil producing cellulosic ethanol from a fraction of sugarcane bagasse. The industrial processes of the adjacent plant would be: (i) pretreatment of bagasse through steam explosion and sulfuric acid; (ii) enzymatic hydrolysis to convert cellulose and hemicellulose into fermentable sugars; (iii) biological fermentation of C5 and C6 sugars; (iv) ethanol distillation; (v) liquid–solid (stillage-lignin) separation; (vi) stillage feed-back to sugarcane field as fertilizer; (vii) lignin burnt to produce steam and electricity.
Feedstock	Sugarcane bagasse
Functional unit	Ton EtOH
Systems boundary	Cradle-to-gate (including first generation sugarcane ethanol production and adjacent cellulosic ethanol production from sugarcane bagasse)
Data sources	Emission factors – USEPA, Raw data – published literatures (two papers)
LCA software used	SUMMA
Impact categories	GWP, energy, AC
Outcomes	1. Biorefinery is able to produce 1.57 times more ethanol than Conventional Ethanol Plants (CEP) and 40 times more than IFEES. 2. IFEES had better performance for all other indicators (except transformity index), compared to the biorefinery and CEP. 3. Biorefinery and CEP have similar net energy efficiency, while the biorefinery has better energy efficiency. 4. For all other indices (water, biotic material, GWP, and AC), biorefinery has a better rating than CEP
Reference	**Integrated versus stand-alone second generation ethanol production from sugarcane bagasse and trash (Dias et al. 2012).**
Type of assessment	Comparative LCA, scenario analysis, sensitivity analysis
Aims	This paper aims to evaluate environmental impacts of second generation ethanol production from sugarcane bagasse and trash. Both for stand-alone and integrated (with first generation ethanol production from sugarcane juice) plants were assessed.
Conversion technology	Second generation ethanol production consists of pretreatment, hydrolysis, and fermentation. Different configurations to improve conventional bioethanol conversion are considered including waste recovery, pretreatment, and hydrolysis technologies.
Feedstock	Sugarcane bagasse and trash
Functional unit	Kg EtOH
Systems boundary	Cradle-to-gate
Data sources	NREL process model (1996) Aspen plus simulation Literature
LCA software used	SimaPro and CML2
Impact categories	Abiotic Depletion (ADP); Acidification (AP); Eutrophication (EP); Global Warming (GWP); OzoneLayer Depletion (ODP); Human Toxicity(HTP); Fresh Water Aquatic Ecotoxicity (FWAET); Marine Aquatic Ecotoxicity (MAET); Terrestrial Ecotoxicity (TET); and Photochemical Oxidation (POP)

(Continued)

Table 14.4 (Continued)

Outcomes	1. Integrated plants using current technology presents the best environmental indicators for most categories among all the evaluated alternatives. 2. Integrated plants require less steel and yield more electricity output per unit of ethanol produced than stand-alone plants. 3. Steel used in the industrial equipment has little influence on the ethanol environmental impacts. 4. Future second generation ethanol scenarios has higher environmental impacts due to high sodium hydroxide consumption for alkaline delignification prior to hydrolysis. 5. Standalone plant presented the highest environmental impacts because it produces only second generation ethanol. 6. Integrating plants in Brazil has equivalent impacts as exporting lignocellulosic materials to countries with similar conditions.
Reference	**Environmental life cycle assessment of bioethanol production from wheat straw (Borrion et al. 2012).**
Type of assessment	Attributional LCA
Aims	An LCA was carried out to assess the environmental burdens of ethanol production from wheat straw and its use as ethanol blend fuels. Two ethanol based fuel E15 (a mixture of 15% ethanol and 85% petrol by volume) and E85 (85% ethanol and 15% petrol by volume) were assessed and results were compared to those of conventional petrol (PT) in 1 km driven by an equivalent car.
Conversion technology	NREL process (pretreatment, enzymatic hydrolysis and fermentation)
Feedstock	Wheat straw
Functional unit	Kg ethanol and the amount of fuel for travelling 1 km
Systems boundary	Cradle-to-gate and cradle-to-grave
Data sources	NREL simulation, EcoInvent
LCA software used	SimaPro 7.2
Impact categories	Global warming, ozone depletion, photochemical oxidant formation, acidification, ecotoxicity, eutrophication, water depletion, and fossil depletion
Outcomes	1. Compared to petrol, life cycle GHG emissions are lower for ethanol blends, with a 73% reduction for an E85-fuelled car and 13% reduction with E15. 2. A modest savings of 40% and 15% in fossil depletion was also reported when using E85 and E15 respectively. 3. Similar results are also observed for ozone depletion. 4. Ethanol blend use does not offer any advantages compared with petrol, in terms of acidification, eutrophication, ecotoxicity, and water depletion.
Reference	**Exploring impacts of process technology development and regional factors on life cycle greenhouse gas emissions of corn stover ethanol (McKechnie et al. 2015).**
Type of assessment	Life cycle GHG emissions, mass allocation, system expansion allocation
Aims	This paper aims to examine impacts of regional factors affecting biomass and process input supply chains and ongoing technology development on the life cycle GHG emissions of corn stover ethanol production.
Conversion technology	NREL process model (2002, 2011)
Feedstock	Corn stover

(Continued)

Table 14.4 (Continued)

Functional unit	1 MJ of ethanol (E100)
Systems boundary	Cradle to-gate (corn cultivation and bio refinery)
Data sources	Published literature National Agricultural Statistics Service (NASS) survey (fertilizer and yield) NREL process model (2002, 2011)
LCA software used	Spreadsheet-based model
Impact categories	GHGs (CO_2, CH_4, and N_2O) based on 100-year global warming potentials
Outcomes	1. GHG emissions from corn stover supply ranges from −6 gCO_2eq./MJ ethanol to 13 gCO_2eq./MJ ethanol. This reflects location-specific soil carbon and N_2O emissions responses to stover removal. 2. GHG emissions from Biorefinery based on the 2011 NREL process model are the highest (18 gCO_2eq./MJ ethanol) and are approximately double those assessed for the 2002 NREL process model. 3. Energy demands of on-site enzyme production included in the 2011 design contribute to reducing the electricity co-product and associated emissions credit, which is also dependent on the GHG-intensity of regional electricity supply. 4. Life cycle GHG emissions vary between 1.5 and 22 gCO_2eq/MJ ethanol (2011 design) depending on production location. 5. Using system expansion for co-product allocation, ethanol production in studied locations meet the Energy Independence and Security Act emissions requirements for cellulosic biofuels; however, regional factors and on-going technology developments significantly influence these results.
Reference	**Impacts of pre-treatment technologies and co-products on greenhouse gas emissions and energy use of lignocellulosic ethanol production (Pourbafrani et al. 2014).**
Type of assessment	Life cycle GHG and energy use; system expansion allocation; sensitivity scenario analysis
Aims	The aim is to understand the life cycle energy use and GHG emissions implications of alternative pre-treatment technologies and co-products through developing a consistent life cycle framework for ethanol production from corn stover.
Conversion technology	The process is based on NREL 2002 design model, but adopting three pretreatment pathways (dilute acid, AFEX, and autohydrolysis) and four potential co-products (electricity, lignin pellets, xylitol, and protein).
Feedstock	Corn stover
Functional unit	1 MJ of fuel produced (E85 or gasoline) and used in a light-duty vehicle
Systems boundary	The E85 pathways include activities associated with corn stover collection and transportation to a biorefinery, conversion of corn stover to ethanol in the biorefinery, blending of the ethanol with gasoline and its distribution, combustion of E85 in a flexible fuel light-duty vehicle and finally, utilizing biorefinery co-products.
Data sources	NREL process model (2002) GREET 1.8d.1
LCA software used	Not specified, but data from GREET 1.8d.1
Impact categories	GHG emission and life cycle energy use

(Continued)

Table 14.4 (Continued)

Outcomes	1. Ethanol yield, life cycle energy use and GHG emissions are affected by the choices of pretreatment technologies. 2. Dilute acid pathways generally exhibit higher ethanol yields and lower net total energy use than the autohydrolysis and ammonia fibre expansion pathways. 3. Similar GHG emissions are found for the pretreatment technologies when producing the same co-product. Xylitol co-production diverts xylose from ethanol production and results in the lowest ethanol yield. 4. Compared to producing only electricity as a co-product, the co-production of pellets and xylitol decreases life cycle GHG emissions associated with the ethanol, while protein production increases emissions. 5. The life cycle GHG emissions of blended ethanol fuel E85 range from −38.5 to 37.2 g CO_2eq/MJ of fuel produced, reducing emissions by 61–141% relative to gasoline. 6. All ethanol pathways result in major reductions of fossil energy use relative to gasoline, at least by 47%.
Reference	**Impact of Cellulase Production on Environmental and Financial Metrics for Lignocellulosic Ethanol (Hong et al. 2013).**
Type of assessment	Life cycle GHG emissions using scenarios and process simulation
Aims	This study evaluates life cycle emissions and cellulase production costs for bioethanol production, considering on-site and off-site production options.
Conversion technology	NREL process model (2012), with a complete enzyme production process simulated using AspenPlus. Both on-site and off-site production for enzymes were considered
Feedstock	Corn stover
Functional Unit	litre ethanol
Systems boundary	Cradle-to-gate
Data sources	NREL process model (2012) Aspenplus simulation – Cellulase production mass and energy balance Greet model
LCA software used	Aspenplus v7.3
Impact categories	GHG emissions
Outcomes	1. GHG emissions for cellulase production range from 10.2 to 16.0 g CO_2 eq/g enzyme protein, depending on on-site or off-site production and the method of transportation. 2. Enzyme GHG emissions are predicted to be 258 g CO_2 eq/l of ethanol for on-site production, versus 403 g CO_2 eq/l for off-site production.
Reference	**Life cycle assessment of hemp hurds use in second generation ethanol production (González-García et al. 2012).**
Type of assessment	CML methodology, scenario analysis, sensitivity analysis, different allocation methods
Aims	This paper aims to assess the environmental performance of ethanol produced from hemp hurds, using the Life Cycle Assessment (LCA) approach. The environmental performance of two ethanol fuel blends, E10 and E85, in a flexi fuel vehicle was analysed and compared.
Conversion technology	NREL process model (2002)

(Continued)

Table 14.4 (Continued)

Feedstock	Hemp hurds
Functional unit	The amount of fuel for the distance travelled by vehicles with the vehicle tank full of conventional petrol
Systems boundary	Hemp cultivation, transport of hurds to the ethanol plant, conversion into ethanol, transport and distribution of ethanol and ethanol blends combustion in FFVs
Data sources	Questionnaires NREL process model (2002) Published literature
LCA software used	Not specified CML methodologies
Impact categories	Fossil fuel use, global warming, acidification, eutrophication, and photochemical oxidant formation
Outcomes	1. Ethanol-based fuels can offer improved environmental performance in impact categories such as global warming, as well as a decrease in fossil fuels use. 2. The use of petrol is the best option in terms of other impact categories, such as photochemical oxidants formation, acidification, and eutrophication. 3. The choice of allocation methods significantly influences the environmental performance.
Reference	**Life cycle assessment of butanol production in sugarcane bio-refineries in Brazil (Pereira et al. 2015).**
Type of assessment	Life cycle assessment, sensitivity analysis and Virtual Sugarcane Biorefinery (VSB)
Aims	1. To compare the different technological configurations or the integrated production of butanol in sugarcane biorefineries from the environment point of view 2. To compare the studied biorefinery route with the petrochemical pathway of butanol based in the Ecoinvent version 2.2 database To assess the potential of butanol as an environmentally friendly fuel by comparing it with the use of gasoline in a gasoline dedicated vehicle
Conversion technology	3. Biochemical pathway
Feedstock	4. Sugarcane
Functional unit	1. To answer first objective: kg of butanol for butanol producing scenarios 2. To answer second objective: US$ earned 3. To answer third objective: km run by a vehicle powered with a gasoline dedicated engine using butanol and gasoline as fuels
Systems boundary	Cradle-to-grave Scenarios include: 1. 1GRS and 1G2GRS used a regular microorganism (*Clostridiium saccahroperbutylaceronium DSM 2152*) 2. Scenarios 1GMS and 1G2GMS used a genetically modified strain (*Clostridium beijerinckii BA 101)* 3. Bases comparison: the microorganisms *S.cerevisae*
Data sources	Ecoinvent v2.2, CML 2 Baseline 2000 v2.05 method
LCA software used	Simapro 7.3.3, CanaSoft Model

(Continued)

Table 14.4 (Continued)

Impact categories	Abiotic Depletion Potential (ADP), Global Warming Potential (GWP), Ozone Depletion Potential (ODP), Human Toxicity Potential (HTP), and Photochemical Oxidation (POP)
Outcomes	1. The results for first objective showed that the first generation scenarios (1G, 1GRS, and 1GMS) displayed the lowest impact for ADP, GWP, ODP, HTP, and POP. On the other hand, second generation (1G2G, 1G2GRS, and 1G2GMS) presented the lowest AP and EP values. 2. Results for second objective showed that the second generation scenarios have lower environmental impacts than first generation per US$ revenue, in particularly, the scenario 1G2GMS Compared to the petrochemical route, the biomass-based butanol has higher values for acidification and eutrophication due to the use of fertilizer on the agricultural stage.

The table provides comparisons of how differences in LCA methodological choices, systems boundaries and functional units impact on outcomes, before considering the impacts of technology and feedstock choices (i.e. first and second generation technologies for various biomass feedstocks in the production of bioethanol).

follows that the details of LCA calculations and reported results must be specified and carried out so that they are appropriate for addressing the production, consumer, policy, techno-economic, or engineering question that is under consideration. In addition, the results of LCA studies ought to be interpreted with a clear understanding of the research questions originally raised (Matthews et al. 2014).

It is quite evident that there is no single answer to a LCA for a given subject. The result depends strongly on the way in which the system is defined and the boundaries are selected. As described above, this is just one basic aspect of the LCA methodology. It follows that other aspects of LCA methodology may also vary and can strongly influence the result generated. It is very easy, therefore, for different LCA studies to produce a variety of results that sometimes may differ. It is, therefore, unsurprising that results for GHG emissions show a wide range that may appear confusing and sometimes contradictory, particularly, with a system as sophisticated as the production and consumption of biofuels and biochemicals, given the many possible sources of feedstocks and possibilities for process chain stages, and the increasing opportunities for by-products and wastes.

14.12 Perspectives: Moving Forward with the LCA Concept

Open-access and commercially available software for carrying out LCA analyses are becoming increasingly sophisticated and, as biofuels systems are developing, more data is becoming available to the LCA practitioner so that LCA outcomes are becoming more robust and rely less on generic data and assumptions. Modelling of impact categories also continues to develop, providing more sophisticated analyses to support the LCA approach. LCA is a powerful but potentially complex tool which should always be implemented according to guidelines that have been set out (e.g. European Commission, Joint Research Centre, Institute for Environment and Sustainability, 2010b). The design of the LCA, data sources and reporting of results should always be as transparent as possible, as this is critical to allow for meaningful comparisons to be made and differences in outcomes to be understood.

The interest in biofuels and sustainable biochemicals as alternatives to fossil resource-based fuels and chemicals has been evolving from theory to practice and from the laboratory to the market. This industry is set to continue its development. The use of biofuels and other bioproducts is well aligned with current trends in circular economy thinking, recycling, and the bioeconomy. The bioeconomy, comprising those economic activities that use renewable biological resources (BIS/15/146 2015), is one of the two cycles considered in circular economy thinking, i.e. the biological cycle, which views the production of feedstocks and conversion to products separately to the technical cycle, which considers the minerals and mined materials used in manufacture and maintenance of factories and infrastructure. Both cycles are inter-related through the end user/consumer and the management of waste generated by both systems (EMF 2013). There are various pathways for achieving circular biological material cycles, including adding value to biomass, by-products and waste materials through further bioenergy or biochemical extraction, with the aim of creating a zero waste circular system. LCA, as a tool, can support decision-making in pathways to go forward and help answer questions on the comparative sustainability of a circular process. Many of the LCA challenges discussed in this chapter also apply when assessing circular biological material pathways. Therefore, this approach requires ongoing understanding of supply chain LCA, system boundaries, displacement effects, direct and indirect effects and the impacts these have on wider consideration of sustainability.

References

Aden, A. (2008). *Biochemical Production of Ethanol from Corn Stover: 2007 State of Technology Model.* CO: National Renewable Energy Laboratory Golden.

Agostinho, F. and Ortega, E. (2013). Energetic-environmental assessment of a scenario for Brazilian cellulosic ethanol. *Journal of Cleaner Production* 47: 474–489. https://doi.org/10.1016/j.jclepro.2012.05.025.

Almodares, A. and Hadi, M.R. (2009). Production of bioethanol from sweet sorghum: a review. *African Journal of Agricultural Research* 4: 772–780.

Biofuels Digest [WWW Document], n.d. URL http://www.biofuelsdigest.com (accessed 12.19.16).

BIS/15/146, 2015. Supporting growth of the UK bioeconomy: opportunities from waste - Publications - GOV.UK [WWW Document]. URL https://www.gov.uk/government/publications/supporting-growth-of-the-uk-bioeconomy-opportunities-from-waste (accessed 12.20.16).

Black, M.J., Whittaker, C., Hosseini, S.A. et al. (2011). Life cycle assessment and sustainability methodologies for assessing industrial crops, processes and end products. *Industrial Crops and Products*, The next generation of industrial crops, processes and products AAIC 2009 34: 1332–1339. https://doi.org/10.1016/j.indcrop.2010.12.002.

Borrion, A.L., McManus, M.C., and Hammond, G.P., web-support@bath.ac.uk, (2012). Environmental life cycle assessment of bioethanol production from wheat straw. *Biomass and Bioenergy* 47: 9–19.

Boustead, I. and Hancock, G.F. (1979). *Handbook of Industrial Energy Analysis* (ed. E. Horwood). Chichester, England\New York: Halsted Press.

Brander, M., Tipper, R., Hutchison, C., Davis, G., 2009. Technical Paper: Consequential and attributional approaches to LCA: a Guide to policy makers with specific reference to greenhouse gas LCA of biofuels. Ecometrica Press. Date correction required in reference section

Buchspies, B. and Kaltschmitt, M. (2016). Life cycle assessment of bioethanol from wheat and sugar beet discussing environmental impacts of multiple concepts of co-product processing in the context of the European Renewable Energy Directive. *Biofuels* 7: 141–153. https://doi.org/10.1080/17597269.2015.1122472.

Cardona, C.A., Quintero, J.A., and Paz, I.C. (2010). Production of bioethanol from sugarcane bagasse: status and perspectives. *Bioresource Technology* 101: 4754–4766. https://doi.org/10.1016/j.biortech.2009.10.097.

Cooker, B., 2016. Recent Biomass-Fed Chemical Process Projects: A Review.

Curran, M.A. (1996). *Environmental Life-Cycle Assessment*. New York: McGraw-Hill Professional Publishing.

Curran, M.A. (ed.) (2012). *Life Cycle Assessment Handbook: A Guide for Environmentally Sustainable Products*, 1e. Hoboken, N.J.: Wiley.

Curran, M.A., Mann, M., and Norris, G. (2005). The international workshop on electricity data for life cycle inventories. *Journal of Cleaner Production* 13: 853–862. https://doi.org/10.1016/j.jclepro.2002.03.001.

Dangol, N., Shrestha, D.S., and Duffield, J.A. (2015). Life cycle analysis and production potential of camelina biodiesel in the Pacific Northwest. *Transactions of the ASABE* 58: 465–475. https://doi.org/10.13031/trans.58.10771.

Delucchi, M. (2011). A conceptual framework for estimating the climate impacts of land-use change due to energy crop programs. *Biomass and Bioenergy* 35: 2337–2360. https://doi.org/10.1016/j.biombioe.2010.11.028.

Delucchi, M.A. (2013). Estimating the climate impact of transportation fuels: moving beyond conventional lifecycle analysis toward integrated modeling systems scenario analysis. Washington Academy of Sciences. *Journal of the Washington Academy of Sciences* 99: 43.

Demirbaş, A. (2001). Biomass resource facilities and biomass conversion processing for fuels and chemicals. *Energy Conversion and Management* 42: 1357–1378. https://doi.org/10.1016/S0196-8904(00)00137-0.

Dias, M.O.S., Junqueira, T.L., Cavalett, O. et al. (2012). Integrated versus stand-alone second generation ethanol production from sugarcane bagasse and trash. *Bioresource Technology* 103: 152–161. https://doi.org/10.1016/j.biortech.2011.09.120.

EMF (2013). *Towards the Circular Economy*. Ellen McCarthy Foundation http://circularfoundation.org/sites/default/files/tce_report1_2012.pdf.

European Commission, Joint Research Centre, Institute for Environment and Sustainability (2010a). *International Reference Life Cycle Data System (ILCD) Handbook: Framework and Requirements for Life Cycle Impact Assessment Models and Indicators*. Luxembourg: Publications Office.

European Commission, Joint Research Centre, Institute for Environment and Sustainability (2010b). *International Reference Life Cycle Data System (ILCD) Handbook – General guide for Life Cycle Assessment – Detailed guidance*. Luxembourg: Publications Office.

Foteinis, S., Kouloumpis, V., and Tsoutsos, T. (2011). Life cycle analysis for bioethanol production from sugar beet crops in Greece. *Energy Policy* 39: 4834–4841. https://doi.org/10.1016/j.enpol.2011.06.036.

Gerbrandt, K., Chu, P.L., Simmonds, A. et al. (2016). Life cycle assessment of lignocellulosic ethanol: a review of key factors and methods affecting calculated GHG emissions and energy use. *Current Opinion in Biotechnology* 38: 63–70. https://doi.org/10.1016/j.copbio.2015.12.021.

Gibreel, A., Sandercock, J.R., Lan, J. et al. (2009). Fermentation of barley by using *Saccharomyces cerevisiae*: examination of barley as a feedstock for bioethanol production and value-added products. *Applied and Environmental Microbiology* 75: 1363–1372. https://doi.org/10.1128/AEM.01512-08.

Gnansounou, E., Dauriat, A., Villegas, J., and Panichelli, L. (2009). Life cycle assessment of biofuels: energy and greenhouse gas balances. *Bioresource Technology* 100: 4919–4930. https://doi.org/10.1016/j.biortech.2009.05.067.

Goldemberg, J. (2007). Ethanol for a sustainable energy future. *Science* 315: 808–810. https://doi.org/10.1126/science.1137013.

Gonzalez, R., Treasure, T., Phillips, R. et al. (2011). Economics of cellulosic ethanol production: green liquor pretreatment for softwood and hardwood, greenfield and repurpose scenarios. *BioResources* 6: 2551–2567.

González-García, S., Luo, L., Moreira, M.T. et al. (2012). Life cycle assessment of hemp hurds use in second generation ethanol production. *Biomass and Bioenergy* 36: 268–279. https://doi.org/10.1016/j.biombioe.2011.10.04.

Gumienna, M., Szwengiel, A., Lasik, M. et al. (2016). Effect of corn grain variety on the bioethanol production efficiency. *Fuel* 164: 386–392. https://doi.org/10.1016/j.fuel.2015.10.033.

Hong, Y., Nizami, A.-S., Pour Bafrani, M. et al. (2013). Impact of cellulase production on environmental and financial metrics for lignocellulosic ethanol. *Biofuels Bioproducts and Biorefining* 7: 303–313. https://doi.org/10.1002/bbb.1393.

Socolow, R., Andrews, C., Berkhout, F., and Thomas, V. (eds.) (1994). *Industrial Ecology and Global Change*. Cambridge, U.K: Cambridge University Press. ISBN: 9780521471978.

ISO 14044:2006 - Environmental management -- Life cycle assessment -- Requirements and guidelines [WWW Document], n.d.. ISO. URL http://www.iso.org/iso/catalogue_detail?csnumber=38498 (accessed 12.16.16).

de Jong, E. and Jungmeier, G. (2015). Biorefinery concepts in comparison to petrochemical refineries. In: *Industrial Biorefineries & White Biotechnology* (eds. A. Pandey, R. Höfer, M. Taherzadeh, et al.), 3–33. Elsevier.

Keshwani, D.R. and Cheng, J.J. (2009). Switchgrass for bioethanol and other value-added applications: a review. *Bioresource Technology* 100: 1515–1523. https://doi.org/10.1016/j.biortech.2008.09.035.

Kumar, S., Singh, J., Nanoti, S.M., and Garg, M.O. (2012). A comprehensive life cycle assessment (LCA) of Jatropha biodiesel production in India. *Bioresource Technology* 110: 723–729. https://doi.org/10.1016/j.biortech.2012.01.142.

Lee, W.-C. and Kuan, W.-C. (2015). Miscanthus as cellulosic biomass for bioethanol production. *Biotechnology Journal* 10: 840–854. https://doi.org/10.1002/biot.201400704.

Martinez-Hernandez, E., Ibrahim, M.H., Leach, M. et al. (2013). Environmental sustainability analysis of UK whole-wheat bioethanol and CHP systems. *Biomass and Bioenergy* 50: 52–64. https://doi.org/10.1016/j.biombioe.2013.01.001.

Marx, S. and Nquma, T. (2013). Cassava as feedstock for ethanol production in South Africa. *African Journal of Biotechnology* 12: 4975–4983. https://doi.org/10.5897/AJB12.861.

Matthews, R., Sokka, L., Soimakallio, S. et al. (2014). Review of literature on biogenic carbon and life cycle assessment of forest bioenergy. *Forest Research*: 1–322. https://ec.europa.eu/energy/en/studies/review-literature-biogenic-carbon-and-life-cycle-assessment-forest-bioenergy.

McKechnie, J., Pourbafrani, M., Saville, B.A., and MacLean, H.L. (2015). Exploring impacts of process technology development and regional factors on life cycle greenhouse gas emissions of corn stover ethanol. *Renewable Energy* 76: 726–734. https://doi.org/10.1016/j.renene.2014.11.088.

McKendry, P. (2002). Energy production from biomass (part 2): conversion technologies. *Bioresource technology* 83: 47–54.

Morales, M., Quintero, J., Conejeros, R., and Aroca, G. (2015). Life cycle assessment of lignocellulosic bioethanol: environmental impacts and energy balance. *Renewable and Sustainable Energy Reviews* 42: 1349–1361. https://doi.org/10.1016/j.rser.2014.10.097.

Muñoz, I., Flury, K., Jungbluth, N. et al. (2014). Life cycle assessment of bio-based ethanol produced from different agricultural feedstocks. *International Journal of Life Cycle Assessment* 19: 109–119. https://doi.org/10.1007/s11367-013-0613-1.

Murphy, R., Woods, J., Black, M., and McManus, M. (2011). Global developments in the competition for land from biofuels. *Food Policy* 36 (Supplement 1): S52–S61. https://doi.org/10.1016/j.foodpol.2010.11.014.

Nuffield Council on Bioethics, 2011. Biofuels: Ethical issues. Chapter 1, p. 8.

OECD, FAO (2016). Biofuels. In: *OECD-FAO Agricultural Outlook 2016–2025*. Paris: Organisation for Economic Co-operation and Development.

Panichelli, L., Dauriat, A., and Gnansounou, E. (2009). Life cycle assessment of soybean-based biodiesel in Argentina for export. *Int J Life Cycle Assess* 14: 144–159. https://doi.org/10.1007/s11367-008-0050-8.

Pereira, C.L.F. and Ortega, E. (2010). Sustainability assessment of large-scale ethanol production from sugarcane. *Journal of Cleaner Production* 18: 77–82. https://doi.org/10.1016/j.jclepro.2009.09.007.

Pereira, L.G., Chagas, M.F., Dias, M.O.S. et al. (2015). Life cycle assessment of butanol production in sugarcane biorefineries in Brazil. *Journal of Cleaner Production* 96: 557–568. https://doi.org/10.1016/j.jclepro.2014.01.059.

Pervez, S., Aman, A., Iqbal, S. et al. (2014). Saccharification and liquefaction of cassava starch: an alternative source for the production of bioethanol using amylolytic enzymes by double fermentation process. *BMC Biotechnology* 14 (1).

Pourbafrani, M., McKechnie, J., Shen, T. et al. (2014). Impacts of pre-treatment technologies and co-products on greenhouse gas emissions and energy use of lignocellulosic ethanol production. *Journal of Cleaner Production* 78: 104–111. https://doi.org/10.1016/j.jclepro.2014.04.050.

Ray, M.J., Brereton, N.J.B., Shield, I. et al. (2012). Variation in cell wall composition and accessibility in relation to biofuel potential of short rotation coppice willows. *Bioenergy Research* 5: 685–698. https://doi.org/10.1007/s12155-011-9177-8.

Rebitzer, G., Ekvall, T., Frischknecht, R. et al. (2004). Life cycle assessment part 1: framework, goal and scope definition, inventory analysis, and applications. *Environment International* 30: 701–720. https://doi.org/10.1016/j.envint.2003.11.005.

REN21, 2016. Renewable 2016 Global Status Report. Paris: REN21 Secretariat.

Rettenmaier, N., Köppen, S., Gärtner, S.O., and Reinhardt, G.A. (2010). Life cycle assessment of selected future energy crops for Europe. *Biofuels, Bioproducts and Biorefining* 4: 620–636.

Salazar-Ordóñez, M., Pérez-Hernández, P.P., and Martín-Lozano, J.M. (2013). Sugar beet for bioethanol production: an approach based on environmental agricultural outputs. *Energy Policy* 55: 662–668. https://doi.org/10.1016/j.enpol.2012.12.063.

Sanchez, S.T., Woods, J., Akhurst, M. et al. (2012). Accounting for indirect land-use change in the life cycle assessment of biofuel supply chains. *Journal of the Royal Society, Interface* 9: 1105–1119. https://doi.org/10.1098/rsif.2011.0769.

Saunders, J., Levin, D.B., and Izydorczyk, M. (2011). *Limitations and Challenges for Wheat-Based Bioethanol Production*. INTECH Open Access Publisher.

Shen, X., Kommalapati, R., and Huque, Z. (2015). The comparative life cycle assessment of power generation from lignocellulosic biomass. *Sustainability* 7: 12974–12987. https://doi.org/10.3390/su71012974.

Singh, A., Pant, D., Korres, N.E. et al. (2010). Key issues in life cycle assessment of ethanol production from lignocellulosic biomass: challenges and perspectives. *Bioresource Technology* 101: 5003–5012. https://doi.org/10.1016/j.biortech.2009.11.062.

Songstad, D.D., Lakshmanan, P., Chen, J. et al. (2009). Historical perspective of biofuels: learning from the past to rediscover the future. *In Vitro Cellular & Developmental Biology – Plant* 45: 189–192.

Soraya, D.F., Gheewala, S.G., Bonnet, S., and Tongurai, C. (2014). Life cycle assessment of biodiesel production from palm oil in Indonesia. *Journal of Sustainable Energy & Environment* 5.

Stephenson, A.L., Dennis, J.S., and Scott, S.A. (2008). Improving the sustainability of the production of biodiesel from oilseed rape in the UK. *Process Safety and Environmental Protection* 86: 427–440. https://doi.org/10.1016/j.psep.2008.06.005.

Talebnia, F., Karakashev, D., and Angelidaki, I. (2010). Production of bioethanol from wheat straw: an overview on pretreatment, hydrolysis and fermentation. *Bioresource Technology* 101: 4744–4753. https://doi.org/10.1016/j.biortech.2009.11.080.

The Renewable Transport Fuel Obligations (Amendment) Order 2009 [WWW Document], n.d. URL www.legislation.gov.uk/uksi/2009/843/contents/made (accessed 12.16.16).

The Renewable Transport Fuel Obligations (Amendment) Order 2011 [WWW Document], n.d. URL www.legislation.gov.uk/uksi/2011/2937/contents/made (accessed 12.16.16).

The Renewable Transport Fuel Obligations (Amendment) Order 2013 [WWW Document], n.d. URL www.legislation.gov.uk/uksi/2013/816/contents/made (accessed 12.16.16).

The Renewable Transport Fuel Obligations (Amendment) Order 2015 [WWW Document], n.d. URL www.legislation.gov.uk/uksi/2015/534/contents/made (accessed 12.16.16).

The Renewable Transport Fuel Obligations Order 2007 [WWW Document], n.d. URL www.legislation .gov.uk/uksi/2007/3072/contents/made (accessed 12.16.16).

US EPA, 2007. Renewable Fuel Annual Standards. Available at: https://www.epa.gov/renewable-fuel-standard-program/renewable-fuel-annual-standards. (Accessed: 11th April 2017)

Wang, L., Littlewood, J., and Murphy, R.J. (2013). Environmental sustainability of bioethanol production from wheat straw in the UK. *Renewable and Sustainable Energy Reviews* 28: 715–725. https://doi.org/10.1016/j.rser.2013.08.031.

Watanabe, M.D.B., Chagas, M.F., Cavalett, O. et al. (2016). Hybrid input-output life cycle assessment of first- and second-generation ethanol production technologies in Brazil. *Journal of Industrial Ecology* 20: 764–774. https://doi.org/10.1111/jiec.12325.

Zamagni, A., Guinée, J., Heijungs, R. et al. (2012). Lights and shadows in consequential LCA. *The International Journal of Life Cycle Assessment* 17: 904–918. https://doi.org/10.1007/s11367-012-0423-x.

Zheng, Y., Lee, C., Yu, C. et al. (2013). Dilute acid pretreatment and fermentation of sugar beet pulp to ethanol. *Applied Energy* 105: 1–7. https://doi.org/10.1016/j.apenergy.2012.11.070.

Part III

Hydrogen and Methane

15

Biotechnological Production of Fuel Hydrogen and Its Market Deployment

Carolina Zampol Lazaro[1], *Emrah Sagir*[1] *and Patrick C. Hallenbeck*[1,2]

[1] *Département de Microbiologie, Infectiologie et Immunologie, Université de Montréal, Québec H3C 3J7, Canada*
[2] *Life Sciences Research Center, Department of Biology, United States Air Force Academy, CO 80840-5002, USA*

CHAPTER MENU

15.1 Introduction

The finite nature of fossil fuel reserves, climate change effects, concerns about energy security, and the dangers that the use of petroleum products are driving research and development of sustainable, renewable fuels (Hallenbeck 2012). Overcoming the effects of fossil fuel use on climate change will require the development and deployment of carbon neutral fuels. Hydrogen has long been touted as a clean, green fuel since its combustion gives only water vapour. In addition, it has the highest known gravimetric energy density of any fuel (142 MJ/kg), nearly triple that of any fossil fuel, however, its low volumetric energy density makes its use as a transportation fuel problematic (Das 2009). Although hydrogen powered internal combustion engines have been developed and are already in use, much higher conversion efficiencies can be obtained using fuel cells (~80% versus ~35%), potentially making hydrogen much more attractive than other fuels. Fuel cells are devices that convert the chemical energy of a fuel into electricity through the chemical reaction of positively charged hydrogen ions with oxygen. A large number of companies are making significant R&D efforts to bring to market hydrogen-powered cars, and other necessary parts of a hydrogen support infrastructure. Already, impressive amounts of hydrogen are produced and consumed every year, about 448 billion m^3 per year, with the vast majority, >96%, being made from fossil fuels (4% electrolysis, 18% coal gasification, 30% oil reformation, 48% methane reformation) (Ajayi-Oyakhire 2013). Thus, in order for hydrogen to become a practical and sustainable energy vector, not only must technical challenges around its storage and conversion be solved, but also truly renewable methods for its production must be developed.

Green Energy to Sustainability: Strategies for Global Industries, First Edition.
Edited by Alain A. Vertès, Nasib Qureshi, Hans P. Blaschek and Hideaki Yukawa.

15.2 Hydrogen Production Through Dark Fermentation

Hydrogen production through dark fermentation has been extensively studied. A variety of microorganisms, including facultative and strict anaerobes, are capable of producing hydrogen either in pure cultures, co-cultures, or in mixed microbial consortia (Hallenbeck 2009; Kalia and Purohit 2008; Wang and Wan 2009). While many studies have been undertaken to evaluate hydrogen production from simple sugars, more recently the focus has been placed on processes that make use of various agricultural, industrial, or municipal wastes (Shah et al. 2016; Sydney et al. 2014; Tawfik and El-Qelish 2012; Zhang et al. 2016). Furthermore, several bioreactor configurations are under investigation, including AFBR (anaerobic fluidized-bed reactors) (Lazaro et al. 2015), AnSBBR (anaerobic sequencing batch biofilm reactor) (Lima et al. 2016), CSTR (continuous stirred-tank reactor) (Kim et al. 2005), EGSB (expanded granular sludge bed) (Abreu et al. 2010), APBR (anaerobic packed-bed reactors) (Perna et al. 2013), and UASB (up-flow anaerobic sludge blanket) (Castelló et al. 2009). The economic feasibility of such processes can be enhanced by optimizing several critical process parameters comprising pH, temperature, or reactor operational modes (batch and continuous).

15.2.1 Microorganisms Involved in Dark Fermentative Hydrogen Production

Hydrogen can be produced by diverse microorganisms, including facultative and strict anaerobes, mesophilic and thermophilic strains.

15.2.1.1 Hydrogen Production by Pure Cultures

In the past many laboratory studies on biohydrogen production have used pure cultures and defined substrates to examine some of the basic microbial physiology involved (Hallenbeck 2009; Hallenbeck 2011). Thus, studies of this nature are important in establishing a fundamental understanding of how various parameters might affect metabolic flux, and ultimately hydrogen production rates and yields. However, given the relatively low value product that is made, operating sterile reactors to convert sugars to hydrogen could probably never be commercially viable. Instead, production processes that are robust and that can use complex and varied substrates will probably be required.

15.2.1.2 Mixed Cultures and Inoculum Pre-treatments

As a result, mixed cultures may be preferred and the evaluation of the performance of different mixed microbial consortia has received much attention recently (Castelló et al. 2011; Etchebehere et al. 2016; Li et al. 2011; Wang and Wan 2009). Complex microbial communities are involved in hydrogen formation from mixed wastes, a property that imparts robustness to the process (Valdez-Vazquez and Poggi-Varaldo 2009). The main advantage of this approach is that often those complex bacterial communities comprise microorganisms that exhibit a wide array of metabolic capabilities, ranging from hydrolytic activities to hydrogen production (Kalia and Purohit 2008). As expected, the interactions among the species within these complex microbial communities may have positive or negative effects on hydrogen production (Bundhoo and Mohee 2016; Saady 2013). There are various positive effects to using mixed cultures. One of them is the formation of granules with the advantages of maintaining high biomass concentration inside the reactors which allows the use of short HRT (hydraulic retention time) and high OLR (organic loading rate). Moreover, the consumption of oxygen by strains such as *E. aerogens*, *Klebsiella* sp. and *Bacillus* sp. allows for operational flexibility, and members of the consortia catalyse the breakdown of complex organic substrates such as cellulose and starch (Hung et al. 2011). However, the hydrogen yield using mixed cultures can be decreased by the consumption of hydrogen by many species present in the reaction mixture, including hydrogenotrophic Archaea (Castelló et al. 2011; Saady 2013; Etchebehere et al. 2016). Furthermore, competition for substrate between hydrogen producers and consumers is possible, as well as adverse effects of some mixed culture members, such as *Sporolactobacillus* sp., which can excrete bacteriocins (Hung et al. 2011). For this reason, inocula are very often pre-treated by acid, alkali, heat, or aeration methods to select spore-forming, hydrogen producing microorganisms, and inhibit anaerobic methanogens.

15.2.1.3 Co-cultures Used for Hydrogen Production

In contrast to processes that make use of aeration to prevent the growth of methanogenic bacteria, co-cultures of facultative and strict anaerobes have been investigated since the presence of facultative anaerobes might plausibly eliminate the need for reducing agents or system sparging with inert gases for initiating and maintaining anaerobiosis (Kalia and Purohit 2008; Kotay and Das 2009). While this approach might be cost-effective at an industrial scale since achieving anaerobic conditions in large scale processes is relatively easier than maintaining aerobic ones, it should be noted that facultative microorganisms will also consume the substrate needed to support the growth and activity of the hydrogen producers. Thus, the potential decrease in hydrogen production efficiency should be carefully evaluated and compared to the use of other strategies that keep the reactor headspace oxygen-free. Although suggesting that such a co-culture approach could be interesting for hydrogen production at the industrial scale from sucrose and starch, the results of a laboratory scale study showed that a drop in the hydrogen yield occurs when *Clostridium acetobutylicum* is cultivated with *Escherichia coli* in a glucose based medium (Hassan and Morsy 2015). In addition to the mostly commonly used combination of *Clostridium* and *Enterobacter* species, other co-cultures have also been used; *C. acetobutylicum* and *Desulfovibrio vulgaris* (Barca et al. 2016), *Bacillus* and *Enterobacter* (Patel et al. 2014), *Clostridium beijerinckii* and *Geobacter metallireducens* (Zhang et al. 2013), *C. acetobutylicum* and *Ethanoligenens harbinense* (Wang et al. 2008a). Finally, co-cultures including cellulolytic hydrogen producers, e.g. *Clostridium cellulolyticum* and non-cellulolytic hydrogen-producing bacteria (Zhang et al. 2016), *Clostridium thermocellum* and *Thermoanaerobacterium thermosaccharolyticum* (Liu et al. 2008) could be used to produce hydrogen from cellulosic substrates.

In addition to evaluating different inoculum sources, much research has centred around the use thermophilic and hyperthermophilic strains, rather than mesophilic ones for hydrogen production. Thermophiles are claimed to give higher hydrogen yields than mesophilic strains, utilize a wider range of substrates, and generate fewer by-products (Hallenbeck 2005). Furthermore, operating at high temperatures should enable better contamination control, especially of hydrogenotrophic methanogens (Pawar and Van Niel 2013). The feasibility of any process depends on its cost-effectiveness, this rationale obviously also applies to the hydrogen economy. Thus, the production of hydrogen under thermophilic temperatures needs to properly address the cost of production since overall the energy input in this fermentation mode is higher than when mesophilic species are used. Another serious drawback is the much lower cell densities observed in comparison to mesophilic cultures, which leads to substantially lower volumetric productivities (Pawar and Van Niel 2013). Nonetheless, it has been pointed out that low culture densities might allow higher specific rates of hydrogen production to be achieved due to lower dissolved hydrogen concentrations (Van Niel et al. 2002).

Although in most studies bioreactors are inoculated with different undefined inocula as starter cultures, for example, anaerobic sludge (Han et al. 2010; Huang et al. 2009; Ozmihci and Kargi 2010), compost (Castelló et al. 2011), cattle manure (Gilroyed et al. 2008), soil (Akutsu et al. 2009a), it might be possible to eliminate altogether the addition of any inoculum when the raw material (typically agricultural waste) that is used as substrate exhibits an abundant indigenous microflora (Kim et al. 2009). The microorganisms present in wastes are probably adapted to the specific characteristics of the substrates on which they are found, and therefore might be particularly active. For example, in one study food waste was successfully used as both substrate and inoculum after being subjected to pre-treatment (Kim et al. 2009). In this case, as potentially in others, heat treatment can serve to select the desired microbes while at the same time improving biomass hydrolysis, thus enhancing hydrogen production. However, the costs associated with performing heat treatment should be carefully evaluated.

15.2.2 Operational Factors Influencing Hydrogen Production

Even though hydrogen production by pure cultures and simple sugars in a synthetic medium has been extensively evaluated, its main drawback remains the low hydrogen yield and rates achieved. Therefore, various attempts have been made to determine the best operational conditions with the hopes of improving process efficiency.

In particular, the influence of temperature on biological hydrogen production has been extensively studied, since this parameter is easy to control. There are three temperature ranges in which hydrogen production has been observed to occur: mesophilic (37 °C), thermophilic (55 °C) and hyperthermophilic conditions (70 °C). From an economic and technological point of view, a mesophilic temperature is preferable because less energy is required for heating the reactors; however, if the substrate to be used is produced at a high temperature, for example, sugarcane or vinasse and stillage from distillation columns operated at high temperature (107 °C) (Salomon and Lora 2009), the thermophilic range could be more feasible since there is no need to cool down the substrate prior to its use.

Wang and Wan (2008) studied the effect of temperature (20–55 °C) on hydrogen production by a mixed consortium at pH 7; they observed that the hydrogen production potential (HPP) of the mixed consortium increases with increasing temperature up to 40 °C. In contrast, Lee et al. (2008) observed higher hydrogen production from starch at a mesophilic temperature (37 °C) in comparison to the same process conducted at 55 °C. On the other hand, Karlsson et al. (2008) observed the highest hydrogen production at 55 °C. Akutsu et al. (2009b) suggested that unstable hydrogen production at mesophilic temperature is due to the activity of hydrogen consumers and therefore, thermophilic temperatures played an important role in inhibiting the consumption of hydrogen by homoacetogenic microorganisms. Nonetheless, for Gilroyed et al. (2008) the higher hydrogen production at a thermophilic temperature (52 °C) in comparison to a mesophilic one (36 °C) is related to a change in the metabolic flux. In contrast, Lima et al. (2016) observed highest hydrogen production at 15 °C in comparison with 30 and 45 °C. For these authors, *Clostridium* species were dominant at the lowest temperature because the growth and activity of lactic acid bacteria (LAB), which have been reported to compete with hydrogen-producing species (Etchebehere et al. 2016), was inhibited. Thus, there is no consensus regarding what is the best temperature for hydrogen production, which seems to depend on inoculum source, substrate type, and possible contamination.

pH has been considered to be one of the most important parameters with significant influence on fermentative hydrogen production (FHP) because of its potential effects on metabolic and enzymatic reactions (Cai et al. 2011). Depending on environmental conditions, *Clostridium* species can produce hydrogen by the fermentation of carbohydrates while producing organic acids (acidogenesis) or by producing solvents (solvetogenesis); however, acidogenesis gives higher hydrogen yields. During the fermentation of sugars, organic acids are produced causing a pH drop, which, when severe enough, causes the cells to shift from acidogenesis to solventogenesis (Bundhoo and Mohee 2016; Hallenbeck 2009; Valdez-Vazquez and Poggi-Varaldo 2009). Thus, to achieve higher hydrogen yields, pH control (around 6–7) should be carried out. It has also been reported that adding peptone to the culture medium could avoid an abrupt pH drop, allowing for better hydrogen production yield (Valdez-Vazquez and Poggi-Varaldo 2009). In general, the optimum pH value appears to be in the pH 5–6 range (Mohan et al. 2007; Davila-Vazquez et al. 2008, Lee et al. 2008), however there are exceptions to this rule of thumb (Zhao et al. 2011; Gadhamshetty et al. 2009; Junghare et al. 2012). Thus, it would seem that the best condition for hydrogen production is a function of a combination of various parameters including inoculum source, previous acclimatization, and substrate type and concentration.

During dark FHP there is co-production of acids and alcohols (Wang and Wan 2009; Valdez-Vazquez and Poggi-Varaldo 2009). It has been reported that even with pH control, the accumulation of organic acids by itself can cause a decline in the hydrogen production due to an inhibitory effect on the growth of fermentative hydrogen-producing bacteria (Valdez-Vazquez and Poggi-Varaldo 2009). This phenomenon may be due to (i) an induced cellular physiological imbalance, or (ii) cell lysis. In the first case, the undissociated (uncharged) soluble metabolites could pass through the cell membrane and then dissociate inside the cells, requiring excessive energy expenditure to restore normal conditions. Secondly, these metabolites, when present at high concentration in their dissociated state, could cause an increase in the ionic strength leading to cell lysis (Wang et al. 2008b). In both cases, the accumulation of fermentation coproducts in hydrogen systems would lead to inhibition of cell growth and, consequently, reduced hydrogen production. Indeed, one study found that the higher the concentration of soluble metabolites the greater the inhibitory effect on hydrogen production (Wang et al. 2008b). However, the inhibitory effect of adding ethanol was smaller than that attained by the addition of acetic, butyric, and propionic acids (Wang et al. 2008b). On the other hand, Baghchehsaraee et al. (2009) found that the addition of lactic acid (up to 3 g/l) causes an increase in the hydrogen rare (HPR) and hydrogen yield (HY) due to the shift in the bacterial metabolic pathways.

Since many hydrogen production experiments have been performed using different wastewaters, potentially containing organic acids, as substrate, it is important to evaluate the possible effect of such compounds on hydrogen production by the system under study. In addition to inhibition of hydrogen production to liquid end-products (organic acids) accumulation during fermentation (Valdez-Vazquez and Poggi-Varaldo 2009), inhibition by the gaseous product (H_2) is also possible (Van Niel et al. 2002). This occurs because hydrogen production by hydrogenase is a reversible process and, depending upon the hydrogen partial pressure, bacterial metabolism can shift to the production of more reduced compounds such as ethanol and lactate (Hallenbeck 2005).

In an anaerobic digestion process, in which the complex organic matter can be converted to CH_4, several physiological groups of microorganisms are metabolically coupled. In these systems, the end-products of the previous step are used as substrate by another microbial group. Thus, for example, the hydrogen produced by fermentative bacteria is consumed by hydrogenotrophic methanogenic microorganisms keeping its gas partial pressure low enough to make the acetogenesis step thermodynamically favourable. This process is called inter species hydrogen transfer and is essential to the good performance of anaerobic digestion. However, when the main goal of the study is hydrogen production, interspecies hydrogen consumption should be avoided, allowing the hydrogen to be accumulated and recovered (Angenent et al. 2004). However, lowering the hydrogen partial pressure makes hydrogen production thermodynamically more favourable. This can be achieved by placing the reactor under a vacuum, with obvious increases in the process cost, or the reactor headspace can be sparged with nitrogen, diluting the hydrogen and making its recovery more difficult and costly (Valdez-Vazquez and Poggi-Varaldo 2009). The continuous release of the biogas produced and the use of a bioreactor that enables efficient liquid–gas transfer are other alternatives to lowering the hydrogen partial pressure.

Thus, the effect of hydrogen partial pressure, varied by sparging gases into the reactor, has been investigated (Mizuno et al. 2000; Hussy et al. 2003), as has the effect of varying the liquid/headspace ratio (van Niel et al. 2002; Nguyen et al. 2009; Junghare et al. 2012). Mizuno et al. (2000) and Hussy et al. (2003) achieved an increase in hydrogen yield from 0.85 mol H_2/mol glucose to 1.43 mol H_2/mol glucose and from 1.26 mol H_2/mol hexose to 1.87 mol H_2/mol hexose, respectively, by using a nitrogen sparging technique. The effect of hydrogen partial pressure on hydrogen production was evaluated in experiments running in batch reactors, by varying the ratio of liquid/gas volumes. Nguyen et al. (2009) first evaluated the best ratio of liquid/headspace in hydrogen production by *Thermotoga neapolitana* and then used nitrogen headspace sparging for lowering the hydrogen partial pressure. They achieved the highest hydrogen production using a liquid to gas ratio of 40/80 ml and observed an HY increase from 1.82 mol H_2/mol glucose to 3.24 mol H_2/mol glucose after employing gas sparging. Junghare et al. (2012) varied the liquid to gas ratio from 1 : 1 to 1 : 12 and also observed an increase in hydrogen production and yield by decreasing the hydrogen partial pressure (from 33.9 to 10.13 kPa). Van Niel et al. (2002) performed experiments to evaluate the influence of hydrogen partial pressure on gas production and growth of *Caldicellulosiruptor saccharolyticus* by allowing hydrogen to accumulate inside the reactor at three different incubation periods; lag phase, early and mid-exponential phase. The extent of hydrogen inhibition was related to the culture density and a hydrogen partial pressure of 1–2.10^4 Pa led to a metabolic shift towards the production of lactate. Contrary to these results, Wang et al. (2007) did not observe any significant difference in hydrogen production when the reactor was sparged with H_2 (2.1 l H_2/l culture) or N_2 (2.1 l H_2/l culture); however, there was an inhibition of cell growth and hydrogen production (1.2 l H_2/l culture) when CO_2 was used.

An interesting approach to diminishing the effect of hydrogen partial pressure without using gas sparging was reported by Laurent et al. (2012), who applied an anaerobic biodisc reactor that enables bacterial immobilization as well as efficient gas transfer from the liquid to the headspace. By using this new design bioreactor, these authors observed the same trend as reported in previous studies: the lower the pressure the higher the hydrogen yield (in this particular experiment, up to an increase of 22.5% in the hydrogen yield).

In addition to the parameters already discussed, OLR is another important factor that should be taken into account in establishing the best culture conditions for hydrogen production. However, as yet there is no consensus on the proper OLR to be used since it seems to vary greatly depending on the other parameters as well as on inoculum source, type of substrate, and reactor design (Ferraz Junior et al. 2015a; Pasupuleti et al. 2014; Souza et al. 2015; Wang and Wan 2009).

15.2.3 Bioreactors Used for Dark Fermentative Hydrogen Production

Biological hydrogen production has been investigated for decades, and recently bioreactor designs have been evaluated in order to improve process performance. Different scale batch reactors, varying from millilitres to litres of total reactor capacity, but none at commercial scale, are commonly used. Reactors have also been operated under fed batch (Tien Anh et al. 2011) and in continuous mode (Sivagurunathan et al. 2016). There are some pilot scale hydrogen production reactors operating successfully; however, there are no commercial scale renewable systems in operation (Levin and Lubieniechi 2013) and this is the main bottleneck of the hydrogen economy (Das 2009).

In addition to various feeding strategies, different bioreactor designs have been applied for biological hydrogen production. Many different types have been reported: AFBR (Lazaro et al. 2015),

AnSBR or ASBR (anaerobic sequencing batch reactor) (Mohan et al. 2007; Moreno-Andrade et al. 2015), AnSBBR (Lima et al. 2016), CSTR (Kim et al. 2005), EGSB (Abreu et al. 2010), APBR (Perna et al. 2013), and UASB (Castelló et al. 2009). Specific examples of hydrogen producing experiments performed in different bio-reactor configurations, using different sources of inoculum (with heat-treated sludge being the most common) and feedstocks are given in Table 15.1. Each reactor configuration has its own advantages and disadvantages.

In addition to reactor configuration, inoculation strategy is another important factor that can affect hydrogen production. Some systems are operated with suspended cells, while others use immobilized cells (biofilms and granules). It is known that the main disadvantage of using suspended cells as a strategy of inoculation is the possible failure of the system due to the biomass washout at short retention times, because with suspended cell operation, cell retention time is the same as the HRT. However, the excessive accumulation of biomass has also been reported to negatively impact hydrogen production. The accumulation of biomass can impact system efficiency because of (i) changes in the ratio of substrate to microorganism (S/M) or (ii) biomass ageing. For example, in long-term experiments, over time non-hydrogen producers or hydrogen consumers can prevail over hydrogen mproducers, causing total cessation of hydrogen production or accumulation (Ferraz Junior et al. 2015a). Periodic discharge of biomass and replacement by fresh inocula can be carried out to overcome these issues by controlling the biomass concentration (Fuess et al. 2016).

The use of granules, cell aggregates cells mediated by extracellular polymeric substances, or self-immobilized biomass, constitutes an inoculation strategy that enables the maintenance of a high biomass concentration inside the reactor, consequently allowing operation at shorter HRT and higher OLR (Hulshoff Pol et al. 2004). This is a useful strategy to avoid biomass washout, particularly when using slow growing bacteria. Additionally, this type of system appears to be more resistant to shock-loads and to be appreciably more stable (Kumar et al. 2016a). Based on these characteristics, a semi-pilot scale biofilm reactor (20 l) was used to assess the viability of scaling up hydrogen production at different OLRs (Pasupuleti et al. 2014).

A high cell density and appropriate settling characteristics, that is the ability of granules to settle down at the bottom of the reactor and to not be washed out, can be achieved in an EGSB which uses self-immobilized biomass in the form of granules. The effects of mesophilic, thermophilic, and hyperthermophilic temperatures, as well as different inocula types (pre-treatments and bioaugmentation) on hydrogen production were evaluated using this reactor configuration (Abreu et al. 2010; Abreu et al. 2011; Abreu et al. 2012). The process temperature was observed to play an important role in reactor performance, with hydrogen production stopping at 37 °C whereas substrate consumption continued. A very similar hydrogen production rate was observed in an EGSB inoculated with heat-treated 'methanogenic granules', an aggregate of microbial comprised of diverse microorganisms including methanogens (Abreu et al. 2010), as well as granules subjected to 2-bromoethane-sulfonate (BES) treatment (Abreu et al. 2011), while a threefold greater hydrogen production rate was observed using engineered heat-treated methanogenic granules enriched with hydrogen producing biomass.

UASB is another reactor configuration that has been used for hydrogen production (Castelló et al. 2009). A UASB reactor fed with cheese whey and inoculated with untreated kitchen compost gave only very unstable hydrogen production and yield due to the persistence of methanogenesis and the presence of non-hydrogen producers (Castelló et al. 2009). A similar hydrogen production rate was achieved in a UASB using sucrose synthetic wastewater (Ning et al. 2013), while a very high hydrogen yield was reported by H_2-producing granules in a UASB operating on glucose (Mu and Yu 2006) (Table 15.1).

Table 15.1 Hydrogen production in different bioreactor configurations.

Reactor type	Inoculum source	Substrate	HY	HPR	Reference
UASB	Kitchen waste compost	Raw cheese whey	<2 mol H_2/mol lactose	1 l H_2/l/d (average)	Castello et al. (2011)
	Sludge from a reactor treating citrate-producing wastewater	Sucrose synthetic wastewater		2.89 l H_2/l/d	Ning et al. (2013)
	H_2-producing granules	Sucrose synthetic wastewater	2.88 mol H_2/mol sucrose	50 ml H_2/l/h	Mu and Yu (2006)
AnSBBR	H_2 producing microflora	Cheese whey	0.8 mol H_2/mol lactose	0.66 l H_2/l/d	Lima et al. (2016)
	Anaerobic sludge	Glucose-based wastewater	0.94 mol H_2/mol glucose	1.36 l H_2/l/d[a)]	Souza et al. (2015)
	Heat-treated anaerobic sludge	Glycerin-based wastewater	21.1 mol H_2/ kg COD	1.5 l H_2/l/d	Bravo et al. (2015)
ASBR	Heat-treated sludge	Alcohol distillery wastewater	172 ml H_2/g $COD_{removed}$	3.31 l H_2/l/d	Searmsirimongkol et al. (2011)
	Anaerobic granular sludge	Food waste	105.3 ml H_2/g$VS_{removed}$	255.4 ml H_2/g$VS_{removed}$.d	Moreno-Andrade et al. (2015)
	Heat-treated anaerobic sludge adapted to cassava wastewater	Cassava wastewater	186 ml H_2/g $COD_{removed}$	3.8 l H_2/l/d	Sreethawong et al. (2010)
CSTR	Mixed microbial culture	Molasses-containing wastewater		5.9 l H_2/l/d	Han et al. (2010)
	Indigenous microflora in the wastewater	Cheese whey	0.78 mol H_2/mol lactose	2.9 l H_2/l/d	Venetsaneas et al. (2009)

EGSB	Anaerobic sludge treated with 2-bromoethane-sulfonate	Glucose/arabinose		0.7 l H_2/l/d	Abreu et al. (2011)
	Engineered heat treated methanogenic granules	Glucose/arabinose		2.7 l H_2/l/d	Abreu et al. (2010)
	Heat-treated metanogenic granules			0.8 l H_2/l/d	
	Anaerobic granular sludge from a citric acid production factory	Fresh leachate originated from municipal solid wastes		2.2 l H_2/1/d	Lie et al. (2011)
APBR	Heat-treated granular sludge	Sugarcane vinasse	3.2 mol H_2/mol carbohydrates total	509.5 ml H_2/l/d	Nunes Ferraz Junior et al. (2015b)
			2.4 mol H_2/mol carbohydrates total	1.1 l H_2/l/d	Djalma Nunes Ferraz Júnior et al. (2014)
AFBR	Heat-treated granular sludge	Glucose	2.15 mol H_2/mol glucose	0.96 l H_2/l/h	Barros et al. (2011)
			1.87 mol H_2/mol glucose	1.07 l H_2/l/h	
		Cheese whey	1.33 mol H_2/ lactose	0.5 l H_2/l/h	Rosa et al. (2014a)
		Cassava processing wastewater/glucose	1.0 mmol H_2/ g COD		Rosa et al. (2014b)
		Sugarcane vinasse	2.23 mmol H_2/ g COD_{added}	1.5 l H_2/l/h	Santos et al. (2014)

a) Converted from original value (using STP, 1 mol equal to 22.4 l).

There are various methods currently used for achieving cell immobilization: adsorption, encapsulation, and entrapment. Among these techniques, adsorption is the most commonly used. It consists of attaching cells on an organic or inorganic matrix by allowing direct contact between the cells and the support microcarrier material inside bioreactors. This can be achieved by keeping a reactor in a closed circuit before initiating operations in a continuous mode (Kumar et al. 2016b). A clear correlation has been demonstrated between the type of support material used and the characteristics of the biofilm formed, which, consequently, impacts reactor performance (Qureshi et al. 2005).

15.2.4 Substrates Used for Dark Fermentative Hydrogen Production

Hydrogen production from simple and pure carbohydrates, e.g. glucose and sucrose, has been extensively studied. However, research is still needed to improve hydrogen yields and production rates from these substrates. As well, in order to become economically feasible for hydrogen production, the challenge of using cheap and abundant substrates, such as domestic, agricultural, and industrial wastes, must be met (Hallenbeck and Liu 2016). Low-value raw materials, such as domestic and industrial wastes, e.g. food waste, kitchen wastewater, sugarcane bagasse, vinasse (stillage), biodiesel waste glycerol, cheese whey, ligno-cellulosic wastes have been recently used as substrates for biological hydrogen production by pure cultures, co-cultures, and mixed consortia (Patel et al. 2010; Im et al. 2012; Maru et al. 2012; Tawfik and El-Qelish 2012; Cheng and Zhu 2013; Sydney et al. 2014) (Table 15.2).

Domestic wastes are potential carbon sources to be used as substrates in addition to agro-industrial residues. They are available everywhere, and their average compositions only vary slightly according to the region, season and people's habits (Alibardi and Cossu 2015). The effect of initial pH and autoclaving on the production of hydrogen from food waste has been examined (Hu et al. 2014). The impact of an enzymatic pre-treatment step of food waste by fungi (*Aspergillus awamori* and *Aspergillus oryzae*) and by bacteria (*Bacillus* sp.) has been assessed on biological hydrogen production (Shah et al. 2016). Aiming to improve hydrogen production efficiency, two CSTRs were operated under thermophilic and hyperthermophilic temperatures to produce hydrogen from food waste (Algapani et al. 2016). In addition to these strategies, a co-digestion system of two residues, which could provide a more balanced nutrient supplementation for the microorganisms, has been performed using municipal food waste and kitchen wastewater to produce hydrogen (Tawfik and El-Qelish 2012).

A promising feedstock for hydrogen production is crude glycerol, a by-product from biodiesel production that is generated during the transesterification reaction (Maru et al. 2012; Lo et al. 2013; Maru et al. 2013). With the increasing demand and, consequently, production of biodiesel, there has also been an increase in the generation of waste glycerol. Crude glycerol has limited commercial value unless an expensive purification process is carried out (Maru et al. 2016). Therefore, as for the other wastes, its use as a carbon source for microbial energy generation is a potentially interesting approach. The use of biodiesel wastes containing glycerol has been evaluated at mesophilic and thermophilic temperatures. At mesophilic temperatures hydrogen production by *Klebsiella pneumoniae* DSM 2026 was increased 5.0-fold after optimization of medium composition by a Plackett–Burman design. Along with hydrogen, fermentation generates 1,3-propanediol, which can be used for polymer production (polyesters, polyethers, polyurethanes) (Liu and Fang 2007). Hydrogen production from glycerol by *Clostridium butyricum*, *Clostridium pasteurianum*, and *Klebsiella* sp. has also been observed at the same temperature range (Lo et al. 2013). Another study evaluated the possibility of producing both hydrogen and ethanol from glycerol by

Table 15.2 Hydrogen production from complex wastes using different inoculum sources.

Substrate	Inoculum source	Hydrogen yield	Experimental conditions	References
Food waste and sewage sludge	Indigenous microorganism (heat treated)	2.26 mol H_2/mol $hexose_{added}$	Batch reactor; 35 °C	Im et al. (2012)
Biodiesel waste containing glycerol	*Klebsiella pneumoniae* DSM 2026	0.53 mol H_2/mol glycerol	Batch reactor	Liu and Fang (2007)
Alcohol distillery wastewater	Heat-treated granular sludge	172 ml H_2/g $COD_{removed}$	ASBR, 37 °C	Searmsirimongkol et al. (2011)
Cheese whey	Heat-treated anaerobic consortium	2.3 mol H_2/mol lactate	AFBR, 30 °C	Gomes et al. (2015)
Cassava wastewater	Heat-treated anaerobic sludge	186 ml H_2/g $COD_{removed}$	ASBR; 37 °C	Sreethawong et al. (2010)
Sugarcane bagasse hydrolysate	*Thermoanaerobacterium aotearoense* SCUT27/Δldh	1.86 mol H_2/mol total sugar	Batch reactor	Lai et al. (2014)
Molasses	*T. neapolitana*	2.95 mol $H_2/mol_{monosaccharide\ consumed}$	Batch reactor; 55 °C	Cappelletti et al. (2012)
Cheese whey		2.50 mol $H_2/mol_{monosaccharide\ consumed}$		

Enterobacter sp. and *Citrobacter freundii* in pure or combined culture (Maru et al. 2013). Hydrogen production from glycerol has also been observed at thermophilic temperatures by cultures of *T. neapolitana* (Ngo and Sim 2012). Thus, it has been reported that a variety of microorganisms can be used for hydrogen production from glycerol. Furthermore, the concomitant generation of valuable products, such as ethanol and 1,3-propanediol, is encouraging for further investigations into the fermentation of this carbon source.

Cheese whey is another substrate that has been extensively used for hydrogen production in batch and continuous culture. It is a liquid waste generated after milk coagulation during cheese manufacturing in the dairy industry. Due to its rich organic and saline content it can cause environmental degradation, such as eutrophication and toxicity in water bodies (Prazeres et al. 2012). The main drawback of using cheese whey for hydrogen production that has been reported is the growth of LAB inside the bioreactors (Castelló et al. 2009; Castello et al. 2011; Perna et al. 2013). These microorganisms can compete for substrate with hydrogen-producers lowering the process yield, and they can also inhibit the growth of certain *Clostridium* species by the production of bacteriocins (Etchebehere et al. 2016). The effect of temperature on the production of hydrogen in an AnSBBR fed with cheese whey has been evaluated (Lima et al. 2016). The best reactor performance was achieved at the lowest temperature (15 °C) in which relatively low abundance and closely related LAB were identified. In another study, the highest efficiency AFBR fed with cheese whey was that where *Clostridium* outnumbered LAB at the beginning of the experimental period and were present at relative equilibrium at the end (Gomes et al. 2015). Therefore, applying proper operational conditions to reactors, e.g. temperature, HRT or OLR, could favour the dominance of hydrogen-producers over LAB, an approach that should be the focus of development efforts in the application of cheese whey as substrate in hydrogen production systems.

In some countries, such as Brazil, distillery effluents, also called sugarcane vinasse or stillage, are produced in large amount by the ethanol industry (Salomon and Lora 2009). This liquid effluent has been used as a fertilizer; however, due to its high COD (chemical oxygen demand), low pH and high potassium content it can contribute to environmental degradation problems due to ground water contamination and land salinization (Wilkie et al. 2000; Salomon et al. 2011). Therefore, researchers have been looking for alternative ways to dispose of stillage, including its utilization as substrate for hydrogen or methane production (España-Gamboa et al. 2012; Sydney et al. 2014). Notably, sugarcane stillage has been applied as a feedstock in several reactor configurations producing hydrogen at mesophilic and thermophilic conditions (Searmsirimongkol et al. 2011; Ferraz Júnior et al. 2014, 2015a,b Santos et al. 2014; Fuess et al. 2016). A possible constraint of using vinasse as substrate was reported to be related to its high potassium and sulfate concentration when an ASBR was fed at high organic loads (40 g COD/l, 60 kg COD/m^3d) (Searmsirimongkol et al. 2011). On the other hand, long-term hydrogen production rates were achieved in an APBR fed with sugarcane stillage by keeping a high and continuous OLR (82.4 kg COD/m^3d); by continuous pH control, and by periodically discharging excess biomass (Fuess et al. 2016). In the latter study, methanogenic activity was observed at an OLR of 50.4 kg COD/m^3d. Thus, the OLR that was observed in one study to be high enough to inhibit methane production (Searmsirimongkol et al. 2011) was considered to be low in another where methanogenic activity took place and de facto decreased the overall hydrogen production (Fuess et al. 2016). This effect might be due to the widely varying characteristics of wastewater, as well as the substrate/microorganism ratio. Moreover, whereas convenient for certain purposes, OLR is a unit that does not take into account the amount of microorganisms present in a system.

Another downside to using raw vinasse as substrate is possible system contamination with LABs, since these Gram positive organisms constitute the most common bacterial contaminants found in ethanol production facilities (Beckner et al. 2011). Along with *Megasphaera* sp. (which is a hydrogen-producer), *Lactobacillus* sp. were found to be present in an AFBR treating sugarcane vinasse (Santos et al. 2014). Therefore, strategies to inhibit both methanogens and LABs are essential for improving process efficiency from ethanol distillery industry effluent.

Lignocellulosic biomass and crop wastes are thought to have great potential as carbon sources for the production of bioenergy. Along with liquid residues, one of the largest agro-industrial cellulosic wastes is sugarcane bagasse, a fibrous residue of cane stalks. It has been used as boiler fuel by the sugar industry (Pandey et al. 2000). However, recent studies have investigated its use to increase ethanol production, also known as secondary ethanol generation (Dias et al. 2011; Wanderley et al. 2013), and also for biohydrogen production (Cheng and Zhu 2013; Lai et al. 2014). Surpassing Brazil, the US is the largest ethanol producer in the world; however, it uses corn as feedstock instead of sugarcane (Sánchez and Cardona 2008). Therefore, the potential crop residue that might be used here as substrate for hydrogen production would be corn stalks. Inevitably, the crop residue to be used for biogas production will vary according to local constraints. The more abundant and available the residue, the lower will be the additional process cost. In addition, biodegradability is another criterion to be taken into account, as a degradation resistant feedstock would increase process costs (Kapdan and Kargi 2006).

The biodegradability of lignocellulosic biomass can be increased by pre-treatment methods that act by breaking the lignin seal, thereby releasing cellulose that can subsequently be further hydrolyzed into fermentable sugars by different processes. Acid and alkali treatments are often used for increasing the yield of reducing sugars for hydrogen production (Argun and Kargi 2011; Rai et al. 2014b; Ren et al. 2016). However, a major disadvantage is the generation of an array of fermentation inhibitors comprising phenolic compounds and furan derivatives (furfural and hydroxymethylfurfural) among others (Bundhoo and Mohee 2016). Biological techniques using fungi or enzymes are an alternative for avoiding the inhibitory effect of these toxic compounds (Ren et al. 2016). Physical and chemical detoxification methods that alleviate inhibition problems include: treating the syrup hydrolysates with NaOH, $Ca(OH)_2$, activated charcoal, and ion-exchange resins; however, these additional steps are not 100% efficient and generate a non-negligible extra cost (Guo et al. 2013).

In general, the main drawback of using complex wastes is the requirement for pre-treatment processes, the main purposes of which are: (i) increasing substrate biodegradability and (ii) suppressing non-hydrogen-producing microorganisms, consequently, increasing the hydrogen yield. In summary, acid/alkali and steam explosion pre-treatments have been applied to increase the yield of fermentable sugars from lignocellulosic biomass. When using food wastes that are highly biodegradable, the aim of the pre-treatment step is rather to select for endospore-forming microorganisms, such as *Clostridium* species, and inhibit indigenous bacteria (Kim et al. 2014). Although methods for selecting endospore-forming bacteria are extensively used, this does not in itself guarantee improvement in process efficiency, especially since some non-spore forming hydrogen-producers, e.g. *Enterobacter* and *Citrobacter*, would be eliminated from the microbial consortium (Kalia and Purohit 2008). However, improvement of hydrogen production has been observed when acid pre-treated food waste was used as substrate (Kim et al. 2014). Notably, the treatment selected for *Clostridium* species, while *Lactobacillus* and *Streptococcus* species were dominant when the food waste was not subjected to acid shock treatment.

In conclusion, a proper inoculum and substrate pre-treatment, and its influence in the hydrogen production seem to be dependent on the combination of three factors: the specific characteristics of the substrate, the dominant indigenous microflora, and the inoculum source. Therefore, using a design of experiments (DOE) approach, which allows the evaluation of the effect and integration of many factors at the same time, can be a powerful tool for improving bioenergy generation from complex substrates. Another point that should be highlighted is that reported units for expressing hydrogen production and yields vary widely, making it difficult to compare and evaluate the economic feasibility of the process (Table 15.2). At the very least, a common variable, the hydrogen yield based on the amount of hydrogen produced per amount of substrate consumed, should be used for making comparisons.

15.3 Hydrogen Production Through Photofermentation

Hydrogen production by photosynthetic bacteria has been proposed as a promising way to replace current hydrogen production methods (Eroglu and Melis 2011, Hallenbeck and Benemann 2002; McKinlay and Harwood 2010). Hydrogen production with photosynthetic bacteria has several potential advantages, including the utilization of solar energy with a variety of agricultural, industrial, and municipal wastes (Eroglu et al. 2008; Keskin and Hallenbeck 2012).

15.3.1 Photo-biological Hydrogen Production by Purple Non-sulfur Bacteria

PNSB (purple non-sulfur photosynthetic bacteria) are versatile and well-characterized organisms that are well-known for their photoheterotrophic growth mode, where organic compounds are used as carbon source, and light as the energy source. Hydrogen production by PNSBs offers many advantages over other processes including high substrate conversion yields, lack of oxygen generation, the ability to use a wide ligh spectrum of light, and the ability to consume a wide range of substrates from organic acids to sugars derived from wastes (Basak and Das 2007, Hallenbeck and Liu 2016). PNSBs have the potential to convert all of the electrons coming from organic substrates into molecular hydrogen including substrates derived from the end products of dark fermentation carried out by mesophilic or thermophilic bacteria (Figure 15.1). Various organic substrates and intermediate products of the citric acid cycle, acetate, lactate, malate, succinate, formate, butyrate, as well as alcohols, sugars, amino acids, glycerol, and aromatic compounds have been shown to be utilized by different PNSBs (Eroglu et al. 1999; Barbosa et al. 2001; Ghosh et al. 2012).

15.4 Hydrogen Production by Combined Systems

Hydrogen production can be realized simultaneously in the same reactor with a combination of dark fermentation and photofermentation. This, as noted elsewhere, has the potential advantage of increasing total hydrogen yields since the dark fermentative bacteria will produce hydrogen from substrates containing sugars and give off organic acids which can then be used by photosynthetic bacteria to produce additional hydrogen with the capture light energy.

15.4.1 Hydrogen Production by Dark and Photofermentation in Co-culture

The theoretical maximum yield of 12 mol H_2/mol glucose could be achieved by using a combined dark and photofermentation process (Figure 15.2). This technique not only enables improved hydrogen production but also reduces fermentation time and other operational costs deriving from the need for a second reactor (Redwood and Macaskie 2006; Singh and Wahid 2015). In addition, the acidification caused by the accumulation of fermenter effluent is also decreased and equilibrated as photofermentation takes place simultaneously (Abo-Hashesh et al. 2011).

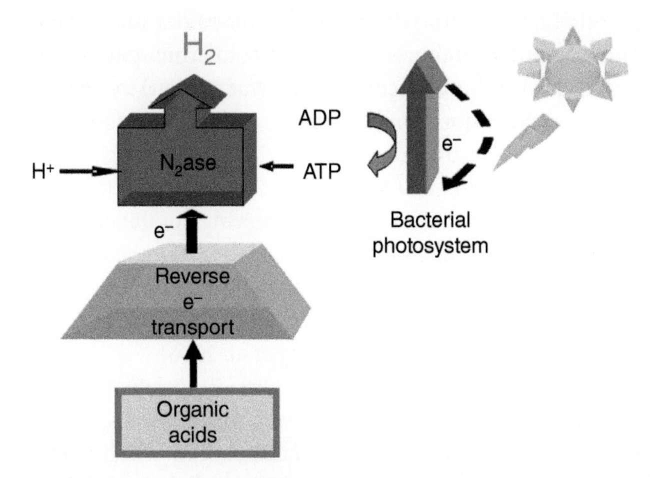

Figure 15.1 Schema of pathways supporting hydrogen production in photosynthetic bacteria. Supply of necessary energy, in the form of ATP and high energy electrons, to nitrogenase is shown. ATP is generated through the stored chemical energy produced by bacterial photosynthesis. Electrons, extracted form organic acids during their catabolism, are made more highly reducing (reverse electron transport), using the energized membrane.

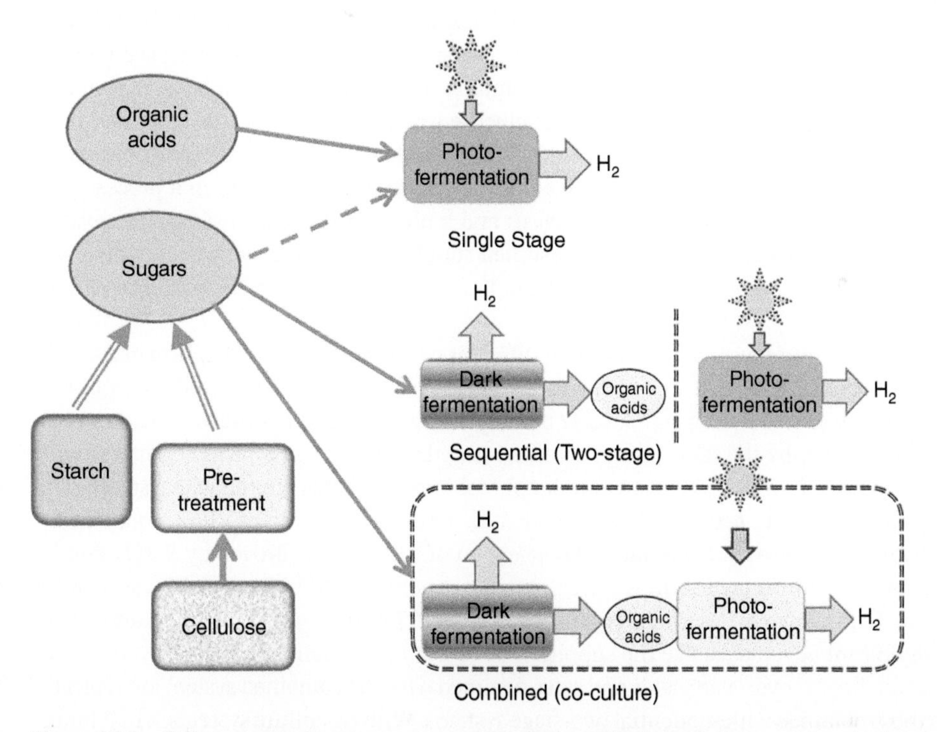

Figure 15.2 Different operational processes using photosynthetic bacteria to produce hydrogen. Organic acids can be used to directly produce hydrogen. Photosynthetic bacteria can also produce hydrogen from sugars, which could be derived from polymers such as cellulose. Some are capable of the direct conversion of sugars to hydrogen in a one step, one stage process. Alternatively, a process could depend upon the initial conversion of the sugar to organic acids and hydrogen, with a subsequent conversion using photofermentation, to produce hydrogen. This can involve a sequential system, or a combined (co-culture) system.

The use of a mixed culture of a photosynthetic bacterium, *Rhodobacter sphaeroides,* and a dark fermentative bacterium, *C. butyricum,* carrying out a combined dark and photofermentation was observed to improve hydrogen yields from 1.9 mol H_2/mol glucose (dark fermentation) to 6.6 mol H_2/mol glucose (Yokoi et al. 1998). Furthermore, the co-immobilization of *Lactobacillus delbrueckii* NBRC13953 and *R. sphaeroides*-RV was observed to result in a yield of 7.1 mol H_2/mol glucose, with the optimal ratio of bacterial cell concentration of *L. delbrueckii* to *R. sphaeroides* being 1 : 5 (Asada et al. 2006). An examination of the relationship between cultures during combined dark and photofermentation has revealed that a syntrophic relation exists between *C. butyricum* and *R. sphaeroides* rather than a competitive one (Fang et al. 2006). This phenomenon was ascribed to a slower consumption of glucose by *R. sphaeroides*, allowing *C. butyricum* to consume glucose, excreting acetic and butyric that can be consumed by *R. sphaeroides* (Fang et al. 2006).

Hydrogen production by the immobilized co-cultures of a dark fermentative (*E. harbinense* B49) bacterium and a photofermentative bacterium (*Rhodopseudomonas faecalis* RLD-53) is another concept that has the advantage of combining both dark and photofermentation when using glucose as substrate. Process optimization experiments were conducted using different control strategies, such as varying glucose and buffer concentrations, or initial pH. The highest hydrogen yield of 3.1 mol H_2/mol glucose was achieved when the reaction medium contained 6 g/l glucose, 50 mmol/l phosphate, and an initial pH of 7.5 (Xie et al. 2010).

Similarly, different combinations of heat-treated anaerobic sludge, dark and photofermentative bacteria including *C. beijerinckii*, and two strains of *R. sphaeroides* (RS-NRRL and RS-RV) were used to enhance hydrogen production from wheat powder starch: the highest hydrogen yield of 0.36 mol H_2/mol glucose was achieved by a mixed culture of heat-treated anaerobic sludge and *R. sphaeroides* RS-NRRL, whereas 0.14 mol H_2/mol glucose was obtained using a pure culture of *C. beijerinckii* (Ozmihci and Kargi 2010). Other feedstocks have also been explored for the production of hydrogen by combined systems. For example, algal biomass starch was used in a process coupling a dark fermentative *Lactobacillus amylovorus* and a photosynthetic bacterium, *Rhodobium marinum* A-501. The lactic acid bacterium consumed starch and excreted lactic acid, which was subsequently used by the photosynthetic bacterium to produce hydrogen in the presence of light (Kawaguchi et al. 2001).

A ground wheat solution was tested as feed at different concentrations for a heat-treated anaerobic sludge and *Rhodobacter* cultures mixed in the same reactor. A hydrogen production of rate 3.44 ml H_2/h, and yield, 63.9 ml/g starch (0.42 mol H_2/mol glucose), were observed (Argun et al. 2009). In another study by the same group, continuous hydrogen production was conducted with *C. beijerinckii* and *R. sphaeroides* RV in an annular photobioreactor using ground wheat starch. The fermentation of starch by *C. beijerinckii* resulted in a high rate of formation of VFAs (volatile fatty acids) and hydrogen. However, lower rates of conversion of VFAs into hydrogen by *R. sphaeroides* RV resulted in their accumulation in the fermentation medium. The highest yield was 90 ml/g starch (0.6 mol H_2/mol glucose) with 5 g/l ground wheat at HRT of six days (Argun and Kargi 2010).

Despite significant advantages like the simplicity of operation, reduction in reactor volume and lower operational costs, hydrogen yields and rates obtained with the combined system are typically lower than those obtained with sequential two-stage systems. With co-culture systems, a maximum hydrogen yield of 7.1 mol H_2/mol hexose has been obtained, significantly less than the theoretical maximum yield that could be obtained, which is 12 mol H_2/mol hexose (Chen et al. 2008).

Some of the key parameters such as the ratio of dark and photofermentative cultures, light intensity, light/dark cycle and effluent dilution ratio were also investigated in a batch experiment by the combined system of *Clostridium butyricum* and immobilized *Rhodopseudomonas faecalis* RLD-53. The hydrogen yields were the maximum when effluent dilution ratio of 1 : 0.5, dark-photo bacteria ratio of 1 : 2, the light intensity of 10.2 W/m^2 were provided, giving a total hydrogen yield of 5.37 mol H_2/mol glucose (Liu et al. 2010).

15.4.2 Two-Stage Dark and Photo-fermentative Hydrogen Production

As commented previously, the hydrogen yield could be raised up to the theoretical maximum value of 12 mol H_2/mol glucose by the metabolism of dark and photo-fermentative bacteria cultivated in co-culture (combine system) or two-stage reactors (sequential system). In either condition, the main aim of the process design is that organic acids produced by dark fermentative bacteria (facultative or strict anaerobes) can be used as substrate for photo-FHP culminating with complete organic matter removal. The PNSBs can utilize a wide range of organic acids, including acetic, butyric, and malic, that are by-products of dark fermentation (Tao et al. 2008). The advantages and disadvantages of co-culturing both microorganisms have already been presented. In this section, the sequential system in which dark and photo-fermentation take place in separate bioreactors are discussed and represented in the biochemical reaction equations below. In Eq. (15.1) is represented hydrogen production through dark fermentation having acetic acid the main end-product that can be further used as substrate for the photo-fermentation step. Butyric acid, another important by-product of dark fermentation, can be also used as carbon source by PNSB for hydrogen generation under the light Eq. (15.2)). By using a dual system (two-stage), it is possible to adopt optimal fermentation conditions (e.g. temperature and pH value) for both dark and photo-fermentative bacteria separately; this forms the basis for achieving processoptimization (Guwy et al. 2011). *Clostridium*, *Caldicellulosiruptor* and mixed consortia were used as inoculum source for dark fermentation; while *Rhodopseudomonas* and *Rhodobacter* were often used in the second stage (Table 15.3).

Dark fermentation step + Photo-fermentation step

$$C_6H_{12}O_6 + 2H_2O \rightarrow 4H_2 + 2CO_2 + 2CH_3COOH; +2CH_3COOH + 4H_2O + \text{“}\textit{light energy}\text{”} \rightarrow 8H_2 + 4CO_2 \quad (15.1)$$

Two-stage overall H_2 yield: $4H_2 + 8H_2 = 12H_2$

$$C_6H_{12}O_6 \rightarrow 2H_2 + 2CO_2 + CH_3CH_2CH_2COOH + CH_3CH_2CH_2COOH + 6H_2O + \text{“}\textit{light energy}\text{”} \rightarrow 10H_2 + 4CO_2 \quad (15.2)$$

Two-stage overall H_2 yield: $2H_2 + 10H_2 = 12H_2$

Hydrogen production from glucose was conducted by sequential cultures of *C. butyricum* and *R. faecalis* RLD-53. A hydrogen yield of 4.1 mol H_2/mol glucose and productivity of 33.85 ml/l/h was obtained with 6 g/l glucose in 100 ml bottle reactors. *R. faecalis* RLD-53 could utilize the acetate that was produced during the dark fermentation step, thus enhancing hydrogen production (Ding et al. 2009). An orthogonal experimental design was used to optimize the culture medium composition to increase hydrogen production and energy conversion efficiency by a combination of *C. butyricum* and *Rhodopseudomonas palustris* growing on glucose. The effluents from dark fermentation medium, mainly acetate and butyrate, was fed together to the *R. palustris* culture to enhance hydrogen production in the photofermentation process. This process design resulted in a significant increase in hydrogen yield from 1.59 to 5.48 mol H_2/mol glucose (Su et al. 2009).

A combination of dark and photofermentation was also used in a sequential two-step process to increase overall hydrogen yields from sucrose by using a dark fermentative (*C. pasteurianum*) and a photofermentative (*R. palustris*) bacterium. The effluents of dark fermentation, mainly acetate and lactate, were given as the feed for the photosynthetic bacteria to use and produce hydrogen.

Table 15.3 Hydrogen production by sequential two-stage dark and photo-fermentation with different substrates and microorganisms.

Dark-fermentative microorganism	Photo-fermentative microorganism	Substrate	Overall hydrogen yield[a)]	References
Clostridium pasteurianum CH_4	*Rhodopseudomonas palustris* WP3–5	Sucrose	10 mol H_2/mol sucrose (~90% COD removal)	Chen et al. (2008)
Clostridium butyricum LS2	*Rhodopseudomonas palustris*	Palm oil mill effluent (POME)	3.1 ml H_2/ml POME (93% COD removal)	Mishra et al. (2016)
Caldicellulosiruptor saccharolyticus	*R. capsulatus* hup^-	Sugar beet molasses	13.7 mol H_2/mol sucrose	Özgür et al. (2010a) Biohydrogen production from beet molasses by sequential
Caldicellulosiruptor saccharolyticus	*Rhodobacter capsulatus* (DSM1710)	Potato steam peels hydrolysate[b)]	3.9 mol H_2/mol hexose	Özgür et al. (2010b) Potential use of thermophilic dark fermentation effluents in
Microbial consortium	*R. sphaeroides* N7	Starch	5.3 mol H_2/mol hexose	Laurinavichene et al. (2012)
Enterobacter aerogenes MTCC 2822	*Rhodopseudomonas* BHU 01 strain	Cheese whey wastewater	58 mmol H_2/l medium	Rai et al. (2014a)
Enterobacter aerogenes MTCC 2822	*Rhodopseudomonas* BHU 01	Hydrolyzed sugarcane bagasse	1755 ml H_2/L	Rai et al. (2014b)

a) Overall theoretical yield: 12 mol H_2/mol hexose.
b) Starch in potato steam peel was hydrolyzed by a-amylase activity.

The resulting total hydrogen yield was increased from 3.8 mol H_2/mol sucrose (dark fermentation) to 10.02 mol H_2/mol sucrose (Chen et al. 2008).

Similarly, a mixed culture of *C. butyricum*, *E. aerogenes* and *Rhodobacter* sp. M-19 was used in a two-step repeated batch culture for hydrogen production from sweet potato starch. In the first step, a mixed culture of *C. butyricum* and *E. aerogenes* produced hydrogen with a yield of 2.4 mol H_2/mol glucose from potato starch residue (Yokoi et al. 2001). Secondly, *Rhodobacter* sp. M-19 produced hydrogen from the supernatant of the culture broth obtained in the repeated batch culture containing *C. butyricum* and *E. aerogenes.* A high yield of hydrogen of 7.0 mol H_2/mol glucose from the starch was attained using sequential two-step repeated batch cultures (Yokoi et al. 2001).

An important process design principle here is that the medium composition needs to be carefully evaluated for the co-culture systems as well as for the sequential one because it must fit the nutritional requirements of both microorganisms. Particularly for the sequential systems, the presence of fixed nitrogen, e.g. ammonium salts, commonly used as a nitrogen source for dark fermentation, can inhibit photo-hydrogen production because it is involved in the control of nitrogenase (Gabrielyan et al. 2010; Hakobyan et al. 2012). The effect of several nitrogen sources, e.g. alanine, asparagine, glutamate, glycine, proline, tyrosine, yeast extract, and ammonia, on the photo-hydrogen production have been evaluated (Gabrielyan et al. 2010; Hakobyan et al. 2012). Glutamate and yeast extract (YE) are largely used; however, the negative effect that even a low level of ammonia can have has been well documented (Kim et al. 2012; Ozgur et al. 2010a; Ozturk et al. 2012). It was also observed that once ammonia has been consumed, hydrogen production resumes (Kim et al. 2012). The influence of both NH_4^+ and YE was evaluated in a two-stage system for hydrogen production from beet molasses. In those dark fermentation experiments, it was observed that both nitrogen sources play an important role in the growth and hydrogen production by the extreme thermophile *C. saccharolyticus*, additionally, the highest hydrogen yield (4.2 mol H_2/mol sucrose) was achieved using YE despite the absence of NH_4^+ in DFE (dark fermentation effluent) making it a very suitable substrate for further hydrogen photo-fermentation. However, in spite of the absence of ammonium, the carbon and nitrogen sources present in DFE were almost 10 times higher than the amounts of those nutrients present in the synthetic medium used to pre-grow the PNSB, albeit it was necessary to dilute DFE three times to achieve the highest hydrogen yield (Ozgur et al. 2010b). The dilution of DFE produced from the fermentation of sucrose by *C. pasteurianum* decreased the ammonia concentration below its inhibitory threshold, making it an adequate substrate for the hydrogen production by *R. palustris* (Chen et al. 2008). Thus, implementing a dilution strategy helps reduce the inhibitory effect of high organic acids and nitrogen concentrations and in parallel reduces the colour of some effluents that can reduce hydrogen production by photo-fermentation.

The concern of using DFE as substrates for PNSBs goes beyond the quality and quantity of carbon, nitrogen sources and its ratio, that can be managed by dilution. Prior to photo-fermentation, DFE has been subjucted to different kinds of pre-treatments that typically involve centrifugation (Chen et al. 2008), and sterilization by autoclaving (Mishra et al. 2016) or filtration (Rai et al. 2014a). All these pre-treatment steps aim at removing contaminant microorganisms and particles that could compete with or otherwise inhibit hydrogen production by PNSBs and hamper the penetration of light. Furthermore, other adjustments such as buffer addition and micronutrient supplementation (iron, molybdenum, nickel, magnesium), have also been explored (Özgür et al. 2010a; Ozgur et al. 2010b; Rai et al. 2014a). Özgür et al. (2010a,b) did not observe photo-FHP from a DFE without buffer supplementation mostly likely due to the pH raising up to 10. Additionally, it was observed that nitrogenase co-factor supplementation (iron and molybdenum) impact the hydrogen production depending on the PNSB strains and the substrates that are used. Rai et al. (2014a) evaluated the hydrogen production from cheese whey wastewater and concluded that nutrient-poor DFE needs to be supplemented with micronutrients to be adequate for photo-fermentation. Thus, it seems that depending upon to the composition of the feedstock and the inoculum source it is necessary to implement macro- and micro-nutrient supplementation. Nevertheless, in order to make the two-stage process less laborious and cost-effective, DFE adjustments should be avoided as much as possible and only carried out if they appear to become critical to improve the process economics.

Many of the recent studies have evaluated the possibility of using DFE as substrate for PNSBs mostly in tests of short duration, typically a few days. Nevertheless, besides achieving an economically feasible yield (around 80% of the theoretical value), process stability is another key issue for hydrogen production at the industrial scale; unfortunately, to date, this aspect has rarely been addressed (Argun and Kargi 2011). One barrier to be overcome in the integration of the dark and photo-bioreactors is related to the different growth rates exhibited by dark and photosynthetic bacteria, either in co-culture or in two-stage processes. In co-culture, buffer supplementation to avoid sharp pH drops and optimum ratios of the species of the microorganisms to balance differences in their respective growth rates have been studied. Whereas in a two-stage process this can be managed more easily, the integration of both systems running in continuous mode requires significant fine tuning. For example, it has been reported that HRT for dark fermentation may vary widely from 0.25 to 30 hours, and for the photo-fermentation this variation is in terms of days (Androga et al. 2011; Argun and Kargi 2011; Fuess et al. 2016; Kim et al. 2005; Tawfik and El-Qelish 2012). Furthermore, it has been reported that DFE dilution is often required in order to obtain satisfactory hydrogen production in the second-stage. Here, a small dark fermentative reactor could provide enough DFE to feed a bigger photo-fermentative one. To summarize, if a reasonable yield is achieved in a dual system, the next task is to achieve the proper integration of both reactors operating in a continuous mode, where the first one provides sufficient substrate for the second stage to be initiated without any time delay. However, based on results reported to date, further incremental improvements still need to be made to achieve an economically sustainable yield while, in parallel, achieving appropriate CO removal. Regarding stability of such dual systems, Chen et al. (2008) reported a relatively stable hydrogen production by photo-fermentation in a CSTR operated at 96 hours HRT. Similarly, long-term hydrogen production (90 days) performed at non-aseptic condition was achieved in repeated batch experiments using starch as a carbon source (Laurinavichene et al. 2012). In that study, hydrogen production in the photo-fermentation step did not seem to be prejudiced by the bacterial contamination present in the DFE used as substrate. However, in another two-stage system operated under non-aseptic conditions, contaminating microorganisms played an important role in inhibiting hydrogen production by PNSBs (Lazaro et al. 2015). The dark fermentation was carried out in AFBR fed with sugarcane stillage, which contains considerable amounts of sulfate. The photo-fermentation was conducted in batch reactors using DFE without any pre-treatment with the exception of a simple dilution and pH adjustment, with the ultimate result that the higher the concentration of DFE, the higher the sulfide production and the lower the hydrogen generation. This observation could be ascribed to the lack of a DFE sterilization pre-treatment thereby allowing sulfate reducers to grow in the photo-fermentation reactors and ultimately competing or inhibiting hydrogen production (Lazaro et al. 2015). In another study, the performance of a pure culture of *R. sphaeroides* AV1b was compared to that of a mixed PNSB consortium in terms of hydrogen production when DFE was used as a substrate. The PNSB consortium-based fermentation resulted in lower yield in comparison to the pure cultures probably due to a too low population of hydrogen-producers in the PSNB consortium, or perhaps due to the presence of opportunist microorganisms in the unsterilized DFE medium (Ghimire et al. 2016). Therefore, further studies on hydrogen production in a two-stage system using real feedstocks should evaluate the impact of many factors in parallel including characteristics of the feedstocks, the inoculum sources and the required DFE pre-treatments. Such a study might best be carried out using a statistical design. This analysis would in turn make it possible to effectively assess the cost-effectiveness of the new approach, including a sensitivity analysis to also take into account the dynamic nature of the cost of goods produced and their market prices.

15.4.3 Hydrogen Production by Multiple Stages (Cellulolytic, Dark Fermentative, and Phototrophic Bacteria)

As discussed previously, the use of abundant and cheap feedstocks, such as lignocellulosic substrates, is a critical factor of economic success of the industrial-scale biotechnological production of hydrogen and large-scale utilization for industry or transportation. However, since these feedstocks that are intrinsically recalcitrant to breakdown and depolymerization into simple sugars cannot be used without a pre-treatment, which furthermore leads to the generation of fermentation inhibitors, one of the critical steps for using these materials is the conversion of cellulose and hemi-cellulose into reducing sugars. The breakdown of these molecules can be achieved by chemical and enzymatic pre-treatments, or by the use of microorganisms that display the necessary hydrolytic capabilities. To our knowledge, the first evaluation of hydrogen production by the use of cellulolytic and photosynthetic bacteria in a co-culture from cellulose was performed almost 30 years ago (Odom and Wall 1983). Much progress has been achieved since, including for example the testing of a wide range of potential agricultural residue feedstocks such as rice straw, corn stalk, water hyacinth, or sugarcane bagasse to name only a few (Cheng et al. 2011; Cheng et al. 2013; Rai et al. 2014a; Yang et al. 2015). For example, pre-treated water hyacinth biomass (microwave-assisted H_2SO_4 before enzymatic hydrolysis) was used as substrate for two-stage hydrogen production by mixed dark- and photo-fermentative consortia (Cheng et al. 2013). In these experiments, the hydrogen yield improved dramatically from 112.3 ml/g TVS to 751.5 ml/g TVS (corresponding to 75.2% of the theoretical H_2 yield). In another study, a microwave-assisted alkali pretreatment was used to improve the enzymatic hydrolysis of rice straw (Chen et al. 2008). The overall yield increases from 155 ml/g TVS to 463 ml/g TVS when the combination process was used (Cheng et al. 2011). Similarly, a delignification process using NaOH was also used to improve the enzymatic hydrolysis with cellulase and hemicellulase from cornstalk (Yang et al. 2015). In the latter experiment, NaOH concentration, cellulase dosage, hydrolysis temperature, hemicellulase, and hydrolysis time were the five most significant parameters, listed in increasing order of influence on sequential dark and photo-FHP (Yang et al. 2015). Both acid and alkali microwave heating are considered efficient methods to aid the disruption of lignocellulosic materials into reducing sugars (RS) and they have been routinely used with a subsequent enzymatic step (Lalaurette et al. 2009; Rai et al. 2014b).

In Figure 15.3, various processes for producing hydrogen from cellulosic biomass are presented. A common characteristic is that the first step is always a hydrolysis step, whether chemical or enzymatic. If the biomass is subjected to acid pre-treatment, the neutralization of the hydrolysis mixture is an important step to be considered before any other step (Figure 15.3a–c). It is at this point that a combination process can be considered. Notably, it is possible to use a sequential dark and photo-fermentation in separate reactors after the cellulose hydrolysis step (Figure 15.3a). There is the possibility of co-culturing cellulolytic microorganisms (bio-hydrolysis) with dark fermentative (Figure 15.3b) and photo-fermentative bacteria (Figure 15.3e) in the same bioreactor to enhance process economics by compressing the reaction time. However, industrial stage culture conditions, starting with medium formulation, in a co-culture system should be better evaluated to optimize the production of hydrogen. Notably, due to the versatility of photosynthetic

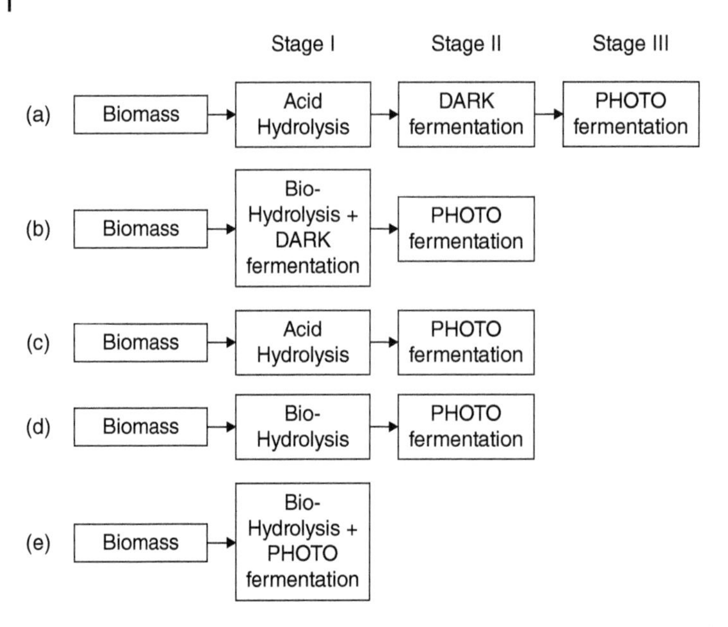

Figure 15.3 Processes for producing hydrogen from cellulosic biomass. Common to all processes is a first step of hydrolysis. Acid pre-treatment requires subsequent neutralization (a–c). Sequential dark and photo-fermentation can be carried out in separate reactors after the cellulose hydrolysis step. Bio-hydrolysis with dark fermentation followed by photofermentation can be carried out (b) or photosynthetic bacterial hydrogen production can be carried out through co-culturing with cellulolytic microorganisms (e). In some cases it might be possible to simply carryout biohydrolysis and photofermentation since some photosynthetic bacteria can directly use sugars derived from cellulose.

bacteria, it is possible to skip the dark-fermentation stage (Figure 15.3c–d), because the sugars released from cellulose can be directly consumed by photosynthetic bacteria (Keskin and Hallenbeck 2012). Further process enhancements can be nonetheless achieved using a co-culture system here since, instead of using the acid/alkali treatments or costly enzymatic hydrolysis to break down the cellulose contained in raw materials such as agricultural waste, it is possible to apply cellulolytic microorganism, e.g. *Cellulomonas fimi*, to achieve this breakdown and release sugars for hydrogen production through dark- and photo-fermentation. Such a system combining the activity of cellulolytic, dark-fermentative, and photo-fermentative bacteria constitutes a multiple-stage process for hydrogen production (Argun and Kargi 2011). Rai et al. (2014b) evaluated hydrogen production from acid-hydrolyzed sugarcane bagasse (SCB), that was subjected to a detoxification pre-treatment aiming at removing furfural and other fermentation inhibitors, with the saccharification conducted by the a relatively recent isolate *C. fimi* (Rai et al. 2014a,b). The subsequent phases involved a dark fermentation step (*E. aerogenes*) followed by a photo-fermentation step (*Rhodopseudomonas* BHU 01). As observed in other studies, the overall yield also increased when this dual system is integrated. Based on results reported to date and considering the vast quantity and quality of biomass feedstocks available, multiple stage processes that combine saccharification with dark-fermentation and photo-fermentation should be exploited more (Balat and Balat 2009; Das 2009).

15.4.4 Hydrogen Production by Combined Dark Fermentation and Microbial Electrolysis

Some anaerobic microorganisms can oxidize organic substrates, releasing electrons to an anode. When that anode is connected to a cathode, the electrons combine with protons and oxygen at the cathode thereby forming water (Hu et al. 2008; Show et al. 2012). In a microbial electrolysis cell (MEC), an external voltage is applied to the cell with the ultimate result of driving hydrogen production at the cathode. As discussed above (Section 15.4.1), dark fermentation alone is not sufficient to completely convert typical substrates to hydrogen, and thus it falls far short of achieving theoretical maximum hydrogen yields and rates. However, fermentation products from the dark stage process could be used in an MEC thereby producing additional hydrogen. In this way, hydrogen production from various substrates, including organic acids (acetate, lactate, and butyrate), glucose, cellulose, and organic wastes using MECs has been reported (Liu et al. 2005; Rozendal et al. 2008; Call and Logan 2008).

Such a combined process has also been used with lignocellulosic materials. For example a hydrogen yield of about 1.6 mol H_2/mol glucose at a rate of 0.25 l H_2/l d with either cellobiose or lignocellulose was obtained in the first dark fermentation step by *C. thermocellum*. MEC was carried out as a second stage to produce additional hydrogen quantities. High electrical efficiencies of 220–230% were obtained with hydrogen rates of 0.96 and 1.0 l H_2/l d with cellobiose and lignocellulose, respectively. Hydrogen yields of 5.5 and 4.1 mol H_2/mol glucose from cellobiose and lignocellulose were achieved in the dark fermentation and MEC processes combined (Lalaurette et al. 2009).

A combined dark and MEC system was furthermore used with the effluent of an ethanol type dark fermentation, giving a hydrogen yield of 1.41 H_2/l/d with a buffered effluent (pH ~7.0) and a hydrogen recovery of 83% in the first step (Lu et al. 2009). The further use of an MEC system at an applied voltage of 0.6 V gave an overall hydrogen recovery of 96% and a hydrogen yield of 2.1 m^3 H_2/m^3/d. The addition of buffer to the MEC system enabled a greater hydrogen production performance as compared to unbuffered effluents (Lu et al. 2009). Another study examined an integrated dark fermentation and an MEC without using an external energy supply, where the electricity needed for the MEC was provided by microbial fuel cells using dark fermentation effluents as substrates (Wang et al. 2011a). This interesting integrated system showed an improved hydrogen with 41% more hydrogen than by fermentation alone with an energy recovery efficiency of 23% and a production rate of 0.24 m^3 H_2/m^3d (Wang et al. 2011a).

Although the advantages of combining dark and microbial electrolysis cells for increased hydrogen production have been demonstrated, the current status of the system is still, to this date, not sufficient to permit its practical industrial scale use, and further studies are required to industrialize such sustainable systems. Similarly, advances in design and construction of suitable cells, electrode material, buffers, and optimization of an external voltage applied are needed in addition to higher hydrogen production yields and rates (Adessi et al. 2012; Hallenbeck et al. 2014).

15.5 Perspectives

The integration of hydrogen production technologies to a fuel bio-hydrogen value chain to serve a subset of the transportation market remains to be achieved at the large scale. A critical factor is to better achieve the integration of hydrogen as an alternative fuel to meet future world energy demands. To this end, it is important to compare and review in detail the HPP of past and future technologies. Hydrogen is an alternative energy carrier to traditional ones and that has the advantages over fossil energy sources of being a clean, non-pollutant fuel with a high energy density and zero emissions when used (Hallenbeck and Benemann 2002; Hallenbeck 2014). At present, hydrogen is mostly produced through the use of fossil fuels, nuclear power, and renewably generated electricity (Ajayi-Oyakhire 2013). In fact, a variety of hydrogen production methods are also available including thermochemical, electrolytic, photolytic, and biological processes though these processes are not currently deployed at the industrial scale (Hallenbeck and Ghosh 2009; Dincer and Acar 2015).

Hydrogen and hydrogen fuel cell technologies are already commercialized and represent many opportunities to generate electricity, heat, and fuel for transportation and power systems (Hallenbeck 2014). However, the majority of the current applications in hydrogen-related commercial industries mostly depend on hydrogen production processes that are non-renewable, energy intensive and high-cost. The US Department of Energy targets a threshold hydrogen production cost of $2.0/gge (USDOE 2016).

The deployment of hydrogen production technologies at the industrial scale offers a number of opportunities for society, economy, and the environment at a time of exponential change when the phenomenon of androgenic climate change will be increasingly generated in the near term. Hydrogen is a high potential fuel that can be used in fuel cells with a variety of purposes, including stationary and mobile power applications; a clear-cut example is the development of fuel cell-powered cars (Trudewind and Wagner 2010). The role and position of hydrogen as a fuel in the energy mix depend less on consumer preference than on logistics and mainly on its production, transportation, and storage. The development and deployment of hydrogen technologies for commercially available long-term applications are one of the ultimate aims in energy policies of numerous governments.

15.5.1 Hydrogen Production Potential and Market Barriers

Global energy demand is positively correlated with economic activities and consequently it has gradually risen annually since the early days of the industrial revolution: this trend has accelerated in the past two decades (Hallenbeck 2014). World energy needs increased by 1.1% in 2014 to reach 13 700 Mtoe in 2014 (IEA 2016) which, when compounded over a decade, represents a phenomenal increase in demand. As a consequence, if the current production and utilization model remain unchanged, a global energy crisis could very well be inevitable since the present energy production model exhibits numerous limitations (e.g. atmospheric pollution exacerbated by coal-fired power stations, peak oil phenomenon), with the first signs of major tensions arising from the first significant stages of global shortages. As a result, making better use of energy, for example by enhancing the energy efficiency of buildings and of engines, as well as investing in technology development and deployment for alternative energy sources, is already presented in the primary policies of numerous governments as a critical global issue, albeit one can lament that a real sense of urgency is still lacking.

The energy content of hydrogen (142 kJ/g) is the highest among the other widely used energy sources such as natural gas and petroleum (Das 2009). The production and use of fuels from non-renewable sources emits greenhouse gasses and ultimately leads to global climate change. In contrast, hydrogen is an environmentally friendly fuel that does not produce any greenhouse gas when consumed. However, regretfully as highlighted earlier in this chapter, most of the hydrogen production technologies currently deployed at the industrial scale rely on fossil fuels. To make matters worse, these processes are highly energy intensive and non-environmentally friendly. At present, hydrogen production is mostly fulfilled from natural gas and oil (70%), coal (18%), and electrolysis (4%) since they are currently still the cheapest sources (Brentner et al. 2010).

Biological hydrogen production has evoked interest as early as the 1970s as an alternative energy carrier. However, the manufacture of hydrogen from renewable sources still needs substantially more research and development for it to become commercially viable. Nevertheless, commercial hydrogen is predicted to provide, by 2050, the energy required to power 700 million fuel cell vehicles (IEA 2007). Biohydrogen has a tremendous potential to compete and replace the current fossil fuel-dependent hydrogen production methods in the market. One of the targets in biohydrogen is to decentralize hydrogen production in certain locations where biowastes are easily accessible, thereby decreasing transportation and economic costs (Adessi and De Philippis 2014).

For it to become a truly sustainable and efficient fuel, numerous hurdles still have to be overcome at every stage of the hydrogen production process. To achieve a sustainable and economically feasible system, major problems in production, storage, transportation, and delivery stages of hydrogen must be dramatically ameliorated or eliminated completely (Hallenbeck 2009). To achieve this transformation, hydrogen production yields must first be enhanced by the development of superior process conditions, bioreactor designs, microbial workhorse, and introducing new biochemical pathways. Another immediate approach is to remove the typical inefficiencies of single production processes by using combined or sequential systems to overcome key issues related to cost and low hydrogen production.

The utilization of solar irradiation, which at the human scale is available for eternity, plays a critical role as green algae, cyanobacteria, and photosynthetic bacteria exhibit various ranges of light absorption. Thus, bioengineering techniques can be applied to modify the light harvesting complexes of the producing organisms or by implementing combination cultures to dramatically improve hydrogen production rates and yields.

Fuel cells are currently mainly considered for use in transportation, stationary, and portable power purposes as they can produce electricity continuously using hydrogen as an energy source (Trudewind and Wagner 2010). As a result, the fuel cell industry has already grown to $2.2 billion sales per year in 2014 (USDOE 2014). An important point, however, remains that hydrogen and fuel cell technologies have to be handled together to achieve low-cost and sustainable energy. The ultimate aim of present fuel cell research and development is to reduce these costs. For this reason, building fuel cells with better and cheaper catalysts and a longer lifetime as well as a proper and efficient cooling system are major objectives in this segment of the energy industry (Trudewind and Wagner 2010).

Hydrogen storage is perhaps one of the remaining vexing issues, as cost-effective and efficient technologies still need to be discovered. To date, there have been many attempts on this front, including physical and chemical storage technologies. Despite these efforts, hydrogen storage still is a major obstacle to many large scale applications (Zannoni et al. 2014).

15.5.2 Hydrogen Generation Market

The reforming of natural gas or biogas for hydrogen generation is the technology most often deployed given its low cost and low CO_2 emissions. This steam-mediated process takes place at high temperatures (700–1000 °C) to produce hydrogen from natural gas, biofuel, or biomass (Hallenbeck 2011). Coal gasification is also a commercially available hydrogen production process in which pulverized coal reacts with pure oxygen at high temperatures (Kothari et al. 2008; Brentner et al. 2010). Biomass gasification is performed following a similar principle to that for coal gasification, but this process remains immature. Therefore, there is still no industrial scale biomass gasification hydrogen plant. One of the earliest techniques explored for hydrogen production is, as highlighted earlier, to use electrolysis, whereby water is split into its elements oxygen and hydrogen (Hallenbeck et al. 2014). Electrolyzers can be built at small or large scale to obtain highly pure hydrogen; however, the cost of this technique is not yet low enough to enable a major commercial impact since large amounts of renewable resources and electricity are required. Notably, it has been assessed that this system costs above the $2.00/kg (USDOE 2015). Photoelectrochemical hydrogen production is another method which requires special semiconductor electrodes to collect the solar energy required to split water into oxygen and hydrogen. This process also needs further investments in research and development before a sustainable hydrogen production can be achieved (Holladay et al. 2009; Dincer and Acar 2015).

No biological hydrogen production method is currently implemented at the industrial scale as more process economic advances are needed for this manufacturing technology to come of age. Photolytic hydrogen production occurs either by photoelectrochemical or photobiological routes. The principle of photolytic processes is based on the lysis of water into hydrogen and oxygen in an enzymatically catalysed reaction mediated by solar energy. However, the industrial use of photosynthetic bacteria to produce hydrogen in this manner still requires significant advances to become industrially viable (Singh and Wahid 2015). All of the production methods and techniques developed to date exhibit various advantages and disadvantages. The crucial challenge in this area is to develop an environmentally friendly, cost-effective, and sustainable hydrogen production system for long-term applications considering the large capital expenses that will be required to develop industrial scale manufacturing plants adapted to these new processes.

15.5.3 Microbial Hydrogen Production: Targets and Future Prospects

Microbial hydrogen production is mainly conducted with photobiological or fermentative processes. Photobiological hydrogen production occurs by exploiting photosynthetic bacteria, cyanobacteria, or algae under certain conditions. These organisms are able to utilize the solar energy for forming hydrogen in biophotolytic or photofermentation processes, and thus have great potential for a sustainable, long-term industrial exploitation in spite of their presently insufficient production rates and light conversion efficiencies (Hallenbeck 2011). Strategies, such as blocking competing pathways, modifying various genes, and decreasing the size of photosynthetic pigments are some examples of paths to explore to improve hydrogen production (Hallenbeck 2014). Currently, an efficient and sustained microbial hydrogen production is being targeted by genetic engineering of algae, cyanobacteria, photofermentative, dark fermentative bacteria and MEC systems. Moreover, improvements in reactor design and processes are still required.

The utilization of hydrogen as a fuel remains limited not only by large scale processing hurdles, but also by logistics constraints including transportation and storage. The cost of biohydrogen production largely depends on four main parameters: the availability and price of the substrates used in dark fermentation, the performance of the photobioreactor and the light conversion efficiency. The current efficiency of incident solar energy to hydrogen by photosynthetic bacteria is less than 1%, which is far from the USDOE 2020 target of 4.5% (USDOE 2015). Therefore, achieving higher light conversion efficiencies is a major barrier for biological hydrogen to play a significant role in the energy mix. Photobioreactor design and the engineering of bacterial photosynthetic pigments using molecular biology techniques exemplify how one could enhance both hydrogen efficiency and targeted economical aspects (Adessi and De Philippis 2014; Basak and Das 2007). Likewise, the photolytic production of hydrogen by green algae and cyanobacteria needs to be improved for these techniques to play a significant economic role. The aim of the US Department of Energy in photolytic biological hydrogen production is to bring this cost to $9.2/gge by 2020 (USDOE 2016).

The major problem that a hydrogen manufacturing plant manager might encounter is that photolytic processes are far from being efficient enough in their light utilization, conversion of solar energy to hydrogen, and production rates. In algae and cyanobacteria, the co-production of hydrogen and oxygen by the photolytic process is the main limiting factor as it typically ultimately results in the inhibition of hydrogenase, thereby limiting the duration of the active hydrogen production phase (Ghirardi 2014). This limitation of course needs to be removed to achieve market competitiveness. It is worth pointing out here that meeting market targets for photolytic hydrogen production will require significant R&D breakthroughs. For example, one of the main 2020 targets cited by the US Department of Energy for photolytic hydrogen is to increase the duration of continuous hydrogen production from 2 minutes (status in 2011) up to 4 hours (USDOE 2015).

Alternatively, biohydrogen production can be implemented via converting biomass using dark fermentation or microbial electrolysis cells. However, the industrial or agricultural wastes used in these processes must be cheap enough and easily obtainable. In dark fermentation, bacteria decompose carbohydrates under anaerobic conditions into various metabolites and hydrogen. A number of different pathways and enzymes are at play in the production of hydrogen by fermentative bacteria. This process has the advantage of using organic wastes and of not requiring any light. However, hydrogen yields remain insufficient due to the incomplete utilization of the organic compounds present in the reaction media of these processes (Hallenbeck 2014; Azwar et al. 2014).

In addition, as discussed above, combining dark fermentation and MEC represents another potential hydrogen production system. However, hydrogen production with MEC has yet to be scaled up and moreover needs more research and development in many areas. Challenges still remain with respect to the use of MEC technology, including the development of low-cost and more efficient electrodes, the reduction of the required electrical input, and the achievement of higher current densities (Adessi et al. 2012; Show et al. 2012). Strategies to increase hydrogen yields and conversion efficiencies, to reduce feedstock cost, and the development of low-cost electrodes are some examples of what is required to enhance hydrogen production in combined dark fermentation and MEC systems in order to be market competitive. Despite obvious limitations, prospects remain bright for the industrial production and use of sustainable fuel hydrogen.

Acknowledgements

This work was supported by grants from the NSERC (Natural Sciences and Engineering Research Council [Canada]) to PCH and by a USAFA contract (FA7000-16-2-0006) to Hallenbeck Associates. The views expressed here are those of the authors and do not reflect the official policy or position of the United States Air Force, the Department of Defense, or the US Government. ES thanks the TUBITAK 2214A International Doctoral Research Fellowship Program, Turkey (Project number: 1059B141500983). CZL was supported by a scholarship from CNPq/Brazil (Process 202426/2014-9).

References

Abo-Hashesh, M., Ghosh, D., Tourigny, A. et al. (2011). Single stage photofermentative hydrogen production from glucose: an attractive alternative to two stage photofermentation or co-culture approaches. *International Journal of Hydrogen Energy* 36: 13889–13895.

Abreu, A.A., Alves, J.I., Pereira, M.A. et al. (2010). Engineered heat treated methanogenic granules: a promising biotechnological approach for extreme thermophilic biohydrogen production. *Bioresource Technology* 101 (24): 9577–9586.

Abreu, A.A., Alves, J.I., Pereira, M.A. et al. (2011). Strategies to suppress hydrogen-consuming microorganisms affect macro and micro scale structure and microbiology of granular sludge. *Biotechnology and Bioengineering* 108 (8): 1766–1775.

Abreu, A.A., Karakashev, D., Angelidaki, I. et al. (2012). Biohydrogen production from arabinose and glucose using extreme thermophilic anaerobic mixed cultures. *Biotechnology for Biofuels* 5: 6–6.

Adessi, A. and De Phillippis, R. (2014). Photobioreactor design and illumination systems for H_2 production with anoxygenic photosynthetic bacteria: a review. *International Journal of Hydrogen Energy* 39: 3127–3147.

Adessi, A., De Philippis, R., and Hallenbeck, P.C. (2012). Combined systems for maximum substrate conversion. In: *Microbial Technologies in Advanced Biofuels Production* (ed. P.C. Hallenbeck), 107–126. Springer, US.

Ajayi-Oyakhire, O. (2013). *Hydrogen – Untapped Energy?* Institution of Gas Engineers and Managers.

Akutsu, Y., Lee, D.Y., Li, Y.Y., and Noike, T. (2009a). Hydrogen production potentials and fermentative characteristics of various substrates with different heat-pretreated natural microflora. *International Journal of Hydrogen Energy* 34 (13): 5365–5372.

Akutsu, Y., Li, Y.Y., Harada, H., and Yu, H.Q. (2009b). Effects of temperature and substrate concentration on biological hydrogen production from starch. *International Journal of Hydrogen Energy* 34 (6): 2558–2566.

Algapani, D.E., Qiao, W., Su, M. et al. (2016). Bio-hydrolysis and bio-hydrogen production from food waste by thermophilic and hyperthermophilic anaerobic process. *Bioresource Technology* 216: 768–777.

Alibardi, L. and Cossu, R. (2015). Composition variability of the organic fraction of municipal solid waste and effects on hydrogen and methane production potentials. *Waste Management* 36: 147–155.

Androga, D.D., Ozgur, E., Eroglu, I. et al. (2011). Significance of carbon to nitrogen ratio on the long-term stability of continuous photofermentative hydrogen production. *International Journal of Hydrogen Energy* 36: 15583–15594.

Angenent, L.T., Karim, K., Al-Dahhan, M.H. et al. (2004). Production of bioenergy and biochemicals from industrial and agricultural wastewater. *Trends in Biotechnology* 22 (9): 477–485.

Argun, H. and Kargi, F. (2010). Bio-hydrogen production from ground wheat starch by continuous combined fermentation using annular-hybrid bioreactor. *International Journal of Hydrogen Energy* 35: 6170–6178.

Argun, H. and Kargi, F. (2011). Bio-hydrogen production by different operational modes of dark and photo-fermentation: an overview. *International Journal of Hydrogen Energy* 36: 7443–7459.

Argun, H., Kargi, F., and Kapdan, I.K. (2009). Effects of the substrate and cell concentration on bio-hydrogen production from ground wheat by combined dark and photo-fermentation. *International Journal of Hydrogen Energy* 34: 6181–6188.

Asada, Y., Tokumoto, M., Aihara, Y. et al. (2006). Hydrogen production by co-cultures of *Lactobacillus* and a photosynthetic bacterium, *Rhodobacter sphaeroides* RV. *International Journal of Hydrogen Energy* 31: 1509–1513.

Azwar, M.Y., Hussain, M.A., and Abdul-Wahab, A.K. (2014). Development of biohydrogen production by photobiological, fermentation and electrochemical processes: a review. *Renewable and Sustainable Energy Reviews* 31: 158–173.

Baghchehsaraee, B., Nakhla, G., Karamanev, D., and Margaritis, A. (2009). Effect of extrinsic lactic acid on fermentative hydrogen production. *International Journal of Hydrogen Energy* 34 (6): 2573–2579.

Balat, M. and Balat, M. (2009). Political, economic and environmental impacts of biomass-based hydrogen. *International Journal of Hydrogen Energy* 34: 3589–3603.

Barbosa, M.J., Rocha, J.M.S., Tramper, J., and Wijffels, R.H. (2001). Acetate as a carbon source for hydrogen production by photosynthetic bacteria. *Journal of Biotechnology* 85: 25–33.

Barca, C., Ranava, D., Bauzan, M. et al. (2016). Fermentative hydrogen production in an up-flow anaerobic biofilm reactor inoculated with a co-culture of *Clostridium acetobutylicum* and *Desulfovibrio vulgaris*. *Bioresource Technology* 221: 526–533.

Barros, A.R., Adorno, M.A.T., Sakamoto, I.K. et al. (2011). Performance evaluation and phylogenetic characterization of anaerobic fluidized bed reactors using ground tire and pet as support materials for biohydrogen production. *Bioresource Technology* 102 (4): 3840–3847.

Basak, N. and Das, D. (2007). The prospect of purple non-sulfur (PNS) photosynthetic bacteria for hydrogen production: the present state of the art. *World Journal of Microbiology and Biotechnology* 23: 31–42.

Beckner, M., Ivey, M.L., and Phister, T.G. (2011). Microbial contamination of fuel ethanol fermentations. *Letters in Applied Microbiology* 53 (4): 387–394.

Bravo, I.S., Lovato, G., Rodrigues, J.A. et al. (2015). Biohydrogen production in an AnSBBR treating glycerin-based wastewater: effects of organic loading, influent concentration, and cycle time. *Applied Biochemistry and Biotechnology* 175 (4): 1892–1914.

Brentner, L.B., Peccia, J., and Zimmerman, J.B. (2010). Challenges in developing biohydrogen as a sustainable energy source: implications for a research agenda. *Environmental Science & Technology* 44: 2243–2254.

Bundhoo, M.A.Z. and Mohee, R. (2016). Inhibition of dark fermentative bio-hydrogen production: a review. *International Journal of Hydrogen Energy* 41 (16): 6713–6733.

Cai, G.Q., Jin, B., Monis, P., and Saint, C. (2011). Metabolic flux network and analysis of fermentative hydrogen production. *Biotechnology Advances* 29 (4): 375–387.

Call, D. and Logan, B.E. (2008). Hydrogen production in a single chamber microbial electrolysis cell lacking a membrane. *Environmental Science & Technology* 42: 3401–3406.

Cappelletti, M., Bucchi, G., De Sousa Mendes, J. et al. (2012). Biohydrogen production from glucose, molasses and cheese whey by suspended and attached cells of four hyperthermophilic Thermotoga strains. *Journal of Chemical Technology and Biotechnology* 87 (9): 1291–1301.

Castelló, E., García y Santos, C., Iglesias, T. et al. (2009). Feasibility of biohydrogen production from cheese whey using a UASB reactor: links between microbial community and reactor performance. *International Journal of Hydrogen Energy* 34 (14): 5674–5682.

Castelló, E., Perna, V., Wenzel, J. et al. (2011). Microbial community composition and reactor performance during hydrogen production in a UASB reactor fed with raw cheese whey inoculated with compost. *Water Science and Technology* 64 (11): 2265–2273.

Chen, C.-Y., Yang, M.-H., Yeh, K.-L. et al. (2008). Biohydrogen production using sequential two-stage dark and photo fermentation processes. *International Journal of Hydrogen Energy* 33 (18): 4755–4762.

Cheng, J. and Zhu, M. (2013). A novel anaerobic co-culture system for bio-hydrogen production from sugarcane bagasse. *Bioresource Technology* 144: 623–631.

Cheng, J., Su, H., Zhou, J. et al. (2011). Microwave-assisted alkali pretreatment of rice straw to promote enzymatic hydrolysis and hydrogen production in dark- and photo-fermentation. *International Journal of Hydrogen Energy* 36 (3): 2093–2101.

Cheng, J., Xia, A., Su, H. et al. (2013). Promotion of H_2 production by microwave-assisted treatment of water hyacinth with dilute H_2SO_4 through combined dark fermentation and photofermentation. *Energy Conversion and Management* 73: 329–334.

Das, D. (2009). Advances in biohydrogen production processes: an approach towards commercialization. *International Journal of Hydrogen Energy* 34: 7349–7357.

Davila-Vazquez, G., Alatriste-Mondragon, F., Leon-Rodriguez, A., and Razo-Flores, E. (2008). Fermentative hydrogen production in batch experiments using lactose, cheese whey and glucose: influence of initial substrate concentration and pH. *International Journal of Hydrogen Energy* 33 (19): 4989–4997.

Dias, M.O.S., Cunha, M.P., Jesus, C.D.F. et al. (2011). Second generation ethanol in Brazil: can it compete with electricity production? *Bioresource Technology* 102: 8964–8971.

Dincer, I. and Acar, C. (2015). Review and evaluation of hydrogen production methods for better sustainability. *International Journal of Hydrogen Energy* 40: 11094–11111.

Ding, J., Liu, B.F., Ren, N.Q. et al. (2009). Hydrogen production from glucose by co-culture of *Clostridium butyricum* and immobilized *Rhodopseudomonas faecalis* RLD-53. *International Journal of Hydrogen Energy* 34: 3647–3652.

Eroglu, E. and Melis, A. (2011). Photobiological hydrogen production: recent advances and state of the art. *Bioresource Technology* 102: 8403–8413.

Eroglu, I., Aslan, K., Gunduz, U. et al. (1999). Substrate consumption rates for hydrogen production by *Rhodobacter sphaeroides* in a column photobioreactor. *Journal of Biotechnology* 70: 103–113.

Eroglu, E., Eroglu, I., and Gunduz, U. (2008). Effect of clay pretreatment on photofermentative hydrogen production from olive mill wastewater. *Bioresource Technology* 99: 6799–6808.

España-Gamboa, E.I., Mijangos-Cortés, J.O., Hernández-Zárate, G. et al. (2012). Methane production by treating vinasses from hydrous ethanol using a modified UASB reactor. *Biotechnology for Biofuels* 5 (1): 1–9.

Etchebehere, C., Castello, E., Wenzel, J. et al. (2016). Microbial communities from 20 different hydrogen-producing reactors studied by 454 pyrosequencing. *Applied Microbiology and Biotechnology* 100 (7): 3371–3384.

Fang, H.H.P., Zhu, H., and Zhang, T. (2006). Phototrophic hydrogen production from glucose by pure and co-cultures of *Clostridium butyricum* and *Rhodobacter sphaeroides*. *International Journal of Hydrogen Energy* 31: 2223–2230.

Ferraz Júnior, A.D.N., Wenzel, J., Etchebehere, C., and Zaiat, M. (2014). Effect of organic loading rate on hydrogen production from sugarcane vinasse in thermophilic acidogenic packed bed reactors. *International Journal of Hydrogen Energy* 39 (30): 16852–16862.

Ferraz Júnior, A.D.N., Etchebehere, C., and Zaiat, M. (2015a). High organic loading rate on thermophilic hydrogen production and metagenomic study at an anaerobic packed-bed reactor treating a residual liquid stream of a Brazilian biorefinery. *Bioresource Technology* 186: 81–88.

Ferraz Júnior, A.D.N., Etchebehere, C., and Zaiat, M. (2015b). Mesophilic hydrogen production in acidogenic packed-bed reactors (APBR) using raw sugarcane vinasse as substrate: influence of support materials. *Anaerobe* 34: 94–105.

Fuess, L.T., Mazine Kiyuna, L.S., Garcia, M.L., and Zaiat, M. (2016). Operational strategies for long-term biohydrogen production from sugarcane stillage in a continuous acidogenic packed-bed reactor. *International Journal of Hydrogen Energy* 41 (19): 8132–8145.

Gabrielyan, L., Torgomyan, H., and Trchounian, A. (2010). Growth characteristics and hydrogen production by *Rhodobacter sphaeroides* using various amino acids as nitrogen sources and their combinations with carbon sources. *International Journal of Hydrogen Energy* 35 (22): 12201–12207.

Gadhamshetty, V., Johnson, D.C., Nirmalakhandan, N. et al. (2009). Dark and acidic conditions for fermentative hydrogen production. *International Journal of Hydrogen Energy* 34 (2): 821–826.

Ghimire, A., Valentino, S., Frunzo, L. et al. (2016). Concomitant biohydrogen and poly-β-hydroxybutyrate production from dark fermentation effluents by adapted *Rhodobacter sphaeroides* and mixed photofermentative cultures. *Bioresource Technology* 217: 157–164.

Ghirardi, M.L. (2014). Hydrogen production by water biophotolysis. In: *Microbial Bioenergy: Hydrogen Production*, Advances in Photosynthesis and Respiration, vol. 38 (eds. D. Zannoni and R. De Philips), 101–135. Springer.

Ghosh, D., Sobro, I.F., and Hallenbeck, P.C. (2012). Optimization of the hydrogen yield from single-stage photofermentation of glucose by *Rhodobacter capsulatus* JP91 using response surface methodology. *Bioresource Technology* 123: 199–206.

Gilroyed, B.H., Chang, C., Chu, A., and Hao, X. (2008). Effect of temperature on anaerobic fermentative hydrogen gas production from feedlot cattle manure using mixed microflora. *International Journal of Hydrogen Energy* 33 (16): 4301–4308.

Gomes, B.C., Rosa, P.R.F., Etchebehere, C. et al. (2015). Role of homo-and heterofermentative lactic acid bacteria on hydrogen-producing reactors operated with cheese whey wastewater. *International Journal of Hydrogen Energy* 40: 8650–8660.

Guo, X., Cavka, A., Jönsson, L.J., and Hong, F. (2013). Comparison of methods for detoxification of spruce hydrolysate for bacterial cellulose production. *Microbial Cell Factories* 12 (1): 93.

Guwy, A.J., Dinsdale, R.M., Kim, J.R. et al. (2011). Fermentative biohydrogen production systems integration. *Bioresource Technology* 102: 8534–8542.

Hakobyan, L., Gabrielyan, L., and Trchounian, A. (2012). Yeast extract as an effective nitrogen source stimulating cell growth and enhancing hydrogen photoproduction by *Rhodobacter sphaeroides* strains from mineral springs. *International Journal of Hydrogen Energy* 37: 6519–6526.

Hallenbeck, P.C. (2005). Fundamentals of the fermentative production of hydrogen. *Water Science and Technology* 52 (1–2): 21–29.

Hallenbeck, P.C. (2009). Fermentative hydrogen production: principles, progress and prognosis. *International Journal of Hydrogen Energy* 34: 7379–7389.

Hallenbeck, P.C. (2011). Microbial paths to renewable hydrogen production. *Biofuels* 2: 285–302.

Hallenbeck, P.C. (2012). Biofuels, the larger context. In: *Microbial Technologies in Advanced Biofuels Production* (ed. P.C. Hallenbeck), 3–12.

Hallenbeck, P.C. (2014). Bioenergy from microorganisms: an overview. In: *Microbial BioEnergy: Hydrogen Production*, Advances in Photosynthesis and Respiration, vol. 38 (eds. D. Zannoni and R. De Philippis), 3–21. Springer.

Hallenbeck, P.C. and Benemann, J.R. (2002). Biological hydrogen production: fundamentals and limiting processes. *International Journal of Hydrogen Energy* 27: 1185–1193.

Hallenbeck, P.C. and Ghosh, D. (2009). Advances in fermentative biohydrogen production: the way forward? *Trends in Biotechnology* 27: 287–297.

Hallenbeck, P.C. and Liu, Y. (2016). Recent advances in hydrogen production by photosynthetic bacteria. *International Journal of Hydrogen Energy* 41 (7): 4446–4454.

Hallenbeck, P.C., Grogger, M., and Veverka, D. (2014). Recent advances in microbial electrocatalysis. *Electrocatalysis* 5: 319–329.

Han, W., Chen, H., Yao, X. et al. (2010). Biohydrogen production with anaerobic sludge immobilized by granular activated carbon in a continuous stirred-tank. *Journal of Forest Research* 21 (4): 509–513.

Hassan, S.H.A. and Morsy, F.M. (2015). Feasibility of installing and maintaining anaerobiosis using *Escherichia coli* HD701 as a facultative anaerobe for hydrogen production by *Clostridium acetobutylicum* ATCC 824 from various carbohydrates. *Enzyme and Microbial Technology* 81: 56–62.

Holladay, J.D., Hu, J., King, D.L., and Wang, Y. (2009). An overview of hydrogen production technologies. *Catalysis Today* 139: 244–260.

Hu, H., Fan, Y., and Liu, H. (2008). Hydrogen production using single-chamber membrane-free microbial electrolysis cells. *Water Research* 42: 4172–4178.

Hu, C.C., Giannis, A., Chen, C.-L., and Wang, J.-Y. (2014). Evaluation of hydrogen producing cultures using pretreated food waste. *International Journal of Hydrogen Energy* 39: 19337–19342.

Huang, C.L., Chen, C.C., Lin, C.Y., and Liu, W.T. (2009). Quantitative fluorescent in-situ hybridization: a hypothesized competition mode between two dominant bacteria groups in hydrogen-producing anaerobic sludge processes. *Water Science and Technology* 59 (10): 1901–1909.

Hulshoff Pol, L.W., de Castro Lopes, S.I., Lettinga, G., and Lens, P.N.L. (2004). Anaerobic sludge granulation. *Water Research* 38: 1376–1389.

Hung, C.-H., Chang, Y.-T., and Chang, Y.-J. (2011). Roles of microorganisms other than *Clostridium* and *Enterobacter* in anaerobic fermentative biohydrogen production systems – a review. *Bioresource Technology* 102 (18): 8437–8444.

Hussy, I., Hawkes, F.R., Dinsdale, R., and Hawkes, D.L. (2003). Continuous fermentative hydrogen production from a wheat starch co-product by mixed microflora. *Biotechnology and Bioengineering* 84 (6): 619–626.

Im, W.-T., Kim, D.-H., Kim, K.-H., and Kim, M.-S. (2012). Bacterial community analyses by pyrosequencing in dark fermentative H_2-producing reactor using organic wastes as a feedstock. *International Journal of Hydrogen Energy* 37 (10): 8330–8337.

International Energy Agency, IEA Energy Technology Essentials: Biofuel Production (2007) available at https://www.iea.org/publications/freepublications/publication/essentials2.pdf

International Energy Agency, Key World Energy Trends, Excerpt from: World energy balances, (2016) available at https://www.iea.org/publications/freepublications/publication/KeyWorldEnergyTrends.pdf

Junghare, M., Subudhi, S., and Lal, B. (2012). Improvement of hydrogen production under decreased partial pressure by newly isolated alkaline tolerant anaerobe, *Clostridium butyricum* TM-9A: optimization of process parameters. *International Journal of Hydrogen Energy* 37 (4): 3160–3168.

Kalia, V.C. and Purohit, H.J. (2008). Microbial diversity and genomics in aid of bioenergy. *Journal of Industrial Microbiology & Biotechnology* 35 (5): 403–419.

Kapdan, I.K. and Kargi, F. (2006). Bio-hydrogen production from waste materials. *Enzyme and Microbial Technology* 38 (5): 569–582.

Karlsson, A., Vallin, L., and Ejlertsson, J. (2008). Effects of temperature, hydraulic retention time and hydrogen extraction rate on hydrogen production from the fermentation of food industry residues and manure. *International Journal of Hydrogen Energy* 33: 953–962.

Kawaguchi, H., Hashimoto, K., Hirata, K., and Miyamoto, K. (2001). H_2 production from algal biomass by a mixed culture of *Rhodobium marinum* A-501 and *Lactobacillus amylovorus*. *Journal of Bioscience and Bioengineering* 91: 277–282.

Keskin, T. and Hallenbeck, P.C. (2012). Hydrogen production from sugar industry wastes using single-stage photofermentation. *Bioresource Technology* 112: 131–136.

Kim, S.H., Han, S.K., and Shin, H.S. (2005). Performance comparison of a continuous flow stirred-tank reactor and an anaerobic sequencing batch reactor for fermentative hydrogen production depending on substrate concentration. *Water Science and Technology* 52: 23–29.

Kim, D.-H., Kim, S.-H., and Shin, H.-S. (2009). Hydrogen fermentation of food waste without inoculum addition. *Enzyme and Microbial Technology* 45 (3): 181–187.

Kim, M.S., Kim, D.H., Cha, J., and Lee, J.K. (2012). Effect of carbon and nitrogen sources on photo-fermentative H_2 production associated with nitrogenase, uptake hydrogenase activity, and PHB accumulation in *Rhodobacter sphaeroides* KD131. *Bioresource Technology* 116: 179–183.

Kim, D.-H., Jang, S., Yun, Y.-M. et al. (2014). Effect of acid-pretreatment on hydrogen fermentation of food waste: microbial community analysis by next generation sequencing. *International Journal of Hydrogen Energy* 39: 16302–16309.

Kotay, S.M. and Das, D. (2009). Novel dark fermentation involving bioaugmentation with constructed bacterial consortium for enhanced biohydrogen production from pretreated sewage sludge. *International Journal of Hydrogen Energy* 34 (17): 7489–7496.

Kothari, R., Buddhi, D., and Sawhney, R.L. (2008). Comparison of environmental and economic aspects of various hydrogen production methods. *Renewable and Sustainable Energy Reviews* 12 (2): 553–563.

Kumar, G., Mudhoo, A., Sivagurunathan, P. et al. (2016a). Recent insights into the cell immobilization technology applied for dark fermentative hydrogen production. *Bioresource Technology* 219: 725–737.

Kumar, G., Sivagurunathan, P., Park, J.-H. et al. (2016b). HRT dependent performance and bacterial community population of granular hydrogen-producing mixed cultures fed with galactose. *Bioresource Technology* 206: 188–194.

Lai, Z., Zhu, M., Yang, X. et al. (2014). Optimization of key factors affecting hydrogen production from sugarcane bagasse by a thermophilic anaerobic pure culture. *Biotechnology for Biofuels* 7: 1–11.

Lalaurette, E., Thammannagowda, S., Mohagheghi, A. et al. (2009). Hydrogen production from cellulose in a two-stage process combining fermentation and electrohydrogenesis. *International Journal of Hydrogen Energy* 34: 6201–6210.

Laurent, B., Serge, H., Julien, M. et al. (2012). Effects of hydrogen partial pressure on fermentative biohydrogen production by a chemotropic *Clostridium* bacterium in a new horizontal rotating cylinder reactor. *Energy Procedia* 29: 34–41.

Laurinavichene, T.V., Belokopytov, B.F., Laurinavichius, K.S. et al. (2012). Towards the integration of dark- and photo-fermentative waste treatment. Repeated batch sequential dark- and photofermentation using starch as substrate. *International Journal of Hydrogen Energy* 37 (10): 8800–8810.

Lazaro, C.Z., Varesche, M.B.A., and Silva, E.L. (2015). Sequential fermentative and phototrophic system for hydrogen production: an approach for Brazilian alcohol distillery wastewater. *International Journal of Hydrogen Energy* 40 (31): 9642–9655.

Lee, K.-S., Hsu, Y.-F., Lo, Y.-C. et al. (2008). Exploring optimal environmental factors for fermentative hydrogen production from starch using mixed anaerobic microflora. *International Journal of Hydrogen Energy* 33 (5): 1565–1572.

Levin, D.B. and Lubieniechi, S. (2013). Patent landscape for biological hydrogen production. *Recent Patents on DNA & Gene Sequences* 7: 207–213.

Li, R.Y., Zhang, T., and Fang, H.H. (2011). Application of molecular techniques on heterotrophic hydrogen production research. *Bioresource Technology* 102 (18): 8445–8456.

Lima, D.M.F., Lazaro, C.Z., Rodrigues, J.A.D. et al. (2016). Optimization performance of an AnSBBR applied to biohydrogen production treating whey. *Journal of Environmental Management* 169: 191–201.

Liu, F. and Fang, B. (2007). Optimization of bio-hydrogen production from biodiesel wastes by *Klebsiella pneumoniae*. *Biotechnology Journal* 2 (3): 374–380.

Liu, H., Grot, S., and Logan, B.E. (2005). Electrochemically assisted microbial production of hydrogen from acetate. *Environmental Science & Technology* 39: 4317–4320.

Liu, Y., Yu, P., Song, X., and Qu, Y.B. (2008). Hydrogen production from cellulose by co-culture of *Clostridium thermocellum* JN4 and *Thermoanaerobacterium thermosaccharolyticum* GD17. *International Journal of Hydrogen Energy* 33: 2927–2933.

Liu, B.F., Ren, N.Q., Xie, G.J. et al. (2010). Enhanced bio-hydrogen production by the combination of dark and photofermentation in batch culture. *Bioresource Technology* 101: 5325–5329.

Liu, Q., Zhang, X., Yu, L., Xia, L., Zhao, A., Tai, J., Jianyong, L., Guangren, Q., Zhi, P.X. (2011). Fermentative hydrogen production from fresh leachate in batch and continuous bioreactors. *Bioresource Technology* 102: 5411–5417.

Lo, Y.-C., Chen, X.-J., Huang, C.-Y. et al. (2013). Dark fermentative hydrogen production with crude glycerol from biodiesel industry using indigenous hydrogen-producing bacteria. *International Journal of Hydrogen Energy* 38 (35): 15815–15822.

Lu, L., Ren, N., Xing, D., and Logan, B.E. (2009). Hydrogen production with effluent from an ethanol-H_2-coproducing fermentation reactor using a single-chamber microbial electrolysis cell. *Biosensors & Bioelectronics* 24: 3055–3060.

Maru, B.T., Bielen, A.A.M., Kengen, S.W.M. et al. (2012). Biohydrogen Production from glycerol using *Thermotoga* spp. *Energy Procedia* 29: 300–307.

Maru, B.T., Constanti, M., Stchigel, A.M. et al. (2013). Biohydrogen production by dark fermentation of glycerol using *Enterobacter* and *Citrobacter* sp. *Biotechnology Progress* 29 (1): 31–38.

Maru, B.T., López, F., Kengen, S.W.M. et al. (2016). Dark fermentative hydrogen and ethanol production from biodiesel waste glycerol using a co-culture of *Escherichia coli* and *Enterobacter* sp. *Fuel* 186: 375–384.

McKinlay, J.B. and Harwood, C.S. (2010). Photobiological production of hydrogen gas as a biofuel. *Current Opinion in Biotechnology* 21: 244–251.

Mishra, P., Thakur, S., Singh, L. et al. (2016). Enhanced hydrogen production from palm oil mill effluent using two stage sequential dark and photo fermentation. *International Journal of Hydrogen Energy* 41: 18431–18440.

Mizuno, O., Dinsdale, R., Hawkes, F.R. et al. (2000). Enhancement of hydrogen production from glucose by nitrogen gas sparging. *Bioresource Technology* 73: 59–65.

Mohan, S.V., Babu, V.L., and Sarma, P.N. (2007). Anaerobic biohydrogen production from dairy wastewater treatment in sequencing batch reactor (AnSBR): effect of organic loading rate. *Enzyme and Microbial Technology* 41 (4): 506–515.

Moreno-Andrade, I., Carrillo-Reyes, J., Santiago, S.G., and Bujanos-Adame, M.C. (2015). Biohydrogen from food waste in a discontinuous process: effect of HRT and microbial community analysis. *International Journal of Hydrogen Energy* 40 (48): 17246–17252.

Mu, Y. and Yu, H.Q. (2006). Biological hydrogen production in a UASB reactor with granules: physicochemical characteristics of hydrogen-producing granules. *Biotechnology and Bioengineering* 94 (5): 980–987.

Ngo, T.A. and Sim, S.J. (2012). Dark fermentation of hydrogen from waste glycerol using hyperthermophilic eubacterium *Thermotoga neapolitana*. *Environmental Progress & Sustainable Energy* 31 (3): 466–473.

Nguyen, T.A.D., Han, S.J., Kim, J.P. et al. (2009). Hydrogen production of the hyperthermophilic eubacterium, *Thermotoga neapolitana* under N_2 sparging condition. *Bioresource Technology* 101 (1): S38–S41.

Ning, Y.-Y., Wang, S.-F., Jin, D.-W. et al. (2013). Formation of hydrogen-producing granules and microbial community analysis in a UASB reactor. *Renewable Energy* 53: 12–17.

Odom, J.M. and Wall, J.D. (1983). Photoproduction of H_2 from cellulose by an anaerobic bacterial co-culture. *Applied and Environmental Microbiology* 45 (4): 1300–1305.

Özgür, E., Mars, A.E., Peksel, B. et al. (2010a). Biohydrogen production from beet molasses by sequential dark and photofermentation. *International Journal of Hydrogen Energy* 35 (2): 511–517.

Özgür, E., Afsar, N., De Vrije, T. et al. (2010b). Potential use of thermophilic dark fermentation effluents in photofermentative hydrogen production by *Rhodobacter capsulatus*. *Journal of Cleaner Production* 18: S23–S28.

Ozmihci, S. and Kargi, F. (2010). Comparison of different mixed cultures for bio-hydrogen production from ground wheat starch by combined dark and light fermentation. *Journal of Industrial Microbiology & Biotechnology* 37: 341–347.

Ozturk, Y., Gokce, A., Peksel, B. et al. (2012). Hydrogen production properties of *Rhodobacter capsulatus* with genetically modified redox balancing pathways. *International Journal of Hydrogen Energy* 37: 2014–2020.

Pandey, A., Soccol, C.R., Nigam, P., and Soccol, V.T. (2000). Biotechnological potential of agro-industrial residues: sugarcane bagasse. *Bioresource Technology* 74: 69–80.

Pasupuleti, S.B., Sarkar, O., and Mohan, S.V. (2014). Upscaling of biohydrogen production process in semi-pilot scale biofilm reactor: evaluation with food waste at variable organic loads. *International Journal of Hydrogen Energy* 39 (14): 7587–7596.

Patel, S.K.S., Purohit, H.J., and Kalia, V.C. (2010). Dark fermentative hydrogen production by defined mixed microbial cultures immobilized on ligno-cellulosic waste materials. *International Journal of Hydrogen Energy* 35 (19): 10674–10681.

Patel, S.K.S., Kumar, P., Mehariya, S. et al. (2014). Enhancement in hydrogen production by co-cultures of *Bacillus* and *Enterobacter*. *International Journal of Hydrogen Energy* 39 (27): 14663–14668.

Pawar, S.S. and Van Niel, E.W. (2013). Thermophilic biohydrogen production: how far are we? *Applied Microbiology and Biotechnology* 97 (18): 7999–8009.

Perna, V., Castello, E., Wenzel, J. et al. (2013). Hydrogen production in an upflow anaerobic packed bed reactor used to treat cheese whey. *International Journal of Hydrogen Energy* 38 (1): 54–62.

Prazeres, A.R., Carvalho, F., and Rivas, J. (2012). Cheese whey management: a review. *Journal of Environmental Management* 110: 48–68.

Qureshi, N., Annous, B.A., Ezeji, T.C. et al. (2005). Biofilm reactors for industrial bioconversion processes: employing potential of enhanced reaction rates. *Microbial Cell Factories* 4: 24–24.

Rai, P.K., Asthana, R.K., and Singh, S.P. (2014a). Optimization of photo-hydrogen production based on cheese whey spent medium. *International Journal of Hydrogen Energy* 39: 7597–7603.

Rai, P.K., Singh, S.P., Asthana, R.K., and Singh, S. (2014b). Biohydrogen production from sugarcane bagasse by integrating dark- and photo-fermentation. *Bioresource Technology* 152: 140–146.

Redwood, M.D. and Macaskie, L.E. (2006). A two-stage, two-organism process for biohydrogen from glucose. *International Journal of Hydrogen Energy* 31: 1514–1521.

Ren, N.Q., Zhao, L., Chen, C. et al. (2016). A review on bioconversion of lignocellulosic biomass to H_2: key challenges and new insights. *Bioresource Technology* 215: 92–99.

Rosa, P.R., Santos, S.C., Sakamoto, I.K. et al. (2014a). Hydrogen production from cheese whey with ethanol-type fermentation: effect of hydraulic retention time on the microbial community composition. *Bioresource Technology* 161: 10–19.

Rosa, P.R.F., Santos, S.C., Sakamoto, I.K. et al. (2014b). The effects of seed sludge and hydraulic retention time on the production of hydrogen from a cassava processing wastewater and glucose mixture in an anaerobic fluidized bed reactor. *International Journal of Hydrogen Energy* 39 (25): 13118–13127.

Rozendal, R.A., Sleutels, T.H.J.A., Hamelers, H.V.M., and Buisman, C.J.N. (2008). Effect of the type of ion exchange membrane on performance, ion transport, and pH in biocatalyzed electrolysis of wastewater. *Water Science and Technology* 57: 1757–1762.

Saady, N.M.C. (2013). Homoacetogenesis during hydrogen production by mixed cultures dark fermentation: unresolved challenge. *International Journal of Hydrogen Energy* 38 (30): 13172–13191.

Salomon, K.R. and Lora, E.E.S. (2009). Estimate of the electric energy generating potential for different sources of biogas in Brazil. *Biomass and Bioenergy* 33 (9): 1101–1107.

Salomon, K.R., Lora, E.E.S., Rocha, M.H., and del Olmo, O.A. (2011). Cost calculations for biogas from vinasse biodigestion and its energy utilization. *Sugar Industry* 136 (4): 217–223.

Sánchez, Ó.J. and Cardona, C.A. (2008). Trends in biotechnological production of fuel ethanol from different feedstocks. *Bioresource Technology* 99 (13): 5270–5295.

Santos, S.C., Rosa, P.R.F., Sakamoto, I.K. et al. (2014). Organic loading rate impact on biohydrogen production and microbial communities at anaerobic fluidized thermophilic bed reactors treating sugarcane stillage. *Bioresource Technology* 159: 55–63.

Searmsirimongkol, P., Rangsunvigit, P., Leethochawalit, M., and Chavadej, S. (2011). Hydrogen production from alcohol distillery wastewater containing high potassium and sulfate using an anaerobic sequencing batch reactor. *International Journal of Hydrogen Energy* 36 (20): 12810–12821.

Shah, A.T., Favaro, L., Alibardi, L. et al. (2016). *Bacillus* sp. strains to produce bio-hydrogen from the organic fraction of municipal solid waste. *Applied Energy* 176: 116–124.

Show, K.Y., Lee, D.J., Tay, J.H. et al. (2012). Biohydrogen production: Current perspectives and the way forward. *International Journal of Hydrogen Energy* 37: 15616–15631.

Singh, L. and Wahid, Z.A. (2015). Methods for enhancing bio-hydrogen production from biological process: a review. *Journal of Industrial and Engineering Chemistry* 21: 70–80.

Sivagurunathan, P., Kumar, G., Bakonyi, P. et al. (2016). A critical review on issues and overcoming strategies for the enhancement of dark fermentative hydrogen production in continuous systems. *International Journal of Hydrogen Energy* 41 (6): 3820–3836.

Souza, L.P., Lullio, T.G., Ratusznei, S.M. et al. (2015). Influence of organic load on biohydrogen production in an AnSBBR treating glucose-based wastewater. *Applied Biochemistry and Biotechnology* 176 (3): 796–816.

Sreethawong, T., Chatsiriwatana, S., Rangsunvigit, P., and Chavadej, S. (2010). Hydrogen production from cassava wastewater using an anaerobic sequencing batch reactor: Effects of operational parameters, COD:N ratio, and organic acid composition. *International Journal of Hydrogen Energy* 35 (9): 4092–4102.

Su, H., Cheng, J., Zhou, J. et al. (2009). Combination of dark- and photo-fermentation to enhance hydrogen production and energy conversion efficiency. *International Journal of Hydrogen Energy* 34: 8846–8853.

Sydney, E.B., Larroche, C., Novak, A.C. et al. (2014). Economic process to produce biohydrogen and volatile fatty acids by a mixed culture using vinasse from sugarcane ethanol industry as nutrient source. *Bioresource Technology* 159: 380–386.

Tao, Y.Z., He, Y.L., Wu, Y.Q. et al. (2008). Characteristics of a new photosynthetic bacterial strain for hydrogen production and its application in wastewater treatment. *International Journal of Hydrogen Energy* 33 (3): 963–973.

Tawfik, A. and El-Qelish, M. (2012). Continuous hydrogen production from co-digestion of municipal food waste and kitchen wastewater in mesophilic anaerobic baffled reactor. *Bioresource Technology* 114: 270–274.

Tien Anh, N., Kim, M.-S., and Sim, S.J. (2011). Thermophilic hydrogen fermentation using *Thermotoga neapolitana* DSM 4359 by fed-batch culture. *International Journal of Hydrogen Energy* 36 (21): 14014–14023.

Trudewind, C.A. and Wagner, H.J. (2010). Hydrogen and fuel cells. In: *Research and Development Targets and Priorities* (ed. D. Stolten), 533–548. Weinheim: Wiley-VCH Verlag.

U.S. Department of Energy, Energy Efficiency & Renewable Energy, Fuel Cell Technologies Market Report, 2014 available at http://energy.gov/sites/prod/files/2015/10/f27/fcto_2014_market_report .pdf

U.S. Department of Energy, Office of Scientific and Technical Information (2016). Final Report: Hydrogen Production Pathways Cost Analysis (2013 – 2016) https://www.osti.gov/servlets/purl/1346418 Available at https://www.osti.gov/biblio/1346418-final-report-hydrogen-production-pathways-cost-analysis

U.S. Department of Energy, Quadrennial Technology Review, Advancing systems and technologies to produce cleaner fuels, 2015 available at http://energy.gov/sites/prod/files/2015/11/f27/QTR2015-7D-Hydrogen-Production-and-Delivery.pdf

Valdez-Vazquez, I. and Poggi-Varaldo, H.M. (2009). Hydrogen production by fermentative consortia. *Renewable and Sustainable Energy Reviews* 13 (5): 1000–1013.

Van Niel, E.W.J., Budde, M.A.W., de Haas, G.G. et al. (2002). Distinctive properties of high hydrogen producing extreme thermophiles, *Caldicellulosiruptor saccharolyticus* and *Thermotoga elfii*. *International Journal of Hydrogen Energy* 27 (11–12): 1391–1398.

Venetsaneas, N., Antonopoulou, G., Stamatelatou, K. et al. (2009). Using cheese whey for hydrogen and methane generation in a two-stage continuous process with alternative pH controlling approaches. *Bioresource Technology* 100 (15): 3713–3717.

Wanderley, M.C.d.A., Martín, C., Rocha, G.J.d.M., and Gouveia, E.R. (2013). Increase in ethanol production from sugarcane bagasse based on combined pretreatments and fed-batch enzymatic hydrolysis. *Bioresource Technology* 128: 448–453.

Wang, J. and Wan, W. (2008). Effect of temperature on fermentative hydrogen production by mixed cultures. *International Journal of Hydrogen Energy* 33 (20): 5392–5397.

Wang, J. and Wan, W. (2009). Factors influencing fermentative hydrogen production: a review. *International Journal of Hydrogen Energy* 34 (2): 799–811.

Wang, X.J., Ren, N.Q., Sheng Xiang, W., and Qian, G.W. (2007). Influence of gaseous end-products inhibition and nutrient limitations on the growth and hydrogen production by hydrogen-producing fermentative bacterial B49. *International Journal of Hydrogen Energy* 32: 748–754.

Wang, A., Ren, N., Shi, Y., and Lee, D.-J. (2008a). Bioaugmented hydrogen production from microcrystalline cellulose using co-culture – *Clostridium acetobutylicum* X-9 and *Etilanoigenens harbinense* B-49. *International Journal of Hydrogen Energy* 33 (2): 912–917.

Wang, B., Wan, W., and Wang, J.L. (2008b). Inhibitory effect of ethanol, acetic acid, propionic acid and butyric acid on fermentative hydrogen production. *International Journal of Hydrogen Energy* 33 (23): 7013–7019.

Wang, A., Sun, D., Cao, G. et al. (2011a). Integrated hydrogen production process from cellulose by combining dark fermentation, microbial fuel cells, and a microbial electrolysis cell. *Bioresource Technology* 102: 4137–4143.

Wilkie, A.C., Riedesel, K.J., and Owens, J.M. (2000). Stillage characterization and anaerobic treatment of ethanol stillage from conventional and cellulosic feedstocks. *Biomass and Bioenergy* 19 (2): 63–102.

Xie, G.J., Feng, L.B., Ren, N.Q. et al. (2010). Control strategies for hydrogen production through co-culture of *Ethanoligenens harbinense* B49 and immobilized *Rhodopseudomonas faecalis* RLD-53. *International Journal of Hydrogen Energy* 35: 1929–1935.

Yang, H., Shi, B., Ma, H., and Guo, L. (2015). Enhanced hydrogen production from cornstalk by dark- and photo-fermentation with diluted alkali-cellulase two-step hydrolysis. *International Journal of Hydrogen Energy* 40 (36): 12193–12200.

Yokoi, H., Mori, S., Hirose, J. et al. (1998). H_2 production from starch by a mixed culture of *Clostridium butyricum* and *Rhodobacter* sp. M-19. *Biotechnology Letters* 20: 890–895.

Yokoi, H., Saitsu, A., Uchida, H. et al. (2001). Microbial hydrogen production from sweet potato starch residue. *Journal of Bioscience and Bioengineering* 91: 58–63.

Zannoni, D., Antonioni, G., Frascari, D., and De Philippis, R. (2014). Hydrogen production and possible impact on global energy demand: open problems and perspectives. In: *Microbial BioEnergy: Hydrogen Production*, Advances in Photosynthesis and Respiration, vol. 38 (eds. D. Zannoni and R. De Philippis), 349–356.

Zhang, X., Ye, X., Guo, B. et al. (2013). Lignocellulosic hydrolysates and extracellular electron shuttles for H_2 production using co-culture fermentation with *Clostridium beijerinckii* and *Geobacter metallireducens*. *Bioresource Technology* 147: 89–95.

Zhang, S.-C., Lai, Q.-H., Lu, Y. et al. (2016). Enhanced biohydrogen production from corn stover by the combination of *Clostridium cellulolyticum* and hydrogen fermentation bacteria. *Journal of Bioscience and Bioengineering* 122 (4): 482–487.

Zhao, X., Xing, D., Fu, N. et al. (2011). Hydrogen production by the newly isolated *Clostridium beijerinckii* RZF-1108. *Bioresource Technology* 102 (18): 8432–8436.

16

Deployment of Biogas Production Technologies in Emerging Countries

Guangyin Zhen[1], Xueqin Lu[1,2], Xiaohui Wang[1], Shaojuan Zheng[1], Jianhui Wang[1], Zhongxiang Zhi[1], Lianghu Su[3], Kaiqin Xu[4], Takuro Kobayashi[4], Gopalakrishnan Kumar[5] and Youcai Zhao[6]

[1] *Shanghai Key Lab for Urban Ecological Processes and Eco-Restoration, School of Ecological and Environmental Sciences, East China Normal University, Shanghai 200241, PR China*
[2] *Department of Civil and Environmental Engineering, Graduate School of Engineering, Tohoku University, Sendai, Miyagi 980-8579, Japan*
[3] *Nanjing Institute of Environmental Sciences of the Ministry of Environmental Protection, Nanjing 210042, PR China*
[4] *Center for Material Cycles and Waste Management Research, National Institute for Environmental Studies, Tsukuba, Ibaraki 305-8506, Japan*
[5] *Department of Environmental Engineering, Daegu University, Gyeongsan, Gyeongbuk, Republic of Korea*
[6] *The State Key Laboratory of Pollution Control and Resource Reuse, Tongji University, Shanghai 200092, PR China*

CHAPTER MENU

16.1 Introduction

According to the United States Energy Information Administration (EIA), global energy needs will increase by 57% during the 2004–2030 period (EIA 2014). Energy consumption and energy depletion in emerging countries has motivated the scientists of the world to search for renewable energy sources that could replace fossil fuels. It is estimated that nearly 60% of all power generation capacity by 2040 will originate from renewable energy (IEA 2016). Bioenergy technologies that use renewable sources for biofuels or chemical commodity production will play a role in energy security. Biogas is, to date, one of the most successful renewable energy alternatives discovered, which can substitute for fossil fuels after purification. It is a methane-rich gas produced from the decomposition of organic wastes under a anaerobic environment. The calorific value of biogas is highly dependent upon methane content; for example, 1 m^3 of biogas with 60% methane has the calorific value of 21.5 MJ, equivalent to 5.97 kWh of electricity under standard conditions45m (Salihu and Alam 2016). Nowadays, one of the most commonly used biogas production technologies is anaerobic digestion (AD), which can utilize a wide range of biomassmes as substrates.

Green Energy to Sustainability: Strategies for Global Industries, First Edition.
Edited by Alain A. Vertès, Nasib Qureshi, Hans P. Blaschek and Hideaki Yukawa.

At present, a myriad of organic wastes have been reported in m AD-related studies for biogas production, such as municipal solid waste (MSW), waste activated sludge (WAS) (Zhen et al. 2017, 2015b; Zhen and Zhao 2017), livestock manure, food waste (Nguyen et al. 2014; Paritosh et al. 2017), microalgae (e.g. *Scenedesmus* sp., *Chlorella* sp., etc.) (Zhen et al. 2016), agricultural waste (rice hull, timber species, willow, switch grass, softwood, rice straw, wheat straw, maize straw) (Khatri et al. 2015; Singh et al. 2014), grass *Egeria densa* (Zhen et al. 2015a), lipid waste (fat, oil, and grease) (Park and Li 2012), smpent coffee waste (Ravindran et al. 2017), and wastewater (e.g. starch, methanol) (Lu et al. 2015a,b).

Anaerobic digestion is a multi-step biological process involving several groups of microorganisms and with different intermediates produced. Generally, the process can be classified into four successive biochemical steps, i.e. hydrolysis, acidogenesis, acetogenesis, and methanogenesis (Figure 16.1) (Appels et al. 2008; Jain et al. 2015). The main obstacle for biogas production has been attributed to the slow hydrolysis of biomass due to the complex components and microstructure. Hence, hydrolysis has been recognized as the rate-limiting step in the whole process. Because of the barrier in hydrolysis, the anaerobic digestion of waste biomass causes a long lag phase before biogas evolution, long solids retention times (SRTs), as well as limited solid degradation efficiencies and biogas yield. In recent years, in order to improve the biodegradability of biomass, different strategies for biomass pretreatment have been developed and employed, including mechanical, ultrasound, microwave, thermal, chemical, and biological, as well as several combinations

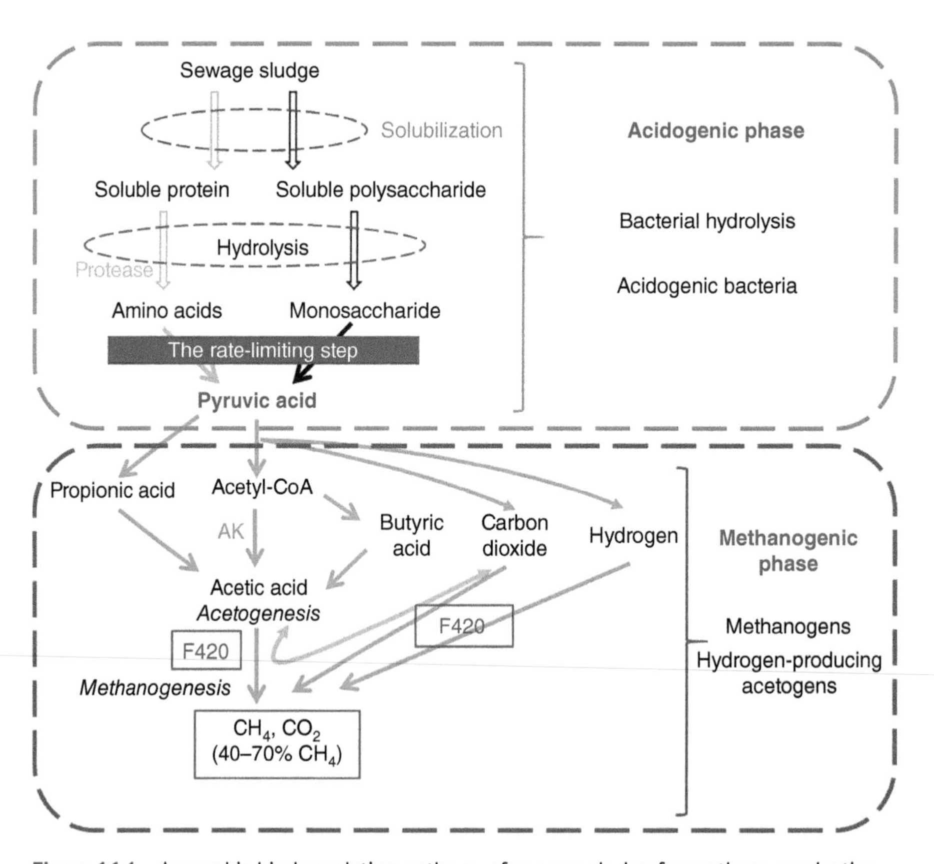

Figure 16.1 Anaerobic biodegradation pathway of sewage sludge for methane production. Source: Modified from Zhen et al. (2014).

of different processes. The pretreatment can accelerate the solubilization and release of organic contents, thereby making them available to subsequent microbial reactions during the anaerobic process. In addition, the high sensitivity of the microbial community to variations in the operating conditions applied can also impair the overall performance of the AD process to a certain degree (Mao et al. 2015).

This chapter will deal with AD technologies for biogas production. Detailed information about the basics and working principles of the AD process, the types and characteristics of various feedstocks, and the commonly adopted pretreatment methods are summarized. In addition, the status of the implementation of AD technologies for biogas production in several emerging countries is also included.

16.2 Types of Feedstock

16.2.1 Waste Activated Sludge

WAS is the part of the activated sludge which is discharged from the secondary sedimentation tank in the activated sludge process (Xiang et al. 2017). The activated sludge process is a type of aerobic biological treatment method that uses suspended microbial flocs. It can remove organic pollutants in the sewage by the biological agglomeration, adsorption, and oxidation of the activated sludge, and then separate the activated sludge from water. Most of the activated sludge is returned to the aeration tank and the excess is discharged from the secondary sedimentation tank. This is called WAS (Suschka and Grübel 2014).

The main constituents of WAS include the microbial population, the microbial self-oxidizing residue, and the undegraded or refractory organic and inorganic substances. Microorganisms mainly include bacteria, protozoa, and algae (Figure 16.2a). WAS from different sewage treatment plants contain different ingredients. For example, WAS discharged from a municipal wastewater treatment plant contains nitrogen, phosphorus and other elements in addition to pathogens and parasite eggs; while WAS discharged from an industrial sewage treatment plant may contain zinc, copper, nickel, lead, chromium and other toxic heavy metals and phenol, aldehydes, cyanides, sulfides, and other non-metallic compounds (Xiang et al. 2017). If WAS is not dealt with in a timely manner or is handled improperly, it will bring secondary pollution because of the large number of organic substances (e.g. toxic and harmful organic matter), significant number of pathogens, parasitic eggs, high content of nitrogen, phosphorus and heavy metals, and high moisture content (over 95%). So, it has a large volume and, if not reduced, will take up a great deal of land and release odour.

Many studies have been conducted to investigate the potential of anaerobic digestion for the treatment of WAS. Anaerobic digestion can reduce WAS volume, reduce pathogenic microorganisms, and produce methane (Table 16.1) (Appels et al. 2010; Ayol 2005; Romano et al. 2009). But there are still some limitations including limited conversion of organic matter, long retention times, and non-biodegradable organic structures (Demirel and Scherer 2008; Tyagi et al. 2014; Yu et al. 2014). Therefore, various cell disruption methods have been explored to enhance biomass hydrolysis rate and methane production. For instance, Charles et al. (2013) studied the potential of electrolytic pretreatment in enhancing methane production in WAS, with methane yield rising from 242 ± 20 ml/g-VS in untreated sludge to 318 ± 15 ml/g-VS in electrolysed sludge. Zhang et al. (2010) found that when the WAS, with a high VS/TS ratio of 0.79, was pretreated by alkali, the highest accumulative methane yield was 398.0 ml/g-VS. Zhen et al. (2014) also employed an

Figure 16.2 (a) Waste activated sludge, (b) lignocellulosic biomass, (c) microalgae cultivation in reactors, (d) food waste, (e) leafy vegetables, and (f) livestock manure.

electrical-alkali process to pretreat WAS and they obtained a 20.3% improvement in methane production at the optimal conditions of 5 V plus pH 9.2. Geng et al. (2016) studied the anaerobic digestion of raw WAS and pre-treated WAS. In this process the methane yield of raw WAS was found to be 123.3 ml/g-VS with a VS reduction of 16.95% and the methane yield was improved to 185.5 ml/g-VS with a VS reduction of up to 24.75%.

16.2.2 Lignocellulosic Biomass

Lignocellulose biomass is composed of lignin (10–20%), cellulose (35–50%), and hemicellulose (20–35%) (Figure 16.2b) (Alonso et al. 2010). These three basic ingredients arrange by the rules, intertwine, gather into a bundle. In addition, there are some gum, protein, and minerals in the lignocellulose biomass (Van Dyk and Pletschke 2012).

The common lignocellulosic biomass includes agricultural residuals, energy crops, MSWs, and forest residues (Akobi et al. 2016; Feng and Lin 2017); it is predicated that annual production of lignocellulosic biomass is about 2.2×10^{11} tons on dry basis, while only 5.5% of it is converted to various forms of fuels and energy (Torres et al. 2007). Most is burnt in an open environment which

Table 16.1 Basic characteristics of different feedstocks.

Feedstocks	Sampling place	Water content (%)	pH	TS (g/l)	VS (g/l)	TCOD (g/l)	Contaminants (organic pollutants, heavy metals, pathogens, etc.)	Treatment methods	References
Waste activated sludge	Sendai, Japan	99.1 ± 0.0	7.5 ± 0.0	7.4 ± 0.3 (TSS)	6.4 ± 0.1 (VSS)	12.7 ± 0.1	—	Electrical-alkali pretreatment + AD	Zhen et al. (2014)
Primary/excess sludge	Hong Kong, China	—	—	—	—	—	Antibiotic resistance genes, human bacterial pathogens (*Collinsella aerofaciens*, *Streptococcus salivarius*, *Gordonia bronchialis*, etc.)	AD	Ju et al. (2016)
Waste activated sludge	Nanjing, China	—	6.8 ± 0.1	27.6 ± 0.3	11.6 ± 0.2	13.5 ± 0.4	—	NaOH + $Mg(OH)_2$ + AD	Huang et al. (2016)
Waste activated sludge	Changsha, China		6.8 ± 0.1	14.5 ± 0.3 (TSS)	10.1 ± 0.2 (VSS)	15.6 ± 0.4	—	Aged refuse+AD	Zhao et al. (2017a)
Waste activated sludge	Yulin, China	98 ± 1.5	7.6 ± 9.5	—	—	7.9	—	Alkaline-modified eggshell and ultrasonic radiation pretreatment + AD	Xiang et al. (2017)
Waste activated sludge	India	—	6.9	8.5	5.2	22.0	—	Thermo-chemical pretreatment + AD	Rani et al. (2012)
Waste activated sludge	Japan	—	—	40.9 ± 0.5	31.3 ± 0.3	47.5 ± 1.6	—	NaOH-MW pretreatment + AD	Chi et al. (2010)
Waste activated sludge	Australia	—	—	10.7	8.9	17.2	—	Electrolysis pretreatment + AD	Charles et al. (2013)
Waste activated sludge	Konya, Turkey	—	7.5 ± 0.1	3.9 ± 0.1	2.5 ± 0.1	3.7 ± 0.1	—	Sono-pretreatment + AD	Şahinkaya and Sevimli (2013)
Waste activated sludge	Bangkok, Thailand	—	6.7 ± 0.2	1.0 ± 0.1 (%)	0.8 ± 0.1 (%)	11.7 ± 0.4	—	Combined chemical-ultrasonic pretreatment + AD	Seng et al. (2010)
Waste activated sludge	Tianjin, China	—	6.8	8.1 (%)	4.5 (%)	—	—	CaO-ultrasonic pretreatment + AD	Geng et al. (2016)
Mesophilic anaerobic digester sludge	Guelph, Ontano, Canada	—	6.9 ± 0.1	17.9 ± 0.5 (%)	11.2 ± 0.3(%)	—	—	Single-stage and two-stage AD	Akobi et al. (2016)

(Continued)

Table 16.1 (Continued)

Feedstocks	Sampling place	Water content (%)	pH	TS (g/l)	VS (g/l)	TCOD (g/l)	Contaminants (organic pollutants, heavy metals, pathogens, etc.)	Treatment methods	References
Miscanthus floridulus substrate	Shandong, China	—	—	95 (%)	92.8 (%)	—	—	Hydrogen peroxide pretreatment + AD	Katukuri et al. (2017)
Maize straw	Lexingtan,IL, USA	—	—	88.2 ± 0.8 (%)	79.7 ± 0.8(%)	—	—	Cartridge design AD	Yang et al. (2017)
Ulva lactuca	Marmara, Turkey	—	—	9.1 (%)	6.8 (%)	—	—	Bioaugmentation pretreatment + AD	Yıldırım et al. (2017)
Corn stover	Wooster, USA	—	—	91.6 ± 0.3 (%)	89.0 ± 0.1(%)	—	—	Liquid AD	Brown et al. (2012)
Fallen tree leaves	Wooster, USA	—	—	93.0 ± 0.1 (%)	86.9 ± 0.2(%)	—	—	Liquid AD	Brown et al. (2012)
Wheat straw	France	—	—	93.5 ± 0.1 (%)	89.4 ± 0.1(%)	—	—	Hydrothermal pretreatment+AD	Eskicioglu et al. (2017)
Sorghum	Montpelli, France	—	—	87.8 ± 0.3 (%)	81.1 ± 1.1(%)	—	—	Hydrothermal pretreatment+AD	Eskicioglu et al. (2017)
Rice straw	France	—	—	91.5 ± 0.4 (%)	77.5 ± 0.1(%)	—	—	Hydrothermal pretreatment+AD	Eskicioglu et al. (2017)
Food waste	Hunan, China	79.83	6.6	28.8 (%)	26.4 (%)	—	—	Anaerobic digestion	Yi et al. (2012)
Food waste	Haerbing, China	—	6.5	—	—	71–90	—	Anaerobic digestion	Li et al. (2016)
Food waste	USA	81.4	—	—	16.3%	50–80	—	Anaerobic digestion	Lopez et al. (2016)
Food waste	—	77.8	5.1	—	20.1%	198 g/kg WW	—	Anaerobic digestion	Fisgativa et al. (2016)
Food waste	China	78.3	6.5	26.1%	19.9%	250–260	—	Anaerobic digestion	Zhao et al. (2017b)
Food waste	—	—	—	40.0%	39.2%	18–26	—	LBR	Yan et al. (2016)
Food waste	China	—	4.68–4.81	12.3–15.8	10.7–14.5	157.0–177.6	—	HVPD pretreatment	Zou et al. (2016)
Food waste	Canada	—	3.4–4.6	39.4–40.5%	35.0–37.5%	—	—	Anaerobic co-digestion	Rajagopal et al. (2017)
Food waste	Italy	—	5.5 ± 0.2	22.6 ± 1.3	22.0 ± 1.2	46.2	—	One- and two-stage anaerobic digestion	De Gioannis et al. (2017)

can further cause environmental pollution (Eskicioglu et al. 2017) or is washed into the wastewater which can hinder the anaerobic digestion of waste sludge.

Lignocellulosic biomass is rich in complex carbohydrates (55–75% in TS) (Wan and Li 2012) and is recognized as the most sustainable energy source all over the world, thus many researchers have studied the potential of applying AD technologies to treat lignocellulosic biomass (Table 16.1). It is predicted that the amount of biogas produced will provide around 38% of the world's direct fuel and 17% of the world's electricity by 2050 (Demirbas 2000). However, due to the stable encapsulation of lignin and the high crystallinity of cellulose, the biodegradability of lignocellulosic biomass is usually low, impairing methane productivity (Chang and Holtzapple 2000). Therefore, in order to improve biodegradability, the structure of lignocellulosic biomass must be 'broken' by certain measures, that is, reducing the degree of polymerization and increasing the porosity (Zhang and Lynd 2004). A work conducted by Khatri et al. (2015) studied the anaerobic digestion of pretreated corn stalks, and reported that methane production was increased by 56.2% with the digestion time reduced from 48 to 7 days in 6% NaOH pretreatment. Song et al. (2013) studied the anaerobic digestion of pretreated rice straw and they observed a 290 ml/g VS of methane yield, 88.0% higher than that without pretreatment. The use of catalysts for the catalytic degradation of lignocellulose before anaerobic digestion has also been carried out by Li et al. (2015), where Ag-AgCl/ZnO was used as catalyst for the photocatalytic degradation of lignin before anaerobic digestion. They observed that the methane yield was 184 ml CH_4/g total organic content (TOC), 10.9% higher than that of the control group.

16.2.3 Algae

Algae is classified into macroalgae and microalgae. Microalgae are unicellular, and usually exist in the ocean and freshwater (Thurman 1997). The size of microalgae ranges from several microns (μm) to several hundred microns, dependent upon the species. They have no root, stems, or leaves. Microalgae are autotrophic organisms and metabolize by using sunlight, water, and atmospheric carbon dioxide (CO_2). During the metabolism, microalgae absorb CO_2 from the atmosphere and convert it into valuable products such as lipids, which can be used as energy sources during photosynthesis (Figure 16.2c) (Saharan et al. 2013). The estimated number of identified species of microalgae is about 100 000, but only about 35 000 of them have been characterized so far (Cardozo et al. 2007). Microalgae contains a high content of proteins, and thus can be used as an important source of single cell proteins. These proteins are high-quality and contain all the essential amino acids required for the human body. The total lipid content in the microalgae is between 1 and 70% of the dry matter, and the majority of it is glycerol fatty acid. Cooney et al. (2011) indicated that the content of lipids has an influence on the algae biomass productivity and it is lower when lipid content is high. In addition, microalgae also contains starch, nucleic acid, vitamins, minerals, and other substances.

Therefore, microalgae can be a good fermentation raw material. Mahdy et al. (2017) studied anaerobic digestion of microalgae where the process can produce about 0.42 LCH_4 /g VS. The main benefit of the microalgae AD is its higher efficiency compared with most other biofuels. The large use of microalgae makes its cultivation necessary. The systems for microalgae cultivation include, open ponds, photo-bioreactors (PBRs), and hybrid systems (Jankowska et al. 2017). The methane production potential of microalgae biomass depends on the digestibility of cell walls, which can be enhanced by chemical, physical, and biological pretreatment. Hence, in order to boost methane potential, microalgae biomass needs pretreatment before the AD process. Proper pretreatment is essential for the enhancement of methane yield. Kinnunen and Rintala (2016) reported that pretreatment increased the biomethane potential of native microalgae biomass by 11–24%. In another investigation, Nuchdang et al. (2017) studied the hydrothermal pretreatment, which increased methane yield to about 200 L_{STP} CH_4/kg VS. Most recently, Cardena et al. (2017) studied anaerobic digestion of microalgae with ozone pretreatment and produced about 432.7 mLCH_4/g VS. In addition, Beltrán et al. (2016) co-digested 25% microalgae with 75% WAS, and obtained the highest methane yield of 0.442 L CH_4/g VS, 22% and 39% higher than when anaerobic digestion of the sole feedstocks, WAS, and microalgae, was performed, respectively.

16.2.4 Food Waste

Food waste (FW) is the leftover food in food processing and production (Figure 16.2d). Specifically, food waste mainly constitutes leftovers, fruit and vegetables, peel, and animal offal which cannot be used. Food waste contains many chemical ingredients, such as water, polysaccharides, salt, fat, and other nutrients, in specific proportions depending on the actual location. For instance, food waste in China contains more fat while in the USA it contains less. If the food waste treatment is not timely or is incomplete, it will have short- or long-term impact on the environment.

With the explosive growth in food consumption, the production of food waste is also greatly increased. In fact, over 1300 billion tons of food waste are produced all over the world every year, emitting over 3.3 billion tons of CO_2 equivalent, which accounts for 9% of global CO_2 emissions. It should be noted that the pollution caused by food waste has gradually become an important problem for urban water pollution and land pollution, affecting all aspects of people's quality of life. How to efficiently deal with food waste has become a critical issue.

The main treatments of food waste are composting, landfill, incineration, anaerobic digestion, and fermentation in batch or integrated reactor systems, etc. The anaerobic digestion of food waste has developed rapidly in recent years (Table 16.1). Considering today's energy shortage, land shortage, and serious food waste pollution, food waste's anaerobic digestion has received much attention. For instance, in Vietnam food waste generation was estimated to be 21 420, 33 264, and 49 920 t/d for 2015, 2020, and 2025, respectively; if food waste was sent to AD plants, the biogas produced could contribute 2.2–4.7% of fuel required for transportation. Alternatively, the total available energy equivalent each day is about 19, 20, and 45 GWh in 2015, 2020, and 2025, respectively, which could contribute approximately 2.4–4.1% of the electricity demand of Vietnam, or double this amount of heat energy (Nguyen et al. 2014).

Liu et al. (2005) studied a wet continuous single stage anaerobic fermentation reactor where the methane yield reached 0.520 m^3/kg VS. Fisgativa et al. (2016) observed that per kilogram volatile solids (VS) of food waste could produce about 460 normal litres (NLs) of CH_4 on average which represents about twice as much as methanogenic cattle manure (270 NL CH_4/kg VS) and sewage sludge (255 NL CH_4/kg VS). Park et al. (2018) studied the effect of feeding mode and dilution in naerobic digestion of food waste where the process can produce about 2.78 L CH_4/l/d. Zhang et al. (2017) co-pretreated FW with WAS, and achieved 24.6% higher methane yield than that of unpretreated substrates and 10.1% more reduction in solids during the digestion. Similarly, the study by Gaur and Suthar (2017) suggested that inoculum type and substrate combination caused a great effect on the biomethanation process. When the ratio of CD (cow dung) : AAGS (acclimatized anaerobic granular sludge) : WAS : FW was 1 : 1 : 1 : 1, the methane yield reached the maximum. What is more, it is suggested that the type and combination of inocula used with FW was the main factor affecting the co-digestion efficiencies. Zhen et al. (2016) also examined the potential of methane production from microalgae (MA) by co-digesting with food waste (FW); the results showed that supplementing food waste significantly improved microalgae digestion performance, with the highest methane yield of 639.8 ± 1.3 ml/g-VS_{added} obtained at a MA : FW ratio of 0.2 : 0.8, which was a fivefold increase with respect to that (106.9 ± 3.2 ml/g-VS_{added}) of the microalgae alone.

16.2.5 Leafy Vegetables

Leafy vegetables are vegetables where the leaves are eaten (Figure 16.2e). With the rapid growth of the world's opulation, there is growth in the volume of wasted leafy vegetables. A large volume of leafy vegetable residues are produced every year, which causes many environmental problems because of their high moisture and organic content. As a result, it is imperative to find an efficient solution to this problem. Biomass resources suitable as AD feedstocks are represented by a large variety of organic materials available on a renewable basis, ranging from simple compounds to complex high-solid matters. Usually, leafy vegetables have a high content of carbohydrates/starch, proteins, and a common feature is their ability to be easily decomposed through AD. This feature makes leafy vegetable residues potential feedstock for reproducible energy via anaerobic digestion. For this reason, in the past decades, many efforts have been devoted to the AD of leafy vegetables. Among the relevant studies, Molinuevo-Salces et al. (2010) and Yao et al. (2014) focused on comparing co-digestion using several vegetables and other feedstocks. Yan et al. (2017) have explored the methane production potential of 20 types of leafy vegetable residues in AD using a standard and unified digestion method. To increase the degradation rate of these substrates, a pretreatment by mechanical, thermal, chemical, or enzymatic processes is usually necessary.

Biogas production from leafy vegetables via anaerobic digestion offers significant advantages over other forms of bioenergy production. Normally, two types of fermentation processes are used to produce biogas, namely wet fermentation and dry fermentation. For dry fermentation, several batch processes with percolation and without mechanical mixing are applied mainly for mono-fermentation of energy crops. Wet digestion processes are operated with TS (total solids) concentrations below 10%, which allows the application of completely stirred tank digesters (Weiland 2010). Leafy vegetable biomass needs to be pretreated before feeding into the digester, which ranges from simple mechanical particle size reduction to more complicated treatments aimed at breaking the lignocellulosic molecules to facilitate the access of anaerobic microorganisms in these structures.

Generally, vegetables with high hemicellulose contents (over 55%) produce higher methane yields, with the exception of Chinese cabbage. The experimental methane yield values of leafy vegetables can be predicted well using an equation, demonstrating which are considered to be more valuable and practical (Yao et al. 2014). After Yao et al. (2014) measured the characteristics of 20 types of leafy vegetable residues and explored their methane production potential with a simple and unified method, a dependent relationship was established between the experimental methane yield and the organic components to predict the AD performance.

16.2.6 Livestock Manure (Chicken, Pig and Swine Manure)

Livestock manure mainly refers to the livestock and poultry industry, a class of rural solid waste, including pig manure, cow dung, sheep manure, chicken manure, and duck manure (Figure 16.2f). Historically, AD has been associated with stabilization treatment of organic wastes. In the same way, anaerobic digestion uses animal manure and slurries. This avoids environmentally polluting waste streams and converts them to valuable resources, such as biogas (as renewable fuel) and digestate (as valuable biofertilizer).

Manures and slurries from a variety of animals can be used as feedstocks for biogas production (pigs, cattle, poultry, horses, mink, and many others). They are characterized by different dry matter contents: solid farmyard manure (10–30% dry matter) or liquid slurry (below 10% dry matter). Their compositions and biogas production potential may vary according to the quality of the animal feed. Hadin and Eriksson (2016) studied the AD of horse manure. They characterized the horse manure which included availability, suitability, digestibility, and impurities and inhibitors, and noticed that fermentation of manure lowers greenhouse gas (GHG) emissions. Nowadays, most of the agricultural biogas plants digest manure from pigs, cows, and chicken with supplementation of other substrates to increase the content of organic material to achieve a higher biogas yield. The high water content of slurry acts as a solvent to ensure proper stirring in the digester and homogeneity of the feedstock mixture. Compared with mono-digestion, co-digestion of manure with organic waste results in a higher stability of the AD process (Angelidaki et al. 2002). Anaerobic digestion of animal manures and slurries is widely applied and is increasingly developing in Europe, Asia, and North America, with multiple objectives including renewable energy recovery, environment protection, and materials recycling etc. AD addresses the environmental problems caused by livestock and poultry manure and recovers bioenergy.

16.3 Pretreatment Technologies of Anaerobic Digestion Feedstocks

16.3.1 Acidic Pretreatment

Acidic pretreatment of lignocellulosic biomass holds great promise for industrial scale development. The advantages and disadvantages for acidic pretreatment of lignocellulosic biomass are summarized in Figure 16.3. The effectiveness of acidic pretreatment is strongly related to the types and characteristics of the biomass substrates. The pretreatment can be carried out with mineral acids (e.g. sulfuric, hydrochloric, hydrofluoric, phosphoric, nitric, and formic acids) in either concentrated or dilute form. The latter is more often used because of its advantages such as less equipment corrosion which greatly reduces operating and maintenance costs. In acidic conditions, the hydrolysis of hemicellulose is the main reaction. During acid pretreatment solubilized lignin

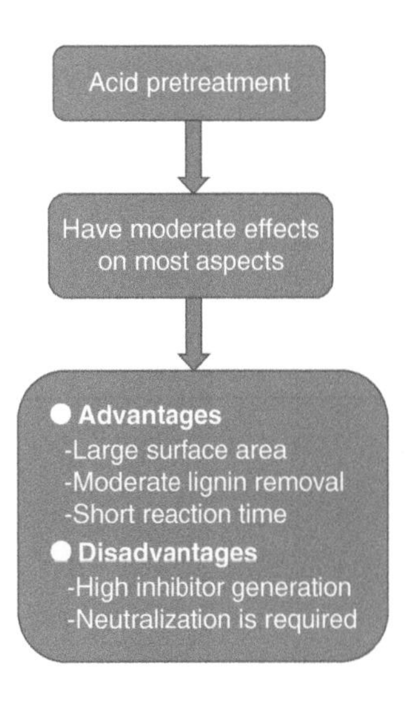

Figure 16.3 Acid pretreatment of lignocellulosic biomass.

quickly condenses and precipitates in acidic environments (Liu and Wyman 2003; Shevchenko et al. 1999). Solubilized hemicellulose (oligomer) can be hydrolysed in an acidic environment to produce monomers, inhibitors (furfural and hydroximethilfurfural (HMF)) and other (volatile) products (Nimz 1984; Ramos 2003). The solubilization of hemicellulose and precipitation of solubilized lignin are more pronounced during strong acid pretreatment compared to dilute acid pretreatment. Xiao and Clarkson (1997) showed that the addition of nitric acid during acid pretreatment has a dramatic effect on the solubilization of lignin in newspaper (Table 16.2).

The acid pretreatments are mainly classified into two kinds: the strong (sulfuric, nitric, or hydrochloric acids) and weak dilute (acetic acid, phosphoric acid, and so on) acids. Dilute sulfuric acid (H_2SO_4) pretreatment is the most commonly used method for converting lignocellulosic biomass into fermentable sugars. In the case of Lloyd and Wyman (2005), they focused on degradation characteristics of hemicellulose during acid pretreatment, and results showed that plant cell walls become more accessible to cellulases (Ishizawa et al. 2007; Sun and Cheng 2005). Devlin et al. (2011) investigated the effects of acid pretreatment (pH 6–1) using HCl on subsequent digestion and dewatering of WAS. The results showed that pretreatment in pH 2 was the most effective, which induced a 14.3% increase in methane yield compared to untreated WAS. Dewatering investigations suggested that the acid pretreated WAS required 40% less cationic polymer addition to achieve the same cake solid content (Table 16.2).

16.3.2 Alkali Pretreatment

Alkali pretreatment is known to be efficient in enhancing the biodegradation of complex materials, such as lignocellulosics. The main benefits of alkaline pretreatment include that it is simple, easy to operate, and highly efficient; Figure 16.4 lists the advantages and disadvantages of alkali pretreatment. In most cases it has demonstrated an increased methane production and a decrease in volatile solids, especially during low-dose alkaline treatment (Lin et al. 2009; Lina et al. 1997; López Torres and Espinosa Lloréns 2008; Navia et al. 2002). In the alkaline pretreatment process, the first reactions are solvation and saponification, which cause the pore size and accessible surface area to increase (Carlsson et al. 2012). An important aspect of the biomass in the alkaline process pretreatment is that the biomass itself consumes some of the alkali (Hendriksa and Zeeman 2009), thus a higher amount of alkali reagents might be required for obtaining the desired AD enhancement. Alkaline pretreatment with NaOH has been widely examined and is very effective in terms of hydrolyzing cellular substances and solubilizing extracellular polymeric substances (EPS). Compared with $Ca(OH)_2$, NaOH is reported to yield greater solubilization efficiency (Kim et al. 2003). Both chemical oxygen demand (COD) and total solids (TS) removal rates increase with the increase in alkaline dosage. According to recent studies, alkaline pretreatment is more efficient than acidic pretreatment for sludge solubilization (Chen et al. 2007; Jie et al. 2014). However, the extremely high pH condition may lead to some biodegradability of sewage sludge (Penauda et al. 1999).

Alkaline pretreatment techniques have been studied by many researchers (Table 16.2). Lin et al. (2009) indicated that alkali/NaOH pretreatment could be an effective method for improving methane yield when using pulp and paper sludge. Wonglertarak and Wichitsathian (2014) reported that alkaline pretreatment increases organic solubility or soluble chemical oxygen demand (SCOD) which improves WAS biodegradability. Alkali was also used to pretreat a large number of feedstocks (agricultural and forest residuals, hardwood, softwood, grass and MSW) at a wide range of temperatures from −15 to 170 °C and that increased methane yield by 3.2–230% in most cases (Zheng et al. 2014).

Table 16.2 Summary of typical studies of mechanical pretreatments on feedstocks and subsequent anaerobic digestion.

	Pretreatment		Anaerobic digestion		
Types of feedstocks	**Conditions**	**Effects**	**Conditions**	**Performances**	**References**
(a) Acidic pretreatment					
Newsprint (20–25% recycled fibre)	35% CH_3COOH + 2% HNO_3	Weight loss: 40%	Batch, 35 °C (60-d)	Increased nearly three times	Xiao and Clarkson (1997)
Rice straw (cellulose 34.63%, hemicellulose 29.74%, and lignin 15.34%)	100–140 °C, 5–15% citric acid	Total weight loss of biomass residue: 18.21%	Batch, 35 °C (45-d)	Sevenfold higher biogas yield	Amnuaycheewa et al. (2016)
Salvinia molesta	30 °C, 500 mL 4% H_2SO_4	Decrease of ash mass: 5.78 g	Batch, 30 °C, 1 atm (30-d)	81.78% increase in biogas yield	Syaichurrozi et al. (2019)
(b) Alkali pretreatment					
Wheat straw substrate	4% NaOH on dry weight basis, 37 °C, 120 h	78–166 l/kg VS	37 °C	111.6% higher methane production	Chandra et al. (2012)
Grass clipping of *Festuca arundinacea* (cellulose 18.98%, hemicellulose 25.88%, and lignin 2.42%)	NaOH: 0.2% (w/v), 100 °C, 10 min	Enzymatic digestibility: 77.65%	SHZ-82A, 150 rpm and 35 °C	252.22% biogas production	Jina et al. (2015)
(c) *Ultrasonication*					
Sludge (18.5 g TS l^{-1})	20 kHz, 750 W, 15000 kJ/kg TS	SCOD: 35%; DD_{COD}: 55%	Batch, 35–37 °C	> +40% methane production	Bougrier et al. (2005)
Food waste (7.5% TS)	20 kHz, 130 W, 30 min	Increase of VS removal: 6.38%	Batch, 50 °C, HRT 30 d, 30 d	+10.12% biogas production	Deepanraj et al. (2017)
Waste activated sludge (25.2 ± 5.8 g TS/l)	20 kHz, 250 W, 80 min	Increase of SCOD: 2406%	Semi-continuous, 35.0 ± 1.0 °C, HRT 20 d, SRT 20 d, over 2 mo	> +60% biogas production	Li et al. (2018)
(d) Microwave irradiation					
Waste activated sludge (11.66 g TS/l)	900 W, 2450 MHz, 70% intensity, 12 min	Increase of COD solubilization: 18.6%	Lab-scale semi-continuous, 37 °C, SRT 15 d	+35% biogas production, +14% SS reduction	Uma Rani et al. (2013)
Dewatered oxidation-ditch sludge (152 g TS/l)	Below 100 W, 2450 MHz, 60 min, 80 °C	Increase of VS removal: 33%	Lab-scale, continuous, 37 °C, SRT 25 d	+27% biogas production	Togari et al. (2016)

Sewage sludge (21.94 ± 0.12 g TS/l)	700 J S^{-1}, 63 s, 20×10^6 J/kg TS	Increase of SCOD from 6.57 ± 0.12 to 7.28 ± 0.05 g/l	Semi-continuous, 35 °C, HRT 35 d, 90 d	+20% methane production, increased biodegradability from 52% to 70%	Gil et al. (2018)
(e) Ozonation					
Waste activated sludge (27 g TS/l)	0.063 g O_3 g^{-1} TSS		Batch, 35 °C, 35 d	+2.0-fold biogas yield, + 8% VS removal	Silvestre et al. (2015)
Anaerobic digested sludge	0.086 g O_3 g^{-1} COD	Increase of SCOD from 1.20 to 2.15 g/l	Batch, 35 °C	+52% methane yield, +94.5% methane production rate	Chacana et al. (2017)
Waste activated sludge	0.200 g O_3 L^{-1}, pH 7.5 ± 0.1	VSS destruction from 47% to 60%	Lab-scale ASBRs, HRT 20 d, SRT 20 d, 350 d	+27.45-fold increase in biogas yield, enhancing sludge biodegradability	Bakhshi et al. (2018)
(f) Thermal pretreatment					
High-solids sludge (15% TS)	70 °C, 30 min	Increase of SCOD: 52.5-fold, DD_{COD}: 13.0%	Pilot-scale, semi-continuous, 35 ± 2 °C, SRT 15–22 d	+11% biogas yield, decrease of digestion time from 22 to 15 d	Liao et al. (2016)
Textile dyeing sludge (2.71% TS)	100 °C, 60 min	Increase of SCOD: 6.2-fold	Batch, 35 ± 1 °C	+ 281.8% methane production	Chen et al. (2017)
Microalgal biomass (2.0–2.5% TS)	130 °C, 1.7 bar, 15 min	Increased of soluble VS: 13–15%	Lab-scale, 37 ± 1 °C, HRT 20 d	+41% methane production	Passos and Ferrer (2015)
(g) Enzymatic pre-treatment					
Chlorella vulgaris	Water bath (50 °C); viscozyme and alcalase	SCOD: 56%	Substrate/inoculums: 0.5 g COD g^{-1}-VS; 35 °C, 3 weeks	+14% methane production	Ahmed et al. (2014)
Waste activated sludge (37991–38 331 TS mg/L)	Endogenous amylase, endogenous protease, combined amylase/protease; 1: 10, 120 rpm, 28 h, 37 °C	78.2%, 29.5%, and 30.2% increase in SCOD; VFAs increased from 133 to 421, 1082, 1128, and 1719 mg/l	Water bath at 37 °C	Biogas production was improved by 18.6–20.2%	Yu et al. (2013)

(Continued)

Table 16.2 (Continued)

Types of feedstocks	Pretreatment		Anaerobic digestion		References
	Conditions	Effects	Conditions	Performances	
Chlorella vulgaris (16.7 ± 0.1 g TS/l, 10.4 ± 2.1 g VS l^{-1})	130 rpm, incubated for 3 h at 50 °C; wet microalgae (16 g/l) were hydrolysed with Alcalase (0.146, 0.293 and 0.585 AU g DW^{-1})	Hydrolysis efficiency > +41.2%	Batch, pH 7–7.5, 35 °C	Methane production 50–70%	Mahdy et al. (2014)
Chlorella vulgaris (32 g TS/l)	Viscozyme (carbohydrase) and Alcalase (protease); 50 °C, water bath	Hydrolysis efficiency of 28.4% ± 3.9 and 54.7% ± 5.6 for samples pretreated with Viscozyme (carbohydrase) and Alcalase (protease)	35 °C, 250 rpm, HRT 20 d	Methane yield increased 5- and 6.3-fold at OLR_1 and OLR_2, respectively	Mahdy et al. (2016)
(h) High-pressure homogenization					
Sewage sludge (19.53 g/l TS)	60 MPa with 0.04 NaOH mol/l	SCOD removal: 73.5%, VS removal: 43.5%	Batch, 100 rpm 35 °C, 15 d	Accelerating sludge digestion and increasing the biogas production	Fang et al. (2014)
Grass clipping of Festuca arundinacea (cellulose 18.98% hemicellulose 25.88%, and lignin 2.42%)	10 MPa, freeze-dried for 48 h	Enzymatic digestibility: 45.25%	SHZ-82A, 150 rpm, 35 °C	+263.48% biogas production	Jina et al. (2015)
Excess activated sludge (19.25 g/l TS)	60 MPa with one homogenization cycle combined with an alkaline dosage of 0.04 mol/l	TCOD, VS removal increased by 24.68%, 18.95%, respectively	35 ± 2 °C	+95.81% cumulative biogas production	Fang et al. (2017)

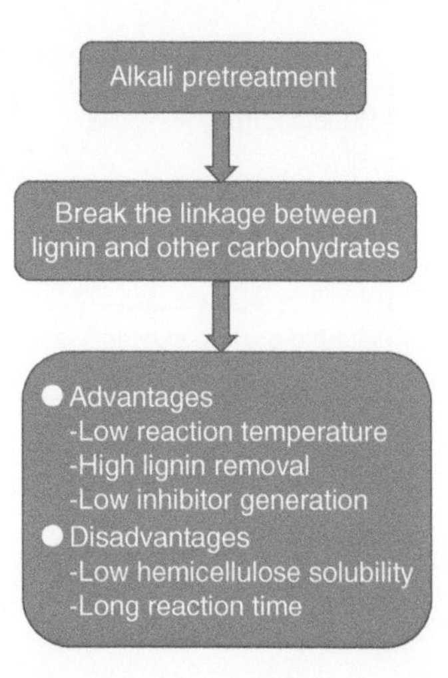

Figure 16.4 Alkali pretreatment of lignocellulosic biomass.

In general, the advantages of alkali treatment include: it is simple, easy to operate, and highly efficient. On the other hand, the use of alkali pretreatment produces some inhibitors that affect anaerobic digestion adversely (Zhen et al. 2014). Both Na^+ and OH^- are anaerobic digestion inhibitors, so the optimal alkali dosage and the negative effects of alkali treatment are yet to be studied.

16.3.3 Ultrasonication

Ultrasound frequencies range from 20 kHz to 20 MHz. Ultrasonic wave technology, a new pretreatment method, is mainly used in sludge treatment. The basic purpose of ultrasonication is to rupture the cell wall and facilitate the release of intracellular substances to increase biodegradability and enhance the anaerobic digestion with lower retention time and higher biogas production (Pavlostathis and Gossett 1986). Cavitation will occur when the local pressure in the aqueous phase falls below the evaporating pressure resulting in small bubbles (Tiehm et al. 1997). The microbubbles formed during this process oscillate in the sound field over several oscillation periods, grow by a process termed rectified diffusion, and collapse in a nonlinear manner (Figure 16.5a). The sudden and violent collapse leads to extreme conditions (a local temperature of around 5000 K and a pressure above 500 bars) (Chu et al. 2002) and initiates powerful hydro-mechanical shear forces and highly reactive radicals (H· and ·OH). Both the hydro-mechanical shear forces and the oxidizing effect of H· and ·OH contribute to the break-up of sludge flocs and the liberation of intercellular material. In comparison with radicals, hydro-mechanical shear forces nonetheless are stronger in sludge rupture. Wang et al. (2005) have evaluated the effect of ·OH and hydromechanical shear forces on sludge disintegration, and the results show that the disintegration of the sludge occurs mainly by hydro-mechanical shear forces produced by cavitation bubbles, and the oxidation effect of the hydroxide radical on sludge solubilization is negligible. The generation of cavitation bubbles and the configurations of ultrasonication are shown in Figure 16.5.

The degree of sludge disintegration depends on the ultrasonic parameters and sludge characteristics. Therefore, the evaluation of the optimum parameters varies with the type of sonicater and sludge to be treated. Generally, the greater the ultrasonic energy density, the better the treatment of the sludge ($\leq$7000 kJ/kg); but when the energy density is higher than 7000 kJ/kg, the methane production capacity of the sludge does not increase (Bougrier et al. 2006). Studies have shown that the ultrasonic frequency should be lower ($\leq$100 kHz) to obtain better sludge pretreatment; the current ultrasonic frequency used for sludge pretreatment is generally about 20 kHz (Hogan et al. 2004).

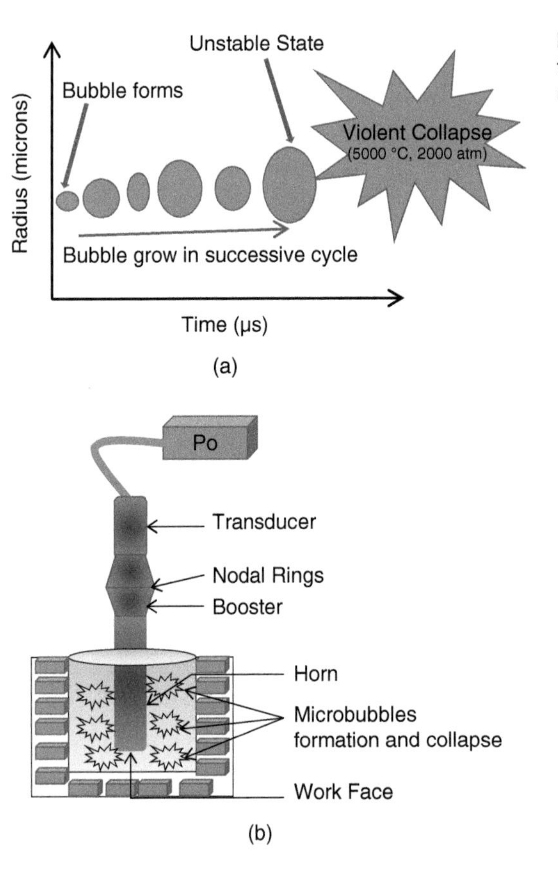

Figure 16.5 (a) Development and collapse of the cavitation bubble; and (b) configurations of ultrasonication.

During the last decades, researchers have conducted extensive studies on the ability of ultrasonic pretreatment to produce methane from sludge; the technology has matured, and there are many successful examples in this field. Bougrier et al. (2005) and Perez-Elvira et al. (2009) evaluated the effect of ultrasonic pretreatment of sludge and the methane produced from it. After ultrasonic pretreatment of the sludge with an energy density of 5.04 kWh/kg methane production increased by 40% (Bougrier et al. 2005). With ultrasonic pretreatment at an energy density of 30 kWh m^{-3}, sludge methane production increased by 42% (Perez-Elvira et al. 2009). In additon, ultrasonic pretreatment has been also adopted using food industry wastes (Luste et al. 2009; Palmowski et al. 2006) and manure as substrates (Elbeshbishy et al. 2011).

Ultrasonic pretreatment has numerous advantages including short processing time, high process efficiency, and improved methane production capacity of sludge. It is advised that in future studies on ultrasonic pretreatment methods, the researchers should focus on the selection of better operating parameters and develop efficient pretreatment devices.

16.3.4 Microwave Irradiation

Microwave (MW) irradiation is considered to be a thermal treatment. MW irradiation operates at wavelengths of $1 \times 10^{-3} - 1$ m with corresponding oscillation frequencies of 0.3–300 GHz. In general, there are two mechanisms for medium heating in the microwave field: (i) ionic conduction by changing dipole orientation of polar molecules (Toreci et al. 2009) and, (ii) the rotation of dipoles (Tang et al. 2010). Microwave cracking is based on microwave heating selectivity, penetration,

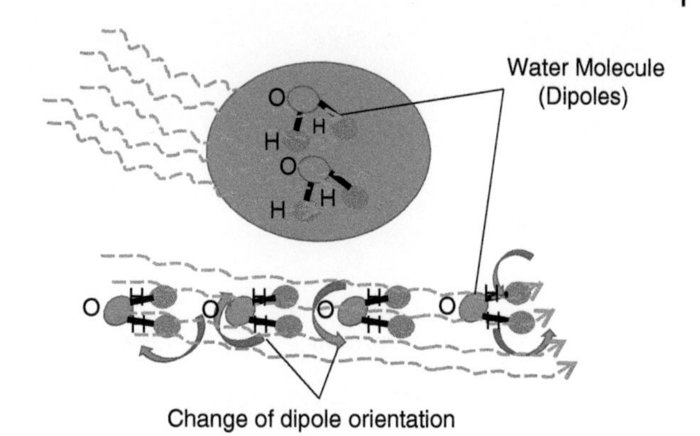

Figure 16.6 Microwave heating.

and efficiency (Figure 16.6). In the real-world application of microwave heating, two kinds of microwave energy dissipation exist at the same time.

(i) *Thermal effect*: The thermal effect of MW irradiation is generated through the rotation of dipoles or polar molecules, which can produce heat, resulting in an elevated temperature of the intracellular liquor to boiling point resulting in the break-up of bacterial cells.
(ii) *Athermal effect*: The athermal effect of MW irradiation is induced by the polarized parts of macromolecules aligning with the poles of the electromagnetic field, resulting in the possible breaking of hydrogen bonds and unfolding and denaturing of complex biological molecules, which kills microorganisms at low temperatures.

Compared to conventional methods, microwave irradiation is faster, more efficient and resource-conserving (Weemaes and Verstraete 1998). Park et al. (2004) found that microwave irradiation increased biogas production and the COD removal rate of the excess sludge by 79% and 64% respectively compared with the untreated sludge. The microwave pretreatment of the sludge could keep the SRT of the anaerobic reactor from the original 15 to 8 days. Microwave irradiation can not only be used independently, but also can be combined with other technologies to form a continuous pre-processing strategy (Beszedes et al. 2009; Sual et al. 2015). The work done by Beszedes et al. (2009) indicated that the combination of microwave and ozone to pretreat sludge could achieve higher biodegradability than ozone treatment alone.

In terms of microwave pretreatment, much effort still needs to be made and several obstacles need to be overcome in future work to push forward its industrialization. The development of more efficient microwave absorbers that can better transfer microwave energy to biomass; microwave pretreatment reactor expansion is also required to meet the large amount of biomass processing.

16.3.5 Ozonation

Ozone is a strong chemical oxidant which is able to oxidize the biological cell wall components (sugar, lipids, protein) into small molecules causing the cell walls to rupture and release organic matter (Figure 16.7). These organic substances continue to be oxidized into small molecules. In general, the greater the amount of ozone dosing the better, but an overdose may give rise to the complete oxidization of cytoplasm substances, which is not conducive to the subsequent anaerobic digestion (Goel et al. 2003).

In recent years, ozonation has been adopted as a pretreatment prior to anaerobic sludge digestion. Ozone pretreatment of sludge can effectively improve the residual sludge solubility and organic matter removal rate. Goel et al. (2003) recorded that the sludge degradation rates were 19% and

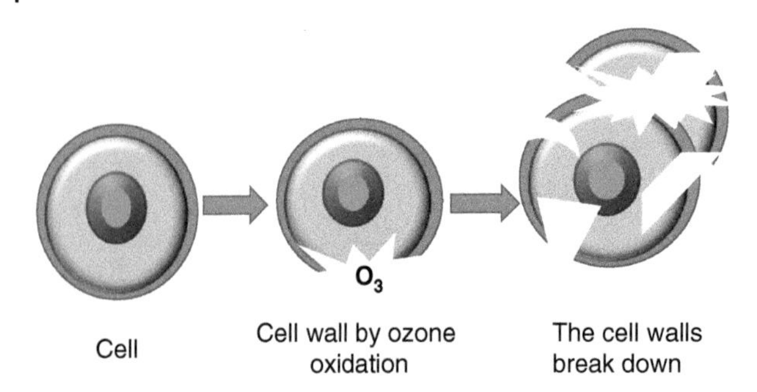

Figure 16.7 Cell decomposition through ozone destruction.

37% respectively when pre-ozonation of 0.015 and 0.05 mg O_3/mg TSS (total suspended solids) was applied. Chacana et al. (2017) evaluated the effect of ozonation using ozone doses ranging from 0 to 192 mg O_3/g-sludge COD on the methane production of anaerobic digested sludge and found that the optimized ozone dose of 86 mg O_3/g-COD increased the methane yield up to 52% and the methane production rate up to 95%. Therefore, ozonation could be used to increase the capacity of anaerobic digesters. But high processing cost hinders its implementation on a large scale.

16.3.6 Thermal Pre-treatment Technique

Thermal pre-treatment is a well-developed and effective pretreatment technology. It is used to improve sludge dewaterability and digestibility. The heat treatment can destroy the colloidal structure of the sludge and release the interstitial water inside the sludge floc and the cells. There are two categories of thermal pretreatment methods: thermal and hydrothermal pre-treatments. At atmospheric pressure, the heating temperature is below 100 °C for thermal pretreatment and above 100 °C for hydrothermal with the subsequent increase of pressure. The efficiency of the thermal pre-treatment process mainly relies upon treatment temperature and time used. González-Fernández et al. (2012) achieved a 2.2-fold methane production after pre-treating microalgae *Scenedesmus* at 90 °C for 120 minutes, and in this study, the soluble organic matter increased 2.4-fold for 90 °C for 30 minutes of pre-treatment and up to 4.4-fold (90 °C) for 180 minutes. However, Carrere et al. (2008) concluded that thermal treatment time had less impact on sludge solubilization in comparison with temperature.

The advantages of thermal pre-treatment include destruction of pathogens, odour removal, reduction of organic waste volume, improved dewaterability with subsequent enhancement of sludge handling as well as positive energy balance.

16.3.7 Enzymatic Pre-treatment

Enzymatic pre-treatment refers to the direct addition of enzyme preparation or the bacteria capable of secreting extracellular enzymes into organic waste. Enzymes can catalyse the decomposition of organic matter, reduce the viscosity of long chain proteins, carbohydrates and lipids, and decompose macromolecules into biodegradable small molecules to improve the biodegradability of organic wastes. There are many studies on the enzymatic pretreatment of organic matter (Table 16.2). Ahmed Mahdya et al. (2014) investigated the effect of enzymatic hydrolysis on microalgae organic matter solubilization and methane production. Their results show that the application of protease prior to anaerobic digestion of *Chlorella vulgaris* could be an effective way to decrease the energetic input required for cell wall disruption.

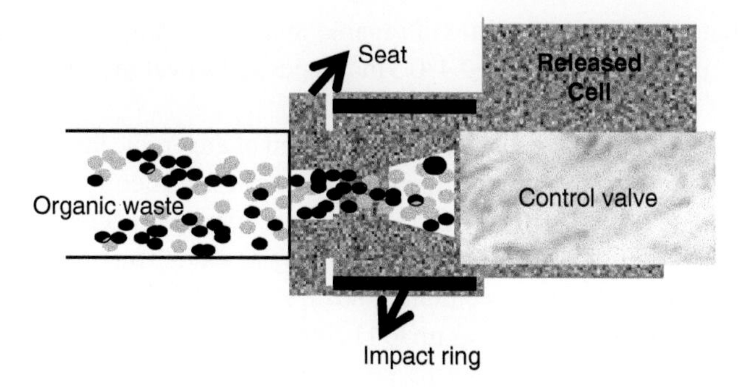

Figure 16.8 Cross-section of a high-pressure homogenizer.

The application of enzymatic pre-treatment technology mainly depends on the development of bio-engineering technology; if the enzyme can be extracted inexpensively, the technology will be able to be applied to a wide range of applications.

16.3.8 High-pressure Homogenization

The high-pressure homogenization pretreatment is applied by pressure gradient, turbulence, cavitation and shear stresses that act on solid surfaces and lead to cell disruption and release of cytoplasm, which gives rise to increased protein concentration and SCOD. The main influencing factors of high-pressure homogenization pretreatment are homogenization pressure and cycle number (Fang et al. 2014). The mechanism of a high-pressure homogenizer is shown in Figure 16.8.

Numerous lines of evidence have demonstrated that the excess sludge pretreatment by a high-pressure homogenizer can significantly increase biogas productivity, as summarized in Table 16.2. The sludge yield is 23% lower than that without pretreatment, and the biogas production is increased by 30% (Onyeche 2004). Fang et al. (2014) demonstrated that the performance of sludge disintegration and anaerobic digestion were markedly enhanced by increasing the homogenization pressure. Recently, high-pressure homogenization has been universally applied to lignocellulose pretreatment and has been shown to be highly efficient. The work by Chen et al. (2010) showed that the enzymatic digestibility of sugarcane bagasse was increased by 101.36% with combined high-pressure homogenization and alkaline pretreatment.

16.4 Full-scale Implementation Status of Anaerobic Digestion in Developing Countries

16.4.1 China

For a long time, the application of anaerobic digestion in China to deal with industrial and agricultural organic wastes has achieved good energy recovery, and environmental and economic benefits. AD as the main technology where biomass energy development and utilization, in the past years, has gradually become a technology that has wide use and has a role in protecting the ecological environment.

According to the survey of existing sewage treatment plants in China, the commonly used sludge stabilization method is mainly anaerobic digestion (38.04%), with aerobic digestion (only 2.81%), and sludge composting (3.45%). Among them, the mesophilic anaerobic digestion process is the most widely used. The results show that the degradation rate of organic matter in sludge could approach higher than 40%, with a biogas production rate of ≥0.4–0.5 m^3/kg-VS and a heat recovery efficiency of ≥35%.

For food waste with high organic, water, oil, and salt content, the application of anaerobic digestion can not only effectively reduce the volume of food waste, but can also recover the energy, and the residue after digestion can be used as agricultural fertilizer. In 2008, the biggest sludge anaerobic treatment project in Asia was put into operation in Shanghai Bailongguan Wastewater Treatment Plant, with the treatment capacity of up to 1020 t/d (water content of 80%). For the biogas project in Harbin Road Area in 2013, its annual treatment capacity was up to 110 000 t. Today, China is still exploring anaerobic digestion processes. High-solids anaerobic digestion which has more applications in Europe is one of the most important processes in the field of anaerobic digestion. In China it is applied mainly in the agricultural field, such as livestock manure, straw, and other agricultural organic waste. In recent years, the direction of research began to turn to the commercial, industrial, and municipal areas, such as food waste, waste from distilleries, municipal sludge, and other organic waste digestion (Duan et al. 2016).

16.4.2 India

Since the beginning of twenty-first century, biomass has been one of the most important focus areas of renewable energy programs in India. The agricultural sector is the main strength of India's biomass resources. It is important to have local biomass databases because crop residue biomasses are distributed resources with variation in their spatial-temporal availability and their characteristics. Overall, India produces 686 Mt gross crop residue biomass on an annual basis, and 234 Mt (34% of gross) are estimated as surplus for bioenergy generation. Amongst all the 28 states, Uttar Pradesh produces the highest amount of crop residue (Hiloidhari et al. 2014). The annual bioenergy potential from the surplus crop residue biomass is estimated to be 4.15 EJ, which is equal to 17% of India's total primary consumption of energy. Variations exist from 679 MJ (West Bengal) to 16 840 MJ (Punjab) of per capita crop residue bioenergy potential amongst the states of India. In terms of overall power, the total installed capacity of electricity generation in India is 2666.64 GW, with 12.83% proportion of biomass power (Kumar et al. 2015).

According to the estimate of Ministry of New and Renewable Energy (MNRE) of India, the energy derived from biomass accounts for about 32% of the total primary energy use in India, and more than 70% of the country's population depends upon biomass for their energy needs. In India, there were under 12 million family size biogas plants installed in 2010, compared to 4 million in 2005 (Rao et al. 2010). The Government of India has realized the essence and potential of biomass energy in India, and over the last 10 years it has initiated many programs for the promotion of efficient biomass conversion technologies. Nevertheless, in spite of the good biomass power potential, many states with agriculture-based economies have not properly utilized biomass for power supplies. Only Uttar Pradesh (in north India) has utilized a large part of the biomass potential, attributed to its sugarcane industry, with cogeneration power plants.

16.4.3 Malaysia

Due to large-scale development and utilization of natural resources in Malaysia, resource consumption has increased dramatically, which unavoidably leads to environmental pollution. With regard to anaerobic digestion, Malaysia mainly uses it to deal with palm oil waste, slaughterhouse wastewater and MSWs. During the past decades, the amount of animal husbandry has grown drastically in Malaysia. A large amount of animal waste including manure, blood, and rumen content is produced annually which provides a considerable source of biogas generation. From the observation of Abdeshahian et al. (2016), there was 4589.49 million m^3 biogas produced from animal waste in 2012, which provides electricity generation of 8.27×10^9 kWh/year.

There has been a rise in the livestock population in Malaysia in response to the increasing demands for farm animal products, in turn, that results in the production of a large amount of organic waste in farms and slaughterhouses. The treatment of farm animal waste by anaerobic digestion or biogas technology can potentially contribute to the generation of huge amounts of renewable energy in the form of methane-rich biogas. In addition, the production of a huge amount of crude palm oil results in even larger amounts of palm oil mill effluent (POME), which causes severe environmental pollution. There are two ways to handle POME. One is a ponding system, and the other is anaerobic digestion. The essential theories of both are the same, but the anaerobic digestion of POME provides the fastest payback of investment because the treatment enables biogas recovery for heat generation and treated effluent for land application (Wu et al. 2010). Anaerobic digestion technology is playing an important role in assisting the Malaysian government to handle the serious environmental problems. It is anticipated that the people and economy of Malaysia will benefit from the emergence of anaerobic digestion.

16.4.4 Vietnam

Vietnam has a long history in the development of anaerobic digestion technology, but until now the contribution of anaerobic digestion to dealing with organic biomass is still very limited (Silva et al. 2016). Moreover, most of the existing biogas plants in Vietnam that have been installed are at a small scale; in 2006, Vietnam had more than 18 000 domestic biogas plants (Nguyen et al. 2014; Nguyen 2011). The rapid urbanization in Vietnam has caused an ever-increasing production of organic waste. The MSW generation was calculated to be 35 700 t/d in 2015, and it will rise to 55 440 t/d in 2020 and then to 83 200 t/d in 2025. In the MSW stream, the organic fraction such as food waste is 21 420, 33 264 and 49 920 t/d in 2015, 2020, and 2025, respectively (Nguyen et al. 2014). Food waste is an ideal feedstock for anaerobic digestion and has great potential for methane production. As further estimated by Nguyen et al. (2014), food waste will generate about 20 and 45 GWh/d in 2020, and 2025, respectively, if it is all sent to anaerobic digestion. This could contribute 2.4–4.1% of the electricity demand, double the amount of energy in the form of heat; or alternatively, contribute approximately 2.2–4.7% of fuel consumption for transportation in Vietnam.

With the rapid economic development, the energy demand in Vietnam will rise fourfold by 2030 and ninefold by 2025 (electricity) compared to 2005. Biogas technology development could be a cost-effective solution to reducing the dependence on fossil fuels and alleviating energy shortage. This has encouraged the Vietnam government to issue many policies for developing renewable energy. However, because of a lack of information and experience in using anaerobic digestion technology, the conversion of biomass-to-energy in Vietnam still has a long way to go.

16.4.5 Thailand

In other emerging countries like Thailand, the biogas technologies have also attracted increasing interest. Thailand has tremendous biomass sources for biogas production, including farm waste (pig and cow dung), agroindustrial waste (POME, cassava starch, ethanol, sugar, and slaughterhouse) and MSW (Intharathirat and Salam 2016). In 2012, they installed around 540 anaerobic digestion systems for food waste digestion, which receives 7884 tons of food waste per year and produces 492 750 m^3 of biogas (Sharp and Sang-Arun 2012). Also, since 2012, the government has initiated an integrated solid waste system in the Nakornratchasima municipality for energy recovery by using the AD process (Intharathirat and Salam 2016). However, inadequate feedstock and a low biodegradable fraction caused inefficiency in the generation of electricity. Similar problems occurred at two other AD plants, located in the Rayong and Chonburi municipalities, owing to insufficient volumes of organic waste.

16.5 Perspectives

Anaerobic digestion is one of the most successful and well-established biogas production technologies, it shows great potential and promise for biomass utilization and simultaneous methane-rich biogas production. The main technical obstacle for producing biogas from biomass is the slow hydrolysis of biomass caused by the complex components and microstructures, and this leads to not only long SRT and but low biodegradation efficiency. Pretreatment using different mechanical, physio-chemical, and biological processes is able to accelerate the solubilization and release of organic contents, thus making them available to subsequent microbial actions during the anaerobic process. However, there are still many barriers limiting the full-scale applications of biogas production technologies in emerging countries, such as high investment costs, large variability in the characteristics of different feedstocks, and limited technical experience. In this regard, more effort should be dedicated to obtaining more information, data, and experience for pushing forward the widespread implementation of anaerobic digestion technology.

References

Abdeshahian, P., Lim, J.S., Ho, W.S. et al. (2016). Potential of biogas production from farm animal waste in Malaysia. *Renewable and Sustainable Energy Reviews* 60: 714–723.

Ahmed Mahdya, B., Mendeza, L., Mercedes Ballesterosa, C., and González-Fernándeza, C. (2014). Enhanced methane production of *Chlorella vulgaris* and *Chlamydomonas reinhardtii* by hydrolytic enzymes addition. *Energy Conversion and Management* 85: 551–557.

Akobi, C., Yeo, H., Hafez, H., and Nakhla, G. (2016). Single-stage and two-stage anaerobic digestion of extruded lignocellulosic biomass. *Applied Energy* 184: 548–559.

Alonso, D., Bond, J., and Dumesic, J. (2010). Catalytic conversion of biomass to biofuels. *Green Chemistry* 12 (9): 1493–1513.

Amnuaycheewa, P., Hengaroonprasan, R., Rattanaporn, K. et al. (2016). Enhancing enzymatic hydrolysis and biogas production from rice straw by pretreatment with organic acids. *Industrial Crops and Products* 87: 247–254.

Angelidaki, I., Ahring, B.K., Deng, H., and Schmidt, J.E. (2002). Anaerobic digestion of olive oil mill effluents together with swine manure in UASB reactors. *Water Science and Technology* 45 (10): 213–218.

Appels, L., Baeyens, J., Degreve, J., and Dewil, R. (2008). Principles and potential of the anaerobic digestion of waste-activated sludge. *Progress in Energy and Combustion Science* 34 (6): 755–781.

Appels, L., Degrève, J., Bruggen, B.V.D. et al. (2010). Influence of low temperature thermal pre-treatment on sludge solubilisation, heavy metal release and anaerobic digestion. *Bioresource Technology* 101 (15): 5743–5748.

Hendriks, A. and Zeeman, G. (2009). Pretreatments to enhance the digestibility of lignocellulosic biomass. *Bioresource Technology* 1: 10–18.

Ayol, A. (2005). Enzymatic treatment effects on dewaterability of anaerobically digested biosolids-I: performance evaluations. *Process Biochemistry* 40 (7): 2427–2434.

Bakhshi, Z., Jauffur, S., and Frigon, D. (2018). Assessing energy benefits of operating anaerobic digesters at low temperature with solids pre-ozonation. *Renewable Energy* 115: 1303–1311.

Beltrán, C., Jeison, D., Fermoso, F.G., and Borja, R. (2016). Batch anaerobic co-digestion of waste activated sludge and microalgae (*Chlorella sorokiniana*) at mesophilic temperature. *Journal of Environmental Science and Health, Part A* 51 (10): 847–850.

Beszedes, S., Kertesz, S., Laszlo, Z. et al. (2009). Biogas production of ozone and/or microwave-pretreated canned maize production sludge. *Ozone: Science and Engineering* 3: 257–261.

Bougrier, C., Carrère, H., and Delgenès, J.P. (2005). Solubilisation of waste-activated sludge by ultrasonic treatment. *Chemical Engineering Journal* 2: 163–169.

Bougrier, C., Albasi, C., Delgenès, J.P., and Carrère, H. (2006). Effect of ultrasonic, thermal and ozone pre-treatments on waste activated sludge solubilisation and anaerobic biodegradability. *Chemical Engineering and Processing* 8: 711–718.

Brown, D., Shi, J., and Li, Y.B. (2012). Comparison of solid-state to liquid anaerobic digestion of lignocellulosic feedstocks for biogas production. *Bioresource Technology* 124: 379–386.

Cardena, R., Moreno, G., Bakonyi, P., and Buitron, G. (2017). Enhancement of methane production from various microalgae cultures via novel ozonation pretreatment. *Chemical Engineering Journal* 307: 948–954.

Cardozo, K.H.M., Guaratini, T., Barros, M.P. et al. (2007). Metabolites from algae with economic impact. *Comparative Biochemistry and Physiology Part C: Toxicology & Pharmacology* 146 (1–2): 60–78.

Carlsson, M., Lagerkvist, A., and Morgan-Sagastume, F. (2012). The effects of substrate pre-treatment on anaerobic digestion systems: a review. *Waste Management* 9: 1634–1650.

Carrere, H., Bougrier, C., Castets, D., and Delgenes, J.P. (2008). Impact of initial biodegradability on sludge anaerobic digestion enhancement by thermal pretreatment. *Journal of Environmental Science Health A Toxic Hazard Substances Environment Engineering* 43 (13): 1551–1555.

Chacana, J., Alizadeh, S., Labelle, M.-A. et al. (2017). Effect of ozonation on anaerobic digestion sludge activity and viability. *Chemosphere* 176: 405–411.

Chandra, R., Takeuchi, H., Hasegawa, T., and Kumar, R. (2012). Improving biodegradability and biogas production of wheat straw substrates using sodium hydroxide and hydrothermal pretreatments. *Energy* 1: 273–282.

Chang, V.S. and Holtzapple, M.T. (2000). Fundamental factors affecting biomass enzymatic reactivity. *Applied Biochemistry and Biotechnology* 84-6: 5–37.

Charles, W., Ng, B., Cord-Ruwisch, R. et al. (2013). Enhancement of waste activated sludge anaerobic digestion by a novel chemical free acid/alkaline pretreatment using electrolysis. *Water Science and Technology* 67 (12): 2827–2831.

Chen, Y., Jiang, S., Yuan, H. et al. (2007). Hydrolysis and acidification of waste activated sludge at different pHs. *Water Research* 3: 683–689.

Chen, D., Guo, Y., Huang, R. et al. (2010). Pretreatment by ultra-high pressure explosion with homogenizer facilitates cellulase digestion of sugarcane bagasses. *Bioresource Technology* 14: 5592–5600.

Chen, X., Xiang, X., Dai, R. et al. (2017). Effect of low temperature of thermal pretreatment on anaerobic digestion of textile dyeing sludge. *Bioresource Technology* 243: 426–432.

Chi, Y.Z., Li, Y.Y., Ji, M. et al. (2010). Use of combined NaOH-microwave pretreatment for enhancing Mesophilic anaerobic digestibility of thickened waste activated sludge. *Advanced Materials Research* 113–116: 459–468.

Chu, C.P., Lee, D.J., Chang, B.-V. et al. (2002). "Weak" ultrasonic pre-treatment on anaerobic digestion of flocculated activated biosolids. *Water Research* 11: 2681–2688.

Cooney, M.J., Young, G., and Pate, R. (2011). Bio-oil from photosynthetic microalgae: case study. *Bioresource Technology* 102 (1): 166–177.

De Gioannis, G., Muntoni, A., Polettini, A. et al. (2017). Energy recovery from one- and two-stage anaerobic digestion of food waste. *Waste Management* 68: 595–602.

Deepanraj, B., Sivasubramanian, V., and Jayaraj, S. (2017). Effect of substrate pretreatment on biogas production through anaerobic digestion of food waste. *International Journal of Hydrogen Energy* 42 (42): 26522–26528.

Demirbas, A. (2000). Biomass resources for energy and chemical industry. *Energy Education Science and Technology* 5: 21–25.

Demirel, B. and Scherer, P. (2008). The roles of acetotrophic and hydrogenotrophic methanogens during anaerobic conversion of biomass to methane: a review. *Reviews in Environmental Science and Bio/Technology* 7 (2): 173–190.

Devlin, D.C., Esteves, S.R.R., Dinsdale, R.M., and Guwy, A.J. (2011). The effect of acid pretreatment on the anaerobic digestion and dewatering of waste activated sludge. *Bioresource Technology* 5: 4076–4082.

Duan, N., Dong, B., Dai, L., and Dai, X.H. (2016). State of the art of high solids anaerobic digestion of organic waste. *Environmental Engineering* 34 (9): 119–124.

EIA (2014). *International Energy Outlook 2014: World Petroleum and Other Liquid Fuels*. Washington, DC.

Elbeshbishy, E., Aldin, S., Hafez, H. et al. (2011). Impact of ultrasonication of hog manure on anaerobic digestability. *Ultrasonics Sonochemistry* 1: 164–171.

Eskicioglu, C., Monlau, F., Barakat, A. et al. (2017). Assessment of hydrothermal pretreatment of various lignocellulosic biomass with CO_2 catalyst for enhanced methane and hydrogen production. *Water Research* 120: 32–42.

Fang, W., Zhang, P., Zhang, G. et al. (2014). Effect of alkaline addition on anaerobic sludge digestion with combined pretreatment of alkaline and high pressure homogenization. *Bioresource Technology* 168: 167–172.

Fang, W., Zhang, P., Shang, R. et al. (2017). Effect of high pressure homogenization on anaerobic digestion of the sludge pretreated by combined alkaline and high pressure homogenization. *Desalination and Water Treatment* 62: 168–174.

Feng, Q.J. and Lin, Y.Q. (2017). Integrated processes of anaerobic digestion and pyrolysis for higher bioenergy recovery from lignocellulosic biomass: a brief review. *Renewable and Sustainable Energy Reviews* 77: 1272–1287.

Fisgativa, H., Tremier, A., and Dabert, P. (2016). Characterizing the variability of food waste quality: a need for efficient valorisation through anaerobic digestion. *Waste Management* 50: 264–274.

Gaur, R. and Suthar, S. (2017). Anaerobic digestion of activated sludge, anaerobic granular sludge and cow dung with food waste for enhanced methane production. *Journal of Cleaner Production* 164: 557–566.

Geng, Y., Zhang, B., Du, L. et al. (2016). Improving methane production during the anaerobic digestion of waste activated sludge: Cao-ultrasonic pretreatment and using different seed sludges. *Procedia Environmental Sciences* 31: 743–752.

Gil, A., Siles, J.A., Martín, M.A. et al. (2018). Effect of microwave pretreatment on semi-continuous anaerobic digestion of sewage sludge. *Renewable Energy* 115: 917–925.

Goel, R., Tokutomi, T., and Yasui, H. (2003). Anaerobic digestion of excess activated sludge with ozone pretreatment. *Water Science and Technology* 47 (12): 207–214.

González-Fernández, C., Sialve, B., Bernet, N., and Steyer, J.P. (2012). Thermal pretreatment to improve methane production of Scenedesmus biomass. *Biomass and Bioenergy* 40: 105–111.

Hadin, A. and Eriksson, O. (2016). Horse manure as feedstock for anaerobic digestion. *Waste Management* 56: 506–518.

Hiloidhari, M., Das, D., and Baruah, D.C. (2014). Bioenergy potential from crop residue biomass in India. *Renewable and Sustainable Energy Reviews* 32: 504–512.

Hogan, F., Mormede, S., Clark, P., and Crane, M. (2004). Ultrasonic sludge treatment for enhanced anaerobic digestion. *Water Science and Technology* 9: 25–32.

Huang, C., Lai, J., Sun, X.Y. et al. (2016). Enhancing anaerobic digestion of waste activated sludge by the combined use of NaOH and Mg(OH)(2): performance evaluation and mechanism study. *Bioresource Technology* 220: 601–608.

IEA (2016) World Energy Outlook 2016 - Executive Summary http://www.iea.org/publications/freepublications; 2016.

Intharathirat, R. and Salam, P.A. (2016). Valorization of MSW-to-energy in Thailand: status, challenges and prospects. *Waste and Biomass Valorization* 7 (1): 31–57.

Ishizawa, C., Davis, M., Schell, D., and Johnson, D. (2007). Porosity and its effect on the digestibility of dilute sulfuric acid pretreated corn stover. *Journal of Agricultural and Food Chemistry* 7: 2575–2581.

Jain, S., Jain, S., Wolf, I.T. et al. (2015). A comprehensive review on operating parameters and different pretreatment methodologies for anaerobic digestion of municipal solid waste. *Renewable and Sustainable Energy Reviews* 52: 142–154.

Jankowska, E., Sahu, A.K., and Oleskowicz-Popiel, P. (2017). Biogas from microalgae: review on microalgae's cultivation, harvesting and pretreatment for anaerobic digestion. *Renewable and Sustainable Energy Reviews* 75: 692–709.

Jie, W.G., Peng, Y.Z., Ren, N.Q., and Li, B.K. (2014). Volatile fatty acids (VFAs) accumulation and microbial community structure of excess sludge (ES) at different pHs. *Bioresource Technology* 152: 124–129.

Jina, S., Zhangb, G., Zhanga, P. et al. (2015). Comparative study of high-pressure homogenization and alkaline-heat pretreatments for enhancing enzymatic hydrolysis and biogas production of grass clipping. *International Biodeterioration and Biodegradation* 104: 477–481.

Ju, F., Li, B., Ma, L.P. et al. (2016). Antibiotic resistance genes and human bacterial pathogens: co-occurrence, removal, and enrichment in municipal sewage sludge digesters. *Water Research* 91: 1–10.

Katukuri, N.R., Fu, S.F., He, S. et al. (2017). Enhanced methane production of *Miscanthus floridulus* by hydrogen peroxide pretreatment. *Fuel* 199: 562–566.

Khatri, S., Wu, S., Kizito, S. et al. (2015). Synergistic effect of alkaline pretreatment and Fe dosing on batch anaerobic digestion of maize straw. *Applied Energy* 158: 55–64.

Kim, J., Park, C., Kim, T.-H. et al. (2003). Effects of various pretreatments for enhanced anaerobic digestion with waste activated sludge. *Journal of Bioscience and Bioengineering* 3: 271–275.

Kinnunen, V. and Rintala, J. (2016). The effect of low-temperature pretreatment on the solubilization and biomethane potential of microalgae biomass grown in synthetic and wastewater media. *Bioresource Technology* 221: 78–84.

Kumar, A., Kumar, N., Baredar, P., and Shukla, A. (2015). A review on biomass energy resources, potential, conversion and policy in India. *Renewable and Sustainable Energy Reviews* 45: 530–539.

Li, H.F., Lei, Z.F., Liu, C.G. et al. (2015). Photocatalytic degradation of lignin on synthesized Ag-AgCl/ZnO nanorods under solar light and preliminary trials for methane fermentation. *Bioresource Technology* 175: 494–501.

Li, H., Tian, Y., Zuo, W. et al. (2016). Electricity generation from food wastes and characteristics of organic matters in microbial fuel cell. *Bioresource Technology* 205: 104–110.

Li, X., Guo, S., Peng, Y. et al. (2018). Anaerobic digestion using ultrasound as pretreatment approach: changes in waste activated sludge, anaerobic digestion performances and digestive microbial populations. *Biochemical Engineering Journal* 139: 139–145.

Liao, X., Li, H., Zhang, Y. et al. (2016). Accelerated high-solids anaerobic digestion of sewage sludge using low-temperature thermal pretreatment. *International Biodeterioration & Biodegradation* 106: 141–149.

Lin, Y., Wang, D., Wu, S., and Wang, C. (2009). Alkali pretreatment enhances biogas production in the anaerobic digestion of pulp and paper sludge. *Journal of Hazardous Materials* 1: 366–373.

Lina, J.-G., Changb, C.-N., and Changa, S.-C. (1997). Enhancement of anaerobic digestion of waste activated sludge by alkaline solubilization. *Bioresource Technology* 3: 85–90.

Liu, C.G. and Wyman, C.E. (2003). The effect of flow rate of compressed hot water on xylan, lignin, and total mass removal from corn stover. *Industrial & Engineering Chemistry Research* 42 (21): 5409–5416.

Liu, Y., Wang, J., and Zhao, D. (2005). Study on the technology of anaerobic digestion treatment of food waste. *Energy Technology* 26: 150–154.

Lloyd, T.A. and Wyman, C.E. (2005). Combined sugar yields for dilute sulfuric acid pretreatment of corn stover followed by enzymatic hydrolysis of the remaining solids. *Bioresource Technology* 18: 1967–1977.

López Torres, M. and Espinosa Lloréns, M.D.C. (2008). Effect of alkaline pretreatment on anaerobic digestion of solid wastes. *Waste Management* 11: 2229–2234.

Lopez, V.M., De la Cruz, F.B., and Barlaz, M.A. (2016). Chemical composition and methane potential of commercial food wastes. *Waste Management* 56: 477–490.

Lu, X., Zhen, G., Chen, M. et al. (2015a). Biocatalysis conversion of methanol to methane in an upflow anaerobic sludge blanket (UASB) reactor: long-term performance and inherent deficiencies. *Bioresource Technology* 198: 691–700.

Lu, X., Zhen, G., Estrada, A.L. et al. (2015b). Operation performance and granule characterization of upflow anaerobic sludge blanket (UASB) reactor treating wastewater with starch as the sole carbon source. *Bioresource Technology* 180: 264–273.

Luste, S., Luostarinen, S., and Sillanpaeae, M. (2009). Effect of pre-treatments on hydrolysis and methane production potentials of by-products from meat-processing industry. *Journal of Hazardous Materials* 1: 247–255.

Mahdy, A., Mendez, L., Blanco, S. et al. (2014). Protease cell wall degradation of *Chlorella vulgaris*: effect on methane production. *Bioresource Technology* 171: 421–427.

Mahdy, A., Ballesteros, M., and González-Fernández, C. (2016). Enzymatic pretreatment of *Chlorella vulgaris* for biogas production: influence of urban wastewater as a sole nutrient source on macromolecular profile and biocatalyst efficiency. *Bioresource Technology* 199: 319–325.

Mahdy, A., Fotidis, I.A., Mancini, E. et al. (2017). Ammonia tolerant inocula provide a good base for anaerobic digestion of microalgae in third generation biogas process. *Bioresource Technology* 225: 272–278.

Mao, C., Feng, Y., Wang, X., and Ren, G. (2015). Review on research achievements of biogas from anaerobic digestion. *Renewable and Sustainable Energy Reviews* 45: 540–555.

Molinuevo-Salces, B., García-González, M.C., González-Fernández, C. et al. (2010). Anaerobic co-digestion of livestock wastes with vegetable processing wastes: a statistical analysis. *Bioresource Technology* 101: 9479–9485.

Navia, R., Soto, M., Vidal, G. et al. (2002). Alkaline pretreatment of kraft mill sludge to improve its anaerobic digestion. *Bulletin of Environmental Contamination and Toxicology* 6: 869–876.

Nguyen, V.C.N. (2011). Small-scale anaerobic digesters in Vietnam - development and challenges. *Journal of Vietnamese Environment* 1 (1): 12–18.

Nguyen, H.H., Heaven, S., and Banks, C. (2014). Energy potential from the anaerobic digestion of food waste in municipal solid waste stream of urban areas in Vietnam. *International Journal of Energy and Environmental Engineering* 5 (4): 365–374.

Nimz, H.H. (1984). Wood-chemistry, ultrastructure, reactions. *Holz als Roh- und Werkstoff* 42 (8): 314–314.

Nuchdang, S., Frigon, J.C., Roy, C. et al. (2017). Hydrothermal post-treatment of digestate to maximize the methane yield from the anaerobic digestion of microalgae. *Waste Management* 71: 683–688.

Onyeche, T.I. (2004). Sludge as source of energy and revenue. *Water Science and Technology* 50 (9): 197–204.

Palmowski, L., Simons, L., and Brooks, R. (2006). Ultrasonic treatment to improve anaerobic digestibility of dairy waste streams. *Water Science and Technology* 8: 281–288.

Paritosh, K., Kushwaha, S., Yadav, M. et al. (2017). Food waste to energy: an overview of sustainable approaches for food waste management and nutrient recycling. *BioMed Research International* 2017 Article ID 2370927.

Park, S. and Li, Y.B. (2012). Evaluation of methane production and macronutrient degradation in the anaerobic co-digestion of algae biomass residue and lipid waste. *Bioresource Technology* 111: 42–48.

Park, B., Ahn, J., Kim, J., and Hwang, S. (2004). Use of microwave pretreatment for enhanced anaerobiosis of secondary sludge. *Water Science and Technology* 9: 17–23.

Park, J.H., Kumar, G., Yun, Y.M. et al. (2018). Effect of feeding mode and dilution on the performance and microbial community population in anaerobic digestion of food waste. *Bioresource Technology* 248: 134–140.

Passos, F. and Ferrer, I. (2015). Influence of hydrothermal pretreatment on microalgal biomass anaerobic digestion and bioenergy production. *Water Research* 68: 364–373.

Pavlostathis, S.G. and Gossett, J.M. (1986). A kinetic model for anaerobic digestion of biological sludge. *Biotechnology and Bioengineering* 10: 1519–1530.

Penauda, V., Delgenès, J.P., and Molettaa, R. (1999). Thermo-chemical pretreatment of a microbial biomass: influence of sodium hydroxide addition on solubilization and anaerobic biodegradability. *Enzyme and Microbial Technology* 3–5: 258–263.

Perez-Elvira, S., Fdz-Polanco, M., Plaza, F.I. et al. (2009). Ultrasound pre-treatment for anaerobic digestion improvement. *Water Science and Technology* 60 (6): 1525–1532.

Rajagopal, R., Bellavance, D., and Rahaman, M.S. (2017). Psychrophilic anaerobic digestion of semi-dry mixed municipal food waste: for North American context. *Process Safety and Environmental Protection* 105: 101–108.

Ramos, L.P. (2003). The chemistry involved in the steam treatment of lignocellulosic materials. *Química Nova* 26 (6): 863–871.

Rani, R.U., Kumar, S.A., Kaliappan, S. et al. (2012). Low temperature thermo-chemical pretreatment of dairy waste activated sludge for anaerobic digestion process. *Bioresource Technology* 103 (1): 415–424.

Rao, P.V., Baral, S.S., Dey, R., and Mutnuri, S. (2010). Biogas generation potential by anaerobic digestion for sustainable energy development in India. *Renewable and Sustainable Energy Reviews* 14 (7): 2086–2094.

Ravindran, R., Jaiswal, S., Abu-Ghannam, N., and Jaiswal, A.K. (2017). Two-step sequential pretreatment for the enhanced enzymatic hydrolysis of coffee spent waste. *Bioresource Technology* 239: 276–284.

Romano, R.T., Zhang, R., Teter, S., and McGarvey, J.A. (2009). The effect of enzyme addition on anaerobic digestion of JoseTall wheat grass. *Bioresource Technology* 100 (20): 4564–4571.

Saharan, B., Sharma, D., Sahu, R. et al. (2013). Towards algal biofuel production: a concept of green bioenergy development. *Innovative Romanian Food Biotechnology* 12: 1–21.

Şahinkaya, S. and Sevimli, M.F. (2013). Sono-thermal pretreatment of waste activated sludge before anaerobic digestion. *Ultrasonics Sonochemistry* 20 (1): 587–594.

Salihu, A. and Alam, M.Z. (2016). Pretreatment methods of organic wastes for biogas production. *Journal of Applied Sciences* 16: 124–137.

Seng, B., Khanal, S.K., and Visvanathan, C. (2010). Anaerobic digestion of waste activated sludge pretreatment by a combined ultrasound and chemical process. *Environmental Technology* 31 (3): 257–265.

Sharp, A. and Sang-Arun, J. (2012) A Guide for Sustainable Urban Organic Waste Management in Thailand: Combining Food, Energy, and Climate Co-Benefit. Asia-Pacific Network for Global Change Research (APN), http://pub.iges.or.jp/modules/envirolib/upload/4130/attach/Attachment4135_Guideline-UOWM-Thailand_Eng.pdf

Shevchenko, S.M., Beatson, R.P., and Saddler, J.N. (1999). The nature of lignin from steam explosion/enzymatic hydrolysis of softwood: structural features and possible uses: scientific note. *Applied Biochemistry and Biotechnology* 77–79: 867–876.

Silva, R., Le, H.A., and Koch, K. (2016). Feasibility assessment of anaerobic digestion technologies for household wastes in Vietnam. *Journal of Vietnamese Environment* 7 (1): 1–8.

Silvestre, G., Ruiz, B., Fiter, M. et al. (2015). Ozonation as a pre-treatment for anaerobic digestion of waste-activated sludge: effect of the ozone doses. *Ozone: Science & Engineering* 37 (4): 316–322.

Singh, R., Shukla, A., Tiwari, S., and Srivastava, M. (2014). A review on delignification of lignocellulosic biomass for enhancement of ethanol production potential. *Renewable and Sustainable Energy Reviews* 32: 713–728.

Song, Z.L., Yag, G.H., Feng, Y.Z. et al. (2013). Pretreatment of rice straw by hydrogen peroxide for enhanced methane yield. *Journal of Integrative Agriculture* 12 (7): 1258–1266.

Sual, H., Liubl, G., Hea, M., and Tana, F. (2015). A biorefining process: sequential, combinational lignocellulose pretreatment procedure for improving biobutanol production from sugarcane bagasse. *Bioresource Technology* 187: 149–160.

Sun, Y. and Cheng, J.J. (2005). Dilute acid pretreatment of rye straw and bermudagrass for ethanol production. *Bioresource Technology* 14: 1599–1606.

Suschka, J. and Grübel, K. (2014). NitrogenI in the process of waste activated sludge anaerobic digestion. *Archives of Environmental Protection* 40 (2): 123–136.

Syaichurrozi, I., Villta, P.K., Nabilah, N., and Rusdi, R. (2019). Effect of sulfuric acid pretreatment on biogas production from Salvinia molesta. *Journal of Environmental Chemical Engineering* 7 (1): 102857.

Tang, B., Yu, L., Huang, S. et al. (2010). Energy efficiency of pre-treating excess sewage sludge with microwave irradiation. *Bioresource Technology* 14: 5092–5097.

Thurman, H.V. (1997). *Introductory Oceanography*, vol. 544. New Jersey: Prentice-Hall.

Tiehm, A., Nickel, K., and Neis, U. (1997). The use of ultrasound to accelerate the anaerobic digestion of sewage sludge. *Water Science and Technology* 11: 121–128.

Togari, T., Yamamoto-Ikemoto, R., Ono, H. et al. (2016). Effects of microwave pretreatment of dewatered sludge from an oxidation-ditch process on the biogas yield in mesophilic anaerobic digestion. *Journal of Water and Environment Technology* 14 (3): 158–165.

Toreci, I., Kennedy, K.J., and Droste, R.L. (2009). Evaluation of continuous mesophilic anaerobic sludge digestion after high temperature microwave pretreatment. *Water Research* 5: 1273–1284.

Torres, W., Pansare, S.S., and Goodwin, J.G. (2007). Hot gas removal of tars, ammonia, and hydrogen sulfide from biomass gasification gas. *Catalysis Reviews-Science and Engineering* 49 (4): 407–456.

Tyagi, V.K., Lo, S.L., Appels, L., and Dewil, R. (2014). Ultrasonic treatment of waste sludge: a review on mechanisms and applications. *Critical Reviews in Environmental Science and Technology* 44: 1220–1288.

Uma Rani, R., Adish Kumar, S., Kaliappan, S. et al. (2013). Impacts of microwave pretreatments on the semi-continuous anaerobic digestion of dairy waste activated sludge. *Waste Management* 33 (5): 1119–1127.

Van Dyk, J.S. and Pletschke, B.I. (2012). A review of lignocellulose bioconversion using enzymatic hydrolysis and synergistic cooperation between enzymes-factors affecting enzymes, conversion and synergy. *Biotechnology Advances* 30 (6): 1458–1480.

Wan, C.X. and Li, Y.B. (2012). Fungal pretreatment of lignocellulosic biomass. *Biotechnology Advances* 30 (6): 1447–1457.

Wang, F., Wang, Y., and Ji, M. (2005). Mechanisms and kinetics models for ultrasonic waste activated sludge disintegration. *Journal of Hazardous Materials* 123 (1–3): 145–150.

Weemaes, M.P.J. and Verstraete, W.H. (1998). Evaluation of current wet sludge disintegration techniques. *Journal of Chemical Technology and Biotechnology* 2: 83–92.

Weiland, P. (2010). Biogas production: current state and perspectives. *Applied Microbiology and Biotechnology* 85 (4): 849–860.

Wonglertarak, W. and Wichitsathian, B. (2014). Alkaline pretreatment of waste activated sludge in anaerobic digestion. *Journal of Clean Energy Technologies* 2: 118–121.

Wu, T.Y., Mohammad, A.W., Jahim, J.M., and Anuar, N. (2010). Pollution control technologies for the treatment of palm oil mill effluent (POME) through end-of-pipe processes. *Journal of Environmental Management* 91 (7): 1467–1490.

Xiang, Y., Xiang, Y., and Wang, L. (2017). Disintegration of waste activated sludge by a combined treatment of alkaline-modi fied eggshell and ultrasonic radiation. *Journal of Environmental Chemical Engineering* 5: 1379–1385.

Xiao, W. and Clarkson, W.W. (1997). Acid solubilization of lignin and bioconversion of treated newsprint to methane. *Biodegradation* 1: 61–66.

Yan, B.H., Selvam, A., and Wong, J.W.C. (2016). Innovative method for increased methane recovery from two-phase anaerobic digestion of food waste through reutilization of acidogenic off-gas in methanogenic reactor. *Bioresource Technology* 217: 3–9.

Yan, H., Zhao, C., Zhang, J. et al. (2017). Study on biomethane production and biodegradability of different leafy vegetables in anaerobic digestion. *AMB Express* 7 (1): 27.

Yang, L.C., Kopsell, D.E., Kottke, A.M., and Johnson, M.Q. (2017). Development of a cartridge design anaerobic digestion system for lignocellulosic biomass. *Biosystems Engineering* 160: 134–139.

Yao, Y.Q., Luo, Y., Yang, Y.X. et al. (2014). Water free anaerobic co-digestion of vegetable processing waste with cattle slurry for methane production at high total solid content. *Energy* 74: 309–313.

Yi, L., Rao, L., Wang, X., and Wang, H. (2012). Analysis of physical and chemical properties of food waste and gas potential of anaerobic fermentation. *Journal of Central South University (Science and Technology)* 43: 1584–1588.

Yıldırım, E., Ince, O., Ince, B., and Aydin, S. (2017). Biomethane production from lignocellulosic biomass enhanced by bioaugmentation with anaerobic rumen fungi. *Process Biochemistry* in press), doi:https://doi.org/10.1016/j.procbio.2017.1002.1026.

Yu, S., Zhang, G., Li, J. et al. (2013). Effect of endogenous hydrolytic enzymes pretreatment on the anaerobic digestion of sludge. *Bioresource Technology* 146: 758–761.

Yu, B., Xu, J., Yuan, H. et al. (2014). Enhancement of anaerobic digestion of waste activated sludge by electrochemical pretreatment. *Fuel* 130: 279–285.

Zhang, Y.H.P. and Lynd, L.R. (2004). Toward an aggregated understanding of enzymatic hydrolysis of cellulose: noncomplexed cellulase systems. *Biotechnology and Bioengineering* 88 (7): 797–824.

Zhang, D., Chen, Y., Zhao, Y., and Zhu, X. (2010). New sludge pretreatment method to improve methane production in waste activated sludge digestion. *Environmental Science & Technology* 44 (12): 4802–4808.

Zhang, J., Li, W., Lee, J. et al. (2017). Enhancement of biogas production in anaerobic co-digestion of food waste and waste activated sludge by biological co-pretreatment. *Energy* 137: 479–486.

Zhao, J., Gui, L., Wang, Q. et al. (2017a). Aged refuse enhances anaerobic digestion of waste activated sludge. *Water Research* 123: 724–733.

Zhao, J., Liu, Y., Wang, D. et al. (2017b). Potential impact of salinity on methane production from food waste anaerobic digestion. *Waste Management* 67: 308–314.

Zhen, G. and Zhao, Y. (2017). *Pollution Control and Resource Recovery: sewage Sludge*. Elsevier Inc https://doi.org/10.1016/B1978-1010-1012-811639-811635.800001-811632.

Zhen, G., Lu, X., Li, Y.-Y., and Zhao, Y. (2014). Combined electrical-alkali pretreatment to increase the anaerobic hydrolysis rate of waste activated sludge during anaerobic digestion. *Applied Energy* 128: 93–102.

Zhen, G., Lu, X., Kobayashi, T.Y.L.Y. et al. (2015a). Mesophilic anaerobic co-digestion of waste activated sludge and *Egeria densa*: performance assessment and kinetic analysis. *Applied Energy* 148: 78–86.

Zhen, G., Lu, X., Li, Y.Y. et al. (2015b). Influence of zero valent scrap iron (ZVSI) supply on methane production from waste activated sludge. *Chemical Engineering Journal* 263: 461–470.

Zhen, G., Lu, X., Kobayashi, T. et al. (2016). Anaerobic co-digestion on improving methane production from mixed microalgae (*Scenedesmus* sp., *Chlorella* sp.) and food waste: kinetic modeling and synergistic impact evaluation. *Chemical Engineering Journal* 299: 332–341.

Zhen, G., Lu, X., Kato, H. et al. (2017). Overview of pretreatment strategies for enhancing sewage sludge disintegration and subsequent anaerobic digestion: current advances, full-scale application and future perspectives. *Renewable and Sustainable Energy Reviews* 69: 559–577.

Zheng, Y., Zhao, J., Xu, F., and Li, Y. (2014). Pretreatment of lignocellulosic biomass for enhanced biogas production. *Progress in Energy and Combustion Science* 42: 35–53.

Zou, L.P., Ma, C.N., Liu, J.Y. et al. (2016). Pretreatment of food waste with high voltage pulse discharge towards methane production enhancement. *Bioresource Technology* 222: 82–88.

17

Hydrogen Production by Algae

Tunc Catal[1] *and Halil Kavakli*[2]

[1] *Department of Molecular Biology and Genetics, Uskudar University, 34662, Istanbul, Turkey*
[2] *Departments of Chemical and Biological Engineering and Molecular Biology and Genetics, Koc University, 34450, Istanbul, Turkey*

CHAPTER MENU

17.1 Importance of Hydrogen Production

Environmental problems such as water shortages, global warming and climate change caused by the combustion of fossil fuels are serious threats to life. This is because the use of fossil fuels, carbon based non-renewable energy, releases greenhouse gases to the atmosphere. Additionally, the fossil fuel reserves are expected to run out in 50 years, which necessitates the generation of alternative energy (Daroch et al. 2013; Praveenkumar et al. 2012; Stuart et al. 2010). Therefore, researchers have put significant efforts into creating alternatives such as energy from sunlight, wind, rain, tides, waves, geothermal heat and biofuels. Among these, biofuels, produced through biological processes using biomass or unicellular organism, are becoming more attractive. Currently, bioethanol and biodiesel are produced on commercial scales from crop plants such as cane, corn, palm, soybean oilseed, and sunflower. The use of crop plants for the production of biofuels raises concerns about the effects on the environment, food security, and prices. Therefore, the production of biofuels from unicellular organisms, and specifically from microalgae, is becoming a popular alternative approach. Since the cultivation of algae does not require fresh water and arable land, its implementation may reduce the dependency on crop plants for the production of biofuels (Ahmad et al. 2011; Chisti 2007; Huang et al. 2010; Lam and Lee 2012). Among the different biofuels, biohydrogen is the more attractive because of its beneficial properties including high gravimetric energy (Argun et al. 2008). In fact, hydrogen has the highest energy content per unit weight of any other known fuel (122 kJ/g) and it possesses 2.75 times more energy density. Currently, the hydrogen produced is mainly used as fertilizer during the production of ammonia and as a key-processing agent

Green Energy to Sustainability: Strategies for Global Industries, First Edition.
Edited by Alain A. Vertès, Nasib Qureshi, Hans P. Blaschek and Hideaki Yukawa.

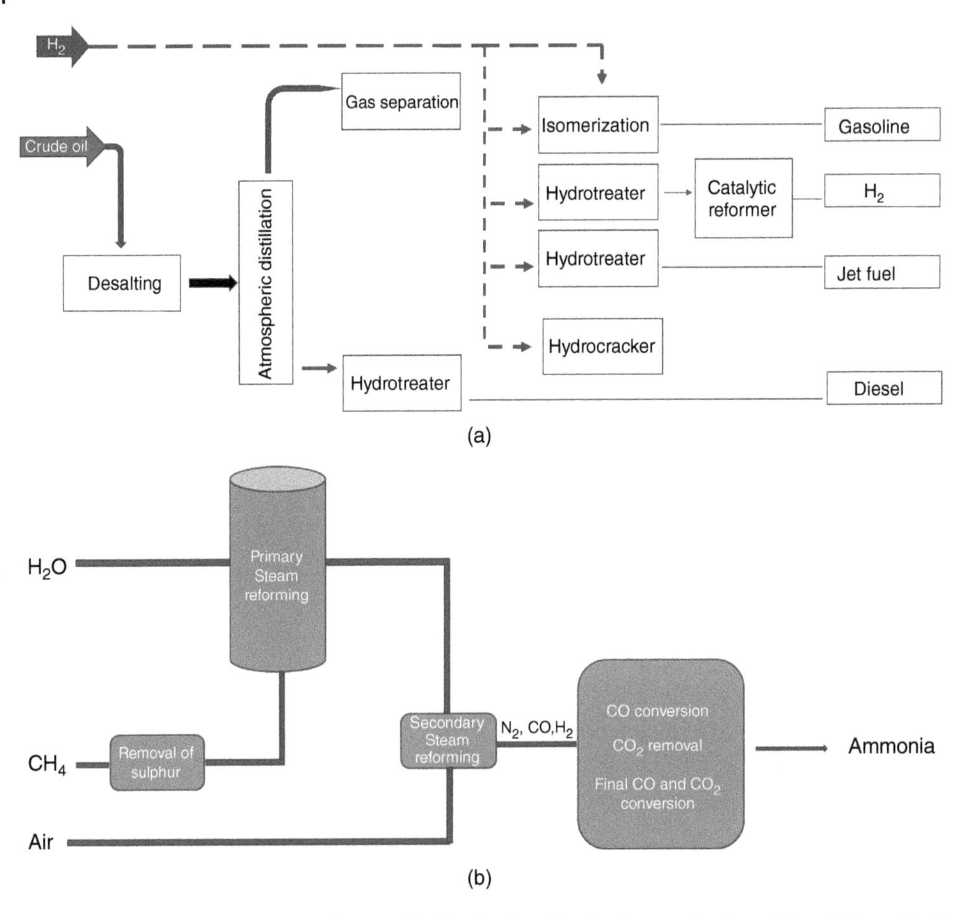

Figure 17.1 Simplified version of H_2 use in (a) oil refineries for the production of the different fuels and (b) production of fertilizer and ammonia from air.

in the petroleum-refining industry. It is also used as hydrotreater, hydrocracker, and isomerization reagent (Figure 17.1a,b). Moreover, hydrogen gas is also used in the chemical, food, transportation, and metal industries (Figure 17.2). Hydrogen also has the potential to be used as fuel because its use result in no production of CO_2 and other toxic gas byproducts (Chang and Lin 2004). In fact, hydrogen is already being used in vehicles, powered by hydrogen fuel cells in many European countries. All this suggests that hydrogen is a clean energy source (Das and Veziroglu 2001; Lay et al. 1999). This indicates that hydrogen is a promising future fuel for cars and other transportation vehicles.

Large amounts of hydrogen are currently produced by steam reforming or oxidation of natural gas and coal gasification and a small quantity is being produced by other routes such as biomass gasification or electrolysis of water (Shobana et al. 2017). Some of the hydrogen is produced as a by-product of naphtha reforming conducted for octane improvement.

As mentioned above, the production of hydrogen from carbon-based non-renewable fossil fuels is not sustainable for the long-term future as they are currently depleting. Therefore, the generation of hydrogen from renewable sources such as production by the fermentation of unicellular organisms is important. Moreover, biological hydrogen production processes have the advantages of being less energy intensive compared with current technologies. Among the various biomass sources used as starting materials for biohydrogen production, algae have attracted particular attention due to

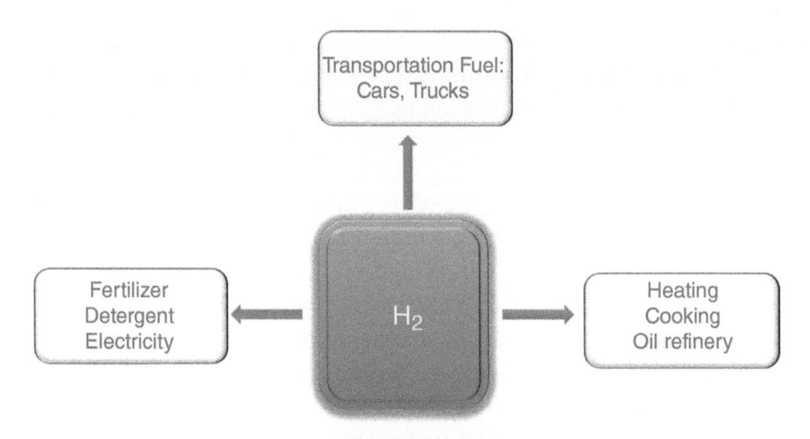

Figure 17.2 The different uses of hydrogen in different industries.

features such as their relatively lower land requirement for cultivation and their remarkable organic matter content (Vardon et al. 2012). Additionally, biohydrogen production from microalgae offers high hydrogen production rates and is capable of converting organic wastes into more valuable energy resources (Zhao and Yu 2008). Furthermore, biohydrogen has the potential to be carbon negative and less energy intensive, as it operates at ambient temperature and pressure (Karthic and Joseph 2012). Although there are obvious environmental advantages to producing biohydrogen through the fermentation of microorganisms the following problems need to be overcome to produce biohydrogen in a cost-effective manner at large scale: the low production rates, low substrate conversion efficiencies, the accumulation of acid-rich intermediate metabolites from the acidogenic process, the lack of cost-efficient photo-bioreactors, and harvesting techniques (Shobana et al. 2017).

17.2 Hydrogen Producing Microorganisms

The traditional synthesis method of hydrogen generation includes chemical conversion of fossil fuels into hydrogen. As an alternative approach, several biological techniques have been examined and possible implementation of biological hydrogen production for commercial purposes is still at the proof of concept stage including microbial electrolysis cell technology and dark fermentation technology. Hydrogen production through dark fermentation from organic materials is one of the most promising technologies for possible commercial hydrogen generation in the future. Biofuels such as starch-derived alcohols, diesel fuel, alkanes, and hydrogen gas can be produced by microorganisms through genetic engineering approaches (Radakovits et al. 2010), and various microorganisms can produce hydrogen gas naturally. For example, diatoms, a major group of algae, can also absorb light and store energy in the form of hydrogen gas (Ghirardi 2006). Hydrogen generating microorganisms such as *Firmicutes*, a phylum of bacteria, enriches among bacterial populations during fermentation (Park et al. 2014). Anaerobic hydrogen fermentation has been investigated using enriched mixed microbial cultures (Sivagurunathan et al. 2016). Dark fermentation hydrogen generation was shown using sludges (Danko et al. 2008). A mixed culture of microorganisms has various advantageous over pure cultures such as easy operation and wide substrate utilization range. However, selecting hydrogen producing microorganisms is a major challenge in this technology (Rafieenia et al. 2017).

Among a diverse array of photosynthetic microorganisms able to produce hydrogen, the green algae *Chlamydomonas reinhardtii* is the model organism and is widely used to study hydrogen production. Despite the well-known fact that the acetate-containing medium enhances hydrogen production in this algae, little is known about the precise role of acetate during this process (Jurado-Oller et al. 2015). Various approaches including biochemical and genetic engineering of algae have been proposed to improve the technological limitations of photosynthetic systems (Srirangan et al. 2011). The combination of photosynthetic microorganisms such as algae/cyanobacteria in natural biofilms has been proposed as a novel bioenergy technology using wastewater to decrease the processing cost of wastewater treatment through hydrogen production (Miranda et al. 2017). Among all eukaryotic organisms, green algae are key organisms, having both an oxygenic photosynthetic metabolism with a hydrogen metabolism. Both hydrogen metabolism and photosynthetic metabolism are closely related to each other. Some genes including the *hydA* gene encoding a type of [Fe]-hydrogenase are induced under anaerobic conditions. Electrons generated through fermentation join the electron transport chain in photosystems through the plastoquinone pool transferring to the hydrogenase helping the algae to live under an anaerobic environment (Happe et al. 2002). This feature of splitting water producing oxygen and hydrogen can be combined with fuel cell technology for electricity generation, and even used as feedstock for more complex fuel production (Murthy and Ghirardi 2013). (http://www.algaeindustrymagazine.com/developing-efficient-hydrogen-production-algae). In this respect, algae have great potential to achieve commercial hydrogen generation.

17.3 Hydrogen Producing Algae (Macro–Micro) Species

Various hydrogen producing algae including macro and micro species have been reported. Twenty species of macroalgae have been studied and fermentation parameters such as methane production have been examined (Machado et al. 2014). Cyanobacterial hydrogen production through light-based photo-physiology shows the importance of biophotolysis which may help to develop a sustainable hydrogen gas production method (Melnicki et al. 2012). *C. reinhardtii* is one of the representative algal species having the ability to generate hydrogen gas via biophotolysis of water. Hydrogen generation occurs under anoxic conditions, which may be imposed metabolically by depriving the algae of sulfur (Tamburic et al. 2012). The green algae can be co-cultured with some other bacteria to improve hydrogen production. In a study, improvement in hydrogen production was achieved by co-culturing *C. reinhardtii* cc849 and *Azotobacter chroococcum* through enhanced hydrogenase activity (Xu et al. 2017). *C. reinhardtii*, a unicellular green algae, produce hydrogen gas under anaerobic conditions, and enhanced hydrogen production ability can be improved without mutations, but not in the presence of sufficient sulfur, oxygen, and moderate light conditions (Yagi et al. 2016). Previously, a potential indicator, cellular nicotinamide adenine dinucleotide phosphate (NADPH) fluorescence, was investigated for biohydrogen production by *C. reinhardtii* (White et al. 2014).

Macroalgae can also be used for fermentative hydrogen production by anaerobic microorganisms. *Laminaria japonica*, a type of macroalgal species, can produce hydrogen gas with mixed cultures, and different biomass pretreatment methods were examined for improved efficiency of saccharification and hydrogen production (Liu and Wang 2014). Bio-gas was produced from *L. japonica* through co-fermentation using biomass hydrolysis and bio-gas production. By using a co-culture of *Clostridium butyricum* and *Erwinia tasmaniensis*, bio-gas generation containing hydrogen and methane was significantly enhanced (Lee et al. 2012). Marine algae, especially *L.*

japonica, can be used for fermentative hydrogen production (Jung et al. 2011). Green algal species utilize hydrogenases to oxidize and/or generate hydrogen gas under anaerobic conditions, and anaerobic metabolism of *C. reinhardtii* has been compared with other strains such as *Chlamydomonas moewusii* and a *Lobochlamys culleus* to understand features of algal hydrogenase activity (Meuser et al. 2009).

Scenedesmus obliquus produces high yield of hydrogen with continuous electron supply under anaerobic condition showing a possibility for industrial application in future (Papazi et al. 2014). Generally, sulfur deprivation is necessary to generate hydrogen under anaerobic conditions, and natural isolates of *C. reinhardtii*, *Chlamydomonas noctigama*, and *Chlamydomonas euryale* are able to generate significant amounts of hydrogen gas (under sulfur-deprivation) (Skjånes et al. 2008). Hydrogen production by algae can be affected not only by sulfur, but also other elements such as phosphorus. For example, sustainable hydrogen gas generation can be achieved under phosphorus-deprived conditions (Batyrova et al. 2012). Macro algae such as *Laminaria digitata* can be used in combination with micro algae, *Chlorella pyrenoidosa* and *Nannochloropsis oceanica* to the optimize carbon/nitrogen (C/N) ratio, since the C/N ratio significantly affects hydrogen production (Ding et al. 2016).

Studies show that [Fe]-hydrogenases and hydrogen metabolism are distributed in green algal species. As pointed out earlier, *C. reinhardtii* generates a significant amount of hydrogen gas in the sulfur-deprive conditions. The photosynthetic oxygen formation rate of *C. reinhardtii* under anoxic conditions decreases below the rate of respiratory oxygen utilization because of a reversible inhibition of photosystem II causing an intracellular anaerobiosis (Winkler et al. 2002). A mixed culture of microalgae from natural isolates (predominantly composed of *Scenedesmus* followed by *Chlorella* species) also demonstrated a potential for hydrogen production (Kumar et al. 2016). Investigators have been looking for novel isolates of microalgae for improved hydrogen production (Figure 17.3).

Hydrogen production by algae can be affected by environmental conditions. Various external factors including the inoculum–substrate ratio, volatile fatty acids and nicotinamide adenine dinucleotide also affects hydrogen production. When the inoculum–substrate ratio decreased, hydrogen generation enhanced by *Chlorella* sp. under anaerobic fermentation conditions (Sun et al. 2011). The effects of external factors such as light/dark conditions on hydrogen generation using

Figure 17.3 A microalgae isolate grown in BG-11 Medium (Catal Laboratory).

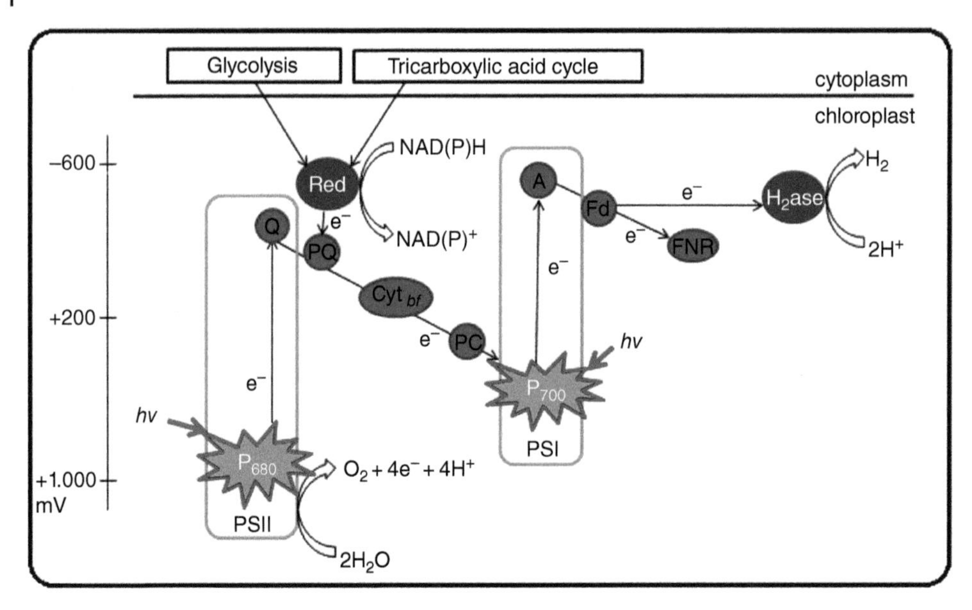

Figure 17.4 Electron transport pathways in green algae. P680, Reaction centre of photosystem II (PS II); P700, reaction centre of photosystem I (PS I); Q, primary electron acceptor of PS II; A, primary electron acceptor of PSI; PQ, plastoquinone; Cyt, cytochrome; PC, plastocyanin; Fd, ferredoxin; Red, NAD(P)H oxido-reductase; H_2ase, hydrogenase; FNR, ferredoxin-NADP+ reductase. Source: Modified from Melis and Happe (2001).

green algae have been examined (Guo et al. 2017). Figure 17.4 shows hydrogenase-related electron transport pathways in green algae. Photosystem II is the original place for electron formation derived from photooxidation of water. Oxidized cellular substrates in the plastoquinone pool may also lead to electron formation. Later on, these electrons are transferred to ferrodoxin through photosystem I, and ferrodoxin transfer electrons to Fe hydrogenase.

Photosynthetic microorganisms show diverse metabolic abilities, and high growth profiles; genetic engineering approaches focus on manipulating the central carbon metabolism (Radakovits et al. 2010). Single-celled microalgae are able to live in both aerobic and anaerobic conditions producing hydrogen gas. The formation of hydrogen is induced by sulfate deprivation that blocks photosystem-II and oxygen in chloroplasts. Under anoxic conditions, algae run anaerobic photosynthetic metabolism to generate ATP evolving hydrogen gas (Melis 2007).

Other algal species with a low C/N ratios such as *Arthrospira platensis* can be used in the fermentation process to generate hydrogen gas after biomass dilute acid pre-treatment for utilization by fermentative microorganisms (Xia et al. 2016). Algal-based hydrogen generation using a gasifier such as a fluidized bed reactor has been reported (Azadi et al. 2014). A three-stage process dark hydrogen fermentation with acid-domesticated hydrogenogens, photohydrogen fermentation, and methanogenesis shows efficient utilization of algal biomass (Cheng et al. 2014). After acid-pretreatment of algal biomass such as *L. digitata,* bioremediation of toxic substances such as chromium, can be achieved (Dittert et al. 2014). As an alternative method to acidic pretreatment, anaerobic digestion of algal biomass can be used for the production of hydrogen gas (Yan et al. 2010). Hydrogen production by *C. reinhardtii*, and photo-electrochemical oxidation has been investigated using hybrid systems (Chatzitakis et al. 2013). Photoproduction of hydrogen gas using *C. reinhardtii* in photobioreactors has been demonstrated (Kosourov et al. 2012). In the photo-fermentation using microorganisms, hydrogen is generated from organic acids, and various parameters such as pH, the type of carbon and the nitrogen sources are important factors for the process (Srikanth et al. 2009). Hydrogenase encoded by the *hyd*A gene in *C. butyricum* is the enzyme responsible for hydrogen generation in dark fermentation (Wang et al. 2008). During wastewater treatment, biomass production, CO_2 sequestration and electricity production can be accomplished using a *C. pyrenoidosa* algal biocathode in microbial carbon capture cells (Jadhav et al. 2017). Different algal species such as *Golenkinia* sp. SDEC-16, *Chlorella vulgaris*, *Selenastrum capricornutum*, *Scenedesmus* SDEC-8 and *Scenedesmus* SDEC-13 were examined for possible energy generation (Hou et al. 2016).

17.4 Production of Biohydrogen Through Fermentation

17.4.1 Biohydrogen Production

Production of biohydrogen using either oxygen-evolving photosynthesis or bacterial fermentation are the most advanced biological H_2 production approaches. Photosynthetic organisms use solar energy to split water into protons (H^+), electrons (e^-) and O_2. Then, either nitrogenase or hydrogenase enzymes convert H^+ and e^- into H_2 (Melis and Happe 2001; Oey et al. 2016; Sharma and Arya 2017). The biohydrogen is produced by two different mechanisms namely light dependent (direct or indirect biophotolysis) and light independent, also known as dark fermentation (Das and Veziroglu 2001; Levin et al. 2006; Oey et al. 2016). Direct biophotolysis, which is observed in microalgae (Volgusheva et al. 2013), requires light to generate O_2 and H_2 from H_2O which is shown in reaction below (Ghirardi et al. 2000; Melis and Happe 2001; Melis et al. 2000) (Figure 17.4).

$$2H_2O + h\nu \rightarrow O_2 + 2H_2$$

Studies with different unicellular organisms have reported that both microalgae and cyanobacteria are able to produce biohydrogen with a light-dependent mechanism (Akkerman et al. 2002; Mudhoo et al. 2011) using either hydrogenase or nitrogenase depending on the species (Bothe et al. 2008; Rogner 2013; Sakurai and Masukawa 2007; Skizim et al. 2012; Srirangan et al. 2011). Although this mechanism seems, at first, to be efficient for the direct conversion of H_2O into H_2, sensitivity of hydrogenase to O_2 inhibition prevents H_2 production at large scales. It is, therefore, required to have both modified cells and production system to obtain a high yield of biohydrogen.

In the indirect biophotolysis mechanism, sunlight energy is first converted into a carbohydrate like starch and then the carbohydrate is utilized for H_2 production (Figure 17.5). The overall reaction of hydrogen production can be given as:

$$C_6H_{12}O_6 + 6H_2O + h\nu \rightarrow 12H_2 + 6CO_2$$

Dark fermentation is a process that requires conversion of organic substrate to biohydrogen by fermentation using a microorganism. This approach is preferred because of the high biohydrogen production efficiency (Ljunggren and Zacchi 2010; Ritimann et al. 2015; Singh and Wahid 2015; van Niel 2016). This approach has not only been studied using algae but also with prokaryotic cells. Many studies with bacteria under anaerobic conditions indicate that biohydrogen production is relatively high compared with other photosynthetic organisms (Duo et al. 2016; Jiang et al. 2014; Shin et al. 2007; Yoshida et al. 2005). Microbial hydrogen production is driven by the heterotrophic mechanism of pyruvate breakdown, formed during the catabolism of various substrates. At the initial stage, glucose is breakdown into the pyruvate and NADH by a pathway known as glycolysis. Pyruvate is then converted into acetyl-CoA catalysed by one of two enzyme systems, which are formate lyase and ferredoxin oxidoreductase (FOR) depending on the type of organism with

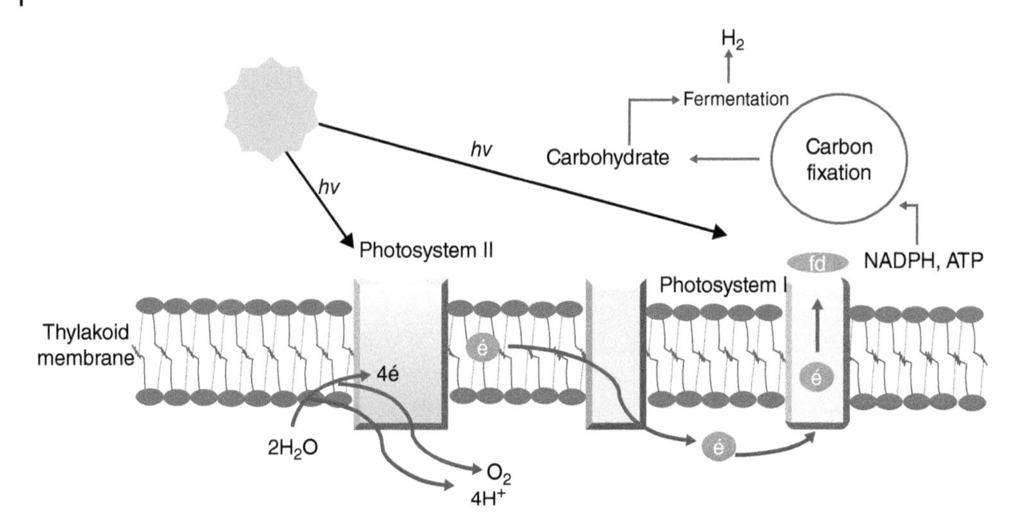

Figure 17.5 Cartoon showing the mechanism of biohydrogen production in indirect photolysis. The process consists of two stages: first, light is utilized for the production of carbohydrate and then accumulated carbohydrate is used for hydrogen production.

following reactions (Akobi et al. 2017; Hallenbeck 2009; Karthic and Joseph 2012; Manish and Banerjee, 2008):

$$\text{pyruvate} + \text{CoA} + 2\text{Fd(ox)} \rightarrow \text{Acetyl} - \text{CoA} + \text{CO}_2 + 2\text{Fd(red)}$$

$$\text{pyruvate} + \text{CoA} \rightarrow \text{Acetyl} - \text{CoA} + \text{formate}$$

Then oxidation of the reduced FOR and NADH Fe–Fe dehydrogenase result in biohydrogen production with following reactions:

$$2\text{H}^+ + \text{Fd(red)} \rightarrow \text{H}_2 + \text{Fd(ox)}$$

$$\text{NADH} + \text{H}^+ \rightarrow \text{H}_2 + \text{NAD}^+$$

Formate is converted into CO_2 and H_2 by the action of either the hydrogen lyase or Ni–Fe dehydrogenase. Dark fermentation processes are carried out using strict or facultative anaerobes (*Clostridium* or *Enterobacter* species) for the biohydrogen production, respectively. Several studies using different types of bioreactor have been reported for the production of biohydrogen using various substrates such as glucose, sucrose, oat straw, dairy waste, fructose, and food waste (Arriaga et al. 2011; Li et al. 2006; Oh et al. 2004; Pasupuleti et al. 2014; Perna et al. 2013; Wu et al. 2007).

17.4.2 Fermentation System for Hydrogen Production

Several photobioreactors have been designed, not only for biohydrogen production, but also for the production of other valuable compounds like pigments for microalgae farming. The selection and design of bioreactor is mainly dependent on the type of microalgae and the method of biohydrogen harvesting (Shobana et al. 2017). At laboratory scale different bioreactors such as a flask, vertical tubular, tubular coiled, stirred tank, bubble column, and flat panel, are used to grow bacteria and microalgae. Each type has advantages and disadvantages in terms of cultivation of the microalgae and harvesting biohydrogen from the system (Skjånes et al. 2016). Harvesting biohydrogen is critical during the fermentation because its accumulation causes inhibition of hydrogen production and instead enhances byproduct synthesis like ethanol and lactate. Therefore, different approaches, mainly membrane technology, have been employed to remove biohydrogen continuously from the system. Different kinds of membranes, non-porous and porous, can be used under the conditions where biohydrogen formation takes place (Bakonyi et al. 2013).

17.5 Technologies (Solar Algae Fuel Cell/Microbial Fuel Cell)

Several technologies including biofuel cells have been proposed in the literature. Biofuel cells employed with algae may offer a cost effective alternative for enhanced productivity (Severes et al. 2017). Hydrogen generation using photosynthetic systems such as manganese/semiconductor catalytic system have been proposed (Hou 2017). Anaerobic digestion technology is also attractive for hydrogen production together with another important energy carrier (methane gas) (Nathoa et al. 2014). The other alternative approach including microbial electrolytic cells (MECs) are novel bioelectrochemical tools to generate hydrogen from renewable sources (Catal et al. 2015, 2017; Catal 2016). The main structural components of MECs are similar to microbial fuel cells (MFCs) which are used to produce electricity through microbial activity from organic wastes and other substrates (Catal et al. 2010, 2011a,b) including lignocellulosic materials (Catal et al. 2008a,b,c, 2018a). Microbial dynamics affect the conversion of wastes or toxic compounds into more environmentally friendly chemicals (Kumru et al. 2012; Bermek et al. 2014). The use of microorganisms and microbial-based technologies for energy production can also help bioremediation of toxic wastes such as heavy metal contamination and sensing of several environmental contaminants (Catal et al. 2009, 2018b, 2019; Abourached et al. 2014; Ozdemir et al. 2019). In this respect, other methods of alternative energy production by microorganisms including hydrogen production by algae may affect not only cost-effective renewable energy production but also have a great impact on the environment. In recent years, hydrogen generation by algae has been studied extensively. Because of their photosynthetic activities, algae consume cheap inorganic compounds for hydrogen production from sunlight. In order to achieve comparable hydrogen production with traditional chemical techniques, the efficiency of algal system-based production techniques must be improved. Microalgae, *C. pyrenoidosa* was successfully introduced into MFCs to generate sustainable electricity (proton leak promoting agents increase the current generation) (Xu et al. 2015). *C. vulgaris* was also used for algal biocathode MFCs for oxygenation for cathodic reaction (Wu et al. 2013).

Hydrogenase enzymes are available in both algae and cyanobacteria. However, each microorganism contains one of two major types of hydrogenase [FeFe] or [NiFe] enzymes showing the same catalytic function (Ghirardi et al. 2007). Hydrogen gas production through light from water may provide an efficient way to generate biofuel in microalgae. HydA which couples to electron transport chains in photosystems eliminate energy waste related to carbohydrate generation. Oey et al. (2013) suggested that the Stm6Glc4L01 strain of *C. reinhardtii* may help to advance the efficiency of photobiological hydrogen generation (Oey et al. 2013). Gadhamshetty et al. (2013) developed two-chamber MFCs using *Laminaria saccharina* for algal-based electron donor, and mixed culture as a catalyst in the anode compartment (Gadhamshetty et al. 2013). Several microorganisms, for example cyanobacteria are capable of generating hydrogen from water, light, and organic compounds. Technologies which focus on hydrogen generation have two major limitations including low yield and high process cost. Different approaches such as reactor modifications and genetic

Figure 17.6 A proposed model of a two chamber self-sustainable microbial fuel cell containing microalgae *Chlorella vulgaris* grown in Bristol Medium (Catal Laboratory).

manipulations have been suggested to overcome this challenge (Gupta et al. 2013). Figure 17.6 shows a novel approach for a self-sustainable MFC model.

Algal hydrogen can be used to enrich natural gas, and *C. reinhardtii* was examined for hydrogen production in the presence of one natural gas component, methane (Antal and Lindblad 2005).

17.6 Possibility of Commercial Production of Hydrogen

Hydrogen fuel is a promising renewable energy source and has potential to be used to generate electricity and for use as fuel in automobiles. These attributes make hydrogen a more sustainable and cleaner energy source. This eliminates the release of emissions of CO_2 into the atmosphere and has potential to generate energy. Development of cost-effective hydrogen processes is required for the success of hydrogen-powered vehicles. Three main processes namely thermal, electrolytic, and photolytic, are currently used for the production of hydrogen. Each main group is further subdivided according to their production technologies (Figure 17.7). Thermochemical processes require heat and chemical reactions to generate hydrogen using organic materials like natural gas, coal, or biomass.

Most hydrogen production is currently achieved by a steam-methane process using two different approaches. First, natural gas (methane) reacts with steam under high pressure and temperature in the presence of an appropriate catalyst to produce hydrogen. Subsequently a water–gas shift reaction is carried out to eliminate the byproduct known as CO, to further purify the hydrogen. The second approach is called partial oxidation, in which natural gas reacts with a limited amount of oxygen to generate hydrogen and CO. To convert CO into CO_2 and enhance hydrogen production, a water–gas shift reaction is performed. Although partial oxidation is much faster and requires small reactors, the amount of hydrogen is quite high in the steam reforming approach. Coal is another starting material for hydrogen production using the thermal process. Coal is converted to hydrogen by gasification. This reaction is carried out under high temperature and pressure, which yields not only hydrogen but also CO and other chemicals. Then hydrogen is further enriched and purified by a water–gas shift reaction. Another thermal process method to obtain hydrogen is to convert biomass to hydrogen by gasification. Hydrogen production from biomass such as crop plants is generated in a fluidized bed reactor by gasification. The gasification reaction is carried out at high temperature in the presence of limited oxygen without combustion. Then hydrogen is purified with a water–gas shift reaction to eliminate CO.

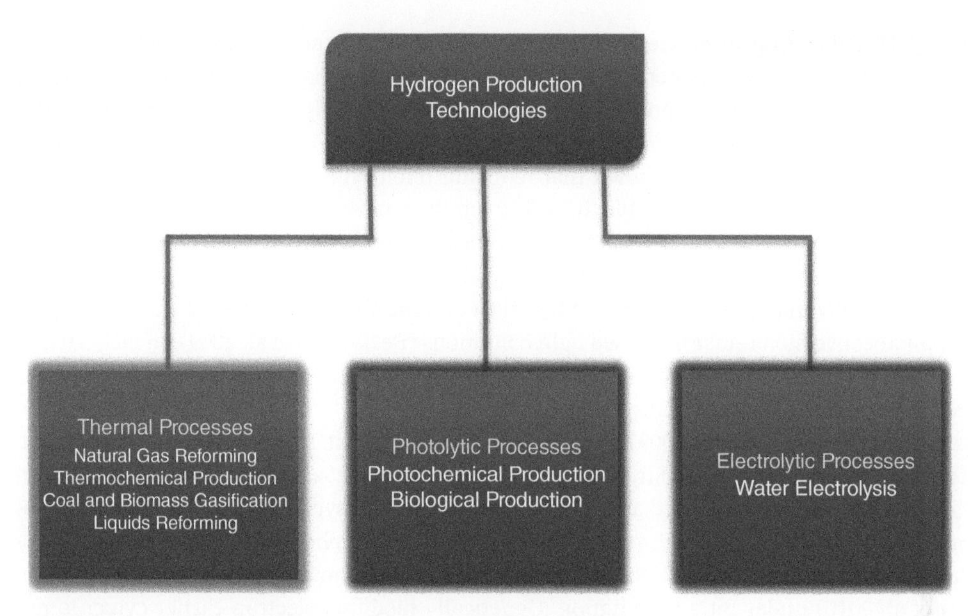

Figure 17.7 Current technologies used for hydrogen production.

Electrolytic processing is an emerging technology that involves direct splitting of water molecules into hydrogen and oxygen by using electricity. Such a reaction occurs in a unit named an electrolyzer. This is an environmentally friendly approach, in which there is no release of CO, SO, or CO_2. An electrolyzer is similar to a fuel cell, consisting of a membrane, cathode, and anodes. The anode and cathode are separated by a membrane and the cathode reacts with water to generate hydrogen ions. Then electrons come from the anode, pass through the membrane and react with hydrogen to form H_2. Like the electrolytic process, photolytic processes are emerging technologies, where hydrogen is generated using solar light. Two processes, the photoelectrochemical and photobiological methods, are being investigated. In the photoelectrochemical process sunlight and semiconductors are directly used to split water into hydrogen and oxygen. In the latter case, hydrogen is produced by microalgae and cyanobacteria using sunlight and water. This approach has great potential for sustainable hydrogen production with almost no environmental impact. In recent years research has been focused to overcome some of the challenges in hydrogen production from microalgae and cyanobacteria at a commercial scale. A number of major challenges need to be overcome. The first is to have nitrogenase or hydrogenase with improved activity, which enables cells to produce more hydrogen. Second, producing strains that have altered capacity of photosynthesis. Finally, designing biophotoreactors that enable cultivation of organisms at large scale for the collection of the hydrogen.

Studies have been carried out to enhance biohydrogen production from both microalgae and cyanobacteria by increasing the capacity of photosynthesis of organisms, by altering the growth media of the organisms, and by using recombinant DNA technology to generate mutants. This work is summarized below. It is possible to enhance hydrogen production to certain degree, but it is still far from possible to generate an organism that can be utilized for biohydrogen production at large scale. Two different types of hydrogenase are required that are based on their cofactors ([FeFe] hydrogenases and [NiFe] hydrogenases), which should be present in microalgae and in cyanobacteria, respectively.

Although [FeFe] hydrogenases are able to couple to the photosynthetic electron transport chain, they generate H_2 directly from water oxidation. The [FeFe] hydrogenases characterized to date are particularly O_2 sensitive, and H_2 photoproduction is only transiently observed prior to the accumulation of O_2 to inhibitory levels under nutrient-depletion conditions in many green microalgae (Cohen et al. 2005). Studies have shown that reducing photosynthetic activity in *C. reinhardtii* under limited nutrient conditions enable the cells to produce more biohydrogen (Kruse et al. 2005; Melis et al. 2000; Volgusheva et al. 2015). Recombinant DNA technology has been used to alter various components of the photosynthetic apparatus to allow algae to absorb more light and, in turn, produce more biohydrogen. For example, PSII of the *C. reinhardtii* is engineered to a small size for efficient photoconversion under increased light conditions (Beckmann et al. 2009). In fact, a transgenic organism was achieved where the repressor of the light harvesting complex II (LHCII) NAB1 is overexpressed. Analysis of the transgenic organism revealed that it showed enhanced biohydrogen production (Beckmann et al. 2009). Analysis of the mutant strain L159I-N230Y of *C. reinhardtii* showed the importance of LHCII in hydrogen production (Scoma et al. 2012). This study also identified the importance of the reduced antenna size and high photosynthesis rates in hydrogen production. In another study, *Rhodobacter sphaeroides* without a peripheral light-harvesting antenna complex deleted by recombinant DNA technology exhibited enhanced biohydrogen production (Eltsova et al. 2010) under nitrate limitation. Genetically modified electron carrier cytochromes lacking the cyt *cbb3* oxidase or the quinol oxidase of *Rhodobacter capsulatus* were generated. Analysis of these various *R. capsulatus* mutants revealed that lack of *cbb3* exhibited elevated biohydrogen production compared with wild type cells (Ozturk et al. 2006). Interestingly, a study with *Nostoc* PCC 7120 using genetic modification tools highlighted important genes in homocitrate in biohydrogen production (Sakurai et al. 2007). It has been shown that disruption of the genes that encode homocitrate synthase, but not nitrogenase, result in prolonged biohydrogen production. Similar approaches are also used in photosynthetic bacteria that result in elevated biohydrogen production, which is reviewed in (Kars and Gunduz 2010). Although these studies advanced our understanding of H_2 metabolism in microalgae, new strains of microalgae need to be developed for sustainable hydrogen production at large scale.

For laboratory scale there are several photobioreactors that have been designed to provide appropriate environmental conditions for biohydrogen production but further improvements in their design are required to be able to produce cost-effective biohydrogen at large scale. Additionally, harvesting of biohydrogen during the algae cultivation still needs to be developed. Designs of the photobioreactor for biohydrogen production depend on the culture conditions and the desired amount of biomass production (Skjånes et al. 2016). All photobioreactors require light input, which comes from either artificial light or sunlight. For an efficient photobioreactor, cells are expected to receive equal amounts of light and an efficient mixing system. There are three different types of photobioreactors that are used for biohydrogen production from microalgae and these are column, tubular, and flat panel. One of the key parameters that should be controlled in all these photobioreactors is to control leakage of biohydrogen produced during cell growth. Therefore, all units should be well sealed. Different tubular photobioreactors are currently used for the production of biohydrogen from algae. The disadvantages of horizontal photobioreactors include an inefficient gas exchange and the requirement for high energy input and sometimes accumulation of biomass in the tubes (Slegers et al. 2013). The flat panel photobioreactor is the one most often used for the production of algae cultivation not only at laboratory scale but also outdoors. The number of parameters, such as light capture efficiency, agitation, and heat exchange, are easily controlled in this system. A study showed that flat panel photobioreactor is more suitable than tubular photobioreactor for biohydrogen production because of the low backpressure generated by hydrogen in the system (Oncel and Kose 2014). Although advances have been made in recent years in the design and construction of photobioreactors for the growth of microalgae, extensive research is still required to discover a suitable photobioreactor to cultivate microalgae for biohydrogen production.

17.7 Perspectives and Future Implications of Algae in Biotechnology

Algae constitute a diverse group of photosynthetic organisms that exist both in fresh and sea water. The microscopic form, so called microalgae, can be found almost everywhere including in plants, for example water springs, ice, damp soils, waste-water, and tree trunks. Algae are the main oxygen producers and the source of the biomass. Moreover, algae are an alternative source for the production of renewable energy (Pittman et al. 2011). Photosynthetic algae use energy from the sun to grow and form biomass that contains proteins, lipids, hydrocarbons, carbohydrates, small molecules and pigments. They also use nitrogen, phosphate, and CO_2 to grow biomass. This makes microalgae attractive for CO_2 mitigation and as bioremediation agents to reduce toxic chemicals and pollutants from the environment. Furthermore, algae biomass has various applications as: food for human and animals, fertilizers to recover soil pH, source of different biomolecules, sewage disposal, in medicine (Correia-da-Silva et al. 2017; Ozakman et al. 2018; Sivakumar et al. 2012; Trentacoste et al. 2015; Wang et al. 2014; Wells et al. 2017) (Figure 17.8).

Research and development on algal biomass cultivation have been increasing for more than 15 years using government research programs, and in company projects. This financial support is mainly used for developing cost-effective methods for the production of renewable biofuels like biodiesel and biohydrogen through the cultivation of microalgae. Biohydrogen is one of the most interesting energy carrier molecules because of its potential to meet energy demands and its renewable nature (Anisha and John 2014). Hydrogen gas has received much attention due to its CO_2-free nature and that it can be obtained from sustainable sources such as photovoltaic (PV), wind, wave, and biological methods to run hydrogen fuel cells (Oey et al. 2016). There is currently substantial interest in utilizing eukaryotic algae for the renewable production of several bioenergy carriers, including starch for alcohols, lipids for diesel fuel, and H_2 for fuel cells. Relative to terrestrial biofuel feedstocks, algae can convert solar energy into fuels at higher photosynthetic efficiencies and can thrive in salt water systems. Recently, there has been considerable progress in identifying relevant bioenergy genes and pathways in microalgae, and powerful genetic techniques have been developed to engineer some strains via the targeted disruption of endogenous genes and/or transgene expression. Collectively, the progress that has been realized in these areas is rapidly advancing our ability to genetically optimize the production of targeted biofuels (Beer et al. 2009), for example, in a previous study, overexpressed *hydA* enhanced H_2 production in *Chlorella* sp. Green algae are able to convert solar energy into biological hydrogen (solar-to-H_2) from H^+ via chloroplast hydrogenase (HydA), which can accept electrons directly from ferredoxin and generates H_2. HydA is nuclear-encoded by the *hydA* gene, and its transcription and activity can only be observed under anaerobic or sulfur-deprived conditions, so various external factors/inhibitors can be examined to understand the limiting part of the system. In future, novel isolates of algae along with an enhanced genetic ability to produce more hydrogen will be a possible way for economical production of this valuable energy carrier.

As shown in Figure 17.8 the production of any product from microalgae consists of different phases: growth and cultivation of algae and extraction of the desired products. Therefore, research is focused on the design of the cost-effective photobioreactors that enhance growth of the

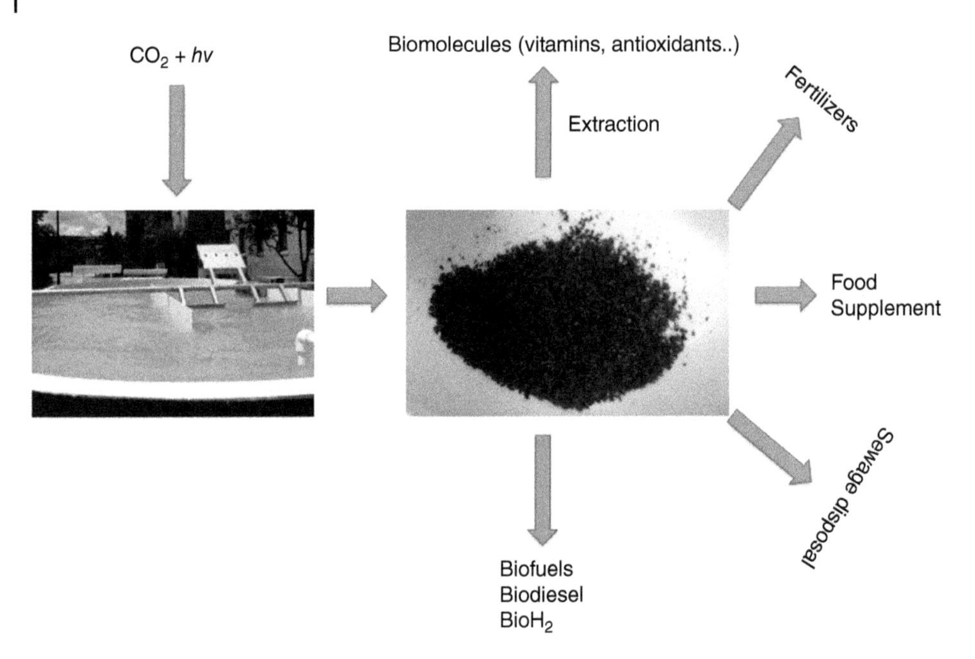

Figure 17.8 Algae capture carbon dioxide and convert into biomass using photons that come from the sun. Biomass can be potentially be used in a number of different industries.

microorganisms and strain improvement for the enhancement of the desired product. Although the production of cost-competitive biofuels from microalgae compared with fossil fuels is still far away, studies are promising and may offer new strains and cost effective photobioreactors for the production of biofuels from algae.

References

Abourached, C., Catal, T., and Liu, H. (2014). Efficacy of single-chamber microbial fuel cells for removal of cadmium and zinc with simultaneous electricity production. *Water Research* 51: 228–233.

Ahmad, A.L., Yasin, N.H.M., Derek, C.J.C., and Lim, J.K. (2011). Microalgae as a sustainable energy source for biodiesel production: a review. *Renewable and Sustainable Energy Reviews* 15 (1): 584–593.

Akkerman, I., Janssen, M., Rocha, J., and Wijffels, R.H. (2002). Photobiological hydrogen production: photochemical efficiency and bioreactor design. *International Journal of Hydrogen Energy* 27 (11–12): 1195–1208.

Akobi, C., Hafez, H., and Nakhla, G. (2017). Impact of furfural on biological hydrogen production kinetics from synthetic lignocellulosic hydrolysate using mesophilic and thermophilic mixed cultures. *International Journal of Hydrogen Energy* 42 (17): 12159–12172.

Anisha, G.S. and John, R.P. (2014). 9 – Bio-engineering algae as a source of hydrogen. *Advances in Hydrogen Production, Storage and Distribution*, 248–262. doi: https://doi.org/10.1533/9780857097736.2.248.

Antal, T.K. and Lindblad, P. (2005). Production of H2 by sulphur-deprived cells of the unicellular cyanobacteria Gloeocapsa alpicola and Synechocystis sp. PCC 6803 during dark incubation with methane or at various extracellular pH. *Journal of Applied Microbiology* 98 (1): 114–120.

Argun, H., Kargi, F., Kapdan, F.K., and Oztekin, R. (2008). Biohydrogen production by dark fermentation of wheat powder solution: effects of C/N and C/P ratio on hydrogen yield and formation rate. *International Journal of Hydrogen Energy* 33 (7): 1813–1819.

Arriaga, S., Rosas, I., Alatriste-Mondragon, F., and Razo-Flores, E. (2011). Continuous production of hydrogen from oat straw hydrolysate in a biotrickling filter. *International Journal of Hydrogen Energy* 36 (5): 3442–3449.

Azadi, P., Brownbridge, G.P.E., Mosbach, S. et al. (2014). Production of biorenewable hydrogen and syngas via algae gasification: a sensitivity analysis. *Energy Procedia* 61: 2767–2770.

Bakonyi, P., Nemestothy, N., and Belafi-Bako, K. (2013). Biohydrogen purification by membranes: an overview on the operational conditions affecting the performance of non-porous, polymeric and ionic liquid based gas separation membranes. *International Journal of Hydrogen Energy* 38 (23): 9673–9687.

Batyrova, K.A., Tsygankov, A.A., and Kosourov, S.N. (2012). Sustained hydrogen photoproduction by phosphorus-deprived *Chlamydomonas reinhardtii* cultures. *International Journal of Hydrogen Energy* 37 (10): 8834–8839.

Beckmann, J., Lehr, F., Finazzi, G. et al. (2009). Improvement of light to biomass conversion by de-regulation of light-harvesting protein translation in *Chlamydomonas reinhardtii*. *Journal of Biotechnology* 142 (1): 70–77.

Beer, L.L., Boyd, E.S., Peters, J.W., and Posewitz, M.C. (2009). Engineering algae for biohydrogen and biofuel production. *Current Opinion in Biotechnology* 20: 264–271.

Bermek, H., Catal, T., Akan, S.S. et al. (2014). Olive mill wastewater treatment in microbial fuel cells. *World Journal of Microbiology and Biotechnology* 30 (4): 1177–1185.

Bothe, H., Winkelmann, S., and Boison, G. (2008). Maximizing hydrogen production by cyanobacteria. *Zeitschrift für Naturforschung. Section C* 63 (3–4): 226–232.

Catal, T. (2016). Comparison of various carbohydrates for hydrogen production in microbial electrolysis cells. *Biotechnology & Biotechnological Equipment* 30 (1): 75–80.

Catal, T., Li, K., Bermek, H., and Liu, H. (2008a). Electricity production from twelve monosaccharides using microbial fuel cells. *Journal of Power Sources* 175 (1): 196–200.

Catal, T., Xu, S., Li, K. et al. (2008b). Electricity generation from polyalcohols in single-chamber microbial fuel cells. *Biosensors and Bioelectronics* 24 (4): 849–854.

Catal, T., Fan, Y., Li, K. et al. (2008c). Effects of furan derivatives and phenolic compounds on electricity generation in microbial fuel cells. *Journal of Power Sources* 180 (1): 162–166.

Catal, T., Bermek, H., and Liu, H. (2009). Removal of selenite from wastewater using microbial fuel cells. *Biotechnology Letters* 31 (8): 1211–1216.

Catal, T., Kavanagh, P., O'Flaherty, V., and Leech, D. (2010). Generation of electricity in microbial fuel cells at sub-ambient temperatures. *Journal of Power Sources* 196 (5): 2676–2681.

Catal, T., Cysneiros, D., O'Flaherty, V., and Leech, D. (2011a). Electricity generation in single-chamber microbial fuel cells using a carbon source sampled from anaerobic reactors utilizing grass silage. *Bioresource Technology* 102 (1): 404–410.

Catal, T., Fan, Y., Li, K. et al. (2011b). Utilization of mixed monosaccharides for power generation in microbial fuel cells. *Journal of Chemical Technology and Biotechnology* 86 (4): 570–574.

Catal, T., Lesnik, K.L., and Liu, H. (2015). Suppression of methanogenesis for hydrogen production in single-chamber microbial electrolysis cells using various antibiotics. *Bioresource Technology* 187: 77–83.

Catal, T., Gover, T., Yaman, B. et al. (2017). Hydrogen production profiles using furans in microbial electrolysis cells. *World Journal of Microbiology and Biotechnology* 33 (6): 115.

Catal, T., Liu, H., Fan, Y., and Bermek, H. (2018a). A novel clean technology to convert sucrose and lignocellulose in microbial electrochemical cells into electricity and hydrogen. *Bioresource Technology Reports* https://doi.org/10.1016/j.biteb.2018.10.002.

Catal, T., Yavaser, S., Atalay, V.E., and Bermek, H. (2018b). Monitoring of neomycin sulfate antibiotic in microbial fuel cells. *Bioresource Technology* 268: 116–120.

Catal, T., Kul, A., Enisoglu-Atalay, V. et al. (2019). Efficacy of microbial fuel cells for sensing of cocaine metabolites in urine-based wastewater. *Journal of Power Sources* 414: 1–7.

Chang, F.Y. and Lin, C.Y. (2004). Biohydrogen production using an up-flow anaerobic sludge blanket reactor. *International Journal of Hydrogen Energy* 29 (1): 33–39.

Chatzitakis, A., Nikolakaki, E., Sotiropoulos, S., and Poulios, I. (2013). Hydrogen production using an algae photoelectrochemical cell. *Applied Catalysis B: Environmental* 142–143: 161–168.

Cheng, J., Liu, Y., Lin, R. et al. (2014). Cogeneration of hydrogen and methane from the pretreated biomass of algae bloom in Taihu Lake. *International Journal of Hydrogen Energy* 39 (33): 18793–18802.

Chisti, Y. (2007). Biodiesel from microalgae. *Biotechnology Advances* 25 (3): 294–306.

Cohen, J., Kim, K., Posewitz, M. et al. (2005). Molecular dynamics and experimental investigation of H-2 and O-2 diffusion in [Fe]-hydrogenase. *Biochemical Society Transactions* 33: 80–82.

Correia-da-Silva, M., Sousa, E., Pinto, M.M., and Kijjoa, A. (2017). Anticancer and cancer preventive compounds from edible marine organisms. *Seminars in Cancer Biology* 46: 55–64.

Danko, A.S., Abreu, A.A., and Alves, M.M. (2008). Effect of arabinose concentration on dark fermentation hydrogen production using different mixed cultures. *International Journal of Hydrogen Energy* 33 (17): 4527–4532.

Daroch, M., Geng, S., and Wang, G.Y. (2013). Recent advances in liquid biofuel production from algal feedstocks. *Applied Energy* 102: 1371–1381.

Das, D. and Veziroglu, T.N. (2001). Hydrogen production by biological processes: a survey of literature. *International Journal of Hydrogen Energy* 26 (1): 13–28.

Ding, L., Cheng, J., Xia, A. et al. (2016). Co-generation of biohydrogen and biomethane through two-stage batch co-fermentation of macro- and micro-algal biomass. *Bioresource Technology* 218: 224–231.

Dittert, I.M., de Lima Brandão, H., Pina, F. et al. (2014). Integrated reduction/oxidation reactions and sorption processes for Cr(VI) removal from aqueous solutions using *Laminaria digitata* macro-algae. *Chemical Engineering Journal* 237: 443–454.

Duo, J., Jin, F.M., Wang, Y.Q. et al. (2016). Na-enhanced hydrogen production from water with Fe and in situ highly efficient and autocatalytic $NaHCO_3$ reduction into formic acid. *Chemical Communications* 52 (16): 3316–3319.

Eltsova, Z.A., Vasilieva, L.G., and Tsygankov, A.A. (2010). Hydrogen production by recombinant strains of *Rhodobacter sphaeroides* using a modified photosynthetic apparatus. *Applied Biochemistry and Microbiology* 46 (5): 487–491.

Gadhamshetty, V., Belanger, D., Gardiner, C.J. et al. (2013). Evaluation of *Laminaria*-based microbial fuel cells (LbMs) for electricity production. *Bioresource Technology* 127: 378–385.

Ghirardi, M.L. (2006). Hydrogen production by photosynthetic green algae. *Indian Journal of Biochemistry & Biophysics* 43 (4): 201–210.

Ghirardi, M.L., Zhang, J.P., Lee, J.W. et al. (2000). Microalgae: a green source of renewable H-2. *Trends in Biotechnology* 18 (12): 506–511.

Ghirardi, M.L., Posewitz, M.C., Maness, P.C. et al. (2007). Hydrogenases and hydrogen photoproduction in oxygenic photosynthetic organisms. *Annual Review of Plant Biology* 58: 71–91.

Guo, Z., Li, Y., and Guo, H. (2017). Effect of light/dark regimens on hydrogen production by tetraselmis subcordiformis coupled with an alkaline fuel cell system. *Applied Biochemistry and Biotechnology* https://doi.org/10.1007/s12010-017-2498-0.

Gupta, S.K., Kumari, S., Reddy, K., and Bux, F. (2013). Trends in biohydrogen production: major challenges and state-of-the-art developments. *Environmental Technology* 34 (13–16): 1653–1670.

Hallenbeck, P.C. (2009). Fermentative hydrogen production: principles, progress, and prognosis. *International Journal of Hydrogen Energy* 34 (17): 7379–7389.

Happe, T., Hemschemeier, A., Winkler, M., and Kaminski, A. (2002). Hydrogenases in green algae: do they save the algae's life and solve our energy problems? *Trends in Plant Science* 7 (6): 246–250.

Hou, H.J.M. (2017). Hydrogen energy production using manganese/semiconductor system inspired by photosynthesis. *International Journal of Hydrogen Energy* 42 (12): 8530–8538.

Hou, Q., Nie, C., Pei, H. et al. (2016). The effect of algae species on the bioelectricity and biodiesel generation through open-air cathode microbial fuel cell with kitchen waste anaerobically digested effluent as substrate. *Bioresource Technology* 218: 902–908.

Huang, G.H., Chen, F., Wei, D. et al. (2010). Biodiesel production by microalgal biotechnology. *Applied Energy* 87 (1): 38–46.

Jadhav, D.A., Jain, S.C., and Ghangrekar, M.M. (2017). Simultaneous wastewater treatment, algal biomass production and electricity generation in clayware microbial carbon capture cells. *Applied Biochemistry and Biotechnology* https://doi.org/10.1007/s12010-017-2485-5.

Jiang, L.J., Long, C.N., Wu, X.B. et al. (2014). Optimization of thermophilic fermentative hydrogen production by the newly isolated Caloranaerobacter azorensis H53214 from deep-sea hydrothermal vent environment. *International Journal of Hydrogen Energy* 39 (26): 14154–14160.

Jung, K.W., Kim, D.H., and Shin, H.S. (2011). Fermentative hydrogen production from *Laminaria japonica* and optimization of thermal pretreatment conditions. *Bioresource Technology* 102 (3): 2745–2750.

Jurado-Oller, J.L., Dubini, A., Galván, A. et al. (2015). Low oxygen levels contribute to improve photohydrogen production in mixotrophic non-stressed *Chlamydomonas* cultures. *Biotechnology for Biofuels* 8: 149. https://doi.org/10.1186/s13068-015-0341-9. eCollection 2015.

Kars, G. and Gunduz, U. (2010). Towards a super H(2) producer: Improvements in photofermentative biohydrogen production by genetic manipulations. *International Journal of Hydrogen Energy* 35 (13): 6646–6656.

Karthic, P. and Joseph, S. (2012). Comparison and limitations of biohydrogen production processes. *Research Journal of Biotechnology* 7 (2): 59–71.

Kosourov, S.N., Batyrova, K.A., Petushkova, E.P. et al. (2012). Maximizing the hydrogen photoproduction yields in *Chlamydomonas reinhardtii* cultures: the effect of the H2partial pressure. *International Journal of Hydrogen Energy* 37 (10): 8850–8858.

Kruse, O., Rupprecht, J., Bader, K.P. et al. (2005). Improved photobiological H-2 production in engineered green algal cells. *Journal of Biological Chemistry* 280 (40): 34170–34177.

Kumar, G., Sivagurunathan, P., Thi, N.B.D. et al. (2016). Evaluation of different pretreatments on organic matter solubilization and hydrogen fermentation of mixed microalgae consortia. *International Journal of Hydrogen Energy* 41 (46): 21628–21640.

Kumru, M., Eren, H., Catal, T. et al. (2012). Study of azo dye decolorization and determination of cathode microorganism profile in air-cathode microbial fuel cells. *Environmental Technology* 33 (16–18): 2167–2175.

Lam, M.K. and Lee, K.T. (2012). Microalgae biofuels: a critical review of issues, problems and the way forward. *Biotechnology Advances* 30 (3): 673–690.

Lay, J.J., Lee, Y.J., and Noike, T. (1999). Feasibility of biological hydrogen production from organic fraction of municipal solid waste. *Water Research* 33 (11): 2579–2586.

Lee, S.M., Kim, G.H., and Lee, J.H. (2012). Bio-gas production by co-fermentation from the brown algae, *Laminaria japonica*. *Journal of Industrial and Engineering Chemistry* 18 (4): 1512–1514.

Levin, D.B., Islam, R., Cicek, N., and Sparling, R. (2006). Hydrogen production by *Clostridium* thermocellum 27405 from cellulosic biomass substrates. *International Journal of Hydrogen Energy* 31 (11): 1496–1503.

Li, C., Zhang, T., and Fang, H.H.P. (2006). Fermentative hydrogen production in packed-bed and packing-free upflow reactors. *Water Science and Technology* 54 (9): 95–103.

Liu, H. and Wang, G. (2014). Fermentative hydrogen production from macro-algae *Laminaria japonica* using anaerobic mixed bacteria. *International Journal of Hydrogen Energy* 39 (17): 9012–9017.

Ljunggren, M. and Zacchi, G. (2010). Techno-economic evaluation of a two-step biological process for hydrogen production. *Biotechnology Progress* 26 (2): 496–504.

Machado, L., Magnusson, M., Paul, N.A. et al. (2014). Effects of marine and freshwater macroalgae on in vitro total gas and methane production. *PLoS One* 9 (1): e85289. https://doi.org/10.1371/journal.pone.0085289. eCollection 2014.

Manish, S. and Banerjee, R. (2008). Comparison of biohydrogen production processes. *International Journal of Hydrogen Energy* 33 (1): 279–286.

Melis, A. (2007). Photosynthetic H2 metabolism in *Chlamydomonas reinhardtii* (unicellular green algae). *Planta* 226 (5): 1075–1086.

Melis, A. and Happe, T. (2001). Hydrogen production. Green algae as a source of energy. *Plant Physiology* 127 (3): 740–748.

Melis, A., Zhang, L.P., Forestier, M. et al. (2000). Sustained photobiological hydrogen gas production upon reversible inactivation of oxygen evolution in the green alga *Chlamydomonas reinhardtii*. *Plant Physiology* 122 (1): 127–135.

Melnicki, M.R., Pinchuk, G.E., Hill, E.A. et al. (2012). Sustained H(2) production driven by photosynthetic water splitting in a unicellular cyanobacterium. *MBio* 3 (4): e00197–e00112. https://doi.org/10.1128/mBio.00197-12.

Meuser, J.E., Ananyev, G., Wittig, L.E. et al. (2009). Phenotypic diversity of hydrogen production in chlorophycean algae reflects distinct anaerobic metabolisms. *Journal of Biotechnology* 142 (1): 21–30.

Miranda, A.F., Ramkumar, N., Andriotis, C. et al. (2017). Applications of microalgal biofilms for wastewater treatment and bioenergy production. *Biotechnology for Biofuels* 10: 120. https://doi.org/10.1186/s13068-017-0798-9. eCollection 2017.

Mudhoo, A., Forster-Carneiro, T., and Sanchez, A. (2011). Biohydrogen production and bioprocess enhancement: a review. *Critical Reviews in Biotechnology* 31 (3): 250–263.

Murthy, U.M.N. and Ghirardi, M.L. (2013). Algal hydrogen production, reference module in biomedical sciences. In: *Encyclopedia of Biological Chemistry* (eds. W. Lennarz and D. Lane), 66–70. Elsevier.

Nathoa, C., Sirisukpoca, U., and Pisutpaisal, N. (2014). Production of hydrogen and methane from banana peel by two phase anaerobic fermentation. *Energy Procedia* 50: 702–710.

van Niel, E.W.J. (2016). Biological processes for hydrogen production. *Anaerobes in Biotechnology* 156: 155–193.

Oey, M., Ross, I.L., Stephens, E. et al. (2013). RNAi knock-down of LHCBM1, 2 and 3 increases photosynthetic H2 production efficiency of the green alga *Chlamydomonas reinhardtii*. *PLoS One* 8 (4): e61375.

Oey, M., Sawyer, A.L., Ross, I.L., and Hankamer, B. (2016). Challenges and opportunities for hydrogen production from microalgae. *Plant Biotechnology Journal* 14 (7): 1487–1499.

Oh, Y.K., Kim, S.H., Kim, M.S., and Park, S. (2004). Thermophilic biohydrogen production from glucose with trickling biofilter. *Biotechnology and Bioengineering* 88 (6): 690–698.

Oncel, S. and Kose, A. (2014). Comparison of tubular and panel type photobioreactors for biohydrogen production utilizing *Chlamydomonas reinhardtii* considering mixing time and light intensity. *Bioresource Technology* 151: 265–270.

Ozakman, G., Yayman, S., Zhmurov, C.S. et al. (2018). The influence of selenium on expression levels of the rbcL gene in *Chlorella vulgaris. 3 Biotech* 8 (4): 189.

Ozdemir, M., Enisoglu-Atalay, V., Bermek, H. et al. (2019). Removal of cannabis metabolite from human urine in microbial fuel cells generating electricity. *Bioresource Technology Reports* 5: 121–126.

Ozturk, Y., Yucel, M., Daldal, F. et al. (2006). Hydrogen production by using *Rhodobacter capsulatus* mutants with genetically modified electron transfer chains. *International Journal of Hydrogen Energy* 31 (11): 1545–1552.

Papazi, A., Gjindali, A.I., Kastanaki, E. et al. (2014). Potassium deficiency, a "smart" cellular switch for sustained high yield hydrogen production by the green alga *Scenedesmus obliquus. International Journal of Hydrogen Energy* 39 (34): 19452–19464.

Park, J.H., Lee, S.H., Yoon, J.J. et al. (2014). Predominance of cluster I *Clostridium* in hydrogen fermentation of galactose seeded with various heat-treated anaerobic sludges. *Bioresource Technology* 157: 98–106.

Pasupuleti, S.B., Sarkar, O., and Mohan, S.V. (2014). Upscaling of biohydrogen production process in semi-pilot scale biofilm reactor: evaluation with food waste at variable organic loads. *International Journal of Hydrogen Energy* 39 (14): 7587–7596.

Perna, V., Castello, E., Wenzel, J. et al. (2013). Hydrogen production in an upflow anaerobic packed bed reactor used to treat cheese whey. *International Journal of Hydrogen Energy* 38 (1): 54–62.

Pittman, J.K., Dean, A.P., and Osundeko, O. (2011). The potential of sustainable algal biofuel production using wastewater resources. *Bioresource Technology* 102 (1): 17–25.

Praveenkumar, R., Shameera, K., Mahalakshmi, G. et al. (2012). Influence of nutrient deprivations on lipid accumulation in a dominant indigenous microalga *Chlorella* sp., BUM11008: evaluation for biodiesel production. *Biomass & Bioenergy* 37: 60–66.

Radakovits, R., Jinkerson, R.E., Darzins, A., and Posewitz, M.C. (2010). Genetic engineering of algae for enhanced biofuel production. *Eukaryotic Cell* 9 (4): 486–501.

Rafieenia, R., Lavagnolo, M.C., and Pivato, A. (2017). Pre-treatment technologies for dark fermentative hydrogen production: current advances and future directions. *Waste Management* pii: S0956-053X(17)30346-X. doi: https://doi.org/10.1016/j.wasman.2017.05.024.

Ritimann, S.K.M.R., Lee, H.S., Lim, J.K. et al. (2015). One-carbon substrate-based biohydrogen production: microbes, mechanism, and productivity. *Biotechnology Advances* 33 (1): 165–177.

Rogner, M. (2013). Metabolic engineering of cyanobacteria for the production of hydrogen from water. *Biochemical Society Transactions* 41 (5): 1254–1259.

Sakurai, H. and Masukawa, H. (2007). Promoting R & D in photobiological hydrogen production utilizing mariculture-raised cyanobacteria. *Marine Biotechnology* 9 (2): 128–145.

Sakurai, H., Masukawa, H., Zhang, X., and Ikeda, H. (2007). Improvement of nitrogenase-based photobiological hydrogen production by cyanobacteria by gene engineering - hydrogenases and homocitrate synthase. *Photosynthesis Research* 91 (2–3): 282–282.

Scoma, A., Krawietz, D., Faraloni, C. et al. (2012). Sustained H-2 production in a *Chlamydomonas reinhardtii* D1 protein mutant. *Journal of Biotechnology* 157 (4): 613–619.

Severes, A., Hegde, S., D'Souza, L., and Hegde, S. (2017). Use of light emitting diodes (LEDs) for enhanced lipid production in micro-algae based biofuels. *Journal of Photochemistry and Photobiology. B* 170: 235–240.

Sharma, A. and Arya, S.K. (2017). Hydrogen from algal biomass: a review of production process. *Biotechnology Reports* 15: 63–69.

Shin, J.H., Yoon, J.H., Ahn, E.K. et al. (2007). Fermentative hydrogen production by the newly isolated *Enterobacter* asburiae SNU-1. *International Journal of Hydrogen Energy* 32 (2): 192–199.

Shobana, S., Kumar, G., Bakonyi, P. et al. (2017). A review on the biomass pretreatment and inhibitor removal methods as key-steps towards efficient macroalgae-based biohydrogen production. *Bioresource Technology* 244: 1341–1348.

Singh, L. and Wahid, Z.A. (2015). Methods for enhancing bio-hydrogen production from biological process: a review. *Journal of Industrial and Engineering Chemistry* 21: 70–80.

Sivagurunathan, P., Kumar, G., Park, J.H. et al. (2016). Feasibility of enriched mixed cultures obtained by repeated batch transfer in continuous hydrogen fermentation. *International Journal of Hydrogen Energy* 41 (7): 4393–4403.

Sivakumar, G., Xu, J., Thompson, R.W. et al. (2012). Integrated green algal technology for bioremediation and biofuel. *Bioresource Technology* 107: 1–9.

Skizim, N.J., Ananyev, G.M., Krishnan, A., and Dismukes, G.C. (2012). Metabolic pathways for photobiological hydrogen production by nitrogenase- and hydrogenase-containing unicellular cyanobacteria Cyanothece. *The Journal of Biological Chemistry* 287 (4): 2777–2786.

Skjånes, K., Knutsen, G., Källqvist, T., and Lindblad, P. (2008). **H2** production from marine and freshwater species of green algae during sulfur deprivation and considerations for bioreactor design. *International Journal of Hydrogen Energy* 33 (2): 511–521.

Skjånes, K., Andersen, U., Heidorn, T., and Borgvang, S.A. (2016). Design and construction of a photobioreactor for hydrogen production, including status in the field. *Journal of Applied Phycology* 28 (4): 2205–2223.

Slegers, P.M., van Beveren, P.J.M., Wijffels, R.H. et al. (2013). Scenario analysis of large scale algae production in tubular photobioreactors. *Applied Energy* 105: 395–406.

Srikanth, S., Venkata Mohan, S., Devi, M.P. et al. (2009). Acetate and butyrate as substrates for hydrogen production through photo-fermentation: Process optimization and combined performance evaluation. *International Journal of Hydrogen Energy* 34 (17): 7513–7522.

Srirangan, K., Pyne, M.E., and Perry Chou, C. (2011). Biochemical and genetic engineering strategies to enhance hydrogen production in photosynthetic algae and cyanobacteria. *Bioresource Technology* 102 (18): 8589–8604.

Stuart, A.S., Matthew, P.D., John, S.D. et al. (2010). Biodiesel from algae: challenges and prospects. *Current Opinion in Biotechnology* 21 (3): 277–286.

Sun, J., Yuan, X., Shi, X. et al. (2011). Fermentation of *Chlorella* sp. for anaerobic bio-hydrogen production: Influences of inoculum–substrate ratio, volatile fatty acids and NADH. *Bioresource Technology* 102 (22): 10480–10485.

Tamburic, B., Zemichael, F.W., Maitland, G.C., and Hellgardt, K. (2012). A novel nutrient control method to deprive green algae of sulphur and initiate spontaneous hydrogen production. *International Journal of Hydrogen Energy* 37 (11): 8988–9001.

Trentacoste, E.M., Martinez, A.M., and Zenk, T. (2015). The place of algae in agriculture: policies for algal biomass production. *Photosynthesis Research* 123 (3): 305–315.

Vardon, D.R., Sharma, B.K., Blazina, G.V. et al. (2012). Thermochemical conversion of raw and defatted algal biomass via hydrothermal liquefaction and slow pyrolysis. *Bioresource Technology* 109: 178–187.

Volgusheva, A., Styring, S., and Mamedov, F. (2013). Increased photosystem II stability promotes H-2 production in sulfur-deprived *Chlamydomonas reinhardtii*. *Proceedings of the National Academy of Sciences of the United States of America* 110 (18): 7223–7228.

Volgusheva, A., Kukarskikh, G., Krendeleva, T. et al. (2015). Hydrogen photoproduction in green algae *Chlamydomonas reinhardtii* under magnesium deprivation. *RSC Advances* 5 (8): 5633–5637.

Wang, M.Y., Tsai, Y.L., Olson, B.H., and Chang, J.S. (2008). Monitoring dark hydrogen fermentation performance of indigenous *Clostridium butyricum* by hydrogenase gene expression using RT-PCR and qPCR. *International Journal of Hydrogen Energy* 33 (18): 4730–4738.

Wang, L., Wang, X., Wu, H., and Liu, R. (2014). Overview on biological activities and molecular characteristics of sulfated polysaccharides from marine green algae in recent years. *Marine Drugs* 12 (9): 4984–5020.

Wells, M.L., Potin, P., Craigie, J.S. et al. (2017). Algae as nutritional and functional food sources: revisiting our understanding. *Journal of Applied Phycology* 29 (2): 949–982.

White, S., Anandraj, A., and Trois, C. (2014). NADPH fluorescence as an indicator of hydrogen production in the green algae *Chlamydomonas reinhardtii*. *International Journal of Hydrogen Energy* 39 (4): 1640–1647.

Winkler, M., Hemschemeier, A., Gotor, C. et al. (2002). [Fe]-hydrogenases in green algae: photo-fermentation and hydrogen evolution under sulfur deprivation. *International Journal of Hydrogen Energy* 27 (11–12): 1431–1439.

Wu, K.J., Chang, C.F., and Chang, J.S. (2007). Simultaneous production of biohydrogen and bioethanol with fluidized-bed and packed-bed bioreactors containing immobilized anaerobic sludge. *Process Biochemistry* 42 (7): 1165–1171.

Wu, X.Y., Song, T.S., Zhu, X.J. et al. (2013). Construction and operation of microbial fuel cell with *Chlorella vulgaris* biocathode for electricity generation. *Applied Biochemistry and Biotechnology* 171 (8): 2082–2092.

Xia, A., Jacob, A., Tabassum, M.R. et al. (2016). Production of hydrogen, ethanol and volatile fatty acids through co-fermentation of macro- and micro-algae. *Bioresource Technology* 205: 118–125.

Xu, C., Poon, K., Choi, M.M., and Wang, R. (2015). Using live algae at the anode of a microbial fuel cell to generate electricity. *Environmental Science and Pollution Research International* 22 (20): 15621–15635.

Xu, L., Cheng, X., Wu, S., and Wang, Q. (2017). Co-cultivation of *Chlamydomonas reinhardtii* with *Azotobacter chroococcum* improved H2 production. *Biotechnology Letters* 39 (5): 731–738.

Yagi, T., Yamashita, K., Okada, N. et al. (2016). Hydrogen photoproduction in green algae *Chlamydomonas reinhardtii* sustainable over 2 weeks with the original cell culture without supply of fresh cells nor exchange of the whole culture medium. *Journal of Plant Research* 129 (4): 771–779.

Yan, Q., Zhao, M., Miao, H. et al. (2010). Coupling of the hydrogen and polyhydroxyalkanoates (PHA) production through anaerobic digestion from Taihu blue algae. *Bioresource Technology* 101 (12): 4508–4512.

Yoshida, A., Nishimura, T., Kawaguchi, H. et al. (2005). Enhanced hydrogen production from formic acid by formate hydrogen lyase-overexpressing *Escherichia coli* strains. *Applied and Environmental Microbiology* 71 (11): 6762–6768.

Zhao, Q.B. and Yu, H.Q. (2008). Fermentative H-2 production in an upflow anaerobic sludge blanket reactor at various pH values. *Bioresource Technology* 99 (5): 1353–1358.

18

Production and Utilization of Methane Biogas as Renewable Fuel

Ganesh Dattatraya Saratale[1], Jeyapraksh Damaraja[2], Sutha Shobana[3], Rijuta Ganesh Saratale[4], Sivagurunathan Periyasamy[5], Gunagyin Zhen[6] and Gopalakrishnan Kumar[7]

[1] *Department of Food Science and Biotechnology, Dongguk University-Seoul, Goyang-si, Gyeonggi-do 10326, Republic of Korea*
[2] *Division of Chemistry, Faculty of Science and Humanities, Sree Sowdambika College of Engineering, Aruppukottai, Tamil Nadu, India*
[3] *Department of Chemistry and Research Centre, Aditanar College of Arts and Science, Tiruchendur, Tamil Nadu, India*
[4] *College of Life Science and Biotechnology, Research Institute of Biotechnology and Medical Converged Science, Dongguk University-Seoul, Goyang-si, Gyeonggi-do 10326, Republic of Korea*
[5] *Center for materials cycles and waste management research, National Institute for Environmental Studies, Tsukuba, Japan*
[6] *School of Ecological and Environmental Sciences, East China Normal University, Shanghai 200241, PR China*
[7] *Institute of Chemistry, Bioscience, and Environmental Engineering, Faculty of Science and Technology, University of Stavanger, Stavanger 4036, Norway*

CHAPTER MENU

18.1 Introduction

The primary energy consumption in the world is documenting a rapid increase in last five decades due to the increasing world population. Fossil fuel resources are widely used in the development of the industrial sector as well as to meet the growing energy demands of the population, however, they are not considered as sustainable from ecological and environmental points of view (Demirbas 2006). In 2010, worldwide primary energy consumption raised to the highest growth percentage in nearly 40 years (Jones and Mayfield 2012; Alam et al. 2012). At this time, 88% of global energy consumption largely resulted from unsustainable energy sources (Fernandes et al. 2007). Therefore, the world is presently facing two major problems, one is an energy crisis and the other is environmental pollution (Gupta and Tuohy 2013). The energy crisis in the past decades was a result of considerable depletion of unsustainable sources like fossil fuels. Wide use of unsustainable sources as transportation fuel, and power generation is considered to be one of the major issues which

Green Energy to Sustainability: Strategies for Global Industries, First Edition.
Edited by Alain A. Vertès, Nasib Qureshi, Hans P. Blaschek and Hideaki Yukawa.

has resulted in huge volumes of carbon dioxide (CO_2) emissions leading to global warming (GW) (Chandra et al. 2012). In the past few years, numerous developed and developing countries in the world have focussed on the development of technologies to enhance energy conversion efficiency and reduce fossil fuel based energy demand (Adenle et al. 2013). At present, biofuels are expected to be potential renewable energy sources and are considered to be sustainable and eco-friendly (Weijermars et al. 2012; Makarevicine et al. 2013).

Algae are sunlight-driven cell factories which have the capability to transform solar energy into chemical energy by photosynthetic mechanisms and have higher growth rates than plants (Brennan and Owende 2010; Ndimba et al. 2013). Additionally, algae can propagate in diverse environments with minimal nutrient requirements. Hence, algae cultivation is possible and easier in areas that are not usually supported by mainstream agriculture (Ndimba et al. 2013; Slade and Bauen 2013). Algae are a huge and diverse group of simple aquatic organisms that are mostly microscopic (Slade and Bauen 2013). Microalgae are unicellular photosynthetic microorganisms, which convert solar energy, CO_2, H_2O, and nutrients to produce biomass (Alam et al. 2012; Demirbas 2010).

Microalgae can be considered as a potential source of biofuel because of their potential advantages such as; efficient photosynthetic mechanisms, higher growth rates, high levels of biomass production and accumulation of biological constituents such as lipids, and carbohydrates(Suali and Sarbatly 2012; Pragya et al. 2013; Andrade-Nascimento et al. 2013). Various aqueous systems such as open ponds, closed ponds, photobioreactors (PBR) or hybrid PBR are widely used for the cultivation of microalgae for biofuel production (Suganya et al. 2016). Microalgae can produce different types of renewable biofuels. The increasing prices of petroleum, depletion of fossil fuel and more significantly, the emerging concern about climate change (global warming) are leading to the use of microalgae as a source of fuel. This chapter summarizes information on anaerobic digestion (AD) using microalgae or extracted microalgal biomass for methane production. The effects of various operational parameters and microalgae characteristics on methane fermentation are included. Lastly, the enhanced methane production from microalgae biomass employing processes such as pretreatments and anaerobic co-digestion is described.

18.2 Anaerobic Digestion

Anaerobic digestion is historically one of the oldest naturally occurring biological process technologies in which microorganisms break down organic matter in an oxygen depleted environment with biogas as the end product. It also has capabilities to recycle nutrients, reduce process volume, and results in sustainable methane biogas production. For this reason, it is a favourable method for the treatment of waste. The produced methane could be used to generate energy, either in the form of heat or electricity (Oliveira et al. 2014). Moreover, low energy costs for operation and the production of less sludge makes AD more advantageous. Golueke et al. (1957) studied AD of two microalgae biomass sources (*Chlorella* and *Scenedesmus*) in which 0.17–0.32 LCH_4 g/VS biogas was produced. Some investigators emphasized that utilization of microalgae for AD and methane production was efficient and cost-effective (Tamilarasan et al. 2018; Ward et al. 2014). In some studies the following microalgae species have been used: *Arthrospora platensis, Dunaliella salina, Chlamydomonas reinhardtii, Chlorella kessleri, Euglena gracilis*, and *Scenedesmus obliquus*, for methane production through AD (Mussgnug et al. 2010). In these studies, they have observed that methane production is dependent on algal species. In other studies, they have utilized freshwater and marine microalgae as a model for large scale production of methane. Similarly, in this study the methane yield was found to dependent on algal species (Frigon et al. 2013). The methanogenic conversion of microalgal biomass residues after oil extraction can be considered as a combination of anaerobic digestion and microalgal cultivation that has been adopted since 1959 (Barontini et al. 2016).This concept promotes energy recovery from the sun in the form of biogas [CH_4 (55–75%) and CO_2 (25–45%)] via photosynthesis. This recycles nitrogen and phosphorous which can be utilized as nutrients for algal cultivation (Sims 2013). However, the kinetics of these anaerobic degradation processes are highly dependent not only on operational and environmental parameters such as organic loading rate (OLR), hydraulic retention time (HRT), solids retention time, pH, temperature, etc., but also on the efficiency of algal species and its cell digestibility (resistance of the cell wall) (Vandenbroucke and Largeau 2007).

18.3 Mechanism of Anaerobic Digestion

The mechanism of anaerobic digestion involves the conversion of the C and N contents of the de-oiled microalgae biomass to methane biogas. The steps include hydrolysis, acidogenesis, acetogenesis, and terminates with methanogenesis (Figure 18.1) in the absence of molecular oxygen (Hartmann et al. 1999); so, the route must take place in a highly redox atmosphere of −200 to −400 mV. The first three steps ferment the complex organic polymeric moieties to acetate/formate/H^+ ions, CO_2 and H_2. This microbial process is shown in Figure 18.2a and also shown in the following reactions (18.1) and (18.2).

$$\text{Glucose} + 2H_2O \rightarrow 2CH_3COO^- + 2H^+ + 2CO_2 + 4H_2 \tag{18.1}$$

$$\text{Glucose} + 2H_2O \rightarrow 2CH_3COO^- + 2HCOO^- + 4H^+ + 2H_2 \tag{18.2}$$

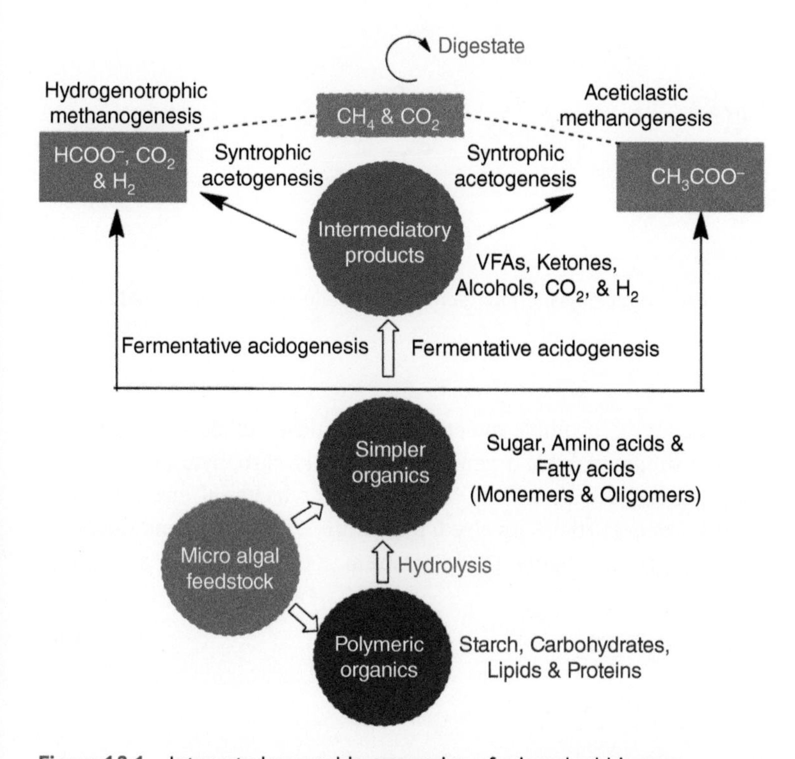

Figure 18.1 Integrated anaerobic conversion of microalgal biomass.

(a) Microbial conversion of sugar moieties in micro algal cellular materials

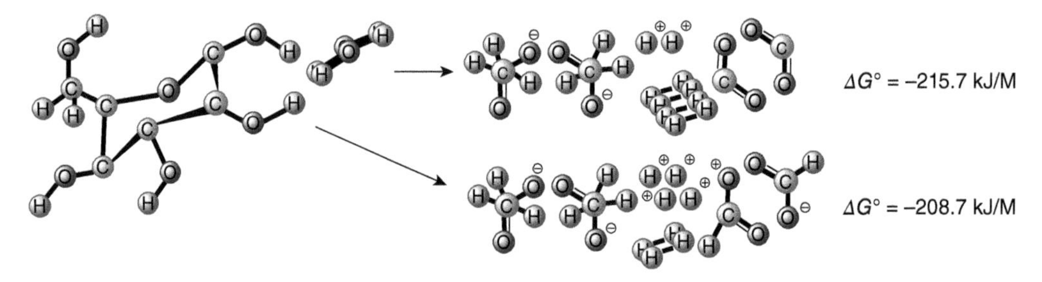

(b) Step 1: Hydrolysis of micro algal cellular biopolymers

(c) Step 2: Formation of intermediatries on acidogenesis

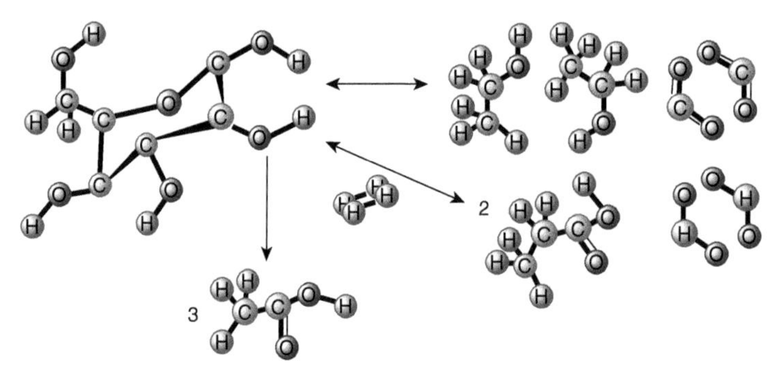

Figure 18.2 Mechanism involved in anaerobic digestion for biomethane production by delipidized microalgal biomass.

Step 1: Hydrolysis

In the first step, the large biopolymeric insoluble moieties like cellulose, hemicellulose, lipids, and proteins are hydrolysed into simpler soluble organic molecules like carbohydrates followed by the formation of hydrogen molecules by means of extracellular bacterial enzymes. The reaction kinetics depend on the size of particles involved, pH, nature of enzymes, and diffusion and absorption of enzymes to particles (Figure 18.2b) (Ostrem 2004).The reaction is given below (18.3)

$$C_6H_{10}O_4 + 2H_2O \rightarrow C_6H_{12}O_6 + 2H_2 \tag{18.3}$$

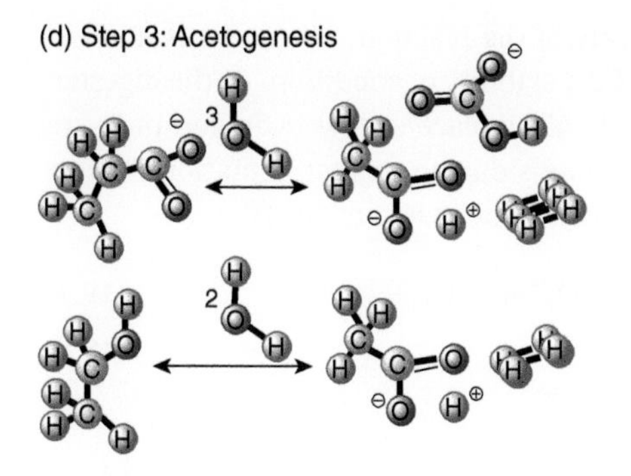

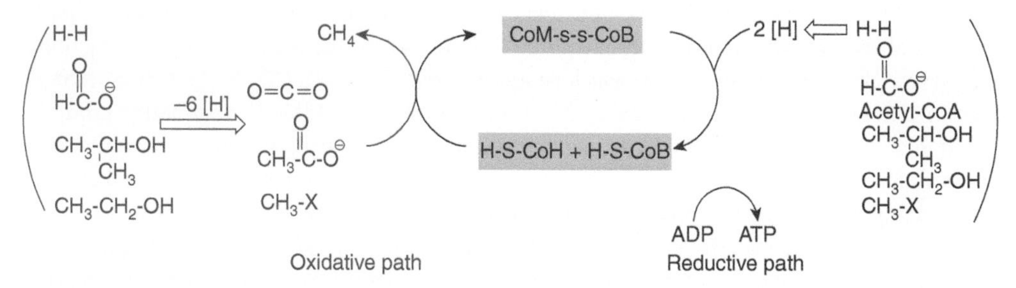

Figure 18.2 (*Continued*)

Step 2: Acidogenesis

This is the fermentative step of anaerobic digestion. It involves the effective conversion of the dissolved organic moieties principally into many intermediate products such as volatile fatty acids (VFAs), ketones, alcohols, hydrogen, and carbon dioxide (Figure 18.2c). The major products formed are C_1-formic [HCO_2H], C_2-acetic [CH_3CO_2H], C_3-propionic and lactic [$CH_3CH_2CO_2H$ and $CH_3(CHOH)CO_2H$] and C_4-butric [$CH_3(CH_2)_2CO_2H$] acids, methanol [CH_3OH] and ethanol [CH_3CH_2OH]. The formed H_2, CO_2, and C_2-acetic acid [CH_3CO_2H] can contribute directly to the last step whereas the other products involved in the third step and the reactions (18.4)–(18.6) are as follows.

$$C_6H_{12}O_6 \leftrightarrow 2CH_3CH_2OH + 2CO_2 \tag{18.4}$$

$$C_6H_{12}O_6 + 2H_2 \leftrightarrow 2CH_3CH_2COOH + 2H_2O \tag{18.5}$$

$$C_6H_{12}O_6 \rightarrow 2CH_3COOH \tag{18.6}$$

This depends on the facts that interspecies hydrogen transfer, pH, HRT and previous acclimation of the anaerobic culture (Seelan and Nallathambi 1997).

Step 3: Acetogenesis

In this step, the acidogenic residues viz C_3-propionic [$CH_3CH_2CO_2H$] and C_4-butric [$CH_3(CH_2)_2CO_2H$] acids and alcohols are transformed into H_2, CO_2, and C_2-acetic acid [CH_3CO_2H] which will in turn be utilized in the fourth step by the methogenic bacteria. The partial pressure of hydrogen plays an intermediary role in the conversion process of all the acids and is low enough for thermodynamical feasibility of the reaction. The concentration of hydrogen in the digester can be used as an indicator of the stability of conditions in the digester. Moreover, the conversion of propionate to acetate ions takes place at low hydrogen pressure and also there is a transformation of ethanol to acetate ions during this fermentation process (Figure 18.2d) and the reactions (18.7) and (18.8) are given below (Ostrem 2004).

$$CH_3CH_2COO^- + 3H_2O \leftrightarrow CH_3COO^- + H^+ + HCO_3{}^- + 3H_2 \tag{18.7}$$

$$CH_3CH_2OH + 2H_2O \leftrightarrow CH_3COO^- + 2H_2 + H^+ \tag{18.8}$$

Step 4: Methanogenesis

The metabolic methanogenic pathway falls into two fractions: the oxidative path which involves the oxidation of the specific co-enzymes M(H-S-CoM,2-thioethanesulfonate) and B(H-S-CoB,7-thioheptanoylthreonine-phosphate) into heterodisulfide; the reductive path which involves the reduction of heterodisulfide (Figure 18.2e). Table 18.1 shows the reactions and enzymes involved in methane biogas formation from acetate and CO_2 in the methanosarcinales with their thermodynamic viability. From Figure 18.2e and Table 18.1, a methyl group is shifted to co-enzyme M (H-S-CoM,2-thioethanesulfonate) in the oxidative path and is coupled with an electro-chemical sodium gradient while in the reductive path, the heterodisulfide bond is coupled with electrogenic proton translocation, these reactions have been coupled with ATP synthesis (Sims 2013). Now the methanogens hold Na^+/H^+ as co-transporter which is in turn capable of the interconversion of the electro-chemical Na^+ and H^+ potentials. The stoichiometry of coupling methanogenesis with ADP phosphorylation depends on the hydrogen concentration. The methanogens can utilize sulfur as a terminal electron acceptor since they can reoxidize the reduced sulfur with CO_2/one of the other methanogenic carbon substrates which in sequence simultaneously reduced to methane biogas.

18.3.1 Theoretical Methane Biogas Production

Buswell and Boruff (1932) used the following stoichiometric relation (18.9) to determine the theoretical methane production when the molar composition of carbon (a), hydrogen (b), oxygen (c) and nitrogen (d) of the substrate is known.

$$C_aH_bO_cN_d + \left(\frac{4a - b - 2c + d}{4}\right) H_2O \rightarrow \left(\frac{4a + b - 2c - 3d}{8}\right) CH_4 + \left(\frac{4a - b + 2c + 3d}{8}\right) CO_2 + dNH_3 \tag{18.9}$$

In addition, the molar volume (Vm) of methane depends on the amount of volatile solids (VSs) present in the substrate (18.10) (Sialve et al. 2009).

$$\text{Methane yield in L/g (VS) destroyed} = \left(\frac{4a + b - 2c - 3d}{12a + b + 16c + 14d}\right) \times Vm \tag{18.10}$$

Literature reveals that the theoretical methane productivity can be calculated for various microalgal biomasses using Eq. (18.2).

Table 18.1 Literature survey of several pretreatment studies for anaerobic digestion from microalgal biomass and methane production.

Type of pretreatment	Pretreatment conditions	Microalgae species	Methane production (LCH_4 g/VS)	References
Thermal	Microwave, until boiling, 600 W, 2450 MHz	*Nannochloropsis salina*	0.487	Schwede et al. (2013)
Thermal	100 °C, 10 h	*Nannochloropsis salina*	0.549	Schwede et al. (2013)
Thermal	Overnight −15 °C	*Nannochloropsis salina*	0.46	Schwede et al. (2013)
Thermal	110 °C, 15 min	*Microspora*	0.413	Alzate et al. (2012)
Thermal	90 °C, 3 h	*Scenedesmus*	170[a)]	González-Fernández et al. (2012)
Thermal	80 °C, 25 min	*Scenedesmus*	0.128[a)]	
Thermal	Low temperature, 95 °C, 10 h	Microalgae mixture[b)]	0.27	Passos et al. (2013a)
Thermal	Microwave, 98 °C, 900 W, 65 400 ASE[b)]	*Scenedesmus* and *Chlorella*	0.307	Passos et al. (2013b)
Thermal	Freeze–thaw −20 °C, 24 h; melted at 20 °C	Microalgae mixture[b)]	0.155	Kinnunen et al. (2014a)
Thermal	60 °C, 24 h	Microalgae mixture[b)]	0.136	Kinnunen and Rintala (2016); Kinnunen et al. (2014b)
Thermal	90 °C, 4 h	*Nannochloropsis oculata*	0.397	Marsolek et al. (2014)
Thermochemical	H_2SO_4, pH 2, 120 °C, 40 min	*Chlorella vulgaris*	0.228	Mendez et al. (2013)
Thermochemical	NaOH, pH 10, 120 °C, 40 min	*Chlorella vulgaris*	0.24	Mendez et al. (2013)
Thermochemical	21 g NaOH/l, autoclaving 121 °C, 10 bar, 30 min	*Nannochloropsis* sp.	0.47	Bohutskyi et al. (2014)
Thermochemical	21 g NaOH/l, autoclaving 121 °C, 10 bar, 30 min	*Chlorella* sp.	0.46	Bohutskyi et al. (2014)
Ultrasound	200 W, 30 kHz, 135 s	*Nannochloropsis salina*	0.274	Schwede et al. (2013)
Ultrasound	E^c = 57 000 kJ kg/TS	*Microspora*	0.314	Alzate et al. (2012)
Ultrasound	E^c = 130 000 kJ kg/TS, 15 min	*Scenedesmus*	0.153[a)]	González-Fernández et al. (2012)
Ultrasound	200 J/ml	*Chlorella vulgaris*	0.418	Park et al. (2013)
Ultrasound	40 J/ml[1]	*Hydrodictyon reticulatum*	0.384	Lee et al. (2014)

(Continued)

Table 18.1 (Continued)

Type of pretreatment	Pretreatment conditions	Microalgae species	Methane production (LCH_4 g/VS)	References
High pressure thermal hydrolysis	Autoclaving 121 °C, 10 bar 30 min	*Nannochloropsis* sp.	0.45	Bohutskyi et al. (2014)
High pressure thermal hydrolysis	Autoclaving 121 °C, 10 bar 30 min	*Thalassiosira weissflogii*	0.41	Bohutskyi et al. (2014)
High pressure thermal hydrolysis	Autoclaving 160 °C, 10 min	*Chlorella vulgaris*	0.256	Mendez et al. (2014)
High pressure thermal hydrolysis	Autoclaving 170 °C, 800 kPa, 30 min	*Scenedesmus*	0.32	Keymer (2013)
Lipid extraction combined with high pressure thermal hydrolysis	Soxhelt, Hexane, 6 h; after Autoclaving 170 °C, 800 kPa, 30 min	*Scenedesmus*	0.38	Keymer (2013)
High pressure homogenization (French press)	Two runs at 10 Mpa	*Nannochloropsis salina*	0.46	Schwede et al. (2013)
Mechanical	Blending (sizes <0.1 mm)	*Rhizoclonium*	0.124	Ehimen et al. (2013)
Biological	Microaerobically digestion, 24 h	*Microspora*	0.266	Alzate et al. (2012)
Biological	Enzimatic digestion (α-amylase, protease, lipase, xylanase (endo-1,4-), cellulase complex), 2 d, pH 7	*Rhizoclonium*	0.145	Ehimen et al. (2013)
Biological	Spontaneous fermentation, storage 15 d	*Mixture of Microcystis, Cyclotella, Cryptomonas, and Scenedesmus)*	0.287	Miao et al. (2013)
Lipid extraction	Soxhelt, hexane, 6 h	*Scenedesmus*	0.24	Keymer (2013)
Lipid extraction	Solvent extraction (acetone and hexane)	*Auxenochlorella protothecoides*	0.25	Bohutskyi et al. (2015)
Lipid extraction	Solvent extraction (chloroform: methanol 2 : 1)	*Tetraselmissuecica*	0.175	Santos-Ballardo et al. (2015)

a) Based on chemical oxygen demand (dm^3/kg).
b) *Chlamydomonas*, *Nitzschia*, *Ankistrodesmus*, *Monorraphidium*, *Chlorella*, and *Scenedesmus*.

18.4 Significant Factors Influencing Anaerobic Digestion

Various physical, chemical, and operational factors affect the process of methanogenesis where the key operating conditions include; temperature, pH, HRT, and OLR, etc. Also, the algal species, their chemical composition along with cell wall resistance can influence the performance of an anaerobic methane production process and these are discussed in detail in the following section.

18.4.1 Effect of Temperature

Anaerobic bacteria are also sensitive to temperature changes, thus anaerobic digestion performance and methane production depends strongly on the temperature. It is essential to maintain a stable temperature for the growth of anaerobic microbes. Biogas formation can occur over a wide range of temperatures at psychrophilic (<20 °C), mesophilic (20–40 °C), and thermophilic (>40 °C) conditions. In most cases, mesophilic methanogenesis occurs at 20–45 °C with an optimum temperature of about 35 °C and thermophilic methanogenesis at 50–65 °C with an optimum temperature around 55 °C. A lowered temperature may cause an increase in volatile acid concentrations that can drop pH which directly affects metabolic rate of methanogens. Thus, a temperature decrease can have drastic repercussions on process operation (Oliveira et al. 2014), whereas at increased temperature higher metabolic growth rate, with higher digestion was observed. However, at thermophilic conditions higher death rate of methanogens was observed as compared to mesophilic condition which suggests that AD process is highly susceptible to change in the environmental and operating conditions. Golueke et al. (1957) using *Scenedesmus* and *Chlorella* biomasses. Samson and Leduy (1983) reported that with the use of *Spirulina maxima*, mesophilic condition is better for both anaerobic digestions as well as for methane production. In contrast, Zamalloa et al. (2012) reported that biogas production by microalgal biomass in anaerobic digestions in a hybrid flow-through reactor, under mesophilic and thermophilic conditions in which there is no significant difference in methane production was observed. Recently, Santos-Ballardo et al. (2015) utilized the residual biomass *Tetraselmis bsuecica* obtained after the biodiesel production for anaerobic digestion using mesophilic and thermophilic conditions. The results suggest that the methane production under mesophilic condition is two times higher than that of the thermophilic condition.

18.4.2 Effect of pH

Similar to other biological processes, pH is another key operating parameter on the performance of anaerobic digestion and methane production. The pH inside the digester influences the growth of anaerobic microbes, particularly methanogens, by impacting enzyme activity. Methanogenesis occurs at a pH between 6.6 and 7.6 with an optimum range of about 7.0–7.2. The distribution of ionized and non-ionized forms of sulfide and ammonia in the digester is also governed by the pH. At lower pH range, non-ionized sulfide (H_2S) predominates in the system, whereas at higher pH non-ionized ammonia (NH_3) dominates over the ionized form NH_4 which inhibits the bacterial activity. Fluctuation in pH influences hydrogen ion concentration which has a direct impact on bacterial growth development and consequently on anaerobic digestion process too. So, it is necessary to regulate pH condition in the digester. Microalgae contains low amounts of sulfonated amino acids, so there is less production of H_2S. However due to algal protein the presence of ammonia in biogas was observed. Thus, pH also influences biogas composition (Santos-Ballardo et al. 2015).

18.4.3 Hydraulic Retention Time (HRT) and Substrate Loading Rate

HRT represents the time that organic matter remains in contact with microbes. Another parameter, the OLR is expressed as the mass of organic matter chemical oxygen demand (COD) per unit reactor volume per unit time. To maintain the optimal conditions, it is essential that both factors are regulated for the anaerobic process and effective conversion of organic matter to biogas. Moreover, it also depends on the composition of substrates and their accessibility to the anaerobic microorganisms (Inglesby and Fisher 2012). Sánchez-Hernández and Travieso-Córdoba (1993) reported AD of microalgae *Chlorella vulgaris* for periods of 64 days in batch mode in which the conversion of biomass to biogas ranged from 70% to 90% in COD was observed. Recently, Alzate et al. (2012), demonstrated that in AD mixtures containing microalgae such as *Chlamydomonas*, *Scenedesmus*, *Nannochloropsis*, *Acutodesmus obliquus*, *Oocystis*, *Phormidium*, *Nitzschia* sp. and *Microspora* can effectively produce biogas. They observed that CH_4 production reached between 52% and 77% of the final productivities within the first five days of culture. They suggested that methane production kinetics vastly depends on microalgae species and recalcitrant algal cell walls which impede the hydrolysis rate and prevent the release of soluble biodegradable compounds.

18.4.4 Microalgae Cell Wall Composition and Degradability

In anaerobic digestion, the most important parameter is the algal cell wall consisting of complex organic matter which determines the methanogenic conversion efficiency of the substrate. The microalgal cell wall accounts for 11–37% of the total dry algal biomass (Domozych et al. 2012). Some microalgae have a protective tri-laminar outer cell wall which is more resistant to degradation that directly influences the performance of the anaerobic digestion (Kwietniewska and Tys 2014). The microalgae cell wall is embedded with glycolipids, glycoproteins, and polysaccharides. It also contains cellulose, hemicellulose, pectin, and glycoprotein (González-Fernández et al. 2012). In addition to this, other biopolymers such as uronic acid, glucosamine, hidroxyproline, proline, sporollenin, alganeans, carotenoids also make the microalgal cell wall more resistant (Ward et al. 2014). The resistance of the cell wall is considered to be one of the main issues for the anaerobic digestion of microalgae (Chen et al. 2008; Kwietniewska and Tys 2014). Mussgnug et al. (2010) studied the effect of cell wall composition by taking six microalgae species for the anaerobic digestion process where they observed that methane production was higher in microalgae without a protein based cell wall whereas carbohydrate based cell walls showed less methane production. In other studies, using *C. vulgaris* they observed that cell walls containing cellulose and a highly resistant outer stratum of alganean directly affects AD performance. (Rodrigues and Da Silva Bon 2011; Ras et al. 2011). Also, the cell wall of *S. obliquus* is considered as a rigid wall constituting glucose, mannose, and galactose (González-Fernández et al. 2012). Thus, due to the cell wall the process of AD is limited. The literature suggests that in order to improve the hydrolysis and the anaerobic digestion of the rigid cell wall various pretreatment steps to disrupt the microalgal cell wall are essential.

18.5 Strategies Applied to Enhance Microalgae Methane Biogas Production

In the microalgal AD process for methane production two major factors, mainly algal cell wall's resistance for disruption and unbalanced nutrients or lower C/N ratio of microalgal biomass, were regarded as important limiting factors. To overcome these bottlenecks some investigators proposed different pretreatment methods for the disruption of cell walls and increasing the accessibility of the biological content for AD. Similarly, increase in the C/N ratio of algal biomass for co-digestion process is also proposed. These are discussed in this section.

18.5.1 Different Pretreatment Techniques

The composition of the organic substrates is one of the most important factors that determine the methane conversion yield. Microalgal cells and biomass are found to be resistant against cell disruption processes and thus reducing the cell biodegradability. Pretreatment can be successfully applied in order to lower the recalcitrant organic fraction and thus increase the methane conversion. Some of the microalgal species like *S. obliquus* produce a very low yield of biogas as it has a rigid cell wall sporopollenin biopolymer. Biodegradability enhancement can be achieved by adopting some pretreatment methods of the delipidized microalgal biomass. This process induces cell wall disruption to raise the bacterial hydrolytic reaction rate before addition to anaerobic digester (Barontini et al. 2016) and those pretreatment methods involve chemical treatment via acids, bases and ozonation, thermal treatment, and ultrasonic lysis. Among them, the literature reveals that 33% of methane biogas production could be reachable by optimal thermal pretreatment at 100 °C for about eight hours which is more effective (Sialve et al. 2009). Gonzalez-Fernandez et al. (2011) investigated the thermal treatment of *Scenedesmus* sp. to increase methane production at 70 °C up to 9% and at 90 °C up to 57% when compared with untreated biomass. Samson and Leduy (1983) investigated thermo-chemical and ultrasonic lysis at 50 and 150 °C and found the increase in the yield from 20% to 43%. Gonzalez-Fernandez et al. (2013) observed 2.9 to 3.4 fold increases in biogas production at 90 °C with the OLRs of 1 and 2.5 kg COD/m^3 day, respectively. Alzate et al. (2012) found 46–62% yield improvement. Keymer (2013) investigated a commercially existing technology of high pressure thermal hydrolysis (HPTH) for pretreating the microalgal cells. The quantity of anaerobic digestate ammonia can be calculated by Eq. (18.9), when the molar composition of carbon (a), hydrogen (b), oxygen (c) and nitrogen (d) of the substrate is known.

Several methods of algal biomass pretreatment including thermal, mechanical, chemical, autoclaving, microwave, sonication for lipids extraction, and biological have been studied to increase its biodegradability, performance of AD, and methane yield and these are summarized in Table 18.1.

18.5.2 Co-Digestion Process

Another important factor is the carbon/nitrogen (C/N) ratio associated with microalgal biomass. The C/N ratio in substrates for anaerobic digestion should range from 20 : 1 to 30 : 1. This can inhibit methane production and results in accumulation of VFAs (Vergara-Fernandez et al. 2008; Sialve et al. 2009; Zhao et al. 2014; Santos-Ballardo et al. 2015). To overcome the problems with low C/N ratios, several researchers have investigated a co-digestion, where microalgae biomass has been co-digested with other carbon-rich wastes (such as waste paper, switch grass or glycerol) to increase the C/N ratio to enhance AD and methane production. Ehimen et al. (2012) found that the lower and upper extremity towards C/N ratio is 15: 1 and 36.4 : 1, respectively. Parkin and Owen (1986) concluded the best C/N ratio to be 20 : 1 and 30 : 1 to yield high methane productivity. Yen and Brune (2007) confirmed that the addition of waste paper improves C/N ratio up to 36.4 for *Scenedesmus* sp. and *Chlorella* sp. with the co-digestion ratio of 50% microalgae and 50% paper and concluded that the best C/N ratio for anaerobic digestion is 20 : 1 and 25 : 1. In a similar manner many researchers tried to blend microalgal biomass with some other co-digesters and achieved higher productivity of CH_4 biogas by improving the C/N ratio, for instance: blending

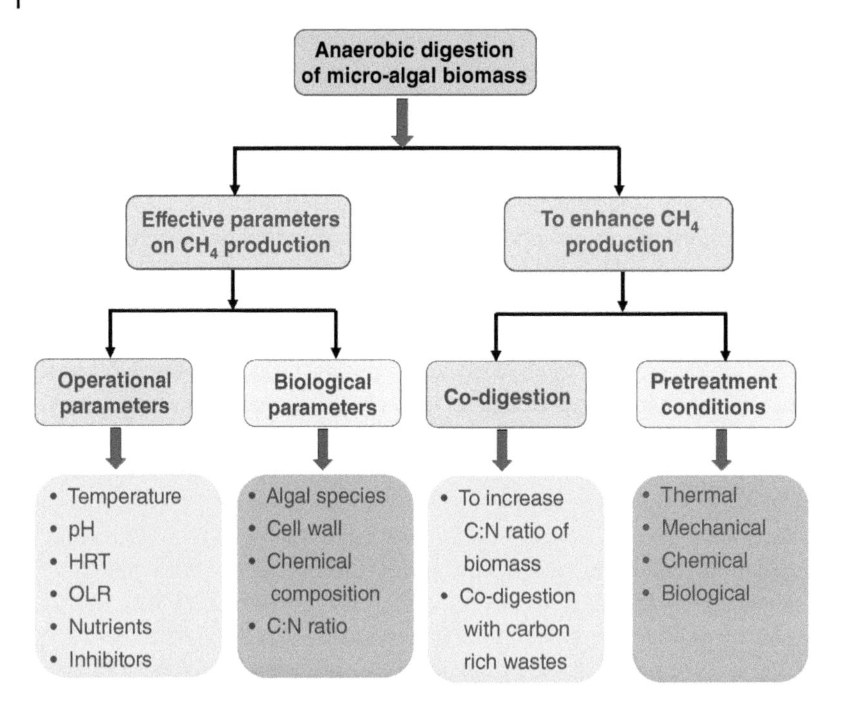

Figure 18.3 Effective parameters and pretreatment techniques for the enhancement of methane production in the AD process of microalgal biomass.

of microalgae with pig manure (Gonzalez-Fernandez et al. 2011; Shouquan et al. 2009), green filamentous microalgae and water hyacinth with cow manure (Saxena et al. 1984), *Zygogonium* sp. with corn stalks (Ramamoorthy and Sulochana 1989), *Spirulina maxima* with sewage sludge (Samson and Leduy 1983), *Chlorella* sp. and *Spirulina platensis* with municipal wastewater solids (Yaun et al. 2012), *Nannochloropsis salina* with lipid rich fats/oil/grease (Park and Li 2012) and microalgae with domestic municipal wastewater (Salerno et al. 2009). Ramos-Suárez et al. (2014) studied anaerobic co-digestion of *Scenedesmus* microalgal biomass and *Opuntia maxima* cladodes at high organic load rates showed that methane yield increases between 63.9 and 66.4% with no ammonia inhibition. Further, the co-digestion reduces the toxicity of certain ingredients involved in this process and retains their reflection in their toxic range (Sialve et al. 2009). Some co-substrates can have an effect on enzymatic stimulation synthesis that can also improve the anaerobic digestion yield. Also, it leads to the reduction of certain toxic compounds (like ammonia). Maintaining them under their toxic threshold allows increased loading rate as well as higher biogas yield (Yen and Brune 2007). Figure 18.3 briefly depicts the effective parameters and pretreatment techniques for the enhancement of methane production in the AD process of microalgal biomass.

18.6 Utilization of Methane Biogas as a Renewable Fuel

Methane is converted to electricity by burning it as a fuel in a gas turbine or steam boiler. Compared to other hydrocarbon fuels, burning methane produces less carbon dioxide for each unit of heat released (Oilgae 2015a). At about 891 kJ per mole, methane's combustion heat is lower than any other hydrocarbon; however, as the simplest hydrocarbon, it produces more heat per unit mass than the more complex hydrocarbons (Oilgae 2015b). In the anaerobic digestion process the resulting biogas mainly consists of methane (about 40–41%) and carbon dioxide (15–60%). In addition to this 2% of other trace gases such as hydrogen sulfide (H_2S), ammonia (NH_3), hydrogen (H_2), and nitrogen (N_2) are also present. Moreover, it also consists of water vapour, and organic silicon compounds. CO_2 reduces the calorific value, water vapour causes corrosion in compressors, gas storage tanks, and engines as well as condensation and/or freezing due to high pressure. H_2S produces oxidation to corrosive gases (e.g. SO_2) causing corrosion in compressors, gas storage tanks, and engines and NH_3 can cause some complications in compressors, gas storage tanks, and engines when combined with water. Moreover, trace amounts of siloxanes can produce abrasive materials (SiO_2 and microcrystalline quartz) causing excessive wear and tear and scaling of spark plugs and other equipment. Damage to microturbine engines due to excessive abrasion caused by silicon oxide have also been widely reported. CO_2 removal is required only when pure methane is desirable and will be used as natural gas standard or for vehicle fuel.

Anaerobic digestion is a practical and cost-effective approach for biogas production and nutrient recovery from manure and other biomass. However, a high level of biogas purification is required when the use of methane is necessary for a range of applications including heating and cooking, power generation, transport fuel, natural gas supply, and even raw materials for the petroleum industry. Thus, the economic value of methane utilization increases substantially with these applications, however, we should consider the cost of biogas purification. Methane gas purification is still a major bottleneck for energy production and other forms of beneficial use. Considering this, research should also be focused on the digestate utilization after AD. Digestate can be used as a bio-fertilizer and liquid fertilizer.

18.7 Perspectives

The anaerobic process has been widely adopted for anaerobic digestion of microalgal biomass and methane production because of several inherent merits. However, this process still lacks scientific and technological advances for large scale application. This may be because of microalgae species characteristics, biochemical composition and resistance to degradation. Moreover, optimizing various operational parameters or applying some strategies such as various pretreatment methods the accessibility of biological components of algal biomass for AD is enhanced. In addition, the co-digestion process was also found to be effective to increase the performance of AD and methane production. If the process utilized extracts algal biomass then it will become cost competitive and practically applicable, however, more research is needed to accomplish this. In addition to this, there is a perpetual need to develop innovative reactor designs that are cost-effective, energy efficient and reliable. Utilization of advanced molecular tools, and research and development efforts should be focused on integrating microbial data in designing, process automation and control as a part of the bioreactor design.

References

Adenle, A.A., Haslam, G.E., and Lee, L. (2013). Global assessment of research and development for algae biofuel production and its potential role for sustainable development in developing countries. *Energy Policy* 61: 182–195.

Alam, F., Date, A., Rasjidin, R. et al. (2012). Biofuel from algae – is it a viable alternative ? *Procedia Engineering* 49: 221–227.

Alzate, M.E., Muñoz, R., Rogalla, F. et al. (2012). Biochemical methane potential of microalgae: influence of substrate to inoculum ratio, biomass concentration and pretreatment. *Bioresource Technology* 123: 488–494.

Andrade-Nascimento, I., Izabel-Marques, S.S., Dominguez-Cabanelas, I.T. et al. (2013). Screening microalgae strains for biodiesel production: lipid productivity and estimation of fuel quality based on fatty acids profile as selective criteria. *Bioenergy Resource* 6: 1–13.

Barontini, F., Biagini, E., Dragoni, F. et al. (2016). Anaerobic digestion and co-digestion of oleaginous microalgae residues for biogas production. *Chemical Engineering Transactions* 50: 91–97.

Bohutskyi, P., Betenbaugh, M.J., and Bouwer, E.J. (2014). The effects of alternative pretreatment strategies on anaerobic digestion and methane production from different algal strains. *Bioresource Technology* 155: 366–372.

Bohutskyi, P., Liu, K., Nasr, L.K. et al. (2015). Bioprospecting of microalgae for integrated biomass production and phytoremediation of unsterilized wastewater and anaerobic digestion centrate. *Applied Microbiology and Biotechnology* 99 (14): 6139–6154.

Brennan, L. and Owende, P. (2010). Biofuels from microalgae a review of technologies for production, processing, and extractions of biofuels and co-products. *Renewable and Sustainable Energy Reviews* 14: 557–577.

Buswell, A.M. and Boruff, C.S. (1932). The relationship between chemical composition of organic matter and the quality and quantity of gas produced during sludge digestion. *Sewage Works Journal* 4: 454–460.

Chandra, R., Takeuchi, H., and Hasegaw, T. (2012). Methane production from lignocellulosic agricultural crop wastes: a review in context to second generation of biofuel production. *Renewable and Sustainable Energy Reviews* 16: 1462–1476.

Chen, Y., Cheng, J.J., and Creamer, K.S. (2008). Inhibition of anaerobic digestions process: a review. *Bioresource Technology* 99: 4044–4064.

Demirbas, M.F. (2006). Current technologies for biomass conversion into chemicals and fuels. *Energy Sources* A28: 1181–1188.

Demirbas, A. (2010). Use of algae as biofuel sources. *Energy Conversion and Management* 51: 2738–2749.

Domozych, D.S., Ciancia, M., Fangel, J.U. et al. (2012). The cell walls of green algae: a journey through evolution and diversity. *Frontiers in Plant Science* 3 (82): 1–7.

Ehimen, E.A., Sun, Z., and Carrington, G.C. (2012). Use of ultrasound and co-solvents to improve the *in-situ* transesterification of microalgae biomass. *Procedia Environmental Sciences* 15: 47–55.

Ehimen, E.A., Holm-nielsen, J.B., Poulsen, M., and Boelsmand, J.E. (2013). Influence of different pre-treatment routes on the anaerobic digestion of a filamentous algae. *Renewable Energy* 50: 476–480.

Fernandes, S.D., Trautmann, N.M., Streets, D.G. et al. (2007). Global biofuel use, 1850–2000. *Global Biogeochemical Cycles* 21: 1–15.

Frigon, J.C., Matteu-Lebrun, F., Abdou, R.H. et al. (2013). Screening microalgae strains for their productivity in methane following anaerobic digestion. *Applied Energy* 108: 100–107.

Golueke, C.G., Oswald, W.J., and Gotaas, H.B. (1957). Anaerobic digestion of algae. *Applied Microbiology* 5: 47–55.

Gonzalez-Fernandez, C., Molinuevo-Salces, B., and Garcia-Gonzalez, M.C. (2011). Evaluation of anaerobic co-digestion of microbial biomass and swine manure via response surface methodology. *Applied Energy* 88: 3448–3453.

González-Fernández, C., Sialve, B., Bernet, N., and Steyer, J.P. (2012). Impact of microalgae characteristics on their conversion to biofuel. Part II: focus on biomethane production. *Biofuels, Bioproducts and Biorefining* 6: 205–218.

Gonzalez-Fernandez, C., Molinuevo-Salces, B., and Garcia-Gonzalez, M.C. (2013). Effect of organic loading rate on anaerobic digestion of thermally pretreated *Scenedesmas* sp. biomass. *Bioresource Technology* 129: 219–223.

Gupta, V.K. and Tuohy, M.G. (2013). *Biofuel Technologies: Recent Developments*. Berlin, Heidelberg: Springer.

Hartmann, H., Angelidaki, I., and Ahring, B.K. (1999). Increase of anaerobic degradation of particulate organic matter in full-scale biogas plants by mechanical maceration. In: *Proceedings of the Second International Symposium on Anaerobic Digestion of Solid Wastes*, vol. 1 (eds. J. Mata-Alvarez, A. Tilche and F. Cecchi), 129–136. Barcelona: IWA.

Inglesby, A.E. and Fisher, A.C. (2012). Enhanced methane yields from anaerobic digestion of *Arthrospira maxima* biomass in an advanced flow-through reactor with an integrated recirculation loop microbial fuel cell. *Energy & Environmental Science* 5: 7996–8006.

Jones, C.S. and Mayfield, S.P. (2012). Algae biofuels: versatility for the future of bioenergy. *Current Opinion in Biotechnology* 23: 346–351.

Keymer, P. (2013). High pressure thermal hydrolysis as pre-treatment to increase the methane yield during anaerobic digestion of microalgae. *Bioresource Technology* 131: 128–133.

Kinnunen, V. and Rintala, J. (2016). The effect of low-temperature pretreatment on the solubilization and biomethane potential of microalgae biomass grown in synthetic and wastewater media. *Bioresource Technology* 221: 78–84.

Kinnunen, H.V., Koskinen, P.E.P., and Rintala, J. (2014a). Mesophilic and thermophilic anaerobic laboratory-scale digestion of *Nannochloropsis* microalga residues. *Bioresource Technology* 155: 314–322.

Kinnunen, V., Craggs, R., and Rintala, J. (2014b). Influence of temperature and pretreatments on the anaerobic digestion of wastewater grown microalgae in a laboratory-scale accumulating-volume reactor. *Water Research* 57: 247–257.

Kwietniewska, E. and Tys, J. (2014). Process characteristic, inhibition factors and methane yields of anaerobic digestion process, with particular focus on microalgal biomass fermentation. *Renewable and Sustainable Energy Reviews* 34: 491–500.

Lee, K., Chantrasakdakul, P., kim, D. et al. (2014). Ultrasound pretreatment of filamentous algal biomass for enhanced biogas production. *Waste Management* 34: 1035–1040.

Makarevicine, V., Skorupskaite, V., and Andruleviciute, V. (2013). Biodiesel fuel from microalgae-promising alternative fuel for the future: a review. *Reviews in Environmental Science and Biotechnology* 12: 119–130.

Marsolek, M.D., Kendall, E., Thompson, P.L., and Shuman, T.R. (2014). Thermal pretreatment of algae for anaerobic digestion. *Bioresource Technology* 151: 373–377.

Mendez, L., Mahdy, A., Timmers, R.A. et al. (2013). Enhancing methane production of *Chlorella vulgaris via* thermochemical pretreatments. *Bioresource Technology* 149: 136–141.

Mendez, L., Mahdy, A., Demuez, M. et al. (2014). Effect of high pressure thermal pretreatment on *Chlorella vulgaris* biomass: organic matter solubilisation and biochemical methane potential. *Fuel* 117 (Part A): 674–679.

Miao, H., Lu, M., Zhao, M. et al. (2013). Enhancement of Taihu blue algae anaerobic digestion efficiency by natural storage. *Bioresource Technology* 149: 359–366.

Mussgnug, J.H., Klassen, V., Schlüter, A., and Kruse, O. (2010). Microalgae as substrates for fermentative biogas production in a combined biorefinery concept. *Journal of Biotechnology* 150: 51–56.

Ndimba, B.K., Ndimba, R.J., Johnson, T.S. et al. (2013). Biofuels as a sustainable energy source: an update of the applications of proteomics in bioenergy crops and algae. *Journal of Proteomics* 93: 234–244.

Oilgae (2015a) Cultivation of Algae in Open ponds. [Online] Available from: http://www.oilgae.com/algae/cult/op/op.html [2015-5/11].

Oilgae (2015b) Cultivation of Algae in Photobioreactor. [Online] Available from: http://www.oilgae.com/algae/cult/pbr/pbr.html [2015-5/12].

Oliveira, J.V., Alves, M.M., and Costa, J.C. (2014). Design of experiments to assess pre-treatment and co-digestion strategies that optimize biogas production from microalgae *Gracilariavermiculophylla*. *Bioresource Technology* 162: 323–330.

Ostrem, K. (2004). *Greening Waste: Anaerobic Digestion for Treating the Organic Fraction of Municipal Solid Wastes*. Earth Engineering Center Columbia University.

Park, S. and Li, Y. (2012). Evaluation of methane production and macronutrient degradation in the anaerobic co-digestion of algae residue and lipid waste. *Bioresource Technology* 111: 42–48.

Park, K.Y., Kweon, J., Chantrasakdakul, P. et al. (2013). Anaerobic digestion of microalgal biomass with ultrasonic disintegration. *International Biodeterioration & Biodegradation* 85: 598–602.

Parkin, G.F. and Owen, W.F. (1986). Fundamentals of anaerobic digestion of wastewater sludges. *Journal of Environmental Engineering* 112: 867–920.

Passos, F., García, J., and Ferrer, I. (2013a). Impact of low temperature pretreatment on the anaerobic digestion of microalgal biomass. *Bioresource Technology* 138: 79–86.

Passos, F., Solé, M., García, J., and Ferrer, I. (2013b). Biogas production from microalgae grown in wastewater: effect of microwave pretreatment. *Applied Energy* 108: 168–175.

Pragya, N., Pandey, K.K., and Sahoo, P.K. (2013). A review on harvesting, oil extraction and biofuels production technologies from microalgae. *Renewable and Sustainable Energy Reviews* 24: 159–171.

Ramamoorthy, S. and Sulochana, N. (1989). Enhancement of biogas production using algae. *Current Science* 58: 646–647.

Ramos-Suárez, J.L., Martínz, A., and Carreras, N. (2014). Optimization of the digestion process of *Scenedesmus* sp. and *Opuntia maxima* for biogas production. *Energy Conversion and Management* 88: 1263–1270.

Ras, M., Lardon, L., Sialve, B. et al. (2011). Experimental study on a coupled process of production and anaerobic digestion of *Chlorella vulgaris*. *Bioresource Technology* 102: 200–206.

Rodrigues, M.A. and Da Silva Bon, E.P. (2011). Evaluation of *Chlorella* (Chlorophyta) as source of fermentable sugars via cell wall enzymatic hydrolysis. *Enzyme Research* 2011: 1–5.

Salerno, M., Nurdogan, Y., and Lundquist, T.J. (2009). *Biogas Production from Algae Biomass Harvested at Wastewater Treatment Ponds, Bioenergy Engineering Conference*. Washington, USA: American Society of Agricultural and Biological Engineers.

Samson, R. and Leduy, A. (1983). Improved performance of anaerobic digestion of Spirulina maxima algal biomass addition of carbon-rich wastes. *Biotechnology Letters* 5: 677–682.

Sánchez-Hernández, E.P. and Travieso-Córdoba, L. (1993). Anaerobic digestion of *Chlorella vulgaris* for energy production. *Resources, Conservation and Recycling* 9: 127–132.

Santos-Ballardo, D.U., Font-Segura, X., Sánchez-Ferrer, A. et al. (2015). Valorization of biodiesel production wastes: anaerobic digestion of residual *Tetraselmissuecica* biomass and co-digestion with glycerol. *Waste Management & Research* 33: 250–257.

Saxena, V.K., Tandon, S.M., and Singh, K.K. (1984). Anaerobic digestion of green filamentous algae and water hyacinth for methane production. *National Academy Science Letters* 7: 283–284.

Schwede, S., Kowalczyk, A., Gerber, M., and Span, R. (2013). Anaerobic co-digestion of the marine microalga *Nannochloropsis salina* with energy crops. *Bioresource Technology* 148C: 428–435.

Seelan, G. and Nallathambi, V. (1997). Anaerobic digestion of biomass for methane production: a review. *Biomass and Bioenergy* 13: 83–114.

Shouquan, W., Qun, Y., Hengfeng, M., and Wenquan, R. (2009). Effect of inoculum to substrate ratios on methane production in mixed anaerobic digestion of pig manure and blue green algae. *Transactions of the Chinese Society of Agricultural Engineering* 25: 172–176.

Sialve, B., Bernet, N., and Bernard, O. (2009). Anaerobic digestion of microalgae as a necessary step to make microalgal biodiesel sustainable. *Biotechnology Advances* 27: 409–416.

Sims KM (2013) Strategies to enhance conversion of lignocellulosic biomass to fermentable sugars and to enhance anaerobic digestion of algal biomass for biogas production: A report

Slade, R. and Bauen, A. (2013). Micro-algae cultivation for biofuels: cost, energy balance environmental impacts and future prospects. *Biomass and Bioenergy* 53: 29–38.

Suali, E. and Sarbatly, R. (2012). Conversion of microalgae to biofuel. *Renewable and Sustainable Energy Reviews* 16: 4316–4342.

Suganya, T., Varman, M., Masjuki, H.H., and Renganathan, S. (2016). Macroalgae and microalgae as a potential source for commercial applications along with biofuels production: a biorefinery approach. *Renewable and Sustainable Energy Reviews* 55: 909–941.

Tamilarasan, K., Kavitha, S., Selvam, A. et al. (2018). Cost-effective, low thermo-chemo disperser pretreatment for biogas production potential of marine macroalgae Chaetomorpha antennina. *Energy* 163: 533–545.

Vandenbroucke, M. and Largeau, C. (2007). Kerogen origin, evolution and structure. *Organic Geochemistry* 38: 719–833.

Vergara-Fernandez, A., Vargas, G., Alarcón, N., and Velasco, A. (2008). Evaluation of marine algae as a source of biogas in a two stage anaerobic reactor system. *Biomass and Bioenergy* 32: 338–344.

Ward, A.J., Lewis, D.M., and Green, F.B. (2014). Anaerobic digestion of algae biomass: a review. *Algal Research* 5: 204–214.

Weijermars, R., Taylor, P., Bahn, O. et al. (2012). Review of models and actors in energy mix optimization- can leader visions and decisions aligh with optimum model strategies for our future energy systems? *Energy Strategy Reviews* 1: 5–18.

Yaun, X., Wang, M., Park, C. et al. (2012). Microalgae growth using high strength wastewater followed by anaerobic digestion. *Water Environment Research* 84: 396–404.

Yen, H.W. and Brune, D. (2007). Anaerobic co-digestion of algal sludge and waste paper to produce methane. *Bioresource Technology* 98: 130–134.

Zamalloa, C., Boon, N., and Verstraete, W. (2012). Anaerobic digestibility of *Scenedesmus obliquus* and *Phaeodactylumtricornutum* under mesophilic and thermophilic conditions. *Applied Energy* 92: 733–738.

Zhao, B., Ma, J., Zhao, Q. et al. (2014). Efficient anaerobic digestion of whole microalgae and lipid extracted microalgae residues for methane residues for methane energy production. *Bioresource Technology* 161: 423–430.

Part IV

Perspectives

19

Integrated Biorefineries for the Production of Bioethanol, Biodiesel, and Other Commodity Chemicals

Pedro F Souza Filho[1,2] *and Mohammad J Taherzadeh*[1]

[1] *Swedish Centre for Resource Recovery, University of Borås, Borås 501 90, Sweden*
[2] *Current Address: Biochemical Engineering Laboratory, Chemical Engineering Department, Federal University of Rio Grande do Norte, Natal/RN 59078-970, Brazil*

CHAPTER MENU

19.1 Introduction

Increasing concerns about the environmental impacts of human activities have been leading the search for a new model of *bioeconomy* based on the production of energy and chemicals from renewable sources (Hasunuma et al. 2013). Climate change, energy security and rural development are pointed out as the main drivers for the change from oil reserves towards biomass technology. In this context, biorefineries are facilities that integrate biomass conversion processes and equipment to exploit all the potential of biological raw materials in a sustainable, environment-, and resource-friendly process. The term biorefinery was coined in the late 1990s to describe the idea of a cluster of facilities, processes, and industries that can use any biomass feedstock to produce a range of products, including fuels, power, heat, chemicals, food, feed, and materials (Figure 19.1).

Several similarities can be observed between oil refineries and biorefineries (Cherubini 2010; Kamm and Kamm 2004a). Both petroleum and biomass have complex compositions, which need, in a first stage to be fractionated into their main components for further processing. Fractionation of the biomass into its core constituents is an important step which has the potential to benefit a wide range of bioprocessing industries. (FitzPatrick et al. 2010).

Biorefinery operation proposed by Ragauskas et al. (2006) starts with the extraction of high-value chemicals already present in the biomass, such as fragrances, flavouring agents, food-related products, and high-value nutraceuticals that provide health and medical benefits.

Green Energy to Sustainability: Strategies for Global Industries, First Edition.
Edited by Alain A. Vertès, Nasib Qureshi, Hans P. Blaschek and Hideaki Yukawa.

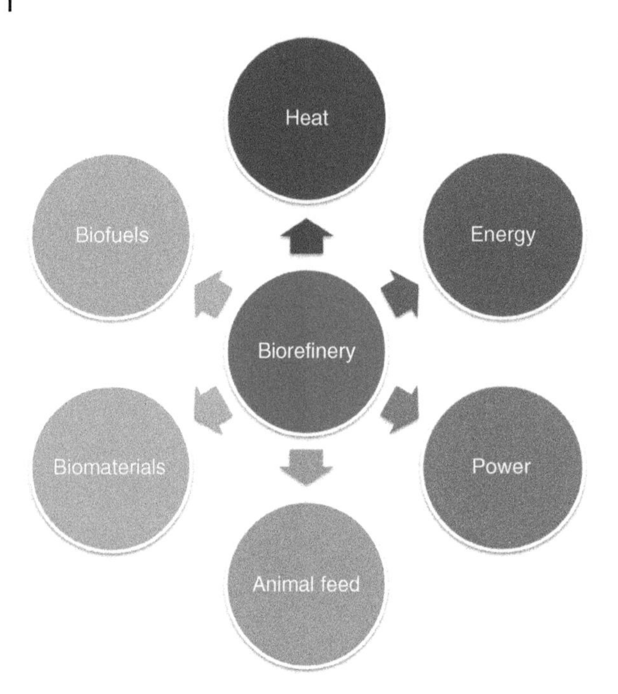

Figure 19.1 Products from a biorefinery.

Once these chemicals are extracted, the biorefinery can focus on processing plant polysaccharides and lignin. Products from a biorefinery can be categorized into high-value low-volume (HVLV) and low-value high-volume (LVHV) products. Biorefineries must be designed to favour HVLV, which also enhances the economical profitability of the industrial unit. LVHV products are generally used as fuel to provide the required energy to the facility. The low energy content of biomass, however, demands that a large amount of feedstock is used to produce energy. Therefore, a limitation in the global demand of energy by biorefineries is caused by constraints in biomass productivity, including seasonality, and soil and water limitations (Cherubini 2010; FitzPatrick et al. 2010; Fernando et al. 2006).

Biorefineries should be developed as dispersed industrial complexes able to revitalize rural areas. Contrary to oil refineries, which are usually very large plants, biorefineries will most probably encompass a whole range of different-sized installations. Several bio-industries can even combine their material flows in order to reach a complete utilization of all biomass components: the residue from one bio-industry becomes an input for other industries, giving rise to integrated bio-industrial systems (Cherubini 2010).

19.2 Types of Biorefineries

Different types of biorefinery can be grouped under certain categories according to their feedstock, flexibility and/or the technology used.

19.2.1 Flexibility

Depending on the flexibility of the operation of a biorefinery, it can be classified as phase I, II, or III. Phase I biorefineries have fixed processing capabilities, use only one feedstock, and have a single main product. Examples of this type of biorefinery are pulp and paper mills, and dry-milling ethanol plant, which convert corn grains into ethanol, feed coproducts, and carbon dioxide. In phase II biorefineries, one single feedstock can be transformed into several end products, depending on the market demand, prices, contract obligations and the plant operating limits. Wet-milling ethanol facilities are examples of phase II biorefineries. They produce ethanol, starch, fructose syrup, corn oil, and gluten feed and meal. Phase I biorefineries can be converted into phase II by identifying ways to upgrade the different side-streams (Kamm and Kamm 2004a; Fernando et al. 2006; Clark and Deswarte 2008).

Biorefineries of phase III, besides producing multiple end-products, can also use different types of feedstocks, including a mixture of them, and processing methods. Phase II are the most advanced type of biorefinery. The flexibility on the choice of feedstock opens various options to achieve profitability and maximize returns, by allowing the plant to easily adapt to market demands (Kamm and Kamm 2004a; Fernando et al. 2006; Clark and Deswarte 2008). Three phase (III) biorefinery systems have been lately the focus of research and development due to their significant potential and are described in the following sections.

19.2.1.1 Lignocellulosic Feedstock (LCF) Biorefinery

It is believed that, due to the optimal situation of the raw material and the good position of the end products within the present petrochemical market and the future bio-based market, lignocellulose feedstock will have the highest success among biorefineries (Kamm and Kamm 2004a). Lignocellulosic material are low cost, large biomass per hectare yield since the whole crop is available as feedstock, and can be grown on land which is not suitable for agricultural crops (Cherubini 2010; Kamm and Kamm 2004a).

Examples of lignocellulose feedstocks which can be used in biorefineries include straw, corn stover, and wood. Lignocellulosic feedstock (LCF) biorefineries can start their development from pulp and paper plants, which are already specialized in collecting and processing large amounts of lignocellulosic biomass. A LCF biorefinery could therefore be built around the mills, either as an extension or as an 'across-the-fence'-type company (Clark and Deswarte 2008).

Lignocellulosic biomass is composed primarily of three fractions called: hemicelluloses, cellulose, and lignin. The carbohydrate components currently have useful applications such as the production of ethanol, acetic acid, butanol, acetone, and succinic acid among other fermentation products from the glucose obtained through the hydrolysis of the cellulose, and furfural from the xylose fraction of hemicellulose. The lignin fraction, although currently underused, has an enormous potential to produce a variety of mono-aromatic hydrocarbons provided efficient forms of isolating and cracking are developed (Fernando et al. 2006; Lora and Glasser 2002).

19.2.1.2 Whole Crop Biorefinery

Whole crop biorefineries will process and consume cereals as, for instance, wheat, maize, and rye to obtain energy, materials, and chemicals. The entire crop will be fed into the unit where, at first, the seed and the straw, which comprise almost the same amount, are separated for the next steps. In order to minimize energy and labour cost, it is ideal that both straw and seed are collected simultaneously and the separation only occurs in the biorefinery. The grain can be treated as presently in wet-milling ethanol facilities and the straw composed mainly by lignocellulosic material, can follow similar processes as the biomass in LCF biorefineries (Fernando et al. 2006; Clark and Deswarte 2008)

19.2.1.3 Green Biorefinery

A green biorefinery is a system that processes natural wet feedstock according to the physiology of the plant, and transforms it into multiple products. Examples of green biorefinery raw materials are grasses, green plants, green crops, and algae. The biomass is firstly refined using wet-fractionation, which produces a fibre-rich press cake and a nutrient-rich green juice (cf. Figure 19.2), preserving the natural forms of the substances. The juice contains numerous valuable chemicals such as proteins, free amino acids, organic acids, enzymes, hormones, dyes, and minerals, while the solid fraction is formed of cellulose, starch, dyes and pigments, crude drugs, and other organics. The press cake can be used as feed or to produce energy, insulation materials, construction panels, biocomposites, etc. The juice is directed to the production of fermentable products, such as lactic acid and ethanol, or for extraction of proteins and amino acids by acidification or heating (Kamm and Kamm 2004a; Fernando et al. 2006; Clark and Deswarte 2008; Starke et al. 2000). This kind of biorefinery has become popular especially in northern European countries (Parajuli et al. 2015).

19.2.2 Feedstock

Biorefineries can be divided into two groups according to the feedstock material used, the biomass-producing type and the waste-material-utilization type. In tropical countries with large areas of arable land such as Australia, Brazil, China, United States and Southeast Asia, the availability of crops will lead to the operation of biorefineries in the first case. On the other hand, Japan and some Europeans countries, which do not have large areas of lands available, neither for crop production nor for garbage landfill, will face biorefineries as a twofold solution. Disposal of organic waste will meet a solution and serve to produce useful products (Ohara 2003).

Municipal solid waste (MSW) represents an important source of carbon which could be useful for biobased technologies (Mohan et al. 2016). It is composed of the discard generated during the quotidian activities from residential and business areas. MSW has also the advantage of being

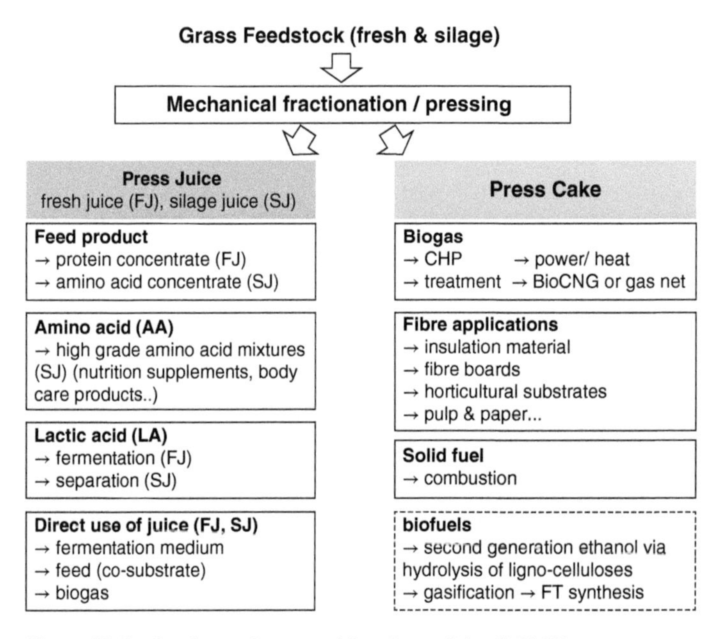

Figure 19.2 Products of a green biorefinery (Mandl 2010).

available during the entire year, unlike crops and their residues which are available seasonaly. The most common destination of MSW is disposition in landfills, a technique with several environmental issues such as ground water contamination and greenhouse gas (GHGs) emissions. Therefore, new technologies such as incineration, composting, and anaerobic digestion have been adopted for the treatment of MSW in order to produce energy, heat, fuel, and bioproducts. Due to its complex nature, it is hard to find a single universal process to exploit MSW. A biorefinery approach is the most appropriate way for producing value added products (Chen et al. 2016; Sawatdeenarunat et al. 2016).

The material (MSW) can also be separated into two groups: carbohydrate-rich material and lipid-rich material. Among the carbohydrate-rich material, one can divide biorefineries into first-generation, based on the easily available carbohydrates, usually used as energy reserve by the plants, like sucrose and starch, and second-generation, processing lignocellulosic material. The first generation of biorefineries is based on biochemical technologies while the second generation biorefineries are the LCF biorefineries previously discussed. Lipid-rich biomass, such as oilseeds, provide a unique opportunity for production of biofuel and high-value fatty acids with the capacity to replace petroleum in the production of lubricants and detergents, as well as biodiesel and glycerol (Octave and Thomas 2009).

19.3 Biorefinery Platforms

Biorefinery technological processes can be divided into four different platforms: thermochemical, biochemical, mechanical/physical, and chemical processes (Figure 19.3). The biochemical platform is focused on fermentation of carbohydrates such as those extracted from LCF. The biomass processing starts with a preparation of the feedstock followed by the conversion of fermentable sugars, bioconversion of the sugars using biocatalysts, and processing to deliver value added chemicals, fuel-grade ethanol and other fuels, heat and/or electricity (Parajuli et al. 2015). Biogas, composed of methane and carbon dioxide, can be produced through digestion of biomass using biochemical platform technology (McKendry 2002).

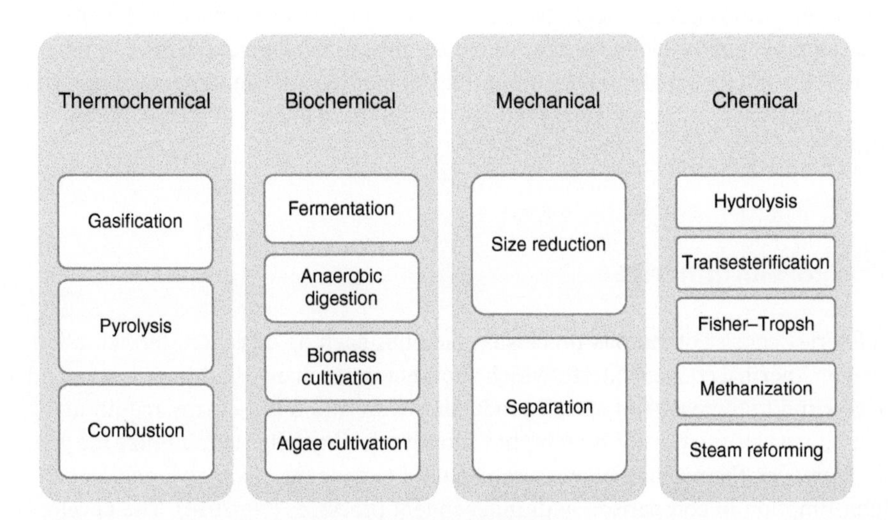

Figure 19.3 Biorefinery platforms and their processes.

Thermochemical platforms convert biomass mainly by gasification and pyrolysis. In the former process, the biomass is heated, usually in the range 800–900 °C, with approximately one third of the oxygen necessary for a total combustion, generating syngas. The gas can be directly burnt generating power, or be used to produce methanol and hydrogen, both having future as fuels for transportation. For the pyrolysis process, the biomass is heated in complete absence of oxygen, producing solid, gas, and liquid phase products. The temperature is generally around 500 °C. The liquid phase obtained, called bio-oil, can be used in engines and turbines (Parajuli et al. 2015; McKendry 2002).

Mechanical platforms involve processes which do not change the state or composition of the feedstock. They are used for size reduction and separation of the components of biomass. Mechanical processes are usually applied in all cases because the use of biomass in reduced size improves the performance of subsequent processes. Separation processes are used to extract and concentrate valuable compounds from biomass (Cherubini 2010).

Hydrolysis and transesterification are commonly used chemical processes in a biorefinery. Hydrolysis reactions are performed by the use of acids, alkalis or enzymes to depolymerize polysaccharides and proteins into their component monomers (e.g. glucose from cellulose) or derive chemicals (e.g. furfural from xylose). Through a transesterification process, vegetable oils can be converted to methyl or ethyl esters of fatty acids and represents the most common method to produce biodiesel currently (Cherubini 2010).

Biochemical processes are advantageous in terms of typically low processing temperature and pressure, and high selectivity and specificity of components targeted and products generated. The disadvantages, on the other hand, include the requirement for a pretreatment step, long processing times, large amounts of space for batch systems, and the downstream processes are energy consuming. Thermochemical platform technologies are usually fast and can be operated in continuous mode. However, they are non-specific, affect all components, and generally require high temperatures, resulting in a reduced energy efficiency and increased capital investment (Budarin et al. 2011).

A strategy using both platforms in the same plant has been studied by Fornell et al. (2013). They took a kraft pulp mill installation as a basis and adapted it to produce ethanol from the obtained pulp and dimethyl ether (DME) from the residue liquor containing lignin through gasification. The results showed heat self-sufficiency although there is electricity deficit. Economically, the process depends on the development of the biofuel prices. The research and development of the conversion platforms will permit the integrated biorefinery to continually increase its diversity and complexity further increasing effectiveness, efficiency, and productivity (Office of Energy Efficiency & Renewable Energy 2016).

19.4 Integrated Biorefineries

Integrated biorefineries consist of various processing facilities such as digestion, fermentation, pyrolysis, gasification, thermal conversion, etc. which allow energy exchange within processes and material reuse/recycling to increase the overall production with minimum energy requirement. Using the waste produced by one facility as an input to another facility helps to minimize the generation of waste. Moreover, the residual biomass can be used to generate electricity, reducing the overall energy consumption in comparison with independent processes (Ng 2010). The development of an integrated biorefinery is unique because of the new technologies involved as well as the integration and bundling of these technologies. A detailed and realistic project scope and expectation is critical to the successful deployment of an integrated biorefinery (Office of Energy Efficiency & Renewable Energy 2016).

Development, design, and operation of chemical processes have had significant successes since the 1970s with the advent of process synthesis. Process synthesis provides a systematic way to identify the types of equipment, flowrates, operating/design conditions and optimal interconnections among different units that create the best total flowsheet. The goal of process synthesis is to develop a methodology for the generation of optimal configurations for transforming a set of given feedstocks into a set of desired final products. It can be considered as an optimal configuration for such a process that provides the highest yield, the highest energy efficiency, or the most sustainable route (Yuan et al. 2013).

The majority of biorefinery processes focus on the production of biofuels based on thermochemical and biochemical technologies. Some of the methods used for the synthesis of petrochemical processes have also been utilized to deal with optimization of biorefining processes in order to search for the total use of feedstock and the incorporation of multiple biomass feedstocks. However, several challenges are still to be overcome in order to design a sustainable, cost-effective, and high carbon conversion efficiency biorefinery process. It might also be noted that sustainability becomes even more important considering the environmental implications of biorenewables (Yuan et al. 2013).

The design of biorefinery configurations commonly started with a core conversion technology such as pretreatment, hydrolysis, fermentation, digestion, gasification, and pyrolysis, which are usually components of a biorefinery. It is then followed by the addition of units for feedstock preparation, product separation and upgrading. It is also common to scale up units developed at laboratory scale and revise the process configuration based on the practical aspects of large-scale production (Pham and El-Halwagi 2012).

Several approaches have been used for the design optimization of biorefineries with many possible scenarios available which can be applied to biomass. Ng (2010) developed an automated procedure targeting finding maximum biofuel production and revenue targets prior to the detail design; Zondervan et al. (2011) proposed a mathematical approach using a mixed integer non-linear program to model a biorefinery producing ethanol, butanol, succinic acid and their blends with gasoline maximizing the yield of products and minimizing the costs and the generation of waste. Bao et al. (2011) introduced a shortcut method for the preliminary synthesis of pathway tracking chemicals on a superstructure representation. Pham and El-Halwagi (2012) presented a forward and backward branching technique which involves forward synthesis of biomass to intermediates and reverse synthesis starting with the desired product and identifying the species and pathways.

Kaparaju et al. (2009) studied a biorefinery model integrating bioethanol, biohydrogen, and biogas production using wheat straw. The solid fraction of wheat straw pretreated by hydrothermal pretreatment (rich in cellulose) was used to produce bioethanol in a process of separate hydrolysis and fermentation using *Saccharomyces cerevisiae* as biological agent. Since baker's yeast can only ferment hexoses, the liquid fraction obtained from pretreatment was used for production of biohydrogen through dark fermentation. The effluents from both fermentation processes were then used for production of methane gas. The authors also investigated the energy obtained from wheat straw through different scenarios, including incineration, biogas, bioethanol, and biohydrogen. Although incineration showed the highest energy output, the production of biofuel is recommended by the authors because of the large demand for liquid fuels in the transportation sector as well as the solid digested material resultant from fermentation which can be used as fertilizer for the cultivation of the same crop.

Microalgae have been the focus of many studies as a promising source of lipids for the production of biodiesel. However, the cultivation of algae aiming only at the production of biofuel was determined not to be profitable. Since lipids are only one of the many algae fractions with commercial interest, production of high value-added coproducts, such as 'nutraceuticals', fertilizer, heat, and energy from waste biomass of algae, is required. Therefore, the idea of a marine biorefinery becomes attractive given the higher growth rate compared to terrestrial biomass sources (Garcia Alba et al. 2011; González et al. 2015).

The transesterification reaction of lipids to produce biodiesel results in the formation of an important residue: glycerol. Due to the recent increase in biodiesel production, an excess of crude glycerol in the market is driving a decrease in glycerol prices. Extra costs associated with the presence of contaminants, like water, salts, methanol, and soap, in the crude glycerol means that it is considered a waste instead of a coproduct. Consequently, it becomes essential to find new applications for the glycerol produced during the biodiesel synthesis. The low price and large availability of glycerol can encourage its use as a primary renewable building block within the biorefinery concept analogous to those of the petrochemical industry (methane, ethylene, BTX, etc.) (Almeida et al. 2012; Zheng et al. 2008). The several chemicals which can be obtained from glycerol are presented later in this chapter (Section 19.5.10).

Huang et al. (2010) simulated a model for an integrated forest biorefinery, based on modern day pulp and paper mills, a phase I biorefinery, with prior extraction of hemicellulose and short fibre cellulose, which is used for ethanol production. The new integrated forest biorefinery had an increased pulping capacity, lower emission of GHGs, and additional revenues and profits. Another example of an integrated forest biorefinery model is presented in the work of Moshkelani et al. (2013) and is shown in the Figure 19.4. Bioproducts like ethanol, furfural, and lignin can be produced from the by-products of the kraft pulp mill together with heat and energy from an associated gasification unit.

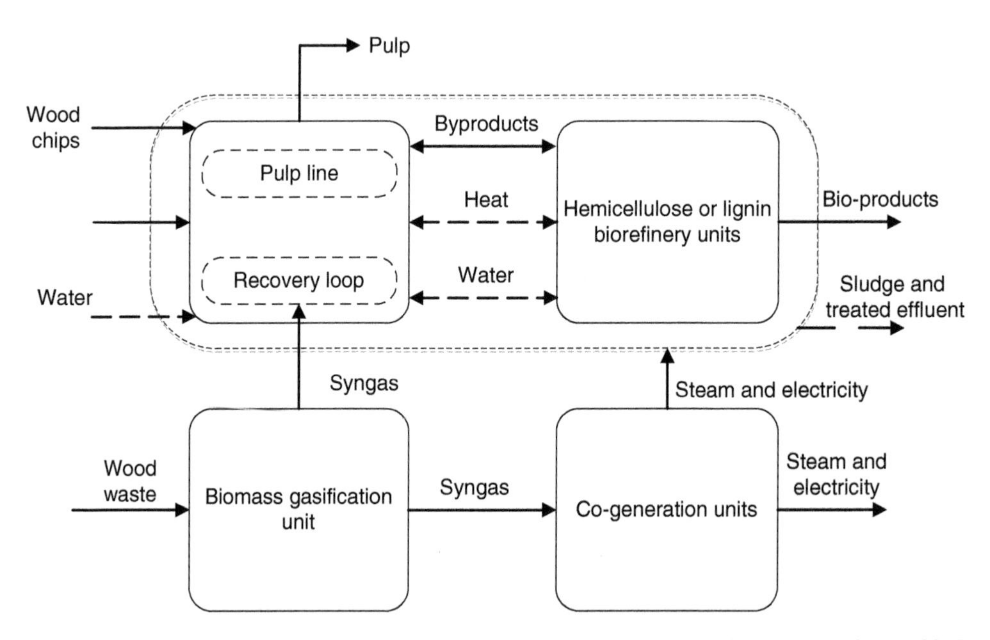

Figure 19.4 Integrated biorefinery concept developed from a Kraft pulp mill (Moshkelani et al. 2013).

The potential of a biorefinery using Zygomycetes and Ascomycetes filamentous fungi as biocatalyst has been reviewed (Ferreira et al. 2013, 2016). The advantages of using these fungi include their ability to grow in a vast range of culture media and the versatility of products they can metabolize, including lactic and fumaric acid and ethanol. Moreover, the biomass of Zygomycetes is composed of proteins with superior amino acid composition, lipids, and chitosan, which makes it eligible for animal feed purposes, such as fish feed, lipid extraction, and chitosan production.

19.5 Coproducts

The replacement of fossil fuels with biomass-derived fuels is considered difficult to achieve due to the extremely large volume of oil utilized and the commodity nature of the pricing. Other products from the refining process, however, can be more easily replaced with renewable sources (Mandalika et al. 2014). The United States Department of Energy (US DOE) identified in 2004 12 building block chemicals that can be produced from sugars via biological or chemical conversions that could serve as an economic driver for a biorefinery. The identification had the purpose of finding common challenges and barriers of associated production technologies. The 12 sugar-based building blocks are 1,4-diacids (succinic, fumaric, and malic), 2,5-furan dicarboxylic acid, 3-hydroxy propionic acid, aspartic acid, glucaric acid, glutamic acid, itaconic acid, levulinic acid, 3-hydroxybutyrolactone, glycerol, sorbitol, and xylitol/arabinitol. The building blocks (Figure 19.5) are formed by molecules with multiple functional groups and can be converted to a variety of high-value bio-based chemicals or materials and the selection factors include cost of feedstock, estimated processing costs, current market volumes and prices, and relevance to current or future biorefinery operations (Werpy and Petersen 2004).

Bozell and Petersen (Bozell and Petersen 2010) reviewed the previous report, delimiting the candidates and including new chemical with the same characteristics such as ethanol, lactic acid, and biohydrocarbons. The criteria used by the authors to evaluate the blocks were based on the market and scientific advances which occurred during the six years after the US DOE publication (Werpy and Petersen 2004).

19.5.1 Four Carbon 1,4-Diacids

Succinic, fumaric, and malic acids were grouped together because of their similar synthetic pathways. Succinic acid, also known as amber acid or butanedioic acid, is widely recognized as a potential platform chemical for the production of various value-added derivatives but its production through a fermentative route is still expensive compared with the traditional chemical production processes (Du et al. 2008). Succinic acid can be used in the production of many chemicals with important industrial applications including adipic acid, 1,4-butanediol, tetrahydrofuran, N-methyl pyrrolidinone, 2-pyrrolidinone, succinate salts, gamma-butyrolactone, biodegradable polymers such as polybutyrate succinate (PBS), polyamides (Nylon®x,4) and various green solvents, which can increase its market demand (Song and Lee 2006). The expectation is that the demand will increase from the current 30 000–50 000 tons/year to 700 000 tons/year by 2020 (Choi et al. 2015).

Fumaric acid is produced by many microorganisms in small amounts as a key intermediate in the citrate cycle. Because of its structure containing two carboxylic groups and a carbon–carbon double bond, and its non-toxic and non-hygroscopic nature, fumaric acid has great potential for industrial applications, e.g. as a starting material for polymerization and esterification reactions. Moreover,

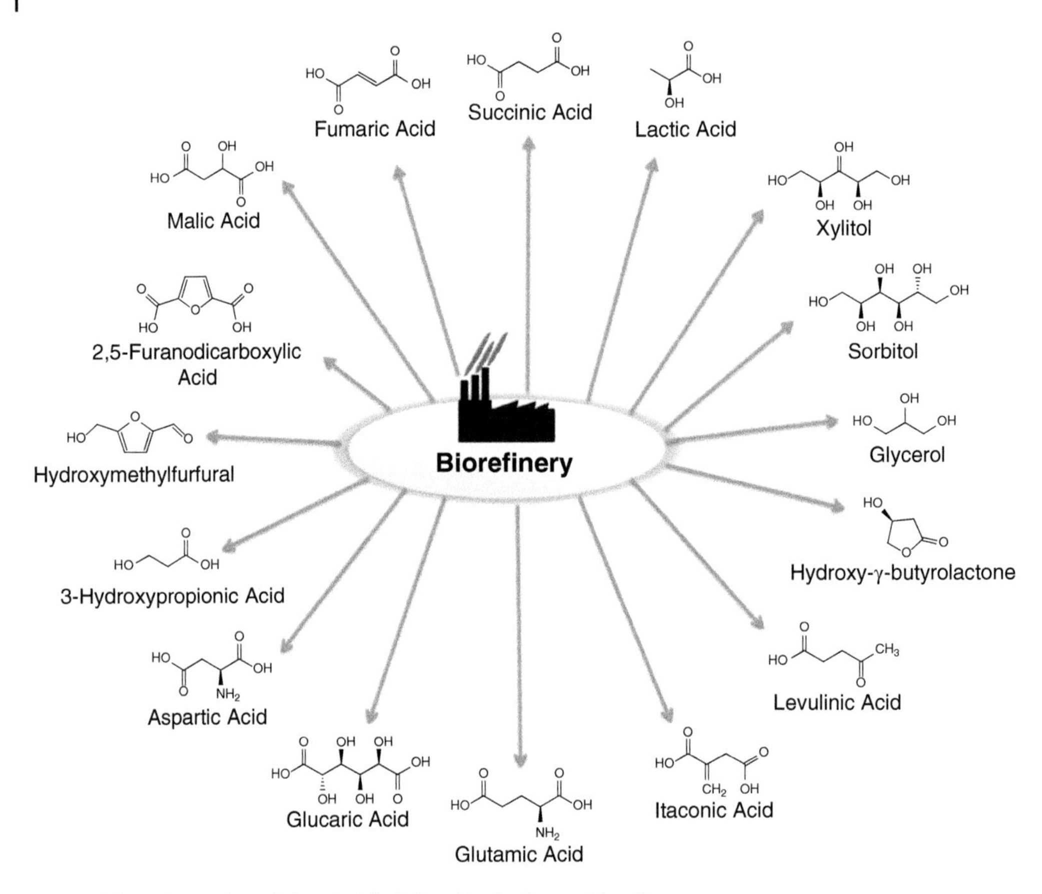

Figure 19.5 Examples of chemical building blocks from a biorefinery.

fumaric acid can be used as a medicine to treat psoriasis, as an acidulant in the food industry, and as a supplement in cattle feed. Recovery and purification of fumaric acid can be achieved by using or employing its low aqueous solubility (7 g/kg at 25 °C; 89 g/kg at 100 °C) and low pK_a values (3.03 and 4.44). Fumaric acid was commercially produced by fermentation until the 1940s, when its fermentation was discontinued and replaced by the petrochemical route, where it is produced by isomerization of maleic acid, which, in turn, is produced from maleic anhydride (Engel et al. 2008; Carta et al. 1999).

Malic acid is also an intermediate of the cell metabolism. It is the second most widely used acidulant in the food industry (10% of the market) and is also used in pharmaceuticals, cosmetics, and medicine. Malic acid is produced chemically by the hydration of fumaric acid (Presečki and Vasić-Rački 2005).

19.5.2 2,5-Furandicarboxylic Acid (FDA) and 5-Hydroximethilfurfural (HMF)

5-Hydroximethilfurfural (HMF) can be obtained by the triple dehydration of glucose. A subsequent hydration reaction or an oxidation can convert HMF into levulinic acid and Furandicarboxylic Acid (FDA), respectively. HMF is rather an unstable molecule which can be found in natural products such as honey and other heat processed food products formed in the thermal decomposition of carbohydrates. The oxidation of HMF is of particular interest. FDA can be used as a replacement for terephthalic acid for the production of polyethylene terephthalate and polybutylene terephthalate. The partially oxidized compounds can also be used as polymer building blocks although these have a more difficult selective production. (de Jong et al. 2012) discussed the substitution of terephthalic acid by FDA for the production of polyesters such as polyethylene terephthalate (PET). FDA is a chemically very stable compound. Its only current uses are in small amounts in fire foams and in medicine to remove kidney stones (Boisen et al. 2009; de Jong et al. 2012).

FDA and its derivatives have not become commercial products yet. However, improvements made in the production of FDA and its derivatives, since the 2004 report was published (US DOE), show that the potential of providing biobased replacements for polymers is the largest segment of the chemical industry. Moreover, the development of more efficient HMF production and one step dehydration/oxidation of sugars to FDA support the potential of FDA as a building block (Bozell and Petersen 2010).

19.5.3 3-Hydroxypropionic Acid (3-HP)

3-Hydroxypropionic acid (3-HP) is a highly reactive C3 molecule, which is a promising building block for the production of polymer materials and other chemical feedstocks. The compound can be used as cross-linking agent for polymer coatings, metal lubricants and antistatic agents for textiles, and for the production of chemicals, such as acrylic acid, 1,3-propanediol (PD), methyl acrylate, acrylamide, ethyl 3-HP, malonic acid, propiolactone, and acrylonitrile. Moreover, 3-HP is a starting material for cyclization and polymerization reactions to produce propiolactone, polyesters, and oligomers. The biodegradable polymers of 3-HP have the potential to replace a variety of traditional petroleum-based polymers and be used in new fields such as surgical biocomposite material and drug release material (Jiang et al. 2009; Kumar et al. 2013).

3-HP is produced by microorganisms as an intermediate or final product. Autotrophic bacteria *Chloroflexus aurantiacus* produces it in the 3-HP cycle for fixation of carbon dioxide. It is also produced by yeast as one of the final products of uracil degradation. However, there is no known organism able to produce 3-HP as a major metabolic end product (Jiang et al. 2009).

19.5.4 Aspartic Acid

Aspartic acid is a C4 amino acid commonly used in the food industry for the production of the sweetener aspartame. It is presently produced, preferably, by the enzymatic conversion of ammonia and fumaric acid. Aspartate is an essential part of metabolism for many species and, therefore, can also be produced by the fermentative route. However, the high cost of fermentative production still favours the enzymatic route. Its derivatives have large application in the polymer and solvent markets (Werpy and Petersen 2004; Leuchtenberger et al. 2005).

19.5.5 Glucaric Acid

Glucaric acid can be found in fruits, vegetables, and mammals and has been the subject of medical studies for cholesterol reduction and cancer chemotherapy. It is presently produced by the oxidation of glucose with nitric acid in a non-selective and costly process. Glucaric acid has the potential to be used as a building block for a number of polymers, including new nylons and hyperbranched polyesters. The chemical is not naturally produced by microorganisms thus a biosynthetic pathway needs to be constructed in an appropriate host organism for their production, like *Escherichia coli* (Moon et al. 2009; Lee et al. 2011).

19.5.6 Glutamic Acid

Glutamic acid is another important amino acid in the food industry. Its salt, monosodium glutamate, is largely used as flavour enhancer. Glutamic acid is presently the amino acid with the highest production capacity and demand. It is produced exclusively by mutants of *Corynebacterium glutamicum*, a Gram-positive soil bacterium; its cellular metabolism has been engineered for more than 40 years (Leuchtenberger et al. 2005; De Graaf et al. 2001).

When it was included in the 2004 report (US DOE), glutamic acid had a status as an existing commercial product which suggested high potential as a source of new derivatives. The amino acid has, however, remained a final product of the chemical industry and research activity in either glutamic acid production; its use as a platform chemical was low (Bozell and Petersen 2010).

19.5.7 Itaconic Acid

Itaconic acid is a white crystalline unsaturated dicarbonic acid with a methylene group conjugated to one carboxyl group. It is used worldwide in the industrial synthesis of resins such as polyesters, plastics, artificial glass, and in the preparation of bioactive compounds in the agriculture, pharmacy, and medicine sectors. Although it can be produced through chemical processes, fermentation by fungi is the most competitive route. *Aspergillus terreus* is the microorganism that is most commonly used in commercial production of itaconic acid. However, the highest yield is achieved when glucose or sucrose is used as the substrate, but the high price of the purified sugars has led to the search for cheaper raw materials, such as starch, molasses, hydrolysates of corn syrup or wood, and other microorganisms that are not as sensitive to particular fermentation conditions (Willke and Vorlop 2001; Okabe et al. 2009).

Polymers of itaconic acid are used in plastics, adhesives, elastomers, and coatings. The reaction of itaconic acid with amines produces pyrrolidones which have applications as thickeners in lubricating grease, shampoos, pharmaceuticals, and herbicides (Willke and Vorlop 2001; Okabe et al. 2009). Poli(acrylamide-co-itaconic acid) is a superabsorbent polymer and represents the most promising market for this building block (Choi et al. 2015).

19.5.8 Levulinic Acid

Levulinic acid can be obtained economically with high yields via acid hydrolysis of cellulosic materials. Levulinic acid has the potential to serve as a platform chemical for the production of a wide range of value-added compounds such as fuel additives, textile dye, antifreeze, animal feed, coating material, solvent, food flavouring agent, pharmaceutical compounds, resin, monomers for plastics and textiles, and other chemicals (Chang et al. 2007). Among its derivative, γ-valerolactone is a potential gasoline additive and 5-aminolevulinic acid can be used as a biological herbicide (Choi et al. 2015).

19.5.9 3-Hydroxybutyrolactone (HBL)

3-Hydroxybutyrolactone (HBL) can be obtained in high yield from carbohydrates such as starch, lactose, arabinose, maltose, and maltodextrin via chemoenzymatic or chemical oxidation. Its chiral properties and the very high enantiomeric purity that can be obtained make it applicable for the synthesis of chiral pharmaceuticals and nutraceuticals such as HIV inhibitors, synthetic statins, and L-carnitine (Choi et al. 2015; Lee and Park 2009; Wang and Hollingsworth 1999).

19.5.10 Glycerol

Currently, there is a large surplus of glycerol in the market. Glycerol has been classically used in the pharmaceutical, cosmetic, and surfactants industries. Nowadays, biodiesel production results in a large production of glycerol as a by-product, which has yet to find new applications in the chemical market. Many microorganisms are able to metabolize glycerol to synthesize a variety of compounds such as 1,3-propanediol, acetate, lactate, formate, succinate, ethanol, butanol, and dihydroxyacetone (the main active ingredient in sunless tanning formulas) (Pagliaro et al. 2007; Amaral et al. 2009).

19.5.11 Sorbitol

Sorbitol is a polyol with applications in the cosmetics, pharmaceuticals, foods, and chemical industries as well as sugar substitute for diabetic diets. The industrial production happens through the hydrogenation of glucose. Sorbitol can be converted into alkanes by repeated cycling of dehydration and hydrogenation reactions in the presence of hydrogen. It has been reported that some strains of the bacterium *Zymomonas mobilis* produce sorbitol from sucrose or glucose-fructose mixtures (Huber et al. 2004; Duvnjak et al. 1991).

19.5.12 Xylitol

Xylitol is a polyol traditionally produced by the catalytic hydrogenation of xylose with applications in food, cosmetic, and pharmaceutical industries. Biotechnological production of xylitol is possible using bacteria, filamentous fungi and yeasts in an economical manner since microorganisms do not need the high pressure and temperature used for the chemical production. Sorbitol and xylitol can be reformed to alkane in a similar process (Kirilin et al. 2012; Sampaio et al. 2008; Santos et al. 2008).

19.5.13 Lactic Acid

Lactic acid is the most widely occurring hydroxycarboxylic acid which can be produced by fermentation or chemical synthesis. The principal application of the lactic acid produced through lactic acid fermentation is in the food industry. It can also be polymerized to form polylactate, which is used in plastic manufacturing. Lactic acid is esterified with ethanol to produce ethyl lactate, a biodegradable solvent used as a replacement for chlorofluorocarbons as the washing solvent for semiconductors (Ohara 2003; Datta and Henry 2006; Kamm and Kamm 2004b).

19.5.14 Biohydrocarbons

Liquid alkanes are important components of fuels and chemicals and attempts have been made to produce them from other organic chemicals like levulinic acid, furanic compounds and cellulose. Bio-oil obtained from pyrolysis can also be upgraded into hydrocarbons or by the Fischer–Tropsch process of aqueous-phase catalytic conversion of methanol. Biohydrocarbons in the range of gasoline (C5–C8) have similar properties to conventional gasoline and are compatible with conventional cars and infrastructure. The quality of the fuel, however, varies depending on the feedstock and production technology used from high-aromatic and high-octane qualities to paraffinic low-octane hydrocarbon mixtures (Aakko-Saksa et al. 2014; Shemfe et al. 2016; Oya Si et al. 2015).

19.5.15 Lignin

Lignin, one of the three major components of lignocellulosic materials, is considered one of the most significant contributors to biomass recalcitrance. It is clear that large amounts of lignin will be produced after the use of the carbohydrate fraction of lignocellulosic biomass. Lignin can be potentially applied to produce power, fuel, syngas, macromolecules, and low-weight aromatic compounds. Combustion, gasification, pyrolysis, and hydroliquefaction are processes capable of converting lignin to power, fuel, and syngas. The use of lignin as macromolecules finds applications in the areas of carbon fibre, polymer modifiers, resins, adhesives, and binders. Moreover, lignin is the most important renewable source of aromatic compounds which are suitable for the production of low-molecular weight aromatic molecules (Yuan et al. 2013; Holladay et al. 2007).

19.6 Integrating Ethanol and Biodiesel Refineries

Biodiesel and bioethanol are considered the most important liquid biofuels used in the transportation sector. The transesterification of vegetable oils to produce biodiesel is carried out preferably using methanol instead of ethanol due to the lower cost of the process since the methanolysis reaction is much more rapid than ethanolysis, and requires smaller reactors. The use of ethanol for biodiesel production, however, can result in a large economic benefit considering a biorefinery plant in which the production of ethanol is integrated with the biodiesel. Ethanol is also advantageous as it is less toxic than methanol and is considered a renewable source, unlike methanol, which is mostly produced from natural gas. The different alcohols result in fuels with little changes in their characteristics, like lower cloud and pour points and moderately larger viscosities for FAEEs (fatty acid ethyl ester) compared to FAMEs (fatty acid methyl ester) (Gutiérrez et al. 2009; Severson et al. 2013; Meneghetti et al. 2006; Pourzolfaghar et al. 2016).

Since the transesterification reaction is carried out with a large excess of alcohol, due to its lower cost, recovering the unreacted alcohol can result in an improvement of the economy of the biodiesel production (Severson et al. 2013). In their article, Gutiérrez et al. (2009) describe an integrated process for production of biodiesel from palm oil using the ethanol obtained from the lignocellulosic residues of the palm oil industry. The strategy resulted in a reduction of energy costs compared to the individual productions of ethanol and biodiesel. Moreover, the approach finds a valuable destination for the solid wastes generated during the palm oil production. The process, however, comes to a bottleneck in the fact that lignocellulosic ethanol production technology is not yet mature enough to economically compete with the ethanol produced from sugars and starch (Gutiérrez et al. 2009).

Figure 19.6 presents the flowsheet of a proposed integrated process for the combined production of ethanol and biodiesel from palm trees. The empty fruit bunches (EFBs) and the palm press fibre (PPF), both lignocellulosic residues, are pretreated and used to produce ethanol, which is purified and mixed with the palm oil in a reactor, resulting in a mixture of biodiesel, ethanol, and glycerol. The chemicals are separated, with the ethanol being recycled to the reactor. In addition to the biodiesel and glycerol produced in the process, a solid side-stream rich in lignin and a syrup stream containing unfermented sugars, like pentoses, are also obtained (Gutiérrez et al. 2009). The plant

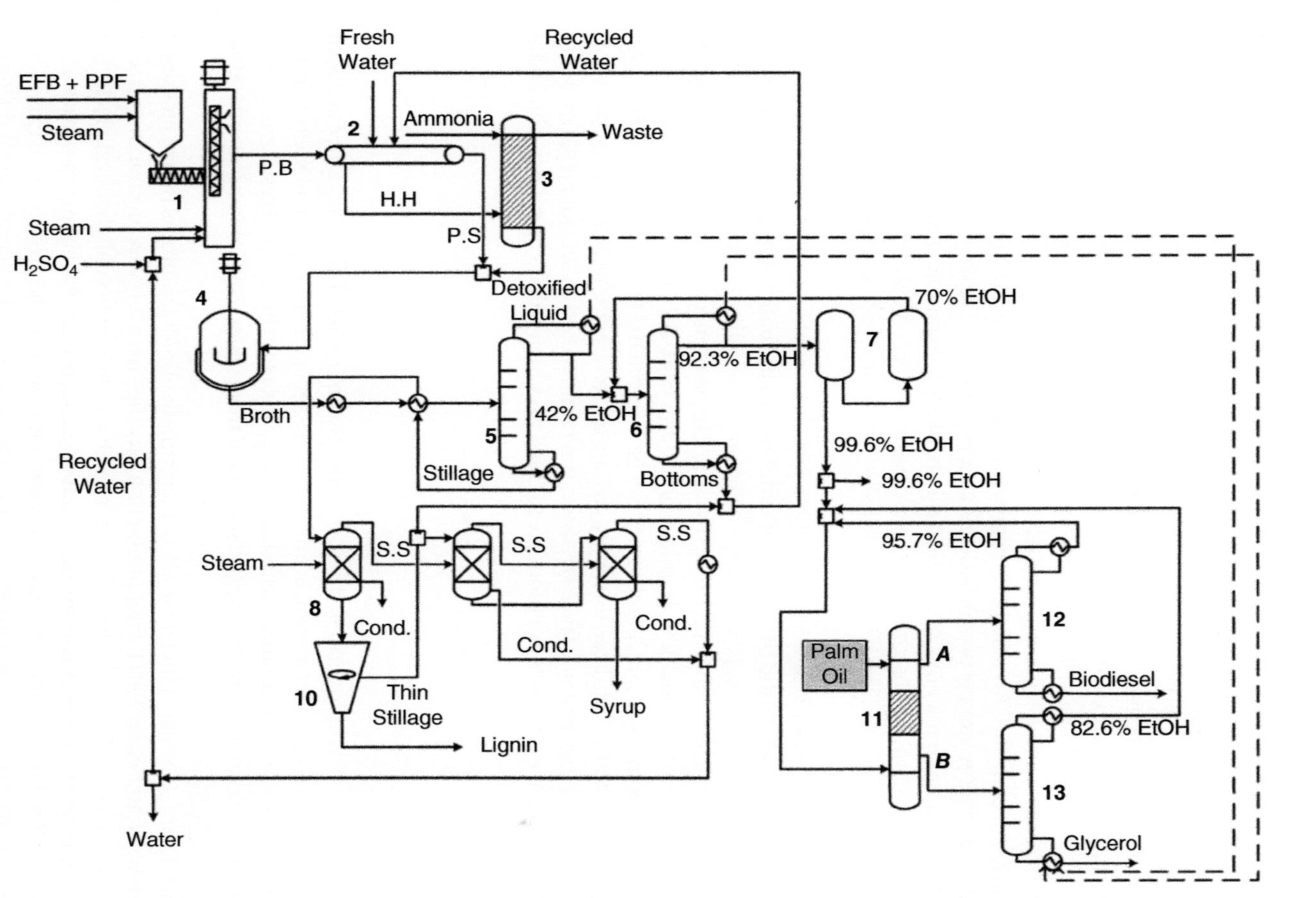

Figure 19.6 Integrated process flowsheet of the combined production of biodiesel and bioethanol. (1): pretreatment reactor, (2): washing, (3): ionic exchange, (4): simultaneous saccharification and cofermentation, (5): concentration column, (6): rectification column, (7): molecular sieves, (8): evaporation train, (10): centrifuge, (11): multi-stage reactor–extractor, (12): distillation column for biodiesel purification, (13): distillation column for glycerol purification. Material streams are represented by continuous lines whereas dashed lines represent heat streams. A: biodiesel-enriched stream, B: glycerol-enriched streams, EFB: empty fruit bunches, PPF: palm press fibre, H.H, hemicellulose hydrolyzate; P.B, pretreated biomass; P.S, pretreated solids, S.S, secondary steam (Gutiérrez et al. 2009).

can be further developed to use these coproducts as raw material to produce new chemicals (as presented in section 19.5) or sent to another facility.

Enzymatic production of biodiesel, using lipase enzymes, instead of the traditional catalysed chemical reaction, would result in an even cleaner process for the production of this biofuel. In the enzymatically catalysed reaction, ethanol has an advantage over methanol because methanol, which is more polar than ethanol and, therefore, less soluble in oil, causes the inactivation of the enzyme. Although the same phenomenon has been observed for ethanol, it happens to a lesser extent (Robles-Medina et al. 2009).

An integrated biorefinery system using algal biomass is interesting because of the many valuable coproducts which can be extracted. Moreover, algae have high growth rates and present the ability to grow in saline and brackish water, which are advantages compared to terrestrial crops. The microalgae biomass can be initially processed to produce recombinant proteins, such as antibodies, antivirals, and enzymes. A second step can be introduced to extract the algal oil. Lipids are produced by algae as energy storage products. Their concentration can be as high as 50–60% of the algal dry weight. Since some fractions of the algal oil are of high economic interest (e.g. omega-3 fatty acids), a separation unit can be added to the process in order to separate them from the rest of the oil. The low value oil can then be used for biodiesel production (Subhadra 2011; Jones and Mayfield 2012).

Macroalgae can also be used for biofuel production. Absolute absence or near absence of lignin in macroalgae makes the enzymatic hydrolysis of cellulose easier than hydrolysis of plant lignocellulosic biomass. The obtained sugars can then be fermented to ethanol in a scheme similar to a LCF biorefinery. Microalgal residue after oil extraction can be used as a source for fermentable sugars as well (Subhadra 2011; Jones and Mayfield 2012; John et al. 2011).

Figure 19.7 shows how the different biorefineries can be integrated. For instance, the dairy/meat industry can produce organic fertilizer to supply the algal biorefinery, which in turn produces enzymes used in the LCF biorefinery. The coproducts can then be used in several industries, contributing to the sustainable production of food, feed, fuel, and high-value coproducts (Subhadra 2011).

19.7 Economical Aspects

One of the major costs for bioproducts is the cost of the raw material. This cost can be reduced by using the whole crop to produce products and coproducts since high coproduct yield reduces the energy demand for the main product. Another strategy for reducing the energy demand is to recycle certain process streams, to minimize the amount of fresh water used (Hahn-Hägerdal et al. 2006).

Making a biorefinery thermally efficient is also a major challenge for the production of bio-based products (Bludowsky and Agar 2009). Fuel production facilities have a low return on investment because of the low value of their products. On the other hand, production of chemicals lacks the potential for a large energy displacement impact. Production of both chemicals and fuel in an integrated biorefinery is a solution to meet energy and economic goals simultaneously, with chemicals contributing to the profitability of the process and fuels impacting on the energy demand (Bozell 2008).

For instance, biodiesel production results in a significant production of glycerol, amounting to 10% (w/w) of the biodiesel produced. Considering the potential for biodiesel production, the associated production of glycerol would be higher than the market demand, generating an oversupply of

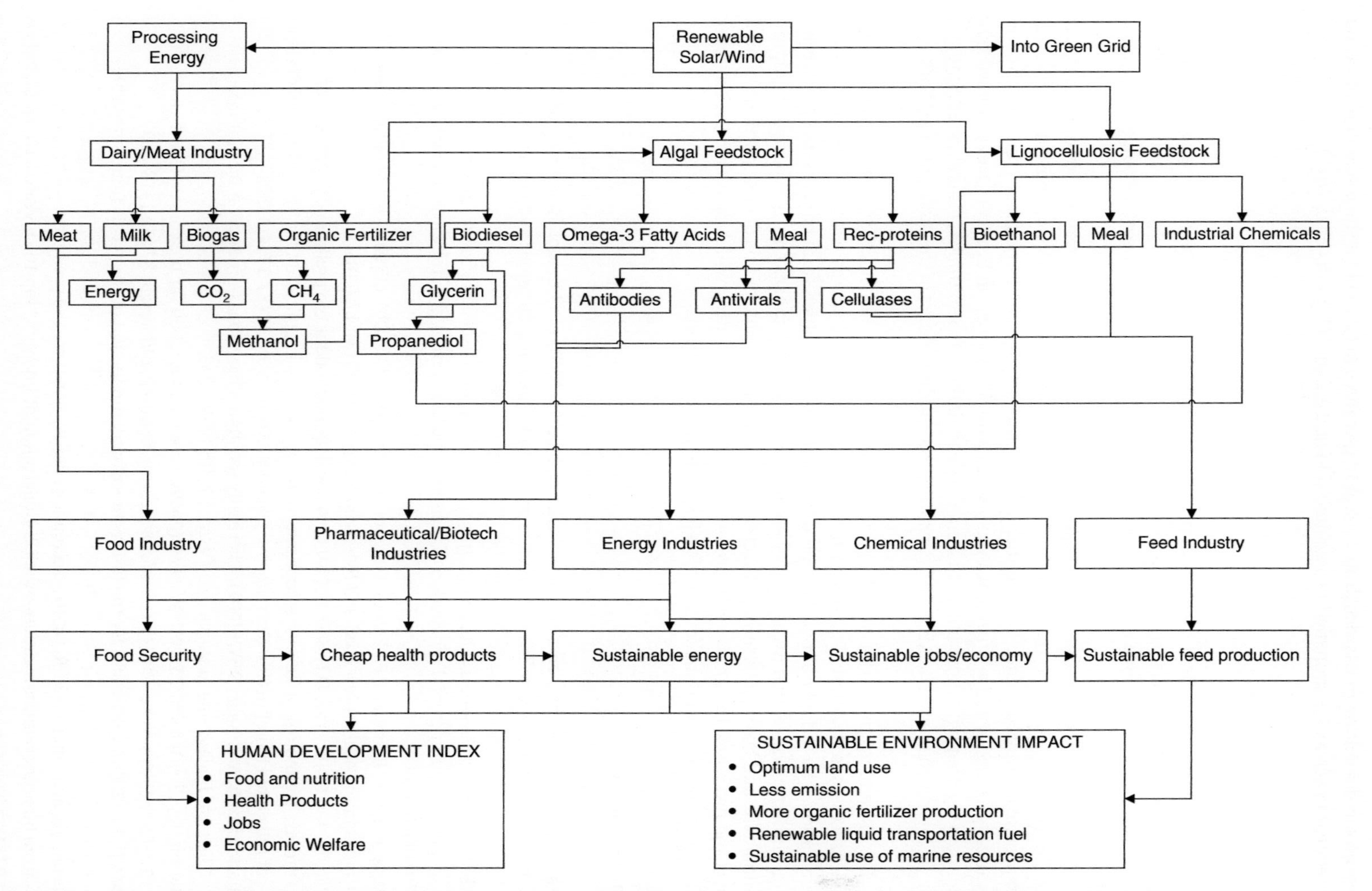

Figure 19.7 Integration of byproducts from bioindustries and its impact on environmental sustainability and human development (Subhadra 2011).

glycerol and resulting in its price drop. However, glycerol can be used for the production of various chemicals, such as 1,3-propanediol, succinic acid and ethanol (Vlysidis et al. 2011).

19.8 Perspectives

For a complete and efficient integration of the biorefineries it is important to set standards for the quality of the bioproducts in order to minimize variability. By achieving this, future research on biorefinery methods will have a focus, given the multidisciplinary nature of the subject (Fernando et al. 2006).

It is equally important that the techniques and methods developed result in minimum impact on the environment since the use of sustainable feedstock per se does not guarantee the lowest impact on the environment. The number of bioproducts which can be obtained from biorefineries is already extensive and will become even greater with further research. Substances extracted from biomass together with those obtained from the degradation of the biomass, either via fermentation or via thermochemical cracking, will constitute the building blocks of the biorefineries (Clark et al. 2009).

The biorefinery aims to harness the maximum of the biomass feedstock capacity, integrating several processes from different platforms in a facility, reducing the energy requirement and the waste generation. The number of potential products that can be obtained from biomass can replace several similar products that are currently obtained from fossil sources. This is expected to improve human life and leave no harmful impact on the environment. Due to this diverse portfolio of products and processes, integration of biorefineries represents a strategy which results in the reduction of wastes and costs, and makes the overall system more efficient.

References

Aakko-Saksa, P.T., Rantanen-Kolehmainen, L., and Skyttä, E. (2014). Ethanol, isobutanol, and biohydrocarbons as gasoline components in relation to gaseous emissions and particulate matter. *Environmental Science & Technology* 48 (17): 10489–10496.

Almeida, J.R.M., Fávaro, L.C.L., and Quirino, B.F. (2012). Biodiesel biorefinery: opportunities and challenges for microbial production of fuels and chemicals from glycerol waste. *Biotechnology for Biofuels* 5 (1): 1.

Amaral, P.F.F., Ferreira, T.F., Fontes, G.C., and Coelho, M.A.Z. (2009). Glycerol valorization: new biotechnological routes. *Food and Bioproducts Processing* 87 (3): 179–186.

Bao, B., Ng, D.K.S., Tay, D.H.S. et al. (2011). A shortcut method for the preliminary synthesis of process-technology pathways: an optimization approach and application for the conceptual design of integrated biorefineries. *Computers & Chemical Engineering* 35 (8): 1374–1383.

Bludowsky, T. and Agar, D.W. (2009). Thermally integrated bio-syngas-production for biorefineries. *Chemical Engineering Research and Design* 87 (9): 1328–1339.

Boisen, A., Christensen, T.B., Fu, W. et al. (2009). Process integration for the conversion of glucose to 2,5-furandicarboxylic acid. *Chemical Engineering Research and Design* 87 (9): 1318–1327.

Bozell, J.J. (2008). Feedstocks for the future–biorefinery production of chemicals from renewable carbon. *CLEAN–Soil, Air, Water* 36 (8): 641–647.

Bozell, J.J. and Petersen, G.R. (2010). Technology development for the production of biobased products from biorefinery carbohydrates—the US Department of Energy's "top 10" revisited. *Green Chemistry* 12 (4): 539–554.

Budarin, V.L., Shuttleworth, P.S., Dodson, J.R. et al. (2011). Use of green chemical technologies in an integrated biorefinery. *Energy & Environmental Science* 4 (2): 471–479.

Carta, F.S., Soccol, C.R., Ramos, L.P., and Fontana, J.D. (1999). Production of fumaric acid by fermentation of enzymatic hydrolysates derived from cassava bagasse. *Bioresource Technology* 68 (1): 23–28.

Chang, C., Cen, P., and Ma, X. (2007). Levulinic acid production from wheat straw. *Bioresource Technology* 98 (7): 1448–1453.

Chen, P., Xie, Q., Addy, M. et al. (2016). Utilization of municipal solid and liquid wastes for bioenergy and bioproducts production. *Bioresource Technology*.

Cherubini, F. (2010). The biorefinery concept: using biomass instead of oil for producing energy and chemicals. *Energy Conversion and Management* 51 (7): 1412–1421.

Choi, S., Song, C.W., Shin, J.H., and Lee, S.Y. (2015). Biorefineries for the production of top building block chemicals and their derivatives. *Metabolic Engineering* 28: 223–239.

Clark, J.H. and Deswarte, F.E.I. (2008). The biorefinery concept-an integrated approach. In: *Introduction to Chemicals from Biomass* (eds. J.H. Clark and F.E.I. Deswarte). Chichester, UK: Wiley.

Clark, J.H., Ei Deswarte, F., and J Farmer, T. (2009). The integration of green chemistry into future biorefineries. *Biofuels, Bioproducts and Biorefining* 3 (1): 72–90.

Datta, R. and Henry, M. (2006). Lactic acid: recent advances in products, processes and technologies—a review. *Journal of Chemical Technology and Biotechnology* 81 (7): 1119–1129.

De Graaf, A.A., Eggeling, L., and Sahm, H. (2001). Metabolic engineering for L-lysine production by *Corynebacterium glutamicum*. In: *Metabolic Engineering* (eds. J. Nielsen, L. Eggeling, J. Dynesen, et al.), 9–29. Springer.

Du, C., Lin, S.K.C., Koutinas, A. et al. (2008). A wheat biorefining strategy based on solid-state fermentation for fermentative production of succinic acid. *Bioresource Technology* 99 (17): 8310–8315.

Duvnjak, Z., Turcotte, G., and Duan, Z.D. (1991). Production of sorbitol and ethanol from Jerusalem artichokes by *Saccharomyces cerevisiae* ATCC 36859. *Applied Microbiology and Biotechnology* 35 (6): 711–715.

Engel, C.A.R., Straathof, A.J.J., Zijlmans, T.W. et al. (2008). Fumaric acid production by fermentation. *Applied Microbiology and Biotechnology* 78 (3): 379–389.

Fernando, S., Adhikari, S., Chandrapal, C., and Murali, N. (2006). Biorefineries: current status, challenges, and future direction. *Energy & Fuels* 20 (4): 1727–1737.

Ferreira, J.A., Lennartsson, P.R., Edebo, L., and Taherzadeh, M.J. (2013). Zygomycetes-based biorefinery: present status and future prospects. *Bioresource Technology* 135: 523–532.

Ferreira, J.A., Mahboubi, A., Lennartsson, P.R., and Taherzadeh, M.J. (2016). Waste biorefineries using filamentous ascomycetes fungi: present status and future prospects. *Bioresource Technology* 215: 334–345.

FitzPatrick, M., Champagne, P., Cunningham, M.F., and Whitney, R.A. (2010). A biorefinery processing perspective: treatment of lignocellulosic materials for the production of value-added products. *Bioresource Technology* 101 (23): 8915–8922.

Fornell, R., Berntsson, T., and Åsblad, A. (2013). Techno-economic analysis of a kraft pulp-mill-based biorefinery producing both ethanol and dimethyl ether. *Energy* 50: 83–92.

Garcia Alba, L., Torri, C., Samorì, C. et al. (2011). Hydrothermal treatment (HTT) of microalgae: evaluation of the process as conversion method in an algae biorefinery concept. *Energy & Fuels* 26 (1): 642–657.

González, L.E., Díaz, G.C., Aranda, D.A.G. et al. (2015). Biodiesel production based in microalgae: a biorefinery approach. *Natural Science* 7 (07): 358.

Gutiérrez, L.F., Sánchez, Ó.J., and Cardona, C.A. (2009). Process integration possibilities for biodiesel production from palm oil using ethanol obtained from lignocellulosic residues of oil palm industry. *Bioresource Technology* 100 (3): 1227–1237.

Hahn-Hägerdal, B., Galbe, M., Gorwa-Grauslund, M.-F. et al. (2006). Bio-ethanol–the fuel of tomorrow from the residues of today. *Trends in Biotechnology* 24 (12): 549–556.

Hasunuma, T., Okazaki, F., Okai, N. et al. (2013). A review of enzymes and microbes for lignocellulosic biorefinery and the possibility of their application to consolidated bioprocessing technology. *Bioresource Technology* 135: 513–522.

Holladay JE, White JF, Bozell JJ, Johnson D. Top Value-Added Chemicals from Biomass-Volume II—Results of Screening for Potential Candidates from Biorefinery Lignin: Pacific Northwest National Laboratory (PNNL), Richland, WA (US); 2007.

Huang, H.-J., Ramaswamy, S., Al-Dajani, W.W., and Tschirner, U. (2010). Process modeling and analysis of pulp mill-based integrated biorefinery with hemicellulose pre-extraction for ethanol production: a comparative study. *Bioresource Technology* 101 (2): 624–631.

Huber, G.W., Cortright, R.D., and Dumesic, J.A. (2004). Renewable alkanes by aqueous-phase reforming of biomass-derived oxygenates. *Angewandte Chemie International Edition* 43 (12): 1549–1551.

Jiang, X., Meng, X., and Xian, M. (2009). Biosynthetic pathways for 3-hydroxypropionic acid production. *Applied Microbiology and Biotechnology* 82 (6): 995–1003.

John, R.P., Anisha, G.S., Nampoothiri, K.M., and Pandey, A. (2011). Micro and macroalgal biomass: a renewable source for bioethanol. *Bioresource Technology* 102 (1): 186–193.

Jones, C.S. and Mayfield, S.P. (2012). Algae biofuels: versatility for the future of bioenergy. *Current Opinion in Biotechnology* 23 (3): 346–351.

de Jong, E., Dam, M.A., Sipos, L., and Gruter, G.J.M. (2012). Furandicarboxylic acid (FDCA), a versatile building block for a very interesting class of polyesters. *Biobased Monomers, Polymers, and Materials* 1105: 1–13.

Kamm, B. and Kamm, M. (2004a). Principles of biorefineries. *Applied Microbiology and Biotechnology* 64 (2): 137–145.

Kamm, B. and Kamm, M. (2004b). Biorefinery-systems. *Chemical and Biochemical Engineering Quarterly* 18 (1): 1–7.

Kaparaju, P., Serrano, M., Thomsen, A.B. et al. (2009). Bioethanol, biohydrogen and biogas production from wheat straw in a biorefinery concept. *Bioresource Technology* 100 (9): 2562–2568.

Kirilin, A.V., Tokarev, A.V., Kustov, L.M. et al. (2012). Aqueous phase reforming of xylitol and sorbitol: comparison and influence of substrate structure. *Applied Catalysis A: General* 435: 172–180.

Kumar, V., Ashok, S., and Park, S. (2013). Recent advances in biological production of 3-hydroxypropionic acid. *Biotechnology Advances* 31 (6): 945–961.

Lee, S.-H. and Park, O.-J. (2009). Uses and production of chiral 3-hydroxy-γ-butyrolactones and structurally related chemicals. *Applied Microbiology and Biotechnology* 84 (5): 817–828.

Lee, J.W., Kim, H.U., Choi, S. et al. (2011). Microbial production of building block chemicals and polymers. *Current Opinion in Biotechnology* 22 (6): 758–767.

Leuchtenberger, W., Huthmacher, K., and Drauz, K. (2005). Biotechnological production of amino acids and derivatives: current status and prospects. *Applied Microbiology and Biotechnology* 69 (1): 1–8.

Lora, J.H. and Glasser, W.G. (2002). Recent industrial applications of lignin: a sustainable alternative to nonrenewable materials. *Journal of Polymers and the Environment* 10 (1–2): 39–48.

Mandalika, A., Qin, L., Sato, T.K., and Runge, T. (2014). Integrated biorefinery model based on production of furans using open-ended high yield processes. *Green Chemistry* 16 (5): 2480–2489.

Mandl, M.G. (2010). Status of green biorefining in Europe. *Biofuels, Bioproducts and Biorefining* 4 (3): 268–274.
McKendry, P. (2002). Energy production from biomass (part 2): conversion technologies. *Bioresource Technology* 83 (1): 47–54.
Meneghetti, S.M.P., Meneghetti, M.R., Wolf, C.R. et al. (2006). Biodiesel from castor oil: a comparison of ethanolysis versus methanolysis. *Energy & Fuels* 20 (5): 2262–2265.
Mohan, S.V., Nikhil, G.N., Chiranjeevi, P. et al. (2016). Waste biorefinery models towards sustainable circular bioeconomy: critical review and future perspectives. *Bioresource Technology*.
Moon, T.S., Yoon, S.-H., Lanza, A.M. et al. (2009). Production of glucaric acid from a synthetic pathway in recombinant *Escherichia coli*. *Applied and Environmental Microbiology* 75 (3): 589–595.
Moshkelani, M., Marinova, M., Perrier, M., and Paris, J. (2013). The forest biorefinery and its implementation in the pulp and paper industry: energy overview. *Applied Thermal Engineering* 50 (2): 1427–1436.
Ng, D.K.S. (2010). Automated targeting for the synthesis of an integrated biorefinery. *Chemical Engineering Journal* 162 (1): 67–74.
Octave, S. and Thomas, D. (2009). Biorefinery: toward an industrial metabolism. *Biochimie* 91 (6): 659–664.
Office of Energy Efficiency & Renewable Energy Development of Integrated Biorefineries. 2016 [cited 2016]; Available from: http://www.energy.gov/eere/bioenergy/development-integrated-biorefineries
Ohara, H. (2003). Biorefinery. *Applied Microbiology and Biotechnology* 62 (5–6): 474–477.
Okabe, M., Lies, D., Kanamasa, S., and Park, E.Y. (2009). Biotechnological production of itaconic acid and its biosynthesis in *Aspergillus terreus*. *Applied Microbiology and Biotechnology* 84 (4): 597–606.
Oya Si, K.D., Watanabe, H., Tamura, M. et al. (2015). Catalytic production of branched small alkanes from biohydrocarbons. *ChemSusChem* 8 (15): 2472–2475.
Pagliaro, M., Ciriminna, R., Kimura, H. et al. (2007). From glycerol to value-added products. *Angewandte Chemie International Edition* 46 (24): 4434–4440.
Parajuli, R., Dalgaard, T., Jørgensen, U. et al. (2015). Biorefining in the prevailing energy and materials crisis: a review of sustainable pathways for biorefinery value chains and sustainability assessment methodologies. *Renewable and Sustainable Energy Reviews* 43: 244–263.
Pham, V. and El-Halwagi, M. (2012). Process synthesis and optimization of biorefinery configurations. *AICHE Journal* 58 (4): 1212–1221.
Pourzolfaghar, H., Abnisa, F., Daud, W.M.A.W., and Aroua, M.K. (2016). A review of the enzymatic hydroesterification process for biodiesel production. *Renewable and Sustainable Energy Reviews* 61: 245–257.
Presečki, A.V. and Vasić-Rački, Đ. (2005). Production of L-malic acid by permeabilized cells of commercial Saccharomyces sp. strains. *Biotechnology Letters* 27 (23–24): 1835–1839.
Ragauskas, A.J., Williams, C.K., Davison, B.H. et al. (2006). The path forward for biofuels and biomaterials. *Science* 311 (5760): 484–489.
Robles-Medina, A., González-Moreno, P.A., Esteban-Cerdán, L., and Molina-Grima, E. (2009). Biocatalysis: towards ever greener biodiesel production. *Biotechnology Advances* 27 (4): 398–408.
Sampaio, F.C., Chaves-Alves, V.M., Converti, A. et al. (2008). Influence of cultivation conditions on xylose-to-xylitol bioconversion by a new isolate of *Debaryomyces hansenii*. *Bioresource Technology* 99 (3): 502–508.
Santos, D.T., Sarrouh, B.F., Rivaldi, J.D. et al. (2008). Use of sugarcane bagasse as biomaterial for cell immobilization for xylitol production. *Journal of Food Engineering* 86 (4): 542–548.
Sawatdeenarunat, C., Nguyen, D., Surendra, K.C. et al. (2016). Anaerobic biorefinery: current status, challenges, and opportunities. *Bioresource Technology*.

Severson, K., Martín, M., and Grossmann, I.E. (2013). Optimal integration for biodiesel production using bioethanol. *AICHE Journal* 59 (3): 834–844.

Shemfe, M.B., Whittaker, C., Gu, S., and Fidalgo, B. (2016). Comparative evaluation of GHG emissions from the use of Miscanthus for bio-hydrocarbon production via fast pyrolysis and bio-oil upgrading. *Applied Energy* 176: 22–33.

Song, H. and Lee, S.Y. (2006). Production of succinic acid by bacterial fermentation. *Enzyme and Microbial Technology* 39 (3): 352–361.

Starke, I., Holzberger, A., Kamm, B., and Kleinpeter, E. (2000). Qualitative and quantitative analysis of carbohydrates in green juices (wild mix grass and alfalfa) from a green biorefinery by gas chromatography/mass spectrometry. *Fresenius' Journal of Analytical Chemistry* 367 (1): 65–72.

Subhadra, B. (2011). Algal biorefinery-based industry: an approach to address fuel and food insecurity for a carbon-smart world. *Journal of the Science of Food and Agriculture* 91 (1): 2–13.

Vlysidis, A., Binns, M., Webb, C., and Theodoropoulos, C. (2011). A techno-economic analysis of biodiesel biorefineries: assessment of integrated designs for the co-production of fuels and chemicals. *Energy* 36 (8): 4671–4683.

Wang, G. and Hollingsworth, R.I. (1999). Synthetic routes to L-carnitine and L-gamma-amino-beta-hydroxybutyric acid from (S)-3-hydroxybutyrolactone by functional group priority switching. *Tetrahedron: Asymmetry* 10 (10): 1895–1901.

Werpy T, Petersen G. Top Value Added Chemicals from Biomass. Volume I—Results of Screening for Potential Candidates from Sugars and Synthesis Gas: United States Department of Energy; 2004.

Willke, T. and Vorlop, K.D. (2001). Biotechnological production of itaconic acid. *Applied Microbiology and Biotechnology* 56 (3–4): 289–295.

Yuan, Z., Chen, B., and Gani, R. (2013). Applications of process synthesis: moving from conventional chemical processes towards biorefinery processes. *Computers & Chemical Engineering* 49: 217–229.

Yuan, T.Q., Xu, F., and Sun, R.C. (2013). Role of lignin in a biorefinery: separation characterization and valorization. *Journal of Chemical Technology and Biotechnology* 88 (3): 346–352.

Zheng, Y., Chen, X., and Shen, Y. (2008). Commodity chemicals derived from glycerol, an important biorefinery feedstock. *Chemical Reviews.*

Zondervan, E., Nawaz, M., de Haan, A.B. et al. (2011). Optimal design of a multi-product biorefinery system. *Computers & Chemical Engineering* 35 (9): 1752–1766.

20

Lignocellulosic Crops as Sustainable Raw Materials for Bioenergy

Emiliano Maletta[1,2] and Carlos Hernández Díaz-Ambrona[2]

[1] *Bioenergy Crops, United Kingdom*
[2] *Polytechnic University of Madrid. School of Agriculture, Spain*

CHAPTER MENU

20.1 Introduction

Debates about the feasibility of biofuels and sustainability started many years ago, chiefly centred on liquid first-generation biofuels based on food crops. The development of new processing technologies able to transform lignocellulosic raw materials into heat, power, gases, or even advanced biofuels, transformed the debate in many ways, especially by enabling new approaches for determining the biomass footprint of dedicated plantations, forestry, or the collection of agricultural residues, in combined energy–food systems (Weiser et al. 2014). Industry and policy makers face complex discussions about sustainability criteria, from food and fuel conflict debates around land availability and food security issues, to impacts and benefits from carbon footprint, indirect land use change effects, emission savings or biodiversity and ecological issues. Can we grow biomass without harming the environment? Are lignocellulosic feedstock options viable to produce energy in the rural sector? Are soil and biodiversity going to be threatened by the expansion of biofuel crops?

While global biomass demand rises, many of the world's ecosystems that support human societies are already overexploited and unsustainable. Climate change could exacerbate these environmental problems by adversely affecting water supplies and agricultural productivity.

The global energy picture is changing rapidly in favour of renewable energy. According to IRENA's (International Renewable Energy Association) global renewable energy roadmap (IRENA 2014) if the realizable potential of all clean energy technologies beyond the business-as-usual scenario are implemented, renewable energy could account for 36% of the global energy mix in 2030. This would be twice the global renewable energy share compared to 2010 levels. Increasing prices and negative environmental impacts of fossil fuels have caused the production

Green Energy to Sustainability: Strategies for Global Industries, First Edition.
Edited by Alain A. Vertès, Nasib Qureshi, Hans P. Blaschek and Hideaki Yukawa.

of biofuels to reach unprecedented volumes over the last 15 years. The use of biomass for energy can significantly displace fossil fuels, and contribute to greenhouse gas (GHG) mitigation, security of energy supply, and rural development. Currently, renewable energy sources contribute 13% to total primary energy supply, of which almost 80% (50 EJ) is supplied by biomass, albeit sometimes with non-sustainable practices (IRENA 2014). The global deployment of energy from biomass is expected to increase to 100–300 EJ by 2050, a two to sixfold increase compared with current figures. According to the International Energy Agency (IEA), global energy demand is increasing, as is the environmental damage due to fossil fuel use. The continued reliance on fossil fuels will make it very difficult to reduce emissions of GHGs that contribute to global warming. Bioenergy currently provides roughly 10% of global supplies and accounts for roughly 80% of the energy derived from renewable sources. Bioenergy was the main source of power and heat prior to the industrial revolution (IEA 2009). Since then, economic development has largely relied on fossil fuels. A major impetus for the development of bioenergy has been the search for alternatives to fossil fuels, particularly those used in transportation (OECD 2009).

Notably, the renewable energy policy of the European Union will generate a considerable increment in lignocellulosic biomass demand by 2030 as suggested in recent reports by the United Nations Food and Agriculture Organization (FAO 2013), the European Environment Agency (EEA 2013) and the European Commission (EC 2012). The increased use of biomass will be driven principally from the energy sector, but also from the industrial and residential sectors. While first-generation biofuel importation and marketing restrictions have been imposed in the European Union, other biomass applications seem to be predominant including domestic heating, biomass power, and advanced (cellulosic) liquid biofuels, which current policies still consider to be simply an option for transport fuels. Most policy makers and organizations involved in bioenergy such as FAO, IRENA, IEA, EEA, and others are promoting more sustainable feedstock alternatives and a multi-product approach in biorefineries using lignocellulosic raw materials (Chundawat et al. 2011). Biorefineries in most cases are conceived as using wet and dry waste streams that can be effectively processed into energy, animal feed or even biofertilizers (Dheeran and Reddy 2018). The key to the future development of world biomass markets resides in regional supply potentials and how well these can mobilize new sources of supply, such as forest residues, agricultural residues and biomass from dedicated bioenergy crops. Recent analysis and scenarios for 2030 by the International Renewable Energy Agency show a clear trend: a growing biomass demand and a major role for dedicated bioenergy crops worldwide, the production of which is expected to grow to 39 EJ by 2030 (IRENA 2014).

It is expected that dedicated biomass plantations will play a substantial role in future energy systems. The scope for their growth is quite substantial. Using FAO statistics, Maletta (2014, 2016) reported that the global area of cropland (arable land plus land with permanent crops) has increased very little since 1961, and almost not at all since 1985–1990, while crop output grew by a factor of 3.5; about 95% of total cumulative growth since 1961 was due to increased productivity per hectare. Alexandratos and Bruinsma (2012), in work sponsored by FAO, estimate that only half the land suitable for rainfed crop production (not counting land covered by forests, built-up, or otherwise strictly protected) is actually used for crops. A total of 2.0 billion hectares (from very suitable to marginally suitable land) could be used for dedicated biomass production without disruptions in food security, and not encroaching onto forests, built-up areas or strictly protected land. About another billion Ha of 'very marginally suitable' land is available and may also be tapped, albeit at low productivity (5–20% of the standard rainfed yield). These data are congruent with findings reported by Souza et al. (2017) and IRENA (2014). The potential area of abandoned agriculture lands that could be used for biomass crops production has been estimated using remote sensing and spatial tools. Campbell et al. (2008) cite a total ranging from 385 to 472 million hectares. Furthermore, Cai et al. (2011) estimated that in Africa, China, Europe, India, South America, and the continental United States, abandoned and degraded cropland and mixed crop and vegetation land represent a total ranging from 320 to 702 million hectares.

Bioenergy options making use of marginal lands are being considered in FAO and IEA bioenergy groups with a focus on the most sustainable alternatives including energy, animal feed, fertilizers and bio-plastics (IEA 2016). Most land is multi-functional, with land needed for food, feed, timber, and fibre productions, in addition to natural resource conservation and climate protection. In this context, a growing proportion of advanced biomass alternatives that promote green covers, perennial agriculture and second-generation biofuels from straw or woody lignocellulosic has been projected (Chum et al. 2011). Electricity generation from biomass, often combined with district heating, would grow by 10% per year to account for nearly a third of global biomass demand by 2030 roughly triple its share in 2010. By 2030, liquid biofuels for transport would grow nearly as fast to 28% of biomass use – also tripling their 2010 share. In parallel, the total energy demand for cooking and heating in industry and buildings would decline to 40% by 2030, compared to its 80% share in 2010 due to growth in the transport and power sectors and substitution of traditional uses. IRENA's recommendations centre on expanding power from biomass and energy crops in the United States are expected to represent a very large share (26–34% of total biomass) using only very marginal lands that are currently not used for food production (IRENA 2014).

When starting a biomass-to-energy project, developers face a challenge regarding feedstock choice. Procurement strategies constitute a rigid framework that must be considered from the earliest developmental phases. In early stages of development, projects should focus on a funding strategy based on specific milestones prior to raising investment in industrial facilities, farm or forestry plantations. This initial period may require tasks such as permitting, licensing, engineering, feasibility studies, environmental and social impact studies, and due diligence regarding contractors and compliance with several standards and legal restrictions. Several solutions and service providers or even biomass suppliers are available in the market and some of them offer planting materials or access to pedigreed varieties, clones, hybrids, or other cultivars.

A number of legal restrictions during the early stages challenge most developers and stakeholders involved in biomass to energy projects. Financial security is often linked to biomass procurement and stable supply chains for industries. The beginning of new plantations often require importation of planting materials, starting trials and demonstrating productivity, or developing native species available in small nurseries to facilitate crop expansion in large scale farming operations. Sometimes forestry management and new plantations may take two to eight years to produce a first harvest making financing strategies complicated. When establishment of new plantations involves sexual propagation with seeds instead of vegetative propagation methods, a faster project escalation and farm or forestry development can be expected. The choice about project site and land management contracts become critical for logistics considering biomass mobility and sources of raw materials that often combine the use of agricultural or forestry residues and biomass from dedicated plantations. Land availability issues considering industrial sites and roads should always be included in feasibility reports conducted prior to funding approvals. Several other environmental and social implications may require life cycle assessments (LCAs) considering all inputs (e.g. fossil energy used in fertilizers, diesel and machinery use for planting, harvesting and agri-chemicals) and other impact categories such as eutrophication, acidification, etc. If applicable, these LCA should consider several end products (e.g. animal feed pellets) and renewable energy (e.g. steam), use of waste materials from industries into biofertilizers for soil amendment and regeneration;

estimated footprint and determination of effects caused by land use changes are often required for certification and permitting. Carbon accountability may include very complex issues such as carbon sequestration by vegetation developed and compared to reference scenarios that often have negative trends (e.g. desertification, erosion, land abandonment). Carbon and total energy balances including net productivity and annual increments (Mg of dry matter or gigajoules per area unit) derived from photosynthesis and biomass obtained can be evaluated in terms of energy content (calorific power). If biomass is fractioned into products for which the final use is animal feed or food (e.g. leaf protein concentrates, starch, oil and other products) and energy, then all inputs shared are considered in equations to determine end product's footprint (for example electricity in the grid or bio-methane blends in pipelines).

Finding economically and environmentally sustainable alternatives may represent a considerable challenge for the companies involved in feedstock procurement. A combination of low environmental footprint of delivered raw material, stable and low economic costs per energy unit of delivered feedstock and farm or forestry management capacities are potential barriers to face. The carbon footprint of dedicated crops has been analysed during the past decade and several studies show reduced chopped grass materials, woody chips, straw bales, and other feedstock options including those that could potentially have negative social implications, soil nutrient depletions risks, deforestation and biodiversity threats and pollution to water streams by agrochemicals. Any of these indicators should be considered when analysing the feedstock footprint. However, most LCA show that a precise accounting may be very difficult to achieve and the scientific community still needs to find a consensus of the appropriate calculation method.

Agricultural and forestry residues available in the region are always considered at the beginning of any biomass project. Nonetheless, those unused resources described here as residues are frequently scarce and policy makers tend to avoid deforestation and unsustainable 'extraction' of biomass that may impact soil organic matter and litter or damage habitats. While conservation and protected areas always need to be considered, vast regions worldwide are occupied with degraded grasslands, invasive bushes or degraded and over-exploited forestry and agriculture. What is a 'waste' and what is not may become a matter of debate (Børresen 1999) and should be approached with a more holistic understanding considering current or baseline scenarios affecting soil, biota and water and the interaction with communities and developers (e.g. farmers, forestry companies, bio-based industries). Residues collected in biomass from energy projects could affect current land use trends and vegetation covers, and determine impacts such as erosion due to excessive biomass residues extraction. When it comes to raw materials from dedicated crops, a similar consideration requires a more complex analysis.

Many bioenergy routes can be used to convert a range of raw biomass feedstocks into a final energy product. Technologies for producing heat and power from biomass are already well-developed and fully commercialized, as are first-generation routes to biofuels for transport (IEA 2009). A wide range of additional conversion technologies are under development, offering prospects of improved efficiencies, lower costs and improved environmental performance.

In this chapter, we discuss realistic cropping systems with a relatively mature commercial readiness level of development such as lignocellulosic raw materials for biobased industries.

20.2 Major Lignocellulosic Industrial Crops

Among several options of lignocellulosic raw materials and dedicated biomass crops, a distinction regarding inputs required for production and farming or logistic chains should consider crop lifetime, rotations, number of harvests and inputs required for regrowth (LogistEC 2012). Several crops have the ability to regrow after being harvested (such as many perennial species) and some other options are annual crops that are planted annually after land ploughing and seedbed preparation. Cultivation and final product characteristics of crops have different features and have been investigated during recent years for several conversion routes.

20.2.1 Annual Crops

Over a number of years the focus of attention for industries were winter and summer cereals, oil crops, fibre and staple crops, tubers, and many other starchy crops. The simplicity of contracting farmers and out-growers for biomass production, with annual crops easy to manage and include in a rotation of their properties, was a major driver for promotion, development, and implementation of these species. Solid biomass from herbaceous annual crops are also similar to straw and other agricultural residues which helped industries to build up a reliable supply chain in surrounding areas minimizing risks, logistic constraints, and promoting dedicated energy crops.

20.2.1.1 Cellulosic Annual Grasses

Several species of winter and summer crops, traditionally cultivated for fodder or grain production have been investigated and even commercially used for energy purposes. This category includes winter cereals such as triticale (*Triticosecale sp.)* and hybrid rye grass (*Secale cereale*), which include new hybrids released specifically for biomass energy. Currently, triticale is grown mainly as a food crop, but increasing attention is being given to it for bio-energy production. In particular, triticale has been tested and commercially used in Europe for ethanol (Lynd et al. 2017), biogas, and power production (LogistEC 2012). Annual rye grass pedigreed hybrids and triticale varieties specifically released for biomass purposes are well described in the scientific literature with above-ground biomass productivity from 5 to 14 Mg/ha reported in most locations in Spain, Italy, and France (Sastre et al. 2014).

Sorghum (*Sorghum bicolor* Monech L.) is a commercially mature option as an energy crop in several countries. Thousands of varieties of sorghum exist including recently developed biomass hybrids. Most hybrids are sorghum × sorghum and sorghum × sudan grass (*S. bicolor x Sorghum sudanense*) that exhibit attributes of particular interest to biobased industries. Many companies offer biomass sorghum hybrids that are suitable options for several processing alternatives including combined heat and power, biogas and cellulosic ethanol. Sorghum is a highly productive C4 grass that has proven its potential for various bioenergy applications; its varieties and hybrids are highly variable in terms of grain and biomass productivity, as well in as leaf/stem ratios; these parameters reflect important biomass quality aspects (Zegada-Lizarazu and Monti 2012). Above-ground biomass productivity from 15 to 30 Mg/ha has been reported in the literature (Olsen et al. 2013; Rocatelli et al. 2012; Tamang et al. 2011).

20.2.1.2 Fibre and Oil Crops

Several annual crops producing seeds with high content of oil, protein or even industrial fibres have been investigated for biomass to energy and other biomass processing options. This category includes canola crops (*Brassica napus*) as well as other related crops such as Ethiopian mustard (*Brassica carinata*), which was found to be particularly suitable in several regions in Europe and the United States. Several studies have highlighted the possibility of introducing this species into commercial cropping systems. Ethiopian mustard was tested extensively in Europe and may occupy fields for approximately five to seven months (from planting to harvesting) and above-ground biomass yields recorded typically range from 6 to 16 Mg/ha when cultivation under rain-fed conditions is considered in areas between 450 and 1000 mm of annual rainfall. Several projects have evaluated potential yields and costs in Europe for biodiesel, bioethanol, heat, and power (Cardone et al. 2003; Copani et al. 2009).

Kenaf is an ancient crop that originates from Africa. A member of the hibiscus family (*Hibiscus cannabinus L.*), it is related to cotton and okra, and grows well in many parts of the United States and many temperate countries. Actually, kenaf is traditionally grown in east-central Africa, west Asia and in several southern states of America for fibre and oil seed (20% oil content) production, whereas it comprises an excellent forage crop, containing 18–30% crude leaf protein, and stalk protein 5.8–12.1%. However, the total above-ground biomass potential productivity in this annual fibre, protein, and oil crop can reach values around 22 Mg/ha as reported by Danalatos and Archontoulis (2004) in irrigated trials in Europe. In the past decade, kenaf has received increased attention in Europe as a high yielding 'non-food crop' for fibre production and particularly for the newsprint paper pulp and other paper products industry (Alexopoulou et al. 2000; Petrini et al. 1994).

Hemp, *Cannabis sativa*, originates from western Asia and India and from there spread around the globe. Hemp is a genetic 'cousin' to marijuana, but contains little to none of the tetrahydrocannabinol (THC), the psychoactive ingredient in marijuana. For centuries, hemp fibres were used to make ropes, sails, cloth and paper, while the seeds were used for oil (including biodiesel), protein-rich food and feed, cosmetic products and construction materials. Interest in hemp declined when other fibres such as sisal and jute replaced hemp in the nineteenth century, however there are some companies that offer a multi-product (food/feed/energy/fibres) at commercial level.[1] Energy use of industrial hemp is today very limited but annual productivity in temperate areas between 9.9 and 14.4 Mg/ha have been reported when the crop is managed for solid fuel or for biogas (Prade et al. 2011). There are few countries in which hemp has been commercialized as an energy crop given the commercial opportunity that its fibres can offer in other markets. Sweden is one such country that has a small commercial production of hemp briquettes. Hemp briquettes are more expensive than wood-based briquettes, but sell reasonably well in regional markets, as exemplified by companies that target animal feed markets, in particular for horses.[2] In 2002, 18 000 ha of land were used for hemp cultivation in the European Union. However, hemp end products were limited, as 70–80% of the hemp fibre produced was used as pulp and cigarette paper, 15% for automotive parts, 5–6% for insulation mats. As for the seeds, 95% were used as animal feed (Karus 2004).

20.2.2 Perennials

There is a growing interest in perennial bioenergy crops which avoid re-planting costs every year, provide several soil benefits due to their long lifetime (often 5–10 years in commercial biomass projects) and relatively low inputs (LogistEC 2012). In this section, we discuss both herbaceous and woody cropping systems with a sound scientific and commercial background.

20.2.2.1 Herbaceous Biomass Crops

Pennisetum purpureum Schum, also known as napier grass, elephant grass or Ugandan grass, is a species of perennial tropical grass native to the African grasslands. *P. purpureum* is a monocot 'warm grass' or C4 species (four carbons in the metabolic photosynthesis pathway) and perennial

---

1 Global Hemp Holding is a Canadian company that has offered a number of products for more than two decades. URL: https://globalhempgroup.com.

2 A relevant example is the company 'Swiss Hanf Production AG' in Switzerland. URL: http://swisshanf-ag.ch.

Figure 20.1 Napier grass in 7000 ha planted in Mexico. Source: Bioenergy Crops Ltd.

grass in the Poaceae family. It is named for the 4-carbon molecule present in the first product of carbon fixation in the small subset of plants known as C4 plants, in contrast to the 3-carbon molecule products in C3 plants also known as 'cool grasses' (Figure 20.1).

There are several varieties of napier grass that are extensively cultivated for fodder, including recently released pearl millet (*Pennisetum glaucum*) × Napier (PMN) non-invasive (no viable flowering) hybrids for which there are several companies offering planting materials for biomass projects in tropical areas worldwide (Jyväskylä Innovation Oy 2009). Among the PMN varieties, there are the recently released first seeded (direct sowing using sexual seeds) pennisetum hybrid alternatives which reduce the costs of establishment by avoiding expensive nursery and vegetative propagation; therefore direct sowing reduces the time required to expand areas during the initial phases in bioenergy project development (Dowling et al. 2013).

While wild napier grass in the United States has been considered invasive in some fragile environments, the crop has evolved into a managed row crop in recent years and is being used extensively in Africa, Latin America, and Asia (Boschma et al. 2010). Some recent experiences in Brazil, report its use in biomass combustion facilities for power production (Isah et al. 2015). With potential for one, two, or even eight harvests per year (depending on biomass quality required) above-ground annual productivity in Napier grass has been reported to reach between 40 and 85 Mg/ha in tropical areas (Filho et al. 2000; Magalhães et al. 2006; Obok et al. 2012; Rueda et al. 2016).

Miscanthus, another group of C4 or warm grass but with relatively good cold tolerance, includes about 20 species. Most of these species are native to South East Asia; however, it is known that there are two Himalayan and four African continent species (Hodkinson et al., 2002). Within the genus Miscanthus, species that are particularly relevant in science and application areas comprise the *Miscanthus sinensis*, *Miscanthus sacchariflorus* and the hybrid M. × giganteus, which is triploid and then sterile (non-viable seeds). Miscanthus is very widespread in the United States, the United Kingdom, and there are several projects in European countries (mainly France and Italy but also Eastern Europe countries) and New Zealand. Miscanthus is highly suitable for temperate areas as it require climatic conditions that are similar to those required for maize production. Miscanthus is often preferred to other crops due to its commercial readiness level as there are several companies[3] that provide planting solutions and pedigreed varieties. In addition, in most areas in Europe and the United States miscanthus becomes senescent during cold winters, even surviving low temperatures (−3 °C) and successfully re-growing the next year. The ash content in miscanthus is probably the lowest compared to similar plants reported in the literature; this observation is probably associated with leaching events that occur during rainy winters when biomass dry content increases; some translocation of nutrients to soil has been verified (Clifton-Brown et al. 2004). The productivities of the miscanthus species and their hybrids are reported to vary between 12 and 33 Mg/ha (Lewandowski et al. 2003).

Giant reed (*Arundo donax*) is a Mediterranean rhizomatous perennial grass that has been developed for several biomass projects in Europe and the United States. The adaptability of this plant to different environments, soils and growing conditions, in combination with its high biomass production and the low input required for its cultivation, give to *A. donax* many advantages when compared to other energy crops. One of the most important and distinguishing characteristics of this plant is its high biomass production per Ha (Angelini et al. 2009). Angelini et al. (2009) reported an average biomass production over a cropping period of 12 years of 30–40 Mg/Ha of dry matter. This biomass is produced with low agronomic input, e.g. low or no use of irrigation, fertilizers, pesticides and low agronomic interventions with machinery (Lewandowski et al. 2003). This translates into both low cost for its cultivation (Soldatos et al. 2004) and low environmental impact. Thanks to these characteristics, giant reed has recently been proposed as an energy crop for producing biogas, and a full field crop campaign started in Italy in 2013 on about 25 farms (Riffaldi et al. 2012).

Another novel bioenergy crop recently developed at commercial scale is switchgrass (*Panicum virgatum* L.). There are two typologies of switchgrass (upland and lowland ecotypes), with many cultivars within each type. While cultivars of both types can be grown in both temperate and warm areas, better adaptability and higher yield potential have made lowland cultivars Alamo and Kanlow, the main focus of biomass research in Europe and the USA (Lewandowski et al. 2003). Upland varieties are desirable when switchgrass is grown as a dual-purpose crop for both forage and biomass or in colder areas in North America and Europe. Lowland ecotypes often produce above-ground biomass yields between 15 and 25 Mg/ha per year, and upland ecotypes have been reported to produce between 5 and 12 Mg/ha (Clifton-Brown et al. 2004; Boehmel et al. 2008; Heaton et al. 2004). Different from napier grass, miscanthus, and giant reeds, switchgrass can be planted using sexual seeds, even with no-tillage methods that may determine lower costs and fossil energy during land preparation (Lewandowski et al. 2003).

While warm grasses represent an opportunity for higher productivity and lower cost of delivered biomass products, these grasses have a four carbon (C4) metabolism pathway that often involves a minimum growing base temperature below which the species may not grow, resist cold effects or even not persist in time (Heaton et al. 2004). Therefore, cool grasses are those species with a three carbon (C3) metabolic photosynthetic pathways that have a very low growth base temperature (often 0 °C). Among them, perennial rye grass (*Lolium perenne* L.), reed canary grass (*Phalaris*

---

3 The following are companies that provide planting materials and/or services for biomass supply chain development: Miscanthus NZ (www.miscanthus.co.nz), Genera Energy (www.generaenergy.com), Terravesta Ltd. (www.terravesta.com), Crops4Energy Ltd. (www.crops4energy.co.uk), Bioenergy Crops Ltd. (www.bioenergycrops.com), and New Energy Farms Ltd. (www.newenergyfarms.com).

arundinacea L.), and tall wheatgrass (*Elytrigia elongata*) have been tested and investigated for bioenergy purposes. These are relatively low input requirement pasture crops that can produce between 5 and 17 Mg/ha in temperate areas in North America and Northern Europe.

When cultivated in areas with limited summer rainfall these crops have less potential to produce high yields because base temperature and drought limit productivity to periods with sufficient soil moisture level which mostly occur during spring and autumn. Summer periods impose a serious limitation to high productivity in warm grasses when rainfall occurs mostly when cold conditions limit plant growth. In contrast, tall wheatgrass is a 'summer dormant' species and more adapted for cultivation in the rainfed conditions occurring in the western USA and in Mediterranean conditions in coastal and continental areas in Europe where its cultivation may deserve more attention as a novel perennial bioenergy crop (Csete et al. 2011; Maletta et al. 2012).

Further adaptation to higher temperature and in particular arid conditions is crassulacean acid metabolism (CAM) photo-synthesis. Among CAM plants, species of maguey (Agavaceae) and opuntia (Cactaceae) have been receiving growing attention as energy crops (Borland et al. 2009; Davis et al. 2011, 2014, 2016; Holtum et al. 2011; Núñez et al. 2011; Escamilla-Treviño 2012; Cushman et al. 2015; Mielenz et al. 2015; Owen and Griffiths 2014; Yang et al. 2015). These plants can achieve high water-use efficiencies and can be grown for bioenergy feedstock in abandoned or degraded agricultural lands, or in drylands where precipitation is too scarce to support traditional C3 or C4 food or even bioenergy crops. Furthermore, some species of Agave and Opuntia have the potential to produce comparable amounts of biomass to C4 and C3 plants but with significantly lower (20–80%, respectively) inputs of water, and under optimal growing conditions. While productivity of these species is under debate due to lack of large areas with reported yields, Nobel et al. (1992) reported average 43 Mg/ha/yr. above-ground dry mass productivity which is comparable to agronomic C4 species (20–80 Mg/ha/yr) and C3 herbaceous species (5–25 Mg/ha/yr) and trees (5–18 Mg/ha/yr). Biomass projects involving CAM plants are frequent in Mexico where the tequila industry has recently started to boost agri-industries with thermal applications and biofuel alternatives, and some biogas projects are reported in Chile near the Atacama desert.[4]

20.2.2.2 Woody Crops

Short rotation forestry (SRF) and short rotation coppice (SRC) biomass systems have been defined by the International Union of Forestry Research Organization as sustainable forestry outside the usual forest areas and for purposes other than sawn timber. These purposes are mainly biomass, with fast-growing species, producing greater than10 Mg/ha/year in significantly fewer years than in conventional forestry, less than 30 years from establishment to harvest and replanting or other land uses. SRC-dedicated energy-woody crops are seen as an option to produce additional woody biomass efficiently in a short time and in a sustainable way without competing with biomass resources from forests. SRC are well suited for biomass production because of the rapid juvenile growth of their trees and their high biomass yields. In temperate areas, fast growing tree species with outstanding performance in marginal areas in Europe, USA, Australia and Northern Asian countries include mainly poplars (*Populus spp.*) and their hybrids, willow (*Salix sp.*), black locust (*Robinia pseudoacacia*), *Eucalyptus sp*, and siberian elm (Ulmus pumila). Most biomass systems developed in Europe and North America are managed with regular harvesting every three to eight years and produce annualized yields between 6 and 18 Mg/ha, with plant populations established between 3000 and 20 000 trees per hectare (Dowell et al. 2009; San Miguel et al. 2015).

---

4 Private Company Elqui Global Energy has been developing biogas units based on cactus in Northern Chile and overseas. A complete report and interview with developers is available at: https://bit.ly/2oN1U7E (last accessed July 2018).

There are several fast-growing nitrogen-fixing species that can be included in biomass supply chains in tropical areas. They often require inoculation with specific nitrogen-fixing microorganisms to use atmospheric nitrogen (N2) via root nodules in exchange for sugar (Hari and Srinivasan 2005). The impact of nitrogen-fixation can be high if nitrogen is the limiting factor for growth or if it needs to be applied to replenish it, and that may depend on application, conservation or availability of phosphate to trees and their symbiotic root fungi (mycorrhiza). Among species with extensive commercial deployment are the most relevant genera studied for wood and fodder including species such as Sesbania, Gliricidia, Robinia, Acacia, Leucaena, Calliandra, and Albizia. With harvests starting from the second or third year from plantation depending on planting density, cultivation and management, fast growing biomass projects can be designed to produce above-ground annualized biomass productivity between 8 and 22 Mg/ha (Castro et al. 2017).

20.3 Social, Economic and Environmental Aspects in Sustainability Criteria

Energy crops are expected to be deployed at a large scale in the very short term, bringing significant social and environmental benefits. Nevertheless, a significant number of studies report a range of very positive to negative environmental implications from growing and processing energy crops, thus great uncertainty still remains.

Large land areas worldwide are classified as wastelands or degraded forests and grasslands where wildfires and desertification process occurs, which could be utilized for growing biomass. This would offer an opportunity to develop a vast extent of wastelands, leading to a greater vegetative cover, thereby protecting such lands from further erosion and degradation. In parallel, the local production of bioenergy can help reduce petroleum imports with clear economic benefits, not forgetting a reduction in the dependency on imported oil thereby resulting in increased energy self-sufficiency.

Increased farm activities linked to bioenergy development, such as growing biofuel and biomass crops, seed collection, briquetting, and transportation of biomass will in addition have a clear positive impact on job creation including several indirect effects notably in rural areas. An increased average income may reduce income disparity between the rich and the poor in rural areas in the poorest countries in the world, and also between rural and urban areas. Higher income levels are positively correlated with rises in literacy rates, medical care and nutrition (IRENA 2017). The traditional use of biomass as domestic fuel for cooking, heating and other purposes causes several health hazards among women and children in rural areas and urban poor areas. Nowadays, several biomass projects in Africa and Asia include the introduction of biopower, biogas and other clean fuels that are expected to drastically improve energy access, rural electrification, reduce health problems related to traditional domestic firewood use, resulting in increased life expectancy and decreased infant mortality (Diaz-Chavez et al. 2015; Souza et al. 2017).

Sharma et al. (2016) published a large and thorough review, which suggests that there are strong multi-dimensional benefits regarding social aspects of agroforestry and perennial systems in sub-Saharian Africa, Latin America and Asia, where land use changes, gender issues, energy poverty, and policy recommendations are covered.

The effect of reduced CO_2 emissions has, however, been questioned by, e.g. Searchinger et al. (2008), who showed that converting grassland into land for producing corn for ethanol can increase total GHG emissions. Crutzen et al. (2008) furthermore pointed out that the application of nitrogen fertilizer for the production of energy crops can increase emissions of nitrous oxide (N_2O), a gas that is characterized by a greenhouse potential that is 298 times higher than that of CO_2 (Solomon et al. 2007), and hence emissions of nitrous oxide counterbalances savings in CO_2 releases in the atmosphere. Moreover, Sastre et al. (2016) showed that LCAs should include nitrogen balance tools to determine a more accurate analysis of emissions during farming energy crops. In addition, Pimentel et al. (2009) argued that the production of bioenergy depends on water resources that in many areas are limited, as well as land and energy that are also necessary for food production.

20.3.1 Annual versus Perennial Options

Several studies have been performed to analyse the various implications of annual and perennial cropping systems and their resulting LCAs, sustainability analysis and socioeconomic implications. Annual crops have been confirmed to hold a major impact in agriculture compared to most perennial alternatives including herbaceous and woody biomass systems. For example, Fazio and Monti (2011) provided evidence and conclusive results that perennial crops exhibit substantially higher environmental benefits than annual crops. In this study, it was shown that significant CO_2 emissions can be avoided through converting arable lands into perennial grasslands. Besides, due to the lack of certain data, soil carbon storage was not included in the calculations, while N_2O emission was considered as an omitted variable bias (1% of N-fertilization). Therefore, especially for perennial grasses, CO_2 savings were reasonably higher that those estimated in this and other studies. For first-generation biodiesel, sunflower showed a lower energy-based impact than rapeseed, while wheat should be preferred over maize for first-generation bioethanol given its lower land-based impact. For second-generation biofuels and thermo-chemical energy, switchgrass provides the highest environmental benefits. With regard to bioenergy systems, first-generation biodiesel had less impact than first-generation bioethanol; similarly bioelectricity had less impact than first-generation biofuels and second-generation bioethanol by thermo-chemical hydrolysis, but significantly more impact than biomass-to-liquid biodiesel and second-generation bioethanol through enzymatic hydrolysis (Fazio and Monti 2011).

Pugesgaard et al. (2015) concluded that willow and grass-clover were superior to winter wheat in terms of low nitrate leaching and high farm gate energy output. As a result, the use of willow or grass as a replacement for annual crops in nitrate sensitive areas will decrease nitrate leaching losses markedly. When removing a perennial crop and replacing it with a new crop there may, however, be a risk of increased nitrate leaching after the conversion, and a long lifetime for the perennial crop may, therefore, be advantageous. Drainage from willow and grass is considerably lower compared to winter wheat which may have a negative effect on the recharge of ground water. The crop nitrogen balances indicate that soil organic matter accumulates below willow and grass, while it is degraded below wheat. The use of pesticides can be considerably reduced by growing willow or grass compared to winter wheat. The results support the conclusion of several other studies that in most cases perennial crops seem to be a more sustainable choice for biomass delivery than annual crops.

Annual crops constitute interesting options when companies seek reliable crop options, however most of them have relatively high costs if only above ground biomass is considered. As suggested by Lalitendu Das et al. (2017) for a kenaf crop, cost analysis should always consider all products from grains and stems or leaves (e.g. protein and ethanol), as well as opportunity costs in case crops are cultivated on suitable or very suitable farms where traditional food or cash crops may offer a higher net profit to farmers.

20.3.2 Soil Issues

Dedicated biomass cropping systems are normally established for specific industrial requirements. From breeding to harvesting solutions, a final raw material will result in a number of tons of organic matter that includes values between 1 and 12% of ash. Minerals are certainly being exported when harvesting operations remove aerial biomass of any biomass cropping system. In general terms, there is an agreement that compared to annual crops, perennial species (both herbaceous or woody species) may have positive effects on soil by reducing erosion, nutrient leaching and increasing organic matter and organic nitrogen pools in the soil (Fazio and Monti 2011; Pugesgaard et al. 2015). In addition, several studies have found that biomass delivered as feedstock for energy production could have a larger footprint in terms of carbon balances and global warming potential when straw removal was analysed in cereals showing a relatively low convenience from environmental perspectives; including energy balances, soil nutrient depletions, impacts on water and air, and greenhouse gases emission savings compared to the cleanest fossil energy sources among several other impact categories (Turley et al. 2003; Saffih-Hdadi and Mary 2008; Tarkalson et al. 2009; Soon and Lupwayi 2012; Nguyen et al. 2013; Weiser et al. 2014).

In a study with several woody species as SRC long term trials, and two perennial herbaceous crops such as giant reed and miscanthus, the conversion of arable land into perennial bioenergy crops provided a substantial soil organic carbon (SOC) sequestration benefit for a seven-year period. In this study, the soil acted as a carbon sink only in the 0.0–0.20 m upper layer. When the hidden cost of industrial nitrogen fertilizers were taken into account, the SOC gain that resulted remained still substantial (Ceotto and Di Candilo 2011). Notably, the SOC increased, and the total nitrogen content in the soil remained fairly constant within the different options for land use, including the nitrogen fixing crop black locust. Herbaceous perennial systems, aimed at producing biomass as feedstock for energy or other purposes, such as miscanthus or reed canary grass were reported to introduce positive effects on earthworm communities (Felten and Emmerling 2011) and more diverse and abundant soil fauna and bacteria pools (Heděnec et al. 2014). Sierra et al. (2013) reported 1.16 Mg/ha/year increments in poplar plantations in Spain and confirmed that such plantations increased the total carbon content in a more effective way than maize does because the duration of use is also correlated with the most recalcitrant carbon forms.

Another relevant aspect to consider about soil is the opportunity that biobased industries provide waste streams that can be used to replace fertilizers, which represent between 40% and 80% of total global warming potential impact of cultivated biomass (Schmer et al. 2015). Working with perennial grasses for biofuel in Greet models to estimate GHG emissions, Schmer et al. (2015) concluded 'Changes in direct soil organic carbon (SOC) can have a major impact on overall greenhouse gas (GHG) emissions from biofuels when using life-cycle assessment (LCA)[…]. Biofuel GHG emissions showed as much as a 154% difference between using near-surface SOC stocks changes only or when accounting for both near- and sub-surface SOC stock changes'. In this regard, any addition of organic matter or soil amendments during cultivation may have a large effect on GHG emissions which are expected to be considered by biofuel certification companies and policy makers. Similar aspects should be considered when planning to replace mineral fertilizers such as nitrogen, phosphorous and potassium, as these fertilizers constitute a considerable share of delivered biomass product feedstock and costs. This includes a positive potential benefit from using ashes from combustion boilers, digestates from anaerobic digesters, vinassc liquid waste application to fields from sugar-mills and solid biochar from gasifiers and pyrolysis units that can be inoculated and blended with organic compost. All these practices have been referenced at the commercial level and have sound scientific backgrounds considering LCAs and soil issues (Smith et al. 2014).

Nonetheless, a key question arises: at the end of the cropping cycle will the carbon remain fixed? If the soil is converted back to arable land, the benefit of carbon sequestration will probably be lost in a few years. The comparison with current land uses (e.g. annual crops, grasslands, etc.) are critical since sometimes the system being replaced also benefits from those systems and not just avoids negative aspects (e.g. wildfires, tillage and desertification, or erosion). In contrast, if perennial energy crops could be rotated, avoiding deep soil tillage, the SOC storage will be assured. The permanence of sequestered carbon is a critical notion, because such permanence implies the commitment to long-term vigilance in the management of the captured carbon. Any policy making decision should consider what the current land use is and what are the realistic alternatives since most food production will determine changes in soil, especially considering that most cash crop options are annual species that would certainly impact the soil.

20.3.3 Biodiversity Issues

Bioenergy and biomass crops are often promoted by environmentalists and government leaders as having the potential to provide tremendous amounts of wildlife habitat and to support biodiversity. Several important studies provide evidence of the negative impacts of first-generation biofuel on wildlife, while others provide evidence that bio-power and second-generation biofuels have positive effects on wildlife and biodiversity. Whether the effects of cropping of biomass or biomass removal has any positive or negative impact on biodiversity depends strongly on specific regional circumstances and the consideration of land use and energy changes at the regional level. While changes at specific plots or sites do not always consider a regional approach (e.g. total diesel or fossil energy used in a given region that also may affect biodiversity and habitats), a more complex understanding of interactions and changes due to enhanced bioeconomy and agri-ecosystems should take place when biomass projects are promoted and implemented (EC 2012).

In other words, the net positive or negative impact will be different in a small island that imports 99% of energy, fertilizers and food that can be partially replaced by a biorefinery, compared to the net change in a country where these inputs come from cleaner and more sustainable alternatives (e.g. electricity mixed with large proportion of renewable energy). Based on the literature and six country studies (Belgium, Denmark, Finland, Netherlands, Sweden, Slovakia), Pedroli et al. (2013) analysed the compatibility of the EU 2020 targets for renewable energy with conservation of biodiversity. These authors concluded that 'increased demand for biomass for bioenergy purposes may lead to a continued conversion of valuable habitats into productive lands and to intensification, which both have negative effects on biodiversity. On the other hand, increased demand for biomass also provides opportunities for biodiversity, both within existing productive lands and in abandoned or degraded lands. Perennial crops may lead to increased diversity in crop patterns, lower input uses, and higher landscape structural diversity which may all have positive effects on biodiversity'.

Finally, in most developing countries with potential for biobased industries, policies to outline environmental standards for bioenergy production are generally lacking, and financial programs for compensating land owners and farmers for habitat- and biodiversity-protecting land practices are also lacking. Most significantly, land conversion could decrease native habitats, reduce biodiversity and decrease ecosystem services. Key issues are habitat loss and fragmentation with expanding acreage dedicated to corn and soybean production; loss of Conservation Reserve Program land; persistence of pesticides in the environment associated with conventional row crops; timing of harvest of perennial crops and forests; and impacts to water quality associated with agricultural run-off.

20.4 Processing Alternatives for Lignocellulosic Bioenergy Crops

Lignocellulosic biomass utilization involves converting the biomass into intermediates and subsequently converting such intermediates into chemical or fuel components. A bioconversion product and process network converts different types of biomass into various fuels and chemicals via a plethora of technologies. Reliable bioconversion processing pathways should be designed considering the volatilities and effects that certain parameters, such as biomass feedstock price and biofuel products demand. The most common intermediates are synthesis gas (also referred to as syngas), pyrolysis oil, sugars, lignin, cooking liquor and biogas. If the intermediate, for example synthesis gas, can be purified to a level suitable for subsequent processes, the same procedures can be applied for intermediates, as in oil refining or the petrochemical industry (Figure 20.2).

Plant biomass covers a wide range of plant and plant-derived materials, including biodegradable wastes such as cereal straw or forestry residues but also lignocellulosic materials from dedicated crops. The chemical compositions of these myriad feedstocks vary greatly; this variability has a crucial impact on process yields, process technical feasibilities, and process economics, and will therefore influence the most suitable type of conversion technology for valorizing a given biomass type. The main conversion routes are: thermal processing (e.g. combustion, gasification) to produce heat or electricity; anaerobic digestion to produce a biogas that may be used for heat or electricity production or as a transport fuel; first-generation biofuel production technologies; bio-refining technologies for the production of chemicals or second-generation biofuels.

Research efforts are ongoing to isolate, identify, characterize, and even tailor microorganisms and enzymes in order to better utilize renewable resources to produce structurally diverse and complex chemicals. Producing higher-value chemicals provides the advantages of high yield and selectivity, as well as minimum waste streams. Nevertheless, there are still problems with current biological transformation technologies including both upstream and downstream processes. Capital costs related to energy requirements, such as pretreatment, sterilization, production, agitation, aeration,

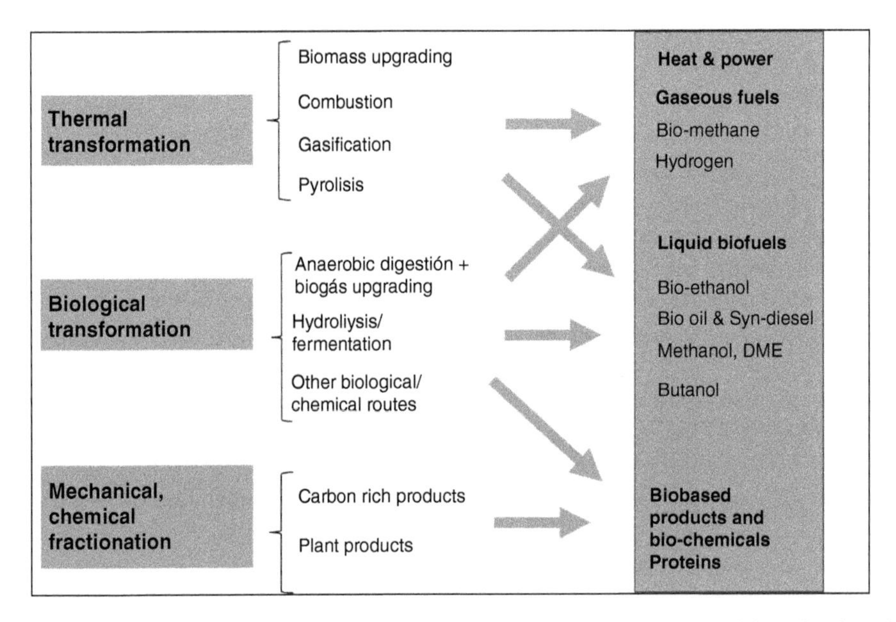

Figure 20.2 Biotechnological routes to biomass conversion. Source: own elaboration based on McMillan (2004).

temperature control, and finally recovery of target products from aqueous systems with low product concentration, result in high-cost processes (Danner and Braun 1999).

Biological conversion or biotransformation is a well-established process and is comprised of fermentation and anaerobic digestion. Sugar and starchy crops provide the main feedstocks for the process of fermentation in which a microorganism converts the sugars into bioethanol. As an economic alternative to costly sugars, lignocellulosic biomass can be used as feedstock after pretreatment which helps to break it down into simple sugars. The pretreatment can be carried out by enzymes or acids. Lignocellulosic conversion would greatly increase the supply of raw materials available for production of various high-value products. The lignin residues could be used as fuel for the energy required and even provide surplus energy, resulting in significantly improved energy balances and resulting potential reductions in GHG emissions (IEA 2013).

The decision for bio-conversion alternatives in a certain project often follows recommendations considering industrial aspects only. However, most projects are designed based on considerations on feedstock options, even after a certain technology has been selected. In waste-to-energy projects based on agricultural residues, municipal solid residue or forestry by-products, uncertainties in the supply chain in the long term make developers consider dedicated plantations. However, while the initial challenge may come from biomass sourcing (e.g. from farm, forest residue collections, etc.), choosing between herbaceous and woody materials often comes after selecting what is expected to be the most suitable technology option considering an initial review of the biomass residues that are available for the new manufacturing plant. The need for risk reduction along a supply chain is a typical first challenge that links the processing plant to farm conditions and biomass procurement logistics and supply chains, with in most cases a lack of reliable sourcing alternatives beyond its designed radius of influence.

20.5 Filling the Gap: From Farm to Industry

As explained in the previous section, there are many dedicated lignocellulosic biomass cropping systems that are already commercially mature delivering fuel or biobased products for industries. Moreover, the sustainability of plantations may secure the stability of the supply chain. Building the bridge between small and medium sized farms to a demanding industry remains, however, a substantial challenge (Figure 20.3).

A deep understanding of that gap enables the practitioners to design appropriate growing conditions, farm management, harvesting techniques, logistic chain development and transportation. A biomass feedstock product needs to meet industrial requirements; however, achieving those requirements should not be carried out without establishing confidence in supply stability and in biomass quality for the specific purpose and processing technologies involved in the target biorefinery, which aims to produce energy or biomaterials and co-products (e.g. animal feed).

The gap between farms and industries is appropriately addressed when a reputable biomass supplier can grow biomass using sustainable practices and farm management techniques suitable for building a sustainable supply chain for industrial needs in the long run. For example, any biomass-to-energy project considering dedicated crops will have to include a reasonable plan, which ensures that farm management (e.g. fertilization), harvesting frequencies (which typically change biomass composition and moisture of harvested biomass), feedstock density (kg/m^3), logistic chain issues with trans-loaders, trailers or in-farm mobility, and the long transportation system that will deliver products to a facility gate. Stable storage systems (e.g. silo bags, piles, sheds, etc.) between farm or plot and final disposal of delivered feedstock may be required to ensure procurement.

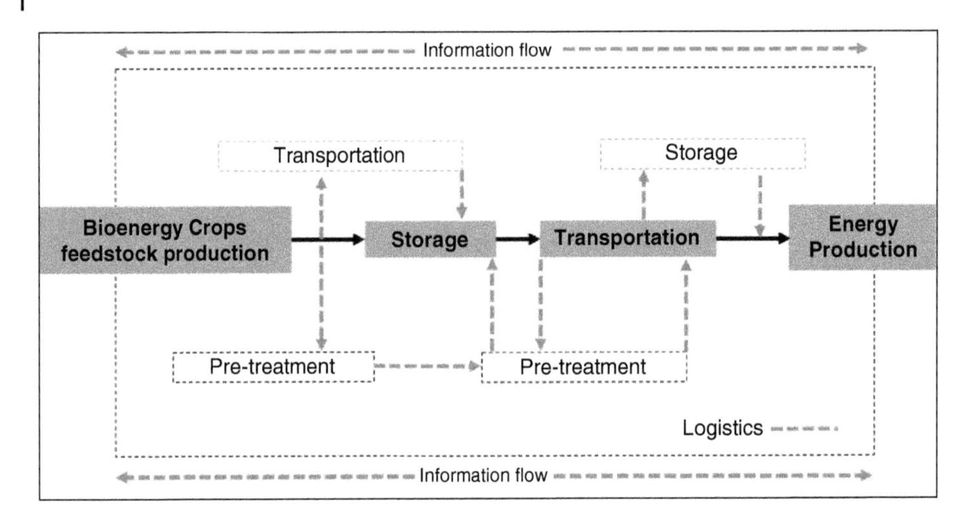

Figure 20.3 Biomass supply chains. Source: own elaboration.

The cost associated with the handling and transporting of biomass is a major concern when using biomass as an energy source. To reduce these costs, low density biomass crops need to be densified by briquetting, pelletizing, or cubing. Eventually, before transportation, a densification unit (e.g. for pelleting, briquetting, torrefaction) may reduce logistic costs by eliminating water and excessive volumes. A typical densification process increases the bulk density of biomass from a density of 40–200 kg/m^3 to a density of 600–800 kg/m^3. The energy consumption of a typical biomass densification step depends on applied compressing pressure and the chosen densification method, the initial moisture content of the biomass material, as well as on a variety of physical properties of the biomass material. Biomass can be densified via two main processes: mechanical densification and pyrolysis. Mechanical densification involves applying pressure to mechanically densify the material. Pyrolysis involves heating biomass in the absence of oxygen. In general, lower temperatures at longer processing times (i.e. slow pyrolysis) favours solid (charcoal) production. Medium temperatures (400–500 °C) at very short times (one to two seconds), known as fast pyrolysis, favours liquid or bio-oil production (Czernik and Bridgwater 2004).

From farm management to end disposal, regional storage systems such as a shed, a patio, a dryer unit, shredders or densification and feeding systems, one has to consider proximate and ultimate analysis for biomass products along the logistic chain. The 'proximate' analysis gives moisture content, volatile content (when heated to 950 °C), the free carbon remaining at that point, the ash content (mineral) of the sample and the high heating value (HHV) based on the complete combustion of the sample to carbon dioxide and liquid water. The low heating value, LHV, gives the heat released when the hydrogen is burned to gaseous water, corresponding to most heating applications and can be calculated from the HHV and the hydrogen fraction. The 'ultimate analysis' gives the composition of the biomass in wt% of carbon, hydrogen and oxygen (the major components) as well as sulfur and nitrogen (if any) (Figure 20.4).

Pellets, cubes and briquettes are just different forms of final product that can be considered in dedicated manufacturing plants that transform lignocellulosic crops. There are several equipment alternatives and sound scientific background on densification systems for dedicated biomass plantations, many including chop length and grinding, physical and chemical pre-treatment options, binding materials, steaming and torrefaction (Deng et al. 2009).

Figure 20.4 Milled grass and pellets examples. Source: Photo: Bioenergy Crops Ltd.

From the point of view of delivered products to be used as feedstocks in biotechnological processes to produce biofuels or sustainable chemicals, biomass dedicated perennial crops can be divided into crops generating wet or dry products. The biomass drying technology has recently been improved; it has now become possible to reduce the moisture content at different parts of the supply chain. Eventually, a partial use of biomass milled into fine powder can be used to feed the thermal energy required for drying the ultimate biomass and fuel products.

A major barrier to using dedicated herbaceous systems, such as perennial grasses, is that woody chips are often preferred mainly because of their lower ash content. Biomass bales with 15% moisture often have higher ash contents, potassium, chlorine, silica, sulphur, nitrogen, calcium or other soluble minerals compared to wood. These materials can be pre-treated to enhance their intrinsic characteristics as process feedstock and meet processing requirements; this is particularly relevant in boilers where the ash composition should not be overlooked during project design. In this regard, several methods have been reviewed including washing systems, screw presses and dewatering alternatives that have been re-considered in biomass projects recently (Kumar et al. 2009). Silage harvesting is becoming more popular in biogas projects and can be more cost effective than mowing, windrowing and baling biomass with farm equipment; and in many cropping systems such as Napier grass in tropical areas or miscanthus in Europe, moisture can be reduced directly in the field, which may allow pick-up choppers to harvest and deliver biomass to a pre-treatment module with a more suitable and uniform moisture content for the processing and conversion pathways involved (Kumar et al. 2009).

These alternatives allow industries to obtain a uniform product during the whole year. The use of screw presses for silage dewatering is energetically more efficient than thermal drying. Tests with two types of screw press showed that at dry matter flow rates of more than 1.0 tDM/h, the specific energy consumption is less than 30 kWh/tDM (Scholz et al. 2009). The water content of the silage is decreased by 5–20% and the share of undesired ingredients by 2–30%. Besides varying press design and plant species, particle size and silage density are essential parameters for the dewatering success. Dryer feedstocks would open markets for the combustion or gasification of the feedstocks, pelleting material for feed, slow release fertilizer or serving as ground cover. At a minimum, producers or processors using a dryer feedstock for feed or land application would be trucking more material and less water. Several studies and even commercial developers are including pre-treatment methods to condition biomass products to be ready for subsequent processing. The biomass ash consists mainly of the elements potassium, chlorine, and silicon. These minerals, together with hemicellulose, may be reduced through a pretreatment step prior to combustion, gasification, pyrolysis or anaerobic digestion process (Kumar et al. 2009). The simplicity of washing and press systems are cost effective alternatives considering the energy requirements and drying costs of other options (Haque and Somerville 2013).

When treating different herbaceous biomass systems, Cui et al. (2015) summarized the impacts of various treatment methods on fuel properties, and provide detailed data on mass and element partitioning between process streams to inform system design. The processed fuels thus obtained exhibited lower ash contents, improved heating values, higher ash deformation temperatures, and higher volatile matter to fixed carbon ratios than the parent materials. The liquid streams generated by the process were characterized for chemical oxygen demand, sugar content, total solids, total suspended solids, and major and trace elements. The pretreatment effectiveness depends on the feedstock type, particle size, and operating conditions, which may include leach water quality and quantity, leaching temperature and duration, as well as the geometric size of the material. Herbaceous biomass supply chains are more frequent in biogas projects and increasing attention by biorefinery managers have been reported since fibre cake and green juice fractions can be utilized in different conversion pathways (Arlabosse et al. 2011). It is worth noting here that grass juice contains useful fractions, including amino acids, organic acids, and dyes (Kromus et al. 2004). Typically, fresh juice contains approximately 35–60% dry matter with a high protein (40–29%) content; moreover, the grass protein concentrate (GPC) can be produced by heat coagulation (Kromus et al. 2004; Grass 2004).

Some recent biorefineries in Europe have been testing and developing processing alternatives to fraction leaf protein concentrate from legume and grass systems; reliable and uniform feedstock for combustion have thus been obtained, as well as biogas or ethanol (see for example www.grassa.nl located in Sint Jansweg 205928 RC Venlo, Netherlands). Ethanol, animal feed protein products and even heat and power supplies are relatively well developed alternatives from the point of view of chemical and physical processing options for a grass based biorefinery (Gnansounou et al. 2017; Hagman et al. 2017; IEA 2016).

To ferment the carbohydrates present in cellulosic feedstocks into ethanol, the first step is to break the long polymers into their component sugars. However, yields from enzymatic hydrolysis are low unless the biomass first undergoes a pretreatment process (Zhang et al. 2015). One promising method to improve the efficiency of the hydrolysis is the ammonia fibre expansion (AFEX) process (Bals et al. 2010). Concentrated ammonia is added to the biomass under high pressure and moderate temperatures, held for a residence time of five minutes, before pressure is rapidly released. This process de-crystallizes the cellulose, hydrolyses hemicellulose, removes and depolymerizes lignin, and increases the sizes of micro-pores on the cellulose surface, thereby, significantly increasing the rate of enzymatic hydrolysis (Mosier et al. 2005; Dale et al. 2009). In addition to coagulation, membrane technologies have been recently significantly improved to achieve reduced costs and improved efficacy when fractioning proteins of legumes and grasses; the deployment of these technologies has been suggested for new biorefineries (Zhang et al. 2015).

20.6 Perspectives

The strategic importance of the bio-economy is linked to those areas in which bio-based products and processes can substitute fossil or synthetic mineral-based products and chemical processes that make use of non-renewable energy sources. Since the overwhelming majority of industrial products and processes are currently based on non-renewable resources and inorganic minerals, such substitution has considerable potential to make various industry sectors more sustainable in the long-run, while also reducing the environmental impacts of the chemical industry in the near-term, especially by reducing GHG emissions and by implementing land disposal requirements. Marginal land-derived biomass can potentially contribute substantially to the renewable energy effort if optimization is implemented at all stages of the process and supply chain. It is likely that the overall feasibility of this industrial transformation will depend upon implementing creative processing to realize the full potential of the emerging processes.

Environmental effects of biofuels should be considered in relation to energy and land- use practices that occur in the absence of their use. The displacement of fossil fuel use can reduce soil subsidence (Morton et al. 2006) and land-use changes associated with exploration and extraction of fossil fuels that significantly negatively impact biodiversity (Finer and Orta-Martinez 2010). Furthermore, the risk an of environmental catastrophe that would affect biodiversity is much lower for biofuels than it is for fossil fuels, as the latter involves exploration and extraction in still pristine environments such as deep seas and arctic regions (Chilingar and Endres 2005; Parish et al. 2013; Butt et al. 2013).

Among the different feedstock options analysed, it seems clear that different sustainability and social issues may relevant when delivering biomass or biofuels to the markets (Diaz-Chavez et al. 2015). An enhanced bioeconomy should consider perennial systems over annual monocultures, at least under large scale agriculture. While farmers are often more reluctant to adopt perennial agriculture, it is well accepted today that perennial culture represents a more sustainable or even regenerative alternative for bioenergy. Perennial cropping systems, both woody and herbaceous species may be integrated in agroforestry schemes, and even in rotational regional approaches with positive impacts on water streams and soil organic matter. Most perennial biomass cropping systems analyzed are believed to annually sequester more than 1 Mg/ha of carbon belowground, even considering the most recalcitrant fractions (Dohleman et al. 2012; Dowell et al. 2009). In this regard, biomass from dedicated perennial cropping systems can have a very low footprint if LCAs consider carbon and nitrogen balances when reporting GHG emissions.

Regarding the implementation and viability of lignocellulosic biomass crops, any operation with dedicated plantations should consider a careful balance between industrial requirements during processing and conversion routes, with logistic chains and feedstock procurement strategies and its products' quality. Management decisions comprising fertilization, harvesting frequency, climatic conditions and genetics may determine cultivation aspects that could endanger or increase risks along the supply chain (e.g. rains during harvesting periods, storage systems required downstream, or higher drying energy costs to meet transportation and feeding system requirements).

The expansion of bioenergy on a global scale also poses important challenges. For example, the potential competition for land and for raw material with other uses for biomass must be carefully managed (Souza et al. 2017). This aspect should always be analysed considering deforestation policy, land grabbing, social ownership schemes, regional legislation, farming and forestry culture, and social stability. To displace and replace fossil reserve-derived fuels, bioenergy must become increasingly competitive with other energy sources. As a result, logistic systems and infrastructure issues must be carefully addressed, and there is a need for further technological innovation leading to more efficient and cleaner conversion of a more diverse range of feedstocks. Further work on these issues is essential so that policies can focus on encouraging sustainable routes and provide confidence in the concept of bioenergy not only to policy makers, but also to the public at large.

References

Alexandratos, N. and Bruinsma, J. (2012). *World Agriculture towards 2030/2050: The 2012 Revision*. Rome: FAO http://www.fao.org/docrep/016/ap106e/ap106e.pdf.

Alexopoulou, E., Christou, M., Mardikis, M., and Chatziathanassiou, A. (2000). Growth and yields of kenaf varieties in central Crecee. *Industrial Crops and Products* 11: 163–172.

Angelini, L.G., Ceccarini, L., o di Nassi, Nasso, N., and Bonari, E. (2009). Comparison of *Arundo donax* L. and Miscanthus x giganteus in a long-term field experiment in Central Italy: analysis of productive characteristics and energy balance. *Biomass and Bioenergy* 33: 635–643.

Arlabosse, P., Blanc, M., Kerfaï, S., and Fernandez, A. (2011). Production of green juice with an intensive thermo-mechanical fractionation process. Part I: effects of processing conditions on the dewatering kinetics. *Chemical Engineering Journal* 168 (2): 586–592.

Bals, B., Rogers, C., Jin, M. et al. (2010). Evaluation of ammonia fibre expansion (AFEX) pretreatment for enzymatic hydrolysis of switchgrass harvested in different seasons and locations. *Biotechnology for Biofuels* 3: 1.

Boehmel, C., Lewandowski, I., and Claupein, W. (2008). Comparing annual and perennial energy cropping systems with different management intensities. *Agricultural Systems* 96: 224–236.

Borland, A.M., Griffiths, H., Hartwell, J., and Smith, J.A.C. (2009). Exploiting the potential of plants with crassulacean acid metabolism for bioenergy production on marginal lands. *Journal of Experimental Botany* 60 (10): 2879–2896.

Børresen, T. (1999). The effect of straw management and reduced tillage on soil properties and crop yields of spring-sown cereals on two loam soils in Norway. *Soil and Tillage Research* 51 (1–2): 91–102, ISSN 0167-1987.

Boschma, S.P., Lodge, G.M., and McCormick, L.H. (2010). Recent tropical perennial grass research and their potential role in maintaining production in a variable and changing climate. In: *NSW* (ed. C. Waters), 85–92.

Butt, N., Beyer, H.L., Bennett, J.R. et al. (2013). Biodiversity risks from fossil fuel extraction. *Science* 342 (6157): 425–426.

Cai, X., Zhang, X., and Wang, D. (2011). Land availability for biofuel production. *Environmental Science and Technology* 45: 334–339.

Campbell, J.E., Lobell, D.B., Genova, R.C., and Field, C.B. (2008). The global potential of bioenergy on abandoned agricultural lands. *Environmental Science and Technology* 42: 5791–5794.

Cardone, M., Mazzoncini, M., Menini, S. et al. (2003). Brassica carinata as an alternative oil crop for the production of biodiesel in Italy: agronomic evaluation, fuel production by transesterification and characterization. *Biomass and Bioenergy* 25: 623–636.

Castro, D., Urzúa, J., Rodriguez-Malebran, M. et al. (2017). Woody leguminous trees: new uses for sustainable development of drylands. *Journal of Sustainable Forestry* 36 (8): 764–786. https://doi.org/10.1080/10549811.2017.1359098.

Ceotto, E. and Di Candilo, M. (2011). Medium-term effect of perennial energy crops on soil organic carbon storage. *Italian Journal of Agronomy* 6 (4): e33. ISSN 2039-6805. Available at: https://bit.ly/2LJdooc. Date accessed: 21 August 2018.

Chilingar, G.V. and Endres, B. (2005). Environmental hazards posed by the Los Angeles Basin urban oil elds: an historical perspective of lessons learned. *Environmental Geology* 47: 302–317.

Chum, H., Faaij, A.P.C., and Moreira, J. (2011). Bioenergy. In: *IPCC Special Report on Renewable Energy Sources and Climate Change Mitigation* (eds. O. Edenhofer, R. Pichs-Madruga, Y. Sokona, et al.), 209–332. Cambridge, United Kingdom/New York, USA: Cambridge University Press.

Chundawat, S.P.S., Beckham, G.T., Himmel, M.E., and Dale, B.E. (2011). Deconstruction of lignocellulosic biomass to fuels and chemicals. *Annual Review of Chemical and Biomolecular Engineering* (2): 121–145.

Clifton-Brown, J.C., Stampfl, P.P., and Jones, M.C. (2004). Miscanthus biomass production for energy in Europe and its potential contribution to decreasing fosil fuel carbon emissions. *Global Change Biology* 10: 509–518.

Copani, V., Cosentino, S.L., Sortino, O. et al. (2009). Agronomic and energetic performance of Brassica carinata a. Braun in southern Italy. In: *Proceedings of the 19th European Biomass Conference and Exhibition*, 166–170. Hamburg, Germany, 29 June – 3 July 2009.

Crutzen, P.J., Mosier, A.R., Smith, K.A., and Winiwarter, W. (2008). N2O release from agro-biofuel production negates global warming reduction by replacing fossil fuels. *Atmospheric Chemistry and Physics* 8: 389–395.

Csete, S., Farkas, Á., Borhidi, A. et al. (2011). *Tall Wheatgrass Cultivar Szarvasi-1 (Elymus elongatus subsp. ponticus cv. Szarvasi-1) as a Potential Energy Crop for Semi-Arid Lands of Eastern Europe.* INTECH Open Access Publisher.

Cui, H., Turn, S.Q., Tran, T., and Rogers, D. (2015). Mechanical dewatering and water leaching pretreatment of fresh banagrass, Guinea grass, energy cane, and sugar cane: characterization of fuel properties and byproduct streams. *Fuel Processing Technology* 139: 159–172.

Cushman, J.C., Davis, S.C., Yang, X., and Borland, A.M. (2015). Development and use of bioenergy feedstocks for semi-arid and arid lands. *Journal of Experimental Botany* 66 (14): 4177–4193.

Czernik, S. and Bridgwater, A.V. (2004). Overview of applications of biomass fast pyrolysis oil. *Energy & Fuels* 18: 590–598.

Dale, B.E., Allen, M.S., Laser, M., and Lynd, L.R. (2009). Protein feeds coproduction in biomass conversion to fuels and chemicals. *Biofuels, Bioproducts and Biorefining* 3: 219–230.

Danalatos N.G. and S.V. Archontoulis, 2004. Potential growth and biomass productivity of kenaf (*Hibiscus cannabinus* l.) Under central greek conditions: I. The influence of fertilization and irrigation. 2nd World Conference on Biomass for Energy, Industry and Climate Protection, 10–14 May 2004, Rome, Italy.

Danner, H. and Braun, R. (1999). Biotechnology for the production of commodity chemicals from biomass. *Chemical Society Reviews* https://doi.org/10.1039/a806968i.

Das, L., Liu, E., Saeed, A. et al. (2017). Industrial hemp as a potential bioenergy crop in comparison with kenaf, switchgrass and biomass sorghum. *Bioresource Technology* 244, Part 1: 641–649, ISSN 0960-8524.

Davis, S.C., Dohleman, F.G., and Long, S.P. (2011). The global potential for agave as a biofuel feedstock. *GCB Bioenergy* 3: 68–78.

Davis, S.C., LeBauer, D.S., and Long, S.P. (2014). Light to liquid fuel: theoretical and realized energy conversion efficiency of plants using Crassulacean acid metabolism (CAM) in arid conditions. *Journal of Experimental Botany* 65 (13): 3471–3478.

Davis, S.C., Kuzmick, E.R., Niechayev, N., and Hunsaker, D.J. (2016). Productivity and water use efficiency of Agave Americana in the first field trial as bioenergy feedstock on arid lands. *GCB Bioenergy* https://doi.org/10.1111/gcbb.12324.

Deng, J., Wang, G.-j., Kuang, J.-h. et al. (2009). Pretreatment of agricultural residues for co-gasification via torrefaction. *Journal of Analytical and Applied Pyrolysis* 86: 331–337.

Dheeran, P. and Reddy, L. (2018). Biorefining of lignocelluloses: an opportunity for sustainable biofuel production. In: *Biorefining of Biomass to Biofuels. Biofuel and Biorefinery Technologies*, vol. 4 (eds. S. Kumar and R. Sani). Springer, Cham.

Diaz-Chavez, R., Morese, M.M., Colangeli, M. et al. (2015). Social considerations. In: *Bioenergy & Sustainability: Bridging the Gaps 72*, 528–552. Paris, France: SCOPE, (ISBN 978-2-9545557-0-6).

Dohleman, F.G., Heaton, E.A., Arundale, R.A., and Long, S.P. (2012). Seasonal dynamics of above- and below-ground biomass and nitrogen partitioning in Miscanthus × giganteus and *Panicum virgatum* across three growing seasons. *Global Change Biology. Bioenergy* 4: 534–544. https://doi.org/10.1111/j.1757-1707.2011.01153.x.

Dowell, R., Gibbins, D., Rhoads, J., and Pallardy, S. (2009). Biomass production physiology and soil carbon dynamics in short rotation grown Populus deltoides and P. deltoids x P. nigra hybrids. *Forest Ecology and Management* 257: 134–142.

Dowling, C.D., Burson, B.L., Foster, J.L. et al. (2013). Confirmation of pearl millet-napiergrass hybrids using EST-derived simple sequence repeat (SSR) markers. *American Journal of Plant Sciences* 4: 1004–1012. https://doi.org/10.4236/ajps.2013.45124. Published Online May 2013 (http://www.scirp.org/journal/ajps).

EC (2012). European Commission, 2012: innovating for sustainable growth: a bioeconomy for Europe. *Industrial Biotechnology* 8 (2): 57–61.

EEA, 2013. EU bioenergy potential from a resource efficiency perspective. Report No 6/2013. European Environment Agency. Available online at: http://bit.ly/1KfDU3o (last access October 2017)

Escamilla-Treviño, L.L. (2012). Potential of plants from the genus agave (Agave sisalana y Agave salmiana) as bioenergy crops. *Bioenergy Research* 5 (1): 1–9.

FAO (2013). *Perennial Crops for Food Security. Proceeding of the FAO Expert Workshop*, 28–30. August, Rome, Italy.

Fazio, S. and Monti, A. (2011). Life cycle assessment of different bioenergy production systems including perennial and annual crops. *Biomass and Bioenergy* 35 (12): 4868–4878, ISSN 0961-9534, https://doi.org/10.1016/j.biombioe.2011.10.014.

Felten, D. and Emmerling, C. (2011). Effects of bioenergy crop cultivation on earthworm communities – a comparative study of perennial (Miscanthus) and annual crops with consideration of graded land-use intensity. *Applied Soil Ecology* 49: 167–177, ISSN 0929-1393.

Filho, Q., De, J.L., Da Silva, D.S., and do Nascimento, I.S. (2000). Dry matter production and quality of elephant grass (*Pennisetum purpureum Schum.*) cultivar Roxo at different cutting ages. *Revista Brasileira de Zootecnia* 29: 69–74.

Finer, M. and Orta-Martinez, M. (2010). A second hydrocarbon boom threatens the Peruvian Amazon: trends, projections, and policy implications. *Environmental Research Letters* 5: 014012.

Gnansounou, E., Alves, C.M., Ruiz Pachón, E., and Vaskan, P. (2017). Comparative assessment of selected sugarcane biorefinery-centered systems in Brazil: a multi-criteria method based on sustainability indicators. *Bioresource Technology* 243: 600–610, ISSN 0960-8524.

Grass, S. (2004). *Utilisation of Grass for Production of Fibres, Protein and Energy*. Paris: OECD Publication Service, September 11.

Hagman, L., Blumenthal, A., Eklund, M., and Svensson, N. (2017). The role of biogas solutions in sustainable biorefineries. *Journal of Cleaner Production*, ISSN 0959-6526.

Haque, N. and Somerville, M. (2013). Techno-economic and environmental evaluation of biomass dryer. *Procedia Engineering* 56: 650–655.

Hari, K. and Srinivasan, T.R. (2005). Response of sugarcane varieties to application of nitrogen fixing bacteria under different nitrogen levels. *Sugar Tech* 7 (2&3): 28–31.

Heaton, E., Voigt, T., and Long, S.P. (2004). A quantitative review comparing the yields of two candidate C4 perennial biomass crops in relation to nitrogen, temperature and water. *Biomass and Bioenergy* 27: 21–30.

Heděnec, P., Novotný, D., Usťak, S. et al. (2014). The effect of native and introduced biofuel crops on the composition of soil biota communities. *Biomass and Bioenergy* 60: 137–146, ISSN 0961-9534.

Hodkinson TR, Chase MW, Renvoize SA (2002) Characterisation of a genetic resource collection for Miscanthus (Saccharinae, Andropogoneae, Poaceae) using AFLP and ISSR PCR. Ann Bot (Lond) 89: 627–636.

Holtum, J.A.M., Chambers, D., Morganz, T., and Tan, D.Y. (2011). Agave as a biofuel feedstock in Australia. *GCB Bioenergy* 3: 58–67.

IEA, 2009. IEA Bioenergy. A sustainable and reliable energy source. Main Report. Paris: International Energy Agency; 2009. Available at: https://bit.ly/2wuWdSj (last accessed date, August 2018).

IEA, 2013. IEA Bioenergy, Task 39. Advanced Biofuels – GHG Emissions and Energy Balances. Report T39-T5. 25 May 2013. Available at: https://bit.ly/2PVvab9 (last accessed, August 2018).

IEA, 2016. The Role of Biomass, Bioenergy and Biorefining in a Circular Economy. Harriëtte Bos, Bert Annevelink & Rene van Ree IEA workshop, Paris, 10 January 2017. Available at: https://bit.ly/2NGZ3dc (last accessed July 2018).

IRENA, 2014. Global Bioenergy. Supply and Demand Projections. A working paper for REmap 2030. International Renewable Energy Association. Masdar City PO Box 236. Abu Dhabi, United Arab Emirates Available at: https://bit.ly/1vtlrcI (last accessed date, 31 August 2018).

IRENA, 2017.Renewable Energy and Jobs. International Renewable Energy Agency (IRENA). Annual review 2017. Available online at (last accessed July 2018): https://bit.ly/2raVDGK

Isah, Y.M., Abakr, Y.A., Kazi, F.K. et al. (2015). Comprehensive characterization of Napier Grass as a feedstock for thermochemical conversion. *Energies* 8: 3403–3417. https://doi.org/10.3390/en8053403.

Jyväskylä Innovation Oy, 2009. Energy from field energy crops – a handbook for energy producers. Jyväskylä Innovation Oy, P.O. Box 27, FI - 40101 JYVÄSKYLÄ, Finland. Available online: https://bit.ly/2PSoGJX (last accessed August 2018).

Karus, M. (2004). European hemp industry 2002: cultivation, processing and product lines. *Journal of Industrial Hemp* 9 (2): 93–101.

Kromus, S., Wachter, B., Koschuh, W. et al. (2004). The green biorefinery Austria – development of an integrated system for green biomass utilization. *Chemical and Biochemical Engineering Quarterly* 18: 7–12.

Kumar, P., Barrett, D.M., Delwiche, M.J., and Stroeve, P. (2009). Methods for Pretreatment of Lignocellulosic biomass for efficient hydrolysis and biofuel production. *Industrial & Engineering Chemistry Research* 48 (8): 3713–3729.

Lewandowski, I., Scurlock, J.M.O., Lindvall, E., and Christou, M. (2003). The development and current status of perennial rhizomatous grasses as energy crops in the US and Europe. *Biomass and Bioenergy* 25: 335–361.

LogistEC, 2012. Logistics for Energy Crops' Biomass. Grant agreement number: FP7–311858. Collaborative project (small or medium-scale focused research project targeted to SMEs) Seventh Framework Programme Priority: Food, Agriculture and Fisheries, and Biotechnology. Deliverable D4.2. Feedstock supply scenarios. Available at: http://www.logistecproject.eu (last accessed October 2017).

Lynd, L.R., Liang, X., Biddy, M.J. et al. (2017). Cellulosic ethanol: status and innovation. *Current Opinion in Biotechnology* 45: 202–211.

Magalhães, J.A., Lopes, E.A., Rodrigues, B.H.N. et al. (2006). Influência da adubação nitrogenada e da idade de corte sobre o rendimento forrageiro do capim-elefante. *Revista Ciência Agronômica* 37: 91–96.

Maletta, H., 2014. Land and Farm Production: Availability, Use, and Productivity of Agricultural Land in the World (October 30, 2014). Available at SSRN: https://ssrn.com/abstract=2484248

Maletta, H. (2016). *Towards the End of Hunger*. Lima, Peru: Universidad del Pacifico.

Maletta, E., Martin-Sastre, C., Ciria, P., del Val, A., Salvadó, A., Rovira, L., Díez, R., Serra, J., González-Arechavala, Y., Carrasco, J.E., 2012. Perennial Energy Crops for Semiarid Lands in the

Mediterranean: *Elytrigia elongata*, A C3 Grass with Summer Dormancy to Produce Electricity in Constraint Environments. 20th European biomass conference and exthibiton. 18–22 June, Milan, Italy.

McMillan, J.D., 2004. Biotechnological Routes to Biomass Conversion. DOE/NASULGC Biomass and Solar Energy Workshops. Aug 3–4, 2004.

Mielenz, J.R., Rodriguez, M. Jr.,, Thompson, O.A. et al. (2015). Development of agave as a dedicated biomass source: production of biofuels from whole plants. *Biotechnology for Biofuels* 8: 79.

Morton, R.A., Bernier, J.C., and Barras, J.A. (2006). Evidence of regional subsidence and associated interior wetland loss induced by hydrocarbon production, Gulf Coast region, USA. *Environmental Geology* 50: 261–274.

Mosier, N., Wyman, C., Dale, B. et al. (2005). *Bioresource Technology* 96: 673–686.

Nguyen, T.L.T., Hermansen, J.E., and Mogensen, L. (2013). Environmental performance of crop residues as an energy source for electricity production: the case of wheat straw in Denmark. *Applied Energy* 104: 633–641, ISSN 0306-2619.

Nobel, P., García-Moya, E., and Quero, E. (1992). High annual productivity of certain agaves and cacti under cultivation. *Plant, Cell and Environment* 15: 339–335.

Núñez, H.M., Rodríguez, L.F., and Khanna, M. (2011). Agave for tequila and biofuels: an economic assessment and potential opportunities. *GCB Bioenergy* 3: 43–57.

Obok, E.E., Aken'Ova, M.E., and Iwo, G.A. (2012). Forage potentials of interspecific hybrids between elephant grass selections and cultivated pearl millet genotypes of Nigerian origin. *Journal of Plant Breeding and Crop Science.* 4: 136–143.

OECD (2009). *The Bioeconomy to 2030: Designing a Policy Agenda*. Main Findings and Policy Conclusions. OECD International Futures Project. Available at: https://bit.ly/2zRWPUJ (last accessed October 2017).

Olsen, S.N., Ritter, K., Medley, J. et al. (2013). Energy sorghum hybrids: functional dynamics of high nitrogen use efficiency. *Biomass and Bioenergy* 56: 307–316.

Owen, N.A. and Griffiths, H. (2014). Marginal land bioethanol yield potential of four crassulacean acid metabolism candidates (Agave fourcroydes, Agave salmiana, Agave tequilana and Opuntia ficus-indica) in Australia. *GCB Bioenergy* 6: 687–703.

Parish, E.S., Kline, K.L., Dale, V.H. et al. (2013). A multi-scale comparison of environmental effects from gasoline and ethanol production. *Environmental Management* 51 (2): 307–338.

Pedroli, B., Elbersen, B., Frederiksen, P. et al. (2013). Is energy cropping in Europe compatible with biodiversity? – opportunities and threats to biodiversity from land-based production of biomass for bioenergy purposes. *Biomass and Bioenergy* 55: 73–86, ISSN 0961-9534.

Petrini, C., Bazzocchi, R., and Montalti, P. (1994). Yield potential and adaptation of kenaf (*Hibiscus cannabinus* L.) in northcentral Italy. *Industrial Crops and Products* 3: 11–15.

Pimentel, D., Marklein, A. Toth, M.A. et al. (2009). Food Versus Biofuels: Environmental and Economic Costs. Human Ecology 37(1): 1–12. DOI: 10.1007/s10745-009-9215-8.

Prade, T., Svensson, S.E., Andersson, A., and Mattsson, J.E. (2011). Biomass and energy yield of industrial hemp grown for biogas and solid fuel. *Biomass and Bioenergy* 35 (7): 3040–3049, ISSN 0961-9534.

Pugesgaard, S., Schelde, K., Larsen, S.U. et al. (2015). Comparing annual and perennial crops for bioenergy production – influence on nitrate leaching and energy balance. *GCB Bioenergy* 7: 1136–1149. https://doi.org/10.1111/gcbb.12215.

Riffaldi, R., Saviozzi, A., Cardelli, R. et al. (2012). Comparison of soil organic matter characteristics under the energy crop giant reed, cropping sequence and natural grass. *Communications in Soil Science and Plant* 41: 173–180.

Rocatelli, A.C., Raper, R.L., Balkcom, K.S. et al. (2012). Biomass sorghum production and components under different irrigation/tillage systems for the southeastern U.S. *Industrial Crops and Products* 36: 589–598.

Rueda, J.A., Ortega-Jiménez, E., Hernández-Garay, A. et al. (2016). Growth, yield, fiber content and lodging resistance in eight varieties of Cenchrus purpureus (Schumach.) Morrone intended as energy crop. *Biomass and Bioenergy* 88: 59–65.

Saffih-Hdadi, K. and Mary, B. (2008). Modeling consequences of straw residues export on soil organic carbon. *Soil Biology and Biochemistry* 40 (3): 594–607.

San Miguel, G., Corona, B., Ruiz, D. et al. (2015). Environmental, energy and economic analysis of a biomass supply chain based on a poplar short rotation coppice in Spain. *Journal of Cleaner Production* 94: 93–101.

Sastre, C.M., Maletta, E., González-Arechavala, Y. et al. (2014). Centralised electricity production from winter cereals biomass grown under Central-Northern Spain conditions: global warming and energy yield assessments. *Applied Energy* 114: 737–748, ISSN 0306-2619, https://doi.org/10.1016/j.apenergy.2013.08.035.

Sastre, C., Carrasco, J., Barro, R. et al. (2016). Improving bioenergy sustainability evaluations by using soil nitrogen balance coupled with life cycle assessment: a case study for electricity generated from rye biomass. *Applied Energy* 179: 847–863, ISSN 0306-2619, https://doi.org/10.1016/j.apenergy.2016.07.022.

Schmer, M.R., Jin, V.L., and Wienhold, B.J. (2015). Sub-surface soil carbon changes affects biofuel greenhouse gas emissions. *Biomass and Bioenergy* 81: 31–34, ISSN 0961-9534.

Scholz, V., Daries, W., and Rinder, R. (2009). Mechanical dewatering of silage. *Landtechnik Agricultural Engineering* 64 (5): 333–335. Available online at: https://bit.ly/2Liqhu7 (last accessed: July 2018).

Searchinger, T., Heimlich, R., Houghton, R.A. et al. (2008). Use of US croplands for biofuels increases greenhouse gases through emissions from land-use change. *Science* 319: 1238–1240.

Sharma, N., Bohra, B., Pragya, N. et al. (2016). Bioenergy from agroforestry can lead to improved food security, climate change, soil quality, and rural development. *Food and Energy Security* 5 (3): 165–183.

Sierra, M., Martínez, F.J., Verde, R. et al. (2013). Soil-carbon sequestration and soil-carbon fractions, comparison between poplar plantations and corn crops in South-Eastern Spain. *Soil and Tillage Research* 130: 1–6, ISSN 0167-1987.

Smith, J., Abegaz, A., Matthews, R.B. et al. (2014). What is the potential for biogas digesters to improve soil fertility and crop production in sub-Saharan Africa? *Biomass and Bioenergy* 70: 58–72, ISSN 0961-9534.

Soldatos PG, Lychnaras V, Asimakis D, Christou M., 2004. BEE – Biomass economic evaluation: a model for the economic analysis of energy crops production. In: Proceedings of 2nd world conference and technology exhibition on biomass for energy, industry and climate protection, Rome; 2004.

Solomon, S., Qin, M., Manning, M. et al. (2007). Technical summary. In: *Climate Change 2007: The Physical Science Basis. Contribution of Working Group I to the Fourth Assessment Report of the Intergovernmental Panel on Climate Change* (eds. S. Solomon, D. Qin, M. Manning, et al.), 19–91. Cambridge, United Kingdom/New York, NY, USA: Cambridge University Press.

Soon, Y.K. and Lupwayi, N.Z. (2012). Straw management in a cold semi-arid region: impact on soil quality and crop productivity. *Field Crops Research* 139: 39–46, ISSN 0378-4290.

Souza, G.M., Ballester, M.V.R., de Brito Cruz, C.H. et al. (2017). The role of bioenergy in a climate-changing world. *Environmental Development* 23: 57–64, ISSN 2211-4645.

Tamang, P.L., Bronson, K.F., Malapati, A. et al. (2011). Nitrogen requirements for ethanol production from sweet and photoperiod sensitive sorghums in the southern High Plains. *Agronomy Journal* 103: 431–440.

Tarkalson, D., Brown, B., Kok, H., and Bjorneberg, D.L. (2009). Impact of removing straw from wheat and barley fields: a literature review. *Better Crops* 93 (3). Available online at: http://bit.ly/1KfE3DR (last accessed 31Aug. 2018).

Turley, D.B., Phillips, M.C., Johnson, P. et al. (2003). Long-term straw management effects on yields of sequential wheat (Triticum aestivum L.) crops in clay and silty clay loam soils in England. *Soil and Tillage Research* 71 (1): 59–69, ISSN 0167-1987.

Weiser, C., Zeller, V., Reinicke, F. et al. (2014). Integrated assessment of sustainable cereal straw potential and different straw-based energy applications in Germany. *Applied Energy* 114: 749–762, ISSN 0306-2619.

Yang, L., Lu, M., Carl, S. et al. (2015). Biomass characterization of agave and Opuntia as potential biofuel feedstocks. *Biomass and Bioenergy* 76: 43–53.

Zegada-Lizarazu, W. and Monti, A. (2012). Are we ready to cultivate sweet sorghum as a bioenergy feedstock? A review on field management practices. *Biomass and Bioenergy* 4: 1–12.

Zhang, W., Grimi, N., Jaffrin, M.Y., and Ding, L. (2015). Leaf protein concentration of alfalfa juice by membrane technology. *Journal of Membrane Science* 489: 183–193, ISSN 0376-7388.

21

Industrial Waste Valorization: Applications to the Case of Liquid Biofuels

Haibo Huang and Qing Jin

Department of Food Science and Technology (DFST), Virginia Polytechnic Institute and State University (Virginia Tech), Blacksburg, VA 24061, USA

CHAPTER MENU

21.1 Introduction

Industrial waste is defined as the waste produced by an industrial activity that is composed of any material that has not been recycled or used during the corresponding manufacturing process. Biofuel production from industrial waste has received wide interest as it is a process with a potentially important impact on the environment, energy, and economics. From an environmental point of view, the disposal of industrial waste can cause severe environmental problems, such as increased green house gas (GHG) emissions, water and soil pollution, and disruption of biogenic cycles due to the presence of toxic contaminants. Recycling industrial waste to biofuels provides an effective way to alleviate the environmental pressure caused by waste disposal while generating a domino effect of savings in other primary raw materials such as food crops that would otherwise be needed to produce the corresponding number of energy units. The world's energy system is still largely based on fossil fuels, an industrial activity that is not sustainable because the burning of fossil fuels leads to an increase of atmospheric CO_2 on the one hand and the continual depletion of fossil resources on the other that push the price of fuel to high levels (Lal and Waldron 2009). Biofuels produced from renewable resources, such as industrial waste, can contribute to reducing both the world's dependence on fossil fuels and CO_2 emission levels (Naik et al. 2010). More importantly, biofuel production from industrial waste can improve society's economic competitiveness by converting waste to valuable resources. In addition, the integration of biofuel production into existing industrial waste streams can provide a low-cost source of carbon, thereby substantially decreasing the feedstock cost of biofuel production.

Green Energy to Sustainability: Strategies for Global Industries, First Edition.
Edited by Alain A. Vertès, Nasib Qureshi, Hans P. Blaschek and Hideaki Yukawa.

In this chapter, technologies for the valorization of industrial waste to liquid biofuels (ethanol, butanol, and biodiesel) are introduced and discussed. Food waste recycling to biofuels is used as the main example of waste to demonstrate the biofuel production potential of this industrial approach. In addition, other industrial wastes suitable for biofuel production are also introduced and discussed.

21.2 Types of Industrial Waste for Biofuel Production

For a typical feedstock to be deemed suitable to produce biofuels, it should contain significant amounts of readily available carbon and energy. Moreover, it should also be converted to biofuel with realistic capital and operating inputs using currently existing or near-future technologies. Meanwhile, its quantity should be large enough to impact the energy mix and support biofuel production at commercial scales. With this in mind, wastes from several industries comprising food processing, wood processing, and bioenergy industries, appear to be suitable feedstocks for biofuel production.

Among several feedstocks considered for biofuel production, food waste has recently triggered a high level of interest because of the large volumes of food waste generated, notably in the most industrialized countries, and its typically high carbohydrate content (Kiran et al. 2014). Food waste includes organic waste discharged from various sources including food processing plants, grocery stores, cafeterias, and restaurants, as well as household kitchens. According to an FAO (Food and Agriculture Organization) report, nearly 1.2 billion tonnes of foods including fresh vegetables, fruits, meats, and grains are lost along the food supply chains, accounting for almost one third of the total food production each year (FAO 2011). In the US, more than 34 million tonnes of food waste is generated each year (EPA 2013). Likewise, the European Union produces around 90 million tonnes of food waste annually (Bio Intelligence Service 2010). Due to its high moisture content, food waste is highly biodegradable. Consequently, the disposal of food waste in a landfill is not only costly but also causes environmental problems, with direct and indirect emissions of GHG. On the other hand, food waste contains abundant amounts of carbohydrates, proteins, fats, and minerals, which makes it a perfect substrate for biofuel production via fermentation or chemical conversion.

The wood processing industry is one of the major industrial domains that produces large amounts of carbohydrate waste. Approximately 131 million m^3 of wood residues are generated every year from the processing of industrial round wood (Lin et al. 2013). Wood residues are primarily produced from the sawing of wood; these rejects from sawmills include sawdust, trimmings, edgings, and veneer rejects. Their main biochemical components are cellulose, hemicellulose, and lignin. This composition makes wood residues a suitable feedstock for lignocellulosic ethanol or butanol production. Another notable use of these materials is the production of wood pellets that are used for domestic heating in replacement of other energy sources.

Another waste worthy of consideration is glycerol. With the fast development of biodiesel production during the past two decades, the volumes of the corresponding byproducts also increased dramatically; glycerol is the main byproduct of biodiesel production using vegetable oil and animal fats as feedstocks. Approximately 1 kg of glycerol is generated per 10 kg of biodiesel produced via a transesterification process (Ardi et al. 2015). It is estimated that around 3 million tonnes of glycerol will be produced from the biodiesel industry (Lin et al. 2013). Glycerol is an important chemical building block for pharmaceuticals, food, plasticizer, and cosmetic production; however, traditional glycerol markets cannot absorb the vast amount of glycerol from the biodiesel industry. Therefore, new applications for this material are greatly needed. New processes have been developed that enable the use of glycerol as carbon sources in various fermentations for the production of biofuels, such as ethanol and butanol; and this will be discussed later in this chapter.

21.3 Ethanol Production

21.3.1 Ethanol and Its Market

Ethanol is a two-carbon alcohol that can be used as an important platform chemical for the production of ethylene and ethylene glycol, two key materials for the further production of polyethylene and other plastics (Koutinas et al. 2014). Another major use of ethanol is as biofuel to partially replace gasoline and thus to make gasoline-ethanol mixtures, such as E15 (15% ethanol and 85% gasoline) or E85 (85% ethanol and 15% gasoline). The worldwide ethanol production reached about 97 billion litres in 2015, and the US and Brazil are the leading countries producing ethanol (Table 21.1). In fact, Brazil has replaced almost 40% of its gasoline needs with ethanol, thanks to its focused commitment to developing a competitive sugarcane industry and the wide adaption of its car fleet to enable the use of hydrous and anhydrous ethanol.

Ethanol can be produced by fermentation of various carbon sources. *Saccharomyces cerevisiae* (baker's yeast) is the most commonly used microorganism for producing ethanol from glucose and fructose, where one hexose molecule is converted into two ethanol molecules and two carbon dioxide molecules:

$$C_6H_{12}O_6 \xrightarrow{\text{yeast}} 2C_2H_5OH + 2CO_2$$

Thus, theoretically, the glucose to ethanol production yield is 0.511 g/g via fermentation. In Brazil, a country well adapted to the culture of sugarcane, sugarcane is the main feedstock used for ethanol production, while corn is the main feedstock in the US and China (Cheng and Timilsina 2011). In the EU, ethanol is generally produced using a combination of sugar beets and wheat (Schnepf 2006). When sugarcane or sugar beets are used as feedstocks, juice is extracted before being sent to the fermenters. The main sugar in extracted juice is sucrose. Since *S. cerevisiae* can hydrolyse sucrose to glucose and fructose for fermentation given its property to secrete invertase (Novick and Schekman 1979), no additional enzymatic treatment is needed. On the other hand, when corn or wheat are used as feedstocks, enzymatic hydrolysis (e.g. amylase and glucoamylase) has to be applied to achieve the breakdown of corn or wheat starch into glucose before fermentation, since the macromolecular starch cannot be directly fermented to ethanol by wild type *S. cerevisiae*, which lacks amylases (Naik et al. 2010; Huang et al. 2012). Besides sugarcane and corn, lignocellulosic materials, such as corn stover and wheat straw, can also be

Table 21.1 Global ethanol production (million litres/year) between 2011 and 2015.

Year	US	Brazil	EU	China	Rest of world
2011	55 967	26 812	5 243	3 073	5 984
2012	54 054	23 398	5 462	2 400	7 560
2013	50 274	23 689	5 182	2 631	6 785
2014	49 964	21 081	4 305	2 098	4 850
2015	52 652	21 066	4 415	2 098	4 275

Source: calculated from Renewable Fuel Association (2016).

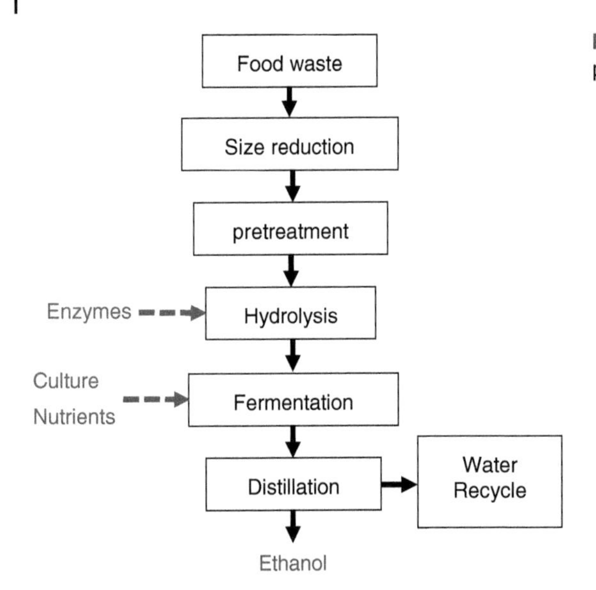

Figure 21.1 Schematic diagrams of ethanol production from food waste.

used as feedstock for ethanol production. Although lignocellulosic biomass is cheaper than cereal crops and sugarcane, the cost of obtaining monomer sugars from such materials for fermentation is, to date, too high to enable the cost-effective production of ethanol (Cheng and Timilsina 2011).

21.3.2 Ethanol from Food Waste

Feedstock cost currently represents the largest portion of ethanol production cost via fermentation. For example, when corn or sugarcane are used as feedstocks for ethanol production, feedstock cost accounts for 60–80% whereas operating cost, including utility and administrative costs, accounts for 20–40% of the ethanol production cost (Quintero et al. 2008; Crago et al. 2010). Recycling food wastes as feedstock can potentially reduce the cost of ethanol production. Various food wastes have been used to produce ethanol, including potato peel waste (Arapoglou et al. 2010), grape pomace (Rodríguez et al. 2010; Zheng et al. 2012a), banana peel (Oberoi et al. 2011b; Manikandan et al. 2008), pineapple peel (Itelima et al. 2013), sugar beet pulp (Zheng et al. 2013; Zheng et al. 2012b), brewer's spent grain (Xiros and Christakopoulos 2009; Xiros et al. 2008), dining food waste (Kim et al. 2008a; Yan et al. 2011; Moon et al. 2009) and household food waste (Matsakas et al. 2014; Ma et al. 2009). The general process to convert food waste to ethanol is shown schematically in Figure 21.1. The process contains several main components: size reduction, pretreatment, hydrolysis, fermentation, and distillation. Hydrolysis and fermentation are sometimes operated together, called simultaneous saccharification and fermentation (SSF) (Kiran et al. 2014).

21.3.3 Pretreatment

For starch-rich food waste, harsh pretreatment may not be necessary because the enzymatic hydrolysis is effective enough to hydrolyse starch to glucose (Kiran et al. 2014). However, for fibre-rich food waste, such as rice hulls and sugar beet pulp, pretreatment is an essential step to break its recalcitrant structure and facilitate the subsequent enzymatic hydrolysis to convert cellulose to glucose. Pretreatment may involve the application of various physical processes (e.g. grinding and milling), physico-chemical processes (e.g. auto-hydrolysis, steam explosion), and chemical processes (e.g. acid and alkali hydrolysis) (Chiaramonti et al. 2012; Alvira et al. 2010).

Acid hydrolysis is a widely used pretreatment method for fibre-rich food waste conversion, and sulfuric acid is the most commonly used acid. The main objective of the acid pretreatment step is to solubilize the hemicellulosic fraction of the food waste and to make the cellulose more accessible to enzymes. For example, Zheng et al. (2012a) pretreated grape pomace with 1 wt % H_2SO_4 at 120 °C for 5 minutes to reduce the recalcitrance of the cellulose structure. The highest ethanol yield was 0.24 g/g grape pomace when the pretreated sample was subsequently subjected to enzymatic hydrolysis and fermented by an engineered *Escherichia coli* expressing the genes *pdc* and *adhB* from *Zymmonas mobilis* (Zheng et al. 2012a). In another study, potato peel waste was pretreated by dilute HCl at 121 °C for 15 minutes, followed by incubation with a mixture of enzymes (beta-glucanase, amylase, cellulase) (Arapoglou et al. 2010). The dilute acid pretreatment and enzymatic hydrolysis released 18.5 g/l reducing sugars and produced 7.6 g/l ethanol after yeast fermentation. One concern of the acid hydrolysis is the formation of an array of potent fermentation inhibitors from the degradation of sugars and lignin, including furfural, and acetic, formic and levulinic acids as well as aromatic compounds (Koutinas et al. 2014). These inhibitors have the potential to reduce enzymatic hydrolysis efficiency and inhibit microbial fermentation. In order to avoid such inhibition, various treatments, such as overliming, for the detoxification of fermentation inhibitors have been investigated (Klinke et al. 2004); however, this extra step increases the process complexity and cost. In another study, Sakai et al. (2007) employed a growth-arrested *Corynebacterium glutamicum* strain to eschew the detoxification steps to keep costs down and reduce process complexity (Sakai et al. 2007). The growth-arrested *C. glutamicum* showed high tolerance to all organic acid, furan, and phenolic inhibitors, retaining 62–100% ethanol productivity compared with inhibitor-free fermentation.

Steam explosion is another widely investigated pretreatment method for fibre-rich food waste. This process consists of a hydrothermal pretreatment in which the food waste is subjected to pressurized steam, typically 0.69–4.83 Mpa, for a period of time ranging from seconds to several minutes, and then suddenly depressurized to atmospheric pressure (Alvira et al. 2010; Sun and Cheng 2002). This pretreatment combines mechanical forces and chemical effects due to the release of acetyl groups presented in hemicellulose. The mechanical effects are caused because the pressure is suddenly reduced and fibres are separated owing to the explosive decompression. Compared to other pretreatments, steam explosion has several advantages such as a lower environmental impact, lower capital investment and potentially higher energy efficiency (Avellar and Glasser 1998). For example, a steam explosion process was applied by Wilkins et al. (2007b) to pretreat citrus peel for ethanol production. The steam explosion process is used for a dual purpose: (i) to increase the enzyme accessibility to the cell wall polysaccharides, and (ii) to remove D-limonene which is an inhibitor for yeast fermentation (Murdock and Allen 1960; Wilkins et al. 2007a). In this experimental process, citrus peel waste was pretreated with steam in a continuous tube reactor at 150–160 °C for 2–4 minutes. The hot material was flashed to a cyclone where most of D-limonene was stripped by steam escape. The pretreated citrus peel waste was subsequently digested with pectinase, cellulase, and beta-galactosidase at different enzyme loadings. After process optimization, the highest ethanol concentration achieved after 24 hours fermentation was 38.7 g/l in broth (Wilkins et al. 2007b).

21.3.4 Enzymatic Hydrolysis

Enzymatic hydrolysis is the most prominent technology for the conversion of macromolecular starch and cellulose to glucose for subsequent fermentation. Although concentrated acid can be used to hydrolyse starch to glucose at an elevated temperature, enzymatic hydrolysis is preferred because it is conducted in mild conditions (pH 5–8, temperature < 90 °C) and does not form inhibitory compounds (Lee 1997).

Starch-rich food waste (e.g. bread, mashed potato, waste noodles) are usually subjected to enzymatic hydrolysis directly, without pretreatment, to convert starch to glucose with starch-hydrolysing enzymes, such as α-amylase, β-amylase, and glucoamylase. For example, Tang et al. (2008) used glucoamylase to convert starch in kitchen waste to glucose. The kitchen waste was collected from a students' dining hall in Japan, mainly containing 59.8% starch sugars, 21.8% protein, and 15.7% lipid based on dry weight. After saccharification using Nagase N-40 glucoamylase at 60 °C, glucose recovery from food waste was as high as 85.5%. Similarly, Yan et al. (2011) used a mixture of α-amylase and glucoamylase to treat food waste collected from a dining room in China. The dining waste contained 63.9% starch sugars, 21.3% protein, and 2.0% cellulose based on dry weight. The highest reducing sugar production was 164.8 g/l at a glucoamylase load of 142.2 μ/g food waste, pH of 4.8, reaction temperature of 55 °C and reaction time of 2.5 hours. In another study, Hong and Yoon (2011), used food waste collected from cafeterias that they treated with a mixture of alpha-amylase, glucoamylase, and protease, resulting in 60 g of reducing sugars per 100 g of food waste. Huang et al. (2015a) furthermore reported the use of novel granular starch-hydrolysing (GSH) enzymes to hydrolyse food waste starch to glucose. These enzymes have high GSH activities and can convert starch into glucose at fermentation temperatures (32 °C); therefore, the use of GSH enzymes does not require heating of the food waste slurry to high temperature, which is usually required by using other enzymes (e.g. alpha-amylase and glucoamylase). This property directly results in an overall reduction of the utility requirements of the hydrolysis process, which contributes to lowering manufacturing costs.

After harsh pretreatments, enzymatic hydrolysis with fibre-rich food waste (e.g. rice hulls, sugar beet pulp, grape pomace) can be carried out by a mixture of highly specific cellulase, hemicellulase, and other assisting enzymes such as esterase and pectinase. For example, in a study on the conversion of rice hulls to ethanol, Saha et al. (2005) applied a combination of cellulase, β-glucosidase, xylanase, and esterase to hydrolyse rice hulls, which were previously pretreated with sulfuric acid. The enzymatic hydrolysis was performed at 45 °C, pH 5.0, for 72 hours. The hydrolysis with the combination of the four enzymes provided the highest yield of 28.7 g sugars per 100 g of rice hulls. Moreover, Foster et al. (2001) performed an enzymatic hydrolysis of ammonia-pretreated sugar beet pulp using three commercially available enzymes (cellulase, cellobiase, and hemicellulase); and the hydrolysis was performed at 40 °C, pH 4.8, for 48 hours. The highest reducing sugar yield was 52 g sugars per 100 g of biomass.

21.3.5 Fermentation

A high ethanol concentration after fermentation is critically important for reducing the cost of ethanol production due to the great energy demand of ethanol distillation (Shihadeh et al. 2013; Thomas et al. 1996; Zhang et al. 2010). While the term 'high ethanol concentration' is ambiguous, the current existing corn-ethanol and sugarcane-ethanol industries can help provide baselines for the final ethanol concentration of food waste fermentation. In the corn-ethanol industry, 110–120 g/l of final ethanol concentrations have been routinely achieved in the past 10 years (Singh et al. 2010). In the sugarcane-ethanol industry, the final ethanol concentrations after juice fermentation are usually between 60 and 90 g/l (Macedo 2010). The final ethanol concentrations in most studies, including those cited above, are below 50 g/l (Table 21.2), which is detrimental to the commercial scale production of ethanol. Low concentrations of ethanol in fermentation broth increase

Table 21.2 Production of ethanol from food wastes from different sources.

Source of food waste	Solid content for fermentation	Final ethanol concentration (g/l)	References
Sugar beet pulp	2.0% (w/w)	8.0	Zheng et al. (2013)
Grape pomace	4.0% (w/w)	~7.3	Zheng et al. (2012a)
Citrus peel	18.6–23.4% (w/w)	24.0–38.7	Wilkins et al. (2007b)
Banana peel	5.0–12.5% (w/v)	6.5–9.8	Manikandan et al. (2008)
potato peel waste	2.0% (w/v)	7.0	Arapoglou et al. (2010)
Cafeteria	12.1% (w/w)	29.1	Moon et al. (2009)
Cafeteria	16.3–19.0% (w/w)	45.0	Kim et al. (2011)
Household	35–45% (w/v)	34.9–42.8	Matsakas et al. (2014)
Dining room	<20.5% (w/w)	75.9–81.5	Yan et al. (2011)
Retail store	35% (w/w)	144.0	Huang et al. (2015a)

the energy consumption of downstream distillation, thereby reducing the energy output-to-input ratio and increasing the ethanol production cost from food wastes.

Higher final ethanol concentrations can be achieved by increasing the dry solid contents of food waste slurries; however, this is limited by the two main factors: high glucose inhibition after hydrolysis and high ethanol concentration after fermentation. High solid substrate results in a high glucose concentration, which negatively affects microbial performance due to high osmotic pressure. Osmotic stress can result in the production of higher concentrations of glycerol as glycerol aids yeast with osmo-adaptation; this reduces the ethanol yield from each unit of glucose. High glucose may also result in catabolite repression of enzymes (Oberoi et al. 2011a, Kiran et al. 2014). This challenge can be mitigated by using an SSF process, which combines enzymatic hydrolysis and ethanol fermentation into a single operation. In an SSF process, glucose is continuously produced by enzymatic hydrolysis and converted to ethanol by microbial metabolism; therefore, its concentration in broth is maintained at a low level (Singh et al. 2010). SSF has been applied in different studies for converting food wastes to ethanol (Hong and Yoon 2011; Kim et al. 2011; Wilkins et al. 2007b). For example Hong and Yoon (2011) studied the production of ethanol from cafeteria food waste in a continuous SSF process where cultivated yeast, enzymes, and nutrients were added simultaneously to autoclaved food wastes to start the SSF process at 33.5 °C. About 36.9 g/l of ethanol were obtained from 100 g/l of food wastes in 48 hours of SSF. In another study, Davis (2008) converted two different types of food waste to ethanol at 27 °C using yeast and starch hydrolysing enzymes in batch SSF experiments. A mathematical model of SSF based on experimental match rate equations for enzyme hydrolysis and yeast fermentation was developed to estimate the key variables for hydrolysis and fermentation. The SSF process has the advantages of being simple by using a single tank for hydrolysis and fermentation, lower energy consumption, and shorter processing times using less enzymes (Öhgren et al. 2007). However, it is important to note that the optimal conditions (pH, temperature) for enzymatic hydrolysis and microbial fermentation are usually different, thus, the optimization of fermentation conditions to maximize the performance of ethanol production is a critical success factor for SSF processes.

High solid fermentation results in a high ethanol concentration in broth, which inhibits microbial activity, causing reduced ethanol yield and fermentation efficiency (Wang et al. 1999). In order

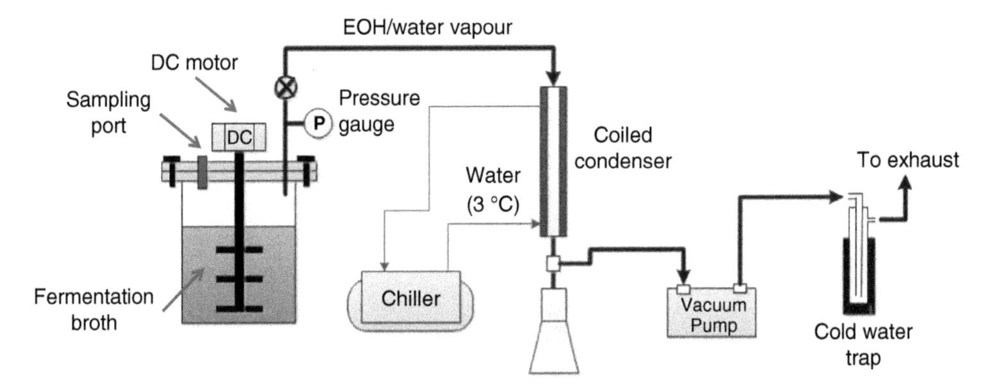

Figure 21.2 Experimental setup for food waste fermentation equipped with a vacuum stripping system. The arrows in the figure show flow direction. Source: Reproduced from H. Huang et al. (2015a), with permission from ACS publications.

to mitigate the ethanol inhibition to microbial culture, the coupling of a downstream process for ethanol recovery with fermentation was found to be useful. For example, a vacuum recovery technology was coupled with ethanol production using mixed food wastes (mashed potatoes, sweet corn and bread) as a substrate of a high solid content (Huang et al. 2015a). The vacuum fermentation system consisted of a fermenter, a vacuum pump, and a condensation unit (Figure 21.2). When the ethanol concentration in broth reached the point of inhibitting yeast, the vacuum pump was turned on to create a vacuum pressure in the fermenter. Fermentation broth in the fermenter boiled at the fermentation temperature generating ethanol vapours; the evaporated ethanol vapours were then condensed by passing them through a coiled condenser. By integrating downstream ethanol recovery with fermentation, the ethanol concentration in the fermentation broth can be maintained at a low level, allowing the high solid fermentation of food waste (35%, w/w) with a high ethanol yield (36 g ethanol/100 g food waste). Gas stripping is another powerful ethanol recovery technology that can be coupled to fermenters to reduce ethanol inhibition to microbial culture. Although not yet tested in food waste fermentation, the gas stripping technique has been successfully coupled with a corn-ethanol fermenter during the fermentation of high solid corn mash (Taylor et al. 2010). In this system, the fermentation broth was continuously recycled through a stripping column, where a non-condensable gas (CO_2) removed ethanol to a condenser. The feasibility of the system was demonstrated at over 40% corn dry solids, using a continuous corn liquefaction system. The system was operated for 60 consecutive days, continuously converting 95% of starch and producing 88% of the maximum theoretical yield of ethanol. The ethanol concentration remained at around 50 g/l in the fermenter while the ethanol concentration in the collected condensate was as high as 200–250 g/l.

21.3.6 Ethanol Production from Other Industrial Wastes

Wood residues (e.g. sawdust, trimmings, edgings) have been investigated for ethanol production in a number of studies (Galbe and Zacchi 2002; Koutinas et al. 2014; Silva et al. 2011; Kim et al. 2013). Wood residues are lignocellulosic materials, from which it is crucial to cost-effectively obtain monomer sugars (C5 and C6) by the hydrolysis of cellulose and hemicellulose. Generally, cellulose and hemicellulose can be hydrolysed to monomer sugars either by direct acid treatment or a combination of pretreatment and enzymatic hydrolysis (Galbe and Zacchi 2002). After hydrolysis, the fermentation of sugars can be carried out in a separate hydrolysis and fermentation (SHF) or SSF mode. For example, Iranmahboob et al. (2002) reported a 78–82% cellulose-to-glucose conversion efficiency when a concentrated sulfuric acid (26 wt %) was used to treat mixed wood chips at 100 °C for 2 hours. Although the direct hydrolysis with concentrated acids avoids the use of expensive enzymes, it causes equipment corrosion and results in high energy requirements for acid recovery (Galbe and Zacchi 2002). In another study of the conversion of sawdust to ethanol, Kim et al. (2013) pretreated the sawdust sample with dilute sulfuric acid, followed by enzymatic hydrolysis (cellulase) and fermentation. Enzymatic hydrolysis tests showed a cellulose digestibility of 67.1%, 70.1%, and 73.6% with 15, 30, and 45 filter paper units per g-cellulose, respectively. In the fermentation test, the maximum ethanol yield was 81.7% of the maximum theoretical value.

Glycerol is the main byproduct from biodiesel production via the transesterification process. The rapidly expanding market for biodiesel has resulted in an increased production and reduced market price of crude glycerol (Taconi et al. 2009). However, this tendency might invert in the near future with the increased sensitivity of citizens and governments alike to the atmospheric pollution generated by diesel powered combustion engines. Since traditional glycerol markets cannot absorb the vast amount of crude glycerol generated nowadays and since the advent of the biofuels industry, the need for advanced development and commercialization of innovative waste conversion technologies is in more demand than ever (Jang et al. 2012). Researchers have found that biodiesel-derived crude glycerol can be used as substrate for ethanol production by isolated *Kluyvera cryocrescens* (Choi et al. 2011). In batch fermentation under a micro-aerobic condition, *K. cryocrescens* produced 27 g/l of ethanol from crude glycerol with high molar yield of 80% and productivity of 0.61 g/l/h. In another study, ethanol was produced from crude glycerol by a mutant strain of *K. cryocrescens* obtained by γ irradiation (Oh et al. 2011). In the fermentation broth of mutant *K. cryocrescens*, the metabolites 2,3-butanediol (2,3-BD), ethanol, lactate, and succinate were significantly increased. The highest ethanol concentration was achieved in a fed-batch fermentation, with an ethanol productivity of 0.78 g/l/h. Hong et al. (2010) studied the enhancement of ethanol production from glycerol by the engineered methylotrophic yeast expressing the pyruvate decarboxylase and aldehyde dehydrogenase genes from *Z. mobilis*. The ethanol production in the engineered yeast was 2.4–3.4 fold greater than that of the wild strain; however, the ethanol concentration in the fermentation broth was still low at 2.4 g/l.

21.4 Butanol

21.4.1 Butanol and Its Market

Butanol is a four carbon alcohol that can be used as transportation biofuel. Using butanol instead of ethanol as a biofuel has several advantages. Firstly, the heating value of butanol is nearly 24% higher than that of ethanol, thus providing a higher mileage/gasoline blend ratio (Pfromm et al. 2010). Secondly, butanol can be blended with gasoline at any ratio, while ethanol can be blended only up to a maximum of 85% gasoline. Thirdly, butanol is less corrosive compared to ethanol, because ethanol can absorb moisture from the atmosphere, which oxidizes most metals. Moreover, some metals, such as magnesium, lead, and aluminium are susceptible to chemical attack by dry ethanol (Jin et al. 2011). Therefore, butanol can be transported through existing pipelines that were built based on the specifications of gasoline transportation. Fourthly, butanol has a lower vapour pressure and is thus safer to handle (Dürre 2007). Butanol has been historically produced from renewable biomass by fermentation at a large scale but its production from the strict anaerobic bacterium *Clostridium acetobutylicum* ceased between the 1950s and 1960s due to its unfavourable economy compared to petrochemically-derived butanol (Qureshi and Ezeji 2008). However, renewed interest

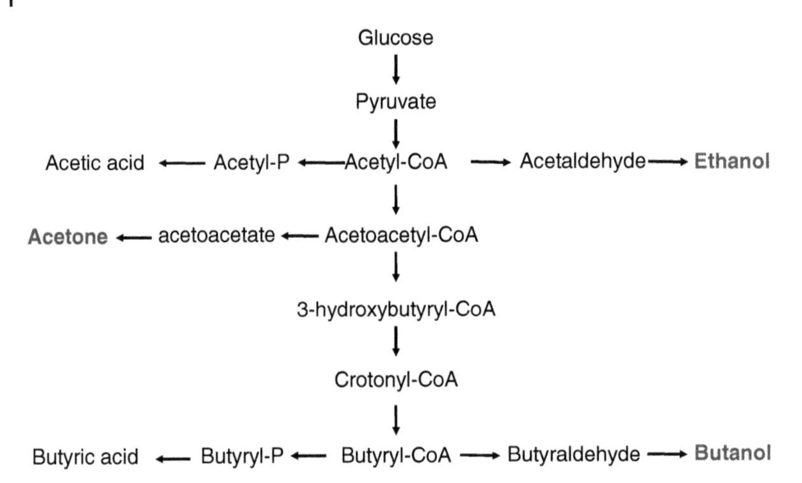

Figure 21.3 Simplified metabolic pathways for ABE synthesis by *C. acetobutylicum*. Source: Jones and Woods (1986).

in the biotechnological production of butanol arose again in recent years over increasing concerns regarding global warming and the peak oil phenomenon and associated oil supply sustainability uncertainty (Lee et al. 2008).

Bio-butanol is conventionally produced by *Clostridium* species via an acetone–butanol–ethanol (ABE) process, which converts carbohydrates into acetone, butanol, and ethanol (Dürre 2007). Several *Clostridium* strains have been identified to produce butanol in significant amounts; these cultures include *C. acetobutylicum*, *C. beijerinckii*, *C. butylicum*, *C. tetanomorphum*,. and *C. saccharoperbutylacetonicum* (Qureshi et al. 2001, 2006; Huang et al. 2015b; Jones and Woods 1986; Becerra et al. 2015). The ABE fermentation process usually proceeds in two phases (Figure 21.3). Glucose is first converted to acetic and butyric acid during the exponential bacterial growth phase, leading to a significant pH drop in the fermentation broth. At the end of the exponential phase of growth, a major metabolic switch takes place in Clostridia. The microorganism slows down acid production, and reuptakes and converts excreted acetic and butyric acids to ABE during the stationary growth phase. The mass ratio of butanol, acetone, and ethanol in a typical ABE process is 6 : 3 : 1, although this ratio varies with different strains. As butanol accumulates in the fermentation broth, it starts to inhibit bacterial growth and ultimately leads to fermentation arrests because butanol damages the cellular membranes and some membrane proteins (Moreira et al. 1981; Bowles and Ellefson 1985). Typically, the ABE process results in butanol production at a concentration lower than 15 g/l (Qureshi and Ezeji 2008).

21.4.2 Butanol Production from Food Waste

Sugar- and starch-based feedstocks have been traditionally used for butanol production. These feedstocks (molasses, sugarcane, sugar beet, and corn) have advantages of being simple to operate and available in bulk amounts; however, their relatively high prices make butanol production no longer economically competitive (Jones and Woods 1986). For example, Xue et al. (2013) reported that the total cost for butanol production from corn is about \$2000 per tonne, in which more than 70% is from feedstock consumption (corn), 20% is from utility consumption in electricity and vapour, and less than 10% is from labour cost, maintenance, and depreciation of facilities.

The use of food waste as feedstock for butanol production is a promising low-cost option for a future commercial process. Although research on butanol production from food waste is not as advanced as that on ethanol production, several studies have been recently published on ABE fermentation of starchy food waste to produce butanol. Notably, Kheyrandish et al. (2015) reported butanol production from potato waste starch using free and immobilized cells of *C. acetobutylicum* NRRL B-591. In that study, potato starch was directly fermented by *C. acetobutylicum* without hydrolysis since the strain used produces amylolytic enzymes to hydrolyse starch to glucose. Batch fermentation of 60 g/l of waste starch with free cells resulted in the production of 9.9 g/l butanol. On the other hand, when bacterial cells were immobilized in calcium alginate-polyvinyl alcohol-boric acid beads, fermentation of 60 g/l of waste starch resulted in the production of 15.3 g/l butanol. In the study by Ujor et al. (2014b), different starchy food waste samples were evaluated for butanol production via ABE fermentation (Ujor et al. 2014b). Similarly, batch fermentations by *C. beijerinckii* using inedible dough, breadings, and batter liquid as substrates showed the highest potential for ABE production among the different 34 food waste samples tested in the study, generated 9.3, 10.5, 10.0 g/l butanol, respectively. Due to low concentration of solvent in the fermentation broth, simultaneous fermentation and product removal techniques, such as vacuum stripping, have been developed and found to be cost-effective by improving both separation efficiency and fermentation yields (Tao et al. 2014). Huang et al. 2015b studied butanol production from mixed food wastes at a high solid content (129 g/l) by ABE fermentation coupled with a vacuum recovery technology. The results show that ABE fermentation coupled with vacuum stripping successfully recovered the ABE from the fermentation broth and controlled the ABE concentration below 10 g/l during fermentation, thereby relieving the butanol inhibition to the culture (*C. beijerinckii*). Under vacuum fermentation conditions, the ABE yield from food waste mixture was 0.36 with an ABE productivity of 0.49 g/l/h. It is worth noting that techno-economic analysis is needed to analyse the economic feasibility of the proposed process by considering both the economic benefits from higher butanol yield and productivity and the added capital and operating costs of vacuum stripping.

Whey permeate is a byproduct of the manufacture of cheese or casein in the dairy industry. The production of 1 kg cheese typically results in the generation of 9 kg of whey permeate (Kosikowski 1979). The total production of whey permeate worldwide is 160 million tonne per year (Stoeberl et al. 2011; Smithers 2008). Lactose makes up a high proportion (>75%) of the total whey solids, and contributes to a large extent to dairy whey being considered one of the most polluting food byproduct effluent streams (Siso 1996). Although glucose is the preferred sugar for *Clostridium* species to produce ethanol, different studies have shown that lactose can be utilized by these organisms when supplied as the sole carbon source (Ujor et al. 2014a). A number of studies have been published on the conversion of whey permeate to butanol via the ABE fermentation, primarily from Maddox, Qureshi, and coworkers (Maddox 1980; Qureshi and Maddox 1987; Ennis and Maddox 1989; Ennis et al. 1986; Qureshi and Maddox 2005; Friedl et al. 1991). In a typical study of Qureshi and Maddox (2005), ABE were produced from whey permeate, supplemented with lactose, in a batch reactor using *C. acetobutylicum*, coupled with ABE removal by perstraction. About 99.0 g/l of ABE (in condensate) were produced from 227 g/l lactose at a yield of 0.44 and productivity of 0.21 g/l/h. It should be noted that although whey permeate has been proven to be a suitable feedstock for butanol production, its low lactose content (due to the low solids present in whey permeate) makes achieving economic production of butanol challenging (Ujor et al. 2014a). The addition of lactose or other sugars may be needed to increase the sugar content in the fermentation broth to increase the final butanol concentration.

21.4.3 Butanol Production from Other Industrial Wastes

Studies have reported that crude glycerol can be used as a fermentation substrate for butanol fermentation (Malaviya et al. 2012; Jang et al. 2012; Taconi et al. 2009; Ahn et al. 2011). Moreover, when used as a fermentation substrate, glycerol can provide a higher reducing power during microbial fermentation compared to the traditionally used hexoses and pentoses (Jang et al. 2012). *Clostridium pasteurianum* is currently the most prominent strain used for converting glycerol to butanol. Notably, Taconi et al. (2009) compared the growth and butanol production of *C. pasteurianum* utilizing both purified and biodiesel-derived crude glycerol as the sole carbon source in batch fermentation. The highest butanol concentration achieved was approximately 7 g/l when purified glycerol was used as a substrate. The maximum yield of butanol from purified glycerol and crude glycerol was 0.36 and 0.30 g/g, respectively. These yields are substantially higher than the 0.15–0.20 g/g butanol yield typically achieved during the ABE fermentation of glucose. However, the butanol productivity in their study was as low as 0.03 g/l/h. To improve the fermentation rate, Malaviya et al. (2012) applied a continuous fermentation process with a hyper producing mutant of *C. pasteurianum* for the conversion of glycerol to butanol (Malaviya et al. 2012). Under the continuous fermentation condition at a dilute rate of 0.9 hours, the butanol productivity was high at 7.8 g/l/h. The continuous fermentation of *C. acetobutylicum* was also conducted by Andrade and Vasconcelos (2003) for converting a mixture of glycerol and glucose for butanol production over a 70 day period. These authors found that low-grade glycerol (65% purity) can be efficiently used as a substrate, resulting in a butanol yield of 0.34 and a butanol productivity of 0.42 g/l/h.

Dried distillers' grains and solubles (DDGS) is a byproduct from the corn-ethanol industry. It contains about 30–40% crude protein, 10–12% crude fat, 5% starch, and 20–30% fibre (neutral detergent fibre), and is usually used as animal feed (Kim et al. 2008b). Although the protein content in DDGS is threefold higher compared to corn, the unbalanced amino acid profile and high fibre content limit the use of DDGS to mainly as a feed for ruminant animals. Other potential uses of DDGS comprise the generation of biogas via anaerobic digestion and the production of biocomposites via thermal compounding of DDGS and poly (lactic acid) (PLA) (Ziganshin et al. 2011; Li and Susan Sun 2011). With an increasing demand for biofuel, DDGS is viewed as a potential bridge feedstock for butanol production from other cellulosic biomass. For example, Ezeji and Blaschek (2008) pretreated DDGS with dilute acid, liquid hot water, or ammonium fibre, followed by enzymatic hydrolysis to produce a mixture of monomer sugars, including glucose, mannose, arabinose, and xylose. The resulting sugar mixture was fermented with solventogenic Clostridia for butanol production. The highest ABE yield reached was 21 g/l, with productivities of 0.28–0.33 g/l/h (Ezeji and Blaschek 2008).

21.4.4 Economic Analysis of Butanol Production

Economic studies have been performed on the production of butanol from different feedstocks, including cellulosic and non-cellulosic biomaterials (Gapes 2000; Kumar et al. 2012; Qureshi et al. 2013; Mariano et al. 2013). These results vary because the studies were conducted at different times, locations, and with different economic assumptions (e.g. feedstock cost, production capacity, interest rate, and inflation). Nevertheless, these economic studies suggested that commercial production of butanol is drawing closer, especially when agricultural byproducts or wastes are used (Ezeji et al. 2007). Feedstock cost is usually one of the largest portions of the total production cost in ABE fermentation (Gapes 2000). Kumar et al. (2012) conducted an economic analysis of ABE fermentation with cellulosic (bagasse, barley straw, wheat straw, corn stover, and switchgrass) and non-cellulosic (glucose, sugarcane, corn, and sago) feedstocks. Their study showed that utilization of glucose required 37% lesser total fixed capital cost than the other cellulosic and non-cellulosic feedstocks for the per year production of 10 000 t of butanol. However, the production cost of butanol from glucose was \$5.32/kg, much higher than that from sugarcane and cellulosic material because of the high cost of glucose. The production cost of butanol from sugarcane and cellulosic materials were estimated at a range of \$0.59–\$0.75/kg.

Qureshi et al. (2013) conducted an economic evaluation of butanol production from wheat straw through ABE fermentation with a capacity of 150 000-t butanol per year. Wheat straw was pretreated with dilute (1% v/v) sulfuric acid at 121 °C for 1 hour followed by separate enzymatic hydrolysis, fermentation, and recovery. The production cost of butanol was estimated to be \$1.30/kg based on the batch fermentation. When an in-situ membrane recovery process was applied to the ABE fermentation, the production cost can be reduced to \$1.00/kg for a plant annexed to an existing distillery. The high capital investment greatly reduced the number of possible investors (Gapes 2000). To solve this problem, Mariano et al. (2013) proposed to integrate butanol production into an existing sugarcane plant. The sugarcane plant was designed to process 2 million-tonne sugarcane per year and utilize 25%, 50%, and 25% of the available sugarcane juice to produce sugar, ethanol, and butanol, respectively. The economic analysis showed that the investment in a sugarcane plant coupled with butanol production was more attractive (internal rate of return = 14.8%) compared with the conventional sugarcane plant (internal rate of return = 13.3%) when butanol is produced by an improved microorganism and traded as a chemical.

21.5 Biodiesel

Biodiesel is an important biofuel that can remarkably be manufactured both by de novo synthesis and recycling.

21.5.1 Biodiesel and Its Market

Biodiesel is generally made by chemically reacting lipids (e.g. vegetable oils, animal fats) with an alcohol (e.g. methanol, ethanol) producing fatty acid esters (Ma and Hanna 1999). The main components of vegetable oils and animal fats are triglycerides that are made up of 1 mol of glycerol and 3 mol of fatty acids (Ma and Hanna 1999). The fatty acids vary in their carbon chain length and in the number of unsaturated bonds, correspondingly leading to varying physical and chemical properties (Enweremadu and Mbarawa 2009). Although vegetable oils were tested for compression diesel engines early in 1900 by Rudolf Diesel (Shay 1993), the direct use of vegetable oils or animal fats as combustible fuels is not suitable for today's diesel engines due to their high kinematic viscosity and low volatility (Lam et al. 2010). Therefore, vegetable oils and animal fats have to be modified to bring their combustion-related properties closer to those of fossil energy-derived diesel fuels.

The transesterification reaction is the dominant process to produce biodiesel from vegetable oils and animal fats. In that reaction, triglycerides are converted into fatty acid alkyl esters in the presence of a short chain alcohol and a catalyst, with glycerol as a byproduct (Shay 1993). A catalyst is usually needed to improve the reaction rate and yield. The transesterification process is shown in Figure 21.4. Stoichiometrically, a 3 : 1 M ratio of alcohol to triglycerides is needed to complete a transesterification reaction. However, because the reaction is reversible, excess alcohol is used to shift the equilibrium to the product side (Shay 1993). Alcohols that can be used in the transesterification reaction are methanol, ethanol, propanol, butanol, and amyl alcohols. Methanol and ethanol are the most commonly used alcohols due to their low cost and their suitable intrinsic

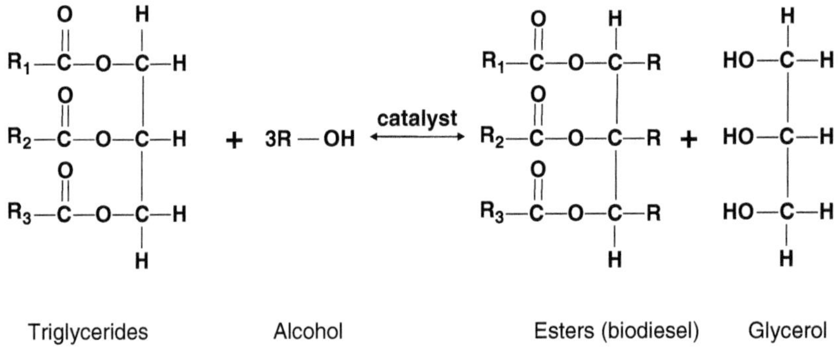

Triglycerides + 3R—OH (Alcohol) —catalyst→ Esters (biodiesel) + Glycerol

Figure 21.4 Transesterification of triglycerides with alcohols to produce biodiesel.

physical and chemical properties (e.g. they are polar alcohols with low boiling points) (Ma and Hanna 1999). The catalyst can be alkalis (e.g. NaOH, KOH, sodium methoxide, and potassium methoxide), acids (e.g. sulfuric acid, hydrochloric acid, and sulfonic acids), or enzymes (lipases) (Ma and Hanna 1999).

Compared to fossil energy-derived diesel fuel, biodiesel has a lower emission profile, and it is renewable (Atabani et al. 2012). The biodiesel production worldwide has increased from 15.2 billion litres in 2008 to 25.0 billion litres in 2012, and is expected to expand to reach 39 billion litres by 2024 (EIA 2012; OECD/FAO 2015). Remarkably, the EU is actively supporting the production of biodiesel from the agriculture sector, and as a result accounts for the largest portion of biodiesel production in the world, in concordance with the higher number of diesel-powered vehicles there (EIA 2012). In the US, the biodiesel production was 4.8 billion litres in 2012, a 1300% increase compared to only 0.34 billion litres in 2004 (DOE 2016). In China, the biodiesel production almost doubled between 2010 and 2015 due to government policies to support biodiesel production (Anderson-Sprecher and Jiang 2014).

21.5.2 Feedstocks for Biodiesel Production

The major feedstocks for biodiesel production are refined vegetable oils, such as soybean oil and canola oil. The high production cost of biodiesel remains one of the main concerns for the further development of biodiesel. For example, the cost of biodiesel production in the US fluctuated between $0.7 and $1.6/l from 2007 to 2014; this number is higher than the US diesel retail prices between $0.5 and 1.2/l during the same time period (Huang et al. 2016). Thus, biodiesel production is still not economically competitive without subsidies from governments. The high cost of biodiesel is mainly due to the high price of refined vegetable oil, because the feedstock cost contributes to more than 80% of the total production cost of biodiesel (Haas et al. 2006; Huang et al. 2016). The short supply of refined vegetable oils is another barrier to widely replacing fossil fuel with biodiesel. In the US, the main feedstock for biodiesel production is vegetable oil derived from soybean. It is estimated that if the entire soybean crop were used for biodiesel production, it would only provide one-tenth of the national use of diesel fuel (Huang et al. 2016).

Due to the high cost and short supply of refined vegetable oils, biodiesel manufacturers and researchers started to use waste cooking oil as feedstock to produce biodiesel. Waste cooking oil is defined as oil that has been used in frying pans or fryers; it was previously discarded as a waste. Compared to refined vegetable oil, waste cooking oil is far less expensive and is generated in large volumes throughout the world (Lam et al. 2010). Zhang et al. 2003b showed that biodiesel

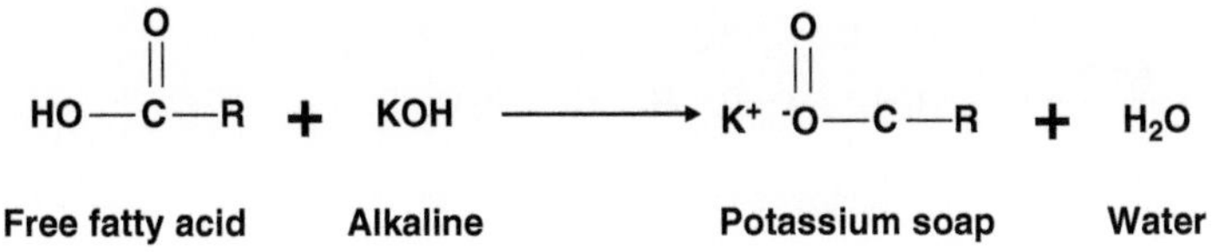

$$\mathrm{HO{-}\overset{\underset{\|}{O}}{C}{-}R} + \mathrm{KOH} \longrightarrow \mathrm{K^+\ {}^-O{-}\overset{\underset{\|}{O}}{C}{-}R} + \mathrm{H_2O}$$

Free fatty acid **Alkaline** **Potassium soap** **Water**

Figure 21.5 Saponification reaction between free fatty acid and alkaline.

production from waste cooking oil was more economically feasible than that from refined vegetable oil, providing a lower total manufacturing cost and a lower biodiesel break-even price.

21.5.3 Biodiesel Production from Waste Cooking Oil with Alkali Catalysts

Alkali catalysts (e.g. NaOH, KOH, CH_3ONa) are commonly used industrially for biodiesel production for several reasons: (i) these catalysts are cheap and readily available; (ii) they are able to catalyse the esterification reaction at low reaction temperature and atmospheric pressure; (iii) they can provide a fast reaction rate compared to other catalysts (Lam et al. 2010). However, the process is highly sensitive to free fatty acids (FFAs) and water content, which are typically high in waste cooking oil and animal fats. Waste cooking oil usually contains 2–7% FFAs, and animal fats contains about 5–30% FFAs; these values are much higher than the 0.5% FFAs typically observed in refined vegetable oil (Van Gerpen 2005; Sharma et al. 2008). High FFA content in waste cooking oil greatly affects the transesterification process catalysed by alkaline, because FFAs react with the most commonly used alkali catalysts (e.g. NaOH, KOH) thereby inducing a saponification reaction (Figure 21.5) and preventing the catalyst from catalysing the transesterification reaction. Furthermore, the high water content present in waste cooking oil can hydrolyse triglycerides into FFAs, further facilitating the saponification process to produce soap (Enweremadu and Mbarawa 2009).

Van Gerpen (2005) reported that the waste cooking oil with an FFA content lower than 5% can still be catalysed with an alkali catalyst, but additional catalyst must be added to compensate for the catalyst lost to soap. The soap thus produced can be removed with glycerol or eliminated during the subsequent water wash step. Felizardo et al. (2006) investigated biodiesel production from waste frying oil from school cafeterias, local restaurants, and local domestic consumers. Methanol was used as alcohol substrate for transesterification reactions and NaOH was used as a catalyst. These researchers evaluated the effects of key operating parameters on biodiesel yield, such as methanol/oil ratio and catalyst/oil ratio. The highest biodiesel yield was achieved when a methanol/oil ratio was 4.8 and a catalyst/oil ratio was 0.6%. Refaat et al. (2008) investigated biodiesel production from waste cooking oil with KOH as a catalyst at different operating conditions. They found that the yield percentage obtained from waste cooking oil was comparable to that obtained from virgin vegetable oil which reached 96.2% under optimum conditions.

21.5.4 Biodiesel Production from Waste Cooking Oil with Acid Catalysts

When the FFA content in waste cooking oil or animal fat is above 5%, an alkali catalyst is no longer suitable, because of the high amount of soap that would be produced through the saponification reaction, in turn inhibiting the separation of the ester (biodiesel) from glycerol and contributing to emulsion formation during the water wash step (Van Gerpen 2005; Lotero et al. 2005). As a result, to treat feedstocks that have an Federal Aviation Agency (FAA) content higher than 5%, acid catalysts (e.g. sulfuric acid and hydrochloric acid) are preferentially used. Acid catalysts are insensitive to

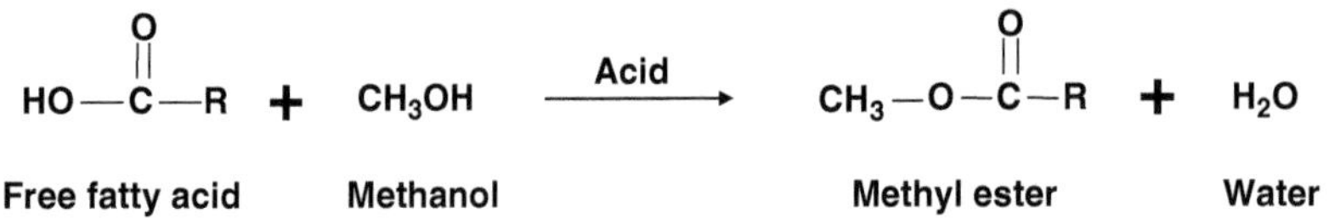

$$HO-\overset{O}{\overset{\|}{C}}-R + CH_3OH \xrightarrow{\text{Acid}} CH_3-O-\overset{O}{\overset{\|}{C}}-R + H_2O$$

Free fatty acid Methanol Methyl ester Water

Figure 21.6 Esterification reaction between free fatty acid and methanol catalysed by acid.

the presence of FFAs in waste cooking oil or animal fats and can catalyse the esterification reaction between FFAs and alcohols to produce esters (Figure 21.6).

Zheng et al. (2006) investigated the production of biodiesel from waste cooking oil using an acid-catalysed transesterification process. They found that the oil/methanol/acid molar ratios and temperature were the most significant operating parameters affecting the yield of biodiesel. The results attained in this study demonstrate that high yields of biodiesel (99%) can be achieved at 70 °C in the presence of a large excess of methanol. In a further study, Al-Widyan and Al-Shyoukh (2002) used ethanol at different excess levels to convert waste palm oil to biodiesel catalysed with H_2SO_4 and HCl. These researchers observed that H_2SO_4 performs better than HCl at 2.25 M as it results in biodiesel with a lower specific gravity. What is more, the optimum conditions for biodiesel production were determined to be 2.25 M H_2SO_4, 100% excess ethanol, 90 °C temperature at three hours of reaction time. Zhang et al. (2003a,b) conducted a technological and economic assessment of four processes for producing biodiesel from waste cooking oil and refined soybean oil. Their results suggest that the alkali-catalysed process using virgin oil had a lower total capital investment but a higher biodiesel production cost. When waste cooling oil was used as feedstock, the acid catalysed process to produce biodiesel had lower break-even price ($644 t^{-1}) than the alkali-catalysed process ($884 t^{-1}).

21.5.5 Biodiesel Production from Waste Cooking Oil with Acid and Alkali Catalysts

Despite their advantage to enable the processing of feedstock with FAA contents greater than 5%, the main limitations of an acid catalysed system are the slow reaction rate, high reaction temperature, high molar ratio of alcohol to oil, and serious environmental pollution and corrosion-related problems (Lam et al. 2010). All these limitations make acid catalysed systems not widely used at a commercial scale. To overcome these limitations, an acid and alkali two-step transesterification process has been developed. In this dual process, an acid catalyst is initially used to convert FFAs to ester through esterification. When the FFA content in the waste cooking oil drops to lower than 0.5–1%, that is when the saponification reaction becomes much less of a problem, an alkali catalyst is then added to catalyse the transesterification reaction. The two-step process has many advantages compared with processes using either acid or alkali catalysis alone, such as higher reaction rate, greater capability to handle high FFA content, and improved conversion efficiency. For example, Wang et al. (2007) studied the production of biodiesel from waste cooking oil via such a two-step catalysis process. In order to avoid the saponification reaction for raw materials with high FFA contents, the FFA was esterified with methanol by ferric sulfate in the first step. When the FFA content was lower than 1.0%, the triglycerides in waste cooking oil were trans-esterified in a second step with methanol and catalysed by NaOH. The results showed that a final product with 97% of biodiesel was obtained after the two-step catalysed process. In another similar study, Canackci and Van Gerpen (2001) demonstrated that the two-step reaction is useful to produce biodiesel with yellow grease with 12% FFAs and brown grease with 33% FFAs (Canakci and Van Gerpen 2001). After reducing the FFA levels of these feedstocks to less than 1% using acid-catalysed pretreatment, the transesterification reaction was completed with different alkali catalysts to produce fuel-grade biodiesel. The highest yield was 82.2% when NaOH was used as an alkali catalyst.

21.6 Perspectives

Much improvement has been achieved during the past two decades regarding the conversion of industrial wastes to valuable liquid fuels. Industrial waste, especially food waste, has been proved to constitute an excellent feedstock for ethanol, butanol, and biodiesel productions. However, challenges associated with the industrial waste reutilizations should be taken into consideration before full-scale commercialization. Firstly, unlike other resources, industrial waste is usually generated at different locations and some waste production is seasonal; therefore, the logistics and storage of industrial waste needs to be further investigated. Secondly, most of the studies regarding the conversion of industrial waste were conducted at the laboratory scale. Pilot scale experiments still need to be conducted in order to assess the reliability and feasibility of the developed technologies. Thirdly, techno-economic analyses are needed to fully assess manufacturability and economic feasibilities before the technology can be deployed to the commercial scale. Last but not least, ethanol, butanol, or biodiesel produced from food waste can only represent a small share of the global market for these commodity chemicals. Consequently, the technology of waste food recycling only provides a partial answer to the supply of these key chemicals. Nonetheless, the biotechnological significance of the emerging food waste recycling is a clear cut contribution to the sustainability goals of the commodity chemicals industry.

References

Ahn, J.-H., Sang, B.-I., and Um, Y. (2011). Butanol production from thin stillage using *Clostridium pasteurianum*. *Bioresource Technology* 102: 4934–4937.

Alvira, P., Tomás-Pejó, E., Ballesteros, M., and Negro, M. (2010). Pretreatment technologies for an efficient bioethanol production process based on enzymatic hydrolysis: a review. *Bioresource Technology* 101: 4851–4861.

Al-Widyan, M.I. and Al-Shyoukh, A.O. (2002). Experimental evaluation of the transesterification of waste palm oil into biodiesel. *Bioresource Technology* 85: 253–256.

Anderson-Sprecher, A. & Jiang, J. 2014. China's 2014 Fuel Ethanol Production is Forecast to Increase Six Percent. Available at: http://gain.fas.usda.gov/Recent GAIN Publications/Biofuels Annual_Beijing_China - Peoples Republic of_11-4-2014.pdf. (Accessed Sept 22, 2016).

Andrade, J.C. and Vasconcelos, I. (2003). Continuous cultures of *Clostridium acetobutylicum*: culture stability and low-grade glycerol utilisation. *Biotechnology Letters* 25: 121–125.

Arapoglou, D., Varzakas, T., Vlyssides, A., and Israilides, C. (2010). Ethanol production from potato peel waste (PPW). *Waste Management* 30: 1898–1902.

Ardi, M., Aroua, M., and Hashim, N.A. (2015). Progress, prospect and challenges in glycerol purification process: a review. *Renewable and Sustainable Energy Reviews* 42: 1164–1173.

Atabani, A.E., Silitonga, A.S., Badruddin, I.A. et al. (2012). A comprehensive review on biodiesel as an alternative energy resource and its characteristics. *Renewable and Sustainable Energy Reviews* 16: 2070–2093.

Avellar, B.K. and Glasser, W.G. (1998). Steam-assisted biomass fractionation. I. Process considerations and economic evaluation. *Biomass and Bioenergy* 14: 205–218.

Becerra, M., Cerdán, M.E., and González-Siso, M.I. (2015). Biobutanol from cheese whey. *Microbial Cell Factories* 14 (1): 200.

Bio Intelligence Service 2010. Preparatory Study on Food Waste across E.U.-27 for the European Commission, 2010, p. 14, Available at http://ec.europa.eu/environment/eussd/pdf/bio_foodwaste_report.pdf, (Accessed July 25, 2016).

Bowles, L.K. and Ellefson, W.L. (1985). Effects of butanol on *Clostridium acetobutylicum*. *Applied and Environmental Microbiology* 50: 1165–1170.

Canakci, M. and Van Gerpen, J. (2001). Biodiesel production from oils and fats with high free fatty acids. *Transactions of the ASAE* 44: 1429.

Cheng, J.J. and Timilsina, G.R. (2011). Status and barriers of advanced biofuel technologies: a review. *Renewable Energy* 36: 3541–3549.

Chiaramonti, D., Prussi, M., Ferrero, S. et al. (2012). Review of pretreatment processes for lignocellulosic ethanol production, and development of an innovative method. *Biomass and Bioenergy* 46: 25–35.

Choi, W.J., Hartono, M.R., Chan, W.H., and Yeo, S.S. (2011). Ethanol production from biodiesel-derived crude glycerol by newly isolated *Kluyvera cryocrescens*. *Applied Microbiology and Biotechnology* 89: 1255–1264.

Crago, C.L., Khanna, M., Barton, J. et al. (2010). Competitiveness of Brazilian sugarcane ethanol compared to US corn ethanol. *Energy Policy* 38: 7404–7415.

Davis, R.A. (2008). Parameter estimation for simultaneous saccharification and fermentation of food waste into ethanol using Matlab Simulink. *Applied Biochemistry and Biotechnology* 147: 11–21.

Dürre, P. (2007). Biobutanol: an attractive biofuel. *Biotechnology Journal* 2: 1525–1534.

Ennis, B. and Maddox, I. (1989). Production of solvents (ABE fermentation) from whey permeate by continuous fermentation in a membrane bioreactor. *Bioprocess Engineering* 4: 27–34.

Ennis, B., Marshall, C., Maddox, I., and Paterson, A. (1986). Continuous product recovery by in-situ gas stripping/condensation during solvent production from whey permeate using *Clostridium acetobutylicum*. *Biotechnology Letters* 8: 725–730.

Enweremadu, C.C. and Mbarawa, M. (2009). Technical aspects of production and analysis of biodiesel from used cooking oil - a review. *Renewable and Sustainable Energy Reviews* 13: 2205–2224.

EPA 2013. Food Waste Basics; U.S. EPA: Washington, D.C., 2014; Available at: http://www.epa.gov/foodrecovery (Accessed Sept 22, 2016).

Ezeji, T. and Blaschek, H.P. (2008). Fermentation of dried distillers' grains and solubles (DDGS) hydrolysates to solvents and value-added products by solventogenic clostridia. *Bioresource Technology* 99: 5232–5242.

Ezeji, T.C., Qureshi, N., and Blaschek, H.P. (2007). Bioproduction of butanol from biomass: from genes to bioreactors. *Current Opinion in Biotechnology* 18: 220–227.

Felizardo, P., Correia, M.J.N., Raposo, I. et al. (2006). Production of biodiesel from waste frying oils. *Waste Management* 26: 487–494.

Food and Agriculture Organization (FAO). 2011. Global food losses and food waste. Available at: http://www.fao.org/docrep/014/mb060e/mb060e.pdf. (Accessed December 2, 2016)

Foster, B.L., Dale, B.E., and Doran-Peterson, J.B. (2001). Enzymatic hydrolysis of ammonia-treated sugar beet pulp. *Applied Biochemistry and Biotechnology* 91: 269–282.

Friedl, A., Qureshi, N., and Maddox, I.S. (1991). Continuous acetone-butanol-ethanol (ABE) fermentation using immobilized cells of *Clostridium acetobutylicum* in a packed bed reactor and integration with product removal by pervaporation. *Biotechnology and Bioengineering* 38: 518–527.

Galbe, M. and Zacchi, G. (2002). A review of the production of ethanol from softwood. *Applied Microbiology and Biotechnology* 59: 618–628.

Gapes, J. (2000). The economics of acetone-butanol fermentation: theoretical and market considerations. *Journal of Molecular Microbiology and Biotechnology* 2: 27–32.

Haas, M.J., Mcaloon, A.J., Yee, W.C., and Foglia, T.A. (2006). A process model to estimate biodiesel production costs. *Bioresource Technology* 97: 671–678.

Hong, Y.S. and Yoon, H.H. (2011). Ethanol production from food residues. *Biomass and Bioenergy* 35: 3271–3275.

Hong, W.-K., Kim, C.-H., Heo, S.-Y. et al. (2010). Enhanced production of ethanol from glycerol by engineered *Hansenula polymorpha* expressing pyruvate decarboxylase and aldehyde dehydrogenase genes from *Zymomonas mobilis*. *Biotechnology Letters* 32: 1077–1082.

Huang, H., Liu, W., Singh, V. et al. (2012). Effect of harvest moisture content on selected yellow dent corn: dry-grind fermentation characteristics and DDGS composition. *Cereal Chemistry* 89: 217–221.

Huang, H., Qureshi, N., Chen, M.-H. et al. (2015a). Ethanol production from food waste at high solids content with vacuum recovery technology. *Journal of Agricultural and Food Chemistry* 63: 2760–2766.

Huang, H., Singh, V., and Qureshi, N. (2015b). Butanol production from food waste: a novel process for producing sustainable energy and reducing environmental pollution. *Biotechnology for Biofuels* 8 (1): 1–12.

Huang, H., Long, S., and Singh, V. (2016). Techno-economic analysis of biodiesel and ethanol co-production from lipid-producing sugarcane. *Biofuels, Bioproducts and Biorefining* 8: 299–315.

Iranmahboob, J., Nadim, F., and Monemi, S. (2002). Optimizing acid-hydrolysis: a critical step for production of ethanol from mixed wood chips. *Biomass and Bioenergy* 22: 401–404.

Itelima, J., Onwuliri, F., Onwuliri, E. et al. (2013). Bio-ethanol production from banana, plantain and pineapple peels by simultaneous saccharification and fermentation process. *International Journal of Environmental Science and Development* 4: 213.

Jang, Y.-S., Malaviya, A., Cho, C. et al. (2012). Butanol production from renewable biomass by clostridia. *Bioresource Technology* 123: 653–663.

Jin, C., Yao, M., Liu, H. et al. (2011). Progress in the production and application of n-butanol as a biofuel. *Renewable and Sustainable Energy Reviews* 15: 4080–4106.

Jones, D.T. and Woods, D.R. (1986). Acetone-butanol fermentation revisited. *Microbiological Reviews* 50: 484.

Kheyrandish, M., Asadollahi, M.A., Jeihanipour, A. et al. (2015). Direct production of acetone–butanol–ethanol from waste starch by free and immobilized *Clostridium acetobutylicum*. *Fuel* 142: 129–133.

Kim, J.K., Oh, B.R., Shin, H.-J. et al. (2008a). Statistical optimization of enzymatic saccharification and ethanol fermentation using food waste. *Process Biochemistry* 43: 1308–1312.

Kim, Y., Mosier, N.S., Hendrickson, R. et al. (2008b). Composition of corn dry-grind ethanol by-products: DDGS, wet cake, and thin stillage. *Bioresource Technology* 99: 5165–5176.

Kim, J.H., Lee, J.C., and Pak, D. (2011). Feasibility of producing ethanol from food waste. *Waste Management* 31: 2121–2125.

Kim, T.H., Choi, C.H., and Oh, K.K. (2013). Bioconversion of sawdust into ethanol using dilute sulfuric acid-assisted continuous twin screw-driven reactor pretreatment and fed-batch simultaneous saccharification and fermentation. *Bioresource Technology* 130: 306–313.

Kiran, E.U., Trzcinski, A.P., Ng, W.J., and Liu, Y. (2014). Bioconversion of food waste to energy: a review. *Fuel* 134: 389–399.

Klinke, H.B., Thomsen, A., and Ahring, B.K. (2004). Inhibition of ethanol-producing yeast and bacteria by degradation products produced during pre-treatment of biomass. *Applied Microbiology and Biotechnology* 66: 10–26.

Kosikowski, F.V. (1979). Whey utilization and whey products. *Journal of Dairy Science* 62: 1149–1160.

Koutinas, A.A., Vlysidis, A., Pleissner, D. et al. (2014). Valorization of industrial waste and by-product streams via fermentation for the production of chemicals and biopolymers. *Chemical Society Reviews* 43: 2587–2627.

Kumar, M., Goyal, Y., Sarkar, A., and Gayen, K. (2012). Comparative economic assessment of ABE fermentation based on cellulosic and non-cellulosic feedstocks. *Applied Energy* 93: 193–204.

Lal, R. and Waldron, K. (2009). Use of crop residues in the production of biofuel. In: *Handbook of Waste Management and Co-Product Recovery in Food Processing*, vol. 2 (ed. K. Waldron), 455–478.

Lam, M.K., Lee, K.T., and Mohamed, A.R. (2010). Homogeneous, heterogeneous and enzymatic catalysis for transesterification of high free fatty acid oil (waste cooking oil) to biodiesel: a review. *Biotechnology Advances* 28: 500–518.

Lee, J. (1997). Biological conversion of lignocellulosic biomass to ethanol. *Journal of Biotechnology* 56: 1–24.

Lee, S.Y., Park, J.H., Jang, S.H. et al. (2008). Fermentative butanol production by Clostridia. *Biotechnology and Bioengineering* 101: 209–228.

Li, Y. and Susan Sun, X. (2011). Mechanical and thermal properties of biocomposites from poly (lactic acid) and DDGS. *Journal of Applied Polymer Science* 121: 589–597.

Lin, C.S.K., Pfaltzgraff, L.A., Herrero-Davila, L. et al. (2013). Food waste as a valuable resource for the production of chemicals, materials and fuels. Current situation and global perspective. *Energy & Environmental Science* 6: 426–464.

Lotero, E., Liu, Y., Lopez, D.E. et al. (2005). Synthesis of biodiesel via acid catalysis. *Industrial & Engineering Chemistry Research* 44: 5353–5363.

Ma, F. and Hanna, M.A. (1999). Biodiesel production: a review. *Bioresource Technology* 70: 1–15.

Ma, H., Wang, Q., Qian, D. et al. (2009). The utilization of acid-tolerant bacteria on ethanol production from kitchen garbage. *Renewable Energy* 34: 1466–1470.

Macedo, E. S. I. 2010. Ethanol and Bioelectricity: Sugarcane in the Future of the Energy Matrix. Available at: http://sugarcane.org/resource-library/books/Ethanol and Bioelectricity book.pdf (Accessed December 12, 2016).

Maddox, I. (1980). Production of n-butanol from whey filtrate using *Clostridium acetobutylicum* NCIB 2951. *Biotechnology Letters* 2: 493–498.

Malaviya, A., Jang, Y.-S., and Lee, S.Y. (2012). Continuous butanol production with reduced byproducts formation from glycerol by a hyper producing mutant of *Clostridium pasteurianum*. *Applied Microbiology and Biotechnology* 93: 1485–1494.

Manikandan, K., Saravanan, V., and Viruthagiri, T. (2008). Kinetics studies on ethanol production from banana peel waste using mutant strain of *Saccharomyces cerevisiae*. *Indian Journal of Biotechnology* 7: 83.

Mariano, A.P., Dias, M.O., Junqueira, T.L. et al. (2013). Butanol production in a first-generation Brazilian sugarcane biorefinery: technical aspects and economics of greenfield projects. *Bioresource Technology* 135: 316–323.

Matsakas, L., Kekos, D., Loizidou, M., and Christakopoulos, P. (2014). Utilization of household food waste for the production of ethanol at high dry material content. *Biotechnology for Biofuels* 7 (1): 4.

Moon, H.C., Song, I.S., Kim, J.C. et al. (2009). Enzymatic hydrolysis of food waste and ethanol fermentation. *International Journal of Energy Research* 33: 164–172.

Moreira, A., Ulmer, D. & Linden, J. Butanol toxicity in the butylic fermentation. Biotechnol. Bioeng. Symp.;(United States), 1981. Colorado State Univ., Fort Collins.

Murdock, D. and Allen, W. (1960). Germicidal effect of orange peel oil and D-limonene in water and orange juice. *Food Technology* 14: 441–445.

Naik, S.N., Goud, V.V., Rout, P.K., and Dalai, A.K. (2010). Production of first and second generation biofuels: a comprehensive review. *Renewable and Sustainable Energy Reviews* 14: 578–597.

Novick, P. and Schekman, R. (1979). Secretion and cell-surface growth are blocked in a temperature-sensitive mutant of *Saccharomyces cerevisiae*. *Proceedings of the National Academy of Sciences* 76: 1858–1862.

Oberoi, H.S., Vadlani, P.V., Nanjundaswamy, A. et al. (2011a). Enhanced ethanol production from Kinnow mandarin (*Citrus reticulata*) waste via a statistically optimized simultaneous saccharification and fermentation process. *Bioresource Technology* 102: 1593–1601.

Oberoi, H.S., Vadlani, P.V., Saida, L. et al. (2011b). Ethanol production from banana peels using statistically optimized simultaneous saccharification and fermentation process. *Waste Management* 31: 1576–1584.

OECD/FAO 2015. "Biofuels," OECD-FAO Agricultural Outlook 2015, OECD Publishing, Paris. DOI: http://dx.doi.org/10.1787/agr_outlook-2015-13-en. (Accessed December 2, 2016).

Oh, B.-R., Seo, J.-W., Heo, S.-Y. et al. (2011). Efficient production of ethanol from crude glycerol by a *Klebsiella pneumoniae* mutant strain. *Bioresource Technology* 102: 3918–3922.

Öhgren, K., Bura, R., Lesnicki, G. et al. (2007). A comparison between simultaneous saccharification and fermentation and separate hydrolysis and fermentation using steam-pretreated corn stover. *Process Biochemistry* 42: 834–839.

Pfromm, P.H., Amanor-Boadu, V., Nelson, R. et al. (2010). Bio-butanol vs. bio-ethanol: a technical and economic assessment for corn and switchgrass fermented by yeast or *Clostridium acetobutylicum*. *Biomass and Bioenergy* 34: 515–524.

Quintero, J.A., Montoya, M.I., Sanchez, O.J. et al. (2008). Fuel ethanol production from sugarcane and corn: comparative analysis for a Colombian case. *Energy* 33: 385–399.

Qureshi, N. and Ezeji, T.C. (2008). Butanol,'a superior biofuel' production from agricultural residues (renewable biomass): recent progress in technology. *Biofuels, Bioproducts and Biorefining* 2: 319–330.

Qureshi, N. and Maddox, I. (1987). Continuous solvent production from whey permeate using cells of *Clostridium acetobutylicum* immobilized by adsorption onto bonechar. *Enzyme and Microbial Technology* 9: 668–671.

Qureshi, N. and Maddox, I. (2005). Reduction in butanol inhibition by perstraction: utilization of concentrated lactose/whey permeate by *Clostridium acetobutylicum* to enhance butanol fermentation economics. *Food and Bioproducts Processing* 83: 43–52.

Qureshi, N., Meagher, M., Huang, J., and Hutkins, R.W. (2001). Acetone butanol ethanol (ABE) recovery by pervaporation using silicalite–silicone composite membrane from fed-batch reactor of *Clostridium acetobutylicum*. *Journal of Membrane Science* 187: 93–102.

Qureshi, N., Li, X.L., Hughes, S. et al. (2006). Butanol production from corn fiber xylan using *Clostridium acetobutylicum*. *Biotechnology Progress* 22: 673–680.

Qureshi, N., Saha, B., Cotta, M., and Singh, V. (2013). An economic evaluation of biological conversion of wheat straw to butanol: a biofuel. *Energy Conversion and Management* 65: 456–462.

Refaat, A., Attia, N., Sibak, H.A. et al. (2008). Production optimization and quality assessment of biodiesel from waste vegetable oil. *International Journal of Environmental Science and Technology* 5: 75–82.

Renewable Fuels Association. 2016. Annual World Fuel Ethanol Production. Available at: https://ethanolrfa.org/statistics/annual-ethanol-production/ Accessed on October 10, 2019.

Rodríguez, L., Toro, M., Vazquez, F. et al. (2010). Bioethanol production from grape and sugar beet pomaces by solid-state fermentation. *International Journal of Hydrogen Energy* 35: 5914–5917.

Saha, B.C., Iten, L.B., Cotta, M.A., and Wu, Y.V. (2005). Dilute acid pretreatment, enzymatic saccharification, and fermentation of rice hulls to ethanol. *Biotechnology Progress* 21: 816–822.

Sakai, S., Tsuchida, Y., Okino, S. et al. (2007). Effect of lignocellulose-derived inhibitors on growth of and ethanol production by growth-arrested *Corynebacterium glutamicum* R. *Applied and Environmental Microbiology* 73: 2349–2353.

Schnepf, R. D. European Union biofuels policy and agriculture: An overview. 2006. Congressional Research Service, Library of Congress. Available at: http://research.policyarchive.org/4344.pdf. (Accessed December 2, 2016).

Sharma, Y., Singh, B., and Upadhyay, S. (2008). Advancements in development and characterization of biodiesel: a review. *Fuel* 87: 2355–2373.

Shay, E.G. (1993). Diesel fuel from vegetable oils: status and opportunities. *Biomass and Bioenergy* 4: 227–242.

Shihadeh, J., Huang, H., Rausch, K. et al. (2013). Design of a vacuum flashing system for high-solids fermentation of corn. *Transactions of the ASABE* 56: 1441–1447.

Silva, N.L.C., Betancur, G.J.V., Vasquez, M.P. et al. (2011). Ethanol production from residual wood chips of cellulose industry: acid pretreatment investigation, hemicellulosic hydrolysate fermentation, and remaining solid fraction fermentation by SSF process. *Applied Biochemistry and Biotechnology* 163: 928–936.

Singh, V., Johnston, D.B., Rausch, K.D., and Tumbleson, M. (2010). Improvements in corn to ethanol production technology using *Saccharomyces cerevisiae*. In: *Biomass to Biofuels: Strategies for Global Industries*, 1e (eds. A.A. Vertés, N. Qureshi, H.P. Blaschek and H. Yukawa), 187–198. United Kingdom: Wiley.

Siso, M.G. (1996). The biotechnological utilization of cheese whey: a review. *Bioresource Technology* 57: 1–11.

Smithers, G.W. (2008). Whey and whey proteins—from 'gutter-to-gold'. *International Dairy Journal* 18: 695–704.

Stoeberl, M., Werkmeister, R., Faulstich, M., and Russ, W. (2011). Biobutanol from food wastes–fermentative production, use as biofuel an the influence on the emissions. *Procedia Food Science* 1: 1867–1874.

Sun, Y. and Cheng, J. (2002). Hydrolysis of lignocellulosic materials for ethanol production: a review. *Bioresource Technology* 83: 1–11.

Taconi, K.A., Venkataramanan, K.P., and Johnson, D.T. (2009). Growth and solvent production by *Clostridium pasteurianum* ATCC 6013 utilizing biodiesel-derived crude glycerol as the sole carbon source. *Environmental Progress & Sustainable Energy* 28: 100–110.

Tang, Y.-Q., Koike, Y., Liu, K. et al. (2008). Ethanol production from kitchen waste using the flocculating yeast *Saccharomyces cerevisiae* strain KF-7. *Biomass and Bioenergy* 32: 1037–1045.

Tao, L., Tan, E.C., Mccormick, R. et al. (2014). Techno-economic analysis and life-cycle assessment of cellulosic isobutanol and comparison with cellulosic ethanol and n-butanol. *Biofuels, Bioproducts and Biorefining* 8: 30–48.

Taylor, F., Marquez, M.A., Johnston, D.B. et al. (2010). Continuous high-solids corn liquefaction and fermentation with stripping of ethanol. *Bioresource Technology* 101: 4403–4408.

Thomas, K., Hynes, S., and Ingledew, W. (1996). Practical and theoretical considerations in the production of high concentrations of alcohol by fermentation. *Process Biochemistry* 31: 321–331.

Ujor, V., Bharathidasan, A.K., Cornish, K., and Ezeji, T.C. (2014a). Evaluation of industrial dairy waste (milk dust powder) for acetone-butanol-ethanol production by solventogenic *Clostridium* species. *SpringerPlus* 3 (1): 387.

Ujor, V., Bharathidasan, A.K., Cornish, K., and Ezeji, T.C. (2014b). Feasibility of producing butanol from industrial starchy food wastes. *Applied Energy* 136: 590–598.

US Department of Energy (DOE) 2016. U.S. Biodiesel Production, Exports, and Consumption. Available at: http://www.afdc.energy.gov/data/.(Accessed July 25, 2016).

US Energy Information Administration (EIA) 2012. International Energy Statistics. Available at: https://www.eia.gov/cfapps/ipdbproject/iedindex3.cfm?tid=79&pid=81&aid=1&cid=ww,&syid = 2008&eyid = 2012&unit = TBPD. (Accessed July 25, 2016).

Van Gerpen, J. (2005). Biodiesel processing and production. *Fuel Processing Technology* 86: 1097–1107.

Wang, S., Ingledew, W., Thomas, K. et al. (1999). Optimization of fermentation temperature and mash specific gravity for fuel alcohol production. *Cereal Chemistry* 76: 82–86.

Wang, Y., Ou, S., Liu, P., and Zhang, Z. (2007). Preparation of biodiesel from waste cooking oil via two-step catalyzed process. *Energy Conversion and Management* 48: 184–188.

Wilkins, M.R., Suryawati, L., Maness, N.O., and Chrz, D. (2007a). Ethanol production by *Saccharomyces cerevisiae* and *Kluyveromyces marxianus* in the presence of orange-peel oil. *World Journal of Microbiology and Biotechnology* 23: 1161–1168.

Wilkins, M.R., Widmer, W.W., and Grohmann, K. (2007b). Simultaneous saccharification and fermentation of citrus peel waste by *Saccharomyces cerevisiae* to produce ethanol. *Process Biochemistry* 42: 1614–1619.

Xiros, C. and Christakopoulos, P. (2009). Enhanced ethanol production from brewer's spent grain by a *Fusarium oxysporum* consolidated system. *Biotechnology for Biofuels* 2 (1): 4.

Xiros, C., Topakas, E., Katapodis, P., and Christakopoulos, P. (2008). Evaluation of *Fusarium oxysporum* as an enzyme factory for the hydrolysis of brewer's spent grain with improved biodegradability for ethanol production. *Industrial Crops and Products* 28: 213–224.

Xue, C., Zhao, X.-Q., Liu, C.-G. et al. (2013. . I.). Prospective and development of butanol as an advanced biofuel. *Biotechnology Advances* 31: 1575–1584.

Yan, S., Li, J., Chen, X. et al. (2011). Enzymatical hydrolysis of food waste and ethanol production from the hydrolysate. *Renewable Energy* 36: 1259–1265.

Zhang, Y., Dube, M., Mclean, D., and Kates, M. (2003a). Biodiesel production from waste cooking oil: 1. Process design and technological assessment. *Bioresource Technology* 89: 1–16.

Zhang, Y., Dube, M., Mclean, D., and Kates, M. (2003b). Biodiesel production from waste cooking oil: 2. Economic assessment and sensitivity analysis. *Bioresource Technology* 90: 229–240.

Zhang, J., Chu, D., Huang, J. et al. (2010). Simultaneous saccharification and ethanol fermentation at high corn stover solids loading in a helical stirring bioreactor. *Biotechnology and Bioengineering* 105: 718–728.

Zheng, S., Kates, M., Dube, M., and Mclean, D. (2006). Acid-catalyzed production of biodiesel from waste frying oil. *Biomass and Bioenergy* 30: 267–272.

Zheng, Y., Lee, C., Yu, C. et al. (2012a). Ensilage and bioconversion of grape pomace into fuel ethanol. *Journal of Agricultural and Food Chemistry* 60: 11128–11134.

Zheng, Y., Yu, C., Cheng, Y.-S. et al. (2012b). Integrating sugar beet pulp storage, hydrolysis and fermentation for fuel ethanol production. *Applied Energy* 93: 168–175.

Zheng, Y., Lee, C., Yu, C. et al. (2013). Dilute acid pretreatment and fermentation of sugar beet pulp to ethanol. *Applied Energy* 105: 1–7.

Ziganshin, A.M., Schmidt, T., Scholwin, F. et al. (2011). Bacteria and archaea involved in anaerobic digestion of distillers grains with solubles. *Applied Microbiology and Biotechnology* 89: 2039–2052.

22

The Environmental Impact of Pollution Prevention, Sustainable Energy Generation, and Other Sustainable Development Strategies Implemented by the Food Manufacturing Sector

Sandra D. Gaona[1], T.J. Pepping[2], Cheryl Keenan[3] and Stephen C. DeVito[1]

[1] *Toxics Release Inventory Program (mail code 7410M), United States Environmental Protection Agency, Washington, DC, 20460*
[2] *Abt Associates Inc., Rockville, MD, 20852*
[3] *Eastern Research Group Inc., Lexington, MA, 02421*

CHAPTER MENU

22.1 Introduction

The US food manufacturing industry is a diverse sector representing $217 billion in value added (about 10% of all manufacturing value added) in 2016 and supporting approximately 1.5 million jobs. This chapter characterizes chemical release and other waste management quantities as well as pollution prevention activities reported by the food manufacturing industry in the United States (US) over the 2007–2017 timeframe.[1] Analysis of information available from federal databases such as the US Environmental Protection Agency's (EPA's) Toxics Release Inventory (TRI) and industry reports reveal the corresponding environmental impacts, and identify opportunities for continued progress. Throughout this chapter several terms are used that may not be familiar to the reader. These terms are defined below.

> A 'TRI chemical' is a chemical that is included on the Toxics Release Inventory (TRI) list of chemicals, as established under Section 313(d)(2) of the *Emergency Planning and Community Right-to-Know Act*. Chemicals included on the TRI list are those that as a result of continuous, or frequently recurring releases are known to cause or can reasonably be anticipated to cause (1) significant adverse acute human health effects at concentration levels reasonably likely to exist beyond facility site boundaries; (2) cancer or teratogenic

1 At the time of the analysis and writing of this chapter, the 2017 reporting year was the year for which the most recent TRI data were available.

Green Energy to Sustainability: Strategies for Global Industries, First Edition.
Edited by Alain A. Vertès, Nasib Qureshi, Hans P. Blaschek and Hideaki Yukawa.

effects or serious or irreversible reproductive dysfunctions, neurological disorders, heritable genetic mutations, or other chronic health effects; or (3) a significant adverse effect on the environment of sufficient seriousness to warrant reporting due to the chemical's toxicity, its toxicity and persistence in the environment, or its toxicity and tendency to bioaccumulate in the environment.

A 'TRI-reported chemical' refers to chemicals on the TRI list of chemicals for which facilities in the US have submitted reports to the US Environmental Protection Agency (EPA) TRI Program indicating releases to the environment or otherwise managed as waste.

'TRI-reported chemical waste' or 'TRI-reported waste' refers to the quantity of the TRI chemical(s) contained in waste and reported to EPA by facilities as released to the environment or otherwise managed as waste, such as through recycling, treatment, or combustion for energy recovery.

Beyond TRI-reported chemical waste management, this chapter also reviews data and literature on a range of pollution prevention and sustainability strategies in the industry such as those related to water conservation, improving energy efficiency and material use. This chapter does not discuss, in detail, the environmental impacts from resource extraction or depletion such as water consumption or energy consumption.

22.2 Overview of the Food Manufacturing Industry

The food manufacturing industry sector consists of any facility that produces foods or food ingredients, from slaughterhouses to bakeries. The industry applies many different industrial processes, varying from preparation (e.g. slaughtering, milling) to processing (e.g. cooking, freezing, fermenting) to packaging. The industry, as discussed in this chapter, consists of facilities classified under the North American Industry Classification System (NAICS) code 311 (Food Manufacturing). The sector consists of nine subsectors: Animal Food; Grain/Oilseed Milling; Sugar and Confectionery; Fruit and Vegetable; Dairy; Meats; Seafood; Bakeries and Tortilla; and Other Food, as described in Table 22.1 (US Census Bureau n.d.-a).[2] The shortened descriptions in Table 22.1 are used throughout the rest of this chapter.

Depending on the subsector and processes involved, TRI-reported chemicals may be manufactured, processed, or otherwise used when manufacturing food products. Some of these chemicals are released into the environment. It is difficult to generalize chemicals used in industrial processes across the entire industry because each subsector has unique processes and techniques. For example, many processes require or produce chemicals during disinfection, cleaning, and waste management, and may also produce large amounts of wastewater. Additionally, most preservation processes have high energy requirements in order to ensure safety of the food product. As a result, combusting fuel for energy leads to releases of TRI chemicals and greenhouse gases (GHGs). Refrigerants are another important class of chemicals for many subsectors in food manufacturing due to the need to prevent microbial growth. As another example, many chemical additives are used for processes including preservation, nutrition supplements, flavouring, and colouring.

2 Although the food manufacturing industry is also known as the food processing industry, the name for NAICS code 311 is food manufacturing and is used throughout this chapter (US Census Bureau n.d.).

Table 22.1 Food manufacturing (NAICS[a]) 311) subsectors.

NAICS	NAICS description	Shortened description	Activities included
3111	Animal food manufacturing	Animal food	Dog, cat, and other animal food manufacturing
3112	Grain and oilseed milling	Grain/oilseed milling	Flour and rice milling; malt manufacturing; soybean and other oilseed processing; fats and oils refining; breakfast cereal manufacturing
3113	Sugar and confectionery product manufacturing	Sugar and confectionery	Sugar manufacturing; chocolate manufacturing; other confectionery manufacturing
3114	Fruit and vegetable preserving and specialty food manufacturing	Fruit and vegetable	Frozen fruit, juice, and vegetable manufacturing; fruit and vegetable canning, pickling, and drying
3115	Dairy product manufacturing	Dairy	Milk, butter, cheese, and ice cream manufacturing
3116	Animal slaughtering and processing	Meats	Animal slaughtering; meat processing; meat byproduct processing
3117	Seafood product preparation and packaging	Seafood	Seafood and seafood products manufacturing
3118	Bakeries and tortilla manufacturing	Bakeries and tortilla	Retail and commercial bakeries; frozen pastries manufacturing, cookie and cracker manufacturing; pasta and dough manufacturing, tortilla manufacturing
3119	Other food manufacturing	Other food	Nuts, peanut butter, coffee and tea, flavouring syrup, prepared sauce, spice and extract, and other miscellaneous food manufacturing

a) North American Industry Classification System.

The food manufacturing industry as discussed in this chapter does not include the following (which are not covered by TRI reporting requirements):

- agriculture (NAICS 111),
- livestock (NAICS 112),
- fishing (NAICS 114),
- support activities for agriculture (NAICS 115),
- supermarkets (NAICS 44-45),
- food warehousing and storage (NAICS 4931), or
- food service such as restaurants (NAICS 722).

It also does not include beverage and tobacco product manufacturing (NAICS 312), although food, beverage, and tobacco manufacturing are frequently combined in other analyses and datasets. Researchers may want to consider these related activities in follow up analyses.

According to recent Statistics of US Businesses (SUSB) data (US Census Bureau 2018), in 2015 there were 26 819 food manufacturing establishments in the US, making up 9% of all manufacturing establishments. During the same year, these establishments employed 1.5 million people, which made up 13% of all employment in all manufacturing industries. Almost half of these establishments employ fewer than 10 full-time employees. Of those food manufacturing establishments with 10 or more employees, 33% were in the Bakeries and Tortilla subsector, which includes retail bakeries. Figure 22.1 presents the number of establishments with 10 or more employees by subsector, representing a total of 13 113 establishments (49% of the total food establishments) during 2015. Establishments in food manufacturing with at least 10 full-time employees (or the equivalent in hours) meet the industry and employment criteria for TRI reporting, and may be required to report to TRI depending on the quantities of TRI chemicals they manufacture, process, or otherwise use annually at the establishment. Of these 13 113 food manufacturing establishments, a relatively small number of facilities (1595) met the TRI reporting requirements during 2015, as evidenced by the fact that each of these facilities submitted at least one TRI reporting form for the 2015 reporting year (RY).

Value added is a measure of the contribution of the industry to the Nation's Gross Domestic Product and is published annually by the US Census Bureau. The Census Bureau derives the value added by subtracting the cost of materials, supplies, containers, fuel, purchased electricity, and contract work from the value of products manufactured plus receipts for services rendered.

As food manufacturing operations have used technological innovations to improve on economies of scale and achieve higher productivity with lower costs, greater industry competition and lower prices have favoured larger companies and resulted in even larger food operations for farms, processing plants, and retailers (Shields 2010). In 2007, the top four companies in the food manufacturing industry held 15% of the market share, while in certain subsectors the market share of the top four firms was much larger, such as soybean processing (80%) and non-poultry slaughterhouses (60%) (US Census Bureau 2011). Vertical integration, where the parts of the supply chain are under a common owner, has also increased, especially in the poultry industry (Martinez 1999). For example, a vertically integrated poultry company may own the feed mill, the breeding stock, the hatchery, and the processing plant, giving them control over the quality and quantity of their supply chain's production. Over a third of the top 50 global food manufacturing companies are headquartered in the US, and American companies such as Tyson Foods, Archer Daniels Midland (ADM), and Kraft are among the largest (Food Engineering 2014).

22.2.1 Production and Economic Trends

Although food manufacturing in the US comprises a significant portion of shipments from all manufacturing industries, due to the necessity and, therewith, inflexible nature of food products as essential commodities, the industry is less impacted by fluctuating economic conditions (US Department of Commerce 2008). In the global recession during 2008–2009, the large impacts felt by many other components of the economy were not as significant in the food manufacturing industry (Ramde 2010). In this analysis, trends in the sector's production levels were based on the annual 'value added' to the overall economy. The annual value added was used as proxy for annual production output. The results are presented in Figure 22.2. Total value added for food manufacturing was $217 billion in 2016, or about 10% of the total value added for all manufacturing industries. As shown in Figure 22.2, estimates of the value added from the food manufacturing subsectors (Bureau of Economic Analysis 2018a, 2018b; US Census Bureau n.d.-b) indicate that the annual value added remained relatively stable over the timeframe, which includes the global recession. Similarly, evaluation of the Federal Reserve Board's production index for food manufacturing (2017), as shown in Figure 22.3, also indicates that this industry was less affected by the recession than other industries and has even expanded in recent years.

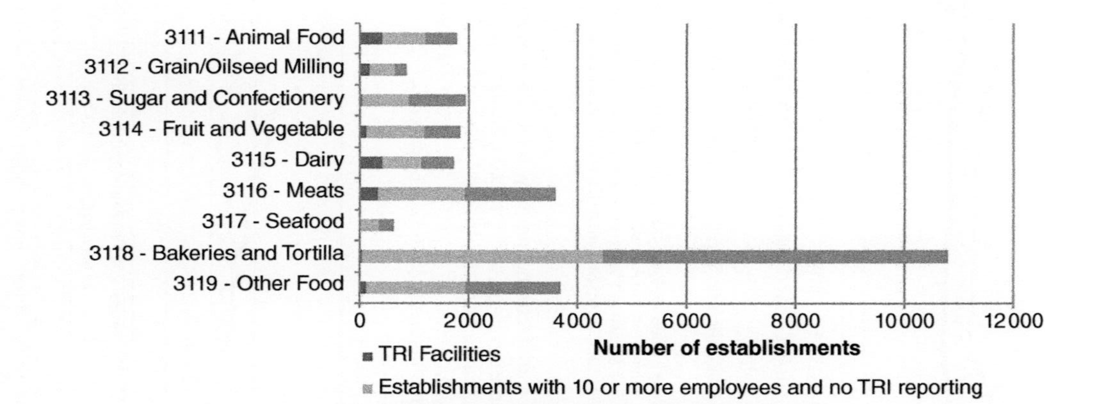

Figure 22.1 Food manufacturing subsectors by TRI reporting status and number of employees, 2015. Note: US Census Bureau data uses the term 'establishment' to denote single physical locations. This analysis assumes that TRI reporting facilities are a subset of the SUSB establishments with 10 or more employees. Sources: US Census Bureau 2018; US EPA Toxics Release Inventory – 2017 National Analysis Dataset.

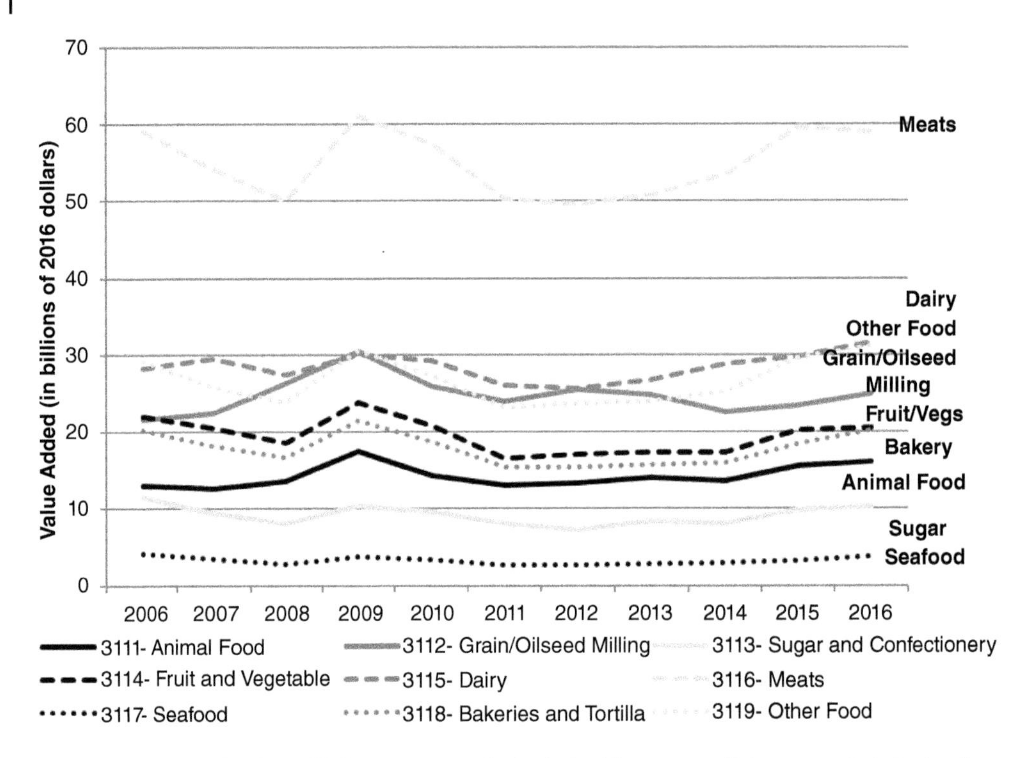

Figure 22.2 Annual value added by the food manufacturing industry, 2006–2016. Note: Value added data is only provided as an aggregated value for food, beverage, and tobacco manufacturing together, so the distribution by subsector of Product Shipments Value from the Annual Survey of Manufacturers was used as a proxy for how to break down value added by subsector as shown in the figure. Data was not available for 2017. Sources: US Census Annual Survey of Manufacturers, 2006–2016; Bureau of Economic Analysis 2018a.

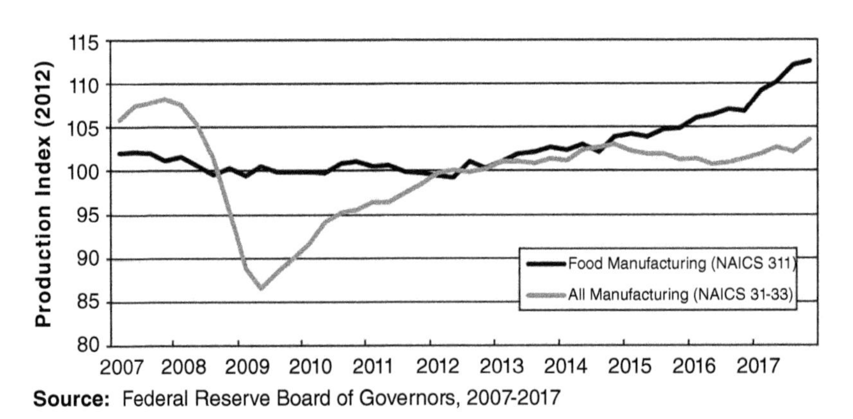

Figure 22.3 Food manufacturing production index – seasonally adjusted, 2007–2017. Source: Federal Reserve Board of Governors, 2007–2017.

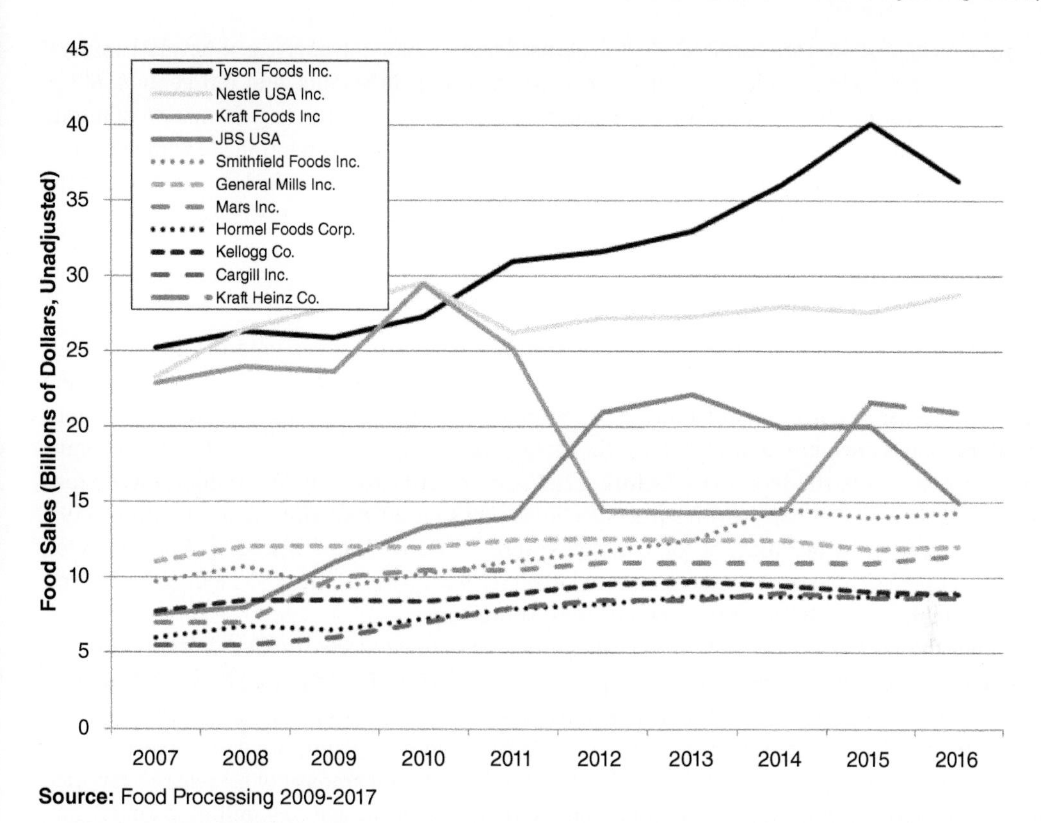

Figure 22.4 Top 10 US and Canadian food manufacturers by food sales, 2007–2016. Source: Food Processing, 2009–2017.

22.2.2 Key Players

Food sales from 2007 to 2016 for the 10 largest US and Canadian food manufacturing companies in 2016 are shown in Figure 22.4. Many of these companies produce a variety of foods and brands and frequently undergo corporate reorganization, restructuring, buyouts, and mergers. For example, the large dip in food sales from Kraft Foods Inc. in 2012 is the result of the reorganization of the company into the separate entities of Kraft Foods and Mondelez International (Food Processing 2018), followed by a merger of Kraft Foods with Heinz in 2015 to form the Kraft Heinz Company. Note that some parent companies may own facilities which are not included in the food manufacturing industry. For example, Nestle USA Inc. also manufactures bottled water and metal can products in addition to food products. The sales figures included in Figure 22.4 represent food sales of the top 10 US and Canadian food manufacturers rather than total sales.

22.3 Chemicals and Chemical Wastes in the Food Manufacturing Industry

Food manufacturing involves processing raw food products, such as produce, grains, meats, and dairy, into more complex and portable products that can be preserved until purchase and used by consumers. In order to create and process these types of food products, various industrial chemicals and chemical processes are required. Chemicals used may include solvents (e.g. *n*-hexane), mineral acids (e.g. nitric acid), artificial additives or preservatives, disinfectants, and refrigerants. Moreover, chemicals may be formed during the food manufacturing processes (e.g. formation of nitrate compounds during wastewater treatment processes when nitric acid is neutralized with alkali), or during activities that support the manufacture of foods (e.g. combustion of fuels for energy, and the resultant creation of GHGs)

22.3.1 Greenhouse Gas Emissions

In regard to the food manufacturing industry, the GHG emissions reported to the EPA's Greenhouse Gas Reporting Program (GHGRP) are mostly generated by the stationary combustion of fuels in order to power manufacturing operations. These are characterized as direct emissions since the GHGs are emitted directly from the facility following combustion of a fuel. Direct emissions reported from the food manufacturing industry for 2017 totalled 28.1 million metric tons of CO2-equivalent (MMCO2-eq) from 336 facilities (US EPA 2018a). More than two-thirds (19.5 MMCO2-eq from 98 facilities) of the emissions were from the Grain/Oilseed Milling subsector, with additional sizable contributions from the Meats (2.9 MMCO2-eq from 99 facilities) and Fruit and Vegetable (2.3 MMCO2-eq from 60 facilities) subsectors.

Among the 336 facilities in the food manufacturing sector that reported to EPA's GHGRP, 207 also reported to EPA's TRI Program for reporting year (RY) 2017 (US EPA 2018c). Of these, 180 facilities reported a total of 25 million pounds of air releases of TRI chemicals as part of their waste management portfolio. Facilities in the Grain/Oilseed Milling subsector contributed the largest amount of GHG emissions reported to GHGRP, as well as the largest total amount of air releases reported to TRI, primarily as *n*-hexane (17.5 million pounds), hydrochloric acid (0.9 million pounds), and acetaldehyde (0.8 million pounds). Air releases of ammonia from facilities in the Meats subsector (2.0 million pounds) were also notable. In 2017, food manufacturers accounted for 1% of total GHG emissions reported to GHGRP and 7% of total air releases reported to TRI.

22.3.2 Conventional Water Pollutants

Water pollutants discharged by industrial facilities are regulated by EPA's National Pollutant Discharge Elimination System (NPDES) Permit Program by authority under the Clean Water Act. Industrial, municipal, and other facilities must obtain permits before they can discharge pollutants into surface waters. Information on pollutant discharges as required under the NPDES Program is reported to EPA and available via the Discharge Monitoring Report (DMR) Pollutant Loading Tool (US EPA n.d.). Note that there is some overlap between facilities which report to EPA's TRI Program and those that are regulated by EPA's NPDES Program; TRI-reported releases, such as nitrate compounds discharged to surface waters, for example, are also reported to the NPDES Program.

In the food manufacturing industry, biological oxygen demand and solids (both dissolved and suspended) are two of the primary concerns associated with wastewater discharges (UN Industrial Development Organization 2004). Based on DMR classifications via Standard Industrial Classification (SIC) codes that correspond to relevant NAICS codes, facilities in the Meats and Seafood subsectors discharged the largest amounts of total dissolved and suspended solids in wastewater over the nine-year timeframe from 2007 to 2015. The large amounts of solids are likely to be associated with processes inherent to these subsectors, such as euthanization, rendering, and bleeding (UN Industrial Development Organization 2004).

22.3.3 Refrigerants

The Montreal Protocol, an international agreement finalized in 1987 to phase out the consumption and production of ozone-depleting substances, significantly impacted the use of refrigerants in the food manufacturing industry. Prior to the Montreal Protocol, chlorofluorocarbons (CFCs) and hydrochlorofluorocarbons (HCFCs) were once widely used refrigerants because of their low toxicity and low flammability. At that time, the three most commonly used refrigerants in the food manufacturing industry were R12 (dichlorodifluoromethane, a CFC), R22 (chlorodifluoromethane, an HCFC also known as HCFC-22), and R502 (a mixture of R22 and R115 [chloropentafluoroethane]). Several CFC and HCFC refrigerants are on the TRI list of chemicals, including R12, HCFC-22, and R115 (also called CFC-115).

Use of CFCs have been almost entirely eliminated in the United States, and industry is aiming to phase out use of HCFCs by 2030. Common substitutes for CFCs and HCFCs are hydrofluorocarbon (HFC) mixtures such as R404A (a mixture of pentafluoroethane, 1,1,1-trifluoroethane, and 1,1,1,2-tetrafluoroethane) and R407C (a mixture of difluoromethane, pentafluoroethane, and 1,1,1,2-tetrafluoroethane), neither of which contain TRI-listed chemicals.

Other substitutes for CFCs and HCFCs include ammonia and propane (James and James 2014). Ammonia in particular is becoming the preferred refrigerant for the food manufacturing industry, largely due to its efficiency in large systems, smaller environmental impact relative to other refrigerants, and strong odour that allows for detection of equipment leaks and serves as a warning in case of accidental releases (Food Manufacturing 2015). In addition, CO2 has been used as a substitute refrigerant. However, due to high operating pressure requirements, applications of CO2 refrigeration are currently limited in scale. For example, CO2 is used as a refrigerant in vending machines (Bhatkar, Kriplana, and Awari 2013).

The common refrigerants mentioned in this section are summarized in Table 22.2. As an example of how releases have changed in recent years, air releases for HCFC-22 as reported to EPA's TRI Program have decreased by 84%, from 72 625 lb in 2007 to 11 625 lb in 2017 (US EPA 2018c).

Beyond industrial use within facilities, refrigerants are also necessary in order to properly and safely transport the items to their intended destination. In 2016, the US Food and Drug Administration (FDA) finalized a rule as part of the Food Safety Modernization Act (FSMA) related to the sanitary transportation of human and animal food products that became effective in June 2016 (US FDA 2016). Among other requirements, adequate temperature controls must be maintained for shipments related to food products; this includes a myriad of industry sectors, such as agriculture

Table 22.2 Examples of chemicals used as refrigerants.

Name (Chemical Abstracts Service [CAS] Number)	Alternative names	Type of refrigerant	TRI-listed chemical?
Ammonia [7664-41-7]		Inorganic	Yes
Chlorodifluoromethane [75-45-6]	HCFC22, R22	HCFC	Yes
R404A[a)]	HFC-404A	CFC + HCFC Mixture	No
R407C[b)]	HFC-407C	CFC + HCFC Mixture	No

a) R404A is a mixture of: pentafluoroethane (HFC125; CAS# 354-33-6), 1,1,1-trifluoroethane (HFC143a; CAS# 420-46-2), and 1,1,1,2-tetrafluoroethane (HFC134a; CAS# 811-97-2). None of these substances are on the TRI chemical list.

b) R407C is a mixture of: difluoromethane (HFC32; CAS# 75-10-5), pentafluoroethane (HFC125; CAS# 354-33-6), and 1,1,1,2-tetrafluoroethane (HFC134a; CAS# 811-97-2). None of these substances are on the TRI chemical list.

and grocery retailers, but also includes shipments from food manufacturing facilities. Based on data from the 2012 Commodity Flow Survey (US Census Bureau 2015), the average miles per temperature controlled shipment for food manufacturing facilities was 214 mi. By weight, most of the shipments occurred via truck, and 35% of the shipments were less than 100 mi.

22.3.4 TRI-Reported Chemical Waste Management

EPA's TRI Program publicly tracks quantities of chemicals included on the TRI chemical list that are released on-site to air, water, and land, transferred off-site to other facilities, or otherwise managed as waste by facilities throughout the United States. Release quantities presented in this chapter include on-site release and disposal, and off-site transfer for further waste management (TRI Form R Sections 5 and 6). Production-related waste management quantities, hereafter referred to simply as 'waste managed', consists of quantities of a TRI chemical released to the environment, as well as recycled, burned for energy recovery, or treated (TRI Form R Sections 8.1–8.7), and excludes any quantity reported as catastrophic or one-time releases (TRI Form R Section 8.8). An overview of TRI's data reporting requirements is available (US EPA 2019b).

For 2017 (the most recent TRI reporting year available at the time of writing), 1585 facilities from the food manufacturing industry reported to the EPA's TRI Program (as illustrated in Figure 22.5) (US EPA 2018c). From 2007 to 2017, the number of food manufacturing facilities reporting to TRI increased by 2%, from 1558 to 1585. Over the same 11-year period, the number of forms for TRI chemicals reported by the food manufacturing industry increased by 7%, from 3478 to 3722. In comparison to the other industry sectors that report to EPA's TRI Program, food manufacturing moved in the opposite direction regarding facilities reporting and forms submitted – across all other industries, both the number of facilities reporting to TRI and the number of forms submitted decreased by approximately 10%. In 2017, food manufacturing accounted for 7% of all facilities, 5% of all waste managed, and 3% of all releases reported to EPA's TRI Program. The food manufacturing industry ranked sixth in waste managed quantities in 2017, after the following industries: chemical manufacturing, primary metals, petroleum, paper manufacturing, and metal mining. In terms of TRI-reported release quantities, food manufacturing was the seventh largest industry contributor to overall TRI-reported release quantities, but was the largest contributor (36%) to the quantities of releases of TRI chemicals (the majority of which was nitrate compounds) to surface waters (Figure 22.5).

A summary of the food manufacturing industry's TRI reporting is included in Table 22.3 and is discussed in more detail throughout the remainder of this chapter. The table also includes the number of facilities reporting source reduction activities. The subsectors with low releases (e.g. Animal Food) tend to have low rates of source reduction reporting. Subsectors that report larger quantities of releases and other waste management quantities (e.g. Dairy) tend to have higher source reduction reporting rates, as there may be more known opportunities to implement source reduction. The type and amount of TRI-reported chemicals handled also depends on the specific subsector and is discussed in the next section.

22.3.5 Trends in TRI-Reported Chemical Waste Management

Between 2007 and 2017, total TRI-reported chemical waste managed by the food manufacturing industry increased by 45%. Most of the increase in waste managed came from the Grain/Oilseed Milling subsector, where increased on-site recycling of *n*-hexane (from 571 million pounds in 2007 to 828 million pounds in 2017) was the key driver. However, a majority of the quantities reported

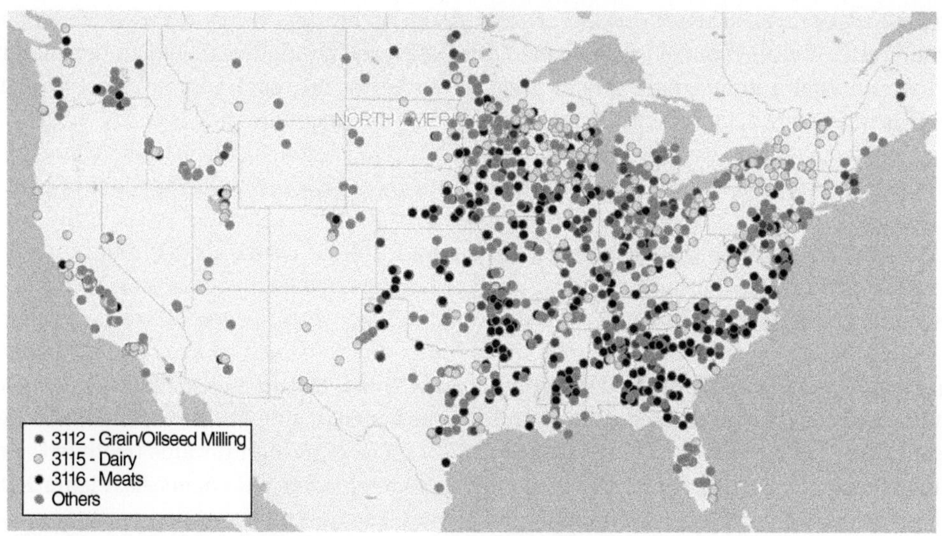

Note: This map does not display any food manufacturing facilities located in Alaska, Hawaii or any of the U.S. territories that filed TRI reports for the 2017 reporting year.

Figure 22.5 Food manufacturing facilities that reported to the TRI program for 2017. Note: This map does not display any food manufacturing facilities located in Alaska, Hawaii or any of the US territories that filed TRI reports for the 2017 reporting year.

Table 22.3 TRI reporting overview for food manufacturing, 2017.

Sector	Number of facilities	Facilities reporting source reduction	Total waste managed		Releases	
			Pounds	% of sector total	Pounds	% of sector total
311 – Food manufacturing (all subsectors)	1585	101	1 450 898 040	100%	127 297 917	100%
3111 – Animal food	406	5	15 491 786	1%	857 944	1%
3112 – Grain/oilseed milling	167	10	987 494 050	68%	34 381 923	27%
3113 – Sugar and confectionery	34	4	10 844 734	1%	7 305 541	6%
3114 – Fruit and vegetable	92	7	17 984 115	1%	8 680 693	7%
3115 – Dairy	416	28	137 929 752	10%	8 611 508	7%
3116 – Meats	334	33	265 179 184	18%	64 839 756	51%
3117 – Seafood	8	0	402 778	<1%	145 472	<1%
3118 – Bakeries and Tortilla	23	2	1 145 543	<1%	766 862	1%
3119 – Other food	107	12	14 426 099	1%	1 708 218	1%

Note: One facility in the Grain/Oilseed Milling subsector reported that 822 million pounds of *n*-hexane were recycled on-site. EPA's TRI Program contacted the facility about this quantity in previous years, and the facility claimed previous values in this range were valid.

Source: US EPA Toxics Release Inventory – 2017 National Analysis Dataset.

are from n-hexane recycled at a single soybean processing facility; this facility reported more than half of all wastes managed in the food manufacturing industry in recent years. When the quantities of n-hexane and methanol that were recycled by this facility are removed from the aggregate, as illustrated in Figure 22.6, total waste managed by the food manufacturing industry increased by 43%. Even without that single facility, the overall trend in waste management is similar, and facilities in the Grain/Oilseed Milling subsector reported increases of 66 million pounds of waste managed across this time period. Facilities in a few other subsectors, such as Meats, Dairy, and Fruit and Vegetable, also reported increases in the quantities of TRI chemicals managed as waste.

As shown in Figure 22.7, the primary waste management method for TRI chemicals in the food manufacturing industry is treatment; 60% of waste reported was treated in 2007, increasing to 66% in 2017. Note that one soybean processing facility was removed from these analyses due to its overwhelming impact on the results. It predominantly recycles both n-hexane and methanol, which are used in the vegetable oil extraction process. Interesting and encouraging TRI-reported waste management trends that are readily apparent in Figure 22.7 include the shift towards treatment and recycling of wastes, and the corresponding shift away from releasing TRI chemicals. Figure 22.8 illustrates the 10 chemicals for which the largest aggregated quantities of waste managed were reported to EPA's TRI Program over the 2007–2017 timeframe.

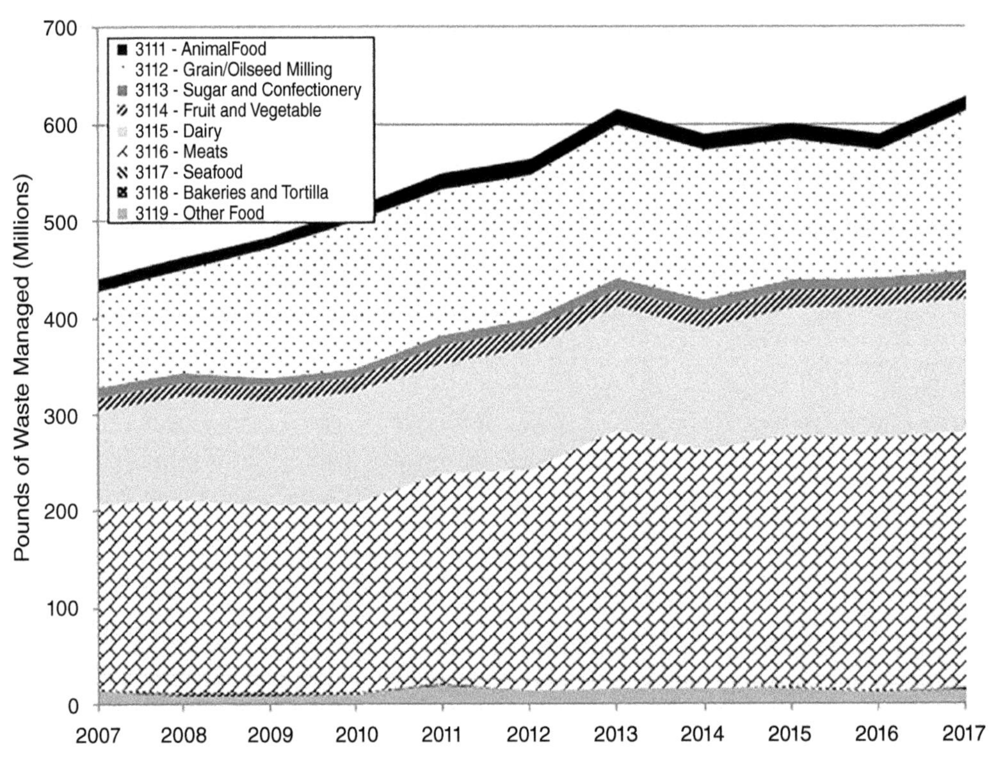

Figure 22.6 Food manufacturing waste managed by subsector, 2007–2017. Note: This figure does not include one soybean processing facility in the Grain/Oilseed Milling subsector. Source: US EPA Toxics Release Inventory – 2017 National Analysis Dataset.

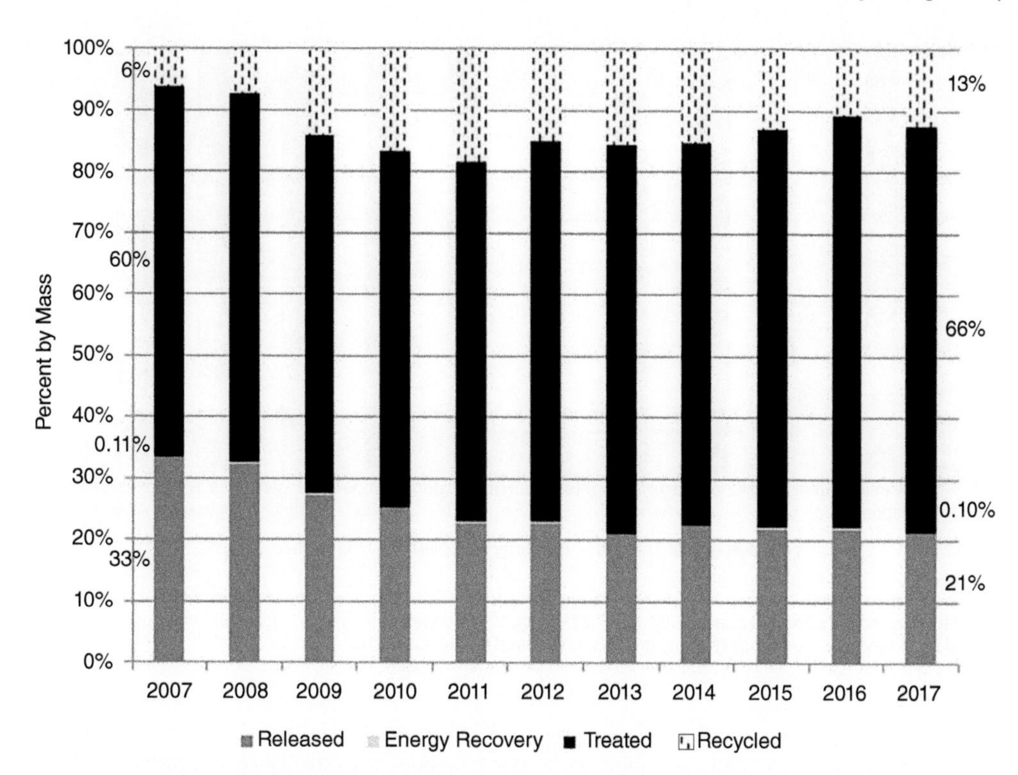

Figure 22.7 Food manufacturing waste management methods, 2007–2017. Notes: (i) Less than 0.2% of TRI chemical waste was used for energy recovery each year, therefore it is barely visible in this figure. (ii) These results do not include one soybean processing facility in the Grain/Oilseed Milling subsector. Source: US EPA Toxics Release Inventory – 2017 National Analysis Dataset.

Treatment is the principal waste management method for TRI chemicals in food manufacturing. The manufacture, processing or otherwise use (and the waste management methods chosen for them) are largely associated with specific subsectors:

- *Nitrate compounds* are manufactured as a result of nitrification during aerobic biological treatment of ammonia in wastewater. They are also manufactured when nitric acid, used for cleaning and sanitation, is treated (neutralized) prior to wastewater discharge. For nitrate compounds, the biggest contributors of waste managed quantities specific to food manufacturing facilities are releases and treatment by the Dairy and Meats subsectors. Management of nitrate compounds in wastewater are regulated under environmental laws such as the Clean Water Act.
- *Ammonia* is often used to sanitize food manufacturing equipment and may also be formed by mineralization of organic nitrogen-containing waste. Releases and treatment of ammonia are largely dominated by the Meats subsector. However, more than half of ammonia waste is managed through recycling, and mostly reported by Grain/Oilseed Milling facilities, where ammonia is used during pre-treatment in the wet corn milling process to improve yields. Recycling of ammonia has increased within the Grain/Oilseed Milling subsector, from 9 million pounds recycled in 2007 to 57 million pounds recycled in 2017. Note that ammonia may also be used in

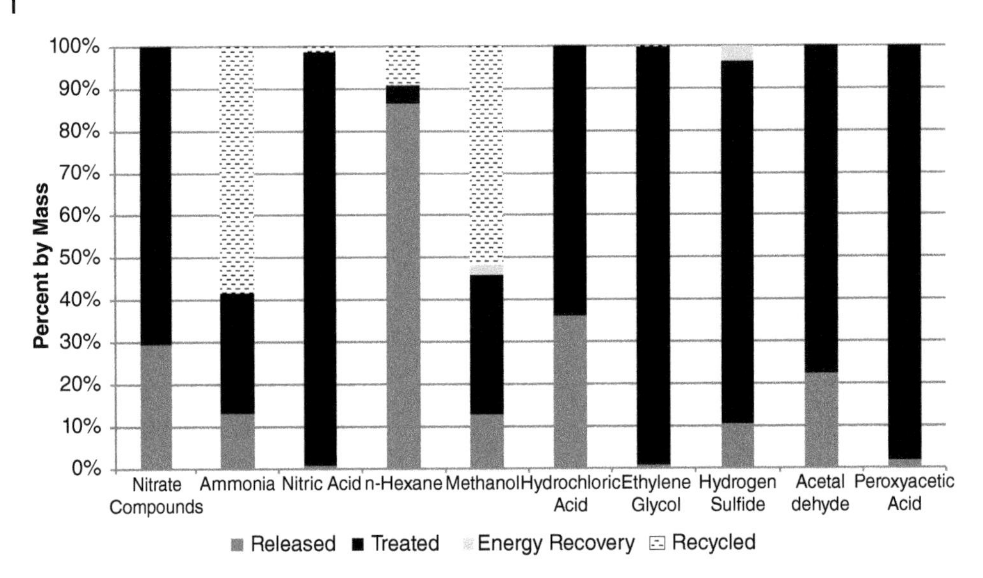

Figure 22.8 Waste management methods for top food manufacturing chemicals during 2007–2017. Notes: (i) The 10 chemicals shown had the highest aggregated quantities of waste managed reported to TRI by the food manufacturing industry over the period 2007–2017. (ii) Hydrogen sulfide reporting began in 2012. (iii) This figure does not include one soybean processing facility in the Grain/Oilseed Milling subsector. Source: US EPA Toxics Release Inventory - 2017 National Analysis Dataset.

refrigeration systems, as discussed previously in Section 22.3.3, but only the quantity used to recharge the system in a given year is reportable to TRI (not the total quantity used for refrigeration). In many cases, this quantity may be below the 10 000 lb. otherwise use threshold for TRI reporting.

- *n-Hexane* is used within the Grain/Oilseed Milling subsector to extract and process a variety of oils from raw agricultural commodities. Most of the *n*-hexane waste reported to EPA's TRI Program from the food manufacturing industry is reported by this subsector, with other reports from the Other Food subsectors including Spice and Extract Manufacturing as well as Flavouring Syrup and Concentrate Manufacturing. As a volatile chemical, *n*-hexane releases are almost entirely to air. TRI reporting forms also indicate a small number of facilities treat *n*-hexane waste through methods such as using absorbers and flaring.

22.3.5.1 Trends in Releases

Figure 22.9 shows total release quantities of TRI chemicals from the food manufacturing industry by subsector and includes quantities released on-site to air, water, and land as well as transferred to an off-site location for disposal. From 2007 to 2017, releases decreased by 10%. Figure 22.9 presents total releases across the 11-year time period from each subsector. These graphs indicate that the quantities of TRI chemicals released largely depend on the subsector. For example, the largest quantity of releases are from the Meats subsector, which are often in the form of nitrate compounds formed in wastewater and are subsequently discharged to surface water. Facilities in the

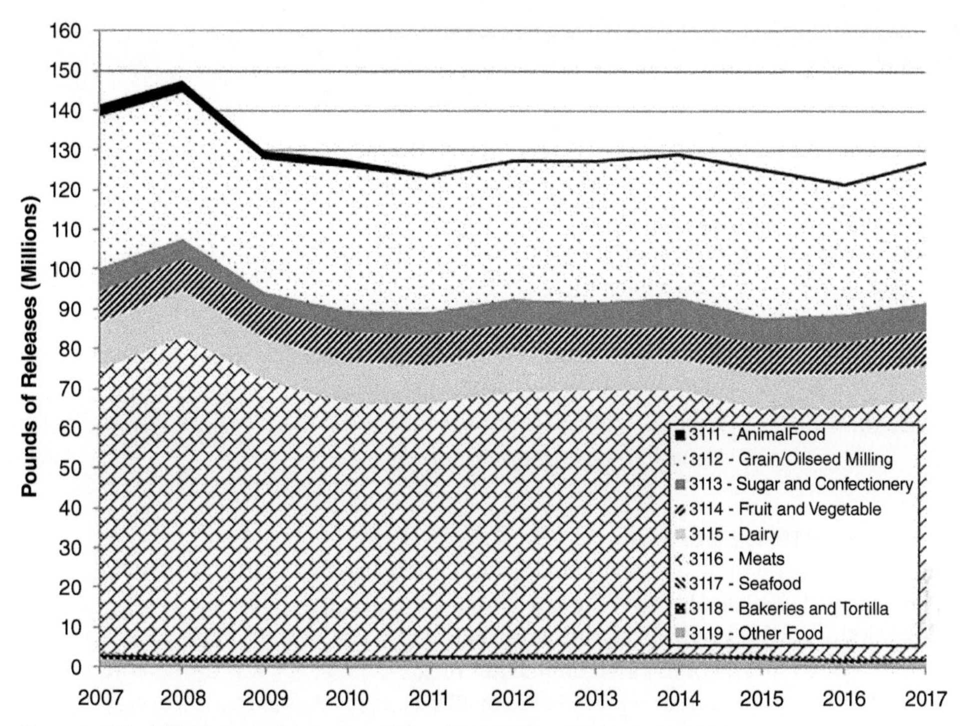

Figure 22.9 Food manufacturing releases by subsector for 2007 through 2017. Source: US EPA Toxics Release Inventory – 2017 National Analysis Dataset.

Grain/Oilseed Milling subsector are responsible for the second greatest quantities of TRI chemicals released to the environment; these releases are often in the form of volatile chemicals such as *n*-hexane, which are released to air. Note that the single soybean processing facility removed from analyses in the previous section on waste managed is not removed from the releases analyses in this or subsequent sections, as it does not have a significant influence on the results.

Facilities in the food manufacturing industry have reported releases for 108 TRI chemicals listed individually or in chemical categories (e.g. nitrate compounds). Nitrate compounds, *n*-hexane, and ammonia accounted for the majority of the quantities of TRI chemicals reported by the food manufacturing industry. These are the same as the top chemicals based on quantities of chemical waste managed. From 2007 to 2017, the top 10 chemicals in terms of quantities released have remained fairly constant, and the following seven chemicals have remained at the top of the list: nitrate compounds, *n*-hexane, ammonia, hydrochloric acid, barium and barium compounds, methanol, and acetaldehyde. The quantities of releases for top chemicals, other than nitrate compounds, have remained relatively constant since 2007; nitrate compound releases decreased by 11% over this time period, though release quantities of nitrate compounds are still large enough for this class of substances to remain at the top of the list.

The three chemicals released in the largest quantities in 2017 for each subsector are shown in Table 22.4. Nitrate compounds ranks in the top three chemicals for every subsector except Bakeries and Tortilla. Ammonia ranks in the top three chemicals for every subsector except Grain/Oilseed Milling.

Table 22.4 The three chemicals released in the largest quantities by food manufacturing subsector, 2017.

Subsector	Top chemicals released
Animal food	Nitrate compounds, ammonia, zinc compounds
Grain/oilseed milling	*n*-Hexane, nitrate compounds, barium compounds
Sugar and confectionery	Ammonia, methanol, nitrate compounds
Fruit and vegetable	Nitrate compounds, ammonia, methanol
Dairy	Nitrate compounds, toluene, ammonia
Meats	Nitrate compounds, ammonia, sodium nitrite
Seafood	Ammonia, nitrate compounds, formaldehyde
Bakeries and tortilla	Ammonia, sulfuryl fluoride, sulfuric acid
Other food	Methanol, nitrate compounds, ammonia

22.3.5.2 Summary of TRI Reporting

The number of facilities within the food manufacturing industry that report to EPA's TRI Program has not changed considerably over the last 11 years. In 2017, the industry accounted for 7% of all facilities, 5% of all TRI chemical quantities managed as waste, and 3% of all release quantities reported to EPA's TRI Program. Based on releases in 2017, the sector was the seventh largest contributor to TRI release quantities reported among all sectors (as defined by three-digit NAICS code) and fourth largest among manufacturing industries, after chemical manufacturing, primary metals, and paper manufacturing.

TRI-reported chemical waste managed in the food manufacturing industry is largely driven by the Grain/Oilseed Milling, Dairy, and Meats subsectors and has increased over the past 11 years. Overall, releases from the food manufacturing industry have decreased by 10% over the same time period, but specific chemical trends are dependent on the subsector and closely tied to the processes involved in that subsector. For example, most releases came from the Meats subsector, which were largely in the form of surface water discharges of nitrate compounds (authorized under the National Pollutant Discharge Elimination System program) due to the organic content of wastewaters resulting from harvesting (i.e. processes such as euthanizing, bleeding, and rendering). Similarly, many of the parent companies whose facilities therein report the largest quantities of TRI chemical releases in aggregate are involved in either the Grain/Oilseed Milling or Meats subsectors.[3] Geographically, the largest quantities of chemicals released are reported from southern and midwestern states, and driven largely by facilities in the Meats and Grain/Oilseed Milling subsectors.

22.4 Pollution Prevention in Food Manufacturing

The prevention of pollution in food manufacturing constitutes a major factor of industrial efficiency, and tackling it answers both economic and environmental goals.

22.4.1 Sustainability Trends in Food Manufacturing

This section discusses the activities, research, and developments that are believed to have contributed to the food industry's reduction in TRI chemicals entering the environment. These

3 As will be discussed in Section 22.4, there are many opportunities for recycling organic materials produced in the harvest process, including through animal rendering. Organic materials that would otherwise be wasted can be processed into a broad range of useful products, which serves to enhance sustainability in the sector.

methods and technologies are compared to TRI-reported (implemented) source reduction activities (Section 22.4.5) to assess their effectiveness and to provide insights as to where additional source reduction activities could be implemented.

Pollution prevention is an essential component of sustainable manufacturing practices. In the US the Pollution Prevention Act of 1990 established a national policy that pollution should be prevented or reduced at source whenever feasible; pollution that cannot be prevented should be recycled in an environmentally safe manner, whenever feasible; pollution that cannot be prevented or recycled should be treated in an environmentally safe manner whenever feasible; and disposal or other release into the environment should be employed only as a last resort and should be conducted in an environmentally safe manner. This hierarchy is illustrated in Figure 22.10 While not mentioned in the Pollution Prevention Act of 1990, energy recovery is a preferred practice over treatment and disposal, and hence, is included in the hierarchy illustrated in Figure 22.10.[4]

As established by the Act, source reduction is more desirable than waste management or pollution control. Source reduction refers to practices that reduce hazardous substances from being released into the environment prior to recycling, or treatment. These practices include equipment or technology modifications, process or procedure modifications, reformulation or redesign of products, substitution of raw materials, and improvements in housekeeping, maintenance, training, or inventory control (Pollution Prevention Act 1990).

Along with source reduction, further sustainability gains are achieved by implementing preferred waste management practices such as recycling and treatment. Pollution prevention and other waste management practices in food manufacturing discussed in this section include:

- **process and technology modifications** implemented to reduce energy consumption and the amount of raw materials used;
- **recycling** of food packaging and organic waste to reduce environmental impacts after the useful life of a material; and
- **wastewater treatment** to destroy TRI chemicals (except for TRI chemicals that are metals or metal compounds).

22.4.1.1 Corporate Sustainability Reports

Corporate sustainability reports provide an overview of sustainability goals, efforts, and accomplishments for some of the largest food manufacturing companies and may include progress

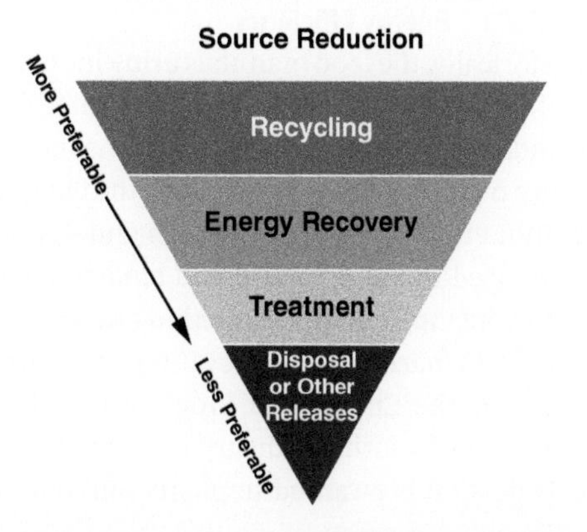

Figure 22.10 Waste management hierarchy.

4 Prior to reporting year 1991, incineration for destruction or energy recovery were not distinguished on the TRI reporting Form R. Beginning with reporting year 1991, the TRI Form R was revised to accommodate the TRI-related mandates of the Pollution Prevention Act. The revisions included a separate section for reporting incineration on-site for energy recovery (Section 8.2) and quantities sent off-site for incineration for energy recovery (Section 8.3).

in reducing CO_2 emissions, water use, energy use, and/or waste generation. Several companies highlight recent innovations in pollution prevention and other improved environmental practices, such as implementing anaerobic wastewater treatment systems that produce biogas and using it to offset their natural gas consumption (Hilmar Cheese Company 2015). Additionally, they provide specific examples of sustainability achievements such as: Cargill's 38 worldwide sites that utilize energy-saving combined heat and power (CHP) systems (Cargill 2015); ADM's sweetener facility that eliminated landfill waste by directing wastewater to local beekeepers and other waste to an energy recovery plant (ADM 2014); and Tyson Food's reduction of almost 1 million pounds of packaging through design modifications (Tyson Foods, Inc. 2015).

22.4.1.2 Eco-Efficiency

In a recent study of sustainability across many manufacturing sectors (Egilmez et al. 2013), sectors were ranked by an *eco-efficiency score*, a measure of environmental performance against economic output. The score accounts for GHG emissions, energy use, hazardous waste, TRI releases, and water use as input categories in comparison to total economic output. The higher the score, the more eco-efficient a sector is said to be. Food manufacturing (NAICS 311) ranked among the highest in *eco-efficiency score* at 100%[5] indicating that the industry is efficiently producing economic output based on the environmental performance indicators included in the model. However, the authors of the study note that this is largely due to the significant economic output of the industry despite its large environmental footprint.

In contrast, when on-site food manufacturing is considered along with its off-site supply chain (including extraction and processing of raw materials), the industry's upstream supply chain is responsible for 81.7% of TRI releases and 97.3% of hazardous waste generation associated with food manufacturing and its inputs. In other words, the production of materials used by the industry to manufacture foods generates significantly more pollution than food manufacturing alone and has a larger environmental impact. Some of the highest-contributing supply chain sectors include: other basic organic chemical manufacturing, petroleum refineries, and plastics material and resin manufacturing (Egilmez et al. 2013).

The following section describes the food manufacturing industry's progress towards reducing TRI chemical waste managed including releases.

22.4.2 Process and Technology Modifications

22.4.2.1 Energy Efficiency

Historically, the food manufacturing industry has been slow to research or adopt new energy-saving technologies. This is due to low profit margins, the need to comply with federal safety and sanitation standards, and competition within the industry that hinders collaboration. As a result, the industry relies on innovations from within the chemical and biotechnology sectors, which are adopted later without the high cost of research and development (Lung et al. 2006). Examples of processes first developed in the chemical and biotechnology sectors and later adopted for use in food manufacturing include oil seed extraction and wet corn milling.

Due to barriers to innovation, government programs may incentivize improvements. For example, the Energy Star Program recognizes facilities with the best energy performance in the food manufacturing industry through an energy performance indicator (Boyd 2011). Energy Star certification is awarded to plants ranking in the top 25% for energy efficiency nationwide. Many

---

5 Eco-efficiency scores are a comparison of environmental impacts and economic output relative to other manufacturing sectors. A score of 100% does not mean that the industry cannot improve upon its eco-efficiency.

large food manufacturing companies, such as General Mills and Cargill, are Energy Star partners (ENERGY STAR n.d.). Energy Star also produced reports containing recommendations for energy-efficiency measures for the Fruits and Vegetables and Bakeries and Tortilla subsectors, providing detailed technical direction, associated cost-savings, and payback times. Some specific recommendations include:

- for the Fruits and Vegetables subsector – adopting technologies in efficient steam production, CHP generation, piping insulation, and reuse of produce washing water;
- for the Bakeries and Tortilla subsector – placing ovens in well-ventilated areas away from processes that require cooler environments, using radio frequency ovens, and reducing the need to open freezer doors (Masanet et al. 2008).

Improving energy efficiency reduces the quantity of fuels burned to produce energy and therefore reduces the quantities of TRI chemicals formed as by-products of combustion or that are contained in the fuel and released to the environment during combustion. If energy is generated during combustion of fossil fuels on-site at a facility, and practices that improve energy efficiencies are implemented by the facility, the corresponding reductions in the releases of TRI chemicals during energy production will likely be reflected in the facility's TRI reporting.

In order to help facilities realize the many possible energy efficiency improvement options, the US Department of Energy's Advanced Manufacturing Office sponsors the development of emerging green technologies to help industry invest in broadly-applicable manufacturing processes that reduce energy consumption and improve efficiency (US DOE 2016). For many facilities in the food manufacturing industry, their processes involve food preservation, which historically requires thermal treatments such as cooking, pasteurization, and drying. These processes account for approximately half of the sector's energy use. To incentivize energy reduction in the sector, the Advanced Manufacturing Office sponsored the development of green technologies, including energy-efficient blanching, pulsed electric field pasteurization, radio frequency drying, and evaporator fan controls for refrigerated storage, all of which provide additional benefits such as reduced wastewater and improved product quality. For some of these technologies, high up-front capital cost may be a barrier to implementation (Lung et al. 2006).

22.4.2.2 Chemical Substitutes

Switching to safer chemicals (i.e. chemicals not included on the TRI list of toxic chemicals) where feasible allows facilities to decrease the generation of wastes that contain TRI chemicals. For example, oilseed processing plants commonly use *n*-hexane (a TRI chemical) for oil extraction, and efforts are underway to find a less toxic (i.e. non-TRI chemical) alternative. Recent laboratory research has identified alternative green extraction methods, such as supercritical CO_2 (a technique used for many years to decaffeinate coffee beans and extract soluble hops), but these methods are expensive to implement on a larger scale, and have not yet replaced *n*-hexane extraction in the industry (Jokic et al. 2012). But, as discussed in Section 22.4.5.3 of this chapter, facilities also face many barriers to source reduction, particularly a lack of adequate substitutes or concerns about product quality.

Food packaging is another research focus area looking into chemical substitutes. Increasingly, biodegradable and compostable materials are being developed for food packaging. For example, polylactic acid, a biodegradable polymer derived from lactic acid, is used in wraps for bakery and confectionery products (Shin and Selke 2014). Several EPA Green Chemistry award nominations have focused on the opportunities to replace conventional food packaging materials with more sustainable options, such as compostable food packaging or eliminating bisphenol A (a TRI-listed chemical) from the interior coatings of aluminium cans (US EPA 2016a).

Chemical substitutes in the food manufacturing industry may also be driven by external forces. For example, one of the largest importers of US poultry is Russia. However, in 2010, the country announced that it was lowering the acceptable level of chlorine residuals on imported poultry. Chlorine was the primary disinfectant used on US poultry products at the time, and the announcement effectively banned US imports of poultry to Russia. In response, the two countries brokered an agreement to allow for the use of alternative disinfectants, such as peroxyacetic acid, that are considered to be acceptable and less toxic than chlorine (Bottemiller 2010). Since 2007, peroxyacetic acid waste managed by the poultry processing subsector (NAICS 311615) has continually increased, while chlorine waste managed has decreased.

22.4.3 Recycling

While strictly speaking it is not considered pollution prevention and thus not the focus of this chapter, recycling continues to play a crucial role in achieving sustainability in the food manufacturing industry both during the manufacturing process and for using food waste. The following text is not meant to be comprehensive but instead provides examples of recycling practices in food manufacturing.

22.4.3.1 Packaging

Almost all food manufacturing facilities are involved in some form of food packaging, either for transportation to another facility or distribution to consumers. Common materials used in food packaging are paper, paperboard, glass, aluminium, and plastic. In the food manufacturing industry, paper is the most recycled material, and plastics are the least recycled (Shin and Selke 2014). Because of risk of contamination, facilities often cannot use recycled paper or plastic for their packaging. The US Food and Drug Administration (FDA) evaluates all recycled plastic on a case-by-case basis and requires food packaging manufacturers to submit a description of the recycled plastic they wish to use, as well as the results of testing to ensure that they removed all contaminants (US FDA 2015).

22.4.3.2 Food Waste

Excess unsaleable food may be donated to those in need and is considered the best alternative after source reduction measures, as explained in EPA's food waste hierarchy shown in Figure 22.11 (US EPA 2019a). As shown in a survey conducted by the Food Waste Reduction Alliance, medium-sized manufacturers are donating food at higher rates than large and small manufacturers, and food donation is being considered as an investment area by over half of the survey respondents (Food Waste Reduction Alliance 2016). Nonetheless, large amounts of organic waste are generated in the food manufacturing industry, and many of these materials may be recycled. According to the survey responses, two types of recycling – land application (60%) and animal feed (34%) – dominated the downstream uses of food waste based on pounds diverted away from landfills (Food Waste Reduction Alliance 2016). The survey results also noted that the most commonly cited barrier to donating or recycling food wastes were transportation constraints.

Specific examples of food waste reutilization and the creation of value-added products is observable in various subsectors. In the Seafood subsector, skin, bone, and fin can be reused as fish meal and fertilizer. Research is also focused on reusing shellfish waste for production of chitin, which is used in several industries, especially textile, cosmetic, and pharmaceutical manufacturing. Fruit and vegetable waste is a source of pectin, which is used in processing jams, jellies, pastries, and confectioneries, and increasingly in pharmaceuticals and cosmetics as well. Other examples are using beet waste as a source of red food colouring and using collagen byproducts from the Meats subsector in leather production (Kosseva 2009).

Figure 22.11 Food recovery hierarchy.

Moreover, according to a report from the National Renderers Association, every year the rendering industry recycles approximately 59 billion pounds of perishable material generated by the livestock and poultry meat/poultry processing, food processing, supermarket and restaurant industries. These materials are then rendered into ingredients for soaps, paints, cosmetics, explosives, toothpaste, pharmaceuticals, leather, textiles, and lubricants (National Renderers Association, Inc. n.d.).

22.4.3.3 Energy Recovery

While energy recovery from combustion of TRI chemical wastes is not typical among large food manufacturers (less than 0.2% of TRI chemical waste was used for energy recovery each year from 2007 to 2017, as shown in Figure 22.7), there is potential for energy recovery from food wastes at these facilities, as evidenced by the municipal solid waste stream where approximately 19% of food waste is combusted for energy recovery (US EPA 2018b). There are also case studies which focus on adopting large-scale energy recovery systems to use municipal food waste as a source for renewable energy generation such as a pilot facility at the East Bay Municipal Utility District, which converts food waste into electric power (East Bay Municipal Utility District n.d.). These energy recovery technologies could represent an alternative for companies seeking to increase value from the food waste and divert it from entering the waste stream to less preferred food recovery practices such as composting, incineration, and landfill disposal.

22.4.4 Wastewater Treatment

Food manufacturing uses a significant amount of water in processing operations (e.g. heating) and for cleaning and sanitation where water use is essential in protecting people from foodborne illnesses. Consequently, the sector generates considerable wastewater. Researchers have investigated if a portion of this wastewater can be reused. However, high levels of organic compounds and bacterial contaminants lead to low rates of wastewater recycling (Visvanathan and Asano 2009). Pressure-driven membrane separation techniques, such as ultrafiltration and reverse osmosis, use permeable membranes to separate and remove dissolved solids and microbes. These processes are already used for reusing wastewater across the food manufacturing industry, including in the fish, poultry, and soybean subsectors (Vourch et al. 2008). Membrane bioreactor technology combines biological treatment with membrane filtration and potentially allows significant reuse of wastewater in food manufacturing plants. However, membrane bioreactor systems are expensive and require a lot of energy to operate (Cicek 2003). Finally, as seen in the corporate sustainability reports, many facilities are using their organic waste to offset their fuel use by producing natural biogas during treatment of wastewater. This process takes advantage of anaerobic bacterial activity that converts organic waste to methane gas and carbon dioxide (CO_2) (Hall and Howe 2012).

Any wastewater that is discharged from facilities must be treated in some manner before release. Some facilities are transitioning away from the use of traditional disinfectants such as chlorine and using alternative technologies for treating wastewater, such as installing ultraviolet (UV) treatment systems. In addition to reducing the chlorine releases reported to TRI by switching to UV treatment, facilities may also reduce the amount of total chlorine residuals reported to EPA's NPDES Program (see Section 22.3.2).

Some food manufacturing facilities are also able to adapt their wastewater treatment plants to collect and handle residual food wastes. For example, a sweet potato processing plant uses anaerobic digestion to generate biogas from its wastewater treatment operations, and was designed to treat both the wastewater stream as well as leftover sweet potato peels (Food Waste Reduction Alliance 2014). The facility is able to offset about 20% of its annual energy demands and divert approximately 10 000 tons of food waste from landfills per year.

22.4.5 Pollution Prevention Activities Reported to TRI

As discussed earlier, quantities of TRI chemicals released by the food manufacturing industry as reported to EPA's TRI Program have decreased between 2007 and 2017, while the amount of waste managed has increased (though there is some variation within subsectors). This section identifies chemicals targeted for pollution prevention and the types of activities implemented to reduce TRI chemical waste quantities. Additionally, it describes the food manufacturing industry's overall progress towards reducing TRI chemical waste managed quantities, including their release.

Facilities that are subject to the TRI reporting requirements, such as those within the food manufacturing industry, are required to report any source reduction activities (e.g. process modifications, substitution of raw materials) that were newly implemented at the facility during the year for which they are reporting. Specific data fields exist in the TRI Reporting Form R for these required data elements. For example, in Section 8.10 of Form R, facilities select source reduction activities[6] from a list of coded processes or improvements (US EPA 2019b). In addition, facilities may voluntarily include specific details on their source reduction activities or other environmental management practices, in the form of text, in Section 8.11 of the TRI Form. Including this additional text provides facilities with a unique opportunity to showcase their achievements in preventing pollution to the public and other users of TRI data and information.

For the 2017 reporting year, 101 facilities in the food manufacturing industry reported 167 source reduction activity codes for 23 chemicals and chemical categories. This represents about 6% of the

6 The terms 'source reduction' and 'pollution prevention' are used interchangeably in this chapter. However, in this section, the term 'source reduction' is used in lieu of pollution prevention to be consistent with TRI reporting terminology.

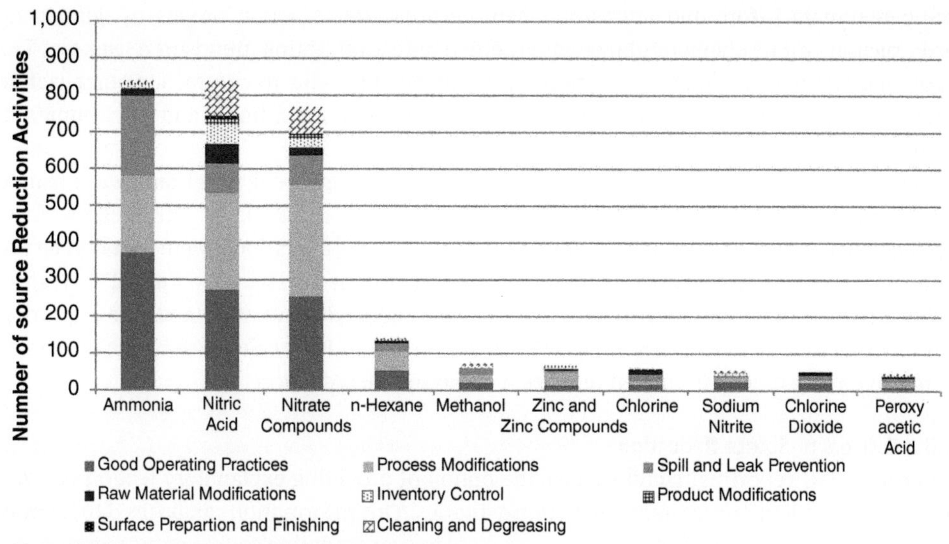

Figure 22.12 Chemicals manufactured, processed or otherwise used in food manufacturing with the greatest number of source reduction activities reported from 2007 to 2017. Note: Metals are grouped with their metal compounds. Source: US EPA Toxics Release Inventory – 2017 National Analysis Dataset.

1585 food manufacturing facilities that reported to EPA's TRI Program for 2017, which is slightly lower than the rate of 7% across all facilities that reported for 2017. 'Good operating practices' was one of the most commonly reported source reduction categories, most often through improved maintenance, record-keeping, or procedures. 'Process modifications' were also frequently reported, most often comprised of optimized reaction conditions or modified equipment, layout, or piping. Figure 22.12 shows that most of the activities were implemented for ammonia, nitric acid, and nitrate compounds, which are also the most commonly reported chemicals.

22.4.5.1 Examples of Source Reduction Activities Reported to EPA's TRI Program

Facilities that report to EPA's TRI Program have the option of describing their source reduction activities in explanatory text. In addition to optional explanatory text, facilities also report specific source reduction activity codes to further characterize the type of source reduction activity implemented within each of the eight categories. Six of these source reduction activity codes are associated with green chemistry practices. Among food manufacturing facilities, the most common green chemistry practice reported is 'W50: Optimized reaction conditions or otherwise increased efficiency of synthesis' and is most commonly reported for nitrate compounds by facilities in the Meats and Sugar and Confectionery subsectors. Other green chemistry practices reported include 'W15: Introduced in-line product quality monitoring or other process analysis system' and 'W43: Substituted a feedstock or reagent chemical with a different chemical'.

22.4.5.2 TRI Pollution Prevention Analysis – Effectiveness of Source Reduction Activities

A study published in 2015 estimated how source reduction activities affect quantities of TRI chemical releases reported by facilities that are subject to the TRI reporting requirements (Ranson et al. 2015). The study used a common economics research technique known as a 'differences-in-differences' analysis. This method estimates how releases of a TRI chemical from a facility change in the years before and after implementing a source reduction project by comparing trends in releases of TRI chemicals targeted by source reduction against trends in releases of TRI chemicals not targeted for source reduction. This comparison helps to control for other factors that affect releases, such as changes in production, economic conditions, and environmental regulations.

Applying this technique to the food manufacturing industry showed that the *average* source reduction project in the industry resulted in a 10% decrease in facility-level TRI releases of the targeted chemical. 'Raw material modifications' and 'product modifications' were found to be the most effective types of source reduction with average decreases in releases of 30% and 23%, respectively. Between 1991 and 2014, source reduction may have reduced cumulative food manufacturing releases reported to EPA's TRI Program by 227–336 million pounds, as calculated by the difference between actual releases and simulated releases with no source reduction.

22.4.5.3 Barriers to Source Reduction

As part of their TRI reporting, facilities have the option of providing explanatory text on barriers that they face in implementing source reduction activities. This information can be used to identify challenges that the food manufacturing industry faces in implementing source reduction.

From 2014 through 2017, 8% of forms (1200 forms) submitted by facilities in the food manufacturing industry included specific barrier text entries. Of these entries, the most commonly reported barrier category was facilities' concerns that source reduction activities would affect product quality. The second most commonly reported barrier category was a lack of substitutes for a specific chemical, and the third most commonly reported barrier category included general concerns such as customer demand. Barriers were primarily reported for nitric acid, nitrate compounds, and ammonia, which are also the chemicals reported most frequently.

The barrier categories reported suggest that some chemicals reported by the food manufacturing industry are integral to the operations and processes, such as the use of nitric acid in cleaning-in-place operations in the Dairy subsector. However, there may still be opportunities for introducing efficiencies in these situations, such as optimizing chemical blends or recovering and reusing cleaning solutions (Solid and Hazardous Waste Education Center 2016). Barriers reported reveal some challenges to source reduction, such as the necessity for sanitation in dairy product manufacturing facilities; however, for some chemicals there is no evidence that source reduction was attempted.

22.4.5.4 Summary of Pollution Prevention Activities Reported to TRI

In the food manufacturing industry, many of the reported source reduction activities are in either the 'good operating practices' or 'process modifications' categories and are most frequently reported for ammonia, nitrate compounds, and nitric acid. For other chemicals with few reported source reduction activities, further research is needed to identify if potential pollution prevention opportunities are available. An analysis of the source reduction activities reported by food manufacturing facilities and releases reported by those facilities suggests that some types of categories, particularly raw material modifications and product modifications, are associated with decreases in reported release quantities of TRI chemicals. The source reduction activity information reported to TRI may be able to provide insight or ideas for facilities that have not yet implemented source reduction activities for particular chemicals or that have additional potential pollution prevention opportunities.

22.5 Perspectives

From 2007 through 2017, overall quantities of TRI chemical wastes managed by the food manufacturing industry as reported to EPA's TRI Program have increased while releases of TRI chemicals have decreased. Trends in chemical releases and waste management are specific to processes within the subsectors included in food manufacturing and can greatly vary by subsector. For example, *n*-hexane is largely associated with soybean processing whereas ammonia and nitrate compounds are often reported by meat and dairy product processing facilities due to the organic content of wastewater.

The TRI chemical waste managed within the industry is not only dominated by a few particular processes, including wet corn milling and meat processing, but also by a few large parent companies due to the economies of scale inherent in the industry and the resulting trend towards mergers. Chemical wastes are also concentrated geographically, with most releases reported to the TRI Program occurring in regions associated with inputs to the food manufacturing industry, primarily southern and midwestern states.

Many companies within the food manufacturing industry include sustainability efforts as part of their corporate reports and highlight progress in reducing GHG emissions, improving energy efficiency, and reducing wastes. A review of the literature identified very little information on research or activities describing chemical substitutes to TRI chemicals in the sector, though there have been successful case studies such as the replacement of chlorine (a TRI chemical) in poultry disinfection systems with the less toxic peroxyacetic acid (also a TRI chemical). Notably, 'no substitutes' is also one of the most common barriers reported to EPA's TRI Program by food manufacturers, indicating that for some chemicals or processes, research into identifying safer chemical substitutes is needed. In other cases, there may not yet be a viable, safer substitute, such as for the use of *n*-hexane in soybean oil extraction, and pollution prevention strategies could focus on identifying alternative substitutes or technologies that obviate the need for TRI chemicals.

The chemicals that facilities most frequently report in the context of source reduction activities are the same chemicals which are also most frequently included in the reported barriers to pollution prevention. These chemicals include nitrate compounds, nitric acid, ammonia, and *n*-hexane. This reporting of both implemented source reduction activities and barriers indicates that there may be opportunities to increase the sector's source reduction activities through information dissemination among facilities in the sector. An analysis of source reduction activities reported to the TRI Program indicates that source reduction can reduce releases by 10% across the food manufacturing industry. Methods for pursuing source reduction that have proven to be effective in the food manufacturing industry include raw material modifications, and process modification. Due to the nature of specific uses of chemicals and associated releases in the food manufacturing industry, pollution prevention activities often need to be both sector- and process-specific.

Companies are seeking synergistic opportunities for reducing food loss and waste including diversion efforts to keep food from being disposed of in a landfill. Such efforts include creation of other value-added products from food wastes, land application, animal feed generation, conversion of food to fuel, and food as an energy recovery source. While energy recovery from combustion of TRI chemical wastes is not as common among food manufacturers (less than 0.2% of TRI chemical waste was used for energy recovery each year from 2007 to 2017), there is a potential for energy recovery from food waste at these facilities, as evidenced by the municipal solid waste stream where approximately 19% of food waste is combusted for energy recovery. Energy recovery technologies documented in case studies could represent an alternative for companies seeking to divert food waste from entering the waste stream who have identified insufficient recycling or industrial use options for optimizing food waste as a resource.

22.5.1 Next Steps

Future efforts to promote pollution prevention activities in this industry sector would have the most impact on subsectors with the larger amounts of waste generation and releases, such as Grain/Oilseed Milling, and Meats. Meat and poultry processing, as well as soybean processing and wet corn milling, are some examples of specific processes where large amounts of TRI chemicals are used and released. These processes may potentially benefit from an increased focus on promoting source reduction activities, depending on the specific chemical and process involved.

A more targeted focus on chemicals involved in upstream or downstream activities (e.g. transportation, distribution, agriculture, and livestock) in the food manufacturing supply chain may also have a larger impact on the overall environmental impact of the industry and warrants additional research. Other efforts related to sustainability include the continued pursuit of more energy efficient operations and elimination of ozone-depleting substances. Resources related to these impact categories, such as energy efficient recommendations for specific subsectors provided by Energy Star, are already available for food manufacturing facilities examining opportunities for pollution prevention.

Disclaimer

The views, statements, opinions, and conclusions expressed in this chapter are entirely those of the authors, and do not necessarily reflect those of the United States Environmental Protection Agency, Abt Associates Incorporated, or the Eastern Research Group, nor does mention of any chemical substance, commercial product, or company constitute an endorsement by these organizations.

References

ADM. (2014). *Corporate Sustainability Report*. Retrieved from http://www.adm.com/en-US/responsibility/2014CRReport/Documents/2014-Corporate-Responsibility-Report.pdf

Bhatkar, V.W., Kriplana, V.M., and Awari, G.K. (2013). Alternative refrigerants in vapour compression refrigeration cycle for sustainable environment: a review of recent research. *International Journal of Environmental Science and Technology* 10: 871–880.

Bottemiller, H. (2010, June 25). Russia Agrees to Lift Ban on U.S. Poultry Imports. Retrieved from http://www.foodsafetynews.com/2010/06/russia-agrees-to-lift-ban-on-us-poultry-imports/#.WL910m8rKM8

Boyd, G. A. (2011). *Development of Performance-based Industrial Energy Efficiency Indicators for Food Processing Plants*. Retrieved from https://www.energystar.gov/ia/business/industry/downloads/Food_EPI_Documentation_final.pdf

Bureau of Economic Analysis. (2018a, November 1). Interactive Access to Industry Economic Accounts Data: GDP by Industry. Retrieved February 14, 2019, from https://apps.bea.gov/iTable/index_industry_gdpIndy.cfm

Bureau of Economic Analysis. (2018b, December 21). Table 2BUI. Implicit Price Deflators for Manufacturing and Trade Sales. Retrieved January 4, 2017, from https://apps.bea.gov/iTable/iTable.cfm?reqid=19&step=2#reqid=19&step=2&isuri=1&1921=underlying

Cargill. (2015). *Cargill 2015 Corporate Sustainability Report*.

US Census Bureau. (2015, February). 2012 Commodity Flow Survey. Retrieved from http://www.census.gov/econ/cfs/2012/ec12tcf-us.pdf

Cicek, N. (2003). A review of membrane bioreactors and their potential application in the treatment of agricultural wastewater. *Canadian Biosystems Engineering*.

East Bay Municipal Utility District. (n.d.). Food scraps recycling. Retrieved March 31, 2019, from https://www.ebmud.com/wastewater/recycling-water-and-energy/food-scraps-recycling

Egilmez, G., Kucukvar, M., and Tatari, O. (2013). Sustainability assessment of US manufacturing sectors: an economic input output-based frontier approach. *Journal of Cleaner Production* 53: 91–102.

ENERGY STAR. (n.d.). ENERGY STAR Focus on Energy Efficiency in Food Processing. Retrieved from https://www.energystar.gov/buildings/facility-owners-and-managers/industrial-plants/measure-track-and-benchmark/energy-star-energy-4

Federal Reserve Board of Governors. (2017). Production Index G.17 – Industrial Production and Capacity Utilization 2005-2015. Retrieved from http://www.federalreserve.gov/datadownload/Build.aspx?rel=G17

Food Engineering. (2014). Top 100 Food and Beverage Companies. Retrieved from http://www.foodengineeringmag.com/global-top-100-food-&-beverage-companies

Food Processing. (2018). Food Processing's Top 100. Retrieved from http://www.foodprocessing.com/top100/top-100-2017/

Food Waste Reduction Alliance. (2014). *FWRA Toolkit: Diversion Beyond Donation Resource: ConAgra LambWeston Case Study*. Retrieved from http://www.foodwastealliance.org/wp-content/uploads/2014/02/ConAgra-LambWeston-Anaerobic-Digestion-With-Header.docx

Food Waste Reduction Alliance. (2016). *Analysis of US Food Waste Among Food Manufacturers, Retailers, and Restaurants*.

Hall, G.M. and Howe, J. (2012). Energy from waste and the food processing industry. *Process Safety and Environmental Protection* 90: 203–212.

Hilmar Cheese Company. (2015). *Committment to Sustainability Report 2014*. Retrieved from http://www.hilmarcheese.com/wp-content/uploads/2016/01/2014Hilmar_Cheese_Company_Sustainability_Report.pdf

James, S.J. and James, C. (2014). Chilling and freezing of foods. In: *Food Processing: Principles and Applications*, 79–105. Wiley.

Jokic, S., Nagy, B., Zekovic, Z. et al. (2012). Effects of supercritical CO_2 extraction parameters on soybean oil yield. *Food and Byproducts Processing* 90 (4): 693–699.

Kosseva, M.R. (2009). Processing of food wastes. *Advances in Food and Nutrition Research* 58: 57–136.

Lung, R. B., Masanet, E., & McKane, A. (2006). *The Role of Emerging Technologies in Improving Energy Efficiency: Examples from the Food Processing Industry*. Retrieved from http://escholarship.org/uc/item/43c841xs

Food Manufacturing. (2015, November 11). The Basics of Ammonia Refrigeration. Retrieved from http://www.foodmanufacturing.com/article/2015/11/basics-ammonia-refrigeration

Martinez, S.W. (1999). *Vertical Coordination in the Pork and Broiler Industries: Implications for Pork and Chicken Products*. Washington, D.C.: USDA Economic Research Service.

Masanet, E., Worrell, E., Graus, W., & Galitsky, C. (2008). *Energy Efficiency Improvement and Cost Saving Opportunities for the Fruit and Vegetable Processing Industry*. Retrieved from https://www.energystar.gov/ia/business/industry/Food-Guide.pdf

National Renderers Association, Inc. (n.d.). *North American Rendering*. Retrieved from http://assets.nationalrenderers.org/north_american_rendering_v2.pdf

Pollution Prevention Act, 42 USC. §13101 et seq. § (1990).

Ramde, D. (2010, November 15). *Food Production a Rare Growth Industry in Gloomy Economy*. Associated Press.

Ranson, M., Cox, B., Keenan, C., and Teitelbaum, D. (2015). The impact of pollution prevention on toxic environmental releases from US manufacturing facilities. *Environmental Science & Technology* 49 (21): 12951–12957.

Shields, D. A. (2010). *Consolidation and Concentration in the US Dairy Industry*. Congressional Research Service.

Shin, J. and Selke, S.E. (2014). Food Packaging. In: *Food Processing: Principles and Applications* (eds. S. Clark, S. Jung and B. Lamsal), 249–273. John Wiley & Sons, Ltd.

Solid & Hazardous Waste Education Center. (2016). *Optimizing CIP to Save Money and Reduce Waste*. Retrieved from http://shwec.engr.wisc.edu/wp-uploads/2015/08/CIP-Fact-Sheet-fin-3-11-16.pdf

Tyson Foods, Inc. (2015). Tyson Foods Sustainability Highlights Fiscal Year 2014.

UN Industrial Development Organization. (2004). Pollution from Food Processing Factories and Environmental Protection. Retrieved from http://www.unido.org/fileadmin/import/32129_25PollutionfromFoodProcessing.7.pdf

US Census Bureau. (2011). Share of Value Shipments Accounted for by the 4, 8, 20, and 50 Largest Companies for Industries: 2007. Retrieved from American Fact Finder website: http://factfinder.census.gov/faces/tableservices/jsf/pages/productview.xhtml?pid=ECN_2007_US_31SR12&prodType=table

US Census Bureau. (2018). 2015 SUSB Annual Data by Establishment Industry. Retrieved December 11, 2018, from http://www.census.gov/data/tables/2015/econ/susb/2015-susb-annual.html

US Census Bureau. (n.d.-a). 2012 NAICS. Retrieved from https://www.census.gov/cgi-bin/sssd/naics/naicsrch

US Census Bureau. (n.d.-b). Historical Data – Annual Survey of Manufacturers. Retrieved from http://www.census.gov/manufacturing/asm/historical_data/index.html

US Department of Commerce. (2008). Industry Report: Food Manufacturing NAICS 311.

US DOE. (2016). Advanced Manufacturing Office. Retrieved March 11, 2016, from http://energy.gov/eere/amo/advanced-manufacturing-office

US EPA. (2016a, March 25). Green Chemistry Program Nomination Table. Retrieved from https://www.epa.gov/greenchemistry/green-chemistry-program-nomination-table

US EPA. (2018a). GHG Reporting Program Data Sets. Retrieved from http://www2.epa.gov/ghgreporting/ghg-reporting-program-data-sets

US EPA. (2018b). *Advancing Sustainable Materials Management 2015 Fact Sheet: Assessing Trends in Material Generation, Recycling, Composting, Combustion with Energy Recovery and Landfilling in the United States.*

US EPA. (2018c). Toxics Release Inventory - National Analysis Dataset 2007–2017. Retrieved December 3, 2018, from https://www.epa.gov/toxics-release-inventory-tri-program/download-trinet

US EPA. (2019a). Sustainable Management of Food. Retrieved April 1, 2019, from https://www.epa.gov/sustainable-management-food

US EPA. (2019b). *Toxic Chemical Release Inventory Reporting Forms and Instructions Revised 2018 Version*. Retrieved from https://ofmpub.epa.gov/apex/guideme_ext/f?p=104:41:0:

US EPA. (n.d.). Discharge Monitoring Report (DMR) Pollutant Loading Tool. Retrieved from https://cfpub.epa.gov/dmr/index.cfm

US FDA. (2015). Recycled Plastics in Food Packaging. Retrieved March 24, 2016, from http://www.fda.gov/Food/IngredientsPackagingLabeling/PackagingFCS/RecycledPlastics/ucm093435.htm

US FDA. (2016, April 6). Sanitary Transportation of Human and Animal Food. Retrieved from https://federalregister.gov/a/2016-07330

Visvanathan, C., & Asano, T. (2009). The potential for industrial wastewater reuse. In *Encyclopedia of Life Support Systems*.

Vourch, M., Balannec, B., Chaufer, B., and Dorange, G. (2008). Treatment of dairy industry wastewater by reverse osmosis for water reuse. *Desalination* 219 (1): 190–202.

23

Financing Strategies for Sustainable Bioenergy and the Commodity Chemicals Industry

Praveen V. Vadlani

Saivera Bio LLC, Puttaparthi, Andhra Pradesh 515134, India

CHAPTER MENU

23.1 The Current Financing Scenario at Global Level

23.1.1 Recovery from 2008 Financial Crisis and Global Economic Trends

Financial deregulation in 2008 resulted in a massive upheaval of domestic and global markets and economic conditions, which primarily raised fundamental questions regarding the influence financing has on economy and income distribution (OECD 2012; Cournède et al. 2015). The years following 2008 triggered a substantial correction among countries to mitigate high unemployment, capacity build-up, and budgetary imbalances. Consequently, global economic activity has been on an upswing, commodity prices have picked up, and financial markets have stabilized. Global growth is expected to accelerate beyond 2017 (2017–2019), with an anticipated stronger growth in developed economies and a significant growth trajectory among emerging and developing economies (International Monetary Fund 2017). However, global trade and foreign policies are undergoing a sea-change, particularly with the established global powers such as the United States, China, and Russia, embarking on inward looking national priorities. These disruptive policy statements have influenced current trade partnerships and institutional mechanisms that sustained globalization and stable world markets in the last few decades. Therefore, it is imperative for the policy planners of all major countries to strengthen mutually beneficial cooperative agreements to sustain a stronger global recovery, financial stability, and market integration; avoid unrequired trade barriers and tariffs, allow free flow of capital to localized opportunities, and ascertain appropriate currency exchange mechanisms.

Green Energy to Sustainability: Strategies for Global Industries, First Edition.
Edited by Alain A. Vertès, Nasib Qureshi, Hans P. Blaschek and Hideaki Yukawa.

While most of the major economies have recovered from this great recessionary trend, emerging economies have fared better in terms of growth and productivity. However, the growth difference between OECD (The Organization for Economic Co-operation and Development) countries and the emerging economies will narrow down considerably in the coming decades. China is likely to have the highest growth rate of about 7% among major economies; however, it is conceivable that countries such as India, Indonesia, and even some East European nations could surpass this growth rate (International Monetary Fund 2017). Advanced economies have experienced a prolonged episode of low interest rates and low growth since the global financial crisis. Over the past several decades, the real term interest rates as stipulated by the federal banks have been on a steady decline. The persistent low rates have challenged the business models of financial institutions, particularly the banks, institutional funds, and other lending agencies.

In addition, in several of the OECD countries, such as Japan and the US, the government debt is several times their Gross Domestic Product (GDP), which will have an adverse impact on federal interest rates and economic stability (OECD 2012). Further, some of the countries with high debt do not have a detailed fiscal plan to reign in the debt and stabilize the interest rates. Political interference, lack of bipartisan support, and welfare and subsidy oriented governmental programs have hampered any attempt to reduce debt and ease borrowing for potential economic growth. Lately, China, the second largest economy, has chosen a path of heightened governmental debt to fuel their economic engine to maintain a high growth rate. This may have consequences that could trigger global financial uncertainties.

Most of the developed economies have a tax system oriented to encourage debt, such as loans, over equity to fund companies. These encouragements result in firms having higher debt with enhanced interest payments, reduced growth opportunities and lack of investment opportunities from private and institutional investors. Structured financial reforms would reduce the risk and thereby the cost of capital required for firms to finance new projects or expand existing ones. So, a balance of an appropriate equity–debt ratio will ensure stock market involvements, an important function of the financial sector, and adequate credit lines from the banks (Cournède et al. 2015).

23.1.2 Financial Conditions at Global Level

Financial conditions at domestic and global level typically refer to the environment of easily obtaining financing for various investments, including for start-ups, new projects, mergers, and acquisitions. The prevailing financial conditions are important since they act as a barometer to determine the monetary policy and the direction of the current and future economic activity. Although global markets and financial upheavals tend to significantly impact the domestic financial conditions, individual countries have a mechanism, primarily the monetary policy, to influence their own prevailing domestic financial conditions. Primarily the domestic financial policy ensures the availability of funds at a reasonable cost and risk to local markets. At the same time, due to almost instant connectivity via technology to world markets, local financial conditions tend to react substantially to global economic and political disasters. As such, the domestic financial policy reactions may not be quick enough to adequately address the situation. Particularly, the emerging economies must constantly monitor the abrupt changes in global financial conditions and have a response ready to adequately put in place a strategy to defend their markets and growth. Along with prevailing interest rates and valuations of assets, financial conditions are influenced by investors' tolerance to risk and willingness to hold illiquid assets

(Cournède et al. 2015; International Monetary Fund 2017). Broadly, financial conditions are the outcome of the current monetary policy of countries, set to sustain and grow their economy and productivity. Monetary policy decides the interest rates that influence the consumption and investment decisions in the economy. Other financial variables, such as risk taking, credit restrictions, and external financial factors also influence the financial conditions.

Financial conditions are driven by factors other than just policy. Some of these factors are: the management decisions of companies that have a powerful presence in their market segments; technological changes that may trigger obsoleteness of an existing established product category; changes in financial institutions and other large institutional funds, such as pension funds, retirement accounts, risk perception; and shifts in investors' confidence and sentiments triggered by unusual domestic and global events. These factors heavily influence the access to capital in an economy and particularly due to globalization most of the markets in all continents are interconnected and usually immediately react to the prevailing global financial conditions. The policy makers must be cognizant of the global trends and should effectively integrate them into domestic fiscal policies in an open economy. For a country whose economy is globally integrated, and its business cycle closely mirrors that of the rest of the world economies, it is prudent to tie their domestic interest rates to that of the world interest rates (Schoenmaker 2013). For open and globally integrated economies, financial conditions enable access to funding from both domestic and global institutions; domestic firms rely more on international markets and as such global trends will have direct impact on their growth and function (International Energy Agency Publications 2017). The United States is the primary driver of the global financial conditions and has a powerful influence on international monetary system. While currencies such as the Euro and lately the Chinese Renminbi are likely to play some significant role in future, the US dollar is still the dominant international currency and takes centre stage as an international currency with important roles such as in global trade invoicing, issuance of loans and asset financing, and commodity and equity transactions. Open advanced economies that are closely aligned with global financial conditions have to be more sensitive to the financial activities in the US markets and the accompanying fiscal policy changes. To dampen the shock of global financial upheavals, the open economies encourage their local companies to have a good mix of debt and equity, depend on a local investor base and have diversified global portfolios for their products.

In recent years, the global financial stability has continued to improve with business-friendly financial and monetary conditions. Federal banks are working on increasing the longer-term interest rates, which is likely to boost the earnings of banks, mortgage, and insurance companies. Indices of all major stock markets have made substantial gains in 2017 and 2018 and the trend is expected to continue in early 2019. Along with global recovery, investors' optimism is fuelled by the anticipated tax reforms that are favourable to businesses, governmental infrastructure spending and the easing of stringent financial regulations imposed after the 2008 global crisis. However, new threats to financial stability are emerging from heightened political and policy uncertainties around the globe. A potential turn towards protectionism in developed economies could dampen the global growth and disrupt the trade balances and free market economies, restrict capital inflow and outflow to efficient markets, and significantly affect consumer confidence. In Europe, new challenges such as Brexit and immigration, have led to a shift in political governance, which could affect the anticipated changes to the banking systems and high debt levels, thereby leading to financial stability concerns (Schoenmaker 2013). Emerging market economies, such as India, have made changes to their monetary policies and diversified the global investments and capital structure of companies to reduce external vulnerabilities.

23.2 Ethanol Biofuel Industry – An Overview

23.2.1 Ethanol Industry Market and Growth Perspective

The United States has become a net exporter of natural gas and has attained self-sufficiency in crude oil, primarily due to the emergence of shale oil and natural gas via hydraulic fracturing of bituminous shale rock. However, despite these new energy sources and policy changes, the United States remains the second-largest growth market for renewables. Biofuels remain the main driver of renewables in the transport sector, while the emergence of electric vehicles is becoming more evident. However, due to the availability of low-cost natural gas and crude oil, the market share of renewables in road transport is anticipated to marginally increase from over 4% in 2016 to nearly 5% in 2022. At the same time, biofuels production is expected to grow by over 16% in the same forecast period (International Energy Agency Publications 2017).

Asia, particularly the emerging economies of India, Indonesia, China, and Philippines, have a rising demand for transport fuel, primarily due to the availability of feedstocks and encouraging governmental initiatives and policies. Brazil is a dominant player and has had a matured bioenergy industry based on its efforts starting from the 1980s and is making efforts to increase sustainable biofuels consumption to meet its national target for 2030. In the United States, despite a surge in the natural gas and crude oil markets, ethanol and biodiesel production are expanding as a result of supportive mandates and an emerging export market. Modest growth is anticipated in the European Union due to the availability of crude oil from multiple sources at competitive prices, and policy changes after 2020 with regards to investments in renewables (International Energy Agency Publications 2017). Advanced biofuels (such as cellulosic ethanol) have made important progress in recent years but are not yet competitive with petroleum products (Guragain et al. 2016). Production is expected to increase sevenfold from the current existing level, which is still just over 1% of total biofuels production (International Energy Agency Publications 2017).

23.2.2 Renewable Ethanol Industry from Grain and Cellulosic Feedstocks

In 2016, 200 operating ethanol plants in 28 states in the US produced about 15 billion gallons of biofuel along with about 42 million metric tons (MMT) of high-protein distillers' dried grains and solubles (DDGS) used as animal feed. The falling oil prices in 2016 triggered a spike in record gasoline consumption, which in turn led to increased consumption of ethanol in E10 blends. In addition, higher blends, such as E15 and E85, got a boost with many gas stations including infrastructure to offer these low-cost, clean fuels (Ethanol Industry Outlook 2017). Lately, the ethanol producers faced regulatory barriers for the use of ethanol, particularly in global markets where the US ethanol and DDGS is competitive in terms of price and availability. Further, oil and other allied energy industries actively seek to out-compete ethanol in terms of price and volumes and lobby efforts to tweak the Renewable Fuel Standard (RFS) program to undermine biofuels in general (Ethanol Industry Outlook 2017; US Environmental Protection Agency (EPA) 2017). The RFS program was created by the US congress to mandate the current transportation fuel to contain a minimum percentage of liquid renewable fuel by volume, and is the main driver of the US ethanol production and consumption. This was set up primarily to provide an economic support structure for the nascent biofuels industry and to also reduce dependence on imported oil; to mitigate greenhouse gas (GHG) emissions (US Environmental Protection Agency (EPA) 2017). The RFS program mandates the need to blend renewable fuels into transportation fuel in increasing amounts each year, reaching 36 billion gallons by 2022; however, the renewable fuels from cellulosic feedstock requirement has been reduced since the anticipated production targets of cellulosic biofuels, particularly ethanol, has not happened due to technological and scalability issues.

The total renewable fuel requirement is primarily met by the mature corn-based ethanol industry, and mandates by the Environmental and Pollution Agency (EPA) of the United States that bases the volume requirements on the supply of various categories of renewable fuels (US Environmental Protection Agency (EPA) 2017). Most of the emerging cellulosic ethanol producers, known as second-generation biofuels enterprises, had grossly underestimated the biomass feedstocks logistics, cost, and variability in terms of composition. As a result, the technological blue-print that they licensed from academic and federal labs was found to be inadequate to process multiple feedstocks and come up with process economics that can at least compete with the grain ethanol business (Guragain et al. 2016, 2014; Chen et al. 2016). Multiple feedstock processing is required for these plants to operate throughout the year for economic viability; most of the individual feedstocks are not available in viable quantities to last for the whole year even if adequate biomass storage facilities were designed. Consequently, the deliverables could not meet the Congressional goals for renewable fuels from cellulosics, albeit progress has been made in the microbial system and overall technology development. Other than biomass-based biodiesel (BBD), all the other biomass-based fuel categories fell below the statutory volume targets set by Congress. As a result, the volume requirement for cellulosic biofuel was reduced for 2018 and beyond in the new EPA's proposed 2018 RFS Required Volume Obligation (RVO) (US Environmental Protection Agency (EPA) 2017). The stakeholders and interested groups promoting cellulosic biofuels have countered these new volume requirements from EPA stating it will dampen the US leadership and the development of technologies and businesses related to advanced and cellulosic biofuels. Further, the RFS regulations do not hold for imports and consequently, other countries will get an advantage in the race to commercialize these cellulosic-based technologies.

Potential roadblocks to the commercialization of the cellulosic biofuels industry include: feedstock costs and availability, high production costs, low product yield, high capital requirements, policy uncertainty, and various technical, environmental and social issues (Warner et al. 2017). Use of non-food biomass feedstocks for biofuel production is encouraging and it expands the production footprint of biofuels, and as a result there is renewed interest in using cellulosic feedstocks, such as crop residues, forestry residues, municipal solid waste (MSW), and dedicated energy crops to produce ethanol (Warner et al. 2017; Guragain et al. 2014). Ethanol produced from cellulosic feedstock is identical to that produced from cereal grains and meets the same American Society for Testing and Materials (ASTM) International fuel quality standards with similar vehicular performance. Over the last few decades, the improvements in pretreatment technologies, enzymatic activity and costs, microbial systems to utilize complete biomass-derived sugars, have advanced the cellulosic biofuels technology and is now reaching commercial production. However, the commercialization has been severely hampered by a lack of enthusiasm among equity and debt investors, particularly after the economic downturn that happened after the 2008 financial crisis. This is primarily due to the high capital costs required for setting up a commercial facility with the associated high cost of capital due to high start-up risks for these new technologies (Guragain et al. 2014).

These risks include feedstock availability, collection, and delivery; pretreatment technology costs; higher capital costs; and technology scale-up challenges. Figure 23.1a,b shows how the financing requirements when a single feedstock leading to single product compares to an integrated bioprocess that can simultaneously handle feedstocks from multiple sources generating a broad spectrum of products. A specific product from a particular source of biomass (Figure 23.1a) is tantamount to high risk for investors in a landscape competing against established chemical and energy companies that have been in business for over a century. Obviously, a manufacturing facility that can

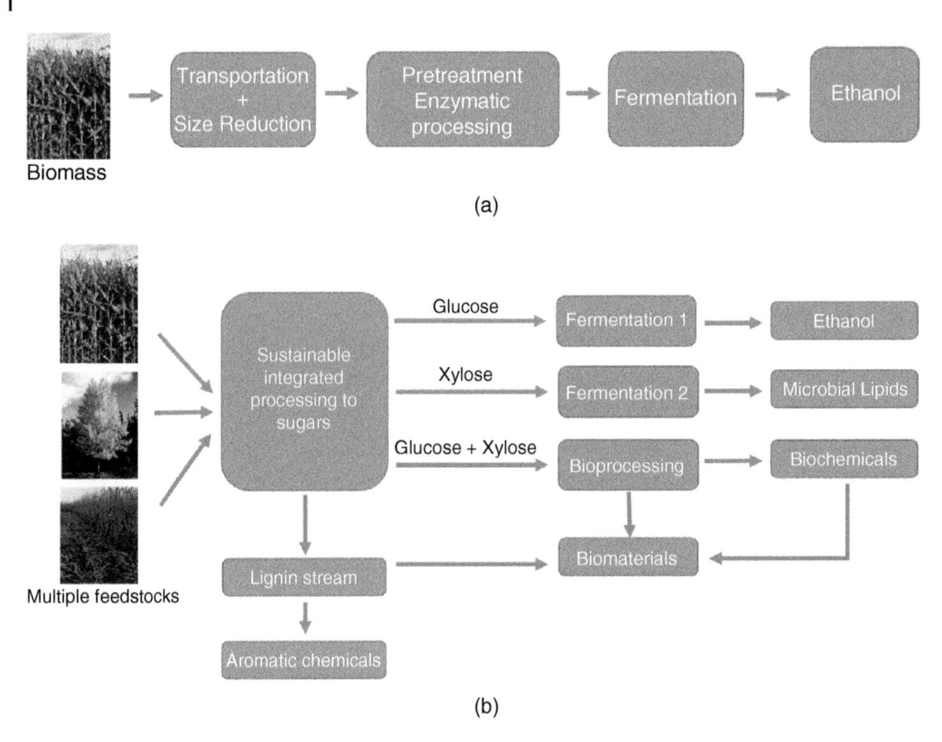

Figure 23.1 (a) Financing for single product bioprocessing. Moderate investment; high risk and high cost of capital; single revenue stream. (b) Financing for integrated bioprocessing. High investment; moderate/low risk and reduced cost of capital; multiple revenue streams.

produce several product categories while utilizing a variety of feedstocks and an integrated sustainable processing method will substantially reduce this risk and the cost of capital to raise funds from both equity and debt investors. Going forward, the technological advances should be geared towards an integrated biorefinery (Figure 23.1b), particularly from a financing perspective. The level of investment for second-generation biofuels and feedstocks has come down considerably, mainly because of the extended timeline in completing the production facility to commercialization. In addition, there is no clear policy statement from world governments regarding the use of cellulosic ethanol as a transport fuel, which has further reduced the flow of investments towards this sector. Large integrated facilities will involve huge investment compared to a typical ethanol production from a single feedstock such as corn stover; it takes considerable expertise and efforts to raise large volumes of capital in competitive open domestic and world markets. This financial dichotomy for the emerging biofuels and bio-based industry has considerably held back the path to commercialization with investors placing less priority on this industry despite the need for a sustainable and clean economy.

In the United States, close to 90% of the grain ethanol is produced from the dry milling process, and the remaining 10% from traditional wet mills. The US ethanol production is predominantly from corn (95%); contribution from other feedstocks is low, e.g. sorghum, barley and wheat (3%), cellulosic biomass (1%), and food/beverage wastes (1%) (Ethanol Industry Outlook 2017). The grain-based ethanol industry is technologically mature and the ethanol yield of about 2.7 gal/bu of grain is close to the theoretical value; a third of the corn goes into DDGS, a valuable animal feed product with large export potential; in wet mills, corn gluten feed and corn gluten meal serve as animal feed products (Guragain et al. 2016). According to the Renewable Fuels Association

(RFA) (Ethanol Industry Outlook 2017), in 2016, the ethanol production of 15.25 billion gallons and 42 MMT of DDGS had supported 74 420 direct jobs in renewable fuel production and agriculture, as well as 264 756 indirect and induced jobs across all sectors of the economy. Further, the RFA claims the ethanol biofuel industry contributed $42 billion to the US GDP in 2016 and about $9 billion in taxes, created jobs and raised household income by $23 billion. At the same time, the US ethanol producers bought $25 billion worth of raw materials and other goods. These numbers are modest compared to the chemical and oil industry; however, the ethanol industry, particularly in the midwest region of the US, has become a significant contributor to the local economy.

The US ethanol plant ownership is spread across 120 organizations; however, four major companies own 29% of plants and 39% of installed capacity: POET (27 plants; 1.6 billion gallons), Archer Daniels Midland (ADM) (8 plants; 1.7 billion gallons), Valero (11 plants; 1.2 billion gallons), and Green Plains Renewable Energy (14 plants; 1.2 billion gallons) (Ethanol Industry Outlook 2017). The United States has consistently been the top producer of ethanol in the world and in 2016 accounted for about 60% of global production. US agriculture has optimized the grain production methods in terms of crop yield, productivity, and modern harvesting techniques; as a result, they have easily met the domestic food and feed requirements, grain export targets, and have enough excess to cater to the biofuel industry. This steady supply of grain, particularly corn, has enabled the US ethanol industry to stay a low-cost producer and emerge as the most reliable supplier to international markets. Further, the entire value-chain: grain producers from individual farmers to cooperatives to the oil industry end users is seamlessly integrated. A significant boost to the US ethanol industry is the rising export market, which in 2016 was about 1 billion gallons. The two major importers of US ethanol are Canada and Brazil, accounting for half of the total exports. Brazil, although being the second largest producer of ethanol had to bear an upheaval in cane sugar raw material prices and to cater for their domestic demand had to import ethanol to make up the requirement. Recently, China has started to import ethanol, mainly to meet their rising demand and to reduce the worsening urban pollution. In 2016 the other major markets for ethanol imports included India, Peru, and South Korea (Ethanol Industry Outlook 2017). Most of the emerging economies, along with advanced economies, have a consistent government-mandated renewable fuels program, and have a blend requirement of mostly ethanol and biodiesel in their domestic transport fuels. Countries, such as India, China, and Philippines, with increasing GDP and standard of living in the last two decades, have better surface transport infrastructure in place with accompanying increase in private and public vehicular traffic. This has led to an enormous surge in transport fuel consumption and to meet the blend requirements they have to source ethanol and biodiesel from outside markets, which is a huge opportunity for the US biofuels industry.

23.2.3 Ethanol Biofuels Industry – Co-products

While the main thrust for the renewable ethanol industry in this century has stemmed from the need for independence from foreign oil and focused national security policies of the US and Western Europe, it had an interesting consequence in the form of the by-product DDGS. The starch component of the grain, such as corn, is fermented to ethanol; the other components including protein, fibre and fat are collected as a wet cake from the distillery column. The solubles are evaporated and added back to the dried cake, which together constitute the DDGS. As mentioned in Section 23.2.2, one-third of every bushel of grain ends up as DDGS, one-third goes to ethanol and the remaining third escapes as carbon dioxide in a dry grind ethanol process (Guragain et al. 2016). DDGS is nutrient-rich and makes an excellent feed product for the livestock, poultry and fish industries in the US, East Asia, and lately in countries such as India, Vietnam and Mexico. As the US ethanol industry ramped up its production to meet the RFS requirements, increasing quantities of DDGS were produced and reached a record output of 42 MMT in 2016 (Ethanol Industry Outlook 2017). Further, by adding front-end fractionators, the dry mill ethanol plants extracted about 2.9 billion pounds of corn distiller's oil (CDO), which is used as a feedstock for biodiesel production along with other applications (Jessen 2013a).

The US grains council and other allied agencies promoted DDGS around the globe and about 11 MMT from the US ethanol producers was exported to around 51 different countries in 2016 (Ethanol Industry Outlook 2017). The lower price of DDGS compared to other feed ingredients, has encouraged feed manufacturers to increase inclusion in livestock, poultry and fish feed across the globe. Countries in Far East Asia with a vibrant animal meat industry, opened their feed markets for DDGS imports, and China, Mexico, and Vietnam were the top three importers, receiving approximately half of the total US DDGS exports. The other leading importers were South Korea, Turkey, Thailand, Canada, and Indonesia, with India actively pursuing DDGS to meet their burgeoning feed industry. While China was the top market for US distillers' grain exports, the Chinese government's recent imposition of anti-dumping and countervailing duties against US distillers' grains led to major disruption in the DDGS export market. As explained in Section 23.1.1, it is imperative that the global trade agreements and policies are stabilized and forward looking rather than being politically motivated to sustain global markets and economies.

23.2.4 Ethanol Biofuels Industry – Cost and Economic Factors

The ethanol industry substantially contributed to the requirements of the US energy policy: independence from foreign oil, boost to rural economy, and sustainability. Particularly, in the corn and sorghum regions of the US, the economic impact of corn grain ethanol has been significant; in the midwest states where ethanol plants are located, the corn growers had a viable opportunity to increase corn grain production to meet the rising demand for ethanol production in the last two decades. Further, an additional boost, in terms of corn processing, marketing, construction of new plants, and research and development, was provided to these regions' economy (Guragain et al. 2016; Hofstrand 2016a).

The estimated production cost, based on an Iowa State University study, considers fixed costs, non-feedstock variable costs (e.g. natural gas, chemicals, and labour), feedstock costs, and revenue contribution from DDGS; the estimated production cost varied from $1.35 to $3.47 /gallon between 2006 and 2015 (Hofstrand 2016a). The single largest cost in the production of ethanol from corn grain is the cost of corn (Guragain et al. 2016). Corn grain prices vary from year to year and have ranged from $2.13 to $3.55 /bu in the last 15 years (from 2001 to 2016) (Hofstrand 2016a; US Department of Agriculture Economic Research Service (USDA-ERS) 2017). Another major production cost contributor is the price of natural gas or other sources of energy needed for the unit operations in the process, including for distillation, evaporation, and starch liquefaction. The recent gains in enzyme costs for starch to glucose conversion and higher fermentation efficiency has brought down the production cost significantly (Guragain et al. 2016). While it is important to consider the production cost of ethanol, the financing for new plants or expansion of existing plants is more complicated and a number of factors, including capital cost, investors' sentiment, availability of funding from banks and equity holders, government policies, and commodity trading, come into play. While the main ethanol producers are in midwest regions, the largest ethanol markets are located on the east and west coasts of the United States. The properties of ethanol prevent it from being pumped through pipelines; therefore, the denatured fuel is shipped on trains to the two coasts' markets. With the crude oil price being low, the transportation costs are marginal; however, with the new safety standards and ethanol competing with the shale oil business, these costs are increasing. Obviously, ethanol prices are typically lowest in the midwest and increase with distance to other domestic markets.

Distillers' grains are sold as livestock feed either as wet cake (46 lb/bu at 65% moisture) to feedlots within 50 mi radius from the ethanol plant or as dry product (18 lb/bu at 10% moisture) to far off domestic and export markets (Hofstrand 2016a). Lately, about 85% of ethanol dry mill plants have added a front-end dry fractionation technology to extract non-edible corn oil at a rate of about 0.5 lb/bu, which can potentially be used as a feedstock for biodiesel plants and in the animal feed industry (Jessen 2013a).

23.3 Bio-Based Industry – Current Status and Future Potential

23.3.1 Emergence of Bio-Based Industry

The bio-based process essentially involves enzyme-based and/or microbial based systems to convert raw materials, preferably renewable resources such as sugars, plant biomass, organic matter present in municipal or industrial waste streams, to a product other than fuels. The food and speciality industry had pioneered several bio-based products, including organic acids, solvents, flavours and fragrances, via fermentation. In the context of large-scale industrial chemicals and bulk drugs, biochemical engineering science, evolved in the last several decades, has developed the appropriate bioprocess technology, molecular biology techniques and procedures for engineered microbial systems for targeted molecules, and dedicated downstream processing for product purification. While the bioprocesses to enable a broad spectrum of chemicals and monomers exist, these processes cannot compete economically with the established fossil fuels based technologies. Further, the established energy and chemical industries have the market presence and financial wherewithal to sustain their paradigm despite calls for sustainable economies based on renewable resources. The world is at a stage where it cannot ignore climate change, water scarcity, over population, and ageing of the work force in developed economies. One of the approaches to mitigate this scenario is to effectively integrate renewable resources with existing or new processing entities to sustain ecological systems, attain food security and ensure sustainability.

The US and the Europe are the two regions in the world that are leading considerable efforts in fostering the bio-based industry. In 2014, the bio-based industry contribution to the US economy was US$393 billion in value addition and sustained 4.2 million jobs. Further, every 1000 jobs in the bio-based industry supported 1760 additional jobs in other parts of the economy (Golden et al. 2016). In Europe the new bioeconomy created 18.3 million jobs with an annual turnover of US$2.8 trillion. The European bio-based industries contributed to a US$826 billion turnover and 3.3 million employees (Bio-based Industries Consortium 2017). With the advent of new bioprocessing methods and the recent gene editing tools to develop efficient microbial systems, the bio-based industries are more resource-effective and environmentally friendly, leading to sustainable and low-cost bio-based chemicals and materials with lower GHG emissions and energy consumption. Broader use of integrated bioprocessing (Figure 23.1b) will significantly reduce the dependence on fossil resources and meet the global climate change targets for 2020 and move towards a low-carbon economy by 2050 (Bio-based Industries Consortium 2017; Golden et al. 2016).

23.3.2 Bio-Based Industry – Policy Landscape and Competitiveness

While the potential of bio-based products as a key driver of the new economy is evident, the existing entities along the value chain of global chemical industry are oblivious to their presence, including wholesale and retail distributors. While the biofuels industry got the required exposure and support from government and the general public; the biobased industry has operated below the radar of public knowledge. The awareness of its existence is crucial, particularly to the substantial private equity holders to successfully integrate this industry into the global economy. More particularly, the federal governments of established economies should have a concrete program to develop and sustain this industry similar to what was done for the early oil and chemical industries in the early twentieth century and in recent times to the biofuels industry. For example, state and federal legislations have promoted biofuels, creating incentives and tax credits; such an initiative has not happened for bio-based products (Golden et al. 2016). One may argue that the computer industry emerged and became a major force in technological markets with minimum support from the federal government. This happened primarily because this was a paradigm shift from industrial economy to information driven economy. Whereas the emerging bio-based industry is not a game changer; it is just a more sustainable and perhaps an economical way of steering the established chemical industry.

For example, the petrochemical supply chain of plastics and industrial chemicals are highly integrated technologically and are seamlessly connected to the global markets and economy. However, the bio-based chemical supply chain is a marriage of the chemical and biotechnology fields; the business models for growth for these two industries are divergent. This is a new development for the industry and consumers involved as stakeholders in this business, and will require a change in mindset and enormous investment both in actual funds and in intellectual efforts. Further, the new bio-based industry will be competing with well entrenched big oil and big chemical entities on commodity products with squeezed margins of profit. This is not an attractive proposition for investors, particularly when it involves huge capital investment to set up the manufacturing base for bio-based chemicals. Seriously, the government has to define the landscape and should create an ecosystem for the nascent bio-based industry to evolve and profitably integrate with the sustainable chemicals business.

In addition, the value proposition of many promising bio-based companies, such as Genomatica (1,4-butanediol and other nylon intermediaries), use the synthetic biology toolbox that effectively transforms their microbial systems as engineered strains, more properly known as genetically modified organisms (GMOs) (Davison and Lievense 2016). While the efficacy of these GMOs is proven, such as with Dupont's Sorona®, there is a general acceptance among consumers to stay away from products derived from GMOs. Similar to RFA for biofuels, the bio-based industry would need a think-tank or interest group to educate the general public and consumers on the merits of using GMOs in non-food applications.

The promising news is that some of the large chemical industries globally are working closely with promising bio-based companies by de-risking the entire value proposition with their substantial investments and managerial support. A few examples include: Dow Chemical initiating and spinning off NatureWorks' bio-based lactic acid venture and DuPont having actively worked with Genencor on 1,3 –propanediol technology via bio-based route (Davison and Lievense 2016). However, production credits, regulatory rules, and financial subsidies require a governmental framework to let this industry take off (Golden et al. 2016). Further, this framework also requires that all stakeholders are made aware that the bio-based industry has to be integrated into the large supply chains of chemicals and materials. The Biomass Crop Assistance Program (BCAP), a part of the Farm Bill, provides assurance to farmers to grow dedicated cellulosic crops in terms of monetary support and defined markets for their produce. Presently, carbon accounting does not treat bio-based carbon feedstocks as neutral; the industry hopes that the governmental framework will provide credit for bio-based feedstocks that are converted into bio-based products (Golden et al. 2016; Lee and Gerald Kutney 2017).

The US bio-based industry has vast experience over the last decade in developing building block monomers to cater to the polymer and plastic industries (Davison and Lievense 2016). Bio-based products such as lactic acid, 1,3-propanediol, and recently butanediol and succinic acid, are good examples. However, the question arises whether these companies, that are at the forefront of developing bioproducts, will seek to locate overseas to alleviate the risk and tap into the low-cost availability of raw materials and lower cost of production. For example, promising bio-based companies, such as Verdezyne, are building facilities in Malaysia, because of the latitude and financial support provided by the host government (Lee and Gerald Kutney 2017). Similarly, Amyris located its commercial plant in Brazil for farnesene production, though the venture failed for reasons other than support from local government. It is a given that bio-based industry will sustain only through effective collaborations along the value-chain. The creativity and tenacity of the emerging bio-based industry leadership lies in forging a unique path forward taking into account all the players that are involved in feedstock procurement, innovative processing and downstream technologies, and large global corporations that are end-users for their products. The scale of operation is also equally important, particularly if you are competing head on with established petrochemical players in commodity markets (Davison and Lievense 2016). Most of the petrochemical companies have been around for several decades; their plants are depreciated, and their technical and managerial teams have enormous experience in tinkering with production outputs based on market response. The new bio-based industry leaders have to consider these factors and innovatively work on their product portfolio and have effective partnership with established players in place. One way to stay in business and thrive is go after new product lines with remarkable properties rather than develop monomers that result in established products that are currently being produced via chemical industry. Another approach is to seek niche speciality products that are of high value with good margins. Particularly, the speciality bio-based products can be produced at a cost lower than the variable cost of related products from known chemical companies (Lee and Gerald Kutney 2017).

23.4 Financing and Investment Strategy for Bio-Based Industries

23.4.1 Financing Challenges for Firms in the New Bioeconomy

The modern capitalistic structure adopted by most of the matured economies is primarily driven by two factors: astute investors and rational markets. Firms need money for capital budgeting purposes, which essentially means seeking projects/new ventures that will generate returns over an acceptable time period and should have positive net present value (NPV) (Berk and DeMarzo 2017). To raise external investments, firms can go to three main sources: bank lending, market debt, and market equity (Campiglio 2016). The bank lending and debt issuance from bonds together constitutes the debt funding; the lenders are mostly interested in interest payments. Whereas the private and institutional investors are interested in having ownership based on their share value and to have a say in the management of the firm. The mix of debt and equity that a company has raised for its capital budgeting and other expenses determines the capital structure (Berk and DeMarzo 2017).

In perfect capital market conditions, the company's capital structure build-up does not affect the value of the company. Equity in a firm that also has debt outstanding is called a levered equity. The firm enhances its value by using leverage to minimize the taxes it pays. That is, the firm pays out more to its investors including interest payments to its debt holders, which enables it to raise more capital. The interest payment is not taxed and is known as a tax shield and is an advantage due to debt financing. Debt as a fraction of firm value has varied in the range from 30% to 50% for average firms (Berk and DeMarzo 2017). Debt issuance has increased recently due to falling interest rates. But firms must have earnings to pay interest payments. This will also enable debt financing at minimal risk cost of debt. The optimal debt to equity ratio is such that interest equals earnings before interest and taxes (EBIT). However, higher debt ratio increases the firm's bankruptcy chances. The debt payments have to be made to avoid bankruptcy, whereas firms have no similar obligations to pay dividends to the equity holders.

New start-ups, mid-size firms interested in renewable bioenergy and biobased products, and dedicated large corporations interested in diversifying into these emerging sectors, must compete for financing with all other opportunities available for investors and lending agencies in rational markets (Table 23.1). The biofuels and bio-based companies could be competing with their allied industries (oil, energy, and chemical industries) for technological prowess and market presence (Astolfi et al. 2008; Campiglio 2016). However, when it needs financing for growth and expansion, they have to compete with every entity seeking investments including high-growth emerging technology and healthcare companies to high-value utilities and retail industry (Hall et al. 2015). Astute individual and institutional investors will base their funding on risk and return on their investments. Similarly, the financial markets react and respond to opportunities on a rational basis, with no inherent bias or favouritism towards any particular industry. In this competitive landscape, the emerging bio-based industries face several challenges to seek financing. Primarily the challenges arise due to the requirement of high initial capital investment, a high proportion of risk and a return that can happen over a timeline of 5 years (Lee and Gerald Kutney 2017; Astolfi et al. 2008; Campiglio 2016). In addition, the bio-based industry is closely allied with chemical and polymer industry sectors, which are mostly in mergers and acquisition mode, and as such do not figure as top contenders for competitive investments. The biofuels and bio-based industry are closely connected since they use the same renewable resources, a common pretreatment and enzymatic process for sugar generation. It is only the final fermentation/bioprocessing step which defines that the product is different (Figure 23.1b). However, strategic planning and infrastructure for interoperability of these product streams for extended value-chain creation is lacking in practice. This is a substantial roadblock for the growth of an integrated bio-based economy (Golden et al. 2016; Davison and Lievense 2016).

Table 23.1 describes the companies by size that are involved in developing bio-based chemicals. Among all the companies that are involved in this sector, the start-ups face immense challenges in

Table 23.1 Growth strategies of companies by size involved in bio-based industry.

Company size	Business model	Business strategy	Financing
Start-ups	Intellectual property	Internal R&D	Private equity/venture capital
Small/medium firms	IP creator process/ technology development	R&D cooperation joint venture	Internal funding private equity debt financing
Large corporations/ multinationals	IP creator commercialization domestic and global markets	Economics of scale mergers and acquisitions	IPO institutional investing optimal debt-equity global financing

moving forward and getting adequate financing to grow (Festel Capital 2010). Since their business model is primarily focused on intellectual property development, it is important that they tie-up with a partner to provide the required financial resources; however, the timing of seeking partnership and concomitant funding is critical to avoid early burn-out or significant dilution of the start-up ownership before it matures. Small/mid-size firms typically have some existing product portfolios that are generating adequate revenues, which reduces the overall risk of the company. Consequently, the ability to raise financing for new projects is more forthcoming compared to the start-ups. Further, the business model and strategy are geared towards technology development that mitigates the process risks and assures deliverability of products (Festel Capital 2010). If the product portfolio is promising, these companies are more than likely to tie-up with large corporations that result in additional internal funding and a real interest from private equity and debt investors. Large corporations/multinationals that are involved in bio-based products are better placed among all firms by size to successfully raise capital from open financial markets (Festel Capital 2010). This is primarily because the large firms will invest a sizable amount of their retained earnings derived from income generated from other established products. Further, these large companies would have done their due diligence on every aspect of commercialization and are closely integrated with global markets. Consequently, the investors' confidence is comparatively high since the overall risk is reduced.

The presence and growth of promising bio-based industries, particularly in a period when the oil prices are historically low, show that the innovative technologies developed by these companies are robust and will deliver to the markets their products at a competitive price. It is imperative that financial institutions are made aware of the merits of this emerging industry. Low margins tend to diminish investment from these institution's in future business, so it is pertinent to take a long-term view of this technology. Government incentives and legislation should be geared to encourage investments in these technologies since it is certain that the outlook for these technologies will be more favourable from a long-term perspective (Lee and Gerald Kutney 2017; Astolfi et al. 2008).

23.4.2 The Financing of Various Steps in the Bio-Based Processes

The financing avenues that are available to a new start-up company depends on the growth strategy followed and also at various stages of technology development and commercialization. Unlike, information technology firms, the ability to generate external investments in bio-based start-ups are intensely challenging, primarily because of more unknowns and lack of assured future returns. So, entrepreneurs, academic researchers and technology developers that have an idea/concept with a large commercial viability should focus on a clear business model and plan to present all the potential risks and defined deliverables to potential investors (Lee and Gerald Kutney 2017; Wustenhagen and Menichetti 2012). This disconnect is obvious since the entrepreneurs are focused on ideas and technology; whereas, the investors have a vague understanding of the technology and their main focus is on risk and returns.

Figure 23.2 shows the financing options that are available at various stages of a company that grows from a fledging ideation to a large commercial corporation. The initial financing for a promising idea/project usually comes from founders and well-wishers. After the firm has been founded and the preliminary results are generated to prove that the idea is practical and workable, government seed funding such as small business programs of federal funding agencies, and non-profit foundations/angel investors are available. With this capital, the new firm establishes a laboratory, hires key personnel, and works on developing proof of concept of the technology. With the

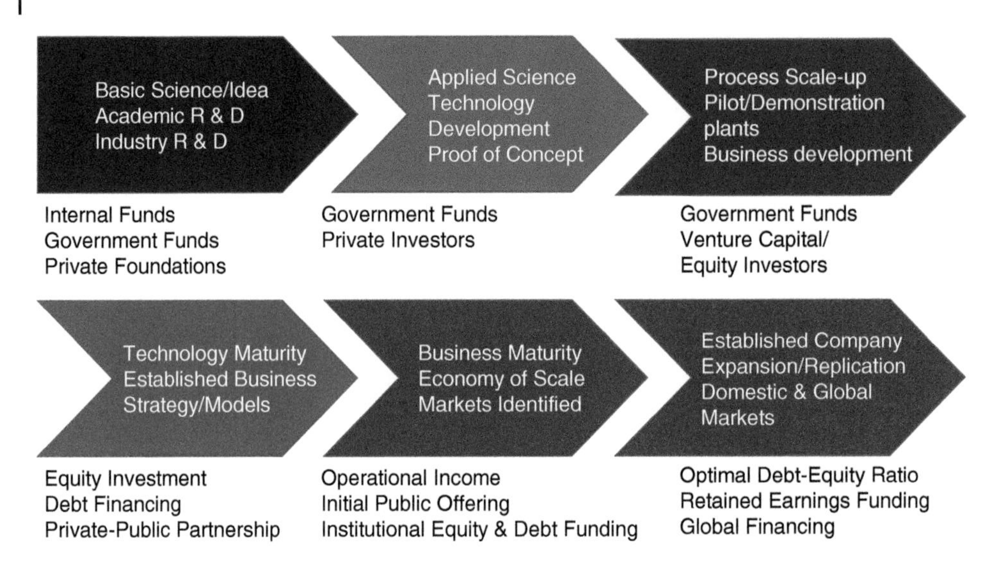

Figure 23.2 Financing avenues at sequential stages of company growth and development.

success of proof of concept, phase II funding from government agencies, early private investors and foundations are available (Lee and Gerald Kutney 2017; Wustenhagen and Menichetti 2012).

After this round of funding, the firm scales up the technology and is sharply focused on proving the technology at pilot/demonstration plants. At this stage, serious investors including venture capitalists (VCs) get involved and the first round of VC funding, known as Series A becomes the established financing method. Some large governmental grants from Departments of Energy and Agriculture in the United States are available at this stage to speed up the commercialization process. Based on VC recommendations, the firm may seek a strategic industry partner that will actively complement their scale-up activities. Since the firm is primarily focused on one aspect of the value chain, such as dedicated biomass feedstock, conversion process technology, or development of a bio-based product, the partnership will be sought at a complementary position in the value-chain. The external financing options open up and the firm can choose the investors it wants in the commercialization plan. Also, the first order techno-economic analysis (TEA) is done at this stage and second round of external financing from VC are obtained. The founders should focus on VC firms that specialize in bio-based industry and/or clean technologies. Once the technology has matured, the founders should focus more on business attributes and hard numbers of interest to the investors; put less emphasis on technology details, social issues and visionary aspects. At this stage, bank loans and other forms of debt financing are available. The firm now grows significantly; having several divisions considering various aspects of business development.

As explained in Section 23.3.2, the emerging bio-based industry has many challenges in terms of high initial capital, long gestation period before returns, changes in government policy, global supply, and demand of crude oil. While the emerging industry is aware of all these factors; most of the cellulosic biofuels industry ignored the feedstock variability factor and it impaired their further commercial efforts. Serious efforts to determine the return on investments are made and the claims of the emerging firm is independently verified by outside validators. After the verification process is completed, the technology has been thoroughly demonstrated with replicable results and productivity, more equity and debt financing are available. These investors include investment banks, and institutional and private equity holders. Several more rounds of independent evaluations of process and business attributes are performed. With the availability of substantial financing, the firm performs due diligence for setting up an economy of scale industrial plant with focus on producing the biobased product and generate early revenue streams by supplying the product to potential large end-users. The firm may choose the option of making an initial public offering (IPO) or retain private ownership. If the VC and other investors want to harvest their investments, the firm will go public with an accompanying large influx of capital. Successful firms will use these financing resources to develop strong value propositions with suppliers and end users of their product. Further, they may tie-up with large global companies to expand their markets and presence. In addition, they may seek other firms to acquire or merge with to become a large multinational with avenues open for both debt and equity financing from all over the world to sustain their high growth and diversification plans. The maturity of the industry goes in parallel with the acceptability of bio-based technologies as a way forward for the new economy.

23.5 Perspectives and Sustainable Financing Approach – Change in Wall Street Mindset in the Valuation of Bio-Based Industries

There is a global convergence on the need to address the climate change that has happened due to the industrial and information growth all over the world in the last two centuries. The world population will reach over 9 billion by 2050 and the current resources from nature that were used to propel the economic engine will be severely stressed; particularly it is predicted that the water scarcity will be evident in large swathes of tropical nations if the global social and environmental issues are not addressed. The modern global economy is predominantly based on Keynesian economic principles, which calls for governmental intervention in the form of fiscal and monetary policies to stabilize the upheavals over the business cycle. While this has helped to prevent events like the Great Depression from happening again; the focus has essentially been on economic stability of markets ignoring the consequences of this model on social and environmental parameters affecting the market regions.

In the last few decades the topic of sustainability has emerged and has been discussed in global climate and economic forums as a crucial factor to be incorporated in markets and economies (McAfee 2016; Jeucken 2001). The triple bottom line of sustainable corporate governance, which includes deliverables incorporating social, environmental and economic metrics, is widely discussed and most of the firms have adopted corporate social responsibility as their core functioning (Jeucken 2001; Sanders and Wood 2015). We have a situation in the world wherein the emerging economies, most of them rich in natural resources, are being nudged by the global climate control agencies to do business in a new ecosystem comprising inclusion of social and environmental metrics, while the old economy thrived in a classical capitalistic framework during their heyday of high growth period. Obviously, the resistance from these emerging economies is palpable; since they are driven by an agenda to maximize their GDP and become dominant players in the global stage. While this is understandable based on the theory that nations place their self-interest ahead of other considerations affecting the global well-being; however, when the same global well-being is under a concrete threat of being destroyed, we have no choice but to take preventive measures.

The awareness of social, environmental and economic issues related to any business and technology has increased; however, the functioning of financial institutions and banking sectors are still based on the capitalistic structure of maximizing the shareholders' wealth and return on investments (Jeucken 2001; Sanders and Wood 2015; Liesen et al. 2013). Though concepts such as green banking and global agencies' green funds are being touted as a mechanism to set the terms for the coming sustainable economy; however, the mind-set of 'Wall Street' investors and bankers are focused on profits and growth of the companies in their portfolios. The metrics, such as price–earnings ratio, return on equity and assets, discounted cash flows, they use to perform valuation of projects and firms are designed for investors to stay or pull from equity portfolios. Similarly, the debt holders are motivated by the interest payments and act as catalyst for premature bankruptcy of firms during challenging times. Even the time-tested NPV method of project evaluation is based on the perceived cost of capital and estimated cash flows over the time period of the project, and these are determined on a purely economic basis with no costs attached to the social and environmental issues involved in the project. Liesen et al. 2013 has proposed a net present sustainable value (NPSV) method to include social and environmental aspects and the project success is closely tied to the sustainable goals of the project and firm. Further, concepts such as sustainable financing and green banking will change the landscape of business perception and evaluation (Jeucken 2001). These are welcome approaches, and unless the methods of valuation of firms change among the investor's community to include concrete sustainable factors, we will be stuck in the current model of a divergent market-driven global economy.

Bioenergy and bio-based industry is certainly poised to augment substantially the current energy and chemical conglomerates in their approaches to streamline their existing and future product lines that will be firmly grounded on sustainability goals. The bioenergy and bio-based industry is hampered by the current method of valuation and financing followed by investment bankers and analysts. If sustainable financing approaches are adopted, the emerging industry will get the deserved investments for its growth to sustain a new bioeconomy. It is understandable that most of the bio-based processes that are using biomass feedstocks involve high-energy chemical-intensive pretreatment and sugar generation steps, which actually make the entire bio-process unsustainable since these steps involve energy generated from non-renewable fuels. An integrated sustainable bioprocessing approach that takes a system approach right from the cropping methods of biomass to the point of how the waste streams are disposed of is the need of the hour.

The perception of the investment bankers and institutional investors in their evaluation of a bio-based technology is rooted in the return on investments. The raw materials cost for a typical bioenergy and bio-based technology can be as high as 50% of the entire product cost; whereas most of the mainstream oil and chemical industries operate with a different cost structure that are significantly lower on raw material costs. As a result, the vagaries of the weather and the demands of the suppliers will dictate the productivity of a bio-based industry, which, from an investment perspective, alludes to elevated risk and drives them to other safer investments. The stakeholders of the entire value-chain of bio-based industry should seriously work on securing a sustainable supply of raw materials at reduced costs. If green banking and sustainable finance methods that factor in environmental risks and social burdens are applied towards commercialization of bio-based technologies, then real and meaningful changes can happen for the new global economy.

Acknowledgements

The author is grateful to his former graduate students Dr. JungEun Lee for her help in getting literature material and Dr. Sai Praneeth Thota for his help in preparing the illustrations. The author also appreciates help from Drs. Nasib Qureshi and Alain Vertes for their constant support and encouragement.

References

Astolfi, P., Baron, S., and Small, M.J. (2008). Financing renewable energy. *Commercial Lending Review* 2008, 23: 3–8.

Berk, J.B. and DeMarzo, P.M. (2017). *Corporate Finance: The Core*. Boston, USA: Pearson Publications. ISBN: 9780134083278.

Campiglio, E. (2016). Beyond carbon pricing: the role of banking and monetary policy in financing the transition to a low-carbon economy. *Ecological Economics* 121: 220–230.

Chen, M., Smith, P.M., and Wolcott, M.P. (2016). U.S. biofuels industry: a critical review of opportunities and challenges. *BioProducts Business* 1 (4): 42–59.

Davison, B.H. and Lievense, J.C. (2016). Technology challenges and opportunities. SBE Supplement: Commercializing Industrial Biotechnology. *Chemical Engineering Progress* 2016: 35–42.

Golden JS, Handfield RB, Daystar J, McConnell TE (2016). An Economic Impact Analysis of the US Bio-based Products Industry. United States Department of Agriculture. Accessed October 2017. https://www.biopreferred.gov/BPResources/files/Bio-basedProductsEconomicAnalysis2016.pdf

Guragain, Y.N., Ganesh, K.M., Bansal, S. et al. (2014). Low-lignin mutant biomass resources: effect of compositional changes on ethanol yield. *Industrial Crops and Products* 61: 1–8.

Guragain, Y.N., Probst, K.V., and Vadlani, P.V. (2016). Fuel alcohol production. In: *Encyclopedia of Food Grains*, 2e (eds. C. Wrigley, H. Corke, K. Seetharaman and J. Faubion), 235–244. Oxford: Academic Press.

Hall, S., Foxon, T.J., and Bolton, R. (2015). Investing in low-carbon transitions: energy finance as an adaptive market. *Climate Policy* https://doi.org/10.1080/14693062.2015.1094731.

Hofstrand, D. 2016a. "Corn-Ethanol Profitability Chart." Agricultural Marketing Resource Center, Iowa State University. http://www.agmrc.org/renewable_energy/ethanol/corn-ethanol-profitability.

International Monetary Fund. Global Financial Stability Report. Getting the Policy Mix Right. World economic and financial surveys, 0258-7440. April 2017. ISBN 978-1-47559-080-7 (PDF)

Jessen, H. 2013a. "Corn Oil Extraction Examined from Several Angles during Webinar." Ethanol Producer Magazine, November 15, 2013. http://ethanolproducer.com/articles/10475/corn-oil-extraction-examined-from-several-angles-during-webinar.

Jeucken, M. (2001). *Sustainable Finance and Banking: The Financial Sector and the Future of the Planet*. Sterling, VA, USA: Earthscan Publications Ltd 20166-2012. ISBN 1-85383-766-0.

Lee W, Gerald Kutney G (2017) Financing Bioeconomy. Accessed November 2017. http://www.biofuelsdigest.com/bdigest/2017/09/25/financing-bioeconomy-ventures/#_ftn1

Liesen, A., Figge, F., and Hahn, T. (2013). Net present sustainable value: a new approach to sustainable investment appraisal. *Strategic Change* 22: 175–189.

McAfee, K. (2016). Green economy and carbon markets for conservation and development: a critical review. *International Environmental Agreements* 16: 333–353.

OECD Medium and long-term scenarios for global growth and imbalances, *OECD Economic Outlook* 2012 2012 1, OECD Publishing, Paris, https://doi.org/10.1787/eco_outlook-v2012-1-44-en.

Sanders, N.R. and Wood, J.D. (2015). *Foundations of Sustainable Business: Theory, Functions, and Strategy*. USA: Wiley. ISBN: 9781118441046.

Schoenmaker, D. (2013). The financial trilemma. *Economic Letters* 111 (1): 57–59.

U.S. Department of Agriculture Economic Research Service (USDA-ERS). 2017. "Feed Grains Database." Accessed October 2017. http://www.ers.usda.gov/data-products/feed-grains-database.aspx.

Wustenhagen, R. and Menichetti, E. (2012). Strategic choices for renewable energy investment: conceptual framework and opportunities for further research. *Energy Policy* 40: 1–10.

U.S. Environmental Protection Agency (EPA). (2017). Renewable Fuel Standard Program. Standards for 2018 and Biomass-Based Diesel Volumes for 2019. Accessed October 12, 2017 https://www.gpo.gov/fdsys/pkg/FR-2017-07-21/pdf/2017-15320.pdf

Bio-based Industries Consortium (2017). The BBI JU – An Institutional PPP supporting the Bioeconomy Strategy. Accessed October 2017. http://biconsortium.eu/sites/biconsortium.eu/files/downloads/BIC_Impact_of_BBI_JU_June_2017.pdf

Cournède B, Denk O, Hoeller P. (2015) Finance and Inclusive Growth, OECD Economic Policy Paper 14, OECD Publishing

Ethanol Industry Outlook (2017) Building Partnerships Growing Markets, 2017. Renewable Fuels Association. http://www.ethanolrfa.org/wp-content/uploads/2017/02/Ethanol-Industry-Outlook-2017.pdf

Festel Capital (2010). Industry Structure and Business Models for Industrial Biotechnology, Research Methodology and Results for Discussion. OECD Workshop on the Outlook on Industrial Biotechnology Vienna, January 14, 2010. Accessed November 2017. https://www.oecd.org/sti/biotech/44776744.pdf

International Energy Agency Publications (2017). Market Report Series: Renewables 2017, Analysis and Forecasts to 2022. ISBN PDF: 978-92-64-28187-5, PRINT: 978-92-64-28185-1

Warner E, Moriarty K, Lewis J, Milbrandt A, Schwab A (2017). 2015 Bioenergy Market Report. United States. doi:https://doi.org/10.2172/1345716. https://www.osti.gov/servlets/purl/1345716

24

Corporate Social Responsibility and Corporate Sustainability as Forces of Change

Asutosh T. Yagnik

AdSidera Ltd., London, SW1Y 4NW, United Kingdom
Institute for Institute for Strategy, Resilience & Security, University College London, London WC1E 6BT, United Kingdom

CHAPTER MENU

24.1 Introduction

Over the last two decades the terms 'corporate social responsibility' (CSR) and 'sustainability/corporate sustainability' (CS) have become increasingly prevalent in discussions in business, academia, and amongst the wider public. In fact, so much so that the popular media and social media platforms all have discussions on these topics on a daily basis, as they relate to companies, governments, and society in general. The two terms have become *en vogue* buzz words businesses use to help manage their brand equity with the public and are often used interchangeably. But different questions remain such as 'what exactly does CSR mean?', 'why and how is it implemented by corporations?' and 'is there any difference between CSR and CS?'. These points are examined in this chapter of the book.

24.2 Corporate Social Responsibility (CSR)

To better understand CSR and best practice in its implementation, the first step is to lay out the fundamentals of the topic, that is, its definition and its key concepts.

24.2.1 What Is CSR?

It is not straightforward to define CSR in the modern context since there are a plethora of different definitions and no commonly accepted international description. In fact, one study in 2006 found that there were at least as many as 37 widely accepted definitions of CSR (Dahlsrud 2006). Some examples are listed in Table 24.1.

Green Energy to Sustainability: Strategies for Global Industries, First Edition.
Edited by Alain A. Vertès, Nasib Qureshi, Hans P. Blaschek and Hideaki Yukawa.

Table 24.1 Some examples of CSR definitions from different institutions.

Institution	Definition of CSR
World Business Council for Sustainable Development, 2000 (Holme and Watts 2000)	'the commitment of business to contribute to sustainable economic development, working with employees, their families, the local community and society at large to improve their quality of life'.
The International Labour Organization (ILO 2009)	'a way in which enterprises give consideration to the impact of their operations on society and affirm their principles and values both in their own internal methods and processes and in their interaction with other actors. CSR is a voluntary, enterprise-driven initiative and refers to activities that are considered to exceed compliance with the law'.
European Commissio, 2011 (European Commission 2011)	'the responsibility of enterprises for their impacts on society' and that 'enterprises should have in place a process to integrate social, environmental, ethical, human rights and consumer concerns into their business operations and core strategy in close collaboration with their stakeholders, with the aim of: • maximizing the creation of shared value for their owners/shareholders and for their other stakeholders and society at large; • identifying, preventing, and mitigating their possible adverse impacts'
International Organization for Standardization (ISO), 2006(ISO 2010)	'responsibility of an organization for the impacts of its decisions and activities on society and the environment, through transparent and ethical behaviour that: • contributes to sustainable development, including health and the welfare of society; • takes into account the expectations of stakeholders; • is in compliance with applicable law and consistent with international norms of behaviour; and • is integrated throughout the organization and practised in its relationships'

The large number of definitions that are currently in use make it difficult to precisely define what CSR is. However, there are clearly common elements that apply to any responsible business:

1. firstly, the business is aware of and carefully considers its impact on society and the environment
2. secondly, the business maintains a transparent and accountable interaction with all its stakeholders
3. finally, the business' core strategy, operations, and organizational culture reflect its commitment to the first two elements, while ensuring it continues to make a profit.

In 1962, Milton Friedman defined CSR in his book *Capitalism and Freedom* as (Friedman 1962):

> There is one and only one social responsibility of business, to use its resources and engage in activities designed to increase its profits so long as it stays within the rules of the game, which is to say engages in open and free competition without deception or fraud.

This outlook of 'profit by any legal means necessary' had, and in some cases continues to have, many proponents. However, it is questionable whether CSR has ever been about just doing only what is required by law, particularly at a time when the consequences of global over-consumption are becoming clear. CSR, as shown by some of the definitions in Table 24.1, is about businesses going above and beyond what is legally required to meet their commitments to the environment and society, whilst making a profit. Since the environmental and societal needs are constantly changing

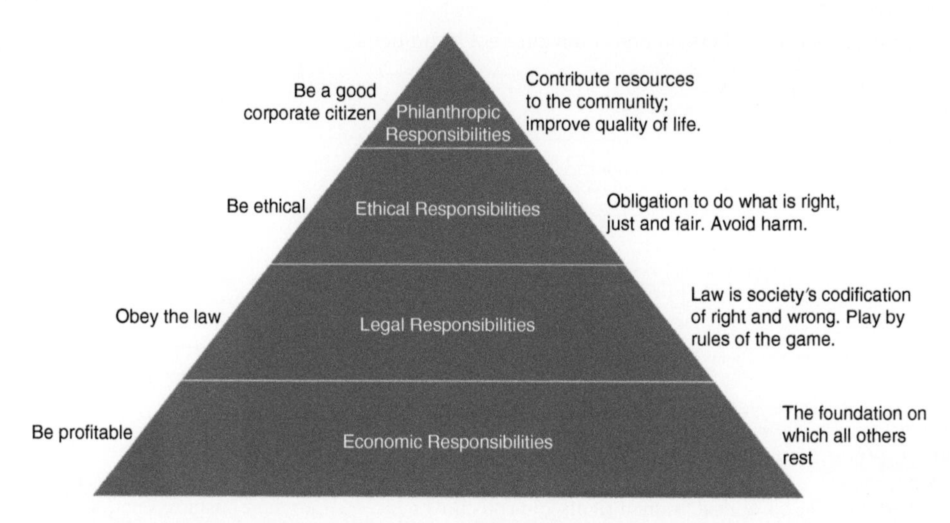

Figure 24.1 Carroll's pyramid of CSR. Source: Adapted from (Carroll 1991) and (Visser 2005) with permission.

over time, CSR is constantly evolving and innovating to preserve the delicate balance between a business' profit and its commitments to the other two elements.

24.2.2 Conceptual Models of CSR

In 1991, during its early origins as an integral part of the business model, Archie Carroll stated that CSR involved the conduct of a business so that it was economically profitable, law-abiding, ethical, and socially supportive. This was depicted as an organization's CSR pyramid (Figure 24.1) (Carroll 1991) (Visser 2005)

Later on, this was refined into a three domain model in the shape of a Venn diagram, where the previous four categories of CSR were reduced to only three: economic, legal, and ethical, with the philanthropic category having been integrated into the ethical category (Figure 24.2) (Schwartz and Carroll 2003).

24.2.3 The History and Evolving Nature of CSR

The constant evolution of CSR since its inception has been in response to the various challenges and commitments businesses face. It is worth understanding the history of CSR in more detail prior to an examination of its future directions.

24.2.3.1 The Four Eras of Early CSR

The recognition of a link between business, society, and the environment is an ancient concept going back to the origins of trade itself. For example, Indian Vedic literatures, some dating back as far as 5000 years and written in the Sanskrit language, such as the Vedas, Upanishads, Mahabharata Valmiki Ramayana and the Puranas, (Radhakrishnan and Moore 1967) (Mani 1998) viewed business as a legitimate and an integral part of society, but which essentially should create wealth for the society through the right means of action (Muniapan and Dass 2008) (Chavan 2012). The phrase *sarva loka hitam* in the Vedic literatures referred to the 'well-being of all peoples [stakeholders]', suggesting an ethical and social responsibility system must have been fundamental and

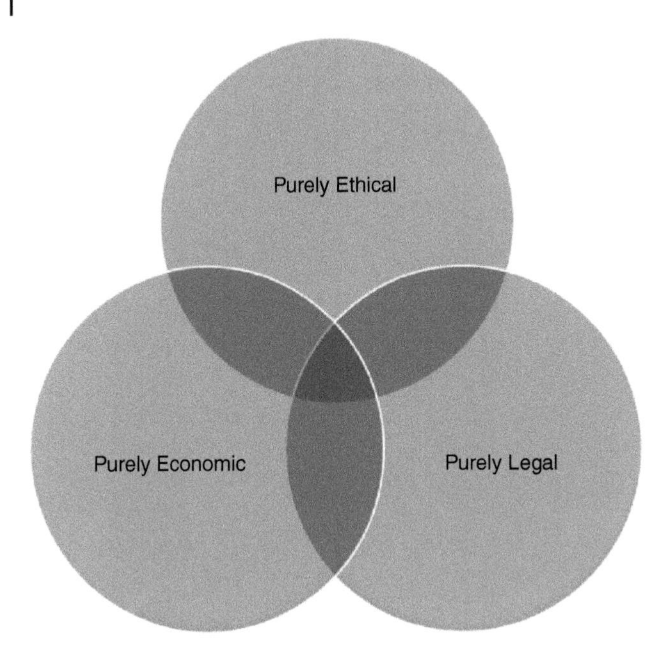

Figure 24.2 The three-domain model of CSR.

functional in business undertakings. In addition, the concept of *dharma*, roughly translating to 'that duty which helps the welfare of all living beings' or in other words, 'social responsibilities', is one of the central tenets of the Bhagavad Gita. Along with *dharma* or simply duty, the Bhagavad Gita also focuses on *karma* which are the actions of a person and it is this duality of the two which leads to the concept of a personal or individual social responsibility, which applies to kings, communities, businesses and individuals, and forms the basis of a wider cultural focus on social responsibility, including CSR. Such concepts of social responsibility are also seen in ancient Mesopotamia around 1700 BCE, where King Hammurabi introduced a code in which builders, innkeepers or farmers were put to death if their negligence caused the deaths of local citizens (Peattie n.d.) (Asongu 2007). Furthermore, in ancient Rome, senators often complained about the failure of local businessmen to contribute sufficient taxes to fund their military campaigns, while in 1622 dissatisfied shareholders in the Dutch East India Company issued pamphlets complaining about management secrecy and 'self-enrichment' (Peattie n.d.) (Asongu 2007).

The idea of the business having some sort of responsibility towards local people and the environment carried on into the eighteenth and nineteenth centuries, when rich industrialists often made huge philanthropic gestures during the growth of their businesses. For example, Sir Titus Salt, who opened a new mill with a model village for his workers in Saltaire, near Bradford in 1853 (Porritt 2012). Similarly, in Bournville and New Earswick, the Cadbury and Rowntree families respectively built parks, hospitals, museums, public baths and reading rooms for the benefit of factory workers. Then, as now, it was sometimes difficult to differentiate what companies were doing for business reasons versus social ones. One way of looking at it is that the companies were doing these things for social consciousness reasons, to help fulfil the needs of employees and make them better contributing members of society. The other, more cynical way of interpreting it is that having all those facilities near to the company's production site was simply a way of ensuring that the workers were more productive and in the long term that benefitted the organization financially. Of course, the two things are not mutually exclusive and this is well demonstrated by the philanthropy of William H. Lever, founder of Lever Brothers, now Unilever (Unilever, London, UK), who used his wealth to build Port Sunlight, a village for Lever Brothers' employees. He started pension schemes, unemployment and sickness benefits, work canteens and the concept of the 8-hour day (Porritt 2012). Similarly, in the late 1800s, major industrialists in the USA such as John H. Patterson, Cornelius Vanderbilt and John D. Rockefeller were making a name for themselves for their social philanthropy (Carroll 2008).

This philosophy of fulfilling social responsibility through philanthropy is not just a Western phenomenon. In 1892 in India, Jamsetji Tata, founder of the now global $100 billion Tata Group (Tata Group, Mumbai, India) business empire, established the JN Tata Endowment to enable Indian students, regardless of caste or creed, to pursue higher studies in England. Although not specifically stated, the obvious quid pro quo was that hopefully the cream of these students would then return to help the growth of the Tata Group businesses.

Notwithstanding these very initial concepts of CSR, it was not until the early 1950s that the modern era of CSR can be said to begin more formally. In his review of CSR, one of the most prolific authors on the topic, Archie B. Carroll (Carroll 2008), states that the modern era of CSR began in 1953 with the publication of the book *Social Responsibilities of the Businessman* by Howard Bowen (Bowen 1953). Bowen is often seen as the 'father of corporate social responsibility' and suggested an initial definition of the social responsibilities of businessmen (businesswomen were not commonplace at the time of the book and not acknowledged in formal writings) as: 'It [Social Responsibility] refers to the obligations of businessmen to pursue those policies, to make those decisions, or to follow those lines of action which are desirable in terms of the objectives and values of our society'. The 1950s was a period of changing attitudes, with business executives learning to get comfortable with CSR talk, but there were very few corporate actions, beyond philanthropy. In his book, Bowen proposed a number of changes for improving business responsiveness to the growing social concern, including changes in the composition of boards of directors, greater representation of the social viewpoint in management, use of the social audit, social education of business managers, development of business codes of conduct, and further research in the social sciences (Bowen 1953).

The early and mid-1960s saw the start of a wider awareness in the business community about CSR and an increased academic interest in the topic with many more scholars writing on the topic. However, philanthropy remained one of the most widely seen manifestations of CSR and it was not until the end of 1960s that it started to manifest in employee working conditions, industrial relations, personnel policies, customer relations and stockholder relations (Heald 1970). The early 1970s saw companies start to address CSR issues such as minority hiring and training, civil rights, pollution control, charities, community affairs, urban renewal and contributions to the arts and education (Eilbirt and Parket 1973) (Holmes 1976). The 1960s and 1970s also saw the rise of an increasing pressure on businesses to be held accountable for their actions which negatively impacted the environment. The seminal book *Silent Spring*, by conservationist Rachel Carson shone a spotlight on the harmful effects of toxic pesticides on nature, wildlife, and humans (Carson 1962). This period also saw the formation and rise of several non-governmental organizations (NGOs) including WWF in 1961 (World Wide Fund for Nature, Gland Switzerland), Friends of the Earth in 1971 (Friends of the Earth International, Amsterdam, Netherlands), and Greenpeace in 1971 (Greenpeace, Vancouver, Canada), as well as a growing activism movement that drew public attention to the impact of businesses on people and the environment. Added to increasing awareness of social and environmental awareness, the pressure from such activist movements resulted in CSR becoming more embedded into mainstream management functions such as board strategy, organizational structure, planning and forecasting, and social performance policies. In the US this resulted in legislative initiatives during the 1970s mandating companies to create organizational mechanisms for complying with federal laws dealing with the environment, product safety, employment discrimination and worker safety (Carroll 1977).

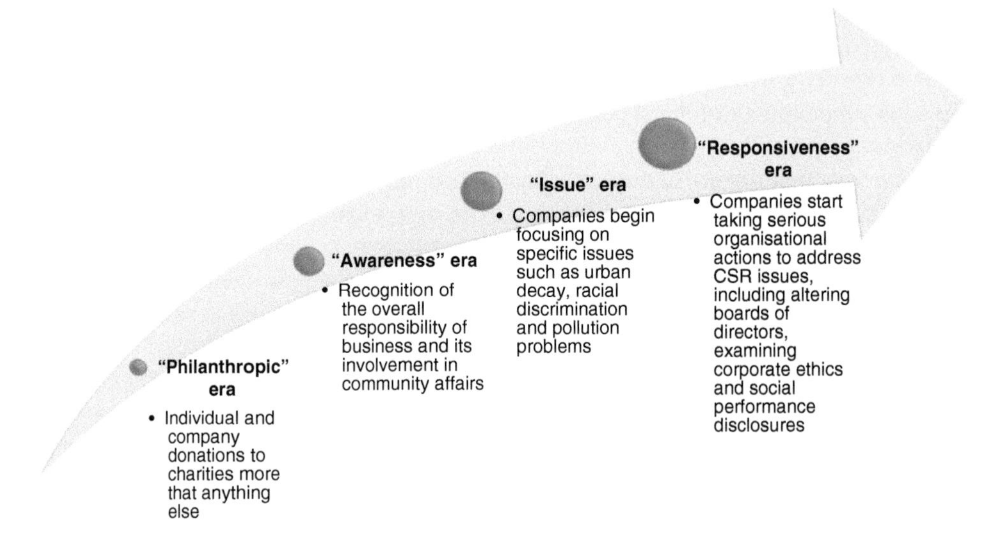

Figure 24.3 Four eras of the evolution of CSR in business, from late 1800s to early 1980s. Source: original figure based on text from source (Murphy 1978).

This period between the late 1800s to mid-1970s was neatly summarized by Patrick Murphy into four eras of CSR (Figure 24.3) (Murphy 1978). From the 1950s on and during those early years, CSR was commonly referred as just social responsibility, mainly because the rise of the modern corporate's dominance in business had not yet fully occurred. Nonetheless, it formed the foundations of modern CSR and it is really from the late 1970s onwards that companies began taking serious management and organizational actions to address CSR issues, including employing and changing appropriate boards of directors, examining corporate ethics and publishing social performance disclosures.

24.2.3.2 The Impact of Environmental Disasters and Globalization in the 1980s to 2000s

The 1970s had already seen an awakening of social awareness of how corporations were starting to negatively affect the environment, wildlife, and humans. This awareness was further increased in the 1980s following a few large-scale, high profile, environmental disasters such as the Bhopal gas tragedy in India in 1984 (Broughton 2005), the Chernobyl nuclear disaster in 1986 (World Nuclear Assocication 2018) and the Exxon Valdez oil spill in 1989 (Holba and Woods 2019), all of which re-emphasized the potentially catastrophic impacts of industry.

In addition to this 'global social awakening', the 1970s also saw high rates of inflation and unemployment all around the globe, particularly in the US and UK. In an effort to combat this 'stagflation', as it became known, US President Ronald Reagan and UK Prime Minister Margaret Thatcher implemented policies in the 1980s to support more private enterprise, reduce the power of labour unions, increase deregulation of industries, reduce taxes, open up freer markets and introduce more incentives for private investment. These policies formed part of 'Reaganomics' and 'Thatcherism', and whilst the benefits are still debated, supporters point to the end of stagflation, stronger gross domestic product (GDP) growth and an entrepreneur revolution in the decades that followed (D'Souza 1997). However, critics of those policies pointed to a widened income gap between the rich and poor and an atmosphere of greed, as well as higher national debts (Weisman 2004). Irrespective of the pros and cons of these policies, one of the results was a push to promote free trade globally, opening up business's access to cheaper resources and new markets around the world and increasing corporate growth. However, one of the negative consequences of this globalization of supply chains was the mass exploitation of people and resources in developing countries: corporations were often found to be paying a pittance to local people, often working in appalling conditions and without regard to environmental impact of their industrial processes. There have been several high-profile examples of such corporate exploitation and subsequent public activism campaigns in the 1990s and 2000s, but probably the 'grandfather' case study of these is the one involving Nike (Nike Inc., Beaverton, Oregon, US). In 1992, Jeffrey Ballinger published an exposé of the labour conditions in Nike's supply chains, specifically in Indonesia, where workers who worked for a Nike subcontractor earned the equivalent for 14 US cents an hour, less than the minimum wage, and documented other abuses (Ballinger 1992). Initially Nike responded that it could not be held responsible for conditions in factories it did not own or for sub-contractors. However, the protests and media reports continued and in 1996 Life magazine published a story headlined 'Six Cents an Hour', with a photo of a 12 year-old Pakistani boy sewing Nike footballs (Schanberg 1996). There followed a relentless cycle of media revelations showing bad CSR: allegations of 'sweatshops' – poor working conditions, low pay and child labour, followed by fervent activism globally, particularly on US college campuses. This cycle of allegations and activism reached a hiatus in 1998 and took its toll: Nike started facing weaker demand for its products, its share price fell while that of its biggest competitor Reebok (Reebok, Boston, Massachusetts, US) increased, and it had to lay off workers.

The company began to realize it needed to change if it was going to reverse the trend and the negative impact on its commercial sales, and CEO Phil Knight promised reform in a 1998 speech: 'the Nike product has become synonymous with slave wages, forced overtime and arbitrary abuse', he said, 'I truly believe that the American consumer does not want to buy products made in abusive conditions' (Cushman Jr 1998). Speaking before the American National Press Club in Washington DC, Knight promised journalists and trade union activists that he would personally ensure six main improvements:

1. all Nike shoe factories would meet US Occupational Safety and Health Administration (OSHA) air quality standards
2. the minimum age would be raised to 18 for workers in Nike shoe factories and 16 for those in for clothing factories
3. Nike would include non-governmental organizations in factory monitoring, and the company would make inspection results public
4. Nike would expand its worker education programme, with free secondary-school equivalent courses
5. a loan programme would be expanded to benefit 4000 families in Vietnam, Indonesia, Pakistan, and Thailand
6. research on responsible business practices would be funded at four universities.

It should be stressed that Nike was not the only or worst company to use sweatshops, but it was the poster boy for activism in that period – the one everybody knew. Many other companies have since been accused of, or found to be, using similar practises, including Adidas (Adidas AG, Herzogenaurach, Germany), H&M (Hennes & Mauritz AB, Stockholm, Sweden), Walmart (Walmart Inc., Bentonville, Arkansas, US), Gap (Gap Inc., San Francisco, California, US) and Zara (Zara S.A., Arteixo, Spain), to name a few.

24.2.3.3 From the 2000s to Today: The Further Evolution of CSR and its Relationship to Brand Management

By the early 2000s it was clear that issues related to global supply-chain management were a real and quantifiable issue for Nike and other manufacturers, with the potential to undermine the value of their brands. Three years after Nike's CEO speech promising six improvements, a report was published on their progress entitled 'Still waiting for Nike to do it' (Connor 2001). This highlighted that although much had been promised, little had changed in the three years on the ground: the company had arranged only one factory audit by one non-profit organization; despite promising to meet the US OSHA air quality standards in all its factories, Nike gave factory owners advance notice of testing giving them the opportunity to change chemical use to minimize toxic emissions on the day the test was conducted; and evidence continued to emerge of young people under the age of 16 employed in Nike contract factories. The report showed many more examples of Nike failing to live up to the promises its CEO made in 1998 and its negative impact on the company's brand and financial bottom line was significant. Although it was slow in coming, Nike eventually set up an extensive and expensive system for monitoring and remedying factory conditions in its supply chain – and the rest of the footwear and apparel industry followed.

Between 2002 and 2004 Nike performed approximately 600 factory audits on its suppliers, including repeat visits to problematic sites (Nisen 2013), and in 2005 it became the first company in its industry to publish details of all the factories it works with as well as a full 108-page report on conditions and pay in its factories, acknowledging widespread issues, particularly in its south Asian factories (Nisen 2013) (Teather 2005). This unprecedented level of transparency and admission of problems, as well as evidence that the company was trying hard to improve on its CSR and much had changed, worked wonders for the company's brand equity. The company slowly started being seen as 'cool' again and as genuinely trying to make a difference on global development issues. The impact on its financial bottom line reflected this, with an increase in sales and profits.

The case of Nike is highlighted as an example because during the 1990s and early 2000s it was one of the most high-profile continuous cases involving CSR – both in terms of shortcomings and ultimately how the organization learnt from its mistakes to become a stronger brand. Today, Nike is a corporate-sustainability leader and its CSR compliance program is accredited by the US Fair Labor Association (FLA), where it makes public all inspection reports from its contract factories (Fair Labor Association 2019). On the FLA website, Nike is quoted as saying, 'Our greatest responsibility as a global company is to play a role in bringing about positive, systemic change for workers within our supply chain and in the industry. We're looking end-to-end, from the first phase of our product creation process to the impacts of our decisions on the lives of workers in the factories that bring our product to life'.

It is not just the fashion or apparel brands that have been found to be using sweatshops. Other examples of companies that have all at one time or another been found to employ child labourers include:

- The chocolate manufacturers such as Cadburys (Cadbury, Uxbridge, UK), Hershey's (The Hershey Company, Derry Township, Pennsylvania, US), Mars (Mars Inc., McLean, Virginia, US) and Nestlé (Nestlé S.A., Vevey, Switzerland), in their cocoa production (Porritt 2012) (Sandler Clarke 2015) (Nieburg 2018) (BBC Panorama 2010). The problem was highlighted in a number of documentaries and reports in early 2000, leading to an international agreement, the Harkin–Engel Protocol, sometimes referred to as the Cocoa Protocol, being signed in 2001 (Chocolate Manufacturers Association 2001). This agreement was aimed at ending the worst forms of child labour and trafficking related to the production of cocoa and pledged to reduce child labour in Ivory Coast and Ghana by 70%. The protocol was extended in 2005, 2008, 2010 and, since the target had not been met as of late 2015, the deadline was extended to 2020.
- Electronics giants Apple (Apple Inc., Cupertino, California, US) and Microsoft (Microsoft Corporation, Redmond, Washington, US) were also found to be failing in their CSR during the production of their products. In the case of Apple, an internal audit in 2013, it found over 100 child

labourers were employed at 11 factories making Apple products (Garside 2013), the company then admitted four years later to using child labour to build its latest iPhone X (Richards 2017), specifically students aged between 17 and 19 years-old being forced to participate in 'internships' to help the company meet its manufacturing deadlines. Microsoft on the other hand was accused by Amnesty International (Amnesty International, London, UK) of using child labour in the Democratic Republic of Congo for the extraction of cobalt, a key component of laptops and mobile phones (Phillpott 2019). Microsoft moved quickly to address these concerns, meaning Amnesty International has suspended its action for the time being (Amnesty International 2018).

- Cigarette manufacturers Phillip Morris (Philip Morris International Inc., New York, New York, US), British American Tobacco (BAT) (British American Tobacco Plc., London, UK) and Japan Tobacco International (Japan Tobacco International, Geneva, Switzerland), long seen as public enemies due to the health impacts of their products anyway. In 2010 Phillip Morris admitted that children as young as 10 years-old were forced into labour on its contracted tobacco farms in Kazakhstan, with allegations by the Human Rights Watch (Human Rights Watch, New York, New York, US) suggesting that passports of many migrants had been confiscated in order to prevent them escaping (Walker 2010). BAT, the world's largest listed tobacco company, found itself at the centre of a report in 2016 which found three main problems in its Bangladesh tobacco fields: child labour was widespread, health ramifications from exposure to pesticides and nicotine were common, and many farmers were trapped in over indebtedness (Swedwatch 2016). Despite evidence to the contrary, the company continues to refute the claims and speaking at its AGM in April 2018, chairman Richard Burrows claimed that BAT has 'not got any questions to answer in respect of these issues (Phillpott 2019)'. Meanwhile, unlike its competitors, Japan Tobacco International buys the tobacco leaf directly from the producers, meaning it has a key say in the prices that are set and the conditions that workers operate in. In a June 2018 report, researchers found that children work on the tobacco farms, with the justification that 'handling dried tobacco is not considered as hazardous as working with wet leaves' (Boseley and Levene 2018). These tobacco companies have clearly not reacted enough to past problems relating to CSR and further allegations have arisen about 'rampant' use of child labour in contracted farms in Malawi (Boseley 2018). Moreover, some of them even consider it not only acceptable, but even beneficial, for children between the ages of 13 and 15 to work on the farms, provided it is 'light work, permitted by local law' (Boseley and Levene 2018).

Although the focus of CSR failures in the above examples is predominantly on the social aspects of a company's responsibilities, there are also many cases in the literature highlighting environmental-related failures. Whilst it is beyond the scope of this work to provide a review of these, it is nonetheless worth highlighting one recent high-profile example: the case of Volkswagen (Volkswagen, Wolfsburg, Germany) and its emissions control scandal, or 'Dieselgate' as it became known. In 2014 researchers at the University of West Virginia in the US found certain VW diesel cars emitted up to 35 times the permissible levels of harmful nitrogen oxide when tested on the road (Ramsey 2015). On 18th September 2016, the US Environmental Protection Agency accused VW of cheating diesel emissions tests using 'defeat devices' and four days later Volkswagen admitted installing software in 11 million diesel engines worldwide which was designed to reduce emissions during laboratory tests (France 24 2018). One day later, CEO Martin Winterkorn stepped down, insisting he knew nothing of the scam, and was replaced by Matthias Mueller. VW shares plunged by 40% in two days. After setting aside billions of Euros to cover anticipated costs of litigation related to the scandal, in April 2016 VW announced a net loss for 2015, its first in

20 years. In June of the same year it agreed to pay $14.7 billion in buybacks, compensation and penalties in a mammoth settlement with US authorities. Later that same year in September, the first VW investors filed lawsuits in a German court seeking billions of Euros in damages.

The following year, in January 2017 VW pleaded guilty to three US charges, including fraud, and agreed to pay $4.3 billion in civil and criminal fines. At the same time German prosecutors started investigating the ex-CEO, Winterkorn, on suspicion of fraud, accusing him of knowing about the defeat devices earlier than he had admitted. By then, Winterkorn was already under investigation for suspected market manipulation over the scandal. In August 2017 a Michigan court sentenced VW engineer James Liang to 40 months in prison and a $200 000 fine, after he pleaded guilty to conspiracy to defraud the US and to violating the US Clean Air Act. Later that year, in December, VW executive Oliver Schmidt was sentenced to seven years in jail after pleading guilty to fraud and violating the US Clean Air Act. The company recorded high sales in 2017, leading it to announce profits again in February 2018, but the positive news was short-lived. In April 2018, VW brand chief Herbert Diess replaced Matthias Mueller as CEO after he too was being investigated for his involvement in 'Dieselgate', whilst at the same time a top manager at Porsche (Porsche AG, Stuttgart, Germany), a VW subsidiary, was arrested in Germany as part of the scandal inquiries. In May 2018, Winterkorn was indicted in the US, accused of trying to cover up the cheating and in June VW agreed to pay a €1 billion fine in Germany, admitting its responsibility for the diesel emissions crisis. Also in June 2018, Rupert Stadler, CEO of VW's Audi (Audi AG, Ingolstadt, Germany) subsidiary, was arrested in Germany, accused of fraud and trying to suppress evidence related to the scandal (France 24 2018). Revelations from the scandal and subsequent consequences continue to unfold and most recently, in April 2019, a public prosecutor in Germany charged the ex-CEO, Martin Winterkorn, and four other managers with fraud over their involvement in VW's diesel emissions scandal (BBC 2019). As of April 2019, the scandal and VW's spectacular failure of CSR was estimated to have cost the company €28 billion ($31 bn; £24 bn).

Apart from the obvious economic impact of the emissions scandal on its financial stability, the reputational cost has also been hugely negative. Current figures of the loss in brand value are difficult to come by, but in 2015, just a few months after the emissions scandal unfolded in the public domain, the company was estimated to have lost up to $10 billion in brand value (Brand Finance 2015). This loss of brand value was the largest ever in the automotive industry, eclipsing that of Toyota (Toyota Motor Corporation, Toyota, Aichi Prefecture, Japan) which dropped almost $3 billion in 2012 due to a series of recalls over mechanical issues from 2009 to 2011 (Brand Finance 2015). At first it seemed that the emissions scandal may not impact VW's brand as much as Toyota since the latter's errors resulted in fatal accidents. However, the scale of the deception in the diesel emissions scandal and their deliberate intent to defraud were at complete odds with VW's advertised brand identity of reliability, honesty, efficiency, and environmentally friendly. As a result, the impact on VW's brand and reputation were unprecedented and evidence of this is further borne out by social media advocacy volumes following the emissions scandal. This data showed that positive recommendations of VW, which had been healthily consistent at an above sector average + 80 points, dropped by an average of 67% post the emissions scandal and furthermore, negative recommendations increased by nearly 2000% (Alva 2015). Similarly, according to YouGov's BrandIndex, a measure of a range of metrics including impression, value, quality, and recommendation, VW's score was 29.3 in 2014 before the emissions scandal, falling sharply to 1.6 it reached in 2015 after the scandal broke. Over three years later the score has still not recovered to anywhere near its highest point and as of November 2018 it stood at 17.4 (Hammett 2018). In an interview with the BBC, VW's current CEO, Herbert Diess, said the company still had to win back the confidence of its customers (BBC 2019). Based on the BrandIndex scores, one would agree.

As the examples shared highlight, there have been many failures in CSR over the years, with enormously negative consequences for a company's brand equity and financial bottom line. These types of failures continue to date, though perhaps arguably to a lesser extent.

24.3 From CSR to Corporate Sustainability

In a time of increased public focus on the environment and related issues such as global warming, air pollution and plastics pollution of the land and water systems, there has been a surge in the belief that every business and even every individual needs to think and act in a 'sustainable' way if we are to 'save the planet for future generations'. Indeed, the last decade has seen an exponential rise in the use of the words sustainable and sustainability within both corporate and non-corporate settings, often in conjunction with or replacing CSR, and the interchanging usage of terminology has caused much confusion regarding the two concepts. This section examines how the focus for businesses has shifted from socially focused CSR to 'corporate sustainability' and the relationship between the two is explored.

24.3.1 What Is Sustainability and how Did the Phrase Come about?

The last 20 years have seen abundance of publications on 'sustainability' but the topic remains a fluid concept with a large number of interpretations and context-specific understanding. The 1980s had seen a number of environmental disasters, as previously discussed, and there was an increasing recognition that the overall goals of environmental conservation and economic development were not conflicting but instead could be mutually reinforcing, resulting in the concept of 'environmentally sustainable' economic development, or sustainable development (Barbier 1987). In 1987 the United Nations World Commission on Environment and Development (UN WCED) published its report 'Our Common Future', known as the Brundtland Report (United Nations 1987), which defined sustainable development as the 'development that meets the needs of the present without compromising the ability of future generations to meet their own needs'. Nowadays the Brundtland definition of sustainable development is one of the most commonly used.

In the years following the Brundtland report this concept of sustainable development became one of the dominant paradigms of the environmental movement and saw much greater international cooperation between business and society to address sustainable development challenges. Sustainable development started becoming synonymous with 'sustainability' and in 1990 the UN Secretary-General, Maurice Strong, tasked businessman Stephan Schmidheiny with injecting a business voice into the global conversation on sustainability and environmental issues. He was engaged with spreading the concept of sustainable development among the world's business leaders and companies ahead of the United Nations Rio de Janeiro Earth Summit in 1992 (UN 1997) (WBCSD n.d.). The World Business Council for Sustainable Development (WBCSD) (World Business Council for Sustainable Development, Geneva, Switzerland), a CEO-led global association of businesses, was founded at the time of the Earth Summit to ensure that the business case for sustainable development was properly promoted. Schmidheiny and the WBCSD launched a seminal book at that Earth Summit, titled *Changing Course*, which was hailed as a landmark step in shaping a business point of view on environmental and economic issues (Schmidheiny and WBCSD 1992).

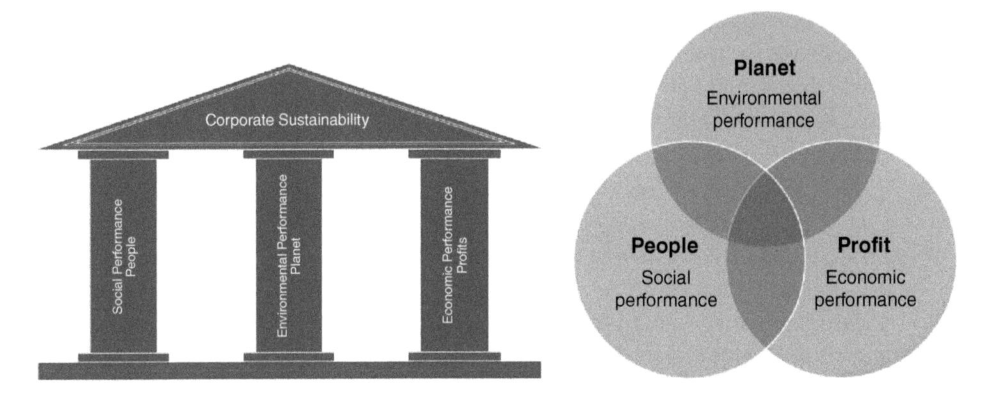

Figure 24.4 Conceptual models of the TBL Corporate Sustainability framework.

24.3.2 Conceptual Models and Frameworks for Corporate Sustainability

Post the Rio Earth Summit there were a large number of research publications in the area of sustainable development, which also became known simply as 'sustainability' or, in a business setting, 'corporate sustainability', with the two words often being interchanged. In 1994 one of the most important frameworks for sustainability was developed by John Elkington, founder of a consultancy called SustainAbility (SustainAbility, London, UK) (Elkington 1994). Elkington proposed that companies should be preparing three different separate bottom lines: the traditional economic bottom line measure of corporate profit; the 'people bottom line' – a measure of how socially responsible an organization has been throughout its operations; and the 'planet bottom line' – a measure of how environmentally responsible it has been. The 'triple bottom line' (TBL), as it become known, therefore consists of three Ps: profit, people, and planet. It aims to measure the financial, social, and environmental performance of a corporation over a period of time, the thinking being that only a company that produces a TBL is taking account of the full cost involved in doing business. This TBL framework has been conceptualized as three pillars underpinning corporate sustainability (Purvis, et al. 2019) and a Venn diagram of intersecting circles (Figure 24.4) (AQA 2017).

Interestingly, the three circle Venn diagram in Figure 24.4 is very similar to Three Domain model previously discussed for CSR (Figure 24.2). The difference between the two models is that in the case of sustainability the ethical and legal aspects of CSR are replaced by the wider reaching and more appropriate areas of environmental and social performance. Nonetheless, since environmental and economic factors are also often considered in modern CSR strategies, it is easy to see why CSR and sustainability/corporate sustainability are often used interchangeably, though wrongly so.

24.3.3 CSR and Corporate Sustainability Are Not the Same Thing

The differences between CSR and corporate sustainability are not necessarily easy to grasp and there are many people who use the term interchangeably. Some people think corporate sustainability is mainly just about environmental issues, whereas others see it in terms of social impacts on developing nations and their populations – and hence the same thing a CSR, and yet even others view it simply from the point of view of the economic bottom line. None of these viewpoints are correct when taken in isolation.

To appreciate the differences between CSR and corporate sustainability it is worth looking at a couple of recent definitions of the latter. In their 2002 publication, Dyllick and Hockerts defined corporate sustainability as 'meeting the needs of a firm's direct and indirect stakeholders (such as shareholders, employees, clients, pressure groups, communities etc.), without compromising its ability to meet the needs of future stakeholders as well' (Dyllick and Hockerts 2002). The Australian government, on the other hand, defines corporate sustainability closer to the 'Daly Rules' (ACWI n.d.) and see it as 'encompassing strategies and practices that aim to meet the needs of the stakeholders today, while seeking to protect, support, and enhance the human and natural resources that will be needed in the future' (Flouris and Yilmaz 2016). Compare these definitions to those previously quoted for CSR and it becomes evident that sustainable development, including the ability to renew natural resources at a rate equal to which it is regenerated, is not officially touched upon by mainstream CSR.

CSR concentrates more on the financial and non-financial societal activities that a company contributes to whereas corporate sustainability also concentrates on both the impact of environmental factors on a company and the company's impact on the environment. The point that businesses need to look both 'inside out', defined as a company's impact on climate, as well as 'outside in', defined as how climate regulatory change may affect the business environment in which the company competes, has been stressed by Michael Porter and Forrest Reinhart (Porter and Reinhardt 2007). They state in their article on the matter that 'companies that persist in treating climate change solely as a corporate social responsibility issue, rather than a business problem will risk the greatest consequences' (Porter and Reinhardt 2007).

One key difference between the CSR and corporate sustainability is the factor of time: if, as per its definition, corporate sustainability must ensure the ability of future generations to meet their own needs is not compromised, then it must have at its heart the need to consider a longer-term view of balancing resource usage and supplies over time. When the resources used match the earth's capacity to regenerate adequate future supply, then the systems remain balanced indefinitely. However, if resources used exceed this capacity, then current demand is being met by borrowing from the future, which eventually leads to an inability to meet society's needs. Whereas CSR focuses on balancing current stakeholder interests, corporate sustainability focuses on taking a longer-term view of ensuring the whole value chain has a future. For example, a socially responsible oil company may build local schools and hospitals to compensate communities for their resource extraction (Bansal and DesJardine 2015). But schools and hospitals require staff and ongoing servicing and so short-term measures do not always acknowledge the long-term impact on the communities. In this way, the shorter-term CSR measures can actually impose long-term liabilities on affected communities, making well-intentioned actions unsustainable (Bansal and DesJardine 2015).

There are many literature studies comparing, contrasting and debating CSR and corporate sustainability and the field of study has mushroomed over the last 15 years. For the purposes of this review, an article by Andy Last is useful to share as an example and is summarized in Table 24.2 (Last 2012). The distinction made in Table 24.2 about how CSR and corporate sustainability tend to be rewarded is an important one. Until recently it has been difficult to show that following a strategy of long-term corporate sustainability is more beneficial to the overall success of a company than other short-term strategies. However, a study by Eccles et al. in 2014 looked at 180 firms, half of which were classified as having high sustainability and half low sustainability, in order to examine issues of governance, culture, and performance (Eccles, et al., 2014). Their study showed that in fact those companies classified as having high sustainability are more likely to have established processes for stakeholder engagement, to be more long-term oriented, and to exhibit higher measurement and disclosure of non-financial information. In addition, high sustainability companies significantly outperformed their low sustainability counterparts over the long-term, both in terms of stock market and accounting performance (Eccles, et al., 2014). This is an important point to

Table 24.2 Six differences between CSR and corporate sustainability.

Corporate social responsibility (CSR)	Corporate sustainability
Vision	
Looks backwards, reporting on what a business has done, typically in the last 12 months, to make a contribution to society.	Looks forward, planning the changes a business might make to secure its future (reducing waste, assuring supply chains, developing new markets, building its brand).
Targets	
Tends to target opinion formers – politicians, pressure groups, media.	Targets the whole value chain – from suppliers to operations to partners to end-consumers.
Business	
Is increasingly becoming about compliance.	Is about how to run a better, more successful business in the longer-term
Management	
Is mainly managed by communications teams.	Is mainly managed by operations and marketing teams
Reward	
Investment is rewarded by the media and politicians	Investment is rewarded by the financial markets due to higher performance
Driver	
Is driven by the need to protect reputations in developed markets.	Is driven by the need to create opportunities in emerging markets.

note when considering how to formulate future business strategy and clearly makes a strong case for incorporating corporate sustainability into an organization's strategy and operations.

Whilst many would argue that CSR and corporate sustainability overlap and are distinct, it would be more accurate to think of the former as a subset of the latter: in a sense CSR may be thought of as the social aspect of corporate sustainability. There is a clear trend of businesses moving on from the simpler CSR approach and towards incorporating corporate sustainability into their strategy.

24.3.4 Corporate Sustainability – The Future of CSR

It is curious to note that Elkington himself has recently suggested a rethinking of the corporate sustainability TBL approach, 25 years after he first introduced the concept (Elkington 2018). The TBL concept was supposed to offer a radical new way for businesses to stop focusing solely on profits and expand their focus to include improving the lives people and the health of the planet. Its stated goal from the outset was a wholescale system change: pushing toward the transformation of capitalism. However, Elkington stated in his 'recall' of the TBL model that this radical goal has been largely forgotten and triple bottom line thinking has been reduced to a mere accounting tool. Moreover, the three Ps of TBL are difficult to quantify as they do not have a common unit of measure: profits are measured in the relevant currency but what unit does one use to measure social capital, or environmental health? The trick with TBL has been to try and measure it in ways that can be compared industry wide. An additional complication is that because it has mainly become about implementing accounting practises, it has at times been used to 'massage' the data in order to present a healthier picture of a company's sustainability. Therefore, one can appreciate Elkington's 'recall' notice for TBL and plea to change the three Ps from accounting practices to creating a system for the fundamental transformation of capitalism, affecting people, planet, and profits – co-existing in balance and harmony.

The massaging of data to portray a rosier CSR/sustainability picture to the media is an issue that is part of a wider problem seen with CSR and, to a lesser extent, with corporate sustainability: namely 'greenwashing' or the practise of communicating unsubstantiated or misleading CSR/sustainability claims. This practise became rampant from the late 1990s onwards and still exists today, albeit perhaps to a slightly lesser extent as it is increasingly difficult to make false claims about one's CSR/sustainability credentials. What has changed, however, is that the greenwashing has become more sophisticated (Watson 2016). For example, in 2013, the energy company Westinghouse Nuclear (Westinghouse Electric Company LLC, Pittsburgh, Pennsylvania, US) launched a new advertisement to put a fresh face on its old claims: 'Did you know that nuclear energy is the largest source of clean air energy in the world?' the advertisement asked viewers right before claiming that its nuclear power plants 'provide cleaner air, create jobs, and help sustain the communities where they operate' (Westinghouse Nuclear 2013). What the commercial failed to mention was that, two years earlier, Westinghouse Nuclear was cited by the US Nuclear Regulatory Commission for concealing flaws in its reactor designs and submitting false information to regulators (Wald 2011). Moreover, in February 2016 another plant that uses Westinghouse Nuclear's reactors, New York's Indian Point, leaked radioactive material into the surrounding area's groundwater (Watson 2016). Another example is that of Nestlé Waters Canada (Nestlé Waters Canada, Puslinch, Ontario, Canada) which ran an advertisement in 2008 claiming 'bottled water is the most environmentally responsible consumer product in the world' (Watson 2016). Several Canadian groups quickly filed a complaint against the company but five years later, during Earth Day 2013, the International Bottled Water Association (International Bottled Water Association, Alexandria, Virginia, US) increased the sustainability claims, announcing that bottled water was 'the face of positive change' because the industry was using less plastic in its bottles and relying more on recycled plastic (Watson 2016). Since only about 31% of plastic bottles end up getting recycled, 'the face of positive change' creates millions of tons of rubbish every year, much of which ends up in landfills or the ocean. These are hardly authentic claims of being sustainable.

The literature is replete with examples of historical and ongoing CSR failures and sophisticated greenwashing, some of which have been shared in this chapter. What is clear from all of these examples is that the initial approach to CSR, or CSR 1.0 as Visser refers to it (Visser 2010), have for the most part failed. Indeed, some people even have gone so far as to say 'CSR is dead: Long live sustainability as a corporate strategy' (Welsh 2018). Whilst this may be an exaggeration – CSR is well and truly alive and kicking in many thousands of organizations around the world – it is true that the tide is shifting to corporate sustainability. For example, Coca-Cola (The Coca-Cola Company, Atlanta, Georgia, US) and Bayer (Bayer AG, Leverkusen, Germany) are part of a movement toward integrated reporting and actual, measurable corporate sustainability goals. Bayer started combining its financial and sustainability reporting in 2013 and made it clear that sustainability was established at the board level, linking it with the Board of Management member responsible for Human Resources, Technology, and Sustainability (Welsh 2018). Coca Cola's objective by 2020 is to have safely returned to nature an amount of water equal to what they use in production. In 2014 it claimed to have returned about 126.7 billion litres of water used in its manufacturing processes back to communities and nature through treated wastewater (Welsh 2018), although it was not clear what percentage of its total water usage that amounted to.

A number of people involved in this tide towards corporate sustainability are focusing on redefining CSR as 'CSR 2.0' and a leading proponent of this movement is Wayne Visser who has proposed that 'CSR 1.0' has failed largely due to 'three curses of modern CSR' (Visser 2010):

1. *Peripheral CSR*: CSR has remained largely restricted to the largest companies, and mostly confined to Public Relations (PR) or other departments, rather than being integrated across the business.
2. *Incremental CSR*: CSR has adopted the quality management model, which results in incremental improvements that do not match the scale and urgency of the problems.
3. *Uneconomic CSR*: CSR does not always make economic sense, as the short-term markets still reward companies that externalize their costs to society.

Visser argues that CSR must be seen for what it is: an outdated, outmoded artefact that was once useful, but whose time has passed. If we admit the failure of CSR, he continues, we may find ourselves on the cusp of a revolution bringing us to CSR 2.0, like the one that transformed the internet from Web 1.0 to Web 2.0. Drawing similarities between Web 2.0 and CSR 2.0 is deliberate: the explosion of social media networks, user-generated web content and open source approaches are a fitting metaphor for the changes CSR must undergo to redefine its contribution and make a serious impact on the social, environmental, and ethical challenges that the world faces (Visser

Table 24.3 The similarities between Web 1.0 and CSR 1.0, and Web 2.0 and CSR 2.0.

Web 1.0	CSR 1.0
A flat world just beginning to connect itself and finding a new medium to push out information and plug advertising.	A vehicle for companies to establish relationships with communities, channel philanthropic contributions and manage their image.
Saw the rise to prominence of innovators like Netscape (Netscape Communications Corporation, Dulles, Virginia, US), but these were quickly out muscled by giants like Microsoft with its Internet Explorer.	Included many start-up pioneers but has ultimately turned into a product for large multinationals.
Focused largely on the standardized hardware and software of the PC as its delivery platform, rather than multi-level applications.	Travelled down the road of 'one size fits all' standardization, through codes, standards, and guidelines to shape its offering.
Web 2.0	**CSR 2.0**
Being defined by watchwords like 'collective intelligence', 'collaborative networks' and 'user participation'.	Being defined by 'global commons', 'innovative partnerships' and 'stakeholder involvement'.
Tools include social media, knowledge syndication and beta testing.	Mechanisms include diverse stakeholder panels, real-time transparent reporting and new-wave social entrepreneurship.
Is as much a state of being as a technical advance – it is a new philosophy or way of seeing the world differently.	Is recognizing a shift in power from centralized to decentralized; a change in scale from few and big to many and small; and a change in application from single and exclusive to multiple and shared.

Source: (Visser 2010) reproduced with permission.

2010). Using this analogy of Web 1.0 and 2.0 also allows the distinctions between CSR 1.0 and CSR 2.0 to be made, as shown in Table 24.3.

Visser suggests businesses that still practice CSR 1.0 will be rapidly left behind like their Web 1.0 counterparts, whereas companies that embrace the CSR 2.0 era will find innovative ways to tackle our global challenges and be rewarded in the marketplace as a result.

To put CSR 2.0 into practise, Visser proposes firstly renaming CSR from Corporate *Social* Responsibility to Corporate *Sustainability* and Responsibility. This reflects the general shift towards corporate sustainability, but the additional thinking behind the name change is that sustainability should be seen as a destination – the challenges, vision, strategy, and goals that organizations should be aiming for, whereas responsibility is about the journey – solutions, responses, management, and action that show how to get there. This subtle change in the definition of CSR 2.0 is quite powerful, though in practise of course the difficulty will be getting large numbers of people to both understand and then also accept the difference if 'CSR' remains part of the acronym.

This CSR 2.0 is seen by Visser as truly transformative and he suggests five principles which lie at the core of the concept:

1. Creativity — In order to succeed in the CSR 2.0 revolution, we will need innovation and creativity.
2. Scalability — The sustainability problems we face, be they climate change or poverty, are at such a massive scale, and are so urgent, that we need corresponding massive scale solutions. For example, when Walmart's former CEO, Lee Scott, decided that all cotton they use will be organic and all fish Marine Stewardship Council (MSC) certified (Marine Stewardship Council 2018).
3. Responsiveness — CSR 2.0 requires uncomfortable, transformative responsiveness, which questions whether the industry or the business model itself is part of the solution or part of the problem.
4. 'Glocality' — In a complex, interconnected a CSR 2.0 world context this is about thinking globally and acting locally; becoming more sophisticated in understanding local contexts and finding the appropriate local solutions they demand, without forsaking universal principles.
5. Circularity — This is about moving from 'cradle-to-grave' thinking to 'cradle-to-cradle' aspirations; creating buildings that produce more energy than they consume and purifying their own waste water; or factories that produce drinking water as effluent, or businesses that constantly feed and replenish their social and human capital, not only through education and training, but also by nourishing community and employee wellbeing.

With these changes, Corporate Social Responsibility (CSR 1.0) would become known as Corporate Sustainability and Responsibility (CSR 2.0) and its performance would be embedded and integrated into the core operations of companies. Visser suggests the whole concept of CSR would lose its Western conceptual and operational dominance, shifting to a more culturally diverse and internationally applied concept. This shift is summarized in Figure 24.5.

24.4 Perspectives

The fields of corporate social responsibility and corporate sustainability clearly have a great deal of overlap in their origins and objectives within organizations. The terms are used interchangeably,

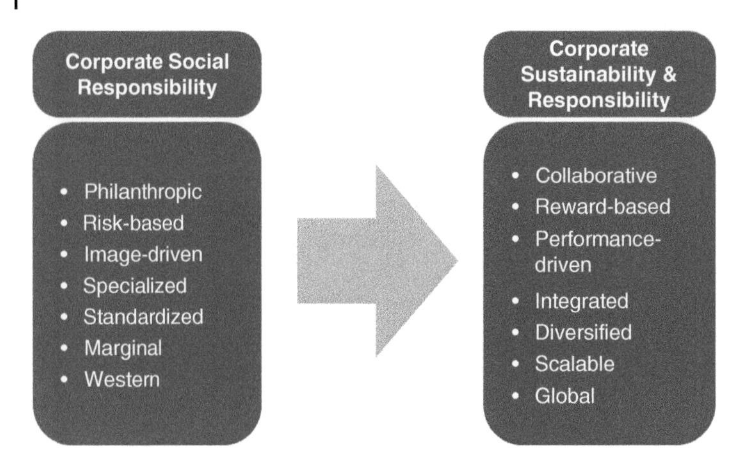

Figure 24.5 Shifts from Corporate Social Responsibility (CSR 1.0) to Corporate Sustainability and Responsibility (CSR 2.0). Source: (Visser 2010) (reproduced with permission).

understandably, partly because of this overlap mentioned but also because there have been many different bodies working in the area, at times collaboratively but also often at cross-purposes. Add to that the self-interest of corporations implementing what suits them in order to portray themselves in the media as being socially responsible and/or sustainable, and one can see why there has traditionally been so much confusion of terminology in the field. That said, there has also been great progress over the last two decades in creating global standards and mobilizing media campaigns against the worst CSR/sustainability offenders, in order to get them to change. However, most of the impacts, important though they have been at times, have not even scratched the surface of what is needed globally if we are to see true transformation of capitalism in terms of environmental, social, and economic change. For that to occur we need a paradigm shift to occur in the thinking of corporate sustainability as well as three key capabilities within organizations: agility, innovation, and resilience (AIR). The word capability is pointedly chosen here as it requires an organization to have both the competency and the capacity, in order to be deemed capable.

24.4.1 Paradigm Shift in Corporate Sustainability Thinking

Corporate sustainability is as much about a business' ability to sustain itself, i.e. long-term strategic viability, as about its external environment

- Sustainability has almost become a synonym how 'green' or 'caring' a company is and much of the focus of corporate sustainability seems to be external to the company – about the impact of the company on the environment and social aspects of what the company is doing to/with its stakeholders.
- Leadership teams must recognize that corporate sustainability is as much about focusing on what is internal to the company as is external. It goes to the heart of 'sustaining' the business i.e. its long-term viability. Therefore, an organization's future survival must include a corporate sustainability strategy.
- The thinking in the boardroom and across the company must change from: 'what are we doing to fulfil our CSR obligations so we can publicise how good we are for the environment?', to 'how do we develop a strategy to sustain our business, taking into account both the external environmental, social, and economic issues which may affect us, and also all factors within our company which may affect those same three issues externally?'

24.4.2 Three Key Capabilities Needed to Support Corporate Sustainability

1. Agility

   Organizational Agility must form part of the DNA of any business.

   - The 'Digital Revolution' technologies such as robotics, virtual reality and artificial intelligence are disrupting and changing every aspect of the way we live. In response, businesses must become Agile: nimbler, more collaborative with, and responsive to, stakeholders, and quicker in their decision making and its implementation.
   - Businesses must draw the distinction between 'being Agile' and 'doing Agile'. Being Agile is about who the organization is – its culture, consciousness and the employees' mindset. Doing Agile, on the other hand, is more about solving an immediate challenge and therefore focuses on way it is implemented – best practises, methodologies and the like. Businesses must 'be Agile' first and deeply embed this culture into all levels of the company.
   - One of the biggest challenges to organizational Agility comes from the mainstay goal of most organizations, namely 'maximizing shareholder value', which in turn requires a top-down mode of management, with all its associated qualities of being slow, bureaucratic, siloed, and risk-averse. However, the primary focus of Agile is not maximizing shareholder value but on delivering value to the customer. Economic success becomes the result, not the goal of Agile. Therefore, developing an organizational Agility will mean moving away from traditional top-down or bottom-up thinking in an organization to one which is outside-in and inside-out: the focus shifts to customers and stakeholders, and as such supports the corporate sustainability view of assessing how a company is impacted from externally and also how it impacts externally.
   - Organizational Agility will still need leadership to set a clear shared purpose and vision across the whole organization, but in addition will require networks of empowered teams, quicker cycles of decision-making and learning, passionate and dynamic people, and leading-edge technologies.

2. Innovation

   The most successful companies in the coming decade will lead with innovation – both in their products and services and approaches to corporate sustainability

   - Within the next 5–10 years most of today's corporate sustainability practises will have become required by law/regulators or other certifying bodies. Companies will not be able to attract and/or retain their customers and investors unless they are seen to be operating sustainably – internally and externally.
   - In addition to innovation in research and development and goods and services, innovation in how a company fulfils its corporate sustainability obligations will become increasingly important and even further linked to brand equity.
   - Customers and investors will be attracted to those companies which not only have great products but are also seen to be 'saving the planet and its people', while being transparent about its sustainability practises.
   - The best corporate sustainability programs will no longer be the remit of a PR or Communications team but rather form part of the executive strategic leadership and perhaps also be implemented via existing specialist functions (e.g. HR, marketing, finance, community engagement, etc.).

3. Resilience

   Corporate sustainability will only be possible in the future for those companies which have true organizational resilience embedded across the leadership and all employees

- The Institute for Strategy, Resilience and Security at University College London, defines resilience as 'the enduring power of a body for transformation, renewal, and recovery through the flux of interactions and flow of events' (ISRS 2017). In today's fast-moving, interconnected world, the sustainability of any organization is inherently linked to its ability to transform and renew itself, and to recover through any events that it may encounter. In short, organizations today need to be resilient.
- Organizational resilience must not be taken to mean the ability to withstand adverse events and survive or recover back to 'business as usual'. That is better described as 'business continuity' which is all about returning a business to where it was before an incident. On the other hand, true resilience is not just about managing the status quo or recovering to a previous state, but rather moving forward past the points of challenge (ISRS 2017). Companies need to be able to anticipate and adapt to radical uncertainty and systemic risk, learn from events and evolve accordingly to thrive.
- Mastering organizational resilience will mean developing and adopting best practises and different habits to change the culture of the company, particularly in three areas:
  - Firstly, building the right capabilities by way of capacity and competencies throughout the business, and cultivating the right values and behaviours so that the way the 'organization thinks', the way it is run and the way it is experienced by its stakeholders, all reflect its resilient nature.
  - Secondly, replacing solely short and medium-term thinking and management of the company which primarily focus on immediate financial gains, with a more balanced approach which includes long-term thinking and strategy. This, then, becomes inherently linked to its corporate sustainability.
  - Thirdly, ensuring that learnings from the collective and individual experiences of incidents, events, exercises and near misses, are captured, evaluated and shared to create a stronger and more adaptable company. These learnings must be continually incorporated into the organization's corporate sustainability strategy.

One hurdle to embracing the paradigm shift in corporate sustainability thinking lies in redefining CSR and then implementing it. Whether the definition is changed to corporate sustainability and responsibility as suggested by Visser (Visser 2010), or perhaps some other working definition which changes the acronym away from CSR, there will no doubt need to be massive global efforts in trying to implement whatever the new frameworks involve in practise. What is clear is that CSR meaning Corporate Social Responsibility is too narrow in terms of current definition and implementation, for it to continue being the mainstay of businesses strategies on how to tackle the environmental, social, and economic problems of the world. Instead, corporate sustainability forms a better basis on which to build a future common definition and implementation approach.

The main challenge for corporate sustainability initiatives remains to have large-scale impact given the environmental and social change urgencies that the world is witnessing in this beginning of the twenty-first century, only 200 years since the start of the Industrial Revolution. This wholescale system change and transformation of capitalism can only occur if all the actors – governments, businesses, NGOs, the media and the general public to name a few – cooperate in building a world where the 'needs of the present are met without compromising the ability of future generations to meet their own needs'.

Incorporating the paradigm shift and capabilities discussed above into corporate sustainability in the future, a new conceptual model emerges (Figure 24.6) (Yagnik 2018). In this model, the organization must have a solid foundation of short, medium, and long-term strategies that fulfil its environmental, social and economic goals and these are delivered through having the right

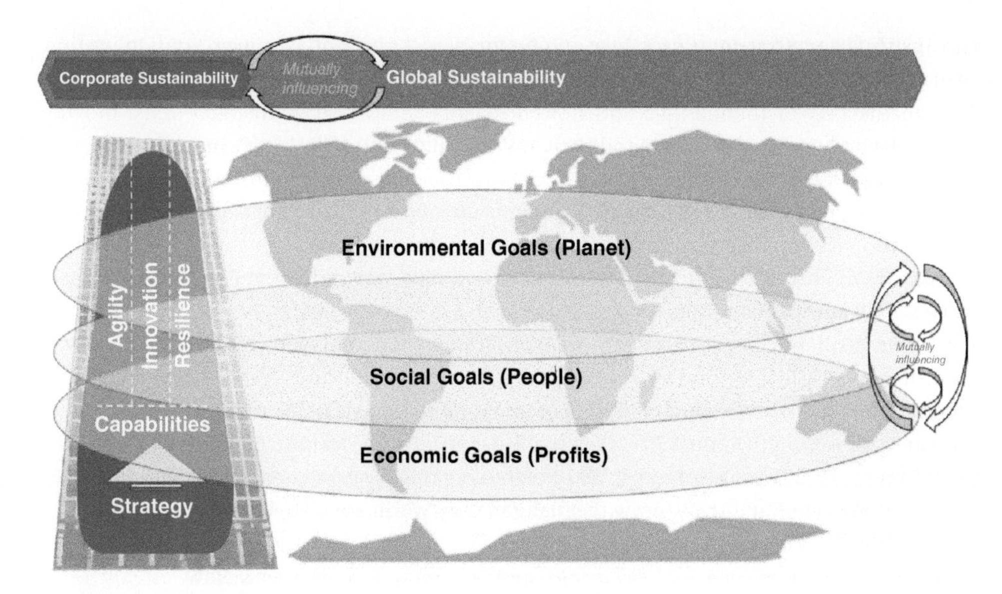

Figure 24.6 A new conceptual model of corporate sustainability that includes a paradigm shift in how it is viewed, as well as the supporting capabilities needed. Source: (Yagnik 2018) (reproduced with permission).

capabilities in Agility, Innovation, and Resilience (AIR), all of which lead to corporate sustainability. The latter is influenced by, and also itself influences, global sustainability. Importantly there must be shared environmental, social, and economic goals globally. It can even be argued that the capabilities of AIR should be developed globally in a wide variety of settings.

This model sees corporate sustainability being more about a company's long-term success, or 'ability to successfully sustain itself', by way of its environmental, social, and economic linked goals. In this approach, in order to be successful, the company must ensure its strategy is defined in such a way so as to fully support these three aspects. Crucial to implementation of said strategy will be the organization's capabilities in AIR. In fact, it may be said that without AIR, corporate sustainability, and hence corporations themselves, may well suffocate and wither in the future..

References

ACWI, n.d.. *What is sustainability?*. [Online] Available at: https://acwi.gov/swrr/whatis-sustainability-wide.pdf [Accessed 01 May 2019].

Alva, 2015. *What is the Reputation Damage of VW's Emissions Admission?*. [Online] Available at: https://www.alva-group.com/reputation-damage-vw-emissions-scandal

Amnesty International, 2018. *Microsoft moves on child labour claims.* [Online] Available at: https://www.amnesty.org.uk/microsoft-moves-child-labour-claims

AQA, 2017. *Teaching guide: Elkington's triple bottom line.* [Online] Available at: https://www.aqa.org.uk/resources/business/as-and-a-level/business-7131-7132/teach/teaching-guide-elkingtons-triple-bottom-line [Accessed 29 April 2019].

Asongu, J.J. (2007). *Strategic Corporate Social Responsibility in Practice.* Lawrenceville: Greenview Publishing Company.

Ballinger, J., 1992. *The new free-trade heel: Nike's profits jump on the back of Asian workers*. [Online] Available at: https://harpers.org/archive/1992/08/the-new-free-trade-heel [Accessed 21 March 2019].

Bansal, T. and DesJardine, M. (2015). Don't confuse sustainability with CSR. *Ivey Business Journal* (January/February): 1–3.

Barbier, E.B. (1987). The concept of sustainable economic development. *Environmental Conservation* 14 (2): 101–110.

BBC Panorama, 2010. *Tracing the bitter truth of chocolate and child labour*. [Online] Available at: http://news.bbc.co.uk/panorama/hi/front_page/newsid_8583000/8583499.stm

BBC, 2019. *Former VW boss charged over diesel emissions scandal*. [Online] Available at: https://www.bbc.co.uk/news/47937141

Boseley, S., 2018. *Child labour rampant in tobacco industry*. [Online] Available at: https://www.theguardian.com/world/2018/jun/25/revealed-child-labor-rampant-in-tobacco-industry

Boseley, S. & Levene, D., 2018. *Special report: The children working the tobacco fields: 'I wanted to be a nurse'*. [Online] Available at: https://www.theguardian.com/world/ng-interactive/2018/jun/25/tobacco-industry-child-labour-malawi-special-report

Bowen, H.R. (1953). *Social Responsibilities of the Businessman*. New York: Harper & Row.

Brand Finance, 2015. *VW Risks its $31 billion Brand and Germany's National Reputation*. [Online] Available at: https://brandfinance.com/news/press-releases/vw-risks-its-31-billion-brand-and-germanys-national-reputation

Broughton, E. (2005). The Bhopal disaster and its aftermath: a review. *Environmental Health* 4 (6).

Carroll, A.B. (1977). *Managing Corporate Social Responsibility*. Boston: Little Brown and Co.

Carroll, A.B. (1991). The pyramid of corporate social responsibility: toward the moral management of organizational stakeholders. *Business Horizons* 34 (4): 39–48.

Carroll, A.B. (2008). A history of corporate social responsibility: concepts and practices. In: *The Oxford Handbook of Corporate Social Responsibility*, 19–46. Oxford: Oxford University Press.

Carson, R. (1962). *Silent Spring*. New York: Houghton Mifflin.

Chavan, S. V., 2012. *A Study on Rethinking of Corporate Social Responsibility (A Holistic Value-Based Societal Perspective)*, Vidyanagari, Jhunjhunu, India: s.n.

Chocolate Manufacturers Association, 2001. *Protocol for the growing and processing of cocoa beans and their derivative products in a manner that complies with ILO Convention 182 concerning the prohibition and immediate action for the elimination of the worst forms of child labor*. [Online] Available at: https://web.archive.org/web/20151208022828/http://www.cocoainitiative.org/en/documents-manager/english/54-harkin-engel-protocol/file

Connor, T. (2001). *Still Waiting for Nike to Do it*. San Francisco: Global Exchange.

Cushman Jr,, J. H., 1998. *International Business; Nike pledges to end child labor and apply U.S. rules abroad*. [Online] Available at: https://www.nytimes.com/1998/05/13/business/international-business-nike-pledges-to-end-child-labor-and-apply-us-rules-abroad.html [Accessed 24 March 2019].

Dahlsrud, A. (2006). How corporate social responsibility is defined: an analysis of 37 definitions. *Corporate Social Responsibility and Environmental Management* 15: 1–13.

D'Souza, D. (1997). *Ronald Reagan: How an Ordinary Man Became an Extraordinary Leader*. New York: Touchstone.

Dyllick, T. and Hockerts, K. (2002). Beyond the business case for corporate sustainability. *Business Strategy and the Environment* 11: 130–141.

Eccles, R.G., Ioannou, I., and Serafeim, G. (2014). The impact of corporate sustainability on organizational process and performance. *Management Science* 60 (11): 2835–2857.

Eilbirt, H. and Parket, I.R. (1973). The current status of corporate social responsibility. *Business Horizons* 16: 5–14.

Elkington, J. (1994). Towards the sustainable corporation: win-win-win business strategies for sustainable development. *California Management Review* 36 (2): 90–100.

Elkington, J., 2018. 25 Years Ago I Coined the Phrase "Triple Bottom Line." Here's Why It's Time to Rethink It. [Online] Available at: https://hbr.org/2018/06/25-years-ago-i-coined-the-phrase-triple-bottom-line-heres-why-im-giving-up-on-it

European Commission (2011). *Communication from the Commission to the European Parliament, the Council, the European Economic and Social Committee and the Committee of the Regions: a Renewed EU Strategy 2011–14 for Corporate Social Responsibility*. Brussels: European Commission.

Fair Labor Association, 2019. *Nike Inc.*. [Online] Available at: http://www.fairlabor.org/affiliate/nike-inc

Flouris, T.G. and Yilmaz, A.K. (2016). *Risk Management and Corporate Sustainability in Aviation*. Abingdon, Oxfordshire: Routledge.

France 24, 2018. *VW 'dieselgate' fraud: Timeline of a scandal*. [Online] Available at: https://www.france24.com/en/20180910-vw-dieselgate-fraud-timeline-scandal

Friedman, M. (1962). *Capitalism and Freedom*. Chicago: University of Chicago Press.

Garside, J., 2013. *Child labour uncovered in Apple's supply chain*. [Online] Available at: https://www.theguardian.com/technology/2013/jan/25/apple-child-labour-supply

Hammett, E., 2018. *Volkswagen should be focusing on effectiveness not efficiency*. [Online] Available at: https://www.marketingweek.com/2018/11/28/volkswagen-marketing-efficiency

Heald, M. (1970). *The Social Responsibilities of Business: Company and Community, 1900–1960*. Cleveland: The Press of Case Western Reserve University.

Holba, C. & Woods, H., 2019. *Exxon Valdez Oil Spill: FAQs, Links, and Unique Resources at ARLIS*. [Online] Available at: https://www.arlis.org/docs/vol2/a/EVOS_FAQs.pdf [Accessed 14 May 2019].

Holme, R. and Watts, P. (2000). *Corporate Social Responsibility: Making Good Business Sense*. Geneva: World Business Council for Sustainable Development.

Holmes, S. (1976). Executive perceptions of corporate social responsibility. *Business Horizons* 19: 34–40.

ILO (2009). *The ILO and Corporate Social Responsibility*. Geneva: International Labour Organization.

ISO (2010). *ISO 26000:2010 Guidance on Social Responsibility*. Geneva: International Standards Organisation.

ISRS, 2017. *About ISRS*. [Online] Available at: www.isrs.org.uk/about [Accessed 1 May 2019].

Last, A., 2012. *Six differences between CSR and sustainability*. [Online] Available at: https://mullenlowesalt.com/blog/2012/10/differences/ [Accessed 1 May 2019].

Mani, V. (1998). *Puranic Encyclopaedia: a Comprehensive Dictionary with Special Reference to the Epic and Puranic Literature*. Delhi: Motilal Banarsidass.

Marine Stewardship Council, 2018. *MSC Fisheries Standard v2.01*. [Online] Available at: https://www.msc.org/docs/default-source/default-document-library/for-business/program-documents/fisheries-program-documents/msc-fisheries-standard-v2-01.pdf?sfvrsn=8ecb3272_11 [Accessed 19 May 2019].

Muniapan, B. and Dass, M. (2008). Corporate social responsibility: a philosophical approach from an ancient Indian perspective. *International Journal of Indian Culture and Business Management* 1 (4): 408–420.

Murphy, P.E. (1978). *An Evolution: Corporate Social Responsiveness*. University of Michigan Business Review November.

Nieburg, O., 2018. *Cocoa child labor lawsuits against Mars and Hershey filed.* [Online] Available at: https://www.confectionerynews.com/Article/2018/02/28/Cocoa-child-labor-lawsuits-against-Mars-and-Hershey-filed

Nisen, M., 2013. *How Nike Solved Its Sweatshop Problem.* [Online] Available at: https://www.businessinsider.com/how-nike-solved-its-sweatshop-problem-2013-5?r=US&IR=T

Peattie, K. (n.d.). *History of Corporate Social Responsibility and Sustainability.* Cardiff: the ESRC Centre for Business Relationships, Accountability, Sustainability and Society (BRASS).

Phillpott, S., 2019. *10 Companies That Still Use Child Labour.* [Online] Available at: https://www.careeraddict.com/10-companies-that-still-use-child-labor

Porritt, J., 2012. *Corporate sustainability: a brief history.* [Online] Available at: http://richardsandbrooksplace.org/jonathon-porritt/corporate-sustainability-brief-history

Porter, M.E. and Reinhardt, F.L. (2007). A strategic approach to climate change. *Harvard Business Review* 85 (10): 22–26.

Purvis, B., Mao, Y., and Robinson, D. (2019). Three pillars of sustainability: in search of conceptual origins. *Sustainability Science* 14 (3): 681–695.

Radhakrishnan, S. and Moore, C.A. (1967). *A Sourcebook in Indian Philosophy.* Princeton: Princeton University Press.

Ramsey, M., 2015. *Volkswagen Emissions Problem Exposed by Routine University Research.* [Online] Available at: https://www.wsj.com/articles/volkswagen-emissions-problem-exposed-by-routine-university-research-1443023854

Richards, D., 2017. *Apple admits to using child labour to build iPhone X.* [Online] Available at: https://www.channelnews.com.au/apple-admits-to-using-child-labour-to-build-iphone-x

Sandler Clarke, J., 2015. *Child labour on Nestlé farms: chocolate giant's problems continue.* [Online] Available at: https://www.theguardian.com/global-development-professionals-network/2015/sep/02/child-labour-on-nestle-farms-chocolate-giants-problems-continue

Schanberg, S. H., 1996. *Six Cents an Hour (1996 Life Article).* [Online] Available at: https://laborrights.org/in-the-news/six-cents-hour-1996-life-article [Accessed 06 April 2019].

Schmidheiny, S. and WBCSD (1992). *Changing Course: a Global Business Perspective on Development and the Environment,* 1e. Cambridge, MA/London England: MIT Press.

Schwartz, M.S. and Carroll, A.B. (2003). Corporate social responsibility: a three-domain approach. *Business Ethics Quarterly*: 503–530.

Swedwatch (2016). *Smokescreens in the supply chain: the Impacts of the Tobacco Industry on Human Rights and the Environment in Bangladesh.* Stockholm: Kalle Bergbom.

Teather, D., 2005. *Nike lists abuses at Asian factories.* [Online] Available at: https://www.theguardian.com/business/2005/apr/14/ethicalbusiness.money

UN, 1997. *UNCED Conference on Environment and Development (1992).* [Online] Available at: https://www.un.org/geninfo/bp/enviro.html[Accessed 30 April 2019].

United Nations (1987). *Report of the World Commission on Environment and Development: Our Common Future.* Geneva: United Nations.

Visser, W. (2005). Revisiting Carroll's CSR pyramid: an African perspective. In: *Corporate citizenship in a development perspective* (eds. E.R. Pedersen and M. Huniche). Copenhagen: Copenhagen Business School Press.

Visser, W. (2010). The age of responsibility: CSR 2.0 and the new DNA of business. *Journal of Business Systems, Governance and Ethics* 5 (3): 7.

Wald, M. L., 2011. *Regulators Find Design Flaws in New Reactors.* [Online] Available at: https://www.nytimes.com/2011/05/21/business/energy-environment/21nuke.html?_r=0 [Accessed 25 April 2019].

Walker, S., 2010. *Tobacco giant Philip Morris sold cigarettes made using child labour*. [Online] Available at: https://www.independent.co.uk/news/world/asia/tobacco-giant-philip-morris-sold-cigarettes-made-using-child-labour-2026759.html

Watson, B., 2016. *The troubling evolution of corporate greenwashing*. [Online] Available at: https://www.theguardian.com/sustainable-business/2016/aug/20/greenwashing-environmentalism-lies-companies [Accessed 1 May 2019].

WBCSD, n.d.. *The birth of WBCSD*. [Online] Available at: https://www.wbcsd.org/Overview/Our-history [Accessed 30 April 2019].

Weisman, J., 2004. *Reagan Policies Gave Green Light to Red Ink*. [Online] Available at: http://www.washingtonpost.com/wp-dyn/articles/A26402-2004Jun8.html [Accessed 20 March 2019].

Welsh, H., 2018. *CSR is dead: Long live sustainability as corporate strategy*. [Online] Available at: https://www.greenbiz.com/article/csr-dead-long-live-sustainability-corporate-strategy [Accessed 1 May 2019].

Westinghouse Nuclear, 2013. *Westinghouse Nuclear: 30 Commercial HD June 2013,* s.l.: s.n.

World Nuclear Association, 2018 *Chernobyl* Accident 1986. [Online] Available at: http://www.world-nuclear.org/information-library/safety-and-security/safety-of-plants/chernobyl-accident.aspx [Accessed 14 May 2019].

Yagnik, A.T. (2018). *A Paradigm Shift in Corporate Sustainability and the Supporting Capabilities that are Required*. London: AdSidera Ltd.

25

The Industrial World in the Twenty-First Century

Alain A. Vertès

London Business School, NW1 4SA, UK
NxR Biotechnologies GmbH, Basel, Switzerland

CHAPTER MENU

25.1 Introduction: Energy and Sustainability

The neoclassical economic growth model developed in 1956 by Robert Solow and Trevor Swan computes exogenous economic growth by computing changes over time in the level of output of an economy as a result of the deployment of (i) capital, (ii) labour, and (iii) technology (or 'effectiveness in labour') (Solow 1956). Notably, at steady-state, the economic growth rate can only be further increased through innovation. While both fundamental and intrinsically useful, this model does not fully account for a critical parameter, that of energy supply. Recent studies have demonstrated that when energy is scarce (i.e. expensive) it strongly impacts economic growth (Stern 2011). On the other hand, when energy is abundant and thus cheap, it has a much reduced effect on economic growth (Stern 2011). The distortion generated by the parameter energy to the predictions of the Solow model, therefore, decreases to the point of becoming a negligible variable in the model only when energy is abundant and cheap. The Industrial Revolution of the late eighteenth and early nineteenth century was initiated in England via an increased use of fossil resources starting with coal because labour there was then relatively expensive and energy relatively cheap as compared to other countries (Figure 25.1). What is more, as highlighted by Allen (2009), the steam engine, the water frame, the spinning jenny and the coke blast furnace resulted in an increase in the use and efficiency of the use of coal and capital relative to labour. However, it must be pointed

Green Energy to Sustainability: Strategies for Global Industries, First Edition.
Edited by Alain A. Vertès, Nasib Qureshi, Hans P. Blaschek and Hideaki Yukawa.

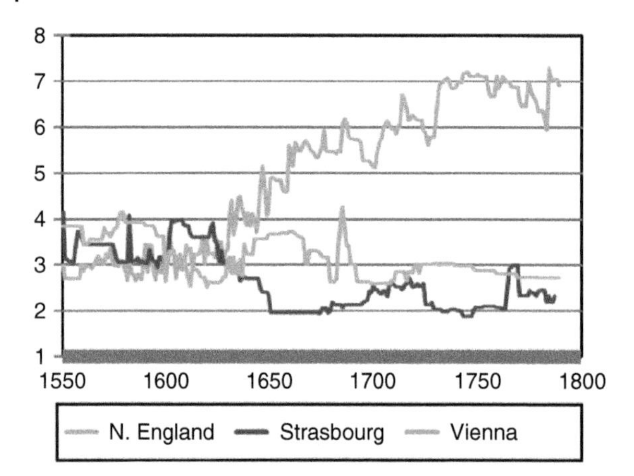

Figure 25.1 Evolution of wages relative to the cost of capital in major European economic centres of the twenty-sixth to twenty-eighth centuries. The ratio of a building labourer's daily wage in northern England, Strasbourg, and Vienna are plotted relative to an index of the rental price of capital in these cities. Remarkably, no particular trend in the evolution of this ratio could be observed between the three cities between 1550 and 1650. However, from the 1650 onwards, a notable divergence occurred and in England labour became increasingly expensive relative to capital. This phenomenon has been ascribed to inflation of nominal British wages after 1650. In contrast, the ratio of wage versus cost of capital decreased in Strasbourg and Vienna from 1650. As a result, there were in those three economic centres different incentives at play: in England, the continuous rise in the cost of labour versus that of capital was the motive force to invent new ways to substitute capital for labour, whereas in Strasbourg and Vienna the incentives were on substituting cheap labour for expensive capital. Source: Figure reproduced with permission (Allen 2009).

out that wood scarcity concerns probably did not play a major role in this transition (Warde 2006; Wrigley 1962). Constraints on economic growth were further decreased by the invention of more efficient methods of extracting the energetic content of coal on the one hand and on the other on the discovery, or rather the technical feasibility of their large scale exploitation, of new fossil fuel resources such as petroleum, which displaced biomass not only for the production of commodity chemicals such as lactic acid or acetone, butanol, and ethanol, but also for the production of transportation fuels (Figure 25.2) (Allen 2009; Benninga 1990; Jones and Woods 1986; Lewis 1981; Stern 2011). Additional time-series analyses further demonstrate from the example of the economy of the United States of America that energy and gross domestic product (GDP) are dependent variables, albeit GDP and energy use can be decoupled particularly through a significant increase in energy use efficiency (Figure 25.2). The twentieth century witnessed three oil shocks: in 1973, 1979, and 2008, all of which strongly encouraged shifts towards cheaper energy sources on the one hand and discouraged energy waste and incentivized more efficient energy use to limit the severity of oil-shock-induced recessions on the other. Nonetheless, this link between energy and economic growth is affected by different mechanisms. Particularly, the energy use per unit of economic output declines due to technological changes and to a shift from poorer quality fuels such as wood and coal to more efficient ones including notably electric power. This relationship is remarkably clear in today's world in developed countries (Stern 2011). Moreover, energy intensity has shown a clear tendency to decrease over the past 40 years in countries that have become richer but not in the others (Csereklyei et al. 2014). One of the major outcomes of the Industrial Revolution is that the living standards of masses of ordinary people underwent sustained growth to the point where they have dramatically improved in the time lapse of only a few generations (Lucas 2002).

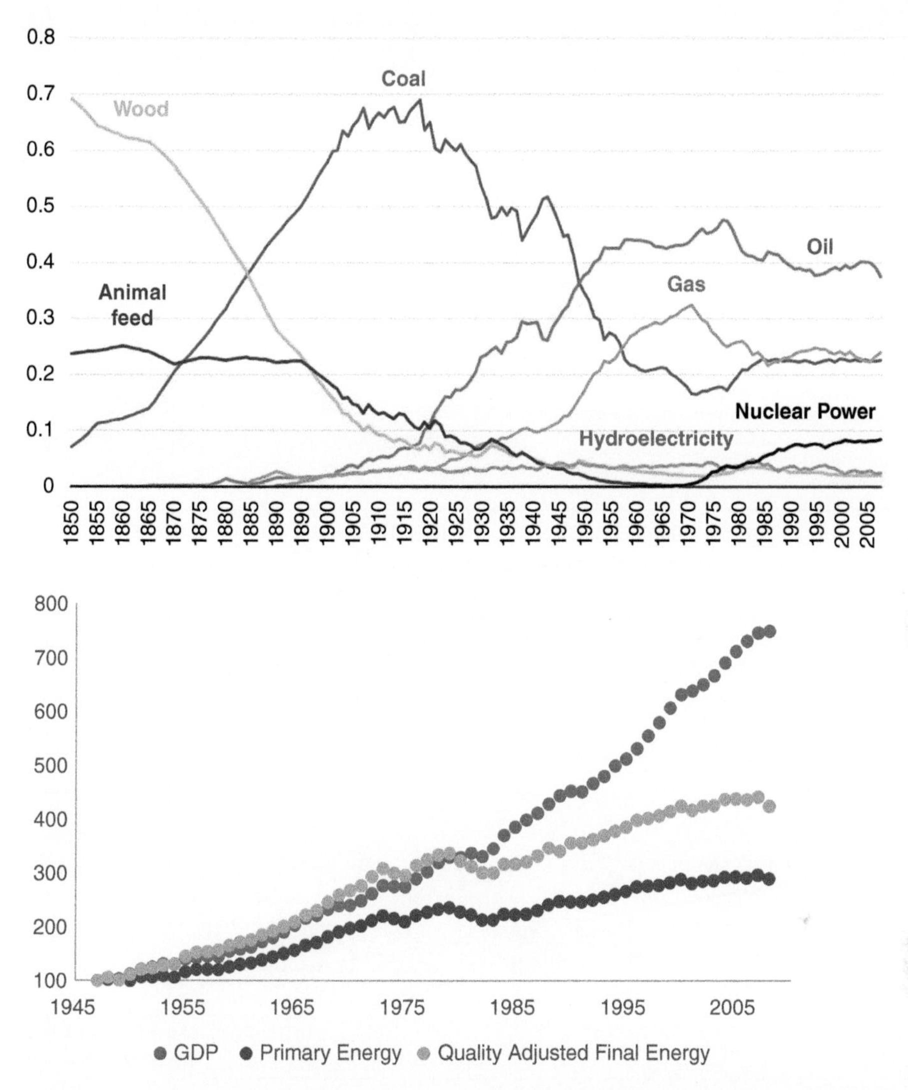

Figure 25.2 The 'energy' and 'economic growth' couple. Upper panel: composition (%) of the US energy input 1850–2005 of: wood, animal feed, coal, oil, gas, hydroelectricity, and nuclear energy. (The global energy mix is presented later in Figure 25.5). Lower panel: US GDP and quality-adjusted final energy use. GDP is in constant dollars i.e. adjusted for inflation. Energy use is a divisia index of the energy content (British thermal units, BTUs) (1947 value is 100) of the principal final energy use categories – including notably oil, natural gas, coal, electricity, and biofuels, and reflecting the energy quality intrinsic to each source. The different fuels are weighted according to their average prices. The final energy consumption is the total energy consumed by end users comprising industry, transport, households, services, and agriculture. Source: Data sourced from US Energy Information Administration and adapted from Stern and Cleveland (Stern and Cleveland 2004).

Overall, as summarized by Csereklyei et al. (2014), during the period spanning the last two centuries, the per capita energy use has tended to rise, energy quality to increase, and the cost share of energy has declined (Figure 25.3). Furthermore, the decline in global energy intensity observed over time has been ascribed to countries becoming richer with a combination of technological or structural changes as engines of this increase in wealth (Csereklyei et al. 2014).

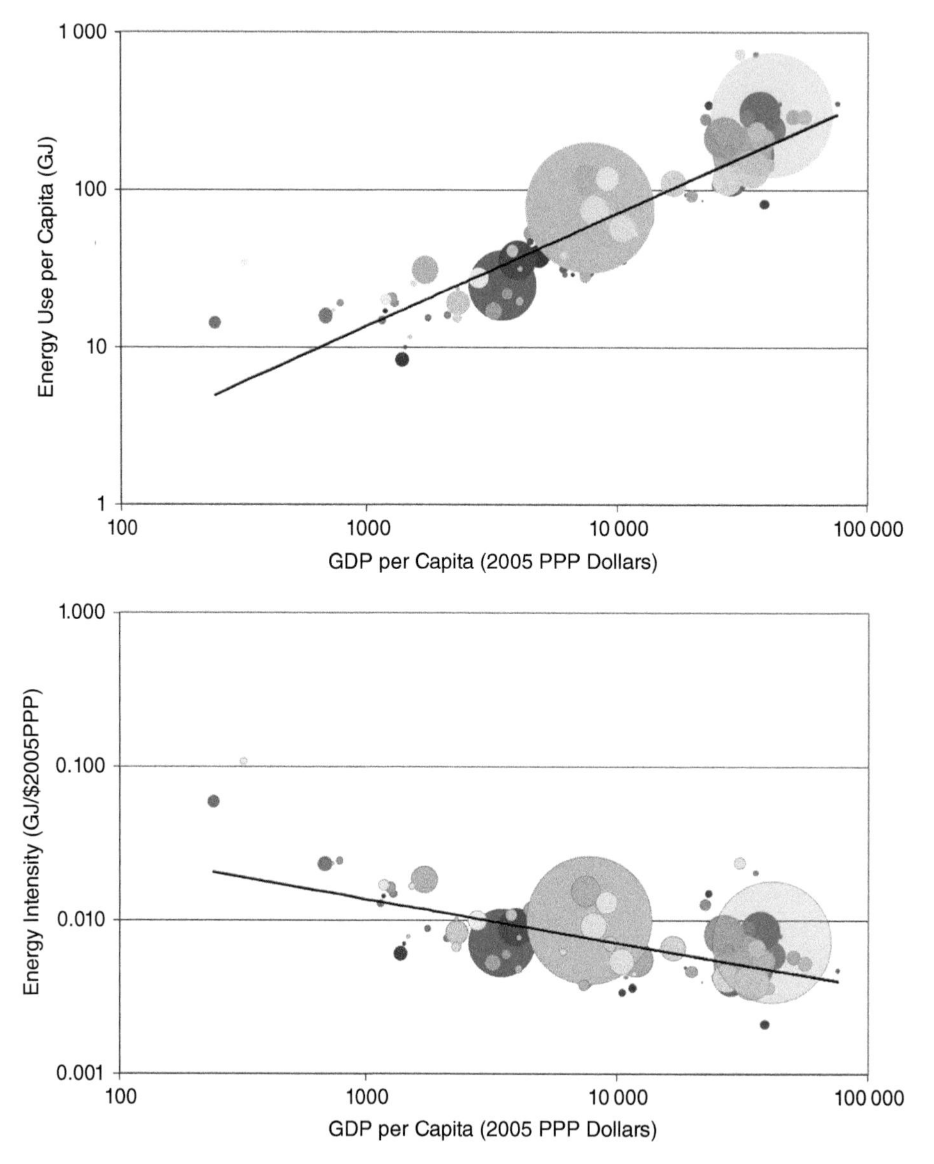

Figure 25.3 Energy use and energy intensity as a function of GDP per capita. Upper panel: the logarithm of energy use per capita increases with the logarithm of income per capita (reference year: 2010). The two largest economies – the United States (top right in blue) and China (middle of the graph in orange) – use more energy per capita than the norm for their income level as predicted by the regression fit. The largest bubble to the left of China is India (dark blue), which uses less energy per capita than the norm for its income level. Lower panel: cross sectional snapshot of energy intensity mapped against real income per capita (year of reference: 2010). A clear-cut tendency is observed: the energy intensity declines with higher income per capita. Interestingly, faster-growing countries also have a faster declining energy intensity, but energy intensity does not decline in the absence of economic growth. Source: Adapted from Csereklyei et al. (2014).

This link between energy demand and economic growth, which is reminiscent of basic thermodynamic principles, is so tight that methods have been developed to forecast global and local energy demands based on economic growth indicators (Suganthi and Samuel 2012).

However, the growth trajectories of economic models to this date do not integrate several considerations that have become critical since the onset of the new millennium. The first one of these fundamental challenges to the global economy is the sustainability of high-quality energy supply. The second of these fundamental challenges is the increasing cost of the consequences of global warming. Peak oil and climate change: these two phenomena of the twenty-first century combine to create a powerful heuristic engine of change of the methods of production (sustainability), the methods of consumption (production and recycling), and ultimately the methods of daily living (resource-consciousness).

Peak oil can be defined as the point in time at which half of all the economically extractable petroleum reserves will have been used up. How fast that point will be reached depends on reserves and consumption rates. Peak oil predictions of the later part of the twentieth century turn out to have been only partially correct as non-conventional oil reserves have been underestimated (Bardi 2019). As petroleum price increases, projects to exploit reserves that are more costly to extract, exemplified by shale oil fields or tar sands, become projects with positive net present value (NPV). What is more, in the absence of suitable alternatives, for example for transportation, petroleum is characterized by its relative price inelasticity that nonetheless can be challenged by biofuels and electric or fuel cell-powered vehicles (Soltani et al. 2015). A concerning trend is that since the 1990s, the global production of oil has exceeded new petroleum reserves discoveries, which suggests that peak oil is a reality at the scale of a few decades (Kuhns and Shaw 2018). Assuming a 'business as usual' scenario, rationing of transportation fuels by only a small percentage value of the market (or a major increase in transportation fuel prices) would be sufficient to generate a dramatic economic impact including societal and protest issues due to car dependence as recently witnessed in the UK and in France (Geroe 2019; Monasterolo and Raberto 2019; Robinson and Mayo 2006; Rocamora 2017). Moreover, the exploitation of non-conventional resources such as tar sand comes at a high environmental impact (Kuhns and Shaw 2018).

Climate change is a global phenomenon initiated by anthropogenic atmospheric CO_2 releases resulting from the burning of fossil fuels including coal (in coal-fired power stations) and petroleum that started at the onset of the Industrial Revolution (Figure 25.4). Given that the earth and its atmosphere constitute a closed system, this phenomenon, driven by the 'selective absorption of the atmosphere' of heat, was predicted as early as 1896 by Svante Arrhenius (1896). Global warming resulting from unbalanced cumulative carbon budgets is exacerbated by the release of biogenic methane, and could be further exacerbated to the point of no-return by the release of methane trapped in the permafrost, a phenomenon that would constitute a climatic 'event horizon' (Hope and Schaefer 2016; Millar et al. 2016; Saunois et al. 2016). The current consequences of climate change are already perceptible notably, but not only, in the poorer countries (Harrington et al. 2016), and its future cost is significant with the calculation that an increase in temperature of only 2 °C, that is, the goal formulated at the close of the Conference of Parties COP21 in Paris in 2015 organized by the United Nations Framework on Climate Change, will already cost several points of global GDP. Critically, climate change constitutes a dynamic force that would slow economic growth rates with long-term and worryingly compounded consequences (Diaz and Moore 2017; Tol 2018). Moreover, climate change impacts also an array of related issues the costs of which have not been factored in current economic models. These include financial and human costs (to take only an anthropocentric view), including the economic cost of more frequent and more severe weather episodes, the availability of water resources, transport, large scale human migration to

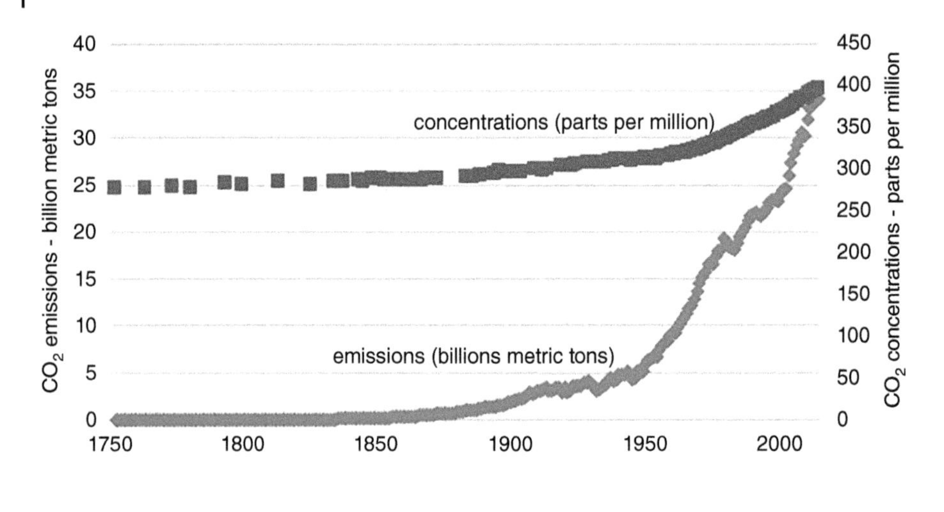

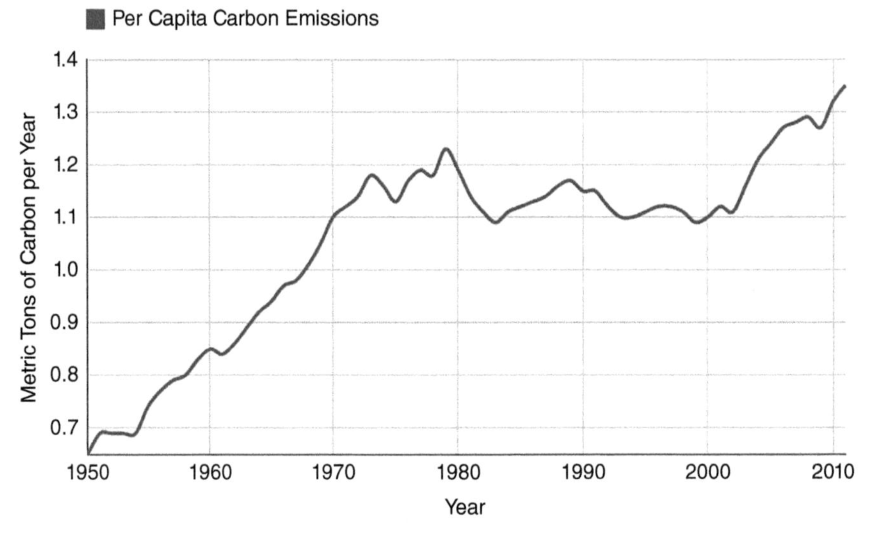

Figure 25.4 World carbon dioxide emissions from fossil fuel combustion and global atmospheric concentrations (1752–2014) and carbon emissions per capita. Since 1751, over 400 billion metric tonnes of carbon have been released into the atmosphere from the consumption of fossil fuels and cement production. Half of these fossil-fuel CO_2 emissions have occurred since the late 1980s. Source: US Energy Information Agency quoting data from Oak Ridge National Laboratory, Carbon Dioxide Information Analysis Centre, Access July 26, 2017 (Anonymous 2018; Boden et al. 2017). Reprinted with permission.

escape the areas that are affected the worst, violent conflicts, energy supply, living space cooling, labour productivity, as well as the tourism and recreation industry (Tol 2018), not to forget unknown (eco)systemic changes of yet unknown consequences (Franklin et al. 2016; Steininger et al. 2015; Stern 2016).

The energy industry, and with it the global economy as a whole, is at a critical crossroad, where decisions with long-term impact accompanied by significant short-term pains must be made. To maintain its course on the projected economic growth trajectory, the global primary energy demand is expected to increase between 2015 and 2040 by approximately 28%, with countries from the *Organisation de Coopération et de Développement Economiques* (OECD) seeing their collective demand decrease by 4% and non-OECD countries, which experience a booming economy, seeing their collective demand increase by 49% (Anonymous 2017). Fossil fuels continue to constitute the greatest share of the global energy mix, with natural gas (up to 47%), petroleum (12%), and coal (2%) use combined being forecast to grow 18% by 2040. In spite of the share of energy demand met by fossil fuels declining, the economy decarbonization index, ID_{50}, of the global economy defined as 50% of the energy being generated by sources other than fossil fuels is not yet in sight as hydrocarbons should still account for 75% of the total global demand in 2040 compared to 81% in 2015 (Anonymous 2017). By comparison, the last time the indicator ID_{50} was met was in 1900 as revealed by available records (Smil 2016). Nevertheless, as a sign of changing times, the International Energy Agency (IEA) estimates that electricity demand will grow much faster (62% global electricity power growth between 2015 and 2040) than energy demand. Interestingly, the use of fuels to generate electricity is expected to grow at a much slower rate (37%) because a large increase in renewable generation will come online. Having said that, the fuel demand is expected to grow 39% in the industrial sector and 27% in both transportation and buildings (Anonymous 2017). In parallel, it is forecast that renewable energy comprising hydropower, Aeolian power, solar power, and biofuels will experience rapid demand growth of approximately 86%. At this rate, renewable energy is expected in 2040 to represent up to 20% of the global energy demand. Remarkably, the developing countries are expected to account for about 70% of the total increase in global renewable energy consumption during the 2015–2040 period. In parallel, nuclear power (for electricity generation) is forecast to grow 49% between 2015 and 2040, with its overall share of global demand increasing to approximately 6%. Under this new policy scenario, CO_2 emissions are expected to rise 10% during the same period in spite of decreasing emissions in developed countries (Anonymous 2017). The pre-industrial value of atmospheric CO_2 was 270–275 ppm; it rose to 310 ppm by 1950, and 310–380 ppm since 1950. This level has reached criticality, with approximately half of this increase occurring in the last 30 years, a phenomenon encompassed in the Anthropocene concept defining humanity as a global geophysical force (Steffen et al. 2007). The impact of this force on animal, plant, and microbial biodiversity and overall on the whole planet Earth itself is so large that whole ecosystems, entire species, wild populations as well as local breeds of domesticated plants and animals are fast shrinking and disappearing with a total of a million species under threat of extinction, which constitutes also a threat to human well-being and life all over the globe, as captured in a report published in May 2019 by the Intergovernmental Science-Policy Platform on Biodiversity and Ecosystem Services (IPBES) (Bonn, Germany), an intergovernmental body that, in response to requests from decision makers, assesses the state of biodiversity and of the ecosystem services it provides to society (Díaz et al. 2019).

Whereas there is a strong macroeconomic need to ensure the supply of ever-increasing levels of energy, in parallel there is a strong need to mitigate the environmental impacts that will be the consequence of this unprecedented increase in global energy demand. As highlighted by Rostand, 'the key challenge for the energy industry in the 21st century – if not for the world – will be to overcome this nexus between energy security and environmental sustainability, and transition the energy system in the most optimal way' (Rostand 2015).

Given technical and infrastructure challenges, including the retrofitting of existing energy production infrastructures, energy transitions take decades (Figure 25.5) (Smil 2010, 2016). It is therefore critical to anticipate the peak oil phenomenon at least two decades in advance of its occurrence as a means to mitigate its negative impacts on the economy (Robinson and Mayo 2006). Likewise, the transition to the ID_{50} point and further to a carbon-free energy operated via the replacement of fossil fuels by renewable energy will take decades, though it could be accelerated by policies (Markard 2018). These practical challenges are further exacerbated by societal and vested interest challenges (Vertès et al. 2006). Those issues are the more acute because the uses of a particular

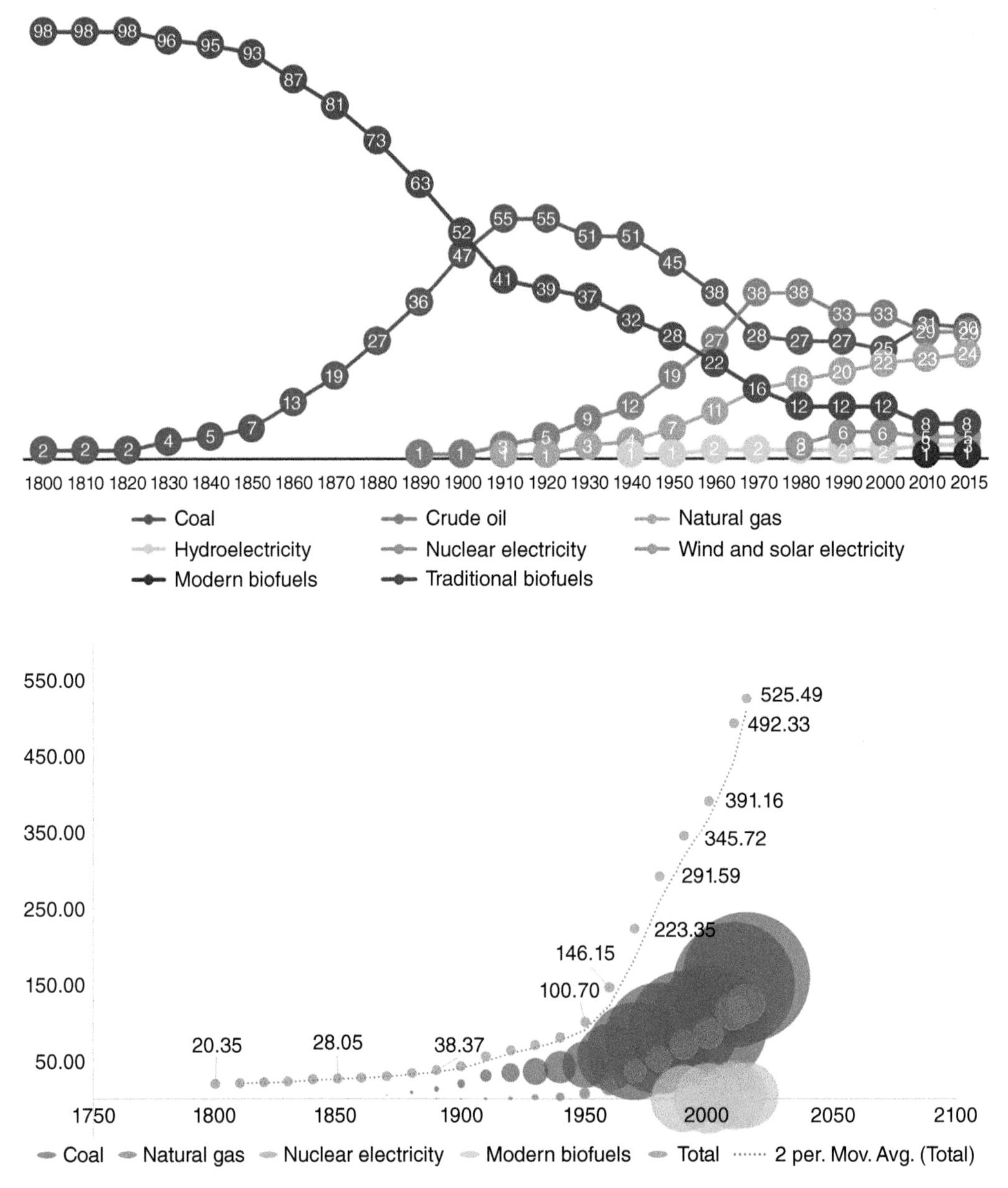

Figure 25.5 Energy transitions take decades. Upper panel global primary energy consumption in percentage. Lower panels: global primary energy consumption in EJ; the bubbles reflect the amounts of energy derived from each main source, the two-year rolling average global consumption is represented by the dotted line. The global energy mix remains heavily carbonized in spite of the emergence of alternative energy sources including nuclear and solar power, and despite a more diverse energy mix today than 50 years ago since when the total amounts of fossil energy use has skyrocketed. Large scale energy transitions all have in common that they take decades, with the time for substitutions depending on the degree of reliance on the dominant energy mix of the previous cycle and how widespread its use is. The forces at play here include vested interest as well as technical and infrastructure requisites. As a result, as emphasized by Smil, 'barring some extraordinary and entirely unprecedented financial commitments and determined actions – none of today's promises for greatly accelerated energy transition from fossil fuels to renewable energy will be realized' (Smil 2010; Smil 2016). Source of data: Vaclav Smil (2016).

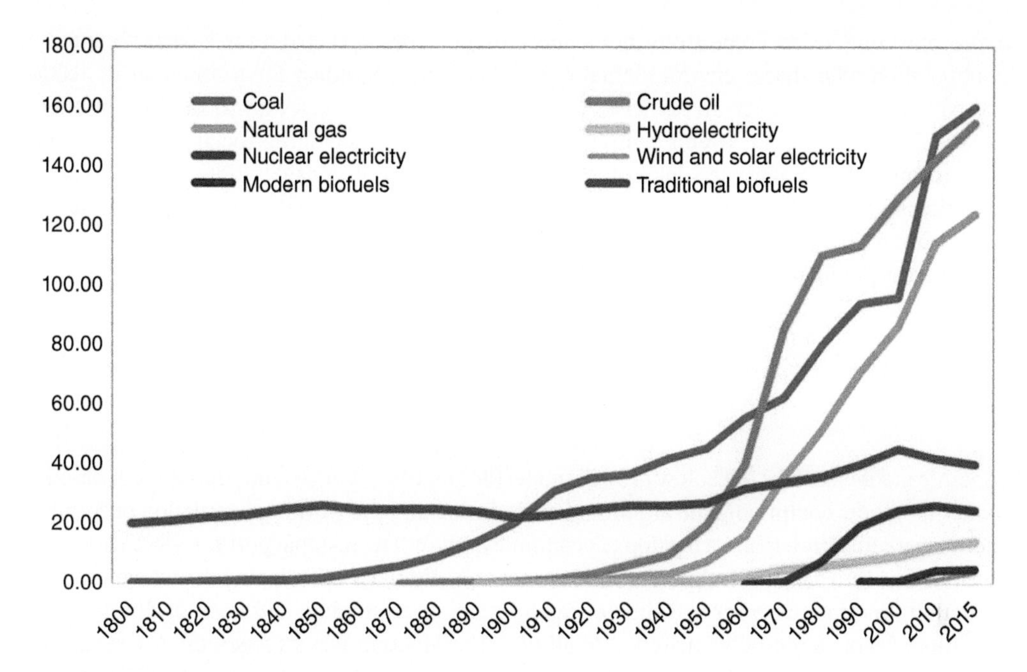

Figure 25.5 *(Continued)*

existing energy source are widespread and end users are locked in an existing technology, such as the combustion engine for transportation for example.

Energy quality is a concept that can be defined as the ability of an energy source to do useful work *in the thermodynamic sense of the term* and to efficiently substitute for other energy forms. Lower quality energy is more dispersed and thus cannot do as much useful work as higher energy quality that is more concentrated, as exemplified by the energy quality of wood as compared to that of petroleum or nuclear fuels. The social cost of carbon (SCC) is a quantity that represents the economic cost caused by the emission of an additional tonne of carbon dioxide or its equivalent. The SCC has been estimated to be \$31 per tonne of CO_2 (in 2010 US\$) in 2015. Under a main assumption, the real SCC is predicted to grow at a compounded annual growth rate (CAGR) of 3% until 2050 (Nordhaus 2017). However, those estimates could be re-evaluated to be well beyond the \$100 per ton depending on the discount rate used in the economic analysis (Nordhaus 2017). Remarkably, the choice of discount rate is a critical one that has the potential to deeply influence the perceived urgency and the choice of policy to mitigate global warming (Stern 2016). To incentivize the reduction of the use of fossil energy, carbon taxes have been implemented in various countries with varying levels of success. Here, Sweden is instrumental in demonstrating that GDP and CO_2 emissions can be successfully decoupled: Swedish GDP increased 75% from 1991 to 2015 whereas Swedish CO_2 emissions decreased 26% during the same period (source: Swedish Environmental Protection Agency, Statistics, Sweden). The tax of SEK 250 (about €24) was introduced in 1991 and levied for each tonne of CO_2 emitted, it has gradually increased to SEK 1180 (€114), its 2019 level. Notably, the carbon tax in Sweden was increased gradually and in a stepwise manner, a method that gave the time for both businesses and households to adapt, thus improving not only the political feasibility but critically the public acceptance of the new tax (Åkerfeldt and Hammar 2015; Dineen et al. 2018). It is particularly remarkable that 95% of current carbon tax revenues in Sweden come from motor fuels, demonstrating that the tax is well accepted by the population who shares environmental concerns (source: Swedish Environmental Protection Agency, Statistics, Sweden). To achieve better sustainability, it is relevant to rank clean energy technologies, according to several complementary dimensions including social, economic, energetic and environmental aspects including raw material life cycle considerations. Based on power generation alone, nuclear energy ranks the highest considering annual generation, capacity factor, mitigation potential, energy requirements, green-house gas emissions, and production costs, whereas solar photovoltaic still lags behind (Dincer and Acar 2015).

25.2 Transportation in the Twenty-First Century: A Carbon Tax Story

Economic models such as the Solow growth model (Solow 1956), which computes growth based on production inputs comprising the amount of capital used and the working population or numbers of hours worked, only explain a portion of economic growth. The residual portion, which is approximated to take into account technological advances contributing to enhancing the efficiency of the production function, whether it is due to 'catching up' by technology diffusion or due to 'novel' innovation, is conventionally attributed to all of the production factors together; it is referred to as total factor productivity (TFP) growth (Nishimizu and Page 1982). As a result, TFP is of critical importance to long-term economic growth, as demonstrated by the observation that over the past century, gains in TFP have accounted for more than half of the growth in measured US labour productivity, that is, in output per hour of work, and as such TFP improvements have contributed more to the measured growth of labour productivity than has growth in the amount of capital per worker (Shackleton 2013). Communication and transportation have constituted fundamental pillars of TFP throughout the history of mankind and trade. Transportation comprises both individual, public, and trade-oriented transportation means, from automobiles to aeroplanes and ships. Like for transportation and the personal cars exemplified by Henry Ford's vision for the model T as a better and cheaper 'motorcar for the great multitude' to democratize access to it (Wells 2007), communication has been revolutionized by the maturity of the computer industry that was achieved by the implementation of Bill Gates' vision of 'a computer on every desk and in every home' coined as early as 1980 (Manes and Andrews 1993). It took pretty much 20 years for the computer vision to be essentially fully realized in the developed world. The information technology (IT) revolution was punctuated in the mid-1990s with the Internet with its wide-ranging applications to create another landmark TFP-related innovation and a new economy (Blinder 2000).

Mapping critical new developments of TFP, one needs to include enhancements to the IT revolution and to communication means, comprising, for example, the coming of age and diffusion of computer-assisted operations and artificial intelligence (AI) technologies including machine learning (Brynjolfsson et al. 2018). Likewise, one needs to consider the continuation and enhancement of mobility, the use of higher quality energy for transportation, such as electricity, and the reduction of air pollution. The development of smart systems to reduce traffic congestion not only helps reduce transportation times, which impacts TFP, but also helps reduce air pollution, which impacts health. Electric vehicles (EVs) (here including both hydrogen fuel-cell vehicles and conventional electric cars) have the potential to bring substantial benefits not only to the environment but also in terms of energy consumption. Long-term environmental benefits largely have been shown to depend on (i) the life-cycle effects with respect to energy, (ii) differences between electric cars and

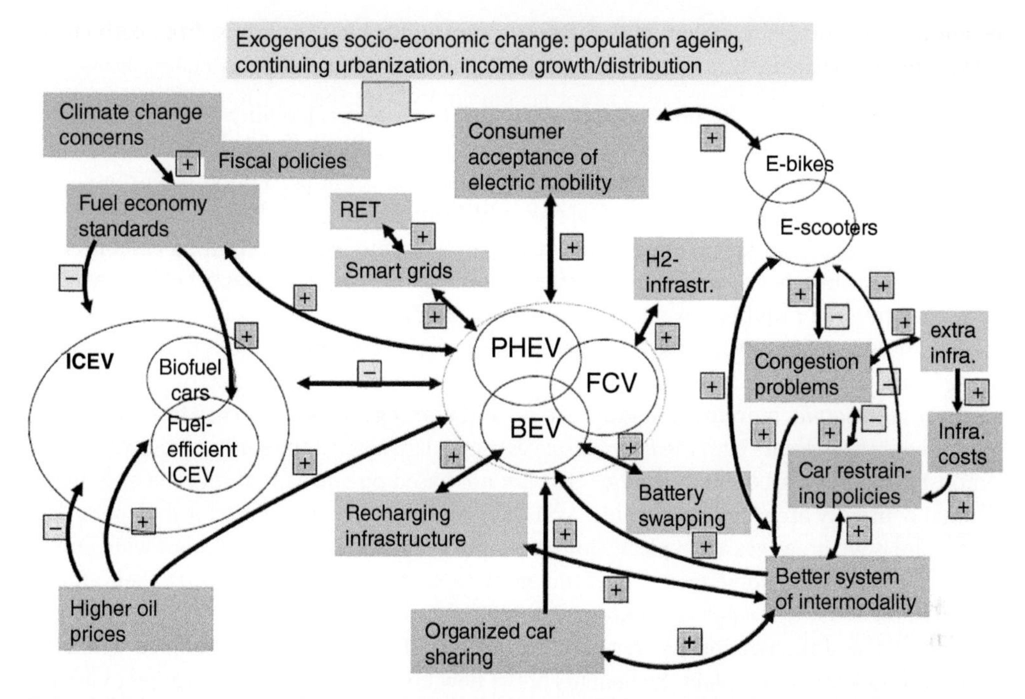

Figure 25.6 Factors promoting (+) or detracting (–) the adoption of different power train technologies and vehicles. Different developments in infrastructure, policy, demand, and congestion may affect and be affected by electric mobility. Electric cars are expected to benefit from the following developments: higher oil prices, better recharging systems, new value propositions such as mobility leasing with battery swapping, urban policies to restrain car traffic for example congestion charges as implemented in central London and to promote clean and silent cars, better systems of inter-modality and the cultural acceptance of electric mobility, and organized car sharing. Some of the developments feed each other: car restraining policies can be expected to stimulate inter-modality that in turn feeds on car sharing and electric mobility. On the other hand, there are balancing developments including the availability of cleaner combustion engine vehicles that will slow down the diffusion of electric vehicles (Dijk et al. 2013). ICEV: internal combustion engine vehicle; BEV: battery electric vehicle; PHEV: plug-in hybrid electric vehicle; FCV: fuel cell vehicle; H_2-infrastr.: hydrogen infrastructure; extra infra.: extra infrastructure. Source: Reproduced with permission (Dijk et al. 2013).

their competitors in 'well-to-wheel' consumption, and (iii) improvements compared to internal combustion engine-powered cars (van Wee et al. 2012). Importantly, policies that intend to promote the diffusion of electric cars need to be adaptive, with for example privileges in central urban areas at the early stages of introduction to provide incentives for adoption, albeit it is clear-cut that these privileges cannot be permanent as the number of electric vehicles on the road increases (van Wee et al. 2012). Perhaps the biggest issue regarding the adoption of electric vehicles is a classical local fitness peak problem, where users are locked into the older (combustion engine) technology; however, local fitness peak escape can be promoted by a variety of policy instruments (cf. Figure 25.6) (Dijk et al. 2013). Considering only cars for 'the great multitude', advances still need to be made regarding battery technologies and obviously the densification of the network of charging stations as well as standardization amongst charging station electricity suppliers (Dijk et al. 2013). As a result, most likely, a combination of 'carrot-and-stick' approaches need to be implemented to promote the deployment of electric cars, with again the carbon tax being a useful instrument as demonstrated in the case of Sweden.

Specifically, regarding the carbon tax, different observations can be derived from carbon tax adoption experiences in different jurisdictions, as follows (Rocamora 2017):

- First of all, the quality of the tax structure is key for its social acceptability and constitutional legality. Cases of failed implementation in France, for example the *gilets jaunes* movement, highlight the importance of (i) tax transparency, (ii) fairness and the perception of fairness, (iii) stability and predictability to limit political risk, and, critically, (iv) affordability. It is worth noting that acceptance can be maximized and costs can be minimized by offsetting carbon taxes with reductions in other taxes, using carbon tax revenues to compensate stakeholders; what is more, investing the resulting revenues in emission reduction activities is a congruent message that enhances tax acceptance. Like the example of Sweden above, jurisdictions that have implemented such carbon taxes have continued to experience strong economic growth. On the other hand, the *gilets jaunes* movement in France, that, as if nothing had been learned from the event, followed only a few years later the *bonnets rouges* movement against the similar *éco-taxe*, clearly demonstrating that carbon tax increases that are not based on substantial revenue and distributional neutrality are simply not viable, as the price of energy, particularly for transportation in rural areas where wages are typically lower, simply cannot be increased without wide public acceptance enabled by some means for compensation and adaptation to enhance their affordability (Geroe 2019; Hourcade 2014; Rocamora 2017).
- Secondly, a sound preparation process and broad stakeholder consultations coupled with efficient communication to explain the benefits of the new tax as an economic instrument with an environmental purpose facilitates the creation of the social consensus is absolutely necessary (Rocamora 2017).
- Importantly, in order to unleash the full potential of a carbon tax as an efficient economic instrument cornerstone to a low-carbon transition, such a carbon tax needs to be placed at the centre of a package of broad economic, social and fiscal reforms. This way, taxpayers will be placed in a position to understand that they are not paying for the environment, but rather that they are accompanying an ongoing change that will affect the whole economy and generate social benefits (Rocamora 2017).

Lithium–ion (Li–ion) batteries are the current predominant design to power electric vehicles. However, the current performance of today's batteries is still insufficient for electric vehicles to replace combustion engine vehicles, even though their displacement has already started, which already translates into significant reduction in CO_2 emissions from transportation (Nykvist and Nilsson 2015; Teixeira and Sodré 2018). Beyond continued reductions in cost due to incremental technology improvements (Nykvist and Nilsson 2015), the widespread deployment of electric vehicles requires high-performance and low-cost energy storage technologies (Cano et al. 2018). Particularly, the long-range, low-cost and high-utilization transportation markets appear not to be well served by current Li–ion-powered electric vehicles. Several fundamental attributes of battery packs need to be improved to fully meet the demands of these vehicle markets, notably: improved specific energy, energy density, cost, safety and grid compatibility. As reviewed by Cano et al. (Cano et al. 2018), lead–acid (particularly lead–carbon) batteries can provide supplementary power for low-cost electric vehicles given their low cost and high specific power, but they have to be paired with a complementary high energy battery due to their low specific energy and energy density. Nickel–metal hydride batteries, in spite of their higher cost and lower specific energy and energy density as compared to Li–ion batteries, may be implemented in place of structural or energy adsorption components in long-range electric vehicles owing to their safer internal chemistry. In parallel, Lithium–sulphur batteries exhibit higher specific energy and lower cost than Li–ion batteries, and are thus preferred in the long-range and low-cost transportation markets. Lithium–air and zinc–air batteries have similarly attractive attributes but they are best suited as range-extenders when coupled with more durable and higher-power batteries. Overall, whereas Li–ion batteries exhibit the best set of attributes for widespread electric mobility applications, the targeted adoption of a diverse mix of battery and fuel-cell-powered EVs would accelerate the full transition to clean, low-carbon transportation (Cano et al. 2018).

The large-scale deployment of battery electric vehicles poses a challenge to the infrastructure of electricity supply as the simultaneous charging of a very large number of electric vehicles in residential distribution grids would result in a significant overload, a problem that would be worsened by a concentrated charging demand before commuting hours or after people return from work, as this would significantly increase the cost for grid operation (Vertès 2010). As a result, coordinated charging of electric vehicles is a crucial problem to be solved for future power systems (Hensley et al. 2009; Sortomme et al. 2011; Vertès 2010). The smart grid vision is that of a high-speed bidirectional communications network that aims at capitalizing on IT and computational intelligence to optimally make use of energy resources in grid operation resulting in enhanced reliability assurance enabled by consumer participation (Gharavi and Ghafurian 2011; Tan et al. 2014). It notably integrates and processes data across electricity generation, transmission, substations, distribution and consumption, with the goal to achieve a system that is clean, safe, secure, reliable, resilient, efficient, and sustainable (Gharavi and Ghafurian 2011). Here, the model of consumer participation proceeds not only from responding to time-based buy-and-sell pricing incentives with the price of electricity varying during peak and off-peak hours, but also to sell back their surplus of energy that they would have generated from their own generators such as solar panels, or simply using their plug-in vehicles as energy storage devices. Remarkably, this seamless integration enables to mitigate the fluctuations of renewable energy resources such as solar and Aeolian power and thus constitutes the basis to be better able to take advantage of these green energy technologies. It is also worth noting that battery storage, as it becomes more efficient and cheaper, constitutes in itself a novel disruptive technology in the power sector since it will facilitate the development of local micro-grids and individual micro-power stations. In turn, the further development of their fundamental attributes and their increased deployment (as a rule of thumb, a 10% market penetration will start having measurable impacts on utility companies' financial bottom lines) will incentivize utility companies to improve their services, incorporate new alternatives for distributed energy, and reduce overall grid-system costs (Frankel and Wagner 2017).

The deployment of electric vehicles is rapidly expanding, with notably the development of electricity generation capacities from the roof-tops of vehicles themselves, which has the added advantage of enabling energy savings from the grid as compared to standard electric vehicles. The use of roof-top solar panels also extends battery lifespan by reducing the battery discharge rate and maximum discharge percentage (Assadian et al. 2016). For example, city buses or delivery trucks can easily be equipped with photovoltaic panels (Figure 25.7). Beyond this simpler innovation to develop more fuel-efficient hybrid vehicles, there is a tremendous impetus to develop novel batteries with a 10-fold higher power as compared to conventional batteries (Evarts 2015).

The deployment on a global scale of electric vehicles is notably impacted by parameters the importance of which is growing over time: (i) high crude oil prices, (ii) carbon constraints, with (iii) changes in fuelling infrastructures, (iv) changes in mobility practices and in the global car market, (v) evolution of energy prices, (vi) climate policies, (vii) and changes in the electric sector influencing innovation in electric engine technology (Dijk et al. 2013). The impact of a global fleet of plug-in electric vehicles on global energy systems remains a matter of debate. In a study, it was calculated that in the short- to mid-term the number of new battery-powered electric vehicles

Figure 25.7 The deployment of e-trucks. Trucks equipped with photovoltaic panels to generate electricity is gaining ground, as exemplified by the Swiss consumer goods company Coop that has started to develop its fleet of hybrid 18-ton trucks equipped with two 120 kWh batteries, with more energy being captured during breaking, resulting in average savings of 22 l of diesel per 100 km as compared to a conventional truck. Source: Photo: courtesy Coop.

would not result in a power-demand shortage crisis albeit this change would be likely to result in a reshaping of the load curve since the model used predicted that when local electric vehicle penetration rates reach the 25% level, peak circuit loads would grow by 30% (Engel et al. 2018). In another study conducted using a large mobility data set collected in Germany, absolute load peaks have been predicted to increase by up to a factor of 8.5 depending on the loading infrastructure, while the load in high load hours is expected to increase by a factor of 3 and the annual electricity demand to double (Fischer et al. 2019). Such uncoordinated recharging would result in significantly increasing peak demand that would require the infrastructure of electricity distribution to be upgraded (Muratori 2018). However, the effect of numerous individual users recharging their vehicles at the same time can notably be mitigated by the deployment of smart grid solutions with dual flow electricity systems where electric vehicle would charge (grid-to-vehicle) and discharge (vehicle-to-grid) in a process optimized for both individuals and utility companies needs and financial incentives. Another impact of the global adoption of battery-powered electric vehicles is a reshuffling in geopolitical and international trade influence in the transportation industry, as evidenced by China's dominant position in Li–ion batteries, both in terms of technology advances and in terms of access to critical raw materials including rare earth elements. In comparison, for example, European manufacturers are lagging behind after having all but lost to China in the race for efficient and cost-effective photovoltaic panels (Goodenough et al. 2018; Kennedy 2013; Knuth 2018; Pitron 2018; Sanderson 2018; Shubbak 2019; Yang and Zhao 2018). What is more, it is worth noting that the integration of renewable energy from fluctuating energy sources such as Aeolian and solar power promotes the development of large battery systems as exemplified in the UK: as the country now sources approximately a quarter of its electricity from renewables, the UK's energy system, National Grid plc (Warwick, UK) was compelled to implement in 2018 a transaction on a total of eight such large battery systems (Clark 2018).

25.3 Cities of Change

Cities are becoming smart on the one hand in the automation of routine functions that serve individuals, buildings, and traffic systems, and on the other in the increasing capability to monitor, understand, analyse and plan a new city or in the majority of cases to retrofit a city with a long history to improve its efficiency, equity and quality of life for its citizens (Batty et al. 2012). As highlighted by Batty et al. (Batty et al. 2012), this vision requires not only 'constellations of instruments across many scales that are connected through multiple networks which provide continuous data regarding the movements of people and materials in terms of the flow of decisions about the physical and social form of the city' but also it requires 'intelligence functions that are able to integrate and synthesise this data to some purpose, ways of improving the efficiency, equity, sustainability and quality of life in cities'.

The IT-driven neo-industrial revolution is enabling via Internet-based applications not only the implementation of smart grids to more efficiently manage power networks and electricity supply and demand, but also this game-changing innovation is driving the development of smart cities. The city of the future is envisaged to be greener, to make more efficient use of scarce resources, to maximize recycling, and to limit CO_2 emissions, thereby offering overall a superior quality of life. Ideally, it will be underlined by improved infrastructures for mobility, for the local economy, and for a sustainable resource management. For example, thanks to information technologies both software and hardware such as fibre networks, the management of services can be optimized via multidirectional information flow, not only from utility companies to consumers and vice versa, but also from devices to other devices including device-to-systems communications, that is, an Internet of Things (IoT) to create a network of sensors and responders, as exemplified in the City of London where thousands of street lights are connected to a mesh network where individual lights play the role of a node in the network that can be leveraged for example to detect pollution (Ejaz and Anpalagan 2019; Rigby 2018). This increased dependence on IT poses issues on its own including ethical ones (Deakin and Reid 2018); however, the increased reliance on IT to optimize the complex system that a megapolis or even a small city represents constitutes a promising resource utilization efficiency maximization tool. Urban areas constitute microcosms of efficiency as transaction costs there are lower, markets are both closer and bigger, and cities benefit from a natural tendency to create clusters of synergistic competence factors via the harnessing of local specialization networks (Clos 2015; van den Berg et al. 2001). As a sign of changing times, urbanization is already massive with more than half of the world population dwelling in concentrated urban areas; this figure is expected to increase up to 75% by 2050 (Clos 2015; Ejaz and Anpalagan 2019). Prosperous urbanization proceeds from (i) an existing strong legal framework, (ii) a carefully thought-out urban design that may include significant retrofit to evolve areas such that better use of the digitalization revolution can be made or sustainable buildings built to maximize resource use, and (iii); all this requires that a financial plan to provide the appropriate level of funding is in place (Clos 2015). The concept of smart cities is already being tested in a number of large cities, for example in Europe including London, Manchester, Amsterdam, Barcelona, and Malmö (Deakin and Reid 2018). Beyond the deployment of technologies, there is a tendency for cities to become greener in the sense that architects now consider more than ever before the use of live foliage to make use of the large surface areas of city buildings and leverage the power of photosynthesis to sequester atmospheric carbon via the construction of biomimetic buildings (Ling and Chiang 2018). Similar to this vertical greening trend, solar panels are being used at a fast increasing rate on essentially all appropriate surface areas such as on the roofs of building to enable the generation of 'city energy' as cheap local power

(Page 2018). In the USA, the installation of solar photovoltaic panels have notably been incentivized by policies from federal, state and local governments; however, as demonstrated from the analysis of a sample of 186 US cities, cities that are deploying local financial incentives harbour 69% more solar photovoltaic capacities than cities without such policies (Li and Yi 2014). In addition, cities that are subject to Renewable Portfolio Standard legislation (RPS) requirements have 295% more solar photovoltaic capacity than cities that are not regulated by state RPS (Li and Yi 2014). There are various important benefits to equipping more urban surface areas with solar panels, since photovoltaic panels not only reduce global warming by minimizing CO_2 emissions for energy generation but also locally they help limit urban heat island effects (Oke 1967). In addition, the production of city energy generates both global and local benefits. First of all, solar panels deployed on urban surface areas help diminish the emission of greenhouse gases by generating sustainable electricity or usable heat. Secondly, photovoltaic panels reduce the urban heat island effect by absorbing solar energy and thus directly modify the energy balance of the urban surface in contact with the atmosphere, in other words, they influence the urban micro-climate. Moreover, solar panels change the levels of energy received by roofs, and thus modify a building's energy balance. In summer, this translates into a decrease by 12% in the energy needed to power air conditioning units, albeit in winter this results in a 3% increase in the need of energy for domestic heating (Masson et al. 2014).

Solar photovoltaic installations have been growing very rapidly in the United States over the last few years, incentivized as discussed above, by policies from federal, state and local governments (Li and Yi 2014), and again as highlighted above, the impact of policies is dramatic regarding the deployment of photovoltaic capacity (Li and Yi 2014). The potential of 'solar cities' thus appears to be nothing short of phenomenal, as exemplified by the suitable rooftop area of Los Angeles and San Francisco being able to generate respectively 60% and 50% of the electricity demand of those cities, and in general rooftop solar power could generate 36.8% of the US electricity demand (Gagnon et al. 2016). Similarly, it was calculated that the city of Mumbai has a solar energy generation potential

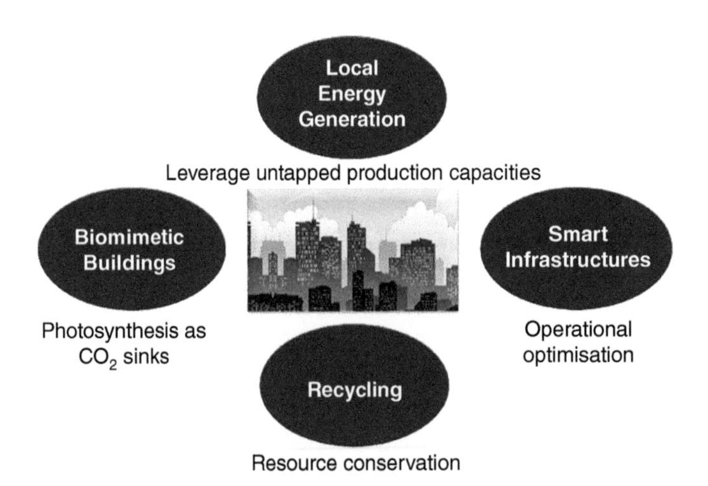

Figure 25.8 Smart cities of the future. Four axes of development are clearly emerging for the cities of the future. The implementation of a dense network of connected devices represents a robust pathway to increase operational efficiency, including particularly energy use. The development of biomimetic buildings that make use of live foliage to harness natural photosynthesis helps reduce pollution while mitigating some of the heat island effects typical of large cities. Resource conservation via increased recycling is a trend that will only increase, which offers the dual benefit of reducing pollution and of opening a sustainable future. A paradigm change, however, is constituted by the local generation and use of energy by leveraging roof surface areas that have to this date been largely untapped.

of 2190 MW using median efficiency panels at an annual average capacity factor of 14.8%, and the large scale deployment of rooftop solar photovoltaic systems could provide 12.8–20% of the average daily demand and 31–60% of the morning peak demand for different months (Singh and Banerjee 2015). Noteworthily, photovoltaic panels could be either for local use or grid-connected. Ideally, grid power should primarily be used to serve industrial needs, while households would primarily use the electricity they generate using their own (photovoltaic) electricity generators.

To sum this up, the efficiency of cities and the quality of life that they offer can be enhanced via four critical pillars: (i) smart infrastructures, (ii) biomimetic buildings, (iii) recycling, and (iv) local energy generation for local or grid use (Figure 25.8).

Defining globalization as global processes increasingly shaping local social relations, another interesting viewpoint is that the emergence of the concept of smart cities signals a departure from the fast forward globalization trend of the early 2000s, where increased focus is now being placed on harnessing the economic potential on local economies that imply, and promote, the local grounding of capital and labour (Lawrence and Almas 2003; Le Heron 2016).

25.4 The Chemical Industry Revisited

The demand for chemicals is predicted to grow at a rate that will double the rate of total energy demand through 2050, whereas the demand for electricity will be twice as large as that for other energy sources, with solar and Aeolian energy representing approximately 80% of the net added capacity until 2050 and 34% of energy generation (Roelofsen et al. 2016). A growing factor of this trend is that light electric vehicles are expected to represent 50% of the new cars sold in the European Union, the USA, and China, and about 30% globally. Here again, policies are important tools to catalyse a transition to a decarbonized economy, as exemplified in Europe by France, joined shortly thereafter by the UK, that have announced in 2017 a ban on the sale of new diesel or new petrol vehicles from 2040 (Marchand 2018). In parallel, the demand for chemicals is expected to decline as the impacts of global plastic recycling increases and as plastic packaging efficiency increases. Notably, assuming global plastic recycling increases from 8% in 2016 to 20% in 2035 and plastic packaging demand decreases by 5%, the demand for liquid hydrocarbon for chemicals manufacturing could be 2.5 million barrels per day lower than under business-as-usual conditions. What is more, gains in energy usage efficiency translate into a decrease in the energy intensity of GDP growth, which could globally fall by 50% by 2050 (Roelofsen et al. 2016). Petroleum price volatility and petrochemicals are intimately linked, with petroleum price shocks exerting three major effects on the chemical industry: (i) crude oil is one of the major cost drivers of the petrochemical industry, (ii) the main drivers of the cost of most commodity chemicals are the production costs and thus both volatilities are linked, (iii) the price of crude oil has a direct impact on demand, which is adjusted upwards or downwards to mitigate price hikes or to benefit from price cuts (Hong et al. 2015). All of these variables that directly impact cost structures and price mechanisms have an obvious direct and significant impact on the profit margins of chemical producers and constantly reshape value creation and competitive dynamics in the petrochemistry industrial sector (Hong et al. 2015).

Nonetheless, the chemical market is expected to continue to grow in the coming decades to pass the $5 trillion bar in 2020; this reflects the current strong capital market performance of this industrial sector as compared to other sectors (Ezekoye et al. 2018). Numerous commodity and specialty chemicals can already be manufactured in a sustainable manner using biomass feedstock, a development that constitutes nothing else but a return to the origin of chemicals manufacturing to the times when lactic acid was produced commercially from biomass (Benninga 1990). Likewise, the electric vehicle is an old idea that is becoming new again, with the first electric vehicle (a small locomotive) having been built in 1835, and the first electric car having been put on the road in 1890. Remarkably, in 1900 electric cars represented a third of vehicles on the road in the USA until the introduction of the Ford Model T in 1912, the rest being essentially horse-driven vehicles (Anonymous 2014). Such green chemicals can be manufactured in a so-called 'biorefinery', that is, a chemical plant where the main feedstock is of biomass origin (Vertès 2014). The production of chemicals from biomass constitutes an opportunity to reduce petroleum use. The biorefinery design aims at mimicking the model of the petrochemical refinery where conventional chemical manufacturing is operating along a product tree of chemical precursors and which proceeds from lower value and higher volume feedstock (comprising petroleum and natural gas), to lower volume but higher value chemicals (comprising polymers, speciality chemicals, and active ingredients), thereby harnessing economies of scale, economies of scope, and economies of experience (Kannegiesser 2008; Philp 2018). However, it is hard to imagine how a biorefinery could cost-effectively leverage economies of scale on a par with the petrochemical industry considering the lower energy density of biomass as compared to crude oil. As reported earlier, the implementation of the biorefinery vision requires several critical success factors: (i) the supply of a subset of existing fuels and chemical building blocks, (ii) the supply of novel chemical building blocks to create novel materials with novel attributes, (iii) the management of some risks associated with the breakdown of the petrochemical supply chain, (iv) the creation of jobs including positions in rural areas that offer suitable logistics, (v) the increased management of global-warming risks, and (vi) the valorization of waste including agricultural and certain types of urban or industrial waste (Vertès 2014). Critically, the roadmap for achieving a sustainable chemical industry is determined by the management of several risk factors, including (i) technological risk, (ii) feedstock price volatility and supply risk, (iii) market risk, (iv) financing risk, and (v) policy risk (Vertès 2014). Ideally, integrated processing should be implemented to manufacture an array of commodity chemicals and high-value-added chemicals from various feedstocks comprising particularly lignocellulosic materials, and full use should be made of all the carbohydrate fractions of the hydrolysates derived from these raw materials. Whereas base chemicals such as lower olefins and benzene–toluene–xylene are not only still too cheap but also require too much carbohydrate to be produced competitively, more oxidized products can already be manufactured cost-efficiently (Straathof and Bampouli 2017). Notably, and making use of the more oxidized redox potential of biomass as compared to petroleum, several products that require multiple conversion steps in petrochemical production, such as adipic acid, acrylic acid, acrylate esters, and 1,4-butanediol, can in theory be produced competitively from carbohydrates (Straathof and Bampouli 2017). A total of 12 commodity chemicals that can be manufactured from biomass has been identified to represent near-term opportunities, including: butadiene (1,3-), butanediol (1,4-), ethyl lactate, fatty alcohols, furfural, glycerine, isoprene, lactic acid, propylene glycol, succinic acid, and xylene (para) (Biddy et al. 2016). The industrial productions of lactic acid or succinic acid are examples of the successful production of sustainable commodity chemicals, with biomass-derived succinic acid having a carbon footprint as determined by life cycle analysis (LCA) that is more than 60% lower than that of fossil-feedstock-derived succinic acid (Benninga 1990; Jiang et al. 2017). At the time of writing, there are four commercial plants manufacturing biomass-derived succinic acid for a consolidated annual capacity of 65 kt (Myriant, Lake Providence, LA, USA; BioAmber, Sarnia, ON, Canada; Succinity, Montemelo, Spain, and Reverdia, Cassano Spinola, Italy); all these companies target specialized markets rather than bulk markets since to produce a price-competitive platform chemical the cost of bio-succinic acid should be near $1/kg (Ferone et al. 2019).

While petroleum-derived chemicals still constitute the greatest share of chemicals produced to this date, it is envisaged that biorefinery products will increasingly displace some of the products of the petrochemistry, starting with speciality chemicals, which are expected to represent a global market of approximately half a trillion USD by 2020 (Anonymous 2015). Numerous microorganisms have been engineered to transform biomass into an array of useful industrial chemicals comprising bio-ethanol, bio-diesel and bio-butanol including yeast, algae, Gram(+) and Gram(−) bacteria (Banner et al. 2011; Vertès 2016). What is more, various processes have been developed including processes where the fermenter biomass production stage and the product production stage are uncoupled as a means to direct more carbon fluxes towards product formation rather than towards biomass generation (Vertès et al. 2008). The application of systems biology techniques to guide metabolic engineering approaches have enabled the practitioner to significantly gain in productivity and yield of microbial processes albeit further enhancements still being required to sufficiently improve process economics for bulk chemical production (Clomburg et al. 2017; Nielsen and Keasling 2016; Vertès et al. 2012). The technique of synthetic biology further complements conventional heterologous gene expression methods to create de novo synthetic pathway or alleviate tight metabolic network control (Figure 25.9) (Chubukov et al. 2016; Nielsen and Keasling 2016).

Biorefineries will play an increasingly important role to supply not only chemicals that are currently produced by petrochemistry, but also novel chemicals, given that biomass has a fundamentally different redox potential to petroleum, which is significantly more reduced (Vertès 2010; Vertès 2014). Moreover, increasingly there will be intersecting synthesis routes where biomass-derived precursors and chemical-derived precursors will be reacted together to generate chemicals. Depending on specific considerations such as the cost of precursor synthesis and redox states, and where products and useful co-products (including not only various chemicals but also energy or recyclable biomass) will be generated in parallel to make maximum us of the carbon feedstock up to the theoretical maximum of the carbon feedstock (Kim and Han 2017; Vertès 2010; Vertès 2014).

Several biorefineries are already in commercial operation, as exemplified in Europe that has several biofuel-driven biorefineries, biobased chemical building blocks biorefineries, and biomaterial-driven biorefineries. The European chemical building blocks biorefineries notably produce azelaic and pelargonic acid (Matrica, Italy); 1,4-butanediol (Novamont, Bottrighe, Italy); 2,5-furandicarboxylic acid (Synvina, Antwerp, Belgium); lactic acid (Glanbia Ingredients, Ireland); levulinic acid (GF Biochemicals, Caserta, Italy); methanol (BioMCN, Delzjil, the Netherlands; Air Liquide, Akzo Nobel Specialty Chemicals, Enerken, the Port of Rotterdam, Rotterdam, Netherlands); 1,3-propanediol and butyric acid (METabolic Explorer, France); and succinic acid (Reverdia, Cassano Spinola, Italy; Succinity, Montmelo, Spain).

Transport biofuels comprising bioethanol, biobutanol, and biodiesel mark a transition from the combustion engine to the electric engine (Vertès 2010). These fuels are predicted to capture a significant market share in the short- to medium-term (15–30 years), but have been modelled in various independent studies to be substituted in the longer term by electric and hydrogen cars (Ahlgren et al. 2017; Sperling 2018). The diffusion of electric and fuel cell-powered cars as well as the diffusion of electricity-powered lorries is already encouraged by policy incentives in several countries combining subsidies, taxes and investment incentives. In addition, as emphasized above France and Britain, but also China, have all signalled that the sale of cars that are not electric will be banned by 2040, (Hensley et al. 2009; Langbroek et al. 2016; Marchand 2018; Meckling and Nahm 2019; Sperling 2018). These initiatives are critically important to circumvent hurdles of vested interest and hurdles to technology adoption (Sperling 2018). The future of the automobile industry is

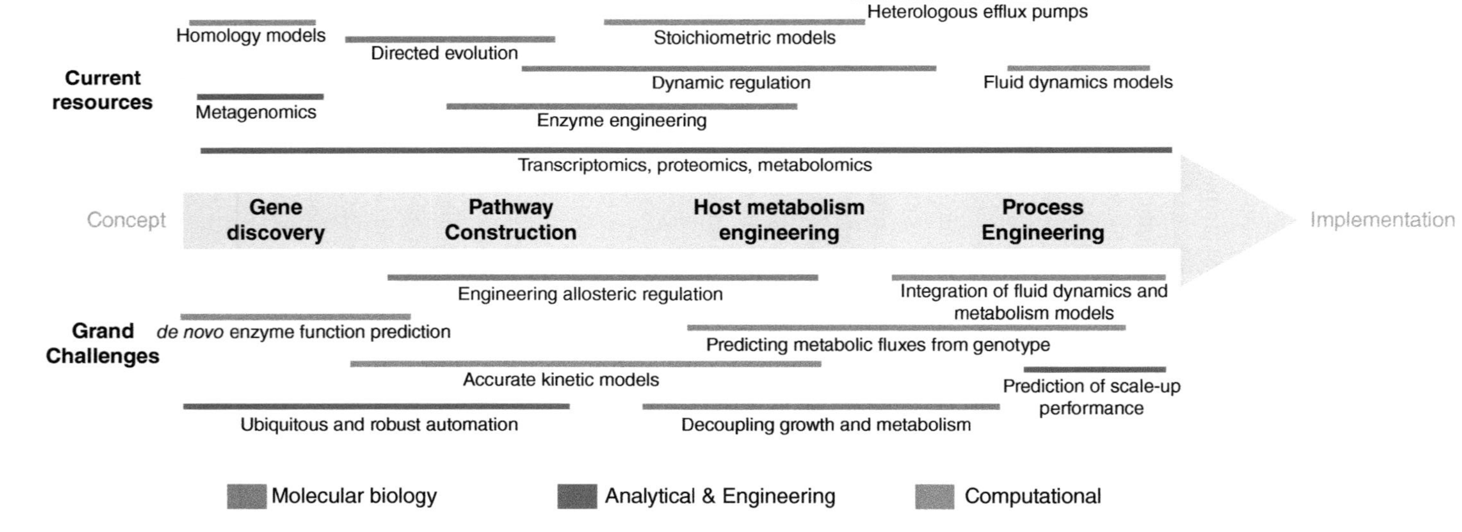

Figure 25.9 Synthetic and systems biology flow chart for the production of commodity chemicals. The process of engineering microbial workhorses for the production of commodity chemicals is represented from initial concept (target molecule selection) to scale up (process engineering and implementation), along with the enabling molecular engineering toolbox applicable to each step and the guiding objective. The lines, coloured according to the type of tool or challenge, indicate which parts of the process the tool or challenge applies to (e.g. dynamic regulation can be used for pathway construction but also for control of toxic intermediates that affect host metabolism). The engineering process is iterative by nature. Source: Reproduced with permission (Chubukov et al. 2016).

electric. This vision was clear in 2010 (Vertès 2010), it is even clearer today, a decade later. This paradigm change will reverberate not only in the car industry but also of course in the biofuel industry as well as in the utility industry and create new markets, notably that of high efficiency and high load batteries (Hensley et al. 2009; Sperling 2018). Notably, for automakers, electric cars represent a major threat of substitution that, if not managed properly, could promote bankruptcies of a similar scale as those experienced by silver halide firms when digital cameras came of age (Habersang et al. 2019). Moreover, this evolution towards electricity-powered transportation will affect the economies of countries deriving most of their revenues from crude oil production, with the coming changes already being signalled by diversification efforts exemplified by the building by 2030 of a series of major solar parks in Saudi Arabia totalling a 200 GW capacity with a budget of US$200 billion (Alnaser and Alnaser 2019; Raval and Inagaki 2018; Salam and Khan 2018). Notably, the acceleration in the use of solar (and Aeolian) energy in Gulf countries are expected to generate numerous benefits that will deeply permeate the economy of the region, including: (i) significant reduction in the solar electricity prices; (ii) increased investments in renewable energy; (iii) innovative house designs and integration of renewable energy devices; (iv) rise of service and maintenance businesses from solar technology companies; (v) training of a skilled workforce by the establishment of new academic programs in renewable energy technologies; (vi) building the experience curve on the connection of renewable energy sources to national grids; (vii) significant reduction in the carbon footprint per capita; (viii) more efficient and low consumption household and industrial devices; (ix) boost in the battery industry for solar electricity storage (Alnaser and Alnaser 2019). Notably, the Saudi government has implemented several plans to greatly diversify its economy, a process that is accelerated by improving the legislative environment, fostering competition, and attracting large multinational firms to enter the Saudi market given that these foreign direct investments will help raise the level of efficiency of the private sector there and thus contribute in Saudi Arabia achieving its main goal of being less dependent on the oil sector (Albassam 2015; Banafea and Ibnrubbian 2018).

25.5 Paradigm Changes in Modes of Consumption

Recycling is critical to sustainability. Waste production is rising fast as the global population and particularly the global population of city dwellers in affluent cities is increasing; reducing the production of waste and converting those wastes that are unavoidable into useful products is increasingly being perceived as being a critical part of the solution (Hoornweg et al. 2013). The global revenue of the plastic, resin, and synthetic fibre industry exceeded $630 billion in 2019 (Anonymous 2019). This industry thus generates very large amounts of waste in the form of plastic solid waste, with packaging representing more than 40% of the European and US demand (Anonymous 2013; Hopewell et al. 2009). A lower range estimate is that 150 million tonnes of solid plastic waste are generated each year, the majority of which ends up in landfills and incineration plants despite its potential for reuse (Rahimi and García 2017). Chemical recycling appears to be a more efficient mode of recycling since the intrinsic properties of most plastics are compromised after several cycles of reuse (Rahimi and García 2017). Treating used plastics as chemical feedstock by chemolysis for the generation of fuels or compounds for the chemical industry is a particularly attractive approach as compared to the technique of mechanical recycling by melting and regranulation, with the cleavage of polymers being achieved by an array of techniques including catalysis, or the action of chemical agents or of temperature (Vasudeo et al. 2016). Whereas today only 12% of plastic waste is recycled (European plastic is recycled at 26%), increasing this rate to 50% by 2030 would represent two thirds of the petrochemical and plastic sector pool growth ($60 billion) (Clark 2015; Hundertmark et al. 2018). This potential represents, as in any major change, both threats and opportunities for the chemical and petrochemical sectors (Hundertmark et al. 2018). This represents also a powerful force of change in oil demand, which under such a scenario is expected to decrease by 30% as compared to a business-as-usual scenario (Hundertmark et al. 2018). Needless to say, implementing the vision of recycling a majority of the plastic used requires an efficient value chain and logistics to drive the collection and revalorization of bulky used plastics. Indeed, petrochemical and plastics companies have to prepare to adapt to another business model as they will have to source plastics-waste supply from a large number of scattered players rather than sourcing their raw materials in bulk from one source (Hundertmark et al. 2018). Here again, policies could help tremendously in implementing this change of paradigm from all polymer synthesis from petroleum to recycling plastic polymers, with for example increased public procurement to prime the pump of the new market, a ban on the incineration of recyclable materials, and interestingly providing specifications on the design of plastic products as a means to decrease the number of different polymers as well as the number and usage of additives, as a means to facilitate recycling and enhance the quality of recycled products (Milios et al. 2018). The implementation of a robust value chain where all its actors are appropriately incentivized is an important part of the implementation of efficient plastic recycling (Mwanza et al. 2018). Moreover, given that virgin plastic manufacturers can gain cost of goods advantages when the price of crude oil is low (Clark 2015), petroleum price volatility has a strong impact on the cost-efficiency of plastic recycling, despite the enormous sustainability and pollution-avoidance impact of recycling. Integrating their incremental cost of pollution into the cost of virgin materials, for example in the form of a commensurate carbon tax, which in the business model of today is inefficiently implemented (Lin and Jia 2019; Philp 2018), would provide a means to compare on the same footing the cost versus benefits of the two product classes, another means of course to reduce the impact of crude oil volatility remains to provide, in a politically and internationally acceptable manner, subsidies to the recycling industry to reduce the profit margin gap.

25.6 International Action for Curbing the Pollution of the Atmosphere Commons: The Case of CFCs and the Ozone Layer

A success story of the industrial world of the twentieth century is the Montréal protocol that was negotiated, signed and enacted in 1987 by 197 countries to gradually eliminate the production and consumption of substances that deplete the ozone layer (Benedick 1998). As a result of this international protocol, chlorofluorocarbon gases (CFCs) that were used as refrigerants in various appliances including air conditioning units, freezers, and refrigerators were banned and replaced by less environment-harmful refrigerant gases. Remarkably, this treaty, was the first treaty in the history of the United Nations to receive universal ratification; it was implemented widely. While it is rightly considered to be the most successful global action for the environment, it importantly signals the feasibility of such global actions and this brings hope to achieve the reduction of deleterious effects of climate change. A global deal on climate change remains the key, albeit national policies help prime the pump to the new industry via bridging gaps in investments and actions to reduce CO_2 emissions. As a result of the Montréal protocol, the ozone hole over Antarctica has been slowly decreasing, and current projections suggest that the ozone layer should return to its 1980 levels between 2050 and 2070 (Douglass et al. 2014). This success, however, is only a first step as research and development continues to identify superior refrigerant materials that have a lower greenhouse effect that those in use today and require lower energy inputs, an important goal since up to 30% of the world's electricity is used for refrigeration purposes (Li et al. 2019). Remarkably, colossal barocaloric effects, i.e. large cooling effects of small pressure-induced phase transitions, have been achieved near room temperature in plastic crystals (Li et al. 2019; Lloveras et al. 2019). These novel materials have thus the potential to achieve the important goal of environment-friendly cooling without sacrificing performance.

25.7 Social Activism as an Engine of Change: Requiem for a Wonderful World

Corporate social responsibility (CSR) is an old concept that has taken major importance in the last two decades, notably in terms of consumer and even investor loyalty. A short definition of CSR is that of an operational concept urging that all stakeholders of the firm must be treated ethically or in a responsible manner (Carroll 1999; Dahlsrud 2008). Remarkably, CSR operates as a driver of sustainable development, with the concept evolving as a subset of 'corporate sustainability', discussed in detail in Chapter 24 of this monograph (Moon 2007). A definition of corporate sustainability responsibility is that of a concept urging a firm to meet the needs of its direct and indirect stakeholders but without doing so on credit and thus compromising its ability to meet the needs of future stakeholders (Dyllick and Hockerts 2002). What is more, CSR has a measurable impact on a firm's financial strength and performance due to enhanced investor trust, reputation and customer perception expressed in terms of social capital and trust. Firms that show proactive support to social responsibility and environmental sustainability have been observed to post significantly higher profit margins relatively to others in the same industrial sector (DiSegni et al. 2015; Lins et al. 2017; Orlitzky et al. 2003; Wang and Sarkis 2017).

Climate change is here to stay. The effects of climate change are already perceptible, notably its impact on agriculture and on water scarcity, or on sea levels, or on biodiversity (Barbier et al. 2018; Campanhola and Pandey 2019; Díaz et al. 2019; Gosling and Arnell 2016; Mantyka-Pringle et al. 2015; Pilkey and Pilkey 2019). The costs of climate change are already significant, an expression of which are increases in the premium of insurances (related to climate change) with the novel tendency to calculate risks using predictive algorithms rather than historic data (Linnerooth-Bayer and Hochrainer-Stigler 2015; Stromberg 2013). The costs of climate change will only worsen, including not only the costs of extreme climatic events but also of decreased productivity and increased poorer health and mortality due to increased summer temperatures (Auffhammer 2018; Tol 2018; Vöhringer et al. 2019). Realistically, it appears unlikely that countries can meet the 2 °C target of international agreements, even if ambitious policies are introduced in the near term (Mehling and Tvinnereim 2018; Nordhaus 2018). From an economic viewpoint only, the SCC and the price of carbon needed to achieve current targets have risen over time as policies have been delayed (Nordhaus 2018) due to short-termism approaches. General short-termism attitudes and practical inaction of various governments in the face of climate change have allowed the future costs of climate change to worryingly compound, which will in time make its consequences absolutely irreversible, especially when methane starts being released from the permafrost at a massive scale (Hope and Schaefer 2016; Paulson 2015; Saunois et al. 2016). Due to natural cognitive biases, it is hard to appreciate the true effects of a compounded or exponential phenomenon (Kurzweil 2004). As a result, as paraphrased from H. Paulson (Paulson 2015), a former United States treasury secretary and chairman and CEO of Goldman Sachs, it is very hard indeed to get governments 'to do anything controversial or difficult unless there is an immediate crisis in immediate sight'.

Adapting to climate change has thus become an inescapable reality. A first major adaptation is to 'decarbonize' electricity (Audoly et al. 2018). This requires that coal-fired power stations be without delay fully replaced by 'carbon-free' technologies, essentially solar power, with its potential to fuel not only systemic-grids but also microgrids and individual users, and nuclear power, which is still necessary until sufficient photovoltaic capacities are reached, which is unlikely to take fewer than half a century to be achieved. A powerful policy instrument is of course novel regulations and legal mandates (Mehling and Tvinnereim 2018). Incentives could be implemented transnationally by including in the SCC a parameter reflecting the energy mix of particular jurisdictions, with bonus coefficients to promote the deployment of 'greener' technologies, for example coal versus nuclear or solar, and nuclear versus solar energy where the costs of pollution and of recycling of spent feedstock are also factored in. Another example of a required major adaptation is to evolve agricultural practices to the new reality, such as by fully defining and implementing a climate-smart agriculture concept that includes appropriate soil management practices with the aim to achieve in parallel a sustainable increase in food production, while improving the resilience of farming systems and contributing in mitigating climate change when possible (Burke and Emerick 2016; Chandra et al. 2018; Hellin and Fisher 2019; Paustian et al. 2016).

Social activism can act as a policy catalyst by giving a sense of urgency to public and private actors, as observed in the pharmaceutical industry to tackle rare or neglected diseases (Siegel and Jakimo 2015). Likewise regarding climate change, activism is an important force that has the potential to contribute to closing the gap between awareness and action, via a mechanism where activists help create the conditions that are necessary for social innovation by catalysing changes in beliefs thereby influencing behavioural changes, but with corporations, and eventually the public sector, taking the lead in developing the new practices (Carberry et al. 2017; Doyle 2016; Gunningham 2018; Swim et al. 2019). An example of how the implementation of the CSR concept bringing both social and corporate benefits including reputational ones is proactively deployed by large corporations is the switch of the pharmaceutical company Novo Nordisk (Bagsværd, Denmark) from fossil fuels to renewable energy (Morsing et al. 2019; Nussbaum 2009). In 2015 Novo joined the global collaborative initiative of businesses RE100 that brings together influential businesses committed to 100% renewable electricity with the purpose of accelerating change towards zero carbon grids at a global scale, and in so doing committed to a 10-year plan to using 100% renewable power for its operations and transportation. Novo announced in 2019 a $70 million investment in a 105-MW solar energy installation to be built in Pender County in North Carolina aiming at providing, by 2020, renewable electricity to all its existing US offices, laboratories and manufacturing facilities, and to achieve the same for all its operations around the world by 2030 using its 'circular-to-zero' environmental strategy to transform Novo from a company running conventional operations to a company running operations with zero environmental impact (Morsing et al. 2019; Palmer 2019).

25.8 Perspectives: A Brave New World

Infinite economic growth in a finite world is not possible, without the equivalent recycling of resources utilized per cycle. With the exacerbation of the consequences of global warming that can no longer be denied, the industrial world has entered a period of radical change. The last period of fundamental change was that of the late 1990s and early 2000s with globalization and the virtual reduction of distances by the quality of communication driven by major advances in IT and more affordable transportation. This resulted in unequalled global convergence and the implementation of comparative advantages by moving manufacturing jobs in areas where labour cost arbitrage could be found. This strategy undoubtedly generated economic benefits to many despite implementation pains for some and created global markets of standardized products, thus enabling corporations to leverage yet unmatched economies of scale and thus granting consumers access to cheaper products (Levitt 1993). However, the role of manufacturing nowadays has evolved, and in advanced economies, manufacturing now acts as a promoter of innovation, productivity and trade to a much greater extent than growth and employment (Maniyika et al. 2012). There is thus a rethinking of international trade where labour cost arbitrage exists; a clear sign is that trade intensity, measuring the ratio of gross exports to gross outputs, has globally decreased by 5.6% between 2017 and 2007 (respectively, 22.5% and 28.1%), with the deceleration in trade intensity being more pronounced in the most complex and most traded value chains, and driven by a change in the geography of global demand (Lund et al. 2019). Particularly, manufacturers with complex needs may very well assess how to maximize their reach to low-cost transportation, consumer insights, and highly skilled workers, which is making higher-cost nations currently consider a return to manufacturing (Maniyika et al. 2012; Marsh 2012). Here, seven critical success factors for the manufacturing of the future have been highlighted: (i) green thinking and the environment imperative, (ii) global supply chains and networked manufacturing, (iii) technological acceleration, (iv) return to the dynamics of economic clusters, (v) worldwide choice of suitable manufacturing jurisdictions, (vi) niche thinking and emphasis on design of world-class manufacturing, (vii) serving customer needs better by conducting personalized production schemes (Marsh 2012). IT advances furthermore enabled not only smart grids that make the consumer's role evolve as one of a full buy-and-sell actor, notably of the energy market, but also local grids that, for example, bring electricity to rural areas and enable a more efficient use of renewable energy (Schleicher-Tappeser 2012; Tollefson 2014; Willemin 2017). Implementing a local model of energy production and distribution alongside a central grid system is fundamental to energy sustainability (Vertès 2010). A central consideration is perhaps that neglecting the SCC constitutes an unsustainable subsidy to the transportation industry, and thus a corrective plan appears urgently needed, but the issue is tremendously complex since it impacts not only global trade but also multiple stakeholders with diverging viewpoints and interests.

Putting all these perspectives together, maximizing quality of life and quality of the living environment that surrounds human societies can be congruent to maximizing economic growth. A lesson from the private sector is that a long-term view generates superior returns (Barton et al. 2017). Neoliberalism, which promotes free market capitalism, enabled globalization and privileged industry in particular by shifting climate change risk and costs to future generations, notably through what can now be perceived as being inadequate regulation – or rather adequate but only on the short-term and inadequate on the long-term – and weak impact assessment (Baldwin et al. 2019). Neoliberalism promoted the capture of economic comparative advantages by re-localizing manufacturing (and agricultural) jobs in jurisdictions where labour cost arbitrage could be found; however, the local social costs of this economic model can be high with local quality of life deteriorating for some (Beumer et al. 2018; Lawrence and Almas 2018), moreover economic comparative advantages are dynamic and changes occur fast as exemplified by the rebound in a single generation of post-war Japan or the economic rise of China in the past two decades (Domar et al. 1964; Euchner 2018; Leitner 1999). A view for a 'hypermodern-capitalism' is thus to proactively maximize economic and social returns by optimizing the production function in parallel in terms of both local and global parameters (Figure 25.10). What is required first of all for this hypermodern-capitalism to fully emerge is long-term thinking to solidly anchor the new economy on solid fundamentals and overcome any phenomenon of local economic optimum. Precursors of the growing awareness of the need for policies that promote both regional and international economies in parallel are perhaps

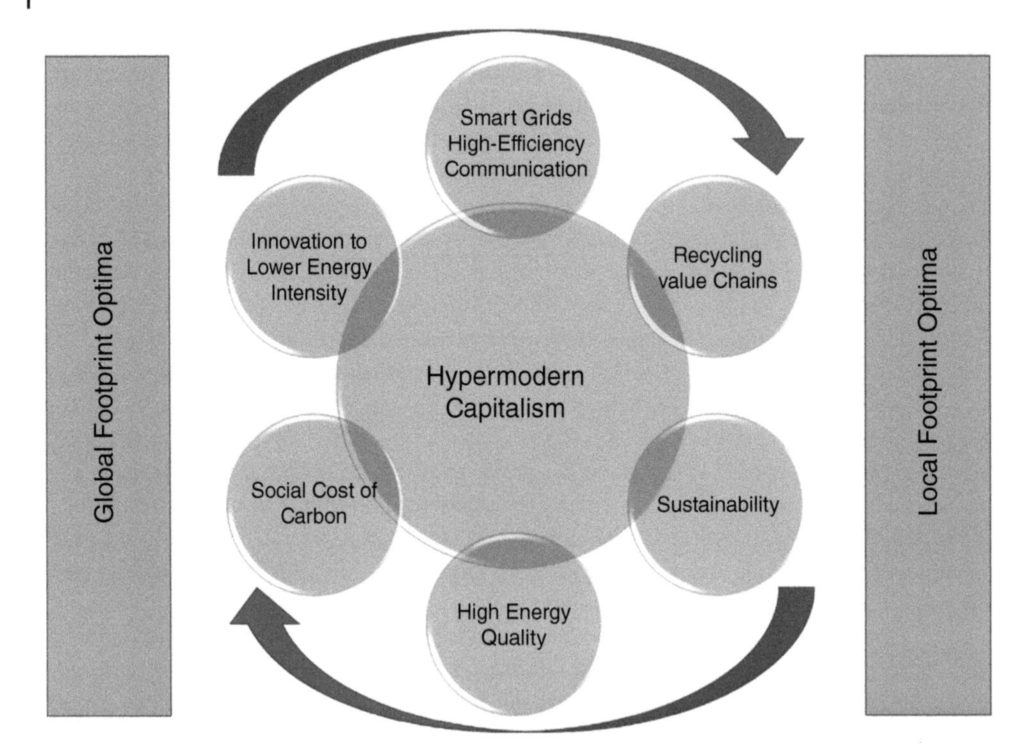

Figure 25.10 Hypermodern-capitalism. Hypermodern-capitalism is a neologism to convey the concept of a hyper flexible economy that senses and responds to challenges quickly by implementing new competitive advantages, with the dimension to build long-term benefits by integrating in parallel both the local and the global levels. It is reminiscent of the hypermodern organization that is characterized by accelerated changes to sense and respond to trends and competition with the objective of maximizing its economic outcomes (Roberts and Armitage 2006). The concept depicted here focuses on energy-related dimensions as described throughout the chapter, though trade and other dimensions related to the economic function could be introduced. The model is congruent with the development in parallel of central grid electricity for powering industries and of distributed generation via micro-grids for cities and towns or individual (photovoltaic) power generators for households, which constitutes a radical departure from the current centralized model in place in OECD countries (Schleicher-Tappeser 2012).

best exemplified by the emergence of micro-currencies, which tend to gain public acceptance as a means to promote the local economy and local benefits by 're-localizing' social and economic relations, including for example by requiring companies to adopt the local culture, to privilege local products, and to adopt a 'green' stance by practicing waste recycling (Anonymous 2014; Kovács et al. 2018; Pacione 1999). Another embodiment of the re-localizing phenomenon described above can perhaps be found in its political exacerbations that can be interpreted as direct consequences of an unsustainable imbalance between local versus global benefits (Flew 2019). A robust way to achieve this vision is by solidly anchoring the economic rationale on long-term horizons (typically beyond a single human generation) and on simple fundamentals. A telling example here could be the prediction of Svante Arrhenius 'on the influence of carbonic acid in the air upon the temperature of the ground' regarding climate change, which might have helped guide industry earlier and better if the simple vision that the Earth is a very large (at the human scale) but nonetheless closed system had been factored in industrial development (Arrhenius 1896). The local economic optimum at the onset of the Industrial Revolution was to consider the atmospheric commons as having an infinite capacity to manage CO_2 emissions, hence free, which more than a century later proves to be totally inaccurate albeit global consequences of man-made carbon imbalances having taken almost 200 years, from the onset of the Industrial Revolution to the beginning of the twenty-first century, to become clear-cut. Having said that, fossil fuels have served mankind tremendously in enabling access to higher quality energy than wood, which ultimately enabled tremendous economic growth and progress in living standards for the multitude. However, the energy model in use since the commoditization of fossil fuels must now change to include a sustainability function.

Decoupling energy demand from economic growth, a phenomenon that would have both economic and environmental benefits, is a function of four forces: (i) a steep decline in energy intensity; (ii) a significant increase in energy efficiency; (iii) the rise of electrification; and (iv) the growing use of renewables (Sharma et al. 2019). As a result, several pillars can already be identified to build the new economy: (i) high energy quality with electricity as the 'universal' energy currency, which translates in the global deployment of photovoltaic and nuclear power, as well as electric vehicles to replace combustion engine-powered vehicles (Vertès 2010); (ii) high efficiency communication to further develop smart cities as well as smart grids and keep virtual distances low while reducing resource use; (iii) sustainability by harnessing renewable energy generation technology and notably solar power the competitiveness of which has been boosted by a sharp decline in the installation costs of photovoltaic systems (Frankel et al. 2014); (iv) integrating the SCC in the cost of goods in an effective and politically acceptable manner for maintaining global trade where it is efficient (the SCC would be subtracted to comparative advantages, which is what the carbon tax fundamentally aims at addressing; it is worth noting that this would make coal-fired power stations economically unsustainable; moreover, it would encourage short trade circuits); (v) innovating and deploying new innovation to decrease energy intensity (a trend that is typically associated with higher incomes); (vi) creating novel value chains to incentivize resource recycling and reduce pollution, and notably making use of used plastics as chemical feedstock alongside of petroleum. Focusing solely on the analysis filter of CO_2 emissions, these evolutions will contribute to mitigating the worsening of climate change. Systematically adapting global efficiency solutions to local realities will tailor the local production functions to their local maxima, with the sum of these local optima integrating to result in superior global long-term returns. Embedded in the application of the concepts of CSR and corporate sustainability, the hope is that such simple changes will green the global economy in a manner such that it can sustain for all a superior quality of life.

References

Ahlgren, E.O., Börjesson Hagberg, M., and Grahn, M. (2017). Transport biofuels in a global energy-economy modelling - a review of comprehensive systems assessment approaches. *GCB Bioenergy* 9: 1168–1180.

Åkerfeldt S, Hammar H. 2015. La taxe carbone en Suède. Revue Projet 7 September https://www.revue-projet.com/articles/2015-09-akerfeldt-hammar-la-taxe-carbone-en-suede/8085

Albassam, B.A. (2015). Economic diversification in Saudi Arabia: myth or reality? *Resources Policy* 44: 112–117.

Allen, R.C. (2009). *The British Industrial Revolution in Global Perspective*. Cambridge, UK: Cambridge University Press.

Alnaser, N.W. and Alnaser, W.E. (2019). The impact of the rise of using solar energy in GCC countries. *Renewable Energy and Environmental Sustainability* 4 https://doi.org/10.1051/rees/2019004.

Anonymous. 2013. Plastic resins in the United States, American Chemistry Council, Washington, D.C., USA

Anonymous. 2014. The history of the electric car, US Department of Energy, Washington, D.C., USA

Anonymous. 2014. Pays Basque. Monnaie locale : l'Eusko, éthique et utile. *Courrier International* 9 July

Anonymous. 2015. Specialty Chemicals Market worth $470 Billion by 2020 Markets & Markets

Anonymous (2017). *World Energy Outlook*. Paris, France: International Energy Agency.

Anonymous. 2018. Greenhouse gas emissions and atmospheric concentrations have increased over the past 150 years. *US Energy Information Administration* https://www.eia.gov/energyexplained/print.php?page=environment_how_ghg_affect_climate

Anonymous. 2019. Plastic resin & synthetic fiber manufacturing industry profile. *Dun & Bradstreet* First Research:February 4th

Arrhenius, S. (1896). On the influence of carbonic acid in the air upon the temperature of the ground. *Philosophical Magazine and Journal of Science* 41: 237–276.

Assadian F, Mallon KR, Fu B. 2016. The impact of vehicle-integrated photovoltaics on heavy-duty electric vehicle battery cost and lifespan. *SAE Technical Paper* 2016-01-1289

Audoly, R., Vogt-Schilb, A., Guivarch, C., and Pfeiffer, A. (2018). Pathways toward zero-carbon electricity required for climate stabilization. *Applied Energy* 225: 884–901.

Auffhammer, M. (2018). Quantifying economic damages from climate change. *Journal of Economic Perspectives* 32: 33–52.

Baldwin, C., Marshall, G., Ross, H. et al. (2019). Hybrid neoliberalism: implications for sustainable development. pp. 1-22. *Society and Natural Resources* 32: 566–587.

Banafea, W. and Ibnrubbian, A. (2018). Assessment of economic diversification in Saudi Arabia through nine development plans. *OPEC Energy Review* 42: 42–54.

Banner, T., Fosmer, A., Jessen, H. et al. (2011). Microbial processes for industrial-scale chemical production. In: *Biocatalysis for Green Chemistry and Chemical Process Development* (eds. J. Tao and R. Kazlauskas). Chichester, UK: Wiley.

Barbier, E.B., Burgess, J.C., and Dean, T.J. (2018). How to pay for saving biodiversity? *Science* 360: 486–488.

Bardi, U. (2019). Peak oil, 20 years later: failed prediction or useful insight? *Energy Research & Social Science* 48: 257–261.

Barton, D., Manyika, J., Kioller, T. et al. (2017). *Where Companies with a Long-Term View Outperform their Peers*. McKinsey & Co.

Batty, M., Axhausen, K.W., Giannotti, F. et al. (2012). Smart cities of the future. *The European Physical Journal Special Topics* 214: 481–518.

Benedick, R.E. (1998). *Ozone Diplomacy*. Cambridge, MA, USA: Harvard University Press.

Benninga, H. (1990). *A History of Lactic Acid Making*. Dordrecth, The Netherlands: Kluwer Academic Publishers.

van den Berg, L., Braun, E., and van Winden, W. (2001). Growth clusters in European cities: an integral approach. *Urban Studies* 38 https://journals.sagepub.com/doi/10.1080/00420980124001.

Beumer, C., Figge, L., and Elliott, J. (2018). The sustainability of globalisation: including the social robustness criterion. *Journal of Cleaner Production* 179: 704–715.

Biddy, M.J., Scarlata, C., and Kinchin, C. (2016). *Chemicals from Biomass: A Market Assessment of Bioproducts with Near-Term Potential*. Golden, CO, USA: National Renewable Energy Laboratory.

Blinder AS. 2000. The Internet and the new economy., The Brookings Institution, Washington D.C., USA

Boden TA, Marland G, Andres RJ. 2017. Global, Regional, and National Fossil-Fuel CO_2 Emissions. *Carbon Dioxide Information Analysis Center, Oak Ridge National Laboratory, U.S. Department of Energy, Oak Ridge, Tenn., U.S.A.* doi https://doi.org/10.3334/CDIAC/00001_V2017

Brynjolfsson, E., Rock, D., and Syverson, C. (2018). Artificial intelligence and the modern productivity paradox: a clash of expectations and statistics. In: *The Economics of Artificial Intelligence: An Agenda* (eds. A. Agrawal, J. Gans and A. Goldfarb). Chicago, IL, USA: University of Chicago Press.

Burke, M. and Emerick, K. (2016). Adaptation to climate change: evidence from US agriculture. *American Economic Journal: Economic Plociy* 8: 106–140.

Campanhola, C. and Pandey, S. (2019). *Sustainable Food and Agriculture*. Academic Press.

Cano, Z.P., Banham, D., Ye, S. et al. (2018). Batteries and fuel cells for emerging electric vehicle markets. *Nature Energy* 3: 279–289.

Carberry, E.J., Bharati, P., Levy, D.L., and Chaudhury, A. (2017). Social movements as catalysts for corporate social innovation: environmental activism and the adoption of green information systems. *Business and Society* https://doi.org/10.1177/0007650317701674.

Carroll, A.B. (1999). Corporate social responsibility: evolution of a definitional construct. *Business and Society* 38: 268–295.

Chandra, A., McNamara, K.E., and Dargusch, P. (2018). Climate-smart agriculture: perspectives and framings. *Climate Policy* 18: 526–541.

Chubukov, V., Mukhopadhyay, A., Petzold, C.J. et al. (2016). Synthetic and systems biology for microbial production of commodity chemicals. *NPJ Systems Biology and Applications* 2: 16009.

Clark P. 2015. Crude woes melt profits for plastic recyclers. *The Financial Times* 1 May:17

Clark P. 2018. Renewables spur National Grid to strike battery deal. *The Financial Times* 27/28 August:2

Clomburg, J.M., Crumbley, A.M., and Gonzalez, R. (2017). Industrial biomanufacturing: the future of chemical production. *Science* 355 https://doi.org/10.1126/science.aag0804.

Clos, J. (2015). *Building Better Cities*. McKinsey & Co.

Csereklyei Z, Rubio M, Stern DI. 2014. Energy and economic growth: the stylized facts. *CCEP Working Paper 1417, November 2014, Crawford School of Public Policy, The Australian National University*

Dahlsrud, A. (2008). How corporate social responsibility is defined: an analysis of 37 definitions. *Corporate Social Responsibility and Environmental Management* 15 (1): 1–13.

Deakin, M. and Reid, A. (2018). Smart cities: under-gridding the sustainability of city-districts as energy efficient-low carbon zones. *Journal of Cleaner Production* 173: 39–48.

Diaz, D. and Moore, F. (2017). Quantifying the economic risks of climate change. *Nature Climate Change* 7: 774–782.

Díaz S, Settele J, Brondízio E, Ngo HT, Guèze M, et al. 2019. Summary for policymakers of the global assessment report on biodiversity and ecosystem services of the Intergovernmental Science-Policy Platform on Biodiversity and Ecosystem Services, Intergovernmental Science-Policy Platform on Biodiversity and Ecosystem Services, Bonn, Germany

Dijk, M., Orsato, R.J., and Kemp, R. (2013). The emergence of an electric mobility trajectory. *Energy Policy* 52: 135–145.

Dincer, I. and Acar, C. (2015). A review on clean energy solutions for better sustainability. *International Journal of Energy Research* 39: 585–606.

Dineen, D., Ryan, L., and Gallachóir, B.Ó. (2018). Vehicle tax policies and new passenger car CO_2 performance in EU member states. *Climate Policy* 18: 396–412.

DiSegni, D.M., Huly, M., and Akron, S. (2015). Corporate social responsibility, environmental leadership and financial performance. *Social Responsibility Journal* 11: 131–148.

Domar, E.D., Eddie, S.M., Herrick, B.H. et al. (1964). Economic growth and productivity in the United States, Canada, United Kingdom, Germany and Japan in the post-war period. *The Review of Economics and Statistics* 46: 33–40.

Douglass, A.R., Newman, P.A., and S, S. (2014). The Antarctic ozone hole: an update. *Physics Today* 67: 42–48.
Doyle, J. (2016). *Mediating Climate Change*. UK: Taylor & Francis Group, Rootledge.
Dyllick, T. and Hockerts, K. (2002). Beyond the business case for corporate sustainability. *Business Strategy and the Environment* 11: 130–141.
Ejaz, W. and Anpalagan, A. (2019). Internet of things for smart cities: overview and key challenges. In: *Internet of Things for Smart Cities* (eds. W. Ejaz and A. Anpalagan), 1–15. Cham, Switzerland: Springer.
Engel, H., Hensley, R., Knupfer, S., and Sahdev, S. (2018). *The Potential Impact of Electric Vehicles on Global Energy Systems*. McKinsey & Co.
Euchner, J. (2018). The emergence of innovation in China. *Research-Technology Management* 61: 9–10.
Evarts, E.C. (2015). Lithium batteries: to the limits of lithium. *Nature* 526: S93–S95.
Ezekoye, O., Milutinovic, A., and Simons, T.J. (2018). *Chemicals and Capital Markets: Back at the Top*. McKinsey & Co.
Ferone, M., Raganati, F., Olivieri, G., and Marzocchella, A. (2019). Bioreactors for succinic acid production processes. *Critical Reviews in Biotechnology* https://doi.org/10.1080/07388551.2019.1592105.
Fischer, D., Harbrecht, A., Surmann, A., and McKenna, R. (2019). Electric vehicles' impacts on residential electric local profiles–a stochastic modelling approach considering socio-economic, behavioural and spatial factors. *Appllied Energy* 233: 644–658.
Flew, T. (2019). Populism and globalization: towards a post-global era? *SSRN* https://doi.org/10.2139/ssrn.3321448.
Frankel, D. and Wagner, A. (2017). *Battery Storage: The Next Disruptive Technology in the Power Sector*. McKinsey & Co.
Frankel D, Ostrowski K, Pinner D. 2014. The disruptive potential of solar power. *McKinsey Quarterly* April
Franklin, J., Serra-Diaz, J.M., Syphard, A.D., and Regan, H.M. (2016). Global change and terrestrial plant community dynamics. *Proceedings of the National Academy of Sciences of the United States of America* 113: 3725–3734.
Gagnon, P., Margolis, R., Melius, J. et al. (2016). *Rooftop Solar Photovoltaic Technical Potential in the United States. A Detailed Assessment*. Golden, CO, USA: National Renewable Energy Laboratory.
Geroe, S. (2019). Addressing climate change through a low-cost, high-impact carbon tax. *Journal of Environment and Development* 15 https://doi.org/10.1177/1070496518821152.
Geroe, S. (2019). Addressing climate change through a low-cost, high-impact carbon tax. *The Journal of Environment and Development* https://doi.org/10.1177/1070496518821152.
Gharavi, H. and Ghafurian, R. (2011). Smart grid: the electric energy system of the future. *Proceedings of the IEEE* 99: 917–921.
Goodenough, K.M., Wall, F., and Merriman, D. (2018). The rare earth elements: demand, global resources, and challenges for resourcing future generations. *Natural Resources Research* 27: 201–216.
Gosling, S.N. and Arnell, N.W. (2016). A global assessment of the impact of climate change on water scarcity. *Climatic Change* 134: 371–385.
Gunningham, N. (2018). Mobilising civil society: can the climate movement achieve transformational social change? *Interface: A Journal on Social Movements* 10: 149–169.
Habersang, S., Küberling-Jost, J., Reihlen, M., and Seckler, C. (2019). A process perspective on organizational failure: a qualitative meta-analysis. *Journal of Management Studies* 56: 19–56.

Harrington, L.J., Frame, D.J., Fischer, E.M. et al. (2016). Poorest countries experience earlier anthropogenic emergence of daily temperature extremes. *Environmental Research Letters* 11: 055007.

Hellin, J. and Fisher, E. (2019). Climate-smart agriculture and non-agricultural livelihood transformation. *Climate* 7: 48.

Hensley, R., Knupfer, S., and Pinner, D. (2009). Electrifying cars: how three industries will evolve. *McKinsey Quarterly* 3: 87–96.

Hong, S., Musso, C., and Simons, T.J. (2015). *Oil-Price Shocks and the Chemical Industry: Preparing for a Volatile Environment*. McKinsey & Co.

Hoornweg, D., Bhada-Tata, P., and Kennedy, C. (2013). Waste production must peak this century. *Nature* 502: 615–617.

Hope, C. and Schaefer, K. (2016). Economic impacts of carbon dioxide and methane released from thawing permafrost. *Nature Climate Change* 6: 56–59.

Hopewell, J., Dvorak, R., and Kosior, E. (2009). Plastics recycling: challenges and opportunities. *Philosophical Transactions of the Royal Society and Biological Science* 364: 2115–2126.

Hourcade, J.C. (2014). L'écotaxe, un combat désespéré? *Revue Projet* 4: 67–74.

Hundertmark, T., Mayer, M., McNally, C. et al. (2018). *How Plastics Waste Recycling Could Transform the Chemical Industry*. McKinsey & Co.

Jiang, M., Ma, J., Wu, M. et al. (2017). Progress of succinic acid production from renewable resources: metabolic and fermentative strategies. *Bioresource Technology* 245: 1710–1717.

Jones, D.T. and Woods, D.R. (1986). Acetone-butanol fermentation revisited. *Microbiology Reviews* 50: 484–525.

Kannegiesser, M. (2008). *Value Chain Management in the Chemical Industry: Global Value Chain Planning of Commodities*. Heidelberg, Germany: Physica-Verlag.

Kennedy, A.B. (2013). China's search for renewable energy: pragmatic techno-nationalism. *Asian Survey* 53: 909–930.

Kim, S. and Han, J. (2017). Enhancement of energy efficiency and economics of process designs for catalytic co-production of bionergy and bio-based products from lignocellulosic biomass. *International Journal of Energy Research* 41: 1553–1562.

Knuth, S. (2018). Breakthroughs for a green economy? Financialization and clean energy transition. *Energy Research & Social Science* 41: 220–229.

Kovács, S.Z., Lakócai, C., and Gál, Z. (2018). Local alternative currencies. New opportunities in expanding local financial services. *Pénzügyi Szemle (Public Finance Quarterly)* 63: 473–489.

Kuhns, R.J. and Shaw, G.H. (2018). Peak oil and petroleum energy resources. In: *Navigating the Energy Maze: The Transition to a Sustainable Future* (eds. R.J. Kuhns and G.H. Shaw), 53–63. Cham, Switzerland: Springer.

Kurzweil, R. (2004). The law of accelerating returns. In: *Alan Turing: Life and Legacy of a Great Thinker* (ed. C. Teuscher), 381–416. Berlin, Heidelberg, Germany: Springer.

Langbroek, J.H.M., Franklin, J.P., and Susilo, Y.O. (2016). The effect of policy incentives on electric vehicle adoption. *Energy Policy* 94: 94–103.

Lawrence, G. and Almas, R. (2003). *Globalisation, Localisation and Sustainable Livelihoods*. Taylor & Francis Group: Routledge.

Lawrence, G. and Almas, R. (2018). *Globalisation, Localisation and Sustainable Livelihoods*. Rootledge: Taylor & Francis.

Le Heron, R. (2016). Globalisation'and local economic development in a globalising world: critical reflections on the theory-practice relation. In: *Theories of Local Economic Development* (ed. J.E. Rowe), 115–134. Routledge: Taylor & Francis Group.

Leitner, P.M. (1999). Japan's post-war economic success: Deming, quality, and contextual realities. *Journal of Management History* 5: 489–505.

Levitt, T. (1993). The globalization of markets. In: *Readings in International Business: A Decision Approach* (eds. R.Z. Aliber and R.W. Click), 249–266. Cambridge, MA, USA: The MIT Press.

Lewis, C.W. (1981). Biomass through the ages. *Biomass* 1: 5–15.

Li, H. and Yi, H. (2014). Multilevel governance and deployment of solar PV panels in US cities. *Energy Policy* 69: 19–27.

Li, B., Kawakita, Y., Ohira-Kawamura, S. et al. (2019). Colossal barocaloric effects in plastic crystals. *Nature* 567: 506–510.

Lin, B. and Jia, Z. (2019). Impacts of carbon price level in carbon emission trading market. *Appllied Energy* 239: 157–170.

Ling, T.Y. and Chiang, Y.C. (2018). Well-being, health and urban coherence-advancing vertical greening approach toward resilience: q design practice consideration. *J. Cleaner Production* 182: 187–197.

Linnerooth-Bayer, J. and Hochrainer-Stigler, S. (2015). Financial instruments for disaster risk management and climate change adaptation. *Climatic Change* 133: 85–100.

Lins, K.V., Servaes, H., and Tamayo, A. (2017). Social capital, trust, and firm performance: the value of corporate social responsibility during the financial crisis. *Journal of Finance* 72: 1785–1824.

Lloveras, P., Aznar, A., Barrio, M. et al. (2019). Colossal barocaloric effects near room temperature in plastic crystals of neopentylglycol. *Nature Communications* 10: 1803.

Lucas RE. 2002. The industrial revolution: Past and future. *Lectures on economic growth*:109–88

Lund, S., Manyika, J., Woetzel, J. et al. (2019). *Globalization in Transition: The Future of Trade and Global Value Chains*. McKinsey & Co.

Manes, S. and Andrews, P. (1993). *Gates: How Microsoft's Mogul Reinvented an Industry-and Made himself the Richest Man in America*. New York City, NY, USA: Simon & Schuster, Inc.

Maniyika, J., Sinclair, J., Dobbs, R. et al. (2012). *Manufacturing the Future: The Next Era of Global Growth and Innovation*. McKinsey & Co.

Mantyka-Pringle, C.S., Visconti, P., Di Marco, M. et al. (2015). Climate change modifies risk of global biodiversity loss due to land-cover change. *Biological Conservation* 187: 103–111.

Marchand L. 2018. Ces pays d'Europe où les voitures diesel ne sont plus les bienvenues. Les Echos 4 March https://www.lesechos.fr/2018/03/ces-pays-deurope-ou-les-voitures-diesel-ne-sont-plus-les-bienvenues-985720

Markard, J. (2018). The next phase of the energy transition and its implications for research and policy. *Nature Energy* 3: 628–633.

Marsh P. 2012. Future factories. *The Financial Times* 11 June:9

Masson, V., Bonhomme, M., Salagnac, J.L. et al. (2014). Solar panels reduce both global warming and urban heat island. *Frontiers in Environmental Science* 2 https://doi.org/10.3389/fenvs.2014.00014.

Meckling, J. and Nahm, J. (2019). The politics of technology bans: industrial policy competition and green goals for the auto industry. *Energy Policy* 126: 470–479.

Mehling, M. and Tvinnereim, E. (2018). Carbon pricing and the 1.5 degree Celsius target: near-term decarbonisation and the importance of an instrument mix. *CCLR* 12: 50.

Milios, L., Holm Christensen, L., McKinnon, D. et al. (2018). Plastic recycling in the Nordics: a value chain market analysis. *Waste Management* 76: 180–189.

Millar, R., Allen, M., Rogelj, J., and Friedlingstein, P. (2016). The cumulative carbon budget and its implications. *Oxford Review of Economic Policy* 32: 323–342.

Monasterolo, I. and Raberto, M. (2019). The impact of phasing out fossil fuel subsidies on the low-carbon transition. *Energy Policy* 124: 355–370.

Moon, J. (2007). The contribution of corporate social responsibility to sustainable development. *Sustainaible Development* 15: 296–306.

Morsing, M., Oswald, D., and Stormer, S. (2019). The ongoing dynamics of integrating sustainability into business practice: the case of Novo Nordisk a/S. In: *Managing Sustainable Business* (eds. G.G. Lenssen and N.C. Smith), 637–669. Dordrecht, The Netherlands: Springer, McMillan International.

Muratori, M. (2018). Impact of uncoordinated plug-in electric vehicle charging on residential power demand. *Nature Energy* 3: 193–201.

Mwanza, B.G., Mbohwa, C., and Telukdarie, A. (2018). Strategies for the recovery and recycling of plastic solid waste (PSW): a focus on plastic manufacturing companies. *Procedia Manufacturing* 21: 686–693.

Nielsen, J. and Keasling, J.D. (2016). Engineering cellular metabolism. *Cell* 6: 1185–1197.

Nishimizu, M. and Page, J.M. (1982). Total factor productivity growth, technological progress and technical efficiency change: dimensions of productivity change in Yugoslavia, 1965-78. *The Economic Journal* 92: 920–936.

Nordhaus, W.D. (2017). Revisiting the social cost of carbon. *Proceedings of the National Academy of Sciences of the United States of America* 114: 1518–1523.

Nordhaus, W.D. (2018). Projections and uncertainties about climate change in an era of minimal climate policies. *American Economic Journal: Economic Plociy* 10: 333–360.

Nussbaum, A.S.K. (2009). Ethical corporate social responsibility (CSR) and the pharmaceutical industry: a happy couple? *J. Journal of Medical Marketing* 91: 67–76.

Nykvist, B. and Nilsson, M. (2015). Rapidly falling costs of battery packs for electric vehicles. *Nature Climate Change* 5: 329–332.

Oke, T.R. (1967). City size and the urban heat island. *Atmospheric Environment* 7: 769–779.

Orlitzky, M., Schmidt, F.L., and Rynes, S.L. (2003). Corporate social and financial performance: a meta-analysis. *Organization Studies* 24: 403–441.

Pacione, M. (1999). The other side of the coin: local currency as a response to the globalization of capital. *Regional Studies* 33: 63–72.

Page, J.M. (2018). A smart future for energy in megacities. In: *Rise of Megacities, the: Challenges, Opportunities and Unique Characteristics* (eds. J. Kleer and K.A. Nawrot), 259–278. Singapore, Singapore: World Scientific.

Palmer E. 2019. Novo manufacturing will be powered by 100% renewable enregy by 2020. *Fierce Pharma* April 30

Paulson, H.M.J. (2015). *Short-Termism and the Threat from Climate Change*. McKinsey & Co.

Paustian, K., Lehmann, J., Ogle, S. et al. (2016). Climate-smart soils. *Nature* 532: 49–57.

Philp, J. (2018). The bioeconomy, the challenge of the century for policy makers. *New Biotechnology* 40: 11–19.

Pilkey, O.H. and Pilkey, K.C. (2019). *Sea Level Rise: A Slow Tsunami on America's Shores*. Durham, N.C., USA: Duke University Press.

Pitron, G. (2018). Voiture électrique: Une aubaine pour la chine. *Le Monde Diplomatique* 773: 1,10–1,11.

Rahimi, A. and García, J.M. (2017). Chemical recycling of waste plastics for new materials production. *Nature Review Chemistry* (6) https://doi.org/10.1038/s41570-017-0046.

Raval A, Inagaki K. 2018. Saudis seal Softbank deal for largest solar project. *The Financial Times* 29 March:15

Rigby R. 2018. Which networks will underpin smart cities? *Financial Times* 10 September:4

Roberts, J. and Armitage, J. (2006). From organization to hypermodern organization: on the accelerated appearance and disappearance of Enron. *Journal Organizational Change Management* 19: 558–577.

Robinson, B. and Mayo, S. (2006). Peak oil and Australia; probable impacts and possible options. In: *Proc. 8th SEGJ International Symposium*, 1–6.

Rocamora AR. 2017. The rise of carbon taxation in France, Institute for Global Environmental Strategies

Roelofsen, O., Sharma, N., Sutorius, R., and Tryggestad, C. (2016). *Is Peak Oil Demand in Sight?* McKinsey & Company.

Rostand A. 2015. Energy transition: oil and gas in a prime position to capitalize. *Schlumberger Business Consulting* http://www.sbc.slb.com/Our_Ideas/Energy_Perspectives/Winter12_Content/Winter12_Energy_Transition.aspx

Salam, M.A. and Khan, S.A. (2018). Transition towards sustainable energy production - a review of the progress for solar energy in Saudi Arabia. *Energy Exploration & Exploitation* 36: 3–27.

Sanderson H. 2018. Europe's fledging battery industry should heed the lessons from solar. *The Financial Times* 10 October:20

Saunois, M., Jackson, R.B., Bousquet, P. et al. (2016). The growing role of methane in anthropogenic climate change. *Environmental Research Letters* 11: 120207.

Schleicher-Tappeser, R. (2012). How renewables will change electricity markets in the next five years. *Energy Policy* 48: 64–75.

Shackleton R. 2013. Total factor productivity growth in historical perspective. *Congressional Budget Office, Washington D.C.* Working paper 1 http://go.usa.gov/ULE

Sharma N, Smeets B, Tryggestad C. 2019. The decoupling of GDP and energy growth: a CEO guide. *McKinsey Quarterly* April

Shubbak, M.H. (2019). The technological system of production and innovation: the case of photovoltaic technology in China. *Research Policy* 48: 993–1015.

Siegel, B. and Jakimo, A.L. (2015). The role of patient advocacy in the clinical translation of regenerative medicine. In: *Stem Cells in Regenerative Medicine: Science, Regulation, and Business Strategies* (eds. A.A. Vertès, N. Qureshi, A.I. Caplan and L. Babiss), 543. Chichester, UK: Elsevier.

Singh, R. and Banerjee, R. (2015). Estimation of rooftop solar photovoltaic potential of a city. *Solar Energy* 115: 589–602.

Smil, V. (2010). *Energy Transitions: History, Requirements, Prospects.* Santa Barbara, CA, USA: Prager: ABC-CLIO.

Smil, V. (2016). *Energy Transitions: Global and National Perspectives (Second Expanded and Updated Edition).* Santa Barbara, CA: USA: Praeger (ABC-CLIO).

Solow, R.M. (1956). A contribution to the theory of economic growth. *The Quarterly Journal of Economics* 70: 65–94.

Soltani, N.Y., Kim, S.-J., and Georgios, B.G. (2015). Real-time load elasticity tracking and pricing for electric vehicle charging. *IEEE Transactions on Smart Grid* 6: 1303–1313.

Sortomme, E., Hindi, M.M., MacPherson, S.D.J., and Venkata, S.S. (2011). Coordinated charging of plug-in hybrid electric vehicles to minimize distribution system losses. *IEEE Transactions on Smart Grid* 2: 198–205.

Sperling, D. (2018). Electric vehicles: approaching the tipping point. In: *Three Revolutions* (ed. D. Sperling), 21–54. Washington, D.C., USA: Island Press.

Steffen, W., Crutzen, P.J., and McNeill, J.R. (2007). The Anthropocene: are humans now overwhelming the great forces of nature. *Ambio: A Journal of the Human Environment* 36: 614–622.

Steininger, K.W., Wagner, G., Watkiss, P., and König, M. (2015). Climate change impacts at the national level: known trends, unknown tails, and unknowables. In: *Economic Evaluation of Climate Change Impacts* (eds. K.W. Steininger, M. König, B. Bednar-Friedl, et al.), 441–459. Cham, Switzerland: Springer.

Stern, D.I. (2011). The role of energy in economic growth. *Annals New York Acadamy of Sciences* 1219: 26–51.

Stern, N. (2016). Economics: current climate models are grossly misleading. *Nature News* 530: 407.

Stern, D.I. and Cleveland, C.J. (2004). Energy and economic growth. *Encyclopedia of Energy* 2: 35–51.

Straathof, A.J. and Bampouli, A. (2017). Potential of commodity chemicals to become bio-based according to maximum yields and petrochemical prices. *Biofuels, Bioproducts and Biorefining* 11: 798–810.

Stromberg J. 2013. How the insurance industry is dealing with climate change. Smithsonian.com 24 September

Suganthi, L. and Samuel, A.A. (2012). Energy models for demand forecasting - a review. *Renewable Sustainable Energy Reviews* 16: 1223–1240.

Swim, J.K., Geiger, N., and Lengieza, M.L. (2019). Climate change marches as motivators for bystander collective action. *Frontiers in Communication* 4 https://doi.org/10.3389/fcomm.2019.00004.

Tan, Z., Yang, P., and Nehorai, A. (2014). An optimal and distributed demand response strategy with electric vehicles in the smart grid. *IEEE Transactions on Smart Grid* 5: 861–869.

Teixeira, A.C.R. and Sodré, J.R. (2018). Impacts of replacement of engine powered vehicles by electric vehicles on energy consumption and CO_2 emissions. *Transportation Research Part D: Transport and Environment* 59: 375–384.

Tol, R.S. (2018). The economic impacts of climate change. *Review of Environmental Economics and Policy* 12: 4–25.

Tollefson, J. (2014). Energy: islands of light. *Nature* 507: 154–156.

Vasudeo, R.A., Abitha, V.K., Vinayak, K. et al. (2016). Sustainable development through feedstock recycling of plastic wastes. *Macromolecular Symposia* 362: 39–51.

Vertès, A.A. (2010). Axes of development in chemical and process engineering for converting biomass to energy. In: *Biomass to Biofuels: Strategies for Global Industries* (eds. A.A. Vertès, N. Qureshi, H.P. Blaschek and H. Yukawa), 491–521. Chichester, UK: Wiley.

Vertès, A.A. (2014). Biorefinery roadmaps. In: *Biorefineries* (eds. N. Qureshi, D. Hodge and A.A. Vertès), 59–71. Elsevier.

Vertès, A.A. (2016). The economic translation of paradigm-changing innovation: a short history of biotechnology. *Technology Transfer and Entrepreneurship* 3: 56–66.

Vertès, A.A., Inui, M., and Yukawa, H. (2006). Implementing biofuels on a global scale. *Nature Biotechnology* 24: 761–764.

Vertès, A.A., Inui, M., and Yukawa, H. (2008). Technological options for biological fuel ethanol. *Journal of Molecular Microbiology and Biotechnology* 15: 16–30.

Vertès, A.A., Inui, M., and Yukawa, H. (2012). Postgenomic approaches to using corynebacteria as biocatalysts. *Annual Review of Microbiology* 66: 521–550.

Vöhringer, F., Vielle, M., Thalmann, P. et al. (2019). *Cost and Benefits of Climate Change in Switzerland.* EPFL Scientific Publications https://doi.org/10.1142/S2010007819500052.

Wang, Z. and Sarkis, J. (2017). Corporate social responsibility governance, outcomes, and financial performance. *Journal of Cleaner Production* 162: 1607–1616.

Warde, P. (2006). Fear of wood shortage and the reality of the woodland in Europe, c. 1450–1850. *History Workshop Journal* 62: 28–57.

van Wee, B., Maat, K., and de Bont, C. (2012). Improving sustainability in urban areas: discussing the potential for transforming conventional car-based travel into electric mobility. *European Planning Studies* 20: 95–110.

Wells, C.W. (2007). The road to the model T: culture, road conditions, and innovation at the dawn of the American motor age. *Technology and Culture* 48: 497–523.

Willemin A. 2017. La tendance est à l'énergie local. *MMS* 20 February:22–7

Wrigley, E.A. (1962). The supply of raw materials in the industrial revolution. *The Economic History Review* 15 (1): 16.

Yang, F.F. and Zhao, X.G. (2018). Policies and economic efficiency of China's distributed photovoltaic and energy storage industry. *Energy* 154: 221–230.

Index

Green Energy to Sustainability: Strategies for Global Industries, First Edition.
Edited by Alain A. Vertès, Nasib Qureshi, Hans P. Blaschek and Hideaki Yukawa.

b

f

g